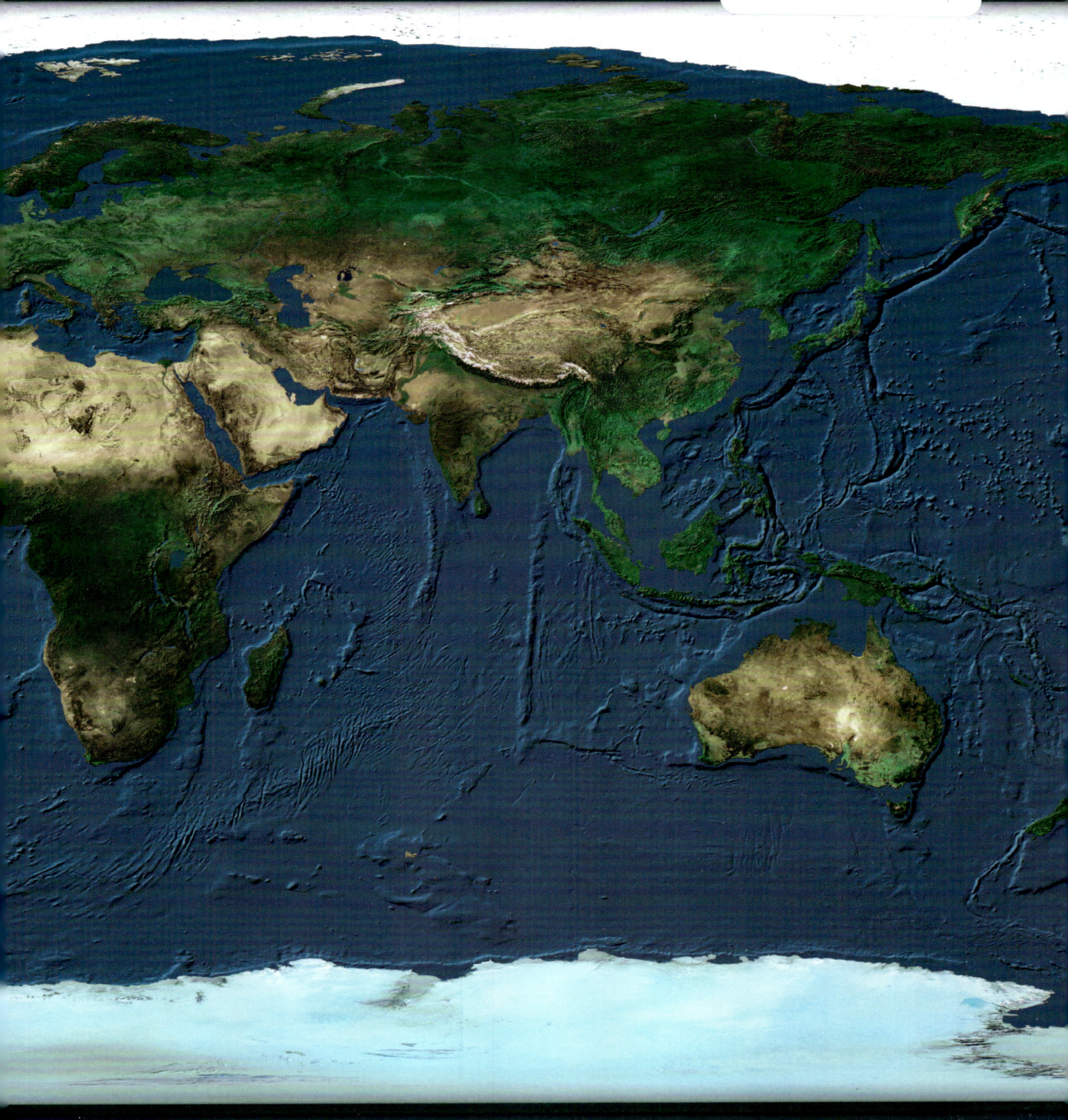

INTERNATIONAL PETROLEUM ENCYCLOPEDIA 2007

International Petroleum Encyclopedia is published annually by:
PennWell Corporation
1421 South Sheridan Road
Tulsa, Oklahoma 74112, USA
Phone: 918-835-3161
Fax: 918-831-9555
www.pennwellbooks.com

1421 South Sheridan Road
Tulsa, Oklahoma 74112-6600 USA

800.752.9764
+1.918.831.9421
sales@pennwell.com
www.pennwellbooks.com
www.pennwell.com

ISBN 978-1-59370-103-1

Printed in the United States of America

1 2 3 4 5 11 10 09 08 07

PennWell

Contents

Key Stats & Tables

Information Sources

Guest Essay

Navigating the Investment Challenge

by Peter J. Robertson, Chevron Corporation, Vice Chairman

Energy—finding it, producing it, transporting it, and using it—is one of the most critical issues facing the global community.

Global energy demand is projected to increase by more than 50% by 2030. Growth in energy demand will be particularly strong in the developing world, which includes several billion people whose rightful first step out of poverty is access to basic energy supplies.

The industry's response to this demand has been to increase investment in new supplies—which we are doing at record levels across the entire energy value chain.

Today's investment cycle: different

We've been through investment cycles before. There were three major investment phases in the 20th century. But there are several factors that differentiate today's investment cycle from those of the past.

The first difference is the sheer *scale* of investment. The International Energy Agency's World Energy Outlook 2006 estimates that through 2030, global investments in oil and gas will require just over $8.0 trillion to meet demand projections.

THE JACK WELL WAS COMPLETED AND TESTED IN 7,000 FT OF WATER, AND MORE THAN 20,000 FT UNDER THE SEA FLOOR—THE DEEPEST SUCCESSFUL WELL TEST IN THE GULF OF MEXICO.

THE DISCOVERER DEEP SEAS DRILL SHIP DRILLED THE JACK WELL.

If investment is added across the entire energy spectrum, including power generation, distribution and infrastructure, the investment target is more than $20 trillion by 2030—an unprecedented scale of investment.

Another characteristic of the current investment phase that is different from the past is its *complexity*. A significant portion of new investment is going toward large, complex projects such as extra-heavy oil and the deep water, which take years to bring onstream and involve multiple networks of partners, sometimes with different objectives. These projects have enormous fixed costs, long lead times, and extended recovery periods for capital investment. They require sophisticated, disciplined, and far-sighted management and very stable investment regimes.

An example of this complexity is Chevron's recent tests at the Jack Field in the deep water of the U.S. Gulf of Mexico, a project Chevron is leading with several partners, including Devon Energy and Statoil.

The Jack well test was the result of several years of active exploration. It required major investments in seismic imaging and drilling technologies to navigate the salt overlay and tight rock formations that characterize the Lower Tertiary formation in that region of the Gulf.

It will take about another year to fully assess likely production from the field, and another three to five years after that to build it out and produce first oil. Industry estimates are that the Lower Tertiary in this region of the Gulf may contain anywhere from 3 billion to 15 billion barrels of recoverable oil (equivalent). This high-profile Jack discovery will require a multibillion dollar investment to develop.

This type of complexity is what is meant when talking about the end of "easy" oil. But it's also ushering in a new and promising frontier for our most advanced technology and exploration methods.

Another point of distinction in the current investment cycle is the *range of risk* that we face: geologic risk, geopolitical risk, price risk, and contract risk. But the history of the energy industry is one of identifying, managing, and overcoming risks. It is what we do. And despite all of the hurdles we face today, the global energy industry is stepping up to the investment challenge.

New investment by the five major international oil companies alone totaled approximately $70 billion in 2005, an increase of nearly 20% over the previous year. Double-digit increases in capital and exploratory spending in 2007 seem likely if current trends hold. Investment on a similar scale is also being made by some of the national oil companies, notably Saudi Aramco and Petrobras.

Even with this historic level of capital spending, it's probably fair to ask the question: are current investments enough?

Constraints on investment

Many argue that investment should be more aggressive—projections of demand growth justify such a posture, and current cash flow in the industry certainly makes it possible. But the realities of constraints in investment today, which are significant, must also be considered.

One constraint is the *rising cost* of almost every investment input. Rig rates, for instance, soared in the past several years, compounded by hurricanes Katrina and Rita. Deepwater rig rates, such as in the U.S. Gulf of Mexico, have more than doubled since 2003–2004. High-spec jack-ups in the North Sea are commanding fees of $300,000 a day. Rates for deepwater, high-end drillships are expected to surpass $520,000 a day for 2008 and 2009 contracts.

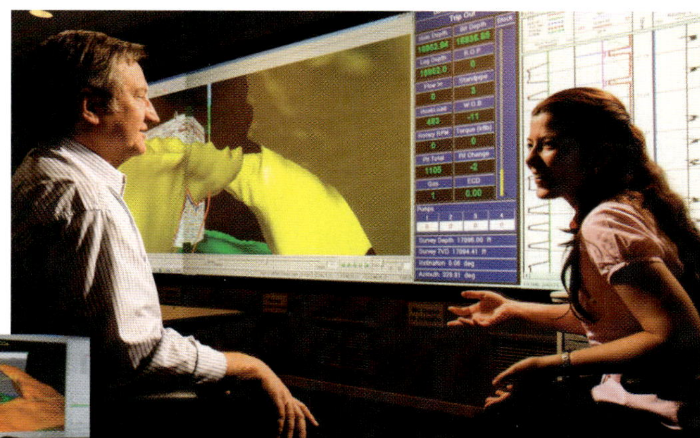

At Chevron's WellDECC facility in Houston, engineers can analyze data as a well is being drilled.

The price of raw materials is also emerging as a bottleneck to investment. Spot prices for steel and iron have nearly quadrupled in some market locations since 2003. Copper prices have increased more than 300 percent since 2002. Nickel has nearly doubled in price over the last year.

While these commodities aren't purchased on the spot market, these trends play out as increased costs for a broad range of fabricated steel structures, valves, compressors, pumps, and other items. Coupled with reduced numbers of contractors and high demand for shop space and engineering resources, the cost of these goods has increased as much as 200 percent over the last few years.

Another constraint is the *availability of skilled people*. After reaching a peak of 11,000 in 1983, the number of petroleum engineering graduates in the United States hit a low of 1,300 in 1997, and then rose slowly to 2,400 in 2006. This is hardly surprising in the wake of the 120,000 positions eliminated by the 25 largest private oil and gas companies worldwide since 1999.

Add to this fact that thousands of the industry's most experienced engineers are planning to retire over the next decade, and it is clear that a labor deficit looms, which will be an investment bottleneck for some time to come—coinciding with the most intensive phase of the investment cycle.

Investment decisions are also clearly impacted by *access*—or lack of access—to new production opportunities. Successful development of energy thrives best in an environment where multiple players apply a broad range of financial, technical and management resources.

It's no coincidence that the U.S. Gulf of Mexico, one of the few offshore areas in the United States open to robust exploration and production, continues to be a prolific source of oil and gas. It is a function of hundreds of oil and gas companies from all over the world, working in a variety of partnerships—and competing with one another—to develop the region to its full potential.

Investing to sustain performance

The energy industry faces a paradox. It is generating strong revenues and cash flow tied to the price of crude, but it also faces a variety of challenges to sustaining that performance. There are a few ways to navigate those challenges and take better advantage of the opportunities in front of us.

First, given the growing demand for energy from the global economy, our industry needs to take a *portfolio approach* to investment and look for sound investment opportunities across the entire energy spectrum. The fact is that investments must be made for the near term, the medium term and the long term. The challenge is to do them all, but to weight them and prioritize them in a way that will deliver predictable, economic returns as well as reliable flows of energy.

Chevron, for example, continues to make significant investments in its *base business*. That ranges from maintenance of existing infrastructure to exploration and production, including enhanced recovery techniques such as the steamflood technology Chevron developed in California and is currently sharing with its Saudi partners.

CHEVRON IS WORKING TO DEVELOP THE AGBAMI FIELD, WITH AN ESTIMATED 900 MILLION BARRELS OF OIL, 113 KM (70 MILES) OFF THE NIGER DELTA'S CENTRAL COAST. IT IS THE LARGEST DEEPWATER FIND IN NIGERIA'S HISTORY.

Chevron is also broadening its portfolio with growing investments in *emerging energy sources*, such as extra-heavy oil in Venezuela, oil sands in Western Canada, and gas-to-liquids in Nigeria and Qatar.

In a world that needs every Btu of energy we can develop, there is a business model emerging for reasonable investments in *alternative energy*. Chevron has put a strategic emphasis on investing in second-generation biofuels based on cellulosic conversion technology.

For example, Chevron is funding a portfolio of research partnerships aimed at developing indigenous feedstocks into fuel products. The Georgia Institute of Technology is focusing on conversion of local forest products in the southern United States, and the University of California at Davis is exploring the use of high-end agricultural waste as a resource. Chevron is also investing in a large-scale facility in Texas that will produce biodiesel from soybeans when it comes onstream in early 2007.

It's clear that oil, gas and coal will continue to anchor the global energy portfolio for decades to come. But new energy sources—as well as focused technology leading to *higher levels of energy efficiency*—can make a material contribution to long-term supplies. In fact, Chevron has a wholly owned operating company dedicated to selling energy efficiency services. This is a profitable business in itself.

Policy factors in investment decisions

Another requirement for sustained and successful investment is *predictability*, based on contract sanctity and increased transparency.

In the current price environment, there is a natural and rightful inclination on the part of governments and resource-holders to maximize the returns on production. But there is a tipping point where those policies can be counterproductive.

The EIA captured this issue in classic understatement: "Governments must strike a balance between short- and long-term objectives," it said in a recent report. "Once investment has occurred, the host government may be inclined to raise taxes and royalty rates to increase revenues quickly, but at the risk of discouraging further investment."

This concept of balance is extremely important. Sudden shifts in contract terms should be carefully considered in light of increased costs, more sophisticated technology and the inevitability of a cyclical downturn in prices at some point in the future. Stable, predictable and reasonable terms are needed to ensure that investments continue to flow.

Another way to increase predictability in the marketplace is through the *support of multilateral organizations* that create a robust environment for global trade.

The energy industry must demand political leadership that promotes inclusive commercial frameworks—and level playing fields—to stimulate trade and investment. That includes expanding the World Trade Organization to welcome global players such as Russia. Enhancing the global network of trade and investment will be one of the most important drivers of stability in the marketplace.

The people equation

The final point on how to effectively navigate the investment challenge is probably obvious. It's *people*—how we employ them, how we serve them and where the benefits of energy flow.

A Chevron employee who is also a student in subsurface geology climbs Colorado's Lookout Mountain to capture a satellite reading.

IPE Guest Essay

Peter J. Robertson
Chevron Corporation
Vice Chairman

Peter J. Robertson is vice chairman of the board of directors of Chevron Corporation—a position he has occupied since January 1, 2002.

Since January 1, 2005, he has had direct responsibility for Chevron's Strategic Planning; Policy, Government, and Public Affairs; and Human Resources. He collaborates closely with the chairman in overseeing the strategies and operations of the company.

In the more than three decades since he joined Chevron, Robertson has served in a variety of financial, operating and affiliate board assignments. In recent years he has been responsible for directing Chevron's North America, Chevron's International and, finally, Chevron's worldwide exploration, production, and global gas businesses.

Robertson is a native of Edinburgh, Scotland. He earned a bachelor's degree in mechanical engineering at Edinburgh University, and he holds a master's degree in business administration from the University of Pennsylvania, Wharton School, where he attended as a Thouron Scholar.

The industry needs smart, skilled people to deploy investments effectively. That means more investment in training and development. It means more partnerships with universities. It means building workforces that are truly global.

In some respects, it means nurturing a new kind of skill set in the industry, particularly among people with hybrid skills—half earth scientist or engineer and half information technologist, for example. People who can function in the interface and see the future.

The people equation also means that all of us—the industry, governments, and communities—must be committed to extending the economic benefits of energy production to all levels of society.

Energy is more than mobility, light, and heat. It's a fundamental driver of economic growth and opportunity for the globe.

Energy can drive such growth and opportunity directly, through the creation of jobs in the industry, and indirectly, through the investment of energy revenues in education, economic development, and other forms of social capital.

The more the energy industry and all other stakeholders work together to create broader prosperity and stability in the communities where they operate, the more secure we will become as long-term, profitable enterprises—and the more secure the global energy supply will be.

The fair price of oil

Another key issue tied directly to investment is that of cheap oil. Many ask if the end of cheap oil has come. But a more relevant question might be: what is the fair price of oil?

What is the price of oil that will allow sustained investment over the long term? What is the price of oil that won't create material demand destruction? What is the price of oil that will continue to generate global economic growth?

The marketplace will ultimately determine that price.

Everyone in the industry must play a role to help ensure that the marketplace works efficiently, predictably, and inclusively. That is the best framework for achieving the fair price of any commodity.

Adapted from Mr. Robertson's keynote address to the Oil and Money Conference, London, September 19, 2006.

Chronology

2006 In Brief

JANUARY

Exploration/Development

- OAO Lukoil discovers giant field in Caspian Sea, southwest of Astrakhan, calls it largest found in a decade in Russia. Recoverable volumes estimated at 600 million bbl of oil, 1.2 tcf of gas.
- Belize announces first commercial oil discovery, 50 miles west-southwest of Belize City.

Production

- Production begins from United Kingdom's largest undeveloped gas discovery, Rhum field. Expected production plateau 300 MMcfd; reserves estimated at 800 bcf.
- Cyclone Clare shuts in oil and gas production in northwestern Australia, temporarily halts 130,000 b/d of production.
- Militants kidnap workers at offshore Nigeria oil platform—the first in a series of similar actions—demanding more oil revenues for local governments. Over the year, an estimated 600,000 b/d of Nigerian oil production is lost.
- First consignment of crude oil shipped from Tiaka field in Tolo Bay off central Sulawesi, Indonesia.

Pipelines/Transportation

- Russia cuts natural gas supplies to Ukraine over pricing dispute. Flow is restored quickly, but concerns rise about European reliance on Russian gas.
- U.S. DOE withdraws 11 million bbl of crude oil from Strategic Petroleum Reserve in aftermath of Hurricanes Katrina and Rita—first emergency drawdown since 1991.
- Russia halts gas exports to Georgia after explosions damage pipelines. Georgia blames sabotage, but Russia denies claim.

Refining/Petrochemicals

- Third of three gas-oil separation plants in Saudi Arabia's Ghawar field comes on stream. Peak capacity: 300,000 b/d of light crude, 140 MMcfd of associated natural gas.

Government

- Iran removes United Nations seals at Natanz uranium enrichment facility in ongoing dispute about Iran's nuclear ambitions. The action unsettles world oil markets.

> *"The restrictions that we faced 10 years ago [regarding access to oil and gas] are the same that we face today."*
>
> *—WesternGeco Schlumberger Pres. Dalton Boutte*

FEBRUARY

Exploration/Development

- WoodMac estimates U.S. Rocky Mountain region gas reserves could exceed 130 tcf by 2030 if growth continues at rate indicated by 30 years of statistics.

Production

- Saudi Arabia begins production from Haradh oilfield. Saudi Aramco says field will produce 300,000 b/d by mid-2006.
- Continuing conflict in Niger Delta forces Royal Dutch Shell to extend force majeure on its crude exports from Nigeria.

Refining/Petrochemicals

- Militants in Saudi Arabia attack Abqaiq oil processing terminal. No serious damage is done, Nymex oil prices jump 4% in a single day.

Government

- International Atomic Energy Agency votes to refer Iran to UN Security Council over that country's nuclear program. Referral could lead to sanctions reducing Iran's oil exports.

MARCH

Production

- Oil production starts from Chinguetti field off Mauritania. Woodside Energy estimates reserves at 120 million bbl and eventual output of 75,000 b/d.

Pipelines/Transportation

- Kinder Morgan Energy Partners and Sempra Pipelines & Storage secure shipper commitments for Rockies Express Pipeline, a $4 billion project to transport gas from Wyoming and Colorado to eastern Ohio.
- BP workers at Prudhoe Bay discover pipeline leak that forces shut-in of 100,000 b/d of crude production. Leak is in a gathering line that brings oil to the Trans-Alaskan Pipeline.
- Russia's environmental regulator approves plans to build 2,500-mile (4,025-km) pipeline from interior to Pacific coast. Transneft plans to begin construction in 2006.
- Enbridge delivers first crude from oil sands in western Canada to Cushing, OK refinery.

> *"The oil industry needs to take a proactive stand on the climate change debate."*
>
> *—OPEC Director of Energy Studies Mohammed Hamel*

Government

- G8 energy ministers call for increasing reliability of world energy supplies and improving transparency in world oil markets. The general G8 meeting in July in St. Petersburg focuses on energy security.

APRIL

Markets

- On April 21, front-month price of West Texas Intermediate (WTI) crude oil on Nymex closes above $75 for first time, driven by rising global demand, geopolitical instability.
- In response to rising U.S. gasoline prices and shortages of some grades, President George Bush authorizes exemptions from environmental rules on gasoline sold in some states.

Exploration/Development

- Chevron Canada Ltd. and partners suspend negotiations with Newfoundland and Labrador on development of Hebron oil field due to disagreements on fiscal terms.

Production

- Constitution and Ticonderoga fields in south Green Canyon area of Gulf of Mexico begin producing oil and gas. Initial flows are 32,000 b/d and 30 MMcf/d.

Natural Gas/LNG

- Planta de Regasificacion de Sagunto SA starts up new LNG terminal in Sagunto, Spain, with a capacity of 3 million tpy.

Pipelines/Transportation

- Sudan opens new 900-mile pipeline to carry oil from Melut Basin to export terminal at Port Sudan. Production from Sudan expected to reach about 500,000 b/d by year-end 2006, up from 363,000 b/d in 2005.

- Crosstex Energy completes North Texas Pipeline (NTPL). Extending more than 140 miles across six Texas counties, it transports gas produced from the Barnett shale.

Refining/Petrochemicals

- BP resumes gasoline production at Texas City refinery. Plant was closed September 2005 in anticipation of Hurricane Rita.
- Esso SAF shuts down Port Jerome/NDG 233,000 b/d refining and petrochemical complex near Le Havre in Normandy for €25 million maintenance program over several weeks.

Government

- State-owned PdVSA takes control of two oil fields in Venezuela operated by Total and Eni because companies do not agree to revised operating contracts under new joint venture structure established by government.
- New Ecuador hydrocarbons law imposes additional royalties on private oil producers. The companies threaten legal action, warn of impact on new oil-sector investment.

MAY

Exploration/Development

- Pioneer Natural Resources Alaska prepares to drill at Oooguruk project, 6 miles offshore Alaska near Colville River Delta. Estimated reserves are 50–90 million boe.

Production

- Nexen announces plans to boost coalbed methane production to more than 30 MMcfd by yearend from the first commercial CBM project in the Upper Mannville coals in central Alberta.
- First commercial offshore oil production begins in Perth basin off Western Australia in Cliff Head oil field. Reserves estimated at 15.4 million bbl.
- Third-phase expansion units came on stream at Syncrude Canada Ltd.'s oil sands plant in Fort McMurray, Alberta, boosting capacity to 350,000 b/d of synthetic crude.

Natural Gas/LNG

- China receives first shipment of imported LNG (from Australia) at its first receiving terminal, in Guangdong. Chevron has 25-year contract to deliver 3.3 million tpy.

Government

- Bolivian Pres. Evo Morales begins nationalization of natural gas industry, demanding that all private companies turn over their production to state-owned YPFB.
- United States restores full diplomatic relations with Libya for first time since 1980. It is final normalization step since Libya said in 2003 it would stop developing weapons of mass destruction. U.S. trade sanctions also end.
- Ecuador revokes contract of Occidental Petroleum, largest private oil producer in the country, alleging illegal sale of some production assets without approval. Petroecuador plans to take control of Occidental's production facilities.

JUNE

Exploration/Development

- Husky Oil China Ltd. announces first deepwater discovery and deepest well offshore China, in Block 26/29, 155 miles (250 km) south of Hong Kong.

Natural Gas/LNG

- FERC approves plans for five LNG projects: construction of Crown Landing (New Jersey), Creole Trail (Louisiana), Port Arthur (Texas), and Vista del Sol (Texas) facilities; and expansion of Sabine Pass project (Louisiana).
- WestPac LNG Corp. begins regulatory review for proposed $350 million LNG terminal on Ridley Island near Prince Rupert, British Columbia.

Pipelines/Transportation

- Kazakhstan, Azerbaijan agree to let Kazakhstan export crude oil through Baku-Tblisi-Ceyhan (BTC) pipeline. Kazakhstan will ship up to 500,000 b/d of crude oil to Baku by tanker across Caspian Sea, where the oil will enter BTC system.

> *"The scarcest resource today is trained, experienced people."*
>
> *—Transocean Inc. Pres. and CEO Robert L. Long*

- Iraq resumes crude exports through Kirkuk-Ceyhan pipeline for first time in four months. Militant attacks have hindered export flows through the pipeline, which normally carries 800,000 b/d.

Refining/Petrochemicals

- Nearly all U.S. refiners have started producing ultralow-sulfur diesel (ULSD) with no more than 15 ppm sulfur.
- Qatar brings $950 million Oryx GTL plant online at Ras Laffan. Called first commercial GTL facility, it will produce 34,000 b/d of product, mostly ULSD, from gas from Qatar's North field.

JULY

Markets

- West Texas Intermediate crude for August delivery hits $78.40/bbl July 14 on Nymex. Record price reflects tensions over Israeli incursion into Lebanon to battle Hezbollah, Iran's continued nuclear standoff with West, militant attacks in Nigeria, falling U.S. crude oil inventories.

Exploration/Development

- API reports high crude oil prices pushed U.S. drilling activity to highest level in 20 years during first half 2006, with 24,729 oil wells, gas wells, and dry holes completed.
- Canada-Newfoundland and Labrador Offshore Petroleum Board ups its oil recovery estimates for Hibernia field and Hebron complex to 2.751 billion bbl.

Production

- Husky Energy brings fifth well online, increasing production from White Rose oil field off Newfoundland to 110,000 b/d.

Pipelines/Transportation

- Columbia Gas Transmission moves in-service date for proposed Millennium Phase 1 gas pipeline project back one year, to November 1, 2008.
- China receives first crude oil through new 614-mile (988-km) pipeline from Atasu in northern Kazakhstan to Alashankou (Xinjiang region). Initial capacity is 210,000 b/d, growing to 400,000 b/d by 2010.
- Azerbaijan, Georgia, Turkey inaugurate 1,087-mile (1,750-km) Baku-Tblisi-Ceyhan (BTC) oil pipeline, carrying 1 million b/d of crude from Caspian Sea to world markets.

Companies

- Repsol YPF SA agrees to pay BP $2.15 billion for 28% interest in Shenzi oil field in the deepwater Green Canyon area, Gulf of Mexico.
- ConocoPhillips exercises option to acquire 24% interest in proposed $4.4 billion Rockies Express gas pipeline, scheduled for completion in 2009.

Refining/Petrochemicals

- Fire at Venezuela's Paraguaná refining complex shuts down 190,000-b/d distillation unit for 5–7 months. Facility is largest crude oil refining complex in the world (940,000 b/d).

AUGUST

Exploration/Development

- U.S. Minerals Management Service calls Western Gulf of Mexico Lease Sale 200 "best in nine years." High bids from 62 companies total $340.9 million.

Production

- ExxonMobil pumps first crude oil for export at DeKastri terminal at its Sakhalin-1 project. Project is expected to produce up to 250,000 b/d by year-end.

Natural Gas/LNG

- Excelerate Energy's Excelsior LNG regas carrier transfers about 13,000 cu m of LNG to sister ship Excalibur in first-ever full-scale, ship-to-ship LNG transfer. The transshipment is carried out in international waters in Gulf of Mexico.
- Royal/Dutch Shell PLC starts up its Terminal de LNG de Altamira, near Tampico, Mexico.

> *"Rising global average temperature and growing atmospheric concentrations of carbon dioxide—however incomplete science may be about the correlation of the phenomena—cannot be ignored."*
>
> *—BP PLC VP, Group Technology Ahmed Hashmi*

- Woodside Natural Gas applies for permits to build and operate deepwater OceanWay LNG project, with two delivery buoys moored 20 miles offshore near Los Angeles.

Pipelines/Transportation

- BP shuts in production at 400,000-b/d Prudhoe Bay oil field when internal corrosion causes leaks in a transit line. Workers quickly restore 200,000 b/d of production in western half of field where no pipeline damage was found.

Companies

- Russian court declares OAO Yukos bankrupt in the oil giant's ongoing legal battle with government of Russia regarding back taxes.

Government

- DOE report says hurricanes Katrina and Rita (2005) caused biggest disruption ever to oil and gas operations in Gulf of Mexico. One or both storms hit 3,050 platforms and 22,000 miles of pipelines, damaged 47 major gas processing plants and 17 NGL sites.
- United Nations brokers ceasefire between Israel and Hezbollah militants in southern Lebanon. Month-long conflict threatened to disrupt oil supplies from Middle East.
- Chad orders Chevron and Malaysia's Petronas to leave country, alleging failure to pay nearly $500 million in taxes. Chad later establishes commission to resolve dispute.

SEPTEMBER

Markets

- Price of West Texas Intermediate for October delivery drops to $60.46/bbl, a 6-month low. Fall reflects increases in distillate stocks, continuing negotiations with Iran on its nuclear program, end of summer driving, and increases in non-OPEC crude supply.

Exploration/Development

- Chevron-led consortium announces promising production tests at Jack #2 well in Gulf of Mexico deepwater, 270 miles southwest of New Orleans. Chevron says Lower Tertiary tapped by the well could hold 3–15 billion barrels of recoverable reserves.

Production

- Petro-Canada announces plan to resume production from Terra Nova oil field off Newfoundland and Labrador after maintenance work that began in May.

Pipelines/Transportation

- BP announces plans to increase production from eastern section of Prudhoe Bay oil field to 150,000 b/d. That will bring total production from the field to 400,000 b/d, highest rate since August, when workers found oil leaking from corroded transit line.
- Pipeline explosion near Bazargan, Iran halts natural gas shipments from Iran to Turkey. Officials from both nations blame sabotage. Iran is contracted to supply 777 MMcfd to Turkey through the pipeline.

Government

- European Union says Bulgaria and Romania are sufficiently prepared to become EU members in January 2007.

OCTOBER

Markets

- Crude futures prices rally from 10-month low, buoyed by reports of first decline in U.S. distillate inventories since early August. November contract for benchmark U.S. sweet crudes hits 2006 low $57.22/bbl in overnight Nymex trading October 12.

Exploration/Development

- Gazprom says it will develop $20 billion Shtokman natural gas field alone, rather than award a Production Sharing Agreement to consortium of foreign companies.

Production

- OPEC members agree to reduce production by 1.2 million b/d (to 26.3 million b/d) effective November 1, 2006. First official production cut since April 2004.

> *"In a peak oil world, we need everything, everywhere, all the time."*
>
> *—Peabody Energy Corp. Sr. VP, Government Affairs Fred Palmer*

- Royal Dutch Shell resumes pumping 18,000 b/d at Cawthorne Channel station in Nigeria, which was closed after militant attack in June.
- China confirms media reports that it has begun to fill its first strategic petroleum reserve facility at Zhenhai.

Pipelines/Transportation

- The 540-km (340-mile) southern leg of subsea Langeled gas pipeline opens, carrying gas from Sleipner platform in Norwegian North Sea to Easington, England.

Companies

- Canadian Natural Resources Ltd. plans to buy Anadarko Canada Corp. for $4.24 billion. ACC will keep its Mackenzie Delta and other Canadian arctic frontier properties.
- Royal Dutch Shell PLC offers to buy for $7.7 billion the shares of Shell Canada Ltd. that it does not already own.
- EnCana Corp. plans a $10 billion joint venture with ConocoPhillips to increase output from EnCana's Canadian oil sands eightfold and expand refineries in Texas, Illinois.

Refining/Petrochemicals

- Fire cuts output from Mazeikiu Nafta refinery in Lithuania to less than half nameplate capacity of 197,000 b/d. Facility is only refining complex in Baltic Sea region.
- U.S. DOE begins evaluating bids for first commercial U.S. cellulosic ethanol production facility aiming for first operational plant in 2012.
- Fire temporarily closes Jamnagar oil refining complex in Gujarat, India for four days. It is largest refinery in Asia and third largest in world (660,000 b/d).

Government

- North Korea tests a nuclear device underground, escalating geopolitical tensions.

> *"By the end of the decade, there will be $100 billion of investments planned, under way, or completed in Canadian oil sands."*
>
> *—Syncrude Canada Ltd. CEO Charles Ruigrok*

NOVEMBER

Markets

- In mid-November, the December contract for benchmark U.S. crudes expires at $55.81/bbl on Nymex, lowest close in 17 months. But colder Midwest and Northeast weather drives January crude contract up to $63.13/bbl on November 30.

Exploration/Development

- Husky Energy completes White Rose delineation, increases estimate of recoverable oil by 190 million bbl. Field is 350 km east of St. John's, Newfoundland.
- Shenzi group led by BHP Billiton agrees to buy Anadarko Petroleum's Genghis Khan oil and gas development in deepwater Gulf of Mexico for $1.35 billion.

Production

- Mortar attack strikes crude oil processing plants at Iraq's Kirkuk field, cutting output to 100,000 b/d from 300,000 b/d. Kirkuk crude flows to Ceyhan, Turkey by pipeline.
- Commercial bitumen production begins at Joslyn project in Athabasca region of Alberta. Expected cumulative production is 2 billion bbl.

Pipelines/Transportation

- BP indefinitely postpones resumption of crude exports through Baku-Supsa pipeline between Caspian and Black Seas after finding corrosion damage. Pipeline typically moves 155,000 b/d from Azerbaijan to Supsa.

Refining/Petrochemicals

- BP signs contracts to initiate $3 billion in upgrades to its Whiting, Indiana refinery. Goal is to boost capacity for processing Canadian heavy oil.

Government

- In U.S. Congressional elections, Democrats win 51–49 margin in the Senate and 233–202 margin in the House of Representatives. New Congress will convene January 4, 2007.

DECEMBER

Exploration/Development

- Pres. George Bush signs law that re-opens for oil and gas exploration several tracts in eastern Gulf of Mexico that had been withdrawn in 2001. It also lets coastal states share in federal OCS leasing revenue for first time.
- Royal Dutch Shell, Mitsui, Mitsubishi sign protocol that brings Gazprom into Sakhalin Energy Investment Company, which controls Sakhalin II oil and natural gas project.
- U.S. MMS okays Petrobras subsea development plan for Cascade, Chinook oil fields in Gulf of Mexico. First-ever approval for proposed use of FPSO vessel in gulf.

Production

- At Nigeria meeting, OPEC members vote to cut production by 500,000 bbl/d starting February 1, 2007.
- Second Alpine satellite oil field, Nanuq, comes online on Alaska's North Slope; expected peak production 15,000 b/d. First satellite (Fiord) came online in August.
- Production begins from Shah Deniz gas field in Caspian Sea. Expected initial production plateau is 300 MMcfd. Gas moves through South Caucasus Pipeline.

Pipelines/Transportation

- First commercial delivery of gas begins through Balgzand-Bacton pipeline from the Netherlands to the UK. At full capacity in 2007, it will deliver 15 billion cu m/year.
- Vandalized fuel pipeline in Lagos, Nigeria explodes, killing 269. People had come to site to collect leaking fuel.

Refining/Petrochemicals

- Petro-Canada Oil Sands files application to build, operate heavy oil upgrader northeast of Edmonton. Will eventually convert 340,000 b/d of bitumen from Fort Hills mine and other sources into synthetic crude oil.

Companies

- Statoil and Norsk Hydro, Norway's two largest oil and natural gas companies, agree to merge, pending required approvals. Merged company would be world's largest offshore oil and natural gas operator.

Government

- U.N. Security Council passes resolution aimed at curbing Iran's nuclear program, imposes nuclear-related trade sanctions.

> *"Until a few years ago, the industry did not have the technology in place to find or develop resources found in deeper formations in the ultradeep waters of the gulf."*
>
> *—Devon Energy Corp. Sr. VP, Exploration and Production Steven Hadden*

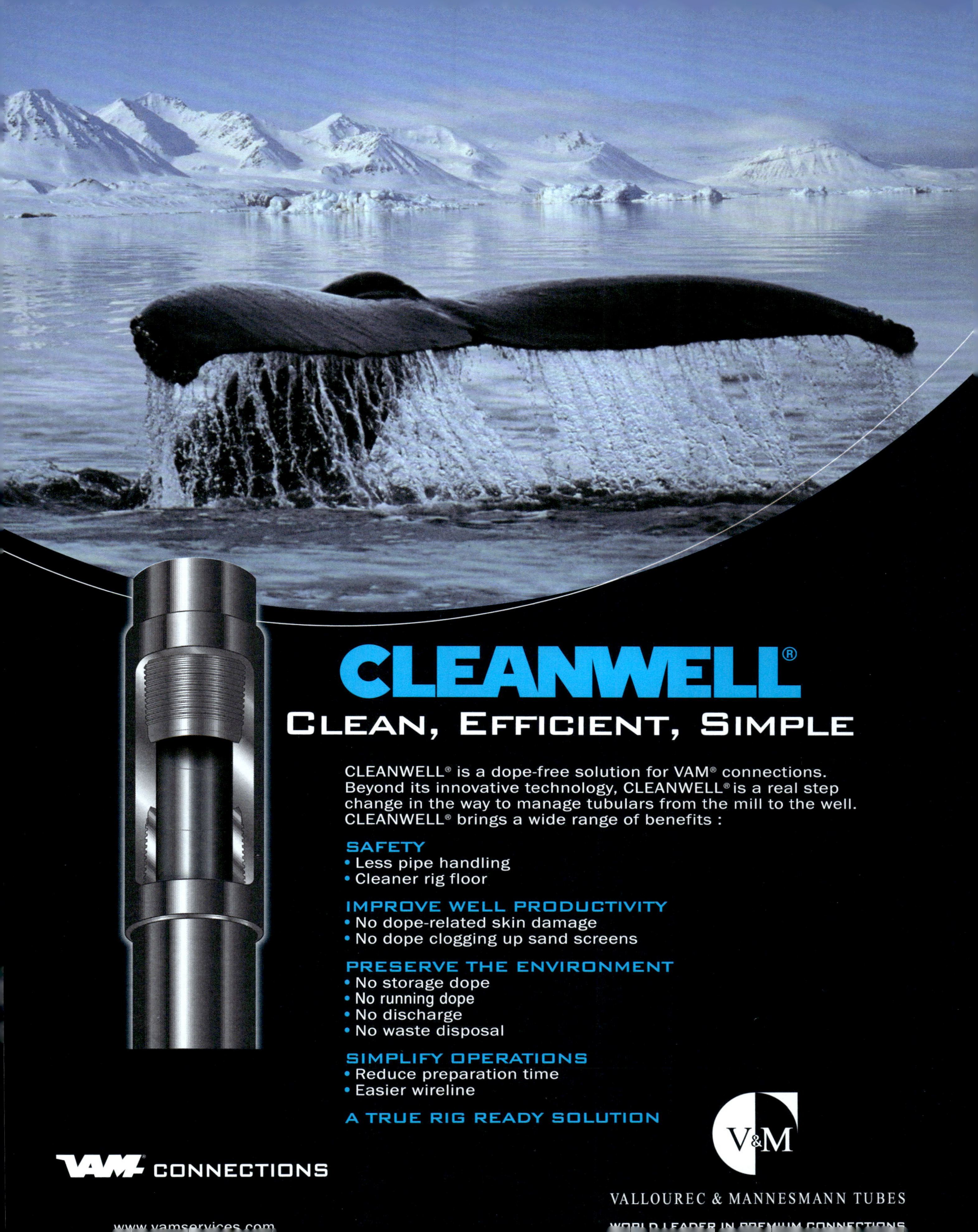
CLEANWELL®
CLEAN, EFFICIENT, SIMPLE
CLEANWELL® is a dope-free solution for VAM® connections. Beyond its innovative technology, CLEANWELL® is a real step change in the way to manage tubulars from the mill to the well. CLEANWELL® brings a wide range of benefits :
SAFETY
• Less pipe handling
• Cleaner rig floor
IMPROVE WELL PRODUCTIVITY
• No dope-related skin damage
• No dope clogging up sand screens
PRESERVE THE ENVIRONMENT
• No storage dope
• No running dope
• No discharge
• No waste disposal
SIMPLIFY OPERATIONS
• Reduce preparation time
• Easier wireline
A TRUE RIG READY SOLUTION
VAM® CONNECTIONS
V&M
VALLOUREC & MANNESMANN TUBES
www.vamservices.com

North America

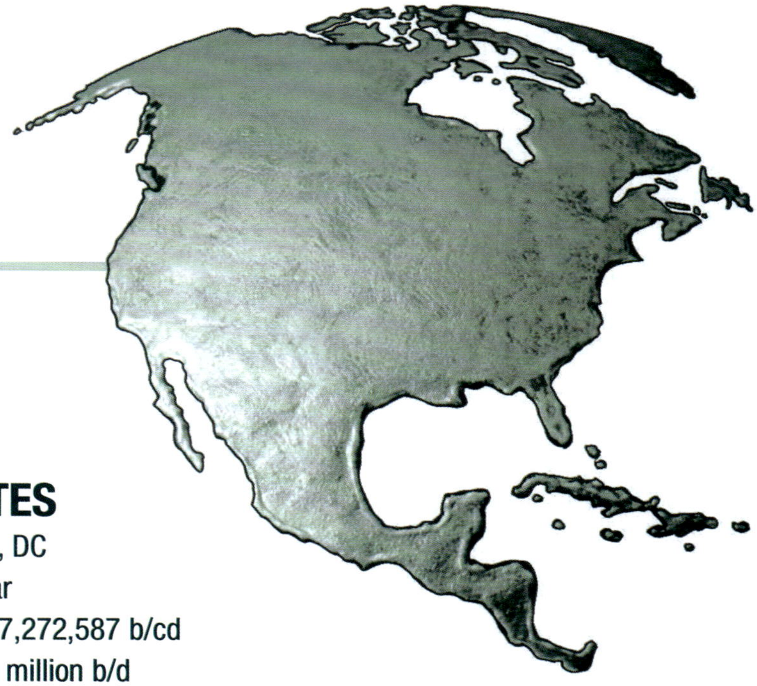

UNITED STATES

Capital: Washington, DC
Monetary unit: Dollar
Refining capacity: 17,272,587 b/cd
Oil production: 5.14 million b/d
Oil reserves: 21.757 billion bbl
Gas reserves: 204.385 tcf

Overview

The year 2006 saw dramatic swings in oil and natural gas prices, driven in large measure by geopolitical unrest in key regions.

Energy markets reflected concerns about continued violence in Iraq, the interest of Iran in uranium enrichment, the abduction of oil-industry workers in Nigeria by militants, conflict between Hezbollah militants and Israel in southern Lebanon, and the underground detonation of a nuclear device by North Korea.

Falling crude oil inventories in the United States also contributed to high prices in world oil markets.

West Texas Intermediate crude, priced at $60.75 on January 6, hit a record $78.40/bbl on July 14, before dropping again due to concerns about slower economic growth and rising inventories.

On November 17, the December contract for benchmark U.S. crudes expired at $55.81/bbl on the New York Mercantile Exchange (Nymex), the lowest closing price in 17 months.

Later in the month, however, arctic cold struck the U.S. Midwest and East Coast. By November 30, the January crude contract was trading at $63.13/bbl and the January gas contract traded as high as $9.05/MMbtu.

The U.S. oil and gas industry made great progress in rebuilding its Gulf Coast infrastructure in 2006, in the aftermath of Hurricanes Katrina and Rita (August 2005). Further, the 2006 hurricane season was much quieter than expected. Although eight to 10 major storms were predicted, only five occurred, none affecting Gulf Coast oil and gas operations.

North America

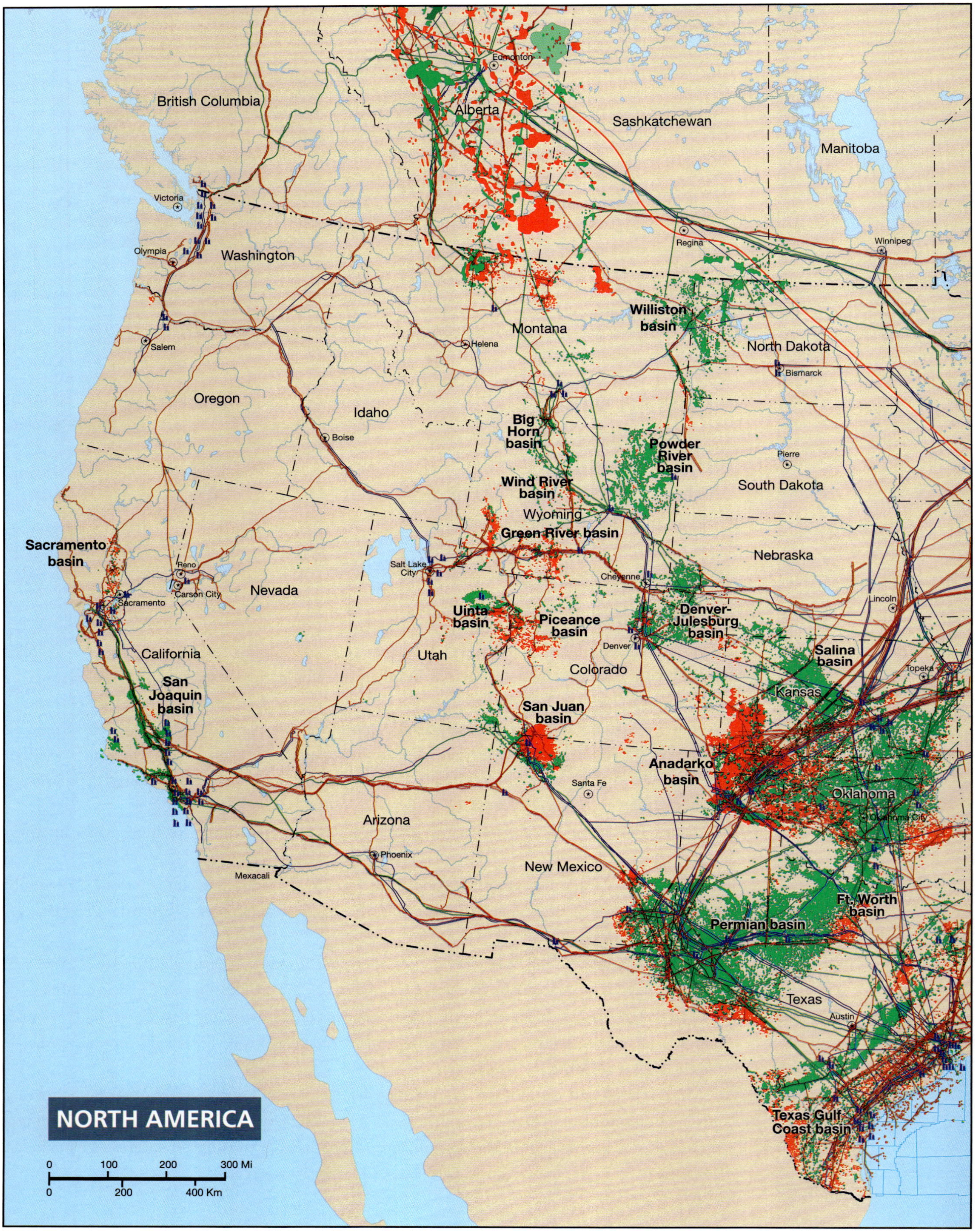

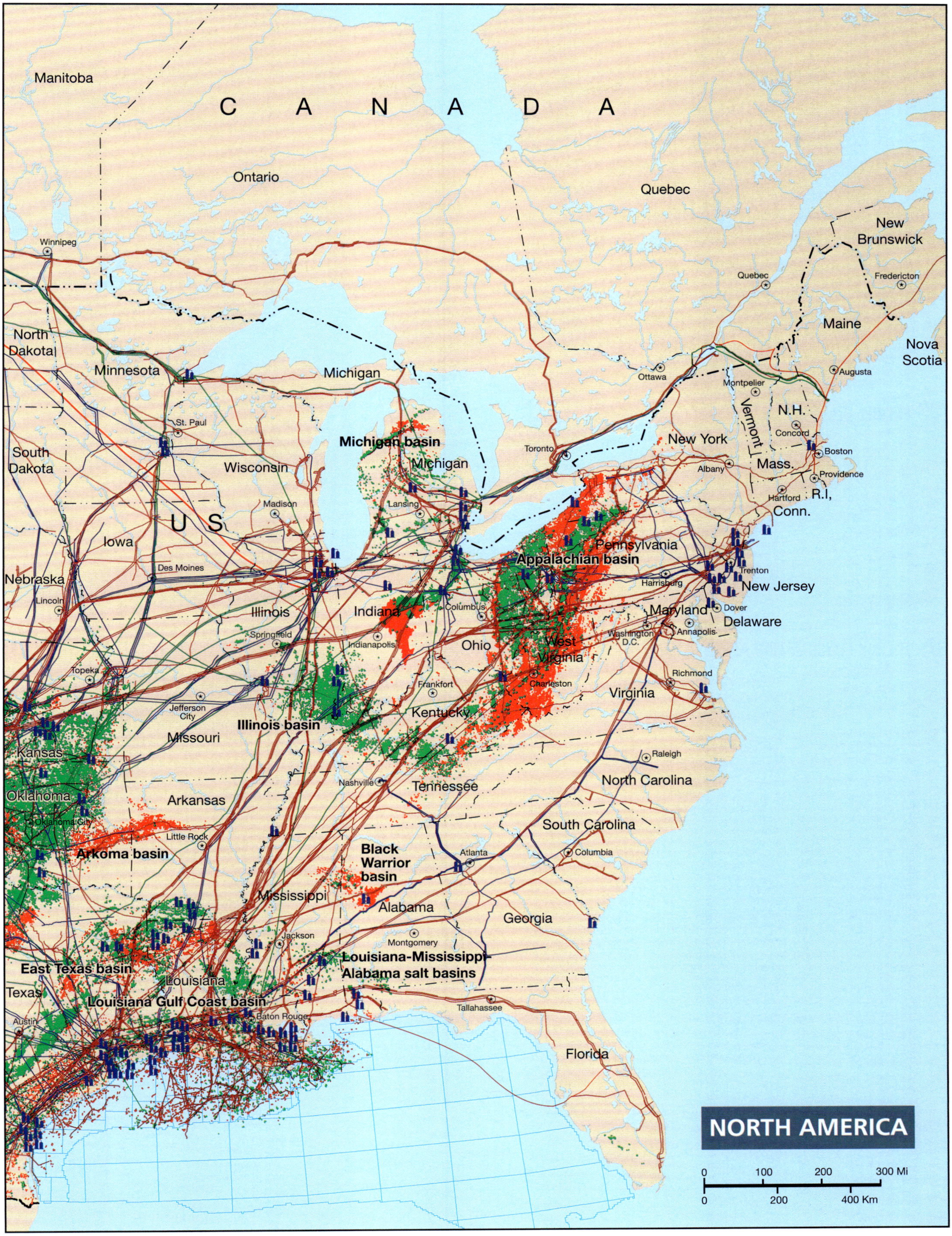
CANADA
Manitoba
Ontario
Quebec
New Brunswick
Nova Scotia
Maine
N.H.
Vermont
Mass.
R.I.
Conn.
New York
Pennsylvania
New Jersey
Delaware
Maryland
West Virginia
Virginia
North Carolina
South Carolina
Georgia
Florida
Alabama
Mississippi
Tennessee
Kentucky
Ohio
Indiana
Illinois
Michigan
Wisconsin
Minnesota
Iowa
Missouri
Arkansas
Louisiana
Texas
Oklahoma
Kansas
Nebraska
North Dakota
South Dakota
US
Michigan basin
Appalachian basin
Illinois basin
Arkoma basin
Black Warrior basin
East Texas basin
Louisiana Gulf Coast basin
Louisiana-Mississippi-Alabama salt basins
Winnipeg
St. Paul
Madison
Lansing
Des Moines
Lincoln
Topeka
Jefferson City
Springfield
Indianapolis
Columbus
Frankfort
Nashville
Little Rock
Oklahoma City
Jackson
Montgomery
Atlanta
Columbia
Raleigh
Richmond
Charleston
Washington D.C.
Annapolis
Dover
Trenton
Harrisburg
Albany
Hartford
Providence
Boston
Concord
Montpelier
Augusta
Fredericton
Quebec
Ottawa
Toronto
Baton Rouge
Austin
Tallahassee
NORTH AMERICA
0 100 200 300 Mi
0 200 400 Km

Short-term U.S. outlook

U.S. energy demand will increase in 2007, said *Oil & Gas Journal* (*OGJ*), as the economy grows at a tempered pace. Winter weather will be milder than normal, requiring less heating, but temperatures in the summer are also expected to be higher than usual, requiring more cooling.

Electric power generation will consume much of the energy used in the United States in 2007. *OGJ* forecasts increases in consumption across all forms of energy. Oil, natural gas, coal, nuclear, and renewable energy each will gain small demand increments from a year ago.

Transportation fuel consumption will push U.S. oil demand in 2007, due to slightly lower prices, economic growth, and growing volumes of ethanol entering the gasoline market.

U.S. production of crude and condensate will move up nearly 2% after declining in 2006. Oil production will increase worldwide.

U.S. natural gas supply will be up just 1% from 2006. Gas demand will be up by less than that, though, despite the need for more electricity generated from gas. Inventories will decline a bit on stronger demand in the summer.

US ENERGY DEMAND

	2005	2006	Change, % 2006/05	2007 Trillion btu	Change, % 2007/06	% share of total energy 2005	2006	2007
	Trillion btu							
Oil	40,735	40,600	-0.3	41,030	1.1	40.6	40.3	40.2
Gas	22,636	22,610	-0.1	22,670	0.3	22.6	22.4	22.2
Coal	22,788	22,780	0.0	23,100	1.4	22.7	22.6	22.7
Nuclear	8,160	8,290	1.6	8,300	0.1	8.1	8.2	8.1
Hydro, other	6,039	6,525	8.0	6,850	5.0	6.0	6.5	6.7
Total	**100,358**	**100,805**	**0.4**	**101,950**	**1.1**	**100.0**	**100.0**	**100.0**

Sources: 2005 US Energy Information Administration; 2006 and 2007 OGJ estimate and forecast

US ENERGY CONSUMPTION AND EFFICIENCY

Year	GDP, billion 2000 $	Energy consumption, trillion btu	Energy consumption per GDP, 2000 $ (Mbtu)	Oil energy consumption, trillion btu	Oil energy consumption per GDP, 2000 $ (Mbtu)	Natural gas energy consumption, trillion btu	Natural gas energy consumption per GDP, 2000 $ (Mbtu)	Total oil and natural gas energy consumption, trillion btu	Total oil and gas energy consumption per GDP, 2000 $ (Mbtu)	Oil and natural gas energy % of total energy
1973	4,341.5	75,808	17.5	34,840	8.0	22,512	5.2	57,352	13.2	75.7
1974	4,319.6	73,991	17.1	33,455	7.7	21,732	5.0	55,187	12.8	74.6
1975	4,311.2	71,999	16.7	32,731	7.6	19,948	4.6	52,679	12.2	73.2
1976	4,540.9	76,012	16.7	35,175	7.7	20,345	4.5	55,520	12.2	73.0
1977	4,750.5	78,000	16.4	37,122	7.8	19,931	4.2	57,053	12.0	73.1
1978	5,015.0	79,986	15.9	37,965	7.6	20,000	4.0	57,965	11.6	72.5
1979	5,173.4	80,903	15.6	37,123	7.2	20,666	4.0	57,789	11.2	71.4
1980	5,161.7	78,289	15.2	34,202	6.6	20,394	4.0	54,596	10.6	69.7
1981	5,291.7	76,335	14.4	31,931	6.0	19,928	3.8	51,859	9.8	67.9
1982	5,189.3	73,234	14.1	30,231	5.8	18,505	3.6	48,736	9.4	66.5
1983	5,423.8	73,066	13.5	30,054	5.5	17,357	3.2	47,411	8.7	64.9
1984	5,813.6	76,693	13.2	31,051	5.3	18,507	3.2	49,558	8.5	64.6
1985	6,053.7	76,417	12.6	30,922	5.1	17,834	2.9	48,756	8.1	63.8
1986	6,263.6	76,722	12.2	32,196	5.1	16,708	2.7	48,904	7.8	63.7
1987	6,475.1	79,156	12.2	32,865	5.1	17,744	2.7	50,609	7.8	63.9
1988	6,742.7	82,774	12.3	34,222	5.1	18,552	2.8	52,774	7.8	63.8
1989	6,981.4	84,886	12.2	34,211	4.9	19,712	2.8	53,923	7.7	63.5
1990	7,112.5	84,605	11.9	33,553	4.7	19,730	2.8	53,283	7.5	63.0
1991	7,100.5	84,522	11.9	32,845	4.6	20,149	2.8	52,994	7.5	62.7
1992	7,336.6	85,866	11.7	33,527	4.6	20,835	2.8	54,362	7.4	63.3
1993	7,532.7	87,579	11.6	33,841	4.5	21,351	2.8	55,192	7.3	63.0
1994	7,835.5	89,248	11.4	34,670	4.4	21,842	2.8	56,512	7.2	63.3
1995	8,031.7	91,200	11.4	34,553	4.3	22,784	2.8	57,337	7.1	62.9
1996	8,328.9	92,446	11.1	35,757	4.3	23,197	2.8	58,954	7.1	63.8
1997	8,703.5	94,800	10.9	36,266	4.2	23,328	2.7	59,594	6.8	62.9
1998	9,066.9	95,200	10.5	36,934	4.1	22,936	2.5	59,870	6.6	62.9
1999	9,470.3	96,837	10.2	37,960	4.0	23,010	2.4	60,970	6.4	63.0
2000	9,817.0	98,976	10.1	38,404	3.9	23,916	2.4	62,320	6.3	63.0
2001	9,890.7	96,453	9.8	38,333	3.9	22,861	2.3	61,194	6.2	63.4
2002	10,048.8	97,967	9.7	38,401	3.8	23,628	2.4	62,029	6.2	63.3
2003	10,301.0	98,273	9.5	39,047	3.8	22,967	2.2	62,014	6.0	63.1
2004	10,703.5	100,415	9.4	40,594	3.8	23,036	2.2	63,630	5.9	63.4
2005	11,048.6	100,358	9.1	40,735	3.7	22,636	2.0	63,371	5.7	63.1
[1]2006	11,400.0	100,805	8.8	40,600	3.6	22,610	2.0	63,210	5.5	62.7
[2]2007	11,650.0	101,950	8.8	41,030	3.5	22,670	1.9	63,700	5.5	62.5

[1]Estimated. [2]Forecast.
Source: US Energy Information Administration

Economy, energy

OGJ forecasts that the U.S. economy will grow in 2007, but more slowly than last year. One big factor is the cooling housing market. Energy prices will be lower, and, according to the U.S. Bureau of Economic Analysis, personal income should continue to rise. Real gross domestic product (GDP) will climb 2.2% after a 3.2% gain in 2006.

U.S. energy demand will grow 1.1%. Consumption of all forms of energy, including oil, gas, coal, nuclear, and renewable energy, will total 101.95 quadrillion btu (quads). Energy efficiency will improve only slightly this year, declining to 8,750 btu/$ of GDP compared to 8,840 btu/$ in 2006.

Oil will remain the largest U.S. energy source. Representing 40.2% of U.S. energy demand, 2007 oil consumption will total 41 quads. This is up 1.1% from 2006, when oil demand contracted 0.3% on strong crude and product prices.

Natural gas consumption will be 22.67 quads, inching up from 2006, when demand recorded a negligible decline.

The largest percentage growth in demand will be for renewable energy, including wind and solar energy and hydroelectric power generation. Energy from renewable sources is expected to climb by 5% in 2007. However, this category will account for only 6.85 quads, or 6.7%, of U.S. energy use.

Oil

U.S. oil production will increase to average 5.25 million b/d, and production of natural gas liquids and liquid refinery gases will move up 1.7% to average 1.775 million b/d. U.S. crude and condensate production last year averaged 5.155 million b/d in 2006.

In 2006, declines in oil production in Alaska, Texas, and California overcame increases in crude and condensate production in Louisiana, New Mexico, and North Dakota. The decline in Alaska's oil production in 2006 was estimated at 11%, one of the largest drops in recent years. Louisiana's oil production recovered somewhat in 2006, reflecting recovery from the impact of Hurricane Katrina.

OGJ expects processing at refineries will remain strong in 2007, with throughputs, operable capacity, and utilization up a bit from last year. U.S. refining capacity is projected to increase marginally to 17.5 million b/d, with a utilization rate of 89.7%.

U.S. demand in 2007 for oil products will increase 1.1%, with consumption of each product type growing or being unchanged from 2006. Total U.S. demand for oil products will average 20.96 million b/d. U.S. exports of products and crude oil will decline 5% to 1.25 million b/d, bringing total demand for oil in the U.S. to 22.21 million b/d.

A weather-driven reduction in demand in January led to a drop in futures prices for crude. In the first two trading days of 2007, the front-month delivery price on the Nymex plunged more than $5/bbl.

Supply and demand for crude in the US

	[1]2006	2005	2004	2003	2002	2001	2000	1999	1998	1997
	1,000 b/d									
SUPPLY										
Crude imports[2]	10,120	10,126	10,088	9,665	9,140	9,328	9,071	8,731	8,706	8,225
Crude production	5,155	5,178	5,419	5,681	5,746	5,801	5,822	5,881	6,252	6,452
Unaccounted for crude	20	76	143	54	110	117	155	191	115	145
Total supply	**15,295**	**15,380**	**15,650**	**15,400**	**14,996**	**15,246**	**15,048**	**14,803**	**15,073**	**14,822**
DEMAND										
Crude refinery runs	15,245	15,220	15,475	15,304	14,947	15,128	15,067	14,804	14,889	14,662
Crude used directly and loss	—	—	—	—	—	—	—	—	—	2
Crude exports	23	32	27	12	9	20	50	118	110	108
Crude into SPR	11	25	102	108	134	26	-73	–11	22	–7
Total demand	**15,279**	**15,277**	**15,604**	**15,424**	**15,090**	**15,174**	**15,044**	**14,911**	**15,021**	**14,765**
Crude stock change (industry)	16	103	46	–24	–94	72	4	–108	52	57
Primary (industry)	330	324	286	269	278	312	286	284	324	305
SPR[3]	695	685	676	638	599	550	541	567	571	563
Total crude stocks (million bbl)	**1,025**	**1,009**	**962**	**907**	**877**	**862**	**827**	**851**	**895**	**868**

[1]Preliminary. [2]Includes imports for the Strategic Petroleum Reserve. [3]Includes Alaskan crude in transit.

North America

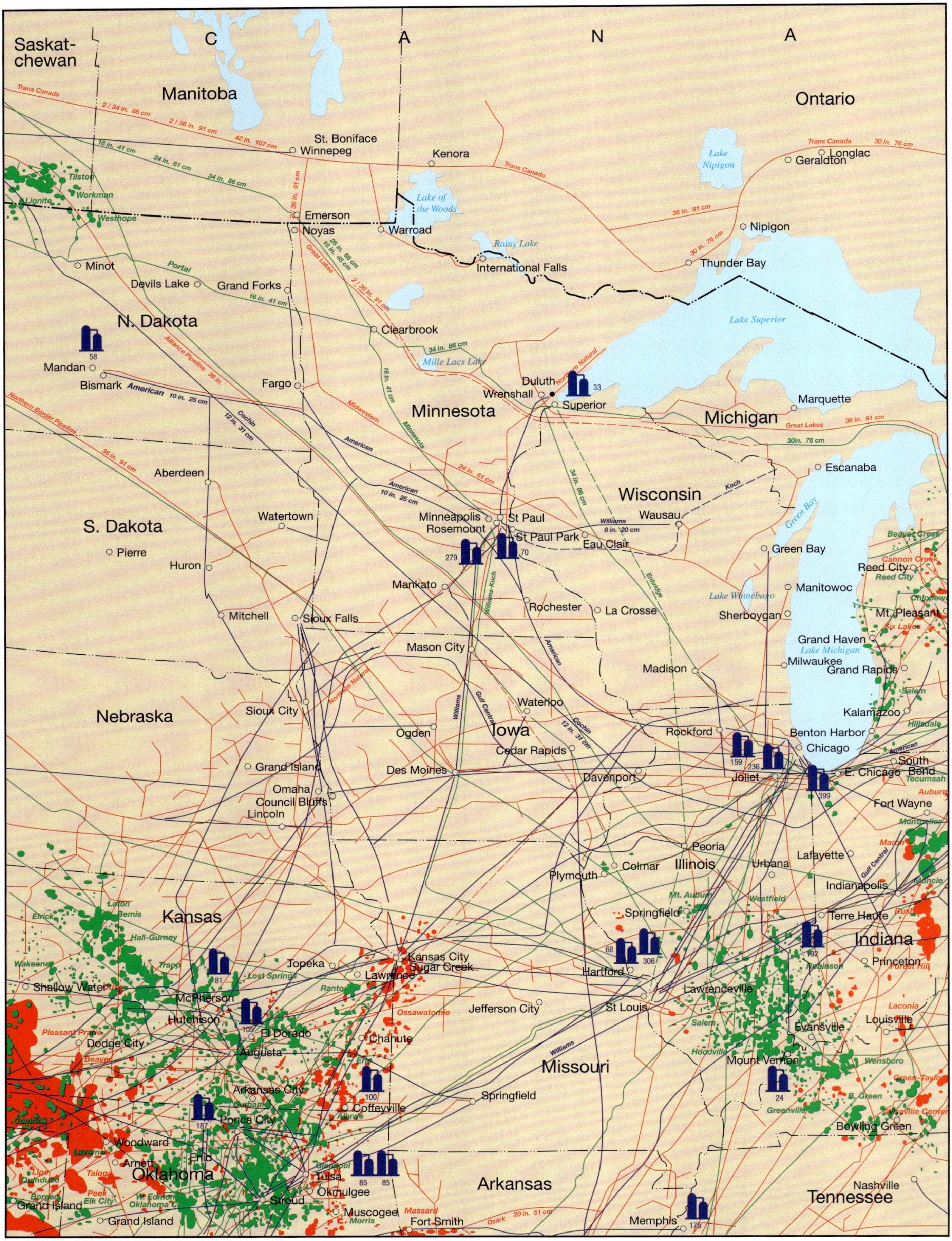

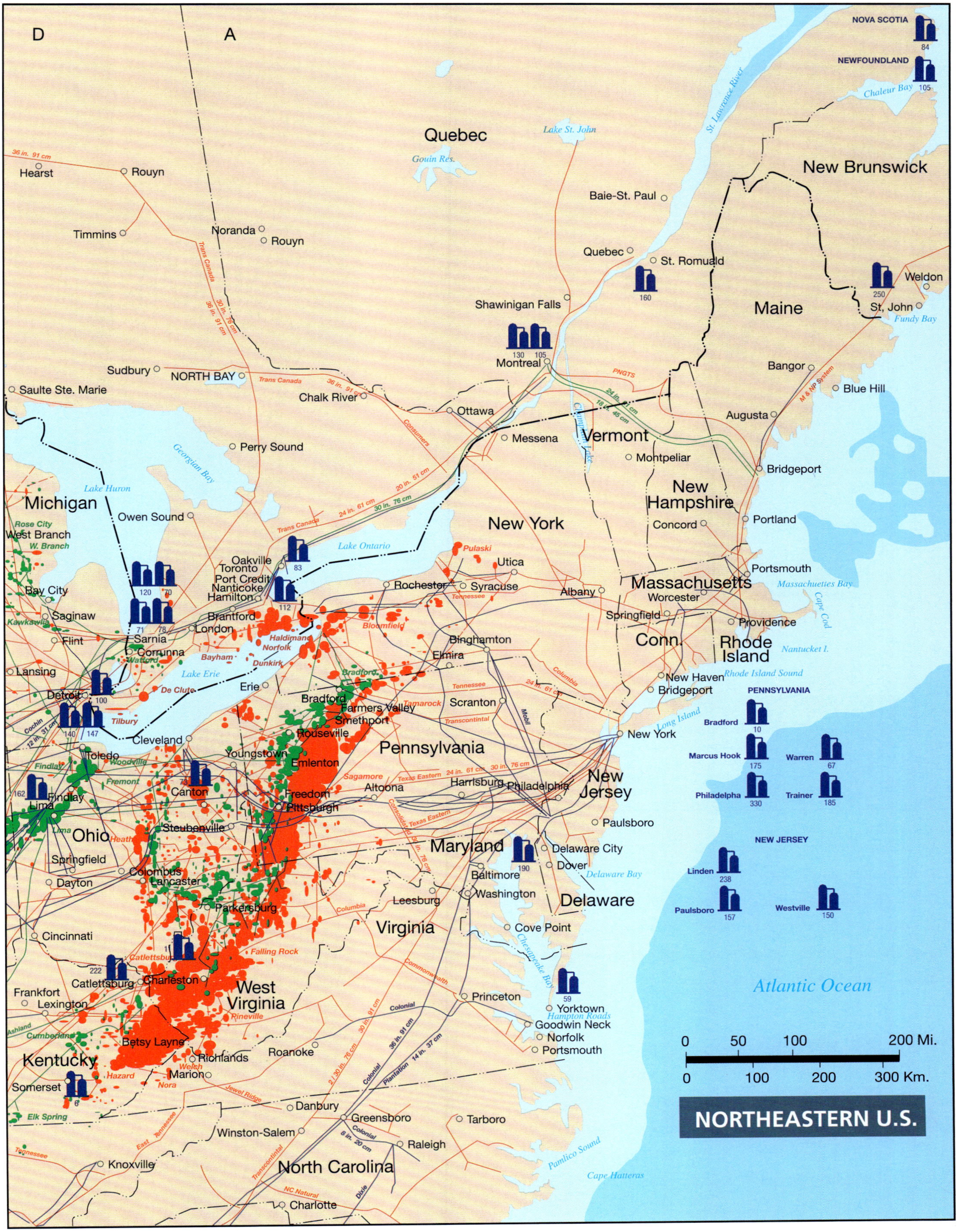
NORTHEASTERN U.S.
Atlantic Ocean
0 50 100 200 Mi.
0 100 200 300 Km.
NOVA SCOTIA
84
NEWFOUNDLAND
105
PENNSYLVANIA
Bradford
10
Marcus Hook
175
Warren
67
Philadelphia
330
Trainer
185
NEW JERSEY
Linden
238
Paulsboro
157
Westville
150
Quebec
New Brunswick
Maine
Vermont
New Hampshire
New York
Massachusetts
Conn.
Rhode Island
Pennsylvania
New Jersey
Maryland
Delaware
Virginia
West Virginia
North Carolina
Kentucky
Ohio
Michigan
Hearst
Rouyn
Timmins
Noranda
Baie-St. Paul
St. Romuald
Shawinigan Falls
Montreal
Sudbury
NORTH BAY
Saulte Ste. Marie
Chalk River
Ottawa
Messena
Montpelier
Augusta
Bangor
Blue Hill
Weldon
St. John
Bridgeport
Portland
Concord
Portsmouth
Worcester
Springfield
Providence
Albany
Utica
Syracuse
Rochester
Binghamton
Elmira
Scranton
New Haven
New York
Harrisburg
Philadelphia
Paulsboro
Delaware City
Dover
Baltimore
Washington
Leesburg
Cove Point
Yorktown
Goodwin Neck
Norfolk
Portsmouth
Princeton
Roanoke
Danbury
Greensboro
Tarboro
Raleigh
Winston-Salem
Charlotte
Knoxville
Perry Sound
Owen Sound
Oakville
Toronto
Port Credit
Nanticoke
Hamilton
Brantford
London
Sarnia
Corrunna
Erie
Bradford
Farmers Valley
Smethport
Rouseville
Emlenton
Youngstown
Cleveland
Canton
Freedom
Pittsburgh
Altoona
Steubenville
Columbus
Lancaster
Parkersburg
Charleston
Catlettsburg
Betsy Layne
Richlands
Marion
Somerset
Frankfort
Lexington
Cincinnati
Dayton
Springfield
Lima
Findlay
Toledo
Detroit
Lansing
Flint
Saginaw
Bay City
West Branch
Lake Huron
Georgian Bay
Lake Ontario
Lake Erie
Lake St. John
Gouin Res.
St. Lawrence River
Chaleur Bay
Fundy Bay
Massachuettes Bay
Cape Cod
Nantucket I.
Rhode Island Sound
Long Island
Delaware Bay
Chesapeake Bay
Hampton Roads
Pamlico Sound
Cape Hatteras
Trans Canada
Consumers
PNGTS
M & NP System
Tennessee
Texas Eastern
Transcontinental
Columbia
Colonial
Plantation
Commonwealth
Dixie
NC Natural
160
250
130
105
83
112
120
70
71
78
100
140
147
162
222
190
59
D
A

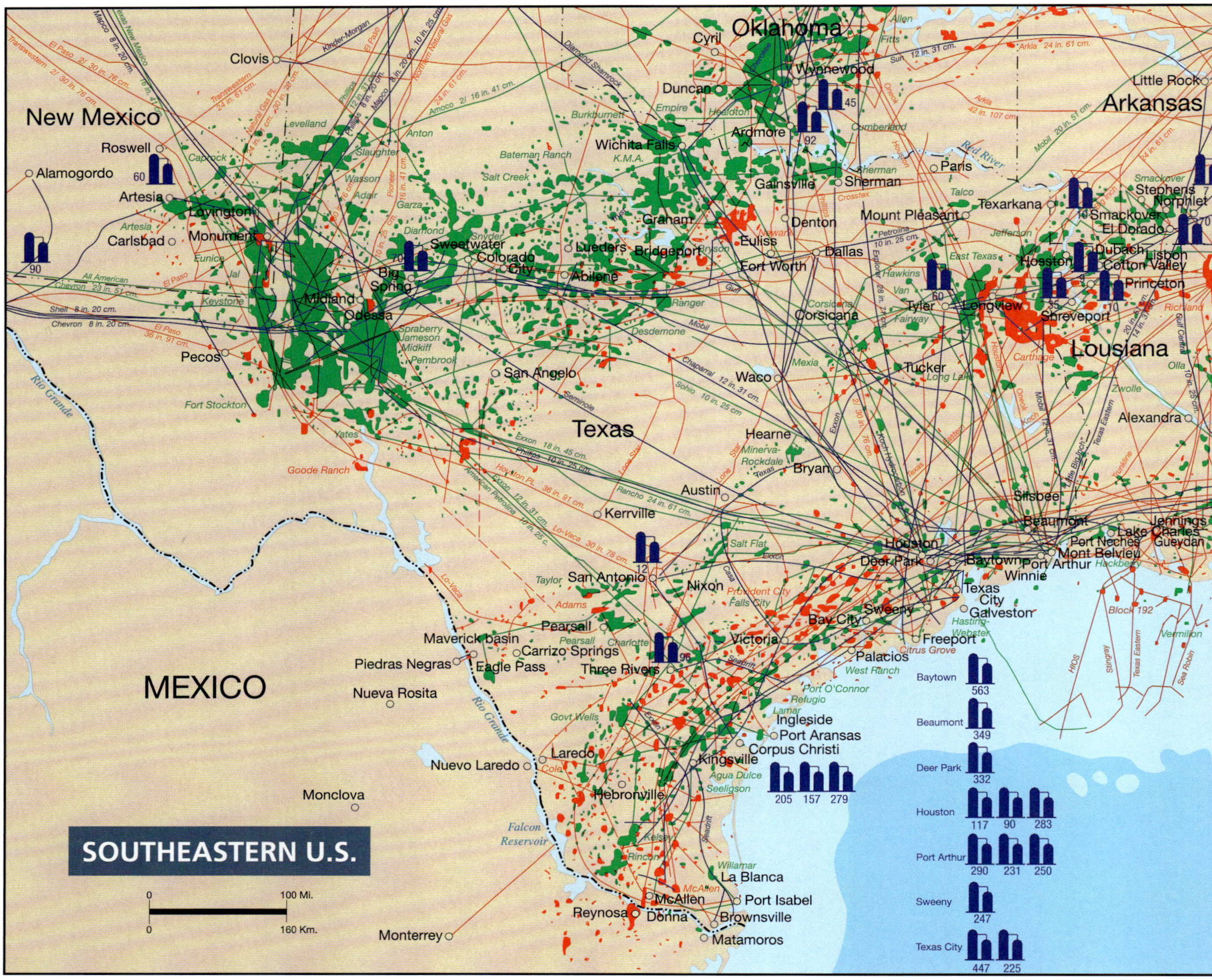

In 2006, demand for motor gasoline, distillate, liquefied petroleum gas, and ethane increased, but demand for jet fuel, residual fuel oil, and other products declined, resulting in a small overall decline for oil product demand. A factor gaining importance in the demand for oil products is the growth of ethanol in the U.S. energy mix.

Natural gas: OGJ view

OGJ forecasts that U.S. production and consumption of natural gas will grow modestly in 2007.

Offshore gas production from federal waters of the Gulf of Mexico will lead growth in U.S. production. New producing fields will slightly outpace decline rates. Gulf of Mexico production is expected to climb almost 3%. Gulf marketed production dropped 2.9% last year due to lingering damage sustained during the 2005 hurricane season.

Production in Texas and Louisiana will increase, though less than during 2006. And collective gas output from all other states will increase by 1.2%.

Through the first half of 2006, production was down in Alabama, Alaska, Florida, Kansas, New Mexico, and the federal offshore area. But big increases in production were posted in Texas, Wyoming, Oklahoma, Montana, Mississippi, and others to offset those declines.

LNG imports will total 570 tcf in 2007, little changed from 2006. U.S. LNG imports plunged 10% in 2006.

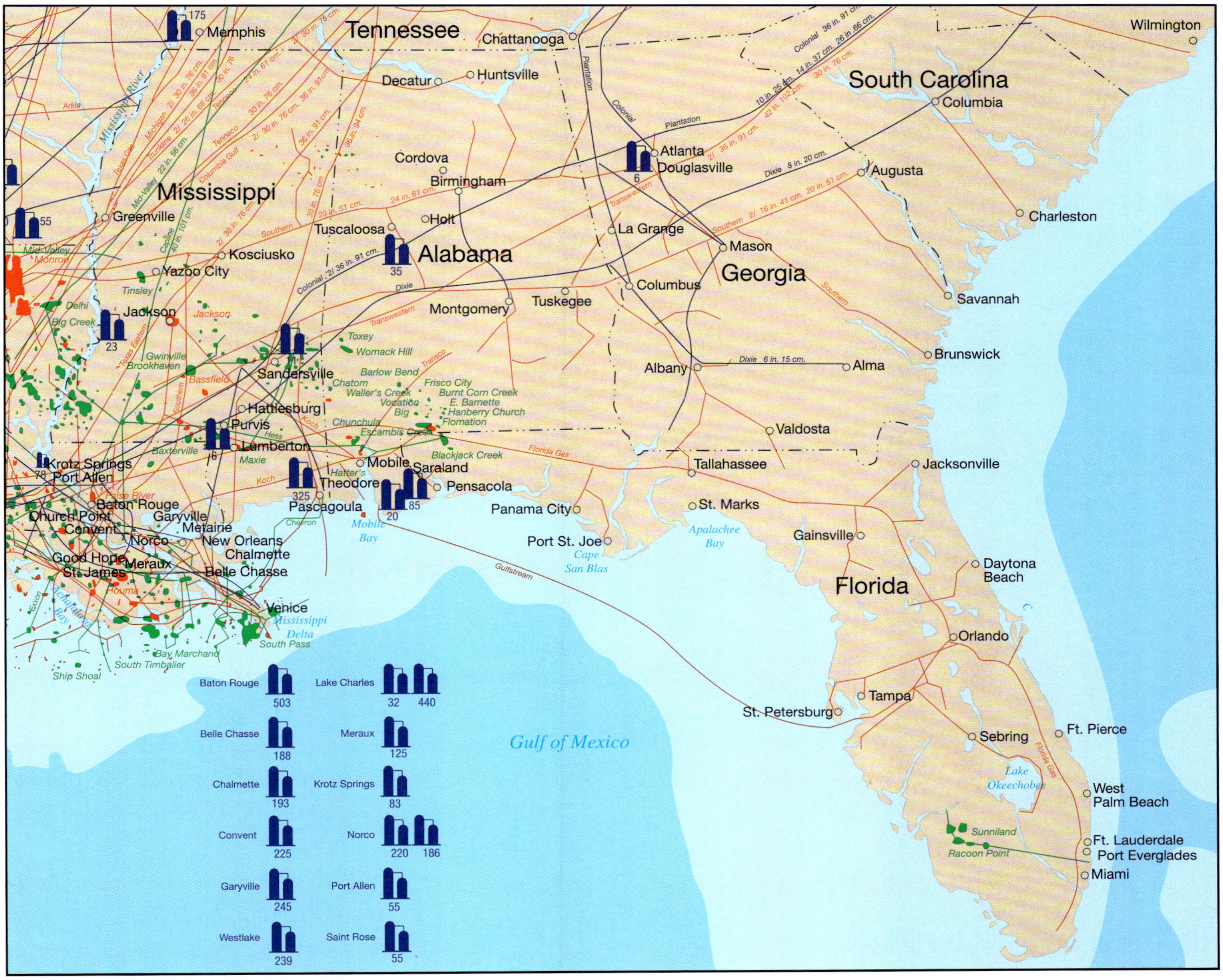

OGJ forecasts that U.S. consumption of gas this year will total 22.288 tcf, up 0.3% from last year. While the amount of gas consumed in the U.S. over the past few years has been fairly steady, gas use for power generation has been increasing, while gas demand by residential, commercial, and industrial consumers has declined.

Natural gas: NGSA, EIA

In late 2006 the Natural Gas Supply Association (NGSA) projected higher gas production and lower prices in the 2006–07 winter heating season. Onshore production—particularly expansions in the Rocky Mountains and Texas before Hurricanes Katrina and Rita hit in late 2005—contributed to the rebound.

The NGSA forecast, prepared by Energy & Environmental Analysis Inc. (Arlington, VA), suggests gas well completions could climb to 30,100/year in 2006 from 27,000/year in 2005.

For 2007, NGSA sees a continued steady increase in production, reflecting high rig counts and the startup of major new deepwater developments. U.S. gas production could climb to an average 50.7 bcfd from 40.9 bcfd, while imports from Canada decline to 8.4 bcfd from 9.2 bcfd. NGSA forecasts an increase in LNG imports to 2 bcfd from 1.5 bcfd.

EIA projected U.S. gas production to increase by 1.9% during 2007, according to its January 2007 Short-Term Energy Outlook.

2006 Congressional Elections

Democrats won narrow majorities in both the U.S. Senate and House of Representatives in midterm Congressional elections in November 2006. As the 110th Congress convened in January 2007, several energy-related issues were being discussed: a windfall profits tax on oil companies, changes to certain tax treatments and accounting procedures, deepwater royalty payments, access to federal lands, the Outer Continental Shelf, and the Arctic National Wildlife Refuge coastal plain for exploration, certain environmental regulations, reauthorization of the 2002 Pipeline Safety Act, and programs to support alternative energy resource development or energy research.

Democrats took over as chairs of the House Energy and Commerce Committee, the Senate Energy and Natural Resources Committee, and the Senate Environment and Public Works Committee.

US Natural Gas Supply and Demand

	2004	2005 bcf	2006	Change, % 06/05	2007 bcf	Change, % 07/06
Production						
Texas	5,067	5,255	5,550	5.6	5,650	1.8
Louisiana	1,353	1,296	1,380	6.5	1,400	1.4
Federal Gulf of Mexico	3,969	3,151	3,060	–2.9	3,150	2.9
Other states	9,128	9,249	9,410	1.7	9,520	1.2
Total production	**19,517**	**18,951**	**19,400**	**2.4**	**19,720**	**1.6**
Imports						
Canada	3,607	3,700	3,550	–4.1	3,200	–9.9
Mexico	0	9	3	–66.7	3	—
LNG	652	631	568	–10.0	570	0.4
Total imports	**4,259**	**4,341**	**4,121**	**–5.1**	**3,773**	**–8.4**
Supplemental gas	60	64	62	–3.1	65	4.8
Losses, etc.*	–479	–437	–390	–10.8	–470	20.5
Total new supply	**23,357**	**22,919**	**23,193**	**1.2**	**23,088**	**–0.5**
Supply from storage	–114	51	–300	—	100	—
Total supply	**23,243**	**22,970**	**22,893**	**–0.3**	**23,188**	**1.3**
Exports	854	729	680	–6.7	900	32.4
Total consumption	**22,389**	**22,241**	**22,213**	**–0.1**	**22,288**	**0.3**

*Extraction losses and unaccounted-for gas.
Sources: 2004 and 2005 EIA; 2006 and 2007 OGJ estimates and forecast

Global energy view: longer-term

U.S. DOE-EIA projections

The U.S. Energy Information Administration (EIA) presented in December an early release of its "Annual Energy Outlook 2007." It projects energy supply, demand, and prices through 2030. The complete document will include alternative scenarios as well as the reference case in the early-release document. All information presented here reflects the reference case analysis.

U.S. GDP grows at 2.9% annually from 2005 to 2030—0.1 percentage point lower than projected for the same period in the AEO2006 reference case.

Average world crude oil price declines slowly in real terms (2005 dollars), from a 2006 average of more than $69/bbl ($11.56/MMBtu) to just under $50/bbl ($8.30/MMBtu) in 2014 as new supplies enter the market. It then rises slowly to about $59/bbl ($9.89/MMBtu) in 2030.

IEA anticipates substantial increases in conventional oil production in several OPEC and non-OPEC countries over the next 10 years, as well as substantial development of unconventional production over the next 25 years. Prices are high enough to trigger market entry of some alternative energy supplies that are expected to be economically viable at $25 to $50/bbl. These include heavy oil sands, gas-to-liquids, and coal-to-liquids.

IEA sees significant long-term supply potential from non-OPEC producers. In several resource-rich regions—with wars ending, new pipelines being built, new exploration and drilling technologies becoming available, and world oil prices rising—access to resources has increased and production has risen. Examples are Angola, Azerbaijan, Kazakhstan, Brazil, and Canada.

World liquids demand is projected to increase from about 84 million b/d in 2005 to 117 million b/d in 2030.

Average U.S. wellhead price for natural gas declines gradually as increased drilling brings on new supplies and new import sources become available.

The average price falls to just under $5/Mcf in 2015 (2005 dollars), then rises gradually to about $6/Mcf in 2030 (equivalent to $9.63/Mcf in nominal dollars).

Imports of LNG, new natural gas production in Alaska, and production from unconventional sources in the lower 48 States are not expected to increase sufficiently to offset the impacts of resource decline and increased demand.

IEA projections

The International Energy Agency (IEA; Paris) released its "World Energy Outlook 2006" in November. Its baseline scenario describes how energy markets are likely to evolve without new government measures to alter underlying energy trends. Under that scenario, the agency predicts that world oil demand will reach 116 million b/d in 2030, compared to 84 million b/d in 2005.

The agency sees nonconventional resources—chiefly oil sands in Canada—playing a growing role in meeting future demand. Also, IEA says that crude prices could exceed $100/bbl by 2030 as countries rapidly consume more energy, particularly emerging economies such as India and China.

IEA also expects that China will overtake the United States as the world's biggest emitter of carbon dioxide before 2010, and called for urgent government action to curb growth in energy demand and greenhouse gas emissions. Global carbon dioxide emissions are anticipated to reach 40 gigatons in 2030, up by 55% compared to 2006.

IEA also developed an alternative scenario describing a "clean, clever, and competitive energy future" based on more efficient energy use and wider use of biofuels, renewables, and nuclear power. Under this scenario, says IEA, energy demand and would expand 10% less than under the baseline scenario.

Reserves

EIA

December 2006 estimates from the U.S. EIA showed increases for both oil and gas reserves during 2005.

Recording a 2% increase, U.S. crude oil reserves moved up for the first time in three years. Reserves declined in Alaska and the Gulf of Mexico federal offshore, and increased in Texas and California. Total U.S. discoveries in 2005 amounted to 1.05 billion bbl, with new field discoveries of 205 million bbl, mostly from the federal Gulf of Mexico. New reservoir discoveries in old fields amounted to 41 million bbl.

A DECADE OF RESERVES CHANGES

Jan. 1	World (1,000 bbl)	OPEC (1,000 bbl)	World gas, bcf
2007	1,317,447,415	902,343,000	6,182,692
2006	1,293,344,534	901,659,000	6,101,158
2005	1,277,701,992	885,188,000	6,040,208
2004	1,265,811,583	869,521,000	6,068,302
2003	1,212,880,852	819,007,000	5,501,424
2002	1,031,100,681	818,842,000	5,451,332
2001	1,028,457,585	814,398,710	5,278,484
2000	1,016,041,221	802,479,710	5,146,207
1999	1,034,264,678	800,479,710	5,144,753
1998	1,019,545,664	797,134,589	5,086,469

Source: OGJ Worldwide Production Reports

While several recent large deepwater discoveries have not yet been booked as proved reserves, crude reserves additions replaced 122% of 2005 production. Texas, Wyoming, and Montana posted the largest increases.

Extensions to existing oil fields in 2005 added 805 million bbl of reserves, mostly from fields in Texas, California, Montana, and Wyoming.

With regard to natural gas, proved U.S. reserves increased 6% during 2005, the largest annual increase since 1970. Additions onshore in the Lower 48 states were large enough to overcome a 10% decline for federal areas of the Gulf of Mexico. Texas and Colorado led reserves additions. Coalbed methane reserves increased 8% and accounted for 10% of U.S. dry gas reserves.

Most gas reserves additions were from extensions of existing fields rather than new discoveries. Field extensions totaled 21.05 tcf in 2005, while new field discoveries were 942 bcf. New reservoir discoveries in old fields were 1.21 tcf.

Other organizations

DOE said the United States holds as much as 430 billion bbl of technically recoverable oil in a resource in place of 1.124 trillion bbl. These volumes include speculative categories such as undiscovered oil, reserve growth, and a "residual oil zone."

According to the American Gas Association (A.G.A.), U.S. natural gas reserves were 197 tcf at the end of 2005, their highest level since 1984. Drilling and revisions to earlier estimates added as much as 35.7 tcf to reserves. Total reserves grew for a seventh consecutive year.

With respect to the U.S. Rocky Mountain region, natural gas reserves could exceed 130 tcf by 2030, said Wood Mackenzie Ltd. (Edinburgh). The resource is dominated by unconventional gas. Rocky Mountain volumes currently account for approximately 31% of Lower 48 proved gas reserves.

Prices

Oil prices

The average price of the front-month crude contract on Nymex in 2006 was about $66.25/bbl, a 17% rise over 2005. According to analysts at Barclays Capital Inc. (London), this was similar in magnitude to the increases of 20%, 32%, and 37% experienced in 2003, 2004, and 2005, respectively.

On December 28, 2006, the February 2007 contract for benchmark U.S. sweet, light crudes was $60.53/bbl. This price level reflected a 20.2 million bbl drop in crude supplies in the preceding five weeks—the biggest five-week drop since the hurricane season in 2005. In addition, forecasts of warm winter weather and fear of a slowing U.S. economy dominated the market.

Natural gas prices

Monthly average spot prices at Henry Hub fell steadily in the first half of 2006, from $9.05/Mcf in January to $6.29/Mcf in June. Record-high U.S. temperatures in July drove up electricity demand as well as gas demand for power generation. That month saw the first-ever drop in summertime natural gas inventories, and the spot gas price jumped to an average of $7.59/Mcf in August.

During the autumn, spot prices softened to about $6/Mcf in October as inventories were rebuilt, then climbed again to $6.83/Mcf in December. EIA projected that high storage levels and the forecast of slightly warmer-than-normal weather would keep natural gas spot prices below $9/Mcf on average through the heating season. The agency also expected a January peak of $8.71/Mcf for the monthly average Henry Hub spot price, and an average of $7.87/Mcf for 2007.

Gasoline prices

U.S. gasoline prices became a highly visible issue in 2006. According to EIA, the average national retail price for regular gasoline spiked at $3.07/gal in mid-September 2005, reflecting the damage that Hurricane Katrina caused to Gulf Coast production and refinery facilities.

At the start of 2006, the average price had dropped to about $2.23/gal, and held at roughly that level until mid-March. However, as the price began to rise again, consumers and members of Congress called for inquiries and action. The price reached $3.04 in early August then fell back to $2.25 in October. By year-end, EIA reported, that the national average price for regular gasoline was $2.38/gal.

Oil industry executives, trade association officials, and others noted the factors contributing to high oil and gasoline prices, including increased global demand for oil, refining capacity lost during the 2005 hurricanes that has not been restored, seasonal growth in demand for gasoline, costs associated with a mandated switch to ethanol from MTBE as a gasoline additive, political instability in some producing countries, and refiners beginning to produce ultralow-sulfur diesel fuel.

The American Petroleum Institute (API) noted that while no new U.S. oil refineries had been built in the last 30 years, domestic capacity at existing refineries has been expanded by the equivalent of 10 to 12 state-of-the-art refineries.

Access

BLM study: resources restricted

Just 3% of federal onshore oil and 13% of onshore federal gas are currently available under standard lease terms, the U.S. Bureau of Land Management (BLM) said in a November 2006 report that is posted online at www.blm.gov.

Access grows to 46% for oil and 60% for gas with additional restrictions, including "no surface occupancy." But its inventory of 99 million acres of federal onshore oil and gas resources found 51% of the oil and 27% of the gas closed to leasing. The study, which was required under the Energy Policy Act of 2005, expands on one published in 2003.

BLM said the 11 areas inventoried in the latest study include six new oil and gas basins in Alaska, the Rocky Mountains, and the eastern United States in addition to five basins studied in 2003. It said the new inventory covers an area containing an estimated 21 billion bbl of oil and 187 tcf of gas, or about 76% of total onshore federal oil and gas resources.

In 2005 the Environmental Protection Agency (EPA) directed that the new inventory consider conditions of approval attached to drilling permits that producers must follow when developing leases—such as, no drilling during migration of sensitive species. The 2003 inventory only considered restrictions in the actual leases.

Proposed 2007–12 OCS lease sales

The U.S. Minerals Management Service released and requested comments on a proposed plan for its 2007–12 OCS leasing program. The plan includes 21 proposed OCS lease sales in 7 of the 25 OCS planning areas—four off Alaska, two in the Gulf of Mexico, and one in the Atlantic.

MMS estimates total undiscovered technically recoverable resources of 67.9 billion bbl of oil and 340.4 tcf of gas from all planning areas where sales are under consideration, though only portions of some of the planning areas are included.

The proposed program includes a special interest sale in selected parts of the Mid-Atlantic Planning Area off Virginia in late 2011. The plan excludes the area within 100 miles of the Florida coast that used to be part of the Eastern Gulf Planning Area.

The program also proposes lease sales in the Beaufort and Chukchi Seas and North Aleutian Basin Planning Areas. A special interest sale may be conducted in the Cook Inlet Planning Area if enough interest is shown by industry. The 2007–2012 OCS Oil and Gas Proposed Final Program and Final Environmental Impact Statement are scheduled for release in spring 2007.

OCS leasing reform enacted

In December, President George W. Bush signed a legislative package that included OCS leasing reform. The Gulf of Mexico Energy Security Act of 2006 opened certain areas of the gulf to leasing for oil and gas exploration.

The OCS leasing legislation focuses on the Sale 181 area and adjacent tracts in deeper water in the eastern Gulf of Mexico. Development of the designated area is expected to bring 1.26 billion bbl of crude oil and 5.8 tcf of gas to U.S. consumers over several years.

One key provision establishes a 125-mile no-lease Florida coastal buffer zone to be in effect through 2022 beyond the Sale 181 area. Another allows the other four gulf states to share 37.5% of future federal lease revenues from the area covered to pay for coastal protection and management.

Other energy-related elements of the bill include tax credits and provisions covering renewable resources, energy efficient home and building construction, photovoltaic devices, fuel cells, coal gasification, and production of methanol and ethanol from coal, and cellulosic biomass ethanol plants.

ERD to tap Pacific OCS?

The U.S. MMS estimates the discovered and undiscovered, conventionally recoverable oil and gas resources of the Pacific OCS at 14–19 billion boe. By comparison, the U.S. Geological Survey (USGS) estimates that the Alaskan area contains 4.3–11.8 billion bbl of technically recoverable oil.

LEASE SALE 181

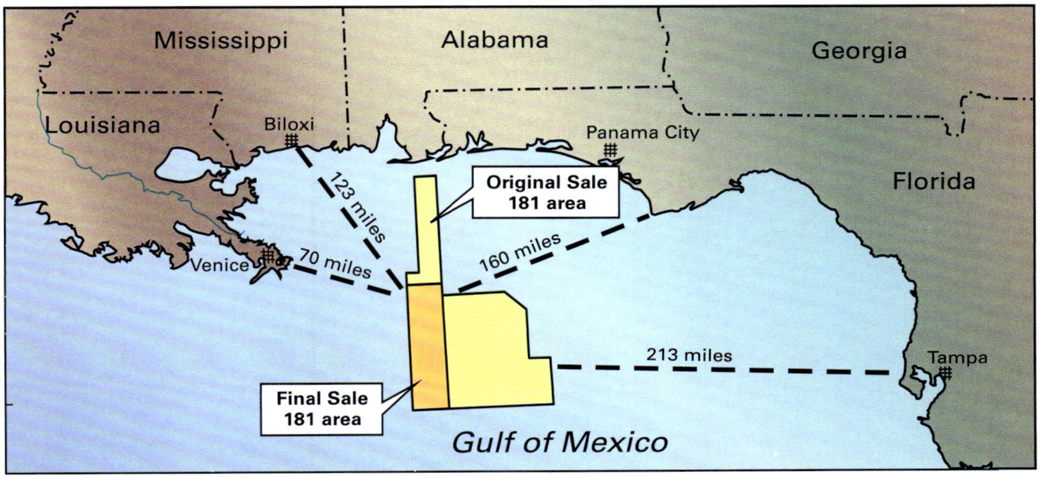

The Pacific OCS resources are located offshore Washington, Oregon, and California, and 80% or more are undeveloped. Thirteen fields offshore of California have produced more than 1.3 billion boe since production began in 1968.

Leasing of the Pacific OCS has been closed by congressional and administrative moratoria. And because virtually all of the resource areas in the Pacific OCS lie within 50 miles of the coast, some parties have concerns about the impact of their development on the environment and tourism.

MMS ASSESSMENT AREAS

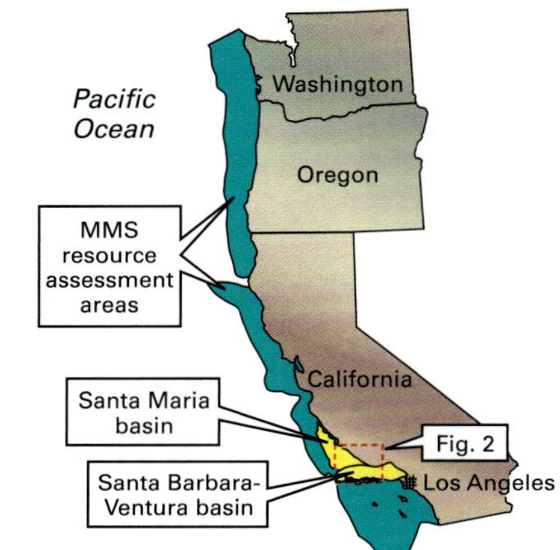

Source: Modified from US Minerals Management Service

In the Santa Maria and Santa Barbara-Ventura basins, 25 offshore fields have been discovered. MMS estimates their unproven reserves to be about 1.3 billion boe.

A 2006 study by the University of Houston said that extended reach drilling (ERD) from onshore locations might provide an economically and environmentally acceptable alternative to offshore platforms for developing near-shore unproven reserves in these two basins.

The study points to successful use of ERD from offshore platforms by ExxonMobil Corp. (at Sakhalin Island and in the Sacate field in the Gulf of Mexico) and by Plains Exploration & Production Co. (Houston) in southern California.

Plains E&P plans to use offshore ERD again to develop the T-Ridge prospect in its Pederenales unit. Looking at the same T-Ridge structure, Sunset Exploration Inc. (Brentwood, CA) has reportedly proposed an ERD project based onshore at Vandenberg Air Force Base to tap resources about four miles off the coast.

Onshore drillsites, the University of Houston study said, could be located within existing coastal oil fields or in undeveloped areas near the coast. Locations of interest include Vandenberg Air Force Base, Point Conception, Capitan oil field, and West Montalvo oil field.

About one-third of the area of unproved offshore reserves (32,000 acres) lies within seven miles of the coast. Based on MMS estimates, ERD wells potentially could produce nearly 500 MMboe from those reserves.

SANTA MARIA, SANTA BARBARA-VENTURA BASIN OIL AND GAS FIELDS

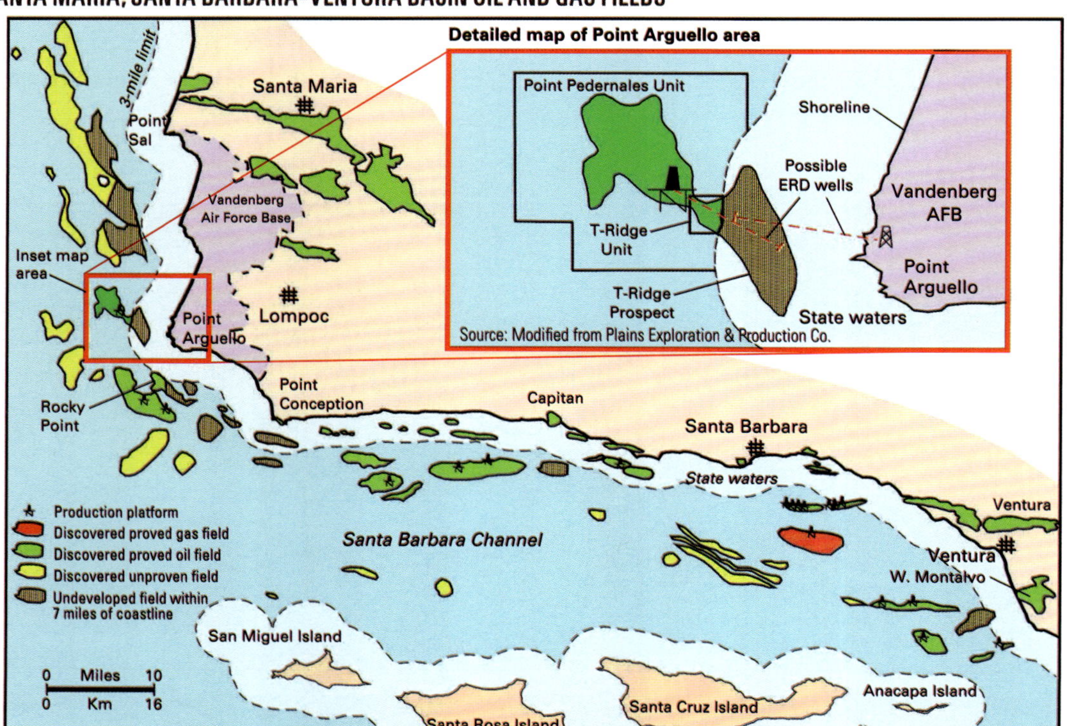

Source: Modified from US Minerals Management Service

MMS reports on inaccessible resources

A report to Congress by the MMS showed big oil and gas potential in areas of the OCS now off-limits to drilling. The report (based on data

Alternative plans to develop T-Ridge prospect

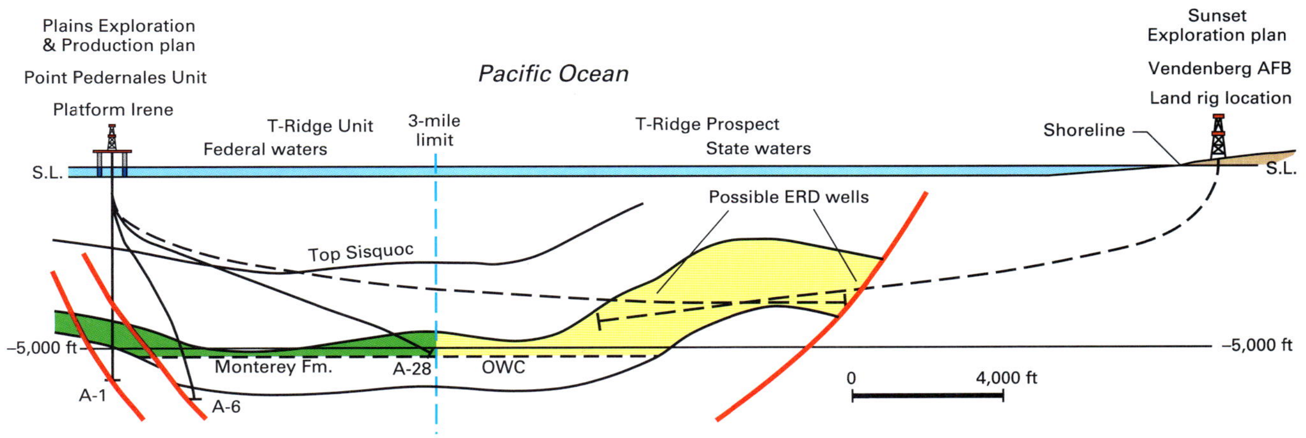

Source: Modified from Plains Exploration & Production Co.

available January 1, 2003) estimated undiscovered, technically recoverable resource (UTRR) in all federal offshore areas at mean values of 85.9 billion bbl of oil and at 419.9 tcf of gas. More than half of the oil-equivalent UTRR is in the Gulf of Mexico.

"Technically recoverable" estimates represent the amount of oil or gas that could be produced with existing or reasonably foreseeable technology. Current technologies include drilling in water deeper than 10,000 ft and subsea depths in excess of 31,000 ft.

The OCS's total hydrocarbon endowment is 115.4 billion bbl of oil and 633.7 tcf of gas. Of the total endowment, about 29.6 billion bbl of oil and 213.8 tcf of gas are in known fields.

The report notes that is has been 20 years or more since any exploration activity has occurred on the Alaska and

OCS undiscovered technically recoverable reserves

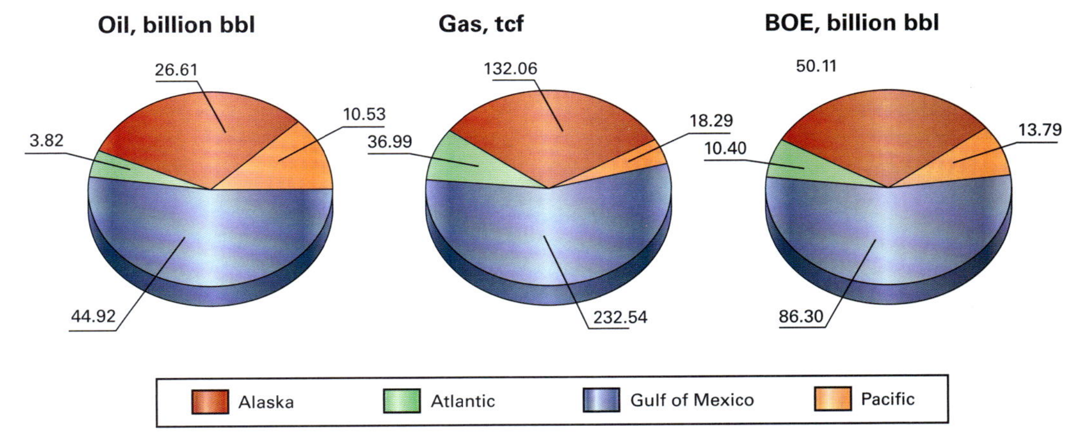

Source: US Minerals Management Service

Distribution of total hydrocarbon endowment

— Resource category —		Alaska	Atlantic	Gulf of Mexico	Pacific	Total OCS
Cumulative production	Oil[1]	0.01	0	13.05	1.06	14.12
	Gas[2]	0	0	152.25	1.32	153.57
	BOE[1]	0.01	0	40.14	1.29	41.45
Reserves	Oil	0.03	0	7.06	1.46	8.55
	Gas	0	0	27.70	1.56	29.26
	BOE	0.03	0	11.98	1.74	13.76
Reserves appreciation	Oil	—	—	6.88	—	6.88
	Gas	—	—	30.91	—	30.91
	BOE	—	—	12.38	—	12.38
UTRR (mean)	Oil	26.61	3.82	44.92	10.53	85.88
	Gas	132.06	36.99	232.54	18.29	419.88
	BOE	50.11	10.40	86.30	13.79	160.60
Total endowment	**Oil**	**26.65**	**3.82**	**71.91**	**13.05**	**115.43**
	Gas	**132.06**	**36.99**	**443.40**	**21.17**	**633.62**
	BOE	**50.15**	**10.40**	**150.81**	**16.82**	**228.18**

[1]Units measured in billion barrels. [2]Units in trillion cubic feet.

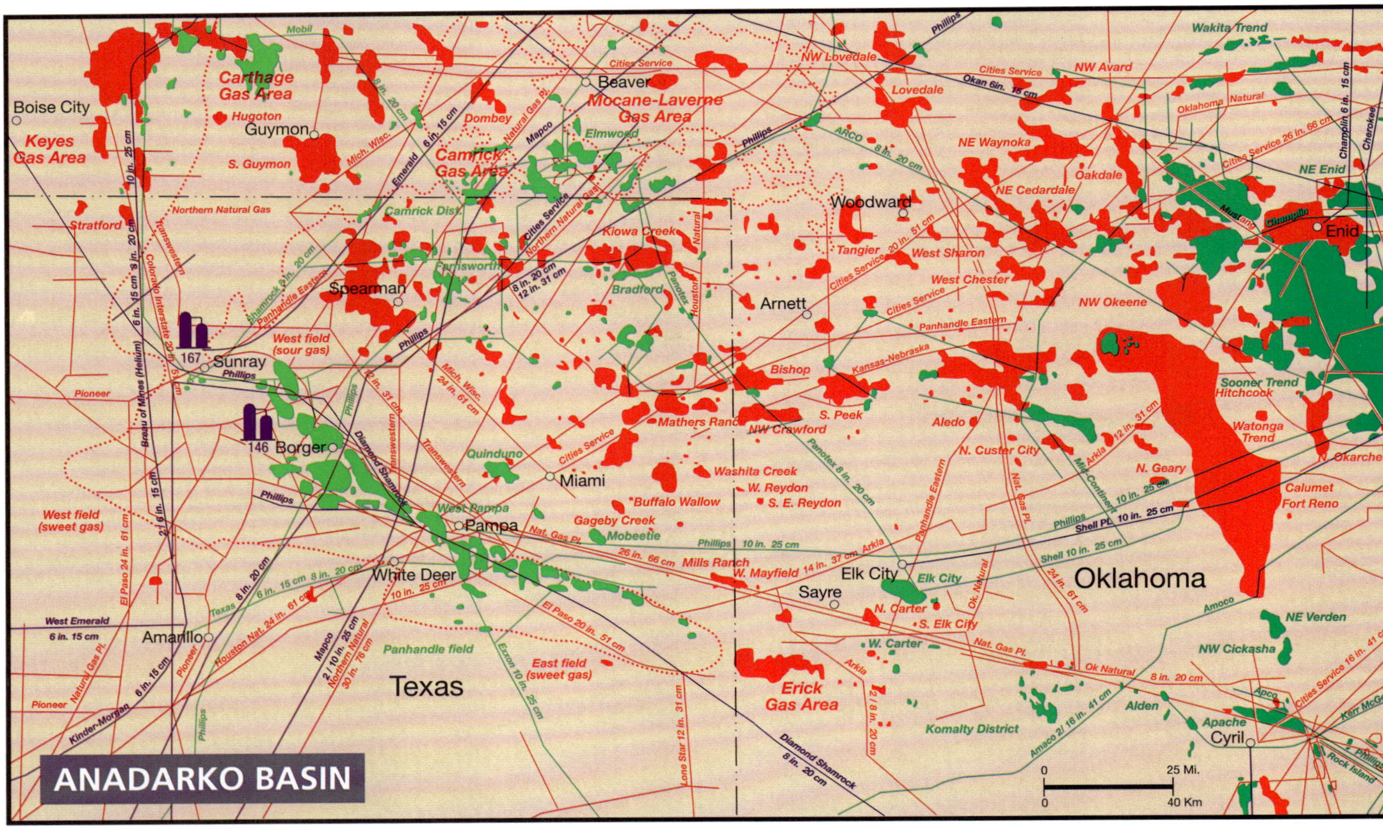

Atlantic OCS. In other areas, such as central and northern California, offshore Oregon and Washington, and the south Florida basin, there has been no meaningful exploration activity since the 1960s.

Exploration and Development

Gas flow milestone in Uinta basin

The giant Greater Natural Buttes area in the Uinta basin reached a milestone in September, averaging a gross 225 MMcfd of gas, said operator Anadarko Petroleum Corp.

Anadarko has more than 1,400 wells and owns and operates a 700-mile gathering system in the basin. The 2006 program called for the drilling of 270 wells, and 4,900 drillsites are in inventory.

The company, which holds 237,000 net acres in the field, is employing 12 rigs, including eight operated rigs. Anadarko has drilled three pilot wells on 20-acre spacing to evaluate recovery potential in the field's more developed areas drilled on 40-acre units.

A slick-water completion pilot program begun in early 2006 has grown to 61 wells and reduced average frac cost by $100,000/well with no material effect on well performance, said Anadarko.

Piceance tight sands could yield 35 tcf

ExxonMobil Gas & Power Marketing Co. signed a 30-year agreement with Enterprise Products Partners LLC for gathering, compression, treating, and conditioning services for natural gas it is developing in tight sands of the Piceance basin of Colorado.

ExxonMobil began a development program in 2005 of about 270,000 gross acres it holds under lease in the area. The company estimated recovery potential at more than 35 tcf and predicted the area will produce gas for more than 50 years. Officials also said gross production could reach 1 bcfd at its peak.

ExxonMobil dedicates to Enterprise the production from 29,000 acres in Rio Blanco County. Enterprise has the option to recover NGL beyond the level needed to make the gas meet pipeline specifications.

Woodford shale play

Newfield Exploration Co. (Houston) signed a 15-year agreement with MarkWest Energy Partners LP (Denver) to construct and operate gas gathering pipelines and related facilities to develop Newfield's Devonian Woodford shale play in the Arkoma basin of southeastern Oklahoma.

The system will include more than 400 miles of large-diameter pipeline and 100,000 hp of compression for a takeaway capacity exceeding 500 MMcfd.

Newfield has drilled more than 100 vertical wells and 30 horizontal wells in the Woodford shale. The company has shifted its drilling program to horizontal wells, which has increased initial production rates and estimated ultimate recovery.

Still acquiring acreage, Newfield expects to employ as many as 20 rigs by late 2007. By year-end 2009, the company expects gross Woodford shale production to reach 350 MMcfd, compared to about 65 MMcfd in late 2006.

Drilling and Production

DOE to support CO_2 EOR in Alabama

DOE plans to share the cost of a project to inject carbon dioxide into Alabama's largest oil field to improve recovery and later store the greenhouse gas. DOE would provide nearly $3 million of the $6 million cost of the enhanced oil recovery project in Citronelle field, in Mobile County, 25 miles northwest of Mobile. The project could recover 64 million bbl of oil, DOE said.

The University of Alabama-Birmingham proposed the project to DOE. After economic oil production ceases, it calls for storage in the reservoir and adjacent formations of CO_2 produced from the combustion of fossil fuels in power plants and other processes.

Project aims to provide oil field operators and CO_2 producers with improved estimates of the oil yields from EOR and the capacity of depleted reservoirs to sequester CO_2.

The Citronelle field is operated by Denbury Resources Inc. (Plano, TX). Other project participants are the University of Alabama-Tuscaloosa; Alabama A&M University (Huntsville); Geological Survey of Alabama; and the University of North Carolina (Charlotte).

NEWFIELD'S ARKOMA WOODFORD ACREAGE

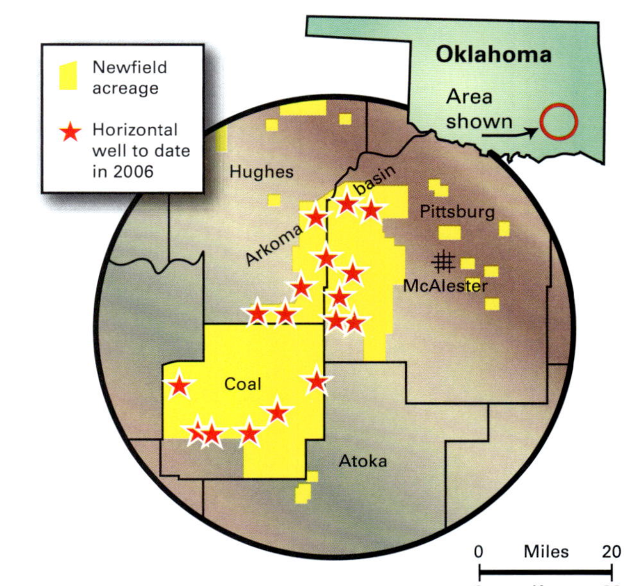

Source: Modified after Newfield Exploration Co.

UNLOCKING TIGHT GAS

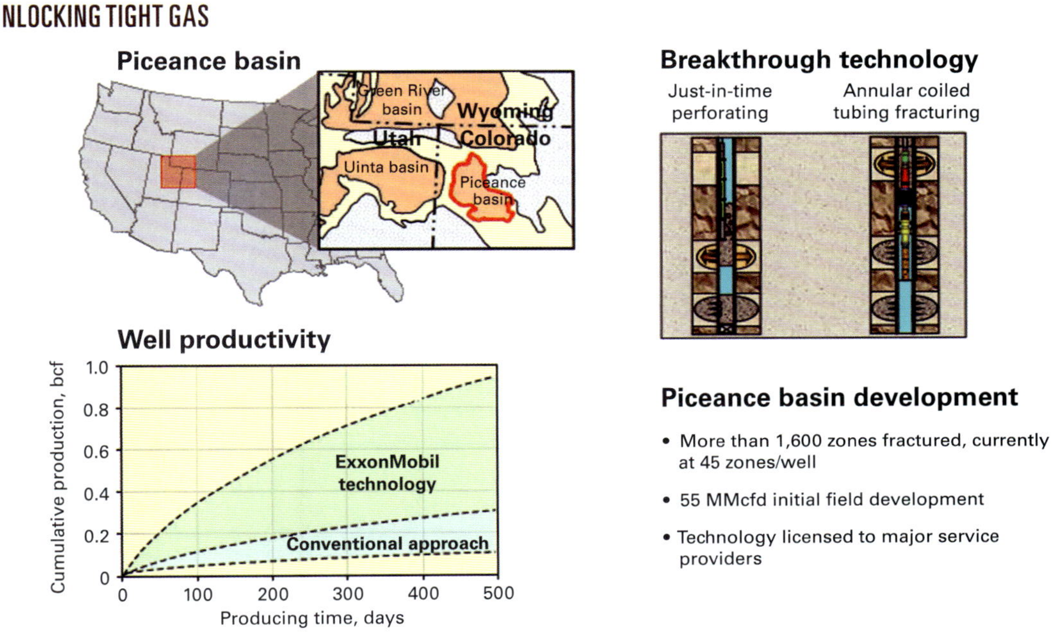

Source: ExxonMobil Corp.

$25 billion in Rockies drilling seen

Wood Mackenzie (Edinburgh) forecasts continued growth in natural gas production the Rocky Mountain region in the next five years, including the drilling of 43,000 wells from 2006 to 2010, requiring capital expenditures of nearly $25 billion.

The company said that, over the past 30 years, Rocky Mountain gas reserves have grown from 19 tcf to more than 57 tcf. Currently, the Rockies region accounts for more than 31% of the total Lower 48 proved gas reserves. If historical trends continue, the region's reserves could exceed 130 tcf by 2030.

Noting that unconventional resources will be particularly important over the next 10 years, the company cautioned that deeper, tighter unconventional plays typically require higher expenditures for advanced drilling and production technologies.

In a mid-2006 evaluation, *OGJ* said technological advances in the United States as well as higher gas prices have made many tight gas, shale gas, and coalbed gas developments feasible, such as the Piceance tight-gas projects in Colorado.

A look at 2005, 2006 drilling

At the start of 2006, *OGJ* looked back as well as forward at U.S. oil and gas drilling. For 2005, year-average rotary rig counts in the United States and Texas were the highest in 20 years. *OGJ* projected that 2006 should be another good year, but said that a duplication of 2005's 16% gain over 2004 was unlikely.

Highlights of *OGJ's* early-year drilling forecast for 2006:

- Operators will drill nearly 43,000 wells in the United States, up 5.3% from an estimated 40,814 wells drilled in 2005. The increase could be higher if prices hold and if operators can make more-productive use of existing units.
- All operators will drill 3,352 exploratory wells of all types, up from an estimated 3,096 last year.
- The Baker Hughes Inc. count of active U.S. rotary rigs will average 1,450 rigs/week in 2006, up from 1,383 rigs/week in 2005.
- Operators will drill 22,695 wells in western Canada, up from an estimated 21,932 wells in 2005.

2005: U.S. rigs break 1985 record

U.S. land drillers fielded an average of 1,383 rigs/week in 2005, the highest since 1,968 rigs/week worked in 1985, according to Baker Hughes. The year 2005 was only the fifth year since 1985, inclusive, that the count has topped 1,000 rigs. The U.S. active rig count rose to 1,471 on December 30, the last reporting period of 2005. The offshore average of 93 rigs/week in 2005 was the lowest since 82 rigs/week in 1993.

A Look at 30 Years of US Well Completions

Year	Total wells[1]	Total footage	Total exploratory wells	Year	Total wells[1]	Total footage	Total exploratory wells
[2]2006	42,948	221,086,000	3,352	1991	28,417	141,391,000	4,399
[2]2005	40,814	210,445,000	3,096	1990	30,615	149,518,000	5,074
2004	36,469	202,146,000	2,848	1989	28,363	134,901,000	5,251
2003	32,767	175,388,000	2,471	1988	32,238	155,164,000	6,350
2002	27,794	145,055,000	2,288	1987	36,253	163,848,000	6,903
2001	36,061	184,462,000	3,136	1986	39,015	177,641,000	7,156
2000	31,261	149,848,000	2,512	1985	70,806	316,778,000	12,208
1999	22,107	109,854,000	2,074	1984	84,983	368,796,000	15,138
1998	25,822	143,625,000	2,690	1983	75,738	316,617,464	13,845
1997	30,208	165,480,000	3,264	1982	83,889	375,382,919	15,882
1996	25,724	138,588,000	3,344	1981	89,234	406,520,453	17,430
1995	23,061	124,426,000	3,401	1980	69,486	311,444,837	12,870
1994	23,324	130,654,000	3,788	1979	51,890	243,685,430	10,735
1993	26,032	138,509,000	3,604	1978	50,061	238,386,248	11,030
1992	23,921	123,456,000	3,494	1977	45,855	215,695,878	10,140

[1]Well counts in most recent years subject to reporting lag. [2]Estimated.
Source: 1977-2004 American Petroleum Institute

The Barnett shale continued as the most active U.S. play in 2005. But drilling is moving away from the North Texas core area counties of Jack and Wise and into several counties in West Central Texas District 7B to the south and west.

As of mid-December 2006, 2,322 drilling permits had been issued for wells in the Barnett shales, a 16% increase over the 2,006 permits issued in 2005.

Companies drilled an estimated 3,650 wells in the state in Wyoming in 2005, compared with 3,588 wells in 2004. Of the 2005 wells, 2,700 were for CBM. At midyear 2005, seven Wyoming basins had CBM wells, of which 14,700 were producing wells in the Powder River basin.

Record well completions in 2006

The American Petroleum Institute (API) reported that an estimated 49,375 oil wells, natural gas wells and dry holes were completed in 2006—a 21-year high. The fourth quarter's estimated 12,439 completions were the highest since first quarter 1986. Other estimates for 2006:

- An all-time high of 29,356 natural gas wells completed
- 15,346 oil wells completed, exceeding every year in this decade and in the 1990s
- 932 exploratory oil wells completed, representing a fourth straight year of increases
- 289.9 million feet of well drilling, the highest level in more than 20 years

Oil & Gas Journal Well Forecast for 2006

	2005 estimate				2006 forecast		
State	**Total comp.**	**Exploratory wells**	**Field wells**	**Total ft (1,000)**	**Total comp.**	**Exploratory wells**	**Field wells**
Alabama	355	15	340	1,124	415	20	395
Alaska	145	5	140	997	157	5	152
Arizona	1	1	0	2	3	3	0
Arkansas	305	12	293	1,763	327	14	313
California land	2,200	55	2,145	5,196	2,300	64	2,236
California offshore	6	0	6	50	9	0	9
Colorado	2,625	105	2,520	14,674	2,800	120	2,680
Florida	1	1	0	12	1	1	0
Illinois	295	27	268	557	320	32	288
Indiana	125	8	117	368	135	8	127
Kansas	2,055	181	1,874	0	2,125	181	1,944
Kentucky	890	27	863	7,029	915	31	884
Louisiana	2,125	192	2,317	18,300	2,355	225	2,130
North	975	88	887	12,968	1,015	95	920
South	490	18	472	861	550	24	526
Offshore	660	86	574	4,472	790	106	684
Michigan	485	92	393	2,149	505	106	399
Mississippi	270	30	240	1,227	300	33	267
Montana	900	138	762	4,090	945	151	794
Nebraska	50	16	34	136	58	17	41
Nevada	8	8	0	61	11	9	2
New Mexico—East	1,025	51	974	4,355	1,090	61	1,029
New Mexico—West	1,360	15	1,345	8,554	1,410	13	1,397
New York	255	8	247	896	275	8	267
North Dakota	240	38	202	839	265	46	219
Ohio	540	70	470	1,041	565	68	497
Oklahoma	3,250	98	3,152	20,524	3,380	112	3,268
Oregon	5	5	0	14	8	8	0
Pennsylvania	3,277	360	2,917	23,781	3,375	351	3,024
South Dakota	21	3	18	158	23	4	19
Tennessee	84	8	76	812	90	9	81
Texas	11,938	1,035	10,903	70,040	12,462	1,112	11,350
Dist. 1	433	34	399	4,046	450	37	413
Dist. 2	635	102	533	2,259	680	112	568
Dist. 3	886	108	778	6,412	925	118	807
Dist. 4	1,225	103	1,122	8,575	1,275	110	1,165
Dist. 5	1,145	34	1,111	6,453	1,200	40	1,160
Dist. 6	1,447	140	1,307	9,731	1,490	140	1,350
Dist. 7-B	725	22	703	5,000	775	27	748
Dist. 7-C	1,375	61	1,314	14,190	1,430	70	1,360
Dist. 8	1,433	99	1,334	7,695	1,500	99	1,401
Dist. 8-A	704	72	632	1,921	732	78	654
Dist. 9	978	23	955	0	1,005	21	984
Dist. 10	840	185	655	3,394	880	202	678
Offshore	112	52	60	363	120	58	62
Utah	717	139	578	3,270	760	153	607
Virginia	425	66	359	7	455	70	385
Washington	1	1	0	14	4	3	1
West Virginia	1,185	178	1,007	4,811	1,215	187	1,028
Wyoming	3,650	102	3,548	13,596	3,900	121	3,779
US total	**40,814**	**3,090**	**37,724**	**210,444**	**42,958**	**3,346**	**39,612**
Western Canada	21,932	5,644	16,288	193,527	22,695	5,960	16,735
Alberta	16,925	4,197	12,728	178,474	17,450	4,397	13,053
Saskatchewan	3,575	997	2,578	8,419	3,750	1,065	2,685
British Columbia	1,155	381	774	5,588	1,200	420	780
Manitoba	277	69	208	1,045	295	78	217
NWT-Yukon	4	4	0	17	9	9	5
Eastern offshore	8	1	7	154	6	2	4
Eastern land	60	5	55	136	65	5	60

Laterals tap Granite Wash gas

Grayhawk Operating Inc. (Plano, Tex.) reported success with what it called the first use of horizontal drilling in the Pennsylvania Granite Wash play in the northeastern Texas Panhandle. Lard Ranch 1-60 yielded 439 MMcf of gas, 9,900 bbl of light oil, and 2,625 bbl of water in its first 11 months on line.

The company estimates that each horizontal well will tap about three times the gas volume of a vertical well, at about twice the cost. It expected to have six horizontal wells producing in 2006.

Continued U.S. drilling growth seen

Raymond James & Associates estimated that the global drilling market would be "pretty healthy" for the next five years," and projected that the U.S. land fleet would continue to grow at 10–12% (net) at least into 2007 and probably into 2009.

California rejects state oil production tax

California voters rejected a state ballot measure on November 7 that would have taxed state oil production at varying rates depending on the price of crude, and tax revenues would have financed the development of alternative energy. Near-final returns indicated Proposition 87 was defeated 55% to 45%.

Oil companies opposed Proposition 87, which, if passed, would have generated an estimated $200–380 million/year in new tax revenue for California. Opponents said it would have caused a reduction in state oil production and a need for increased imports.

At year-end 2006, both oil and gas prices had stopped their rapid upward climb, but rigs were still working. The total number of U.S. oil rigs was up slightly but relatively flat over the last five years. The number of rigs drilling for gas in the United States has doubled in the last three years.

December 2006 contracts on the Nymex for U.S. benchmark crude oils expired November 17 at the lowest closing price in 17 months. In some areas, reduced natural gas prices had caused operators to scale back development drilling. Yet operators were proceeding with major exploration projects and leaving few rigs idle, on or offshore.

Modest softening in gas drilling has caused some companies to shift assets to oil-oriented regions such as the Permian basin and California's San Joaquin basin.

Baker Hughes Inc. reported 1,717 rotary rigs working in the United States the week of December 1, 2006, up 18% (257 rigs) from a year earlier. This includes 1,608 land rigs (up 18% from a year ago), 86 offshore rigs (up 10% from a year ago), and 23 rigs working in inland waters.

Important Texas Panhandle Granite Wash fields

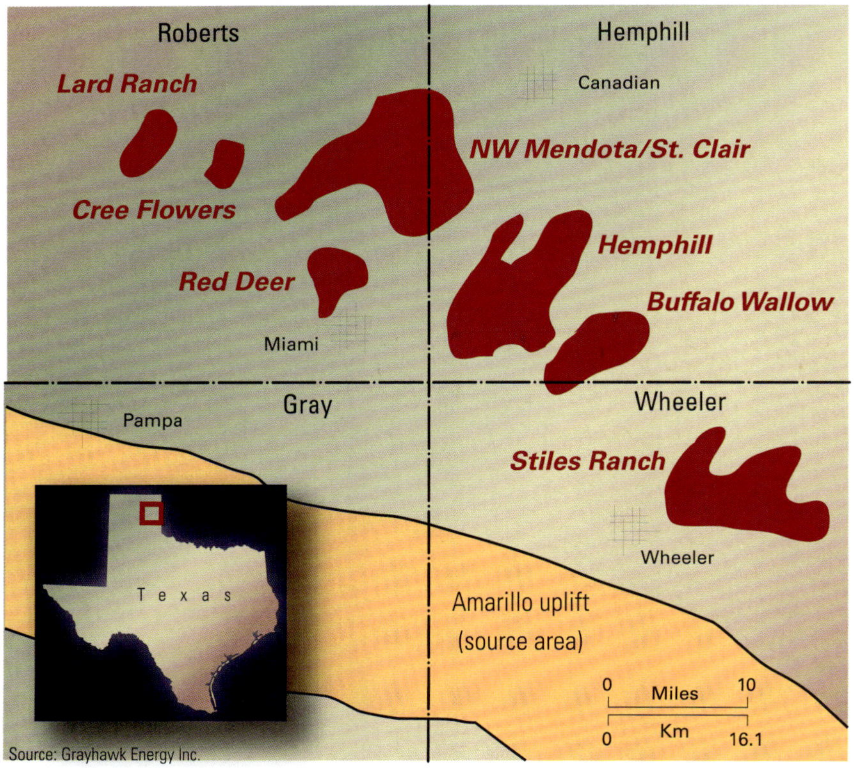

Source: Grayhawk Energy Inc.

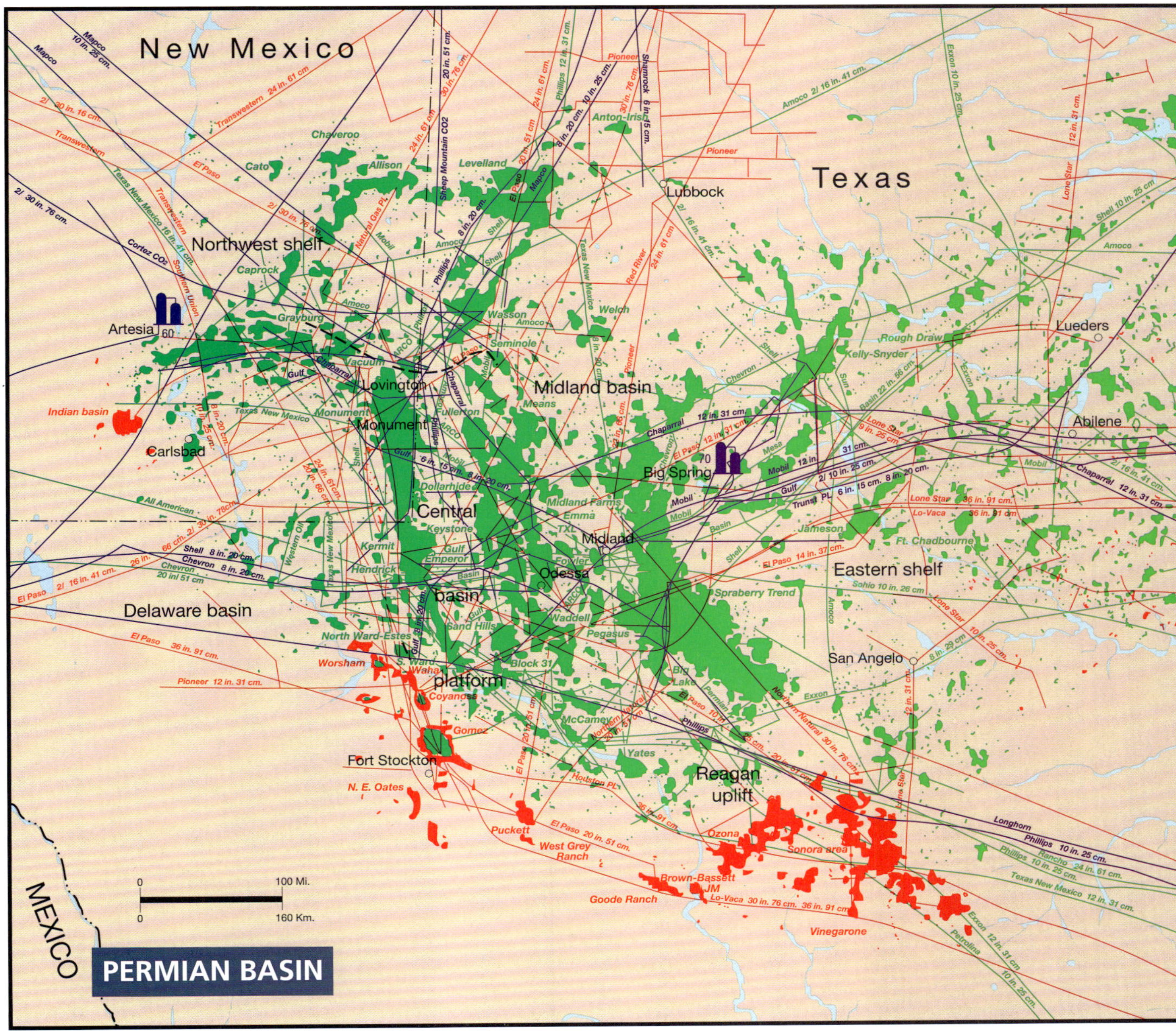

Most rigs (1,423; 83%) were drilling for gas, up 19% from a year ago. Another 289 rigs were drilling for oil, up 9% from a year ago.

Barnett shale

One of the largest U.S. gas resource plays and largest growth areas, the Barnett shale attracted drilling by Devon Energy Corp., XTO Energy Inc., EnCana Oil & Gas, Chesapeake Energy Corp., EOG Resources Inc., Denbury Resources Inc., Quicksilver Resources Inc., Southern California Public Power Authority, Royal Dutch Shell, and DTE Gas Resources, a subsidiary of DTE Energy Co., among others.

As of mid-December, 2,322 drilling permits had been issued for wells in the Barnett, a 16% increase over the 2,006 permits issued in 2005, and 67% increase over the 1,387 issued in 2004. About 165 active rigs were operating in December, and about 200 are likely to be working the play in 2007.

Gulf of Mexico

The offshore fleet working in late 2006 in the Gulf of Mexico was about 112 rigs, with 70 jack ups, 29 semisubmersibles, and 6 drillships. Overall utilization at the end of November was about 80%, according to ODS-Petrodata

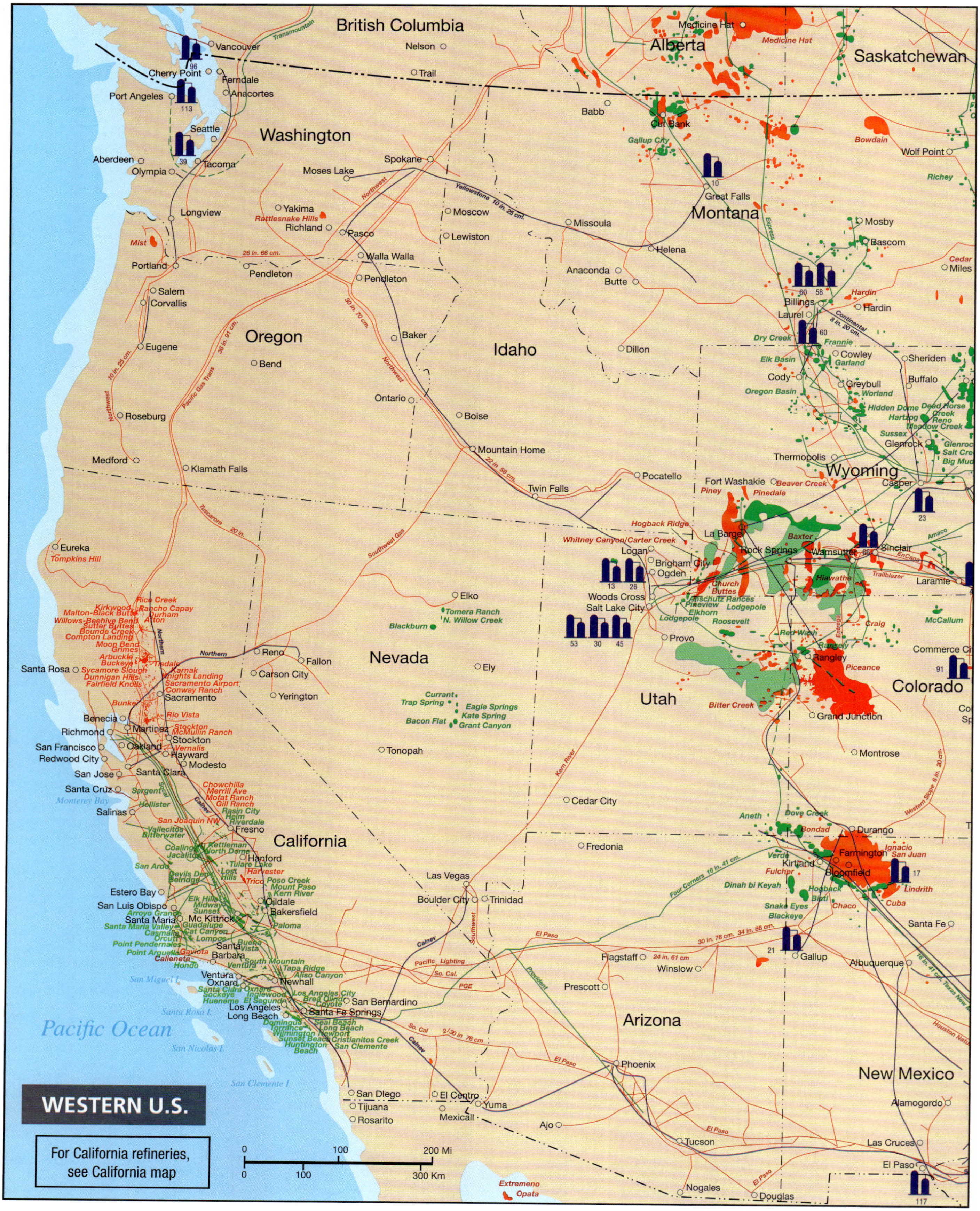
WESTERN U.S.
For California refineries, see California map
British Columbia
Alberta
Saskatchewan
Washington
Montana
Oregon
Idaho
Wyoming
Nevada
Utah
Colorado
California
Arizona
New Mexico
Pacific Ocean
Vancouver
Cherry Point
Ferndale
Anacortes
Port Angeles
Seattle
Tacoma
Aberdeen
Olympia
Longview
Portland
Salem
Corvallis
Eugene
Roseburg
Medford
Klamath Falls
Bend
Pendleton
Yakima
Richland
Pasco
Walla Walla
Moses Lake
Spokane
Moscow
Lewiston
Nelson
Trail
Medicine Hat
Cut Bank
Babb
Great Falls
Missoula
Helena
Anaconda
Butte
Dillon
Billings
Laurel
Hardin
Mosby
Bascom
Wolf Point
Cody
Greybull
Sheriden
Buffalo
Thermopolis
Casper
Glenrock
Fort Washakie
Rock Springs
Wamsutter
Sinclair
Laramie
Baker
Ontario
Boise
Mountain Home
Twin Falls
Pocatello
Logan
Brigham City
Ogden
Woods Cross
Salt Lake City
Provo
Rangley
Grand Junction
Montrose
Durango
Farmington
Bloomfield
Kirtland
Santa Fe
Albuquerque
Gallup
Flagstaff
Winslow
Prescott
Phoenix
Tucson
Nogales
Douglas
Ajo
Yuma
El Centro
Mexicali
Las Vegas
Boulder City
Trinidad
Cedar City
Fredonia
Elko
Ely
Reno
Fallon
Carson City
Yerington
Tonopah
Eureka
Santa Rosa
Sacramento
Benecia
Richmond
Martinez
San Francisco
Oakland
Redwood City
Hayward
Stockton
Modesto
San Jose
Santa Clara
Santa Cruz
Salinas
Fresno
Hanford
Estero Bay
San Luis Obispo
Santa Maria
Mc Kittrick
Oildale
Bakersfield
Santa Barbara
Ventura
Oxnard
Newhall
Los Angeles
Long Beach
Santa Fe Springs
San Bernardino
San Diego
Tijuana
Rosarito
Alamogordo
Las Cruces
El Paso
Commerce City
Piceance
Pacific Gas Trans
Northwest
Southwest Gas
Kern River
El Paso
Transmountain
Yellowstone 10 in. 25 cm.
0 100 200 Mi
0 100 300 Km

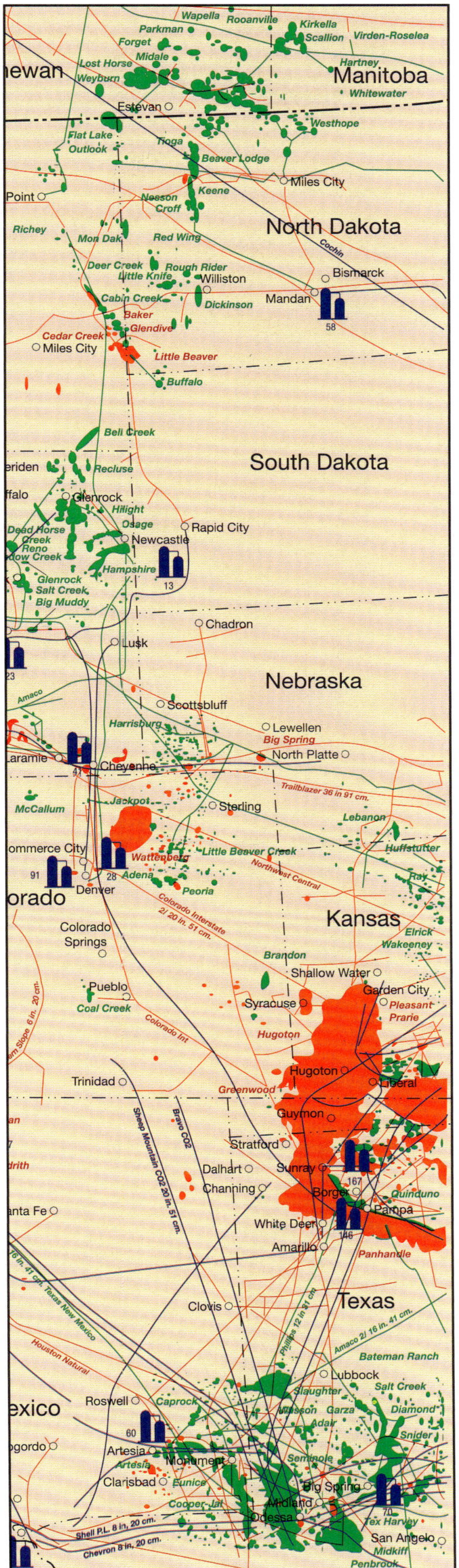

Group. Drillships were 100% engaged and the semisubs 94.5%, but only 73% of the jack ups were working.

The Weatherford-AESC well-service rig count increased to 2,440 rigs for October 2006, up 5% from a year earlier. Rig utilization was at 74%.

Shale

DOE issues leases for oil shale projects

In December the U.S. Dept. of the Interior issued research, development, and demonstration (RD&D) leases for five oil shale projects in Colorado's Piceance basin, said the U.S. Bureau of Land Management. The RD&D leases are for projects proposed in mid-2005 by Chevron USA Inc., EGL Resources Inc., and Shell Frontier Oil & Gas Inc.

The leases grant rights to develop oil shale resources on tracts up to 160 acres for a 10-year initial term. They also contain a preferential right to convert the RD&D acreage, plus as much as 4,960 acres of adjacent land, to a 20-year commercial lease once commercial production levels are achieved and all requirements are met.

The Eocene Green River oil shale formation, which spans parts of Colorado, Utah, and Wyoming, holds the equivalent of 800 billion bbl of recoverable oil, more than 70% of which lies under federal lands.

Palo Duro basin

Bankers Petroleum Ltd. (Calgary) said the right hydraulic fracturing method is the key to unlocking the unconventional gas resources in the Lower Pennsylvanian Bend group sands and shales of the Palo Duro basin in West Texas.

The entire Palo Duro basin covers 22,700 sq miles in the Texas Panhandle, eastern New Mexico, and Oklahoma. It is about 260 miles northwest of the Fort Worth basin Barnett shale play, where improved fracture technologies greatly expanded estimates of recoverable gas. The Palo Duro basin gas play targets shales at 7,000–10,500 ft depths.

Several companies have drilled wells in the basin. The Tyner Stephens well of Tyner Resources Ltd. (Vancouver, BC) is the most successful well to date in the area. The company says that after fracture stimulation the Bend shale produced between 400 Mcfd and 1.5 MMcfd.

Bankers will determine in 2007 or 2008 the viability of the basin for producing gas.

Oil Sands

Ivanhoe Energy upgrades heavy oil bottoms

Ivanhoe Energy Inc. (Vancouver, BC) successfully demonstrated its technology for upgrading vacuum tower bottoms (VTBs) derived from heavy oil

Rotary rig activity by states

	2006	2005	2004	2003	2002	2001	2000	1999	1998	1997
Alabama	4.6	3.6	2.5	2.4	3.0	5.3	4.1	5.5	6.0	3.7
Alaska	8.0	9.3	9.9	9.7	11.2	13.4	8.2	5.0	12.0	10.9
Arizona	0.0	0.0	0.0	0.0	0.3	0.0	0.0	0.0	0.0	0.0
Arkansas	24.0	9.3	6.4	2.1	0.8	1.5	3.9	2.5	6.2	9.8
California	33.3	27.2	23.9	21.1	22.3	36.4	24.1	19.0	27.9	31.7
Land	29.3	23.0	20.4	17.9	19.7	32.5	20.7	17.4	26.1	28.6
Offshore	4.0	4.2	3.5	3.2	2.6	3.9	3.4	1.6	1.8	3.1
Colorado	88.5	73.9	54.2	38.8	27.8	32.3	18.4	12.5	12.8	16.3
Florida	0.3	1.6	1.1	0.7	0.2	0.4	0.2	0.2	0.1	0.4
Idaho	0.3	0.0	0.0	0.0	0.0	0.0	0.0	0.0	0.0	0.0
Illinois	0.0	0.0	0.0	0.0	0.0	0.0	0.0	0.0	0.0	0.0
Indiana	0.0	0.0	0.0	0.0	0.5	0.0	0.0	0.0	0.0	0.1
Kansas	9.6	6.7	6.7	8.7	7.5	22.4	22.0	7.4	13.2	19.0
Kentucky	7.2	4.7	4.4	4.3	4.8	6.4	4.9	5.5	2.4	4.8
Louisiana	188.4	182.1	166.8	157.2	162.8	213.8	194.4	141.1	187.4	193.1
North	57.5	48.4	39.3	28.5	23.2	30.3	24.1	16.2	18.9	20.7
Inland waters	19.2	22.8	18.2	14.3	16.3	20.4	15.8	15.5	21.4	23.2
South	38.5	32.5	30.3	29.6	31.6	44.1	36.7	21.1	40.8	47.8
Offshore	73.2	78.4	79.1	84.8	91.7	119.0	117.9	88.3	106.4	101.4
Michigan	2.2	2.6	3.0	3.1	1.3	1.2	2.4	2.1	5.4	6.2
Mississippi	10.3	10.3	9.8	8.0	7.6	14.2	11.2	7.4	14.0	13.3
Montana	21.3	24.0	19.9	14.0	7.9	10.0	6.5	4.3	8.6	10.8
Nebraska	0.0	0.0	0.8	0.0	0.1	0.2	0.6	0.3	0.5	0.9
Nevada	1.3	1.9	1.5	1.2	0.0	0.0	0.0	0.0	0.0	0.2
New Mexico	93.8	82.8	67.2	64.4	41.5	68.2	67.9	36.0	44.7	52.5
New York	6.4	4.3	4.9	2.8	4.3	5.4	3.3	2.5	2.2	0.4
North Dakota	31.5	20.4	15.0	13.7	10.1	14.3	13.4	5.9	11.2	18.1
Ohio	7.5	9.2	6.7	7.4	8.7	9.6	8.5	10.5	10.1	8.8
Oklahoma	178.7	152.1	158.8	128.2	90.8	130.2	99.4	61.9	84.9	103.4
Pennsylvania	15.7	13.2	8.9	10.1	10.3	10.6	8.7	7.8	10.9	9.7
South Dakota	1.1	2.0	0.5	0.2	0.2	0.6	0.2	0.5	0.1	0.8
Texas	746.4	614.7	505.9	448.5	337.5	462.5	343.4	227.1	302.3	357.4
Gulf Coast	170.3	184.6	156.1	153.0	134.3	168.1	127.1	50.0	63.4	64.7
Offshore & inland waters	14.8	10.5	14.1	20.2	16.2	26.4	16.6	13.8	11.5	15.6
North	33.5	31.8	37.4	39.4	30.1	27.4	14.5	10.5	10.5	12.1
Panhandle	68.2	62.5	47.5	26.0	14.6	21.0	16.7	13.7	20.8	24.8
East	243.3	172.5	131.2	107.2	68.1	106.1	78.0	38.9	54.7	62.7
West Central	79.0	53.0	45.3	28.4	21.9	31.7	17.4	50.1	72.9	68.3
West	137.5	100.0	74.2	74.2	52.5	81.6	73.1	50.1	68.5	109.3
Utah	40.2	27.7	21.5	13.4	13.1	20.8	15.5	8.8	12.5	13.6
West Virginia	26.5	17.4	15.1	15.5	13.2	18.1	14.1	13.7	14.3	14.5
Wyoming	99.0	78.5	73.6	53.6	40.2	55.0	41.0	31.8	38.6	39.3
Others	2.6	3.6	1.5	1.2	2.2	3.6	2.0	5.7	2.6	3.1
Total US	**1,648.7**	**1,383.1**	**1,190.5**	**1,030.3**	**830.2**	**1,156.4**	**918.3**	**624.9**	**830.6**	**943.5**
Land	1,536.6	1,265.9	1,074.0	905.6	699.9	981.4	761.2	502.0	684.7	797.7
Inland waters	22.2	23.7	19.4	16.8	17.7	21.9	17.3	16.6	22.1	24.0
Offshore	89.9	93.4	97.0	107.9	112.6	153.1	139.8	106.3	123.9	121.8
Canada—land	466.5	454.3	361.1	369.8	259.5	336.3	339.7	240.1	255.9	372.7
Canada—offshore	3.6	3.8	3.9	3.8	6.1	5.2	4.7	5.2	3.6	1.3
Grand total	**2,118.8**	**1,841.2**	**1,555.5**	**1,403.9**	**1,095.8**	**1,497.9**	**1,262.7**	**870.1**	**1,090.2**	**1,317.5**

Source: Baker Hughes Inc. Note: May not add due to independent rounding.

US organic shales

	Barnett	Ohio	Antrim	New Albany	Lewis	Fayetteville	Palo Duro
Depth, ft	6,500-8,500	2,000-5,000	600-2,200	500-2,000	3,000-6,000	1,500-6,500	7,000-10,500
Gross thickness, ft	150-700	300-1,000	160	180	500-1,900	50-325	500-1,000
Net thickness, ft	100-600	30-100	70-120	50-180	200-300	20-200	200-?
Bottomhole temperature, °F.	200	100	75	80-105	130-170		180
Total oraganic carbon, %	4.5	0.0-4.7	1-20	1-25	0.45-2.5	4.0-9.5	2.5-4.5
Porosity, %	4.5	4.7	9	10-14	3.0-5.5	2.8	
Reservoir pressure, psi	3,000-4,000	500-2,000	400	300-600	1,000-1,500	600-2,000	3,000-3,250
Gas in place, bcf/section	50-150	5-10	6-15	7-10	8-50	25-60	80-180
Resources, MMcf	500-4,000	150-600	200-1,200	150-600	600-2,000		

Sources: Pickering Partners Inc., Bankers Petroleum Ltd., Vitage Petroleum Inc., and Tyner Resources Ltd.

at its California demonstration facility. VTBs are the heaviest, most viscous component of crude oil. Canada's Athabasca bitumen typically contains 45–55% VTBs, California raw heavy crude about 33%.

The test was significant, Ivanhoe said, because it used a process configuration intended for specific commercial applications—those aimed mainly at viscosity reduction and maximizing liquid volume yields. The company said data from the demonstration and other testing will enable it to start site-specific design and engineering for commercial 10,000–15,000 b/d plants.

Spearhead delivers oil sands crude to Cushing

Enbridge Inc. (Calgary, Alberta) delivered the first crude from oil sands in western Canada to Cushing, OK, through its Spearhead Pipeline in March 2006. The oil entered the Enbridge mainline system at Edmonton, Alta., and traveled 1,565 miles to Chicago before entering the Spearhead Pipeline and traveling 650 miles to Cushing. The initial delivery rate was 80,000 b/d. Enbridge expects throughput to reach Spearhead's 125,000 b/d initial capacity rapidly, and plans to expand capacity to as much as 190,000 b/d.

Pipelines

FERC okays Rocky Mountain gas lines

Three related Rocky Mountain natural gas pipeline projects would be environmentally acceptable with appropriate mitigation measures, Federal Energy Regulatory Commission (FERC) staff concluded in a draft environmental impact statement (EIS).

Known as the Rockies West Phase Project, the three proposed pipelines would transport 1.5 bcfd of gas. The three are the REX-West Project, proposed by Rockies Express Pipeline LLC; the Blanco to Meeker Project, proposed by TransColorado Gas Transmission Co., and the Wamsutter Expansion Project, proposed by Questar Overthrust Pipeline Co.

FERC commissioners will consider the staff recommendations and the final EIS when they issue final decisions on the projects.

CALNV to expand again

Kinder Morgan Energy Partners LP plans to further expand the 550-mile CALNEV pipeline, which transports gasoline, diesel, and jet fuel from California to Nevada. A new 16-in. pipeline from Colton, CA, to Las Vegas, NV would boost system capacity to 200,000 b/d. It would parallel existing utility corridors and will carry product to Nellis Air Force Base. Startup by 2010 is planned.

BEND SHALE POTENTIAL

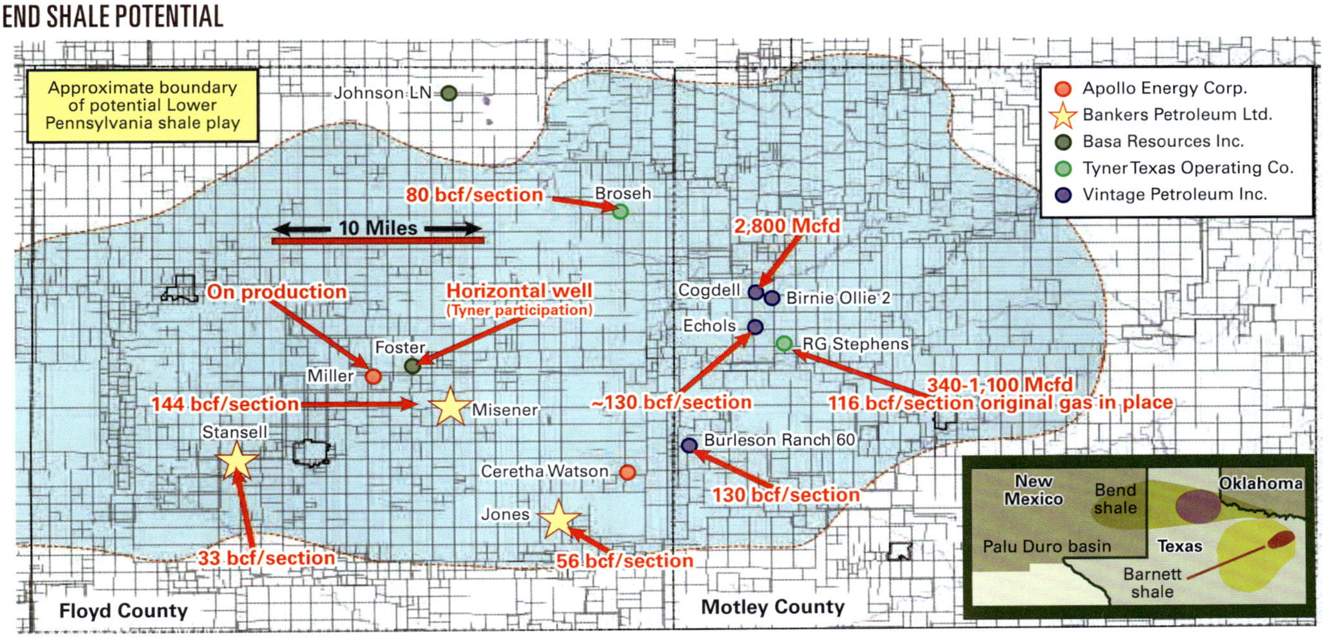

Source: Bankers Petroleum Ltd.

ETNG opens new pipeline lateral

East Tennessee Natural Gas LLC (ETNG), a unit of Duke Energy Gas Transmission, has placed into service its 32-mile Jewell Ridge pipeline, to transport up to 235 MMcfd of natural gas from the Appalachian basin to the U.S. Southeast.

The 20-in. lateral, from a compressor station northwest of Richlands, VA, connects the western Virginia operations of CNX Gas Co. LLC via the existing Cardinal States Gathering System in Tazewell County to ETNG's mainline near Marion in Smyth County. It also will interconnect with the Williams Cos. Inc. Transco interstate pipeline at Cascade Creek, NC.

Final rule on oil spill prevention

The EPA issued a final rule on oil spill prevention, control, and countermeasures (SPCC) that provides compliance alternatives for some facilities. The new rule streamlines options for specifically qualified facilities and exemptions from the regulations for certain vehicle fuel tanks and other onboard bulk oil storage containers.

New pipeline safety law

On December 29, President Bush signed into law a pipeline safety bill that authorizes a 50% increase in the number of federal pipeline inspectors, give additional enforcement authority to the U.S. Office of Pipeline Safety and states to prevent third-party excavation damage, and federally regulate low-stress oil pipelines and local gas distribution systems for the first time.

One major provision mandates use of excess flow valves, which can shut off gas flow when a service line is ruptured, preventing a potential explosion. The Pipeline and Hazardous Materials Safety Administration within the U.S. Dept. of Transportation said it intends to exempt operators of master meter and LPG systems from the requirement, but only because excess flow valves have been shown to be ineffective on such systems.

Millennium gas pipeline delayed

Columbia Gas Transmission Corp. (CGT), a unit of NiSource Inc., and operator of the Millennium natural gas pipeline project, informed the FERC that the anticipated Nov. 1, 2007, in-service date for the proposed Millennium Phase 1 project will be pushed back to Nov. 1, 2008. Phase 1 will transport 500 MMcfd of gas from the Dawn, Ontario, trading hub and several supply and storage basins.

Midcontinent Crossing project

CenterPoint Energy Gas Transmission Co. (Houston) and Duke Energy Gas Transmission (Charlotte, NC) signed a memorandum of understanding to build the Midcontinent Crossing (MCX) natural gas pipeline. The pipeline would run 1,600 miles from the Waha hub, TX, to Oakford-Delmont, PA., carrying 1.5–1.75 bcfd. A 2008 in-service date is estimated.

Plans for 750-mile NGL pipeline

Overland Pass Pipeline Co. LLC (Tulsa, OK) announced plans to build a 750-mile, 110,000 b/d NGL pipeline from Opal, WY to Conway, KS. Construction is slated to begin on the $450 million pipeline in 2007, with completion in 2008.

Rocky Mountain Pipeline System

Rocky Mountain Pipeline System LLC said it is proceeding with the $77 million expansion of its crude oil pipeline system from the terminus of Frontier Pipeline near Evanston, WY to the Salt Lake City, UT refining complex. Parent company is Pacific Energy Partners LP (Long Beach, CA).

The 91-mile, 16-in., 95,000 b/d pipeline will parallel the existing U-Crude pipeline. It will be able to move multiple grades of crude in batches, including heavy crude and synthetic crude. Construction is to be completed in 2007.

North Texas Pipeline

Crosstex Energy LP (Dallas, Tex.) completed construction on the $115 million, 24-in. North Texas Pipeline (NTPL), running about 140 miles from Tarrant County to Lamar County to gather and transport gas produced from the Barnett shale.

The full system became operational in early April. NTPL has an initial capacity of 250 MMcfd, with potential to increase capacity to 375 MMcfd.

Rockies Express Pipeline

Source: Kinder Morgan

Rockies Express Pipeline

Kinder Morgan Energy Partners LP and Sempra Pipelines & Storage secured shipper commitments for the 1,323-mile Rockies Express Pipeline project that will transport natural gas from basins in Wyoming and Colorado to eastern Ohio.

The pipeline, with an estimated cost of more than $4 billion, will have the capability to transport 1.8 bcf/d of natural gas. The first 710-mile segment is projected to be in service by January 2008; the entire project is expected to be completed by 2009.

Louisiana Pipeline

Kinder Morgan also proceeded with plans for its proposed $500 million Louisiana Pipeline. The new interstate natural gas pipeline will provide 3.2 bcfd of take-away capacity from the Cheniere Sabine Pass LNG plant being built in Cameron Parish, LA. The major segment of the new pipeline will consist of a 132-mi, 42-in. line with firm capacity of about 2.0 bcfd, connecting to an existing interstate pipeline in Evangeline Parish. The new pipeline could be in service in 2009.

First market-based gas storage rates approved

FERC in November granted an interstate natural gas storage operator market-based rates for the first time. Under a new approach adopted in Order No. 678 of June 2006, FERC said that Northern Natural Gas Co. may charge market-based rates to initial shippers for firm delivery service from the company's underground storage aquifer field in Redfield, Iowa. Northern plans to expand that facility by about 10 bcf, with a peak withdrawal rate of 175 MMcfd.

Order No. 678 implemented Section 312 of the 2005 energy policy act, which FERC adopted to make gas prices less volatile and to encourage development of more domestic storage capacity.

Refining

Whiting refinery upgrades

BP Products North America Inc. awarded Fluor Corp. packages valued at $3 billion for upgrades to increase Canadian heavy oil processing capacity at its 399,000 b/cd Whiting refinery in northwest Indiana.

The work scope includes engineering, procurement, and fabrication of three major work packages, including a revamped crude distillation unit, a gas oil hydrotreater, and support infrastructure. The project will increase capacity for coking, hydrogen production, hydrotreating, and sulfur recovery. Construction should start in 2007 and be completed by 2011.

BP said the reconfiguration could boost production of motor fuels by about 15% from its current capacity of 4.5 billion gal/year. The upgrade also will boost Canadian heavy crude oil processing capability by about 260,000 b/d.

Further Pascagoula refinery upgrades

Chevron USA, Inc., applied for an environmental permit to build a continuous catalyst regeneration (CCR) unit and several other minor units at its 325,000 b/cd refinery in Pascagoula, MS. The CCR project—to begin in 2008—is expected to further increase gasoline production capacity by 725,000 gal/day.

These projects are to follow the refinery's $150 million fluid catalytic cracking (FCC) project, to be completed by year-end 2006, to boost gasoline production capacity by 10% to 5.5 million gal/day.

Garyville expansion

Marathon Oil Corp. will expand its Garyville, LA refinery. The $3.2 billion project would add 180,000 b/d of crude capacity to the 245,000 b/d refinery. Front-end engineering and design cost estimation is complete and environmental permits are being sought. Construction is planned to begin in mid-2007, with startup in fourth-quarter 2009. Garyville is the last grassroots refinery built in the United States (1976).

CSB report on Texas City 2005 explosion

In October 2006 the U.S. Chemical Safety Board (CSB) released preliminary findings from its investigation into the March 23, 2005 explosion at BP America Inc.'s 460,000-b/d Texas City, TX refinery, which killed 15 workers and injured 180 others. The CSB said that years of budget reductions contributed to conditions leading up to the explosion.

CSB also found that BP was aware of problems at the facility but implemented reforms aimed at improving compliance with procedures and reducing occupational injuries while ignoring potentially catastrophic safety risks. The board does not expect to issue a final report before March 2007.

Other downstream projects planned

Motiva Enterprises LLC said it is in the engineering stages of a 325,000-b/d expansion of its 285,000-b/d refinery in Port Arthur, TX. Construction will begin in 2007, with expanded capacity coming online in 2010. The project would make the refinery the largest in the United States, said Motiva.

Sinclair Oil Corp. said it is in the engineering stage of a project to construct a 30,000-b/d delayed coker at its 60,000-b/d refinery in Tulsa. The unit will begin operating in 2009.

Energy Transfer Partners LP said it is planning a gas processing plant in Johnson County, TX to handle gas produced from the Barnett shale and connect with the partnership's pipeline system. A 115 MMcfd cryogenic gas plant was to begin operating by December 2006, with a second, 170 MMcfd unit to be completed by July 2007. Project cost is $65 million.

GE Energy & Bechtel Power agreed with American Electric Power to proceed with the front-end engineering and design of a 630-Mw coal gasification plant in Mason County, WV. The work was to be completed by year-end 2006.

Catalysts market forecast

The U.S. market for petrochemical and refining catalysts, now about $2.7 billion, will increase 3.0%/year to about $3.2 billion, said Business Communications Company Inc. (BCCI; Norwalk, CT).

Their report classifies petrochemical catalysts as chemical process catalysts and refinery catalysts, which are used in petroleum refineries for the express purpose of producing fuel products.

The report estimates the current chemical catalyst market at about $1.9 billion, growing at 3.2%/year to $2.2 billion in 2010. This includes six major categories: hydrogenation, dehydrogenation, organic syntheses, oxidation, polymerization, and syngas processes.

Most of the market for chemical catalysts comes from polymerization, which in 2005 accounted for $846.5 million or nearly 45% of the total market. It will continue to dominate the market through 2010; demand for polymerization catalysts will grow about 4.5%/year through 2010.

The report estimates refining catalyst sales of $851.4 million in 2005, which will grow at 3.2%/year to about $999 million in 2010. Increased demand for reformulated and other less-polluting gasolines has spurred the growth in petroleum catalysts, said the study.

Refining on tightrope

The July 2006 annual U.S. Refining and Storage Capacity Report of the National Petrochemical & Refiners Association (NPRA) showed the U.S. refining industry contending with high costs for crude oil and natural gas to run their plants, on one hand, and with being suspected of "gouging" consumers, on the other.

NPRA's report said U.S. refining capacity as of Jan. 1, 2006, had increased by 214,000 b/d, or 1.25%, from a year earlier. Capacity utilization was down substantially year-to-year, however, because of lingering effects of Hurricanes Katrina and Rita. Refiners also have faced rapidly increased use of ethanol as a gasoline additive and new requirements for ultralow-sulfur diesel.

US CATALYST MARKET

Catalyst application	– Sales, million $ – 2005	2010	Average growth 2005-10, %/year
Chemical	1,890.4	2,191.0	3.0
Refinery	851.4	999.0	3.2
Total	**2,741.8**	**3,190.0**	**3.0**

Source: BCC Inc.

NEW U.S. DIESEL ERA

A new era dawned in mid-October 2006, said OGJ, for U.S. drivers who use diesel fuel. That's when new rules took effect, restricting the sulfur content of 80% of highway diesel fuel sold in the United States to 15 ppm—a 97% reduction from the 500 ppm allowed previously.

All but the smallest refiners had to start making ultra low sulfur diesel (ULSD) in June 2006. Some are processing to well below the specified level in order to offset contamination by residual sulfur in distribution systems. California's ULSD deadline was September 1, but EPA gave the industry outside that state an extra 45 days to address contamination and other potential transportation problems.

The head of the U.S. Environmental Protection Agency called the transition to ULSD "the single biggest achievement since the removal of lead from gasoline 25 years ago."

EPA set the availability deadline to accommodate exhaust-treatment systems on diesel engines starting with the 2007 model year. ULSD also will yield health benefits, said EPA, through reduced emissions of ozone precursors and soot. Its use by more than 90% of the nation's trucks and buses should cut nitrogen oxide emissions by 2.6 million tons/year and particulate matter emissions by 110,000 tons/year.

At the same time, however, motorists have noticed that the price of diesel is now higher than the price of gasoline—a reversal likely to become more the norm than the exception for several reasons.

- Beyond the estimated $9 billion the refining industry spent on equipment needed for compliance, its operating costs also have increased for the hydrogen inputs, catalysts, processing severities, and other changes made necessary by the EPA diesel program.

— CONTINUED

New U.S. diesel era — cont.

- Indirect costs also arise, such as new pressure on total oil supply as newly intense processing lowers product yields. Overall, the cost of making highway diesel has risen, although a crude-price slump now masks the effects.

- Diesel also faces a market squeeze, says OGJ. Supply of ULSD in the United States now largely depends on domestic and nearby hydrotreating capacity and availability of sweet crude. Because other countries also are lowering diesel sulfur content, competition for on-spec product in international trade will be intense, which means the price will be high.

U.S. demand for highway diesel has a new and growing stimulus: the logistics of fuel ethanol, production of which is geographically diffuse and transportation of which requires trucks.

Further boosts to U.S. demand for ULSD will come from regulations extending the ULSD requirement to 100% of highway diesel by 2010 and to off-road product in 2015.

Higher renewable-fuel market target

The EPA proposed in September 2006 to raise the total share of renewable fuels in the United States to 3.71% in 2007 from 2.78% in 2006 by doubling the use of ethanol, biodiesel, and other renewables.

The renewable fuel standard aims to increase U.S. use of renewable motor fuel to at least 7.5 billion gal in 2012 from about 4.5 billion gal in 2006. It contains compliance tools and a credit and trading system. The proposal could reduce petroleum demand by 3.9 million gal/year and greenhouse gas emissions by 14 billion tons/year by 2012, said EPA.

The National Petrochemical & Refiners Association (NPRA) questioned the wisdom of mandating levels for ethanol, biodiesel, and other renewables. NPRA pointed to disagreement among experts whether any true net reduction of greenhouse gas emissions occurs with renewable fuel use.

Enterprise Products to expand LPG terminal

Enterprise Products Partners LP (Houston) announced plans for a second phase of expansion of its LPG import-export terminal on the Houston Ship Channel. The work will double the import terminal's maximum peak operating rate to 480,000 b/d and increase the export terminal's maximum peak rate to 160,000 b/d from 140,000 b/d.

The terminal, at Oiltanking Houston LP, will be able to simultaneously unload two very large gas carriers, which carry as much as 550,000 bbl of LPG.

Ethanol

Interest in cellulosic ethanol grows

The U.S. Dept. of Energy (DOE) said it is evaluating bids for the first commercial U.S. cellulosic ethanol production facility, aiming for an operational plant by 2012. Cellulosic ethanol—produced from nonfood plant material—is not yet commercial because its production costs are higher than those of ethanol production from grain.

Biobutanol is of interest. It is an alcohol produced from such biomass materials as sugar cane, sugar beets, corn, wheat, cassava, and sorghum.

BP Fuels Management Group said it is collaborating with DuPont to produce biobutanol. BP reported plans to introduce biobutanol as a gasoline component in the UK in 2007. The company said it is working with Associated British Foods PLC subsidiary British Sugar to convert the country's first ethanol fermentation facility to produce biobutanol.

One DOE goal is to drive down the cost of cellulosic ethanol production by 2012 to $1.07/gal from $2.25/gal in 2005. Another is production of enough U.S. ethanol by 2030 to equal 30% of current transportation fuel demand. According to the Renewable Fuels Association, U.S. ethanol production totaled of 318,000 b/d in mid-2006 and demand was 395,000 b/d.

Ethanol imports needed

U.S. demand for ethanol is outstripping supply and raising imports of the product, said Friedman, Billings, Ramsey Group Inc. (FBR; Arlington, VA). Demand jumped to an annualized average of 5.4 billion gal/year in May 2006, said FBR. This was well above the April level of 4.7 billion gal/year average and the comparable year-ago rate.

With domestic annualized U.S. production at only about 4.5 billion gal/year, the shortfall in domestic supply was offset by a draw on inventories and significant imports, explaining a rise in ethanol spot market prices. Although prices returned to more normal levels in August, FBR expressed concern about the impact of a supply/demand mismatch on prices.

Federal officials are debating whether to lift the 54¢/gal tariff on imported ethanol until domestic production increases to meet U.S. demand. EIA estimated that removing the tariff would increase U.S. imports by 10,000–20,000 b/d over the short run from the current 22,000 b/d (chiefly from Brazil).

Lurgi to build ethanol plant

Lurgi AG (Frankfurt, Germany) won a contract worth €66 million to build a bioethanol plant in the United States. The contract was let by Central Illinois Energy LLC for a 100,000-tonne/year bioethanol plant in Canton, IL. The plant is slated to come on stream in 2007 using corn as feedstock.

Air quality issues

While volumes of greenhouse gases emitted in the United States are rising, the rate of increase and amounts emitted in relation to economic growth are falling. In an annual estimate, the U.S. EIA said 2005 emissions of greenhouse gases associated with human activity totaled 7.1 billion tonnes of carbon dioxide equivalent (CO_2e)—up 0.6% from 2004 and 17% from 1990.

Emissions of CO_2, the target of most policy proposals for global-warming remediation, grew in 2005 at a below-average rate of 0.3%. Emissions—adjusted to CO_2e to reflect global warming potential—of nitrogen oxides were up 1.9% and of methane, up 0.9%.

CO_2 represented 84% of U.S. greenhouse-gas emissions in 2005. Nitrogen oxides and methane together made up 14% of the total. The other greenhouse gases are hydrofluorocarbons (HFCs), perfluorocarbons (PFCs), and sulfur hexafluoride (SF6), emissions of which increased by a combined 7.2% in 2005.

Since 1990, U.S. greenhouse gas emissions have increased at 1%/year. EIA notes that the growth rate is below those of population (1.2%/year), primary energy consumption (1.1%/year), electric power generation (1.9%/year), and gross domestic product (3%/year) over the same period.

Greenhouse gas intensity—emissions per unit of real economic output—has fallen by 25% since 1990 at a rate averaging 1.9%/year, EIA said. CO_2 intensity has fallen by 23% over the same period.

The EPA in September announced new air quality standards that would reduce daily allowable fine particulate matter levels by almost 50%. The new National Ambient Air Quality Standards address both fine particles (as small as 2.5 μm) and inhalable coarse particles (2.5–10 μm). The new standards likely will have a bigger impact on electric power generators than on oil and gas producers, according to industry observers.

US greenhouse-gas emissions

Gas	1990	1995	1998	1999	2000	2001	2002	2003	2004	2005*
	Million tonnes of CO_2e									
Carbon dioxide	4,990.6	5,308.5	5,594.0	5,673.9	5,853.4	5,767.0	5,814.7	5,875.3	5,988.7	6,008.6
Methane	701.7	672.5	629.8	616.5	611.2	597.7	600.2	602.2	606.5	611.9
Nitrous oxide	333.5	357.7	348.8	346.8	342.8	337.9	333.6	332.9	359.9	366.6
HFC, PFC, SF_6	87.1	94.9	134.3	133.9	138.0	128.5	137.8	136.6	149.5	160.2
Total	**6,112.8**	**6,433.5**	**6,707.0**	**6,771.1**	**6,945.4**	**6,831.0**	**6,886.3**	**6,946.9**	**7,104.6**	**7,147.2**

*Preliminary.
Source: US Energy Information Administration

LNG

A review of North American LNG import capacity by *OGJ* examined expansions at existing terminals, proposals that advanced to receive regulatory approvals, and all that were under construction on August 1 or about to start.

Five LNG import terminals currently serve U.S. markets. Four are conventional land-based terminals (three on the East Coast); the fifth is 116 miles off Louisiana and is served by new-generation combined LNG shipment and regasification tankers.

Expansion has taken place or is under way at the three land-based terminals in the Lower 48 states that can expand, while owners of the fourth have advanced an offshore project to increase supplies to its markets. Construction is under way at three new U.S. terminals, all on the Gulf Coast.

On Mexico's Gulf coast, a small LNG terminal could start operations by early 2007. On its Pacific coast, another terminal is under construction with one, an offshore concept, on the drawing boards. In Canada, one terminal on the East Coast has cleared ground.

North American LNG terminals: mid-2006[1]

Project name, location	Owner(s)	Baseload sendout capacity, bcfd	Start-up date[2]
Under construction on Aug. 1, 2006			
Altamira, Tamaulipas, Mex.	Shell Gas	0.5	2006
Cameron (Hackberry), La.	Sempra LNG	1.5	2008
Canaport, St. John's, NB	Repsol YPF; Irving Oil	1.0	2008
Cove Point, Md., expansion	Cove Point LNG	0.8	2008
Energía Costa Azul, Baja Calif.	Sempra LNG	1.0	2008
Freeport, Tex.	Freeport LNG; Cheniere	1.5	2008
Sabine Pass, Tex.	Cheniere	2.6	2008
Reg. approval in hand; no construction on Aug. 1			
Bear Head LNG, Point Tupper, NS	US Venture Energy	1.0	2007
Corpus Christi, Tex.	Cheniere	2.6	2010
Creole Trail LNG, Cameron, La.	Cheniere	3.3	2011
Crown Landing, NJ	BP PLC	1.2	2008
Golden Pass, Sabine Pass, Tex.	ExxonMobil	2.0	2009
Gulf Landing (off Louisiana)	Shell	1.0	2010
Ingleside (Corpus Christi, Tex.)	Ingleside Energy; Occidental Energy Ventures	1.0	2010
Kitimat, BC	Galveston LNG	0.61	2009
Port Arthur, Tex.	Sempra LNG	3.0	2010
Port Pelican (off Louisiana)	Chevron	1.6	2007
Sabine Pass, Tex. (expansion)	Cheniere	1.4	2009
Weaver's Cove (Fall River, Mass.)	Weaver's Cove Energy; Hess LNG	0.8	2008

[1]Based on and updated from US Federal Energy Regulatory Commission (http://www.ferc.gov/industries/lng/indus-act/terminals/exist-prop-lng.pdf). [2]According to company-published material or spokesperson.

FERC streamlines rules on gas, LNG

In October the FERC approved a final rule expanding blanket certificate eligibility for projects at storage facilities, pipelines, and facilities to transport gas from LNG receiving terminals.

Blanket certificates, granted to companies already having FERC project approval, allow them to upgrade or build new facilities without case-by-case authorization. The latest ruling raised the cost limits for blanket certificate projects to $9.6 million from $8.2 million for automatic authorizations and to $27.4 million from $22.7 million for prior-notice projects, which require FERC review.

EIA outlook

EIA projected total LNG net imports of 760 bcf in 2006 (up from 630 bcf in 2005), growing to 1 tcf in 2007. High prices for natural gas will accelerate construction of new LNG terminals and result in a large increase in total U.S. LNG receiving capacity by 2015.

Total sendout capacity of U.S. LNG receiving terminals is expected to reach 4.9 tcf in 2015, up from 1.4 tcf in 2004. By 2015, net LNG imports will likely hit 3.1 tcf. Construction of new LNG terminals will slow after 2015. With terminal capacity of 5.8 tcf in 2030, U.S. net LNG imports total 4.4 tcf.

Net U.S. natural gas exports to Mexico will fall through 2011 as new LNG terminals are built in that country. After 2011 U.S. exports to Mexico increase, to 560 bcf in 2030.

Gulf Coast

Mississippi

Construction of a proposed LNG terminal near Pascagoula, MS would have minimal environmental impact, the staff of the FERC said in a final environmental impact statement. The Gulf LNG Clean Energy Project also would include a new pipeline to ship as much as 1.5 bcfd of gas out to three interconnects.

Louisiana

Trunkline LNG's Lake Charles terminal placed its first-phase expansion into service in 2005, increasing baseload sendout capacity to 1.2 bcfd from 630 MMcfd, with an expanded peak capacity of 1.5 bcfd. In April 2006 it added a second ship berth and a new LNG storage tank that increased storage capacity to 9 bcf of natural gas equivalent.

The second-phase expansion started up in July to increase sendout capacity to 1.8 bcfd with peak sendout of 2.1 bcfd.

Construction on two other terminal projects near the Lake Charles terminal is under way or approved. Work started in September 2005 near Hackberry on Sempra LNG's Cameron LNG terminal. In January the company got regulatory approval to expand the terminal's baseload sendout to 2.65 bcfd and add a fourth storage tank. Construction for the $250-million expansion will begin in 2007, with start up in 2010.

One of several projects sponsored wholly or in part by Cheniere Corp. (Houston) is the Creole Trail LNG LP. The project received FERC approval in June to proceed with a new 3.3-bcfd terminal and associated pipelines (Creole Trail Pipeline LP). The terminal will sit at the mouth of the Calcasieu Channel, in Cameron Parish.

The Creole Trail pipeline will interconnect with more than 12 bcfd of interstate and intrastate pipeline capacity in southwest Louisiana and extending northward about 120 miles. Cheniere expects to break ground on the project in 2007and begin operations in 2011.

McMoRan Exploration Co.(New Orleans) said it will seek a change to its Deepwater Port license permit of its proposed $400 million offshore LNG receiving terminal at Main Pass Energy Hub, about 37 miles east of Venice, LA. The terminal would use open rack vaporization (ORV), a technology about which some parties have environmental concerns. ORV uses seawater to warm the LNG.

South of Cameron, Shell U.S. Gas & Power LLC plans to build a gravity-based structure (GBS) terminal—two concrete caissons—to deliver 1 bcfd starting in 2010. Approved in February by the U.S. Maritime Administration, Gulf Landing will deliver to as many as five major interstate pipelines.

Along Sabine Pass where Louisiana and Texas join, three projects are under construction or approved. In June the FERC authorized Cheniere's project, the Sabine Pass LNG LP, to expand capacity at its LNG receiving terminal, already under construction in western Cameron Parish, to 4 bcfd from 2.6 bcfd. The Sabine Pass terminal is just west of Sempra's Cameron LNG and Trunkline's Lake Charles terminal.

Sabine Pass LNG LP, owned by Cheniere, began construction in March 2005 and suffered severely from Hurricane Rita but is back on schedule, said the company. Start-up of the initial phase will be 2008; expansion is to be completed in 2009.

Gas from the terminal will pass through the 16-mile Cheniere Sabine Pass LNG pipeline to connect with several interstate and intrastate pipelines. The pipeline is designed to move 2.6 bcfd.

LNG TRANSSHIPMENT

Excelerate Energy's operational Gulf Gateway Energy Bridge offshore LNG terminal started up in April 2005. An official with the Society of International Gas Tanker and Terminal Operators (SIGTTO) told OGJ in late 2005 that the greater significance of the technologies associated with Energy Bridge may be the evolution toward ship-to-ship LNG transfer, or "transshipment."

A few service companies have worked on technologies, but none has reported full-scale tests or operations. As of August, only Excelerate Energy had reported full-scale field trials using off-the-shelf components. The company's Excelsior transferred about 13,000 cu m of LNG to its sister ship Excalibur in a trial in international waters in the Gulf of Mexico.

SIGTTO is developing procedural guidelines, and Excelerate's goal is to have them ready for the planned year-end 2006 opening of the company's Teesside, UK, offshore LNG terminal. Transshipping might be of great importance for terminals that are constrained from handling large LNG carriers.

Texas

FERC approved in June the Sempra LNG plan to build the Port Arthur LNG project along the Port Arthur ship channel, bout 85 miles east of Houston. The $800 million terminal project would have a baseload sendout capacity of 1.5 bcfd and be capable of expansion to 3 bcfd. First-phase commercial operation would begin in 2010.

The planned Golden Pass project proposes a new LNG import terminal about 10 miles south of Port Arthur. It would deliver the vaporized LNG to 10 interconnections with existing pipeline systems and the ExxonMobil Beaumont refinery.

The proposed LNG facilities would have a baseload sendout capacity of 2 bcfd with a peak capacity of 2.7 bcfd. The facility would be capable of unloading two LNG ships holding up to 250,000 cu m.

Roughly 100 miles southwest of Sabine Pass is the third U.S. Gulf Coast LNG terminal under construction. Freeport LNG Development LP is building and will operate an LNG terminal on Quintana Island, about 70 miles southwest of Houston near Freeport.

The first-phase regasification terminal will have a peak sendout capacity of 1.75 bcfd, and begin operations in early 2008. Phase-2 expansion will add up to 1.15 bcfd of capacity. This second phase will install a second marine berthing dock and additional storage. Phase-2 construction could start in 2007, with operation in 2009.

In December 2005, Cheniere Energy's Corpus Christi LNG LP received FERC approval to begin construction of a terminal near Corpus Christi. Site preparations and construction are under way, but a full construction start date has not been specified. Operations are to begin in 2010.

Initial baseload sendout capacity will be about 2.6 bcfd with peak sendout of 2.9 bcfd. Gas from the terminal will move through the Cheniere Corpus Christi LNG pipeline 24 miles to a connecting pipeline.

Another terminal is planned by a subsidiary of Occidental Petroleum Corp. The Ingleside Energy Center will be adjacent to Occidental Chemical's existing plant west of Ingleside, TX. The terminal would have nominal sendout of 1 bcfd and include a pipeline with up to 12 interconnects. Construction could begin in 2007.

The $600-million Vista del Sol LNG terminal, authorized by the FERC in June, would be located along the northeastern shore of Corpus Christ Bay. Start-up is proposed for 2008–09. It has a planned baseload sendout of 1.1 bcfd and a peak-day sendout of 1.4 bcfd. An associated pipeline would carry gas to an interstate pipeline interconnection and have up to eight interconnects.

Several more projects proposed for the U.S. Gulf Coast had not been approved by FERC as of August. These include the 1-bcfd Calhoun LNG terminal near Port Lavaca, TX, and expansions at Freeport LNG (2.5 bcfd) and Sempra's Cameron terminal (1.15 bcfd).

Also, ConocoPhillips withdrew its application for a deepwater port license for the proposed Compass Port LNG project, 11 miles off Dauphin Island, AL, based on environmental concerns.

West Coast LNG

California

It is not clear whether California will allow installation of LNG receiving terminals of any configuration, but at least three have been advanced. One is land based (Long Beach) and two would be built off the state's shores (Cabrillo Port and Clearwater Port).

Others have also been proposed, but not all will be built.

- ConocoPhillips and Sound Energy Solutions formed SES Terminal LLC in 2005 to jointly build an LNG import terminal at Long Beach by 2009.
- Australia's BHP Billiton Ltd. has applied for permits for the $550 million Cabrillo Port floating LNG project, 21.5 miles off Ventura County. It would involve a permanently moored floating storage and regasification unit (800 MMcfd), LNG storage tanks, and a subsea pipeline to shore,
- Houston-based Crystal Energy LLC's Clearwater Port project would use the existing Grace oil platform as the site of an LNG terminal, approximately 12 miles off Ventura County.
- Esperanza Energy LLC, formed in March by Tidelands Oil & Gas Corp., is evaluating feasibility of a deepwater LNG terminal at an undetermined site as far as 12 miles off California.

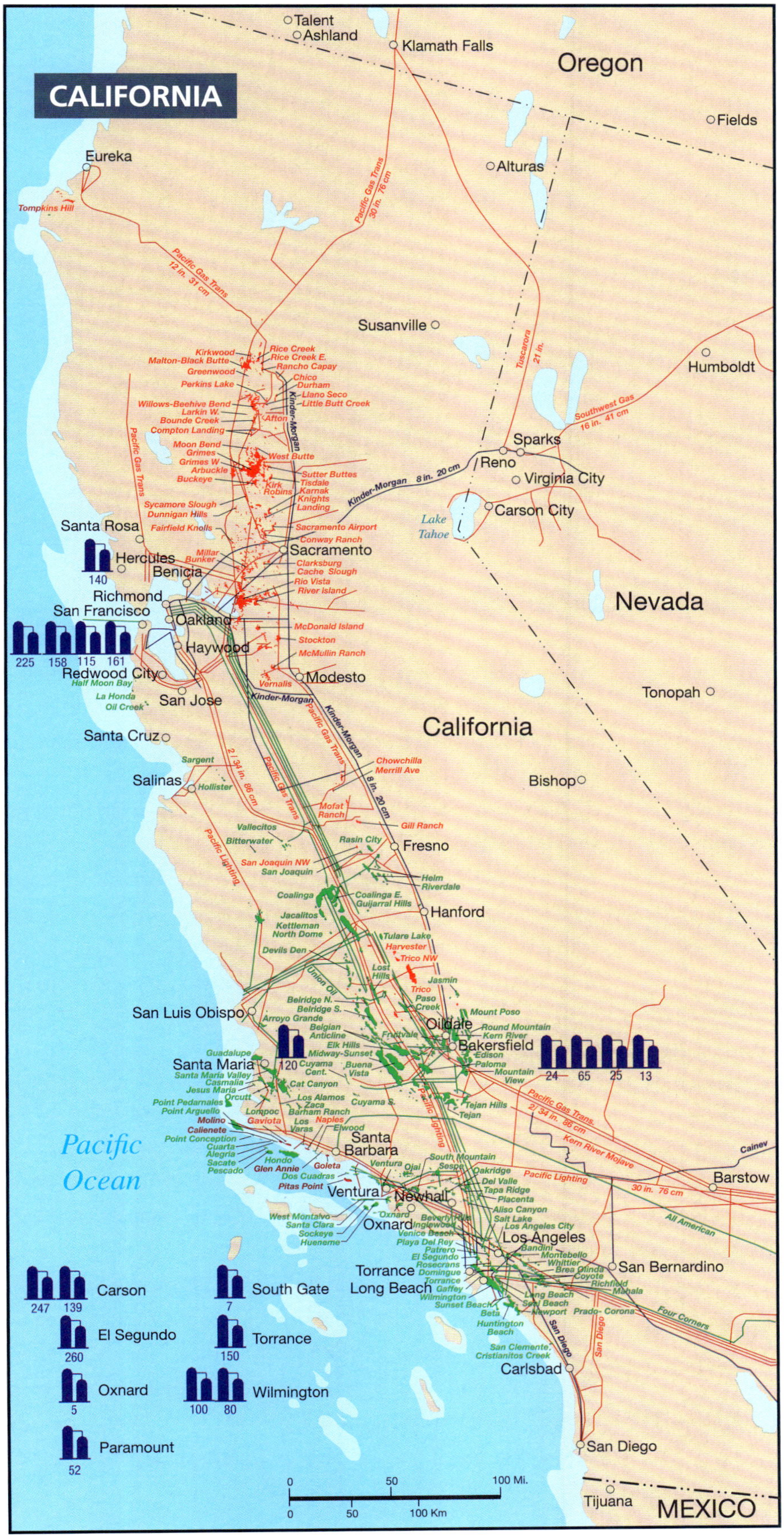

- Woodside Natural Gas Inc. is seeking permits for its OceanWay Secure Energy project about 28 miles off Los Angeles. An underwater buoy system would connect to specially designed vessels that offload regasified LNG into a subsea pipeline.
- Chevron Corp. is looking at sites for LNG import terminals in northern and southern California and planning a terminal eight miles off Tijuana, adjacent the Coronado Islands.

Oregon

The Jordan Cove Energy Project LNG receiving terminal and regasification plant at Coos Bay is proposed by Fort Chicago Energy Partners LP (Calgary, Alberta). The project includes the 250-mile Pacific Connector pipeline to carry gas away from the terminal. That pipeline would be built by 2010 to deliver 1 bcfd to the Pacific Northwest and beyond, including California and northern Nevada.

Bradwood Landing, proposed by Northern Star LNG, Houston, is clearing a site on the southern shore of the Columbia River in Clatsop County. The company filed its application with FERC in June 2006. Installed capacity will be about 1.3 bcfd.

East Coast LNG

New England and South

Suez LNG NA subsidiary Distrigas of Massachusetts LLC owns and operates the LNG import and regasification terminal on the Mystic River near Boston. Rated vaporization capacity is about 1 bcfd, with a sustainable daily throughput capacity of about 715 MMcfd.

The terminal's location prevents significant expansion, so Distrigas parent Suez LNG has proposed an offshore LNG terminal about 22 miles northeast of Boston

and in 250 ft of water 10 miles off the coast. The Neptune offshore project will feed another 400 MMcfd on average and about 700 MMcfd peak-day delivery into New England markets.

LNG carriers will moor to a buoy system and discharge LNG regasified by vaporization equipment aboard ship. The gas will be transported to shore by a subsea pipeline into the existing Algonquin HubLine pipeline. As of third-quarter 2006 regulatory approval had not been obtained. Construction would take about 36 months.

A competing proposal has been submitted by Excelerate Energy LLC. The Northeast Gateway Energy Bridge deep-water port would be built 13 miles southeast of Gloucester, MA, in federal waters. It would use technology similar to that at Excelerate's Gulf Gateway Energy Bridge, off Louisiana. That terminal started up in 2005.

The terminal would move regasified LNG in a new 16-mile pipeline lateral to be built and operated by Algonquin into its HubLine, allowing delivery of 400 MMcfd with peak sendout of 800 MMcfd. Excelerate expects regulatory approvals in 2007 and could have offshore equipment ready to receive gas by 2008.

Another facility proposed for the Northeast received FERC approval in mid-2005, but the $250-million project has been stymied by local opposition, state legislators, and the U.S. Navy. Weaver's Cove Energy LLC envisioned an LNG import terminal near Fall River, MA. The terminal would have baseload sendout capacity of 400 MMcfd and peak sendout capacity of 800 MMcfd.

Also on the U.S. East Coast, the FERC approved in June the construction of the Crown Landing LNG project in Logan Township, NJ. The planned terminal will have a baseload sendout capacity of 1.2 bcfd and peak-day capacity of 1.4 bcfd. Commercial service is planned for 2008, but a court case is pending with regard to state jurisdiction over the project.

Mid-Atlantic LNG

South of Baltimore, MD, is Dominion Cove Point LNG LP's facility, the largest U.S. LNG import terminal. Its sendout capacity is about 1 bcfd. Work began In October to expand sendout capacity to 1.8 bcfd. with completion scheduled for 2008.

Southern LNG Inc. (Birmingham, AL) is conducting Phase 3 expansion at the Elba Island LNG terminal near Savannah, GA, and construction of the 190-mile Elba Express Pipeline. This work will add 8.4 bcf of storage capacity and will increase sendout capacity to 2.1 bcfd by 2010. Total expected capital cost for the expansion and pipeline is $930 million.

Competing plans to base LNG terminals in the Bahamas and send regasified LNG through subsea pipelines to Florida evolved in 2004 into one. Calypso LNG, a subsidiary of Suez Energy NA; Ocean Express, a unit of AES Ocean LNG Ltd.; and units of El Paso Corp., agreed to merge their proposed projects. As of mid-2006, however, prospects for the terminal appeared poor.

Suez Energy NA, unsure of gaining approval for an LNG terminal in the Bahamas, announced plans to replicate its Neptune offshore LNG regasification scheme (discussed above) to waters off Florida.

Subsidiary Calypso LNG LLC will pursue development of a facility off southeastern Florida to be an offshore delivery point for LNG regasification tankers, mirroring the scheme proposed for Massachusetts Bay.

In March the company filed a Deepwater Port License Application with the Coast Guard. Calypso Deepwater Port would be about 10 miles off Port Everglades and consist of a marine offloading buoy system that will reside, when idle, 150 ft below the water's surface. A subsea pipeline would move the gas to shore.

Mexico LNG

The Terminal de LNG de Altamira S de RL de CV, the first LNG terminal in Mexico, began commercial operations in October. A second, Energía Costa Azul, near Enseñada, is under construction. And a third is planned: Terminal GNL Mar Adentro de Baja California. Details are presented in the Mexico section of this book.

Canada LNG

One or more LNG terminals proposed for construction in Canada could supply that country as well as parts of the United States. Information about the Canaport project in New Brunswick, the Bear Head project on Cape Breton Island, and the Kitimat project near Kitimat, British Columbia, are described in the Canada section of this book.

GULF OF MEXICO

Deepwater Developments

According to the U.S. Minerals Management Service (MMS), production from the deepwater Gulf of Mexico (water depths of more than 1,000 ft) accounted for less than 5% of the gulf's total oil and 1% of its total gas output in 1990.

However, in 2004, deepwater gulf production accounted for more than 67% (362 million bbl) of the oil and 37% (1.5 tcf) of the gas from the region.

INDUSTRY RESULTS IN GULF'S LOWER TERTIARY TREND

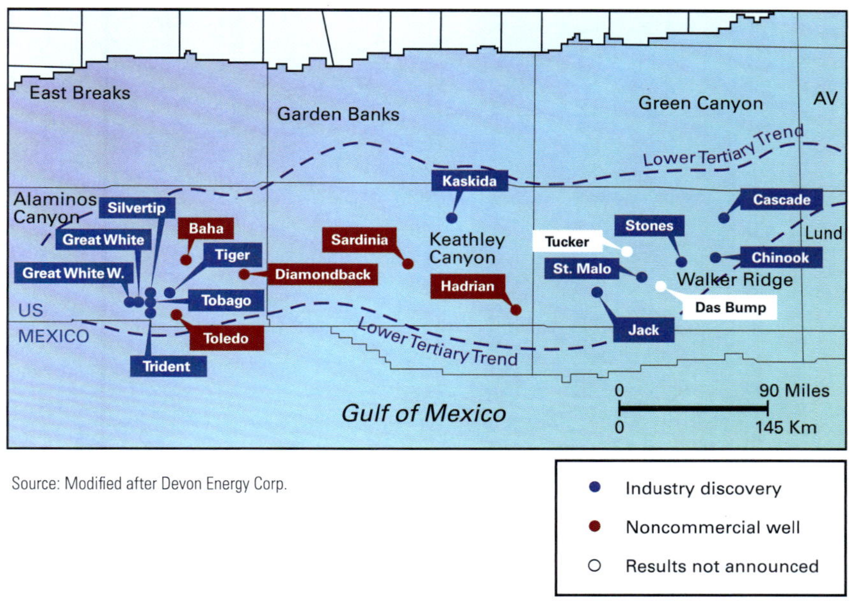

Source: Modified after Devon Energy Corp.

INDUSTRY'S LOWER TERTIARY DISCOVERIES IN THE GULF

Year	Present operator	Discovery	Block	Water depth, ft
2006	BP	Kaskida	Keathley Canyon 292	5,860
2005	BP	Stones	Walker Ridge 508	9,576
2004	Chevron	Jack	Walker Ridge 759	6,965
	Chevron	Silvertip	Alaminos Canyon 815	9,226
	Chevron	Tiger	Alaminos Canyon 818	9,004
	Chevron	Tobago	Alaminos Canyon 859	9,627
2003	BHP Billiton	Chinook	Walker Ridge 469	8,831
	Chevron	St. Malo	Walker Ridge 678	7,036
2002	Devon	Cascade	Walker Ridge 206	8,143
	Shell	Great White*	Alaminos Canyon 857	8,717
2001	Chevron	Trident	Alaminos Canyon 903	9,743

*Information not announced on Great White West.

Deepwater production growth has more than offset declines in shallow water, allowing total U.S. gulf oil production to grow from 283 million bbl in 1990 to 540 million bbl in 2004.

Promising lower tertiary trend

More than 99% of Gulf of Mexico production has come from Tertiary Pleistocene, Pliocene, and Miocene formations. Just five years ago, few observers believed that the Lower Tertiary sands would be present or productive at great depths.

But ultradeepwater discoveries in the gulf extend the Lower Tertiary Trend and represent a resource that could produce oil and gas for decades to come, said MMS.

This trend began with discoveries in the Alaminos Canyon and Walker Ridge areas and now has extended to the Keathley Canyon area. MMS said the area could be as wide as 300 miles and involve as many as 3,000 blocks that MMS administers. The play area covers more than 10 million acres, much of it near Mexican waters.

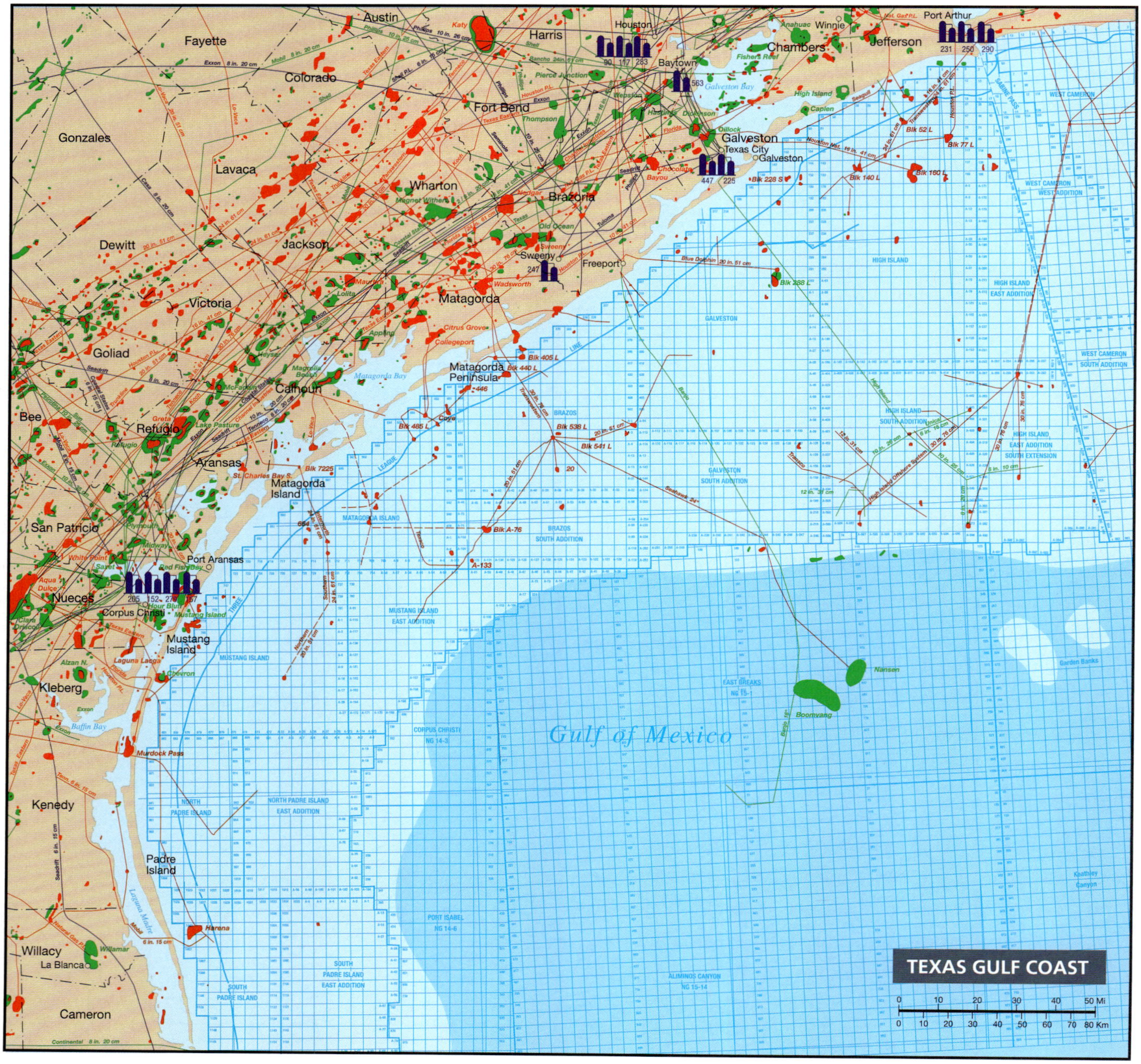

Even with the play still in its infancy, the industry is estimating ultimate recovery at 9–15 billion boe, which could make it the biggest domestic discovery since Alaska's Prudhoe Bay. For comparison, current U.S. proved crude oil reserves are 21.8 billion bbl.

Even if the find does turn out to be as large as estimated, it will take five to seven years and tens of billions of dollars to reach the start of full production.

The industry overall has made 12 discoveries in 19 wells drilled in the trend for a 63% geologic success rate, according to Devon Energy Corp. (Oklahoma City).

Chevron is the largest lease holder in the deepwater Gulf of Mexico and is currently developing the $3.5 billion Tahiti project located in Green Canyon Block 640. That project is scheduled to commence production in 2008. Others with deepwater projects in progress in the Lower Tertiary are Exxon Mobil Corp., BP PLC, and Anadarko Petroleum Corp.

Jack-2 well in the spotlight

In second quarter 2006, Chevron Corp. (San Ramon, CA), Devon Energy Corp. and Statoil ASA (Stavanger, Norway) completed the only extended well test of a Lower Tertiary well to date.

Chevron said the Jack-2 well in Walker Ridge Block 758 sustained flow rates of more than 6,000 b/d of oil from 40% of the total measured net pay of more than 350 ft. Oil and gas came from more than 28,000 ft under the seabed in 7,000 ft of water. The well is 270 miles southwest of New Orleans.

The test set more than a half-dozen world records for test equipment pressure, depth, and duration in deepwater, according to Chevron. It broke the company's 2004 Tahiti well test record as the deepest successful well test in the Gulf of Mexico. A second test well is planned for 2007.

Calls for legislative action

The discovery of the potential new oil reserves in the Lower Tertiary in the gulf prompted calls from some federal lawmakers for more rapid action to open the areas associated with Lease Sale 181 (in the central and eastern gulf) and Lease Sale 181 South for oil and gas development.

Outer Continental Shelf (OCS) leasing-reform legislation was, in fact, signed into law in December, and the MMS has included the region in its proposed five-year OCS leasing plan for 2007–12.

Other Deepwater Activity

Find at Mission Deep

Anadarko Petroleum Corp. discovered oil at its Mission Deep prospect on Green Canyon Block 955 in the Gulf of Mexico. The discovery well found more than 250 ft of net oil pay in the primary middle Miocene objective. The well, in 7,300 ft of water, was drilled to TD 25,000 ft. Anadarko is Mission Deep operator. Devon Energy Corp. is a partner.

Independence Hub nears production

Gas production from the eastern gulf is scheduled to start in 2007 from 10 deepwater fields tied into a floating central processing facility called Independence Hub, moored in Mississippi Canyon Block 920.

Independence Hub will tie together multiple deepwater gas fields lying in 8,000–9,000 ft of water that otherwise would be uneconomical to develop on a standalone basis. Initially, 10 fields will anchor the project and produce about 1 bcfd to the hub platform. Owner of the hub is Independence Hub LLC.

The 132-mile Independence Trail Pipeline, owned by Enterprise Products Partners LP, will transport the gas to an interconnect with the Tennessee Gas Pipeline system in West Delta Block 68.

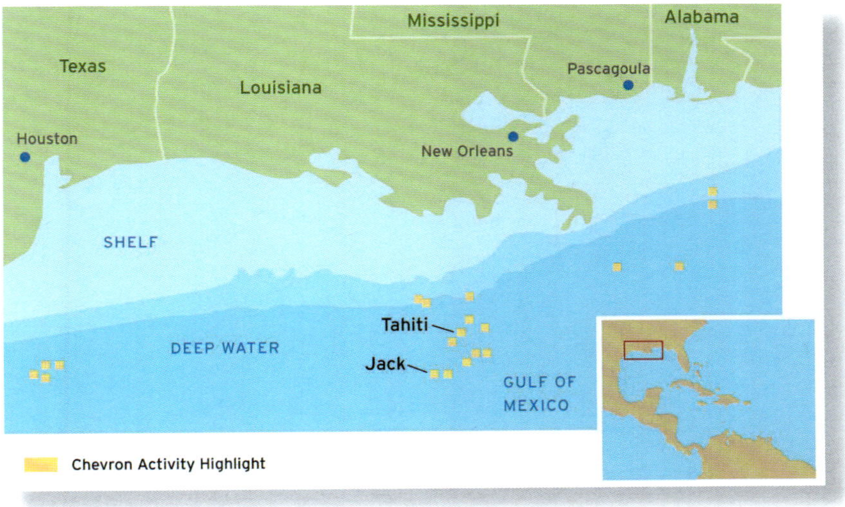

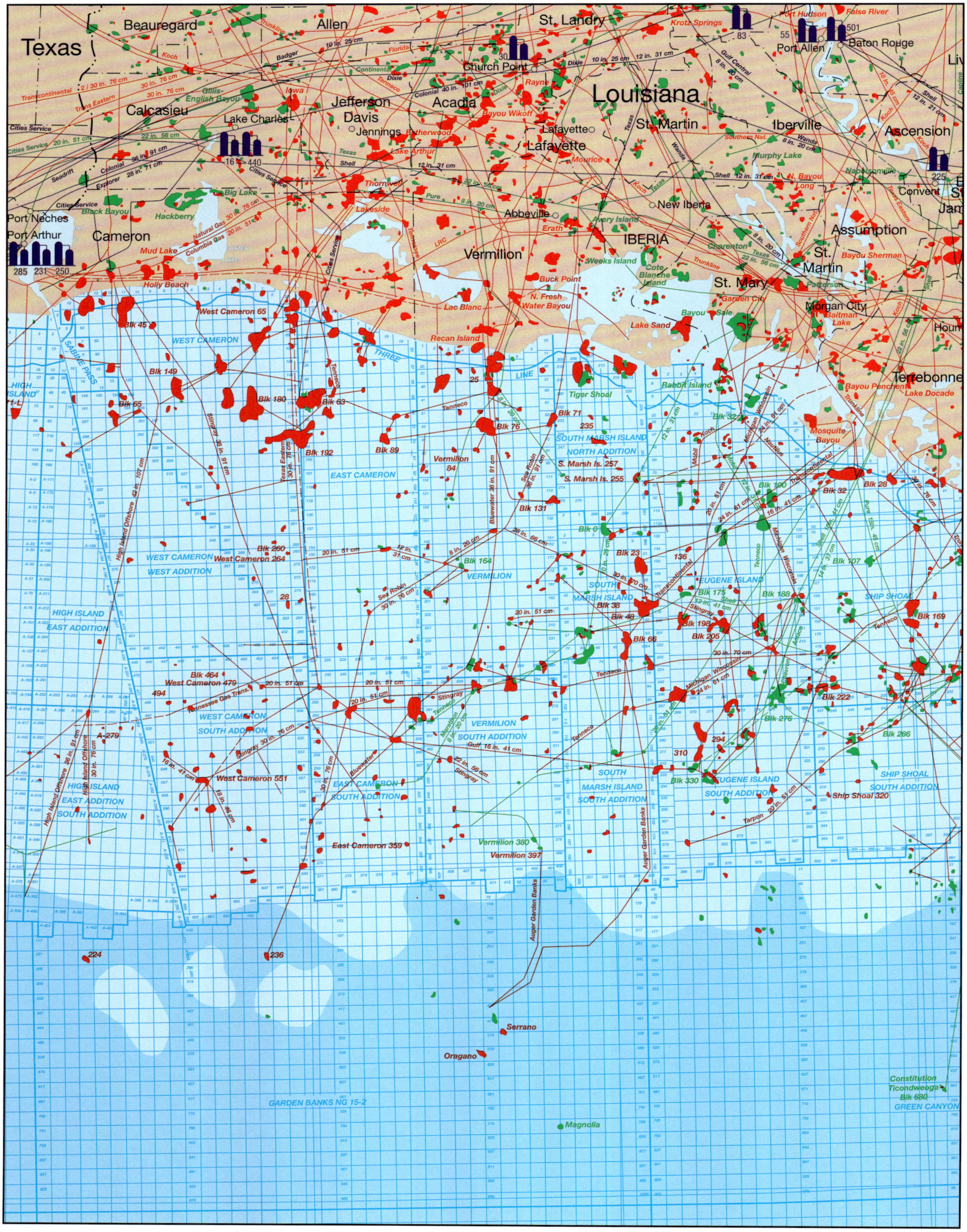

Texas
Louisiana
Beauregard
Allen
St. Landry
Calcasieu
Jefferson Davis
Acadia
Lake Charles
Jennings
Church Point
Lafayette
St. Martin
Iberville
Ascension
Baton Rouge
Port Allen
Port Hudson
False River
Krotz Springs
Port Neches
Port Arthur
Cameron
Vermilion
Abbeville
New Iberia
IBERIA
Assumption
St. Mary
Morgan City
Terrebonne
Convent
Holly Beach
Weeks Island
Avery Island
Lake Arthur
Hackberry
Big Lake
Black Bayou
Mud Lake
Lac Blanc
Recan Island
Buck Point
Lake Sand
Bayou Sale
Lake Docade
Tiger Shoal
Rabbit Island
Mosquito Bayou
WEST CAMERON
WEST CAMERON WEST ADDITION
WEST CAMERON SOUTH ADDITION
EAST CAMERON
EAST CAMERON SOUTH ADDITION
VERMILION
VERMILION SOUTH ADDITION
SOUTH MARSH ISLAND
SOUTH MARSH ISLAND NORTH ADDITION
SOUTH MARSH ISLAND SOUTH ADDITION
EUGENE ISLAND
EUGENE ISLAND SOUTH ADDITION
SHIP SHOAL
SHIP SHOAL SOUTH ADDITION
HIGH ISLAND
HIGH ISLAND EAST ADDITION
HIGH ISLAND EAST ADDITION SOUTH ADDITION
SABINE PASS
THREE LINE
GARDEN BANKS NG 15-2
GREEN CANYON
West Cameron 65
West Cameron 264
West Cameron 479
West Cameron 551
East Cameron 359
Vermilion 380
Vermilion 397
S. Marsh Is. 257
S. Marsh Is. 255
Ship Shoal 320
Serrano
Oragano
Magnolia
Constitution
Ticondweoga
Blk 680
Auger Garden Banks

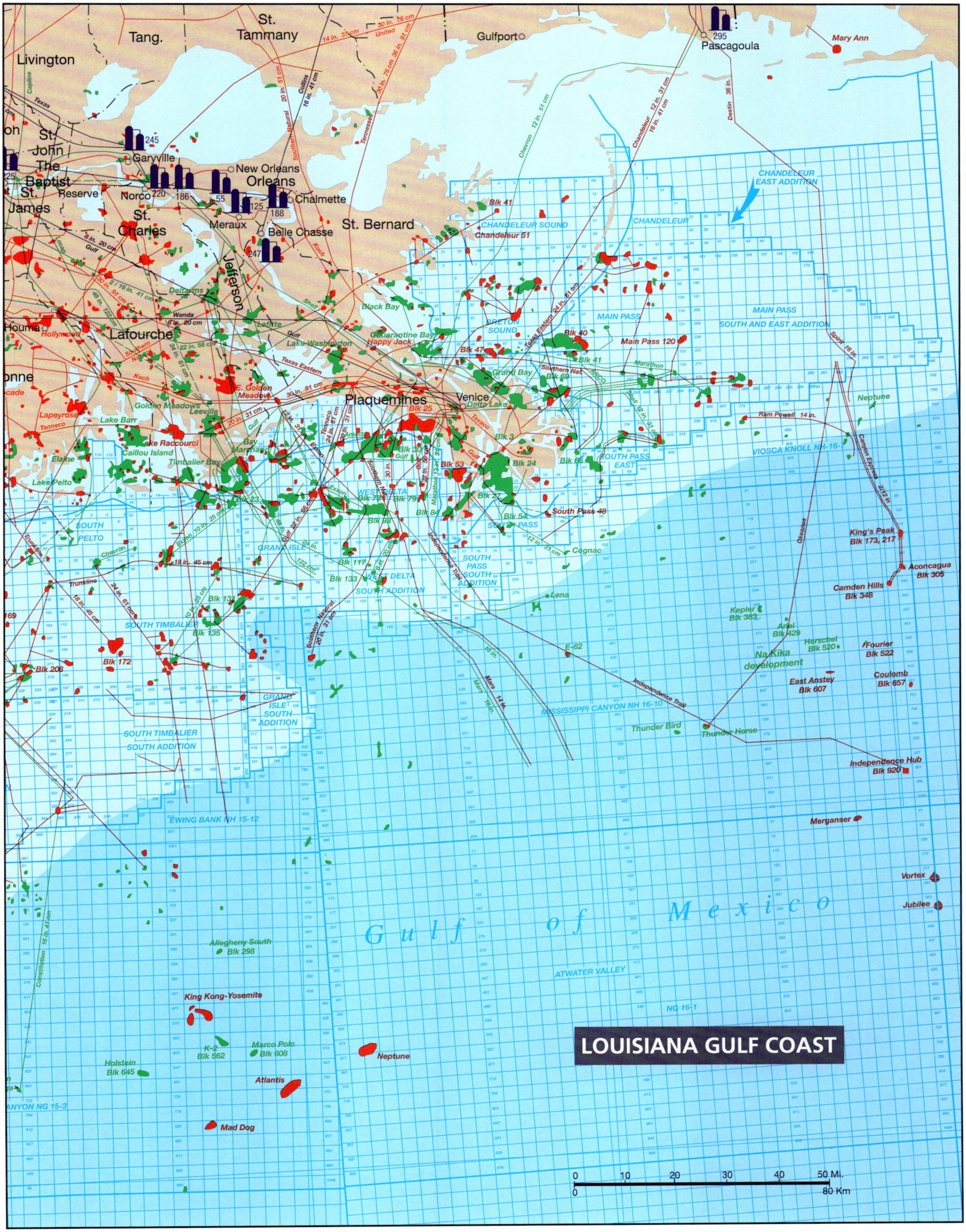
LOUISIANA GULF COAST
Tang.
St. Tammany
Livington
Gulfport
Pascagoula
Mary Ann
St. John The Baptist
St. James
Reserve
Garyville
Norco
New Orleans
Orleans
Chalmette
Meraux
Belle Chasse
St. Charles
St. Bernard
Jefferson
Lafourche
Houma
Plaquemines
Venice
CHANDELEUR EAST ADDITION
CHANDELEUR
CHANDELEUR SOUND
Chandeleur 51
Blk 41
BRETON SOUND
MAIN PASS
Main Pass 120
MAIN PASS SOUTH AND EAST ADDITION
Black Bay
Quarantine Bay
Happy Jack
Lake Washington
E. Golden Meadow
Golden Meadows
Leeville
Lake Barr
Lake Raccourci
Caillou Island
Timbalier Bay
Bay Marchand
Lapeyrose
Elaine
Lake Pelto
SOUTH PELTO
SOUTH TIMBALIER
SOUTH TIMBALIER SOUTH ADDITION
GRAND ISLE
GRAND ISLE SOUTH ADDITION
WEST DELTA
WEST DELTA SOUTH ADDITION
SOUTH PASS
SOUTH PASS SOUTH ADDITION
SOUTH PASS EAST
Grand Bay
Delta Lake
Blk 25
Blk 3
Blk 24
Blk 53
Blk 27
Blk 84
Blk 54
Blk 47
Blk 40
Blk 41
Blk 66
Blk 117
Blk 133
Blk 135
Blk 172
Blk 208
South Pass 48
Cognac
Lena
E-62
Neptune
Ram Powell
VIOSCA KNOLL NH 16-1
King's Peak Blk 173, 217
Aconcagua Blk 305
Camden Hills Blk 348
Kepler Blk 383
Ariel Blk 429
Herschel Blk 520
Fourier Blk 522
Na Kika development
East Anstey Blk 607
Coulomb Blk 657
MISSISSIPPI CANYON NH 16-10
Thunder Bird
Thunder Horse
Independence Hub Blk 920
Merganser
Vortex
Jubilee
EWING BANK NH 15-12
Gulf of Mexico
ATWATER VALLEY
NG 16-1
Allegheny South Blk 298
King Kong-Yosemite
K-2 Blk 562
Marco Polo Blk 608
Holstein Blk 645
Atlantis
Neptune
Mad Dog
Independence Trail
Canyon Express
Destin
Chevron
Texas Eastern
Southern Natural
Tennessee
Trunkline
Tenneco
Koch
Exxon
Marathon
Oceanus
Spirit
0 10 20 30 40 50 Mi.
0 80 Km

Independence Hub Project

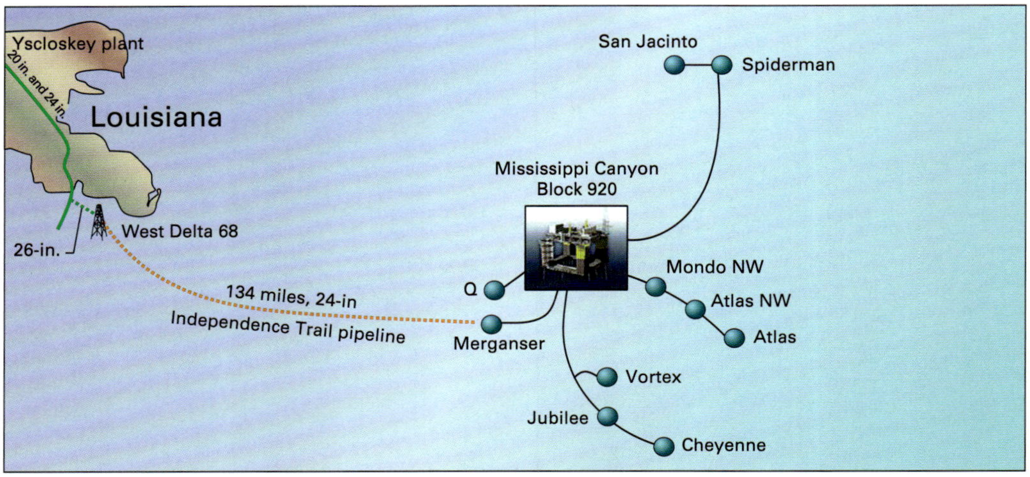

Supercomputer-based seismic

A partnership led by Repsol YPF SA (Madrid, Spain) launched the Kaleidoscope Project to apply advanced computing power to explore deep subsalt prospects under ultradeep water of the Gulf of Mexico. The goal is to commercially image structures at more than 40,000 ft that are impossible to resolve on today's 3D seismic data.

The approach will use computer power not available earlier to perform fast Fourier transforms of 3D data sets. Joining Repsol YPF in the project are 3DGeo Development Inc. (Houston), a private imaging company, and the Barcelona Supercomputing Center (BSC), owned by the government of Spain.

First deployment of RTM in the gulf could come in 2007, and it could eventually have application off West Africa and Brazil, according to Repsol.

Discovery at Bigfoot

Chevron Corp. has made a deepwater oil discovery at the Big Foot prospect on Walker Ridge Block 29, about 225 miles south of New Orleans in the Gulf of Mexico. The discovery well, in 5,000 ft of water, was drilled to a TD of 25,127 ft. It encountered 300 ft or more of net oil pay.

Record-depth spar for Perdido

Shell Offshore Inc. let an engineering, procurement, and construction contract to Technip for a production spar hull and mooring system for the Perdido hub project. Contract value was not disclosed.

Shell and its partners will install the spar in what Shell said will be a world record water depth of 8,000 ft to develop three oil fields along the Perdido fold belt in the Gulf of Mexico. It will be located 200 miles south of Freeport, TX.

Fields producing for Independence Hub

- Atlas: Lloyd Ridge Blocks 49 and 50; Lease Sale 181
- Atlas NW: Lloyd Ridge Block 5; Lease Sale 181
- Jubilee: Atwater Valley Blocks 305 and 349 and Lloyd Ridge Blocks 265 and 309; Lease Sale 166
- Merganser: Atwater Valley Blocks 36 and 37; Lease Sale 175
- San Jacinto: DeSoto Canyon Blocks 618 and 619; Lease Sale 181
- Spiderman: DeSoto Canyon Blocks 620 and 621; Lease Sale 181
- Vortex: Atwater Valley Blocks 217 and 261 and Lloyd Ridge Blocks 177 and 221; Lease Sale 157
- Mondo NW: Lloyd Ridge Block 1; Lease Sale 116
- Cheyenne: Lloyd Ridge Block 399; Lease Sale 181
- Q: (Mississippi Canyon Blocks 960, 961, 1004, and 1005; Lease Sale 175

Subsea layout

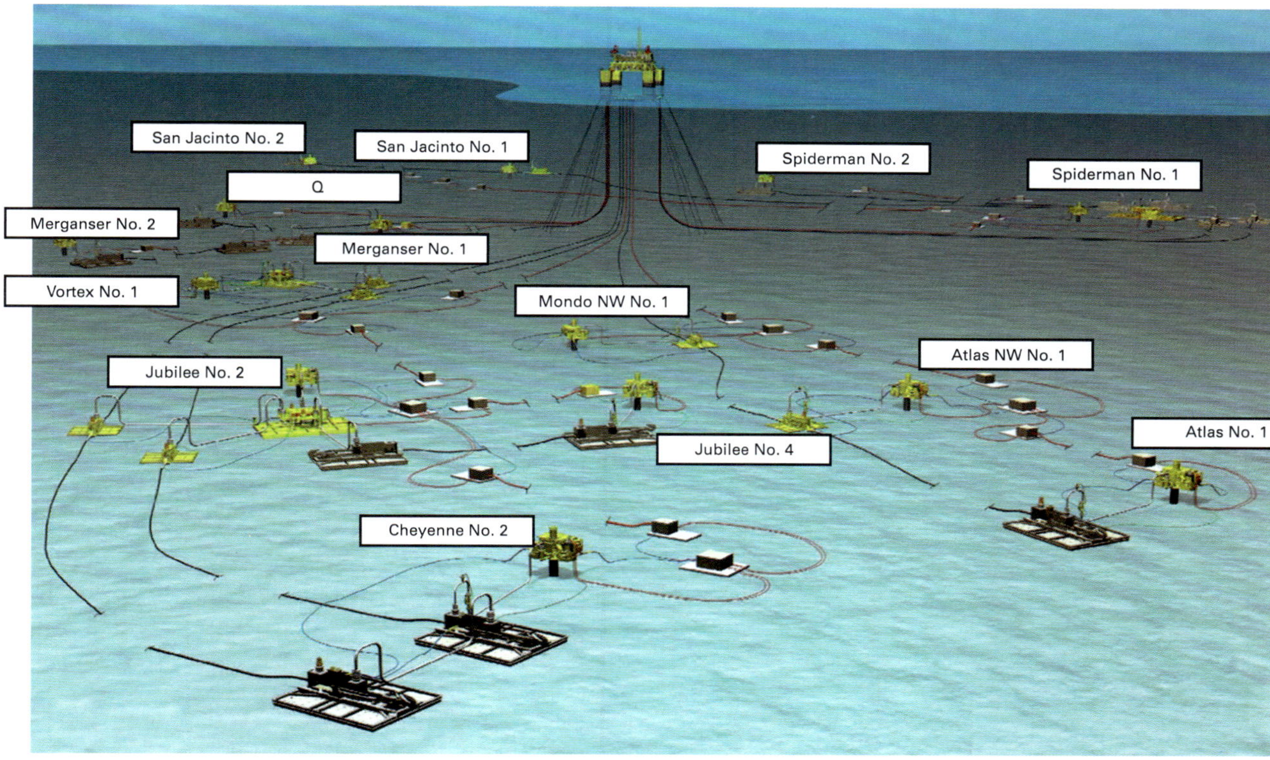

The regional direct-vertical-access spar will handle production from Great White, Tobago, and Silvertip fields in the Alaminos Canyon area. The spar will be able to handle production 100,000 b/d of oil and 200 MMscfd of gas, starting "at the turn of the decade," said Shell. The project will initiate production from the Lower Tertiary play.

Strikes at Caesar, Claymore prospects

Kerr-McGee Corp. reported an oil discovery at its Caesar prospect on Green Canyon Block 683 about 160 miles south of Houma, La., and a gas strike at its Claymore prospect on Atwater Valley Block 140 about 150 miles southeast of New Orleans.

The Caesar-1 well was drilled to 29,721 ft TD in 4,500 ft of water. The Claymore-1 well, drilled to 25,000 ft TD in 3,700 ft of water, encountered more than 150 ft of net pay in multiple zones. Claymore-1 is the first well in the company's 2006 subsalt program in the deepwater gulf.

Well test at Alaminos Canyon

Total SA successfully tested its well at Alaminos Canyon Block 856 No. 1 in deep water 140 miles off Texas. Drilled to 14,600 TD in 7,600 ft of water, the well encountered 290 ft of oil pay in two zones. Results were not reported. The well is west of Shell Offshore Inc.'s Great White discovery.

Mississippi Canyon discovery

Noble Energy Inc. (Houston) reported "high quality pay" in one objective and identified pay in a second objective of its Redrock prospect in Mississippi Canyon Block 204 No. 1.

The well was spudded in 3,334 ft of water on Dec. 30, 2005, and drilled to 23,365 ft TMD.

Noble is operator of the Redrock and adjacent Raton prospects with a 50% interest. Partners are Samson Offshore Co. and Energy Partners Ltd., 25% each.

Other Exploration & Development

FPSO okayed for Gulf use

Petroleo Brasileiro SA (Petrobras) got approval from the MMS for its conceptual subsea development plan of Cascade and Chinook oil fields. This is the first time MMS approved a plan that includes the deployment of a floating, production, storage, and offloading vessel in the Gulf of Mexico. Oil production could start in 2009.

Petrobras proposes to use six technologies new to the U.S. gulf, including a disconnectable turret buoy allowing the FPSO to move offsite during hurricanes and severe weather, crude transportation via shuttle tanker, free-standing hybrid risers, subsea electric submersible pumps, torpedo pile vertical-loaded anchors, and polyester mooring systems.

The development plan consists of the installation and operation of a FPSO in about 8,200 ft of water. It provides for at least two subsea wells in Cascade and one subsea well in Chinook, each drilled to about 27,000 ft and to be tied back to the FPSO. The plan could be expanded to include additional wells on each unit.

Mobile Bay, Main Pass

Petsec Energy Ltd. (Sydney, Australia), reported a gas discovery at the first well of a three-well program in federal waters south of Mobile Bay. The Mobile Block 951-1 well in 67 ft of water went to TD 3,050 ft MD and cut 18 ft of net gas pay.

Separately, Petsec Energy Inc., a U.S. subsidiary of Petsec Energy Ltd., began development planning for Main Pass Block 18 after its Main Pass 18 G-6 well, drilled to 13,018 ft TD, discovered 7.5 bcf of gas and 1.1 million bbl of oil (net to Petsec).

High Island Block

Mariner Energy Inc. (Houston) said it has made a material discovery with its 5ST1 well on High Island Block 116 in the Gulf of Mexico. The well was drilled to a TMD of 14,683 ft and encountered about 540 ft of net vertical pay. Completion operations are under way, and initial production is expected by the end of 2006. Mariner estimates proved and probable gas reserves at 40–60 bcf.

West Cameron

Dominion Exploration & Production (Houston) is completing for production a deeper-pool gas discovery in West Cameron 130 field. Its well No. 3, drilled to 19,365 ft in 40 ft of water 250 miles southwest of New Orleans, cut 250 ft of net pay in multiple sands.

On test, the well flowed at restricted rates of 11.8 MMcfd of gas and 40 b/d of oil.

Dominion E&P, operator, owns a 50% working interest in the well. Other interests are Hydro Gulf of Mexico LLC 35% and Mariner Energy Inc. 15%.

Knotty Head

Nexen Inc. (Calgary) said a sidetrack well confirmed the Knotty Head oil discovery on Green Canyon Block 512 in 3,500 ft of water.

The discovery well was drilled in 2005 to 34,189 ft TD—the deepest well drilled to date in the gulf—and found about 600 ft of net oil pay in multiple zones. Based on drilling results, the company estimates the resource at 200–500 million bbl of oil equivalent.

Transportation

Pipeline for Shenzi oil

Enterprise Products Partners LP (Houston) signed agreements with producers to build, own, and operate an 83-mile pipeline connecting the deepwater Shenzi oil and gas field in Green Canyon Blocks 609, 610, 653, and 654 with existing systems. The pipeline will carry up to 230,000 b/d of crude.

It will connect a tension-leg platform under construction at the field with the Cameron Highway Oil Pipeline and Poseidon Oil Pipeline systems at Enterprise's Ship Shoal 332B junction platform. Production from the Shenzi platform is expected in 2009.

First LNG ship-to-ship transfer in gulf

Exmar NV (Antwerp) completed the first commercial ship-to-ship (STS) transfer—or transshipment—of LNG, in the Gulf of Mexico. Until recently, such transfers were only performed between LNG carriers in emergencies.

A total of 20,650 cu m of LNG was transferred from the Excelsior, a 130,800 cu m vessel, to the Excalibur, a 138,000 cu m vessel. Exmar operates the vessels with Excelerate Energy LLC. The process does not require the construction of an offshore platform.

Drilling and Production

Gomez, Rigel, Seventeen Hands fields

ATP Oil & Gas Corp (Houston) started production from Gomez oil and gas field in 3,000 ft of water on Mississippi Canyon Block 711. Two wells are being produced, and a second development phase was scheduled to start in mid-2006. The field has eight identified and logged oil and gas sands.

Dominion Exploration & Production Inc. brought gas production on line from Rigel field (Block 296) and Seventeen Hands field (Block 299), both 120 miles southeast of New Orleans. Rigel field lies in 5,200 ft of water and Seventeen Hands field in 5,800 ft. Both are subsea developments tied back to Chevron Corp.'s Viosca Knoll 900 platform. The fields' development system is designed to handle a combined production rate of 160 MMcfd of gas.

Lease Sales

Royalties on 1998–99 leases

Five gulf producers agreed in December to pay royalties on their 1998 and 1999 deepwater leases that were originally issued without price thresholds. BP PLC, Shell Oil Co., ConocoPhillips, Marathon Oil Corp., and Houston independent Walter Oil & Gas Corp. will pay royalties for production from October 1 onward.

Leases issued under the Deepwater Royalty Relief Act of 1995 included an incentive that suspended royalties until a specific amount was produced, encouraging producers to explore in areas where costs were high. The incentive did not apply if prices reached a certain level in most years, but it was apparently accidentally omitted for leases issued during 1998–99.

Congress demanded that the MMS correct the error, discovered early in 2006, and recover lost revenues, but some lessees argued that being forced to renegotiate would violate contract law. After legal consultation, MMS asked the leaseholders to voluntarily renegotiate.

MMS OFFSHORE RIK GAS SALE SETS RECORD

MMS sold a record amount of royalty-in-kind (RIK) natural gas from federal offshore leases in a sale concluded in March. The sale calls for delivery of about 118 bcf of RIK gas to 14 offshore pipeline systems originating in the Gulf of Mexico. Records were set for the amount of gas sold, the number of bids received (127), and the number of companies that made offers (21).

Winning bidders included Atlanta Gas Light Co., Virginia Power Co., ConocoPhillips, National Energy & Trade LP, Louis Dreyfus Corp., Shell Trading Co., Chevron USA, Williams Power, and Total SA.

Framework set for environmental studies

MMS identified areas and mitigation measures it will consider in environmental analyses covering 11 proposed oil and gas lease sales in the central and western Gulf of Mexico.

The identification will involve areas in Western Gulf of Mexico sales 204, 207, 210, 215, and 218 and Central Gulf of Mexico sales 205, 206, 208, 213, 216, and 222.

The identification proposes a framework in which potential operating and environmental impacts can be evaluated before MMS releases the related draft multisale environmental impact statement. The sales are included in the MMS draft proposed OCS program for 2007–12.

Lease Sale 200

The August 16, 2006 western Gulf of Mexico Lease Sale 200 was the best in nine years, in terms of number of bids, said MMS.

A federal judge had warned that a pending environmental lawsuit could overturn the sale, but a settlement was reached with the State of Louisiana in October.

Offered were 3,865 unleased blocks covering about 20.9 million acres on the OCS off Texas and in deeper waters off Louisiana. The blocks are 3–210 miles offshore in up to 3,000 m of water. MMS estimated the sale could result in the eventual production of 136–262 million bbl of crude and 810 bcf-1.4 tcf of gas.

Results of Western Gulf of Mexico Lease Sale 200: August 16, 2006

Top Five Companies by Number of High Bids

Company	Total High Bids	Sum of High Bids
Petrobras America Inc.	34	$45,483,774
BP Exploration & Production Inc.	31	$37,469,796
Hess Corporation	30	$16,804,661
Shell Offshore Inc.	28	$35,420,316
Kerr-McGee Oil & Gas Corp.	27	$17,968,470

Top Five Companies by Amount of High Bids

Company	Total High Bids	Sum of High Bids
Petrobras America Inc.	34	$45,483,774
BP Exploration & Production Inc.	31	$37,469,797
Shell Offshore Inc.	28	$35,420,316
Cobalt International Energy, L.P.	24	$33,125,180
Chevron USA Inc.	19	$26,152,108

Source: U.S. Minerals Management Service

MMS reported high bids totaling $340.9 million from 62 companies. The total of all bids submitted was $462.8 million, up 38% from the 2005 lease sale in the western gulf.

More than 50% of the tracts receiving bids were in the Garden Banks and Keathley Canyon areas. Of 445 newly available tracts offered, 130 received bids, including 80 in the Garden Banks and Keathley Canyon areas.

Plan: Expand Lease Sale 205, cancel 201

MMS proposed in November to expand the available acreage in OCS Lease Sale 205, scheduled for September 2007. The proposal seeks to include all available acreage in the Central Gulf of Mexico planning area in Lease Sale 205; originally, only acreage along the eastern boundary of the Central gulf planning area had been included.

The MMS proposal also includes the acreage that would have been offered in Lease Sale 201, which was originally scheduled for March 2007 and included all of the available acreage in the Central gulf planning area as proposed in the five-year program for 2002–07.

As part of a legal settlement with Louisiana, MMS has agreed to prepare an environmental impact statement before conducting any additional lease sales in the gulf. As a result, MMS is cancelling Lease Sale 201.

Lease Sale 198: strong interest

Central Gulf of Mexico Lease Sale 198 on March 15, 2006 drew $588.3 million in high bids from 82 companies, said the MMS. The total of all bids was $978.3 million, 38% higher than the bids submitted during the 2005 Central Gulf lease sale.

MMS received 707 bids on 405 OCS tracts. Water depths of the tracts ranged from less than 650 ft to 8,500 ft. The sale involved 4,040 blocks covering 21.3 million acres off Alabama, Louisiana, and Mississippi.

Amerada Hess Corp. submitted the highest bid with $42.8 million for Green Canyon Block 287. Amerada Hess also had the top high-bid total, offering $62.6 million total in 13 high bids. Woodside Energy (USA) Inc. was second, bidding $47.2 million in 12 high bids.

BP Exploration & Production Inc. submitted the most high bids: 73, for a total of $22.3 million.

Hurricane Aftermath

Hurricanes Katrina and Rita struck the Gulf Coast in August and September 2005, respectively. In its final report issued in September 2006, the U.S. Department of Energy (DOE) said that the two storms caused the biggest disruptions ever experienced by gulf oil and gas operations.

Damage assessment

Katrina destroyed 44 platforms and damaged 20 others. It also damaged at least 100 pipelines in federal water, six of which were 10 in. or larger.

Rita destroyed 69 offshore platforms and damaged 32 others. It also damaged at least 83 offshore pipelines (28 of them large-diameter lines).

MMS estimated that 3,050 of the 4,000 offshore platforms and 22,000 of the 33,000 miles of pipelines in the gulf were in the direct path of either Katrina or Rita.

According to American Global Maritime Inc. (Houston), the magnitudes of both storms exceeded the design criteria of all jack up rigs that were destroyed or floated ashore.

Some 12% of the gulf jack up rig fleet was lost or moved off locations by the two hurricanes, chiefly due to wind and currents, wave inundation, and foundation or soil movement.

Results of Central Gulf of Mexico Lease Sale 198: March 15, 2006

Top Five Companies by Number of High Bids

Company	Total High Bids	Sum of High Bids
BP Exploration & Production Inc.	73	$22,276,118
BHP Billiton Petroleum Inc.	26	$7,661,284
Hunt Oil Company	23	$10,943,280
Dominion Exploration & Production Inc.	16	$41,417,099
Hydro Gulf of Mexico, LLC	14	$20,117,740

Top Five Companies by Amount of High Bids

Company	Total High Bids	Sum of High Bids
Amerada Hess Corporation	13	$62,593,922
Woodside Energy (USA) Inc.	12	$47,162,990
Dominion Exploration & Production Inc.	16	$41,417,099
Newfield Exploration Company	6	$32,208,909
Samson Offshore Company	10	$27,109,326

Source: U.S. Minerals Management Service

The two storms also damaged 47 major gas processing plants and 17 NGL fractionation sites along the coasts of Texas, Louisiana, Mississippi, and Alabama. Those facilities had an aggregate processing capacity of 22.8 bcfd.

Analysts with Wood Mackenzie (Edinburgh) said that lost oil and gas revenue from the hurricanes can be conservatively estimated at $16 billion using annualized 2005 prices and estimated cumulative shut-in volumes. The American Petroleum Institute (API) estimated total offshore damages to rigs, platforms, and pipelines at $18 billion to $31 billion.

The volume of gas production curtailed during the third week of October 2005 from both onshore and offshore areas of the Gulf Coast was 5 bcfd—about one half of pre-hurricane offshore production.

After Katrina hit Louisiana, nearly 2 million b/d of refinery capacity was shut down due to direct damage or power interruption. More than 4.9 million b/d of refinery capacity was shut down after

Katrina-Rita Paths Relative to Oil and Gas Infrastructure

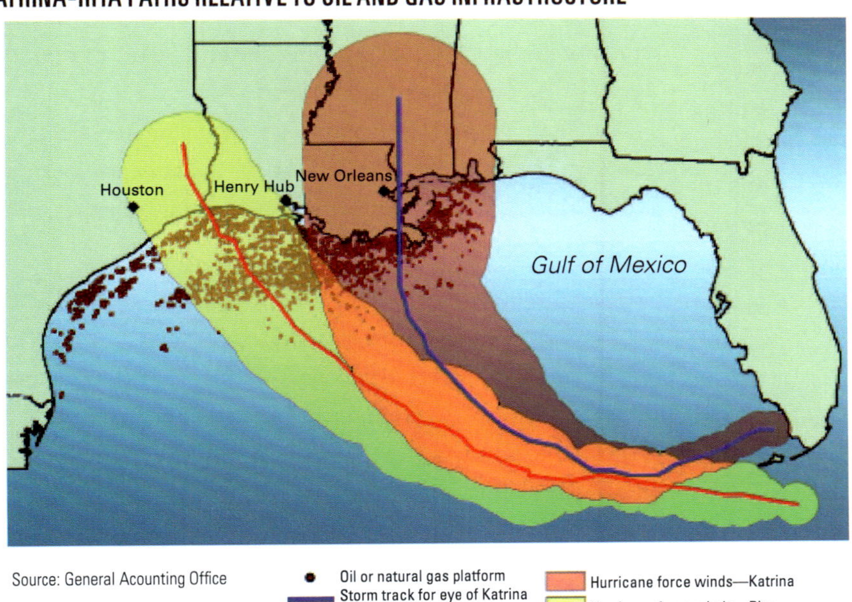

Source: General Acounting Office

Gulf of Mexico Rig Damage, Hurricane Katrina

Major damage, total loss

Diamond Offshore Drilling Inc.'s Ocean Warwick, jack up, drifted 66 miles northeast from Main Pass Block 299; beached on Dauphin Island, Ala.

ENSCO 29, platform rig, severely damaged, future status unclear.

H&P 201, platform rig, on Mars TLP in Mississippi Canyon Block 807, smashed.

Hercules 25, jack up, derrick broke, crushed crew quarters.

Nabors Offshore Corp.'s Super Sundowner XII, platform rig, platform collapsed, rig lost.

Pride 210, platform rig in West Delta Block 73, mast damage, may be scrapped.

Rowan New Orleans, jack up, presumed capsized in Main Pass Block 185.

Damaged

Diamond's Ocean Quest, semisubmersible, minor damage.

Diamond's Ocean Voyager, semisubmersible, drifted 9 miles north from Mississippi Canyon Block 711.

ENSCO 7500, semisub, tow line parted while under tow, listing.

GSF Arctic I, semisub, drifted, grounded near mouth of Mississippi River.

GSF Celtic Sea, semisub, listing.

GSF Development Driller I, semisub, found listing slightly, with water damage to thruster control systems.

GSF Development Driller II, semisub, anchor damage.

Nabors' Dolphin 110, jack up, broken windows, water damage.

Noble Jim Thompson, semisub, drifted 17 miles NNE from Mississippi Canyon Block 935; mooring line damage.

Transocean Inc.'s Deepwater Nautilus, moored semisub, drifted 80 miles; damage to mooring system and thrusters, lost about 3,200 ft of marine riser and part of subsea well control system. 3 months repair required.

Hurricane Rita hit the Texas coast. A Wood Mackenzie study issued in early October 2005 said that Katrina damaged 18 rigs and Rita 15.

Recovery pace

The industry's recovery actions in fourth quarter 2005 returned to service some three-fourths of the supplies shut in by storm damage in Louisiana, and early 2006 saw further improvement. When DOE stopped monitoring recovery efforts in early March 2006, offshore gas production had returned to 9 bcfd, and all but two of the 47 damaged gas processing plants were operating, although at a reduced aggregate level.

MMS in mid-June reported shut-in gas production of 935.67 MMcfd, equivalent to 9.4% of the daily gas production in the gulf. The cumulative shut-in gas production from August 26, 2005, to June 19, 2006, was 803.6 bcf—about 22% of total annual gas production in the gulf (3.65 tcf).

With regard to oil, MMS reported mid-June shut-in production of 179,970 b/d, equivalent to 12% of daily oil production in the gulf. Total oil production in the gulf had been reduced by more than 166 million bbl since Hurricane Rita struck. That amounted to about 30% of a normal year's oil production from federal offshore fields.

Gulf of Mexico Rig Damage, Hurricane Rita

Major damage; total loss

GSF Adriatic VII, jack up, drifted 80 miles from Eugene Island Block 338; ran aground off Louisiana.

GSF High Island III, jack up, ran aground off Louisiana.

Noble Joe Alford, submersible, drifted 8 miles from Vermillion Blcok 52; support members below the hull are bent or broken.

Noble Max Smith, semisubmersible, drifted 123 miles, hole in starboard outboard column and decks damaged; sustained heaviest damage among Noble's rigs.

Rowan-Fort Worth, jack up, drifted from South Marsh Island Block 146; found beached in West Cameron.

Rowan-Halifax, jack up, drifted from East Cameron Block 346; found beached.

Rowan Louisiana, jack up, hull detached in Vermilion Block 338; ran aground near Cameron, La.

Rowan-Odessa, jack up on Ship Shoal Block 250; missing.

Damaged

Diamond Offshore Drilling Inc.'s Ocean Saratoga, semisubmersible,drifted 100 miles northwest, ran aground in Vermilion Block 111.

Diamond's Ocean Star, semisubmersible, drifted 100 miles northwest,ran aground in Eugene Island Block 142.

ENSCO 68, jack up, drill floor shifted.

ENSCO 69, jack up, drilling skid shifted.

ENSCO 90, jack up, listing in South Marsh Block 130

ESV 7500, semisubmersible, moorings broke.

Hercules 21, jack up, listing in Main Pass Block 21.

Nabors Offshore Corp.'s Dolphin 111, jack up, Sabine Pass, windowsblown out, water damage to control systems and quarters.

Nabors's Pool 54, jack up, mast blown over.

Nabors's Rig 300, deep drilling barge, submerged east of Cameron, La., electrical and pump systems damaged.

Noble Lorris Bouzigard, semisubmersible, broke mooring lines, drifted.

Rowan-Louisiana, jack up, legs severed, grounded near Cameron, La.

Transocean Marianas, moored semisub, grounded in shallow water at Eugene Island Block 133; damage to mooring system, thrusters, and hull; one column partially flooded. Under repair in Brownsville, Tex.; return to service first-quarter 2006.

Mitigating future hurricane damage

The industry is considering various precautionary measures to reduce damage from future storms. These include: increasing the space between a jack up hull and water to avoid damage from the waves created by a storm surge, improving mooring reliability, and studying the fatigue life of risers that may be exposed to multiple storms.

In May, when MMS issued its final damage assessment report, operators were still revising assessments and evaluating the economics of repair, abandonment, or expansion.

Regulatory response

Regulatory authorities accelerated progress in gas infrastructure and market developments that had been impeded by slow regulatory responses in recent years.

These include: approving applications for a number of new LNG receiving terminals, approving applications for new pipelines and changes to pipeline tariffs to enable pipeline companies to avoid bottlenecks during periods of supply disruption, and giving more attention to potential gas market manipulation during periods of high prices.

In addition, the U.S. Federal Energy Regulatory Commission (FERC) revised its rules to better monitor and assess the condition of the interstate gas pipeline grid and gas storage infrastructure when service is disrupted by storms, other natural disasters, or acts of terrorism.

FERC also extended its suspension of blanket-certificate rules to help Gulf Coast gas facilities recover from hurricane damage. It extended the deadline from October 31, 2006 to February 28, 2007 to complete construction of projects under the order. The action will allow facilities that can't be put in service by the start of the 2006–07 heating season to be completed during the season.

The commission also temporarily expanded the definition of eligible facilities to include mainline facilities, extension of a mainline, and temporary compression that raises a mainline's capacity.

Company News

Billiton group to buy Khan development

A group led by BHP Billiton agreed to buy Anadarko Petroleum Corp.'s Genghis Khan oil and gas development in the deepwater gulf for $1.35 billion. BHP Billiton, Hess Corp., and Repsol YPF SA already own adjacent Shenzi field. The Genghis Khan sale involves a 100% stake in Green Canyon Block 652 and certain deep rights in Green Canyon Block 608. The sale was expected to close by year-end 2006.

Apache to buy BP gulf shelf assets

Apache Corp. (Houston) announced plans to pay $1.3 billion to acquire BP PLC's interests in 18 oil and gas fields in the gulf. The transaction covers 92 blocks with estimated proved reserves of 27 million bbl of liquid hydrocarbons and 185 bcf of natural gas.

Pioneer deepwater assets to Marubeni

A unit of Pioneer Natural Resources Co. (Dallas) entered into an agreement through which Marubeni Offshore Production (USA) Inc., a subsidiary of Marubeni Corp. (Tokyo), will acquire certain deepwater gulf assets from Pioneer for $1.3 billion.

The transaction includes Pioneer's interest in three producing projects—the Falcon Corridor, Devils Tower, and Canyon Express—two potential development projects—Ozona Deep and Thunder Hawk—and 88 exploration blocks.

IMPACT OF 2005 HURRICANES ON GOM GAS PRODUCTION

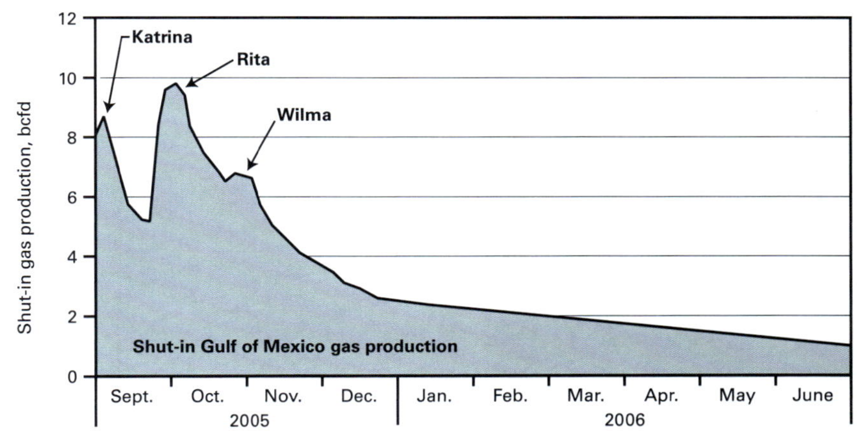

Source: June 2005 US Minerals Management Service data, David Wood & Associates

2005 HURRICANE IMPACT ON HENRY HUB SPOT GAS PRICE

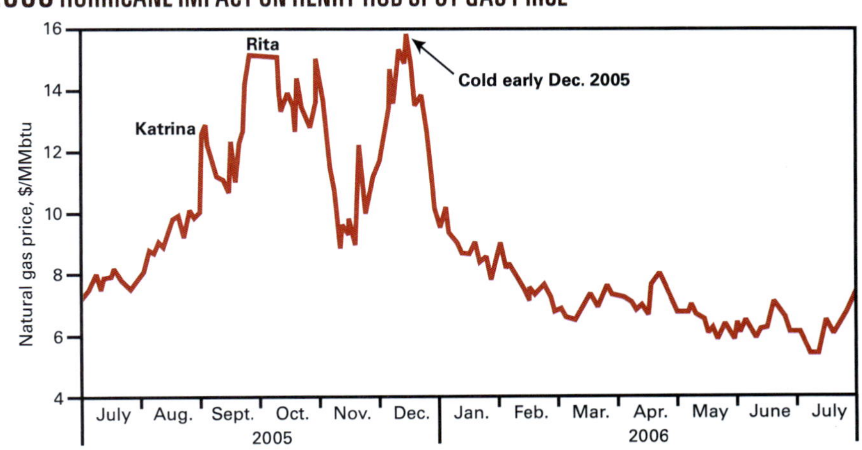

Source: David Wood & Associates

Chukchi Sea
Beaufort Sea
Point Barrow
For detail see Alaska's North Slope map
Barrow
Smith Bay
Simpson
Prudhoe Bay
Endicott
Arctic Plains
Prudhoe Bay
15
15
Camden Bay
National Petroleum Reserve
Umiat
Gubik
E. Umiat
Kavik
Brooks Range
Koakoak
Amauligak
Issungnak
Atkinson
Pitsiulak
Ivik
Mayogiak
Mackenzie Bay
Adgo
Malik
Tuk
Niglintgak
Kumak
Titalik
Ya Ya
Parsons Lake
Kavik
Kotzebue Sound
Arctic Circle
Fort Yukon
Trans Alaskan
Bering Strait
Circle
Tanana
Fairbanks
North Pole
Dawson
Alaska
18
215
Yukon Territory
St. Lawrence I.
Norton Sound
US
Whitehorse
Chugach Mountains
Haines Jct.
Alaskan Range
Ivan River
Anchorage
Beluga River
Granite Point
North Cook Inlet
48
Bethel
Birch Hills
Valdez
Yakataga
Skagway
Trading Bay
72
Gravina Point
Yakutat
Nunivak I.
McArthur River
Swanson River
Montaque I.
Juneau
Kuskokwim Bay
Middle Ground Shoal
Kenai Peninsula
Kenai
Sitka
Cook Inlet
Alaska Peninsula
Bering Sea
Kodiak
Kodiak I.
Gulf of Alaska
ALASKA
0
100
200
300 Mi
0
200
400 Km

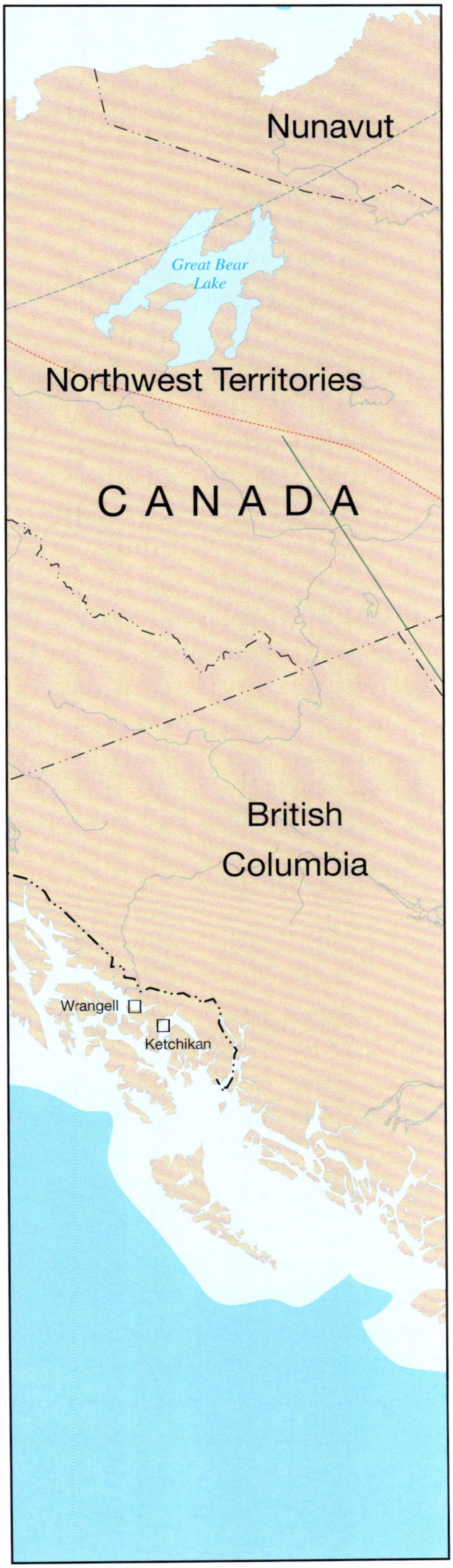

K-M shelf properties sold for $1.34 billion

Kerr-McGee Corp. (Oklahoma City) signed an agreement with W&T Offshore Inc. (Houston) for all of Kerr-McGee's interest in oil and natural gas properties on the gulf shelf for gross proceeds of about $1.34 billion. W&T Offshore also will assume responsibility for abandonment liabilities, which Kerr-McGee recorded at about $135 million at year-end 2005.

After the transaction, W&T will own 100% of the membership interest in the Kerr-McGee subsidiary. Kerr-McGee expects the net after-tax cash proceeds of about $925 million.

The divested properties are located in the gulf in about 100 fields on 249 offshore blocks in an area from Corpus Christi, TX, to Mobile, AL. W&T estimates that there are approximately 362 bcfe of net proved reserves, plus additional probable and possible reserves of 650 bcfe.

ALASKA

Transportation

Alaska Natural Gas Pipeline: slow progress

The proposed Alaska Highway Pipeline Project would build a pipeline of more than 3,000 miles through western Canada into the U.S. Midwest and carry 4 bcfd of gas from the North Slope.

Alaska Gov. Frank Murkowski announced in March that North Slope producers—Alaskan subsidiaries of BP PLC, ExxonMobil Corp., and ConocoPhillips—had reached basic agreement on a gas pipeline contract with the state that calls for partial state ownership of the pipeline.

However, final approval requires resolution of several technical, fiscal and labor issues, plus approval of a tax on natural gas reserves. In several special sessions in 2006, the legislature was unable to reach agreement on contract terms, and voters rejected the gas reserves tax plan in November.

Proposed Alaska Highway Pipeline

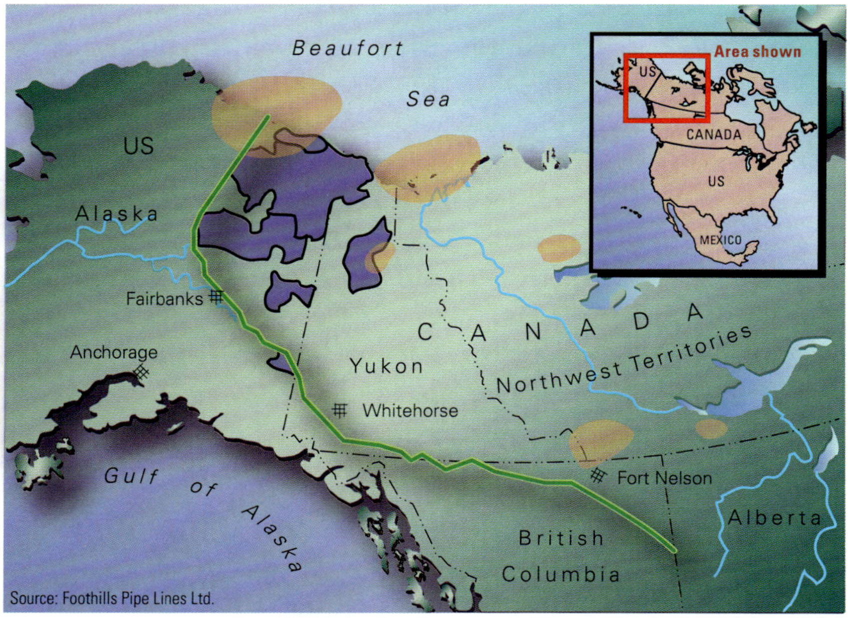

Source: Foothills Pipe Lines Ltd.

Gas reserves tax plan rejected

Alaska voters in November soundly rejected a proposal to tax North Slope natural gas. Ballot Measure 2, also known as the Alaska Gasline Now initiative, aimed to provide an incentive for gas production.

Under that proposal, leaseholders on the North Slope would have paid an annual reserves tax of 3¢/Mcf for undeveloped gas reserves. The payments would have been repealed upon pipeline construction.

Future progress is uncertain because Murkowski lost in the August primary election to incoming governor-elect Sarah Palin. If Alaska's legislature approves the gas pipeline contract, the producer group could make plans and seek permits and a right of way in Alaska and Canada.

In a related development, the state's oil and gas taxation regime was changed just before the August primary. Alaska producers now will pay 22.5% of their net profits to the state, a 0.25% increase for every dollar the price rises above $40/bbl. The law also includes a tax credit of 4–20% for investments in exploration and development. The law will raise an estimated $2.2 billion for the state.

In Washington, 15 federal agencies signed a memorandum of understanding to expedite permitting processes and establish a management framework to reduce bureaucratic delays in construction of a natural gas pipeline from Alaska, according to the DOE.

Point Thompson Unit settlement offered

ExxonMobil Corp. proposed to pay Alaska $20 million and to relinquish 20,000 acres to resolve unmet obligations to develop the 106,200 acre Point Thomson Unit (PTU) gas-condensate leaseholding on Alaska's North Slope.

In October 2005, the state rejected ExxonMobil's development plan and placed the company in default of its 2001 development agreement that called for production by 2008.

The PTU holds estimated reserves of 8 tcf of gas and 200 million bbl of condensate and oil, but there is no PTU production yet because many lessees believe commercial viability requires construction of a gas pipeline from the North Slope to the Lower 48.

BP pipeline leak

In early March, approximately 200,000 gal of oil leaked from a section of a 34-in. buried gathering line operated by BP Exploration (Alaska) Inc. at its Prudhoe Bay production facility on Alaska's North Slope. The leak—confined to less than 2 acres—was attributed to internal pipe corrosion and led to a call by the U.S. Dept. of Transportation for corrosion inspection of other transit lines.

As a result of that additional inspection, BP found severe corrosion in another line and evidence of another small oil leak (4–5 bbl). On August 7 the company began a shutdown of the eastern operating area (EOA), which normally produces half of the 400,000 b/d total output from Prudhoe Bay.

BP decided it could safety continue production from the western operating area (WOA), based on ultrasonic, infrared and visual pipeline inspection data and consultation with federal and state regulators.

In a separate incident in April, a break in a 3-in. BP natural gas line at Prudhoe Bay released about 12 Mcf of gas. BP said that external pipe corrosion probably was the cause.

In late October BP announced that EOA oil production had returned to 200,000 b/d but that work would continue to finish inspecting EOA pipe, start internal inspection in November of WOA pipe, and plan for replacement of 16 miles of corroded transit lines in 2007. BP said it planned to spend $195 million in 2007 on major maintenance at Prudhoe Bay, four times the level spent in 2004.

One significant impact of the Prudhoe Bay incidents was to further delay legislative action regarding access to the Arctic National Wildlife Refuge (ANWR) coastal plain for oil and gas exploration. The U.S. House of Representatives had passed another bill (H.R. 5429) authorizing such leasing in May, but evolving news and Congressional hearings on the Prudhoe Bay situation doomed further progress it.

Exploration and Development

Alaska North Slope plays

A Calgary independent has signed joint operating agreements on seven play areas for exploration mostly west and south of supergiant Prudhoe Bay and giant Kuparuk River oil fields.

Some of the agreements relate to areas near the Cronus-1 exploratory well, which Pioneer Natural Resources Corp. (Dallas) said is an indicated oil discovery. Cronus-1 encountered a Cretaceous Torok sequence similar to that productive at nearby Meltwater.

Cronus is near acreage in the seven play areas now subject to joint operating agreements between TG World Energy Corp. (Calgary) and Brooks Range Petroleum Corp., a subsidiary of Alaska Venture Capital Group LLC (AVCG).

The joint venture has an interest in acreage within 3 km of the Cronus well as well as 31,696 gross acres on the Ocean Point prospect, 24,840 gross acres on the Titania prospect, and 30,651 gross acres at the Whiskey Gulch prospect, all oil prospects on the central North Slope.

The joint venture was identifying drillable prospects in mid-2006 for the coming three winter seasons.

BG Alaska E&P Inc., a subsidiary of BG Group plc (Berkshire, UK), signed an exploration agreement with Anadarko Petroleum Corp. (Houston) to acquire a 40% equity share in 208,000 acres of land along Alaska's Eastern North Slope. The acreage is on the coastal plain near to the Prudhoe Bay field.

Also, BG Group signed a participation agreement with Anadarko and Petro-Canada (Alaska) Inc. for a 33.33% equity share in 2.1 million acres of land in the Foothills area of the North Slope.

Cook Inlet basin

Swift Energy Co. (Houston) entered into an agreement with Aurora Gas LLC (Houston) to acquire a 37.5% working interest in onshore acreage in the Cook Inlet basin. Swift will acquire half of Aurora's working interest in seven areas covering 54,500 acres. The companies plan to spud the Endeavour 1 exploratory commitment well. At least six other prospects have been targeted in the Hemlock-Tyonek oil play and the Beluga-Sterling gas play.

North Slope

Independent oil and gas executives formed a joint venture that may spend $46.9 million in 2006–09 exploring for oil on extensive portions of the North Slope. The venture holds interests in 167,000 acres and plans to drill at least four wells.

Play areas of interest include: Gwydyr Bay, Itkillik, Ocean Point, Slugger, Southland, Titania, and Whiskey Gulch. Gwydyr is just north of and Slugger is east of the Prudhoe Bay field. The others are near Alpine, Nanuq, Tarn, and Meltwater fields. All are prospective for oil, have existing 2D leads or 3D drilling prospects, and are near infrastructure.

Joint venture operator is a subsidiary of Alaska Venture Capital Group LLC (AVCG), owned by eight Kansas independent oil and gas company executives and a former president of ARCO Alaska president. Independent TG World Energy Corp. (Calgary) is to pay $17 million to share interests in the targeted acreage and may elect to participate in each part of the program.

Oooguruk update

Pioneer Natural Resources Alaska Inc. reported progress in setting up a drillsite for the $500 million Oooguruk development project in the Beaufort Sea off Alaska. Oooguruk has 50–90 million boe of reserves, from which Pioneer expects peak production of 15,000–20,000 b/d.

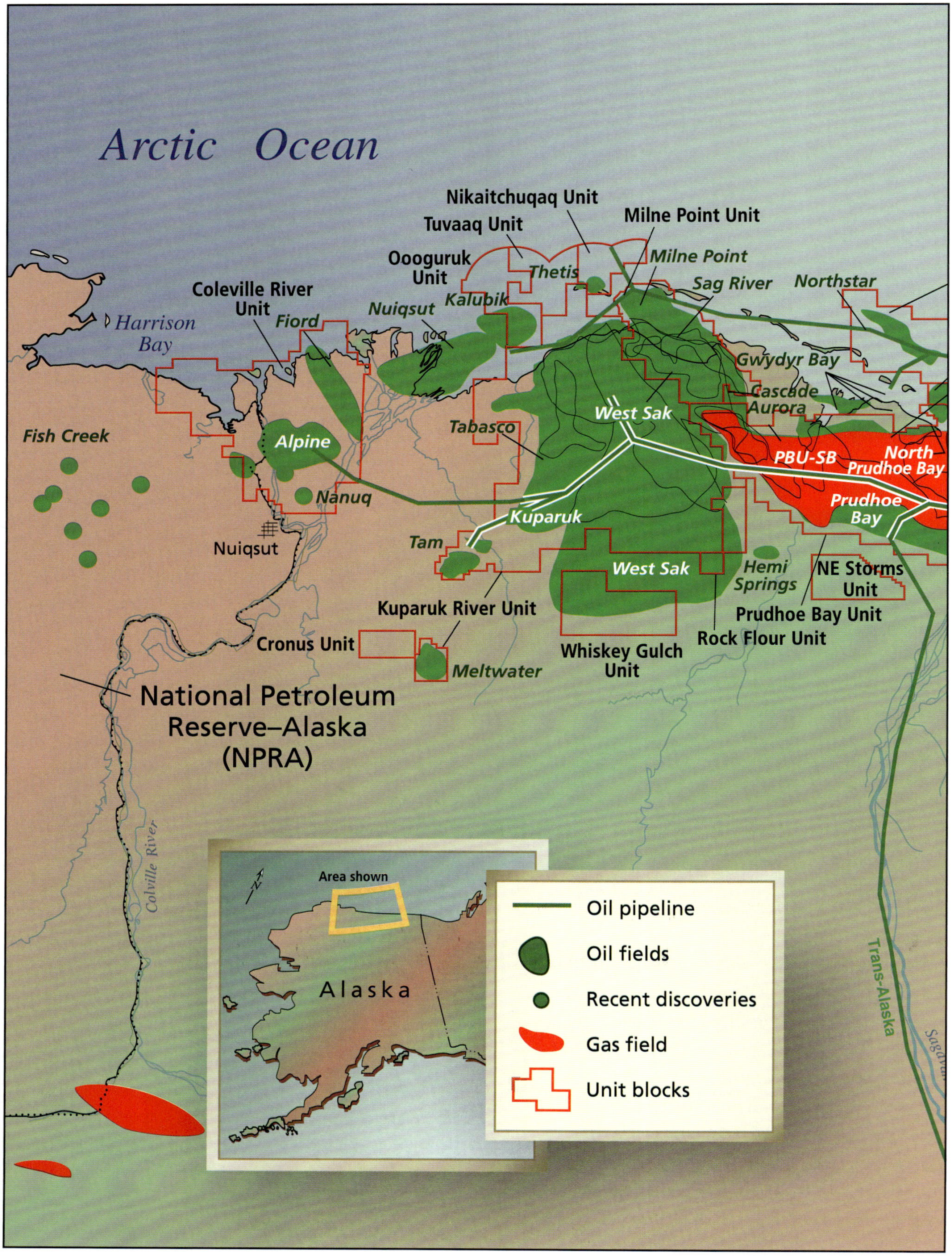
Arctic Ocean
Nikaitchuqaq Unit
Tuvaaq Unit
Milne Point Unit
Milne Point
Ooguruk Unit
Thetis
Sag River
Northstar
Coleville River Unit
Kalubik
Nuiqsut
Harrison Bay
Fiord
Gwydyr Bay
Cascade
Aurora
West Sak
Fish Creek
Alpine
Tabasco
PBU-SB
North Prudhoe Bay
Nanuq
Prudhoe Bay
Kuparuk
Tam
Nuiqsut
West Sak
Hemi Springs
NE Storms Unit
Kuparuk River Unit
Prudhoe Bay Unit
Cronus Unit
Rock Flour Unit
Whiskey Gulch Unit
Meltwater
National Petroleum Reserve–Alaska (NPRA)
Colville River
Area shown
Alaska
Oil pipeline
Oil fields
Recent discoveries
Gas field
Unit blocks
Trans-Alaska

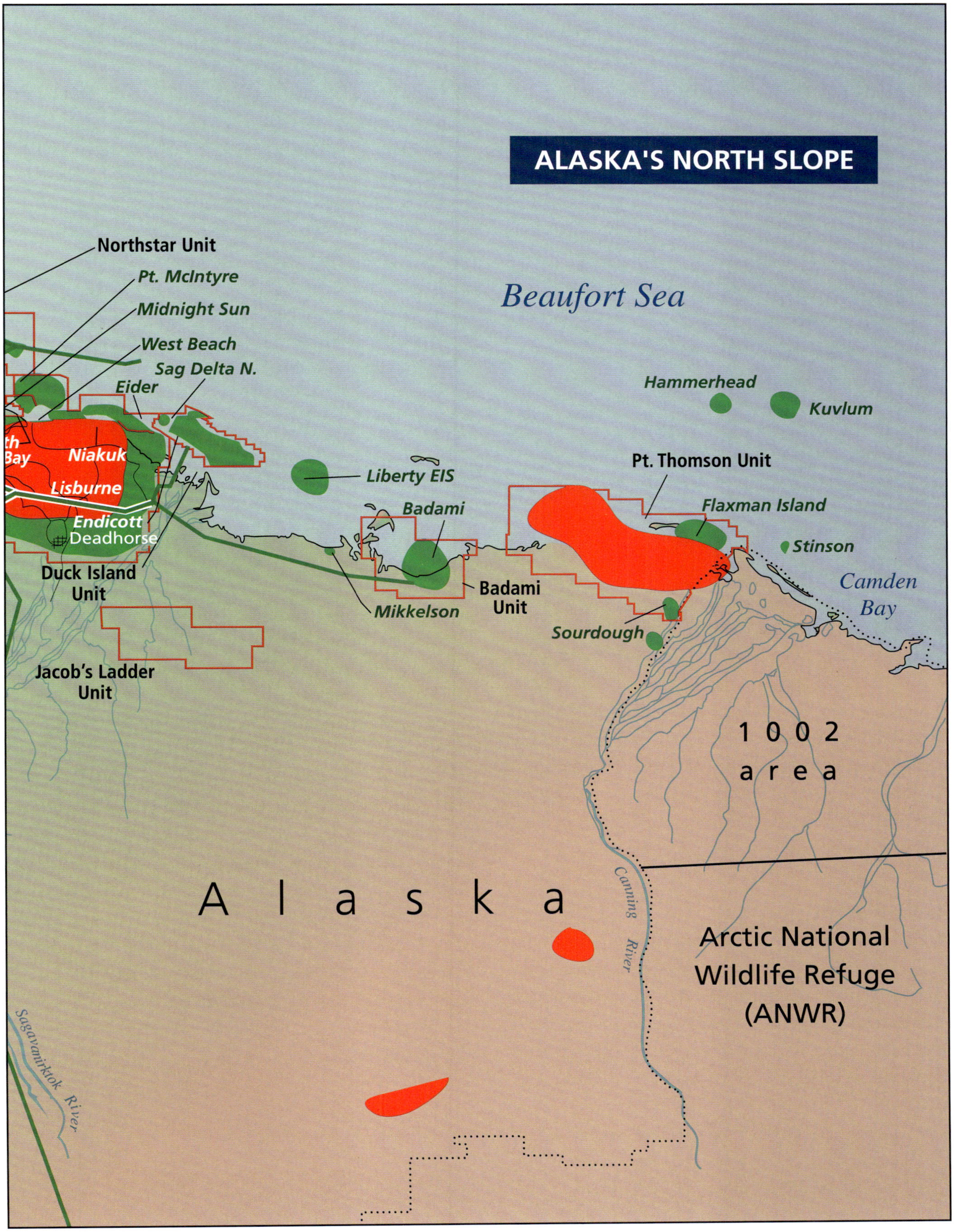
ALASKA'S NORTH SLOPE
Beaufort Sea
Northstar Unit
Pt. McIntyre
Midnight Sun
West Beach
Sag Delta N.
Eider
Niakuk
Lisburne
Endicott
Deadhorse
Duck Island Unit
Jacob's Ladder Unit
Liberty EIS
Badami
Mikkelson
Badami Unit
Hammerhead
Kuvlum
Pt. Thomson Unit
Flaxman Island
Stinson
Sourdough
Camden Bay
1002 area
Alaska
Canning River
Arctic National Wildlife Refuge (ANWR)
Sagavanirktok River

The Oooguruk site is a 6-acre gravel island built in shallow water, about six miles off the coast, near the Colville River delta. The development plans calls for drilling 60–80 wells as deep as 6,000 ft by 2009.

Seismic for Beaufort, Chukchi basins

GX Technology Corp. (Houston) completed data acquisition for the Beaufort East and Chukchi phases of its basin-scale seismic program. The Beaufort East phase will provide high-resolution subsurface images of prospective targets in the Mackenzie Delta region and its contiguous continental margin of the Canada basin offshore Canada. The Chukchi phase will provide a regional 2D seismic framework for the entire Chukchi OCS area.

The seismic program comprises 6,500 km (40 mi) of 2D data along the northern coasts of Alaska and Canada. GXT said the resulting dataset would be ready for analysis in advance of the MMS Chukchi lease sale scheduled for November 2007.

Shell plans Beaufort Sea drilling

Shell acquired $44 million worth of leases in the Beaufort Sea and purchased a drilling rig that is specially designed to work in ice-laden Arctic waters. Drilling could start in mid-2007.

Shell was a major explorer in the Beaufort and Chukchi seas and the Cook Inlet in the 1980's and 1990's, but failed to find any large offshore discoveries. The half million acres it now leases include Hammerhead Block, off Point Thompson, 60 miles east of Prudhoe Bay. It holds an estimated 200 million bbl of oil.

New look at gas hydrates

Geologists have long speculated that naturally occurring hydrates—a crystalline combination of gas and water—contain immense volumes of natural gas. But production of gas from gas hydrates under the world's oceans and in arctic regions remains uneconomic. Development is compared with that of coalbed methane, which was considered uneconomic 20 years ago.

Industry-oriented research and gas hydrate energy assessments accelerated during the past five years as rising gas prices have stimulated interest in the subject.

MMS is conducting a study using a probability-based method to estimate the in-place and technically recoverable gas hydrate energy resource potential for the OCS, including Gulf of Mexico, Pacific, Atlantic, and Alaska. Preliminary results are expected in 2007.

Also, BP Exploration (Alaska) Inc., USGS, and the DOE are working jointly to characterize North Slope hydrate resource potential and assess the commercial viability of gas hydrates in the Milne Point area of northern Alaska. BP tentatively plans a stratigraphic test in the Eileen gas hydrate trend in 2007.

USGS and others suggest that U.S. production of gas from hydrates is likely to start in Alaska and that limited commercial production is possible by 2011.

Lease Sales

MMS 2007–2012 OCS lease plans

The U.S. Minerals Management Service (MMS) released and requested comments on a proposed plan for its 2007–2012 OCS leasing program. The plan includes 21 proposed OCS lease sales in seven of the 25 OCS planning areas, including four off Alaska.

Lease sales are proposed in the Beaufort and Chukchi Seas and North Aleutian Basin Planning Areas. A special interest sale also may be conducted in the Cook Inlet Planning Area if industry interest is sufficient. The 2007–2012 OCS Oil and Gas Proposed Final Program and Final Environmental Impact Statement are scheduled for release in spring 2007.

In a related action, Pres. George Bush lifted the executive ban on oil and gas leasing of an area in the Bristol Bay, North Aleutian basin of Alaska. The North Aleutian basin area was held under a Congress-imposed moratorium from 1990 through 2003. The moratorium was discontinued in 2004.

The Department of the Interior now has the option of offering these areas during the 2007–2012 OCS oil and gas leasing program. The 2007–2012 OCS Oil and Gas Proposed Final Program and Final Environmental Impact Statement are scheduled for release in spring 2007.

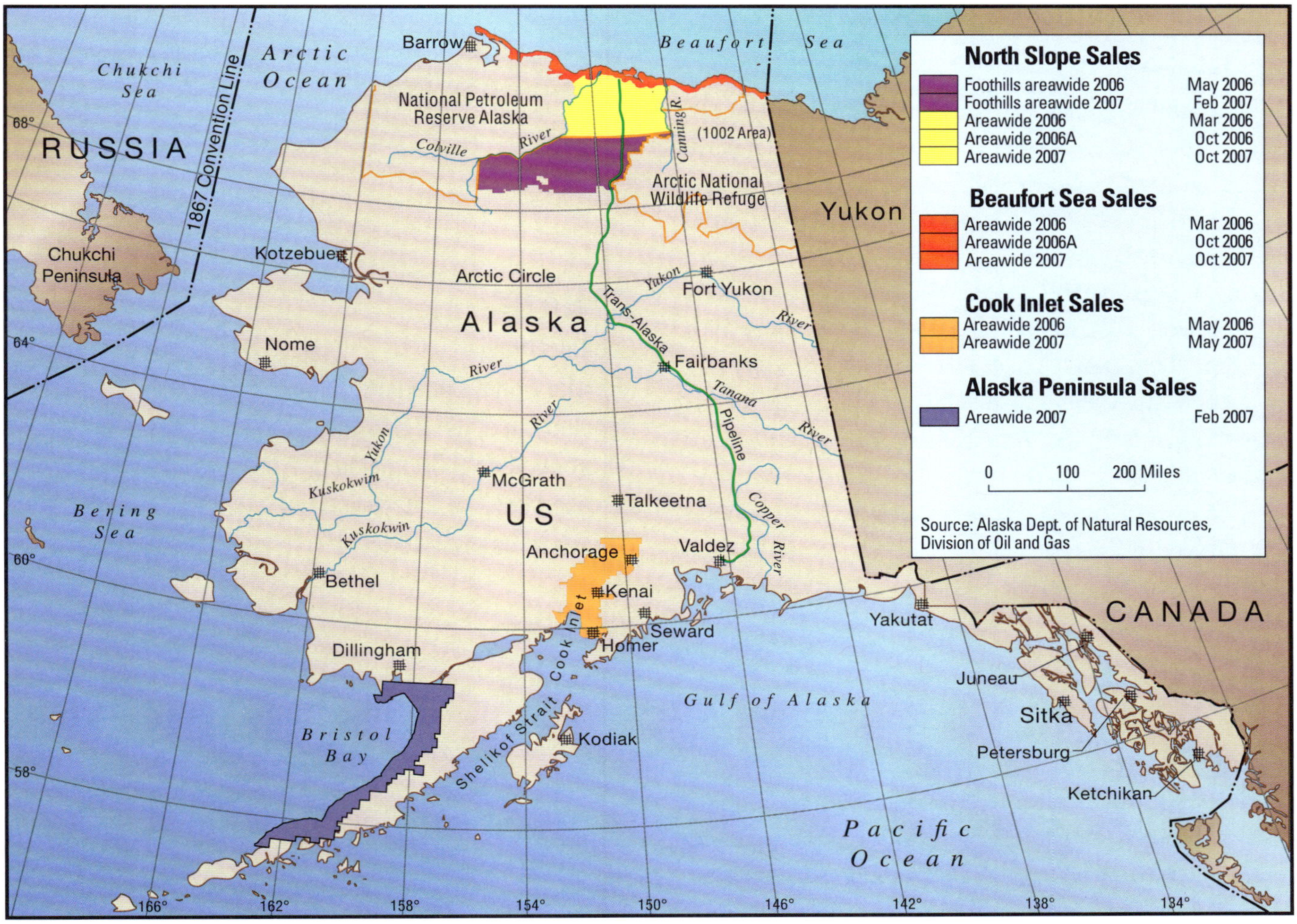

Beaufort Sea: Alaska awards

The Alaska Dept. of Natural Resources announced the winners of the October BS2006A oil and gas lease sale regarding state-owned tracts in the Beaufort Sea. The sale drew 15 bids, and 13 bids were accepted—12 from Alaska Venture Capital Group LLC and one from Ultrastar Exploration LLC.

The sale area encompassed 2 million acres of tide and submerged lands in the Beaufort Sea between the Canadian border and Point Barrow. It is adjacent to both the National Petroleum Reserve-Alaska and the Arctic National Wildlife Refuge. Lease terms vary from five to 10 years.

MMS Beaufort Sea lease sale March 2007

MMS in Anchorage scheduled March 2007 for its next oil and gas lease sale in the Beaufort Sea. The sale area includes 1,877 blocks covering 9.7 million acres. The Beaufort Sea could contain as much as 7 billion bbl of conventionally recoverable oil and 32 tcf of gas, MMS estimates.

Proposed royalty suspensions on oil and condensate production will be included, subject to price thresholds. The sale area, from the Canadian border on the east to a point near Barrow on the west, excludes offshore areas near Barrow and Kaktovik, where the Inupiat conduct bowhead whale subsistence hunts. All offshore oil and gas activity must be coordinated with the whalers during these hunts.

Six other lease stipulations will minimize environmental effects on the Inupiat people from area oil and gas development.

Drilling and Production

Alpine satellite fields

ConocoPhillips brought on stream production from two Alpine satellite oil fields, Fiord and Nanuq. Five miles north of Alpine on Alaska's North Slope, Fiord is expected to have peak production of 22,500 b/d gross in 2008. Development plans envision 17 wells.

Nanuq, 3 miles south of Alpine, was developed exclusively with horizontal wells. Peak production of 12,500 b/d is expected in 2008. Development plans envision 19 wells.

Production from Fiord and Nanuq will be processed through Alpine facilities. Together, the two satellite fields cost $650 million.

Tests on a third Alpine satellite, the Qannik oil accumulation, flowed an average 1,200 b/d. Plans are to develop Qannik from the Alpine CD2 drillsite, with first production in 2008.

Company News

Forest Oil forms Alaska subsidiary

Forest Oil Corp. (Denver) transferred the majority of its Alaska business unit to a new subsidiary, Forest Alaska Operating LLC (Alaska), which will hold and operate the parent's oil and gas interests in Cook Inlet. As of Sept. 30 Forest Alaska had 32 million boe of proved reserves and production of 6,000 boe/d. The company plans to focus on development of McArthur River field.

Canadian crude oil production outlook*

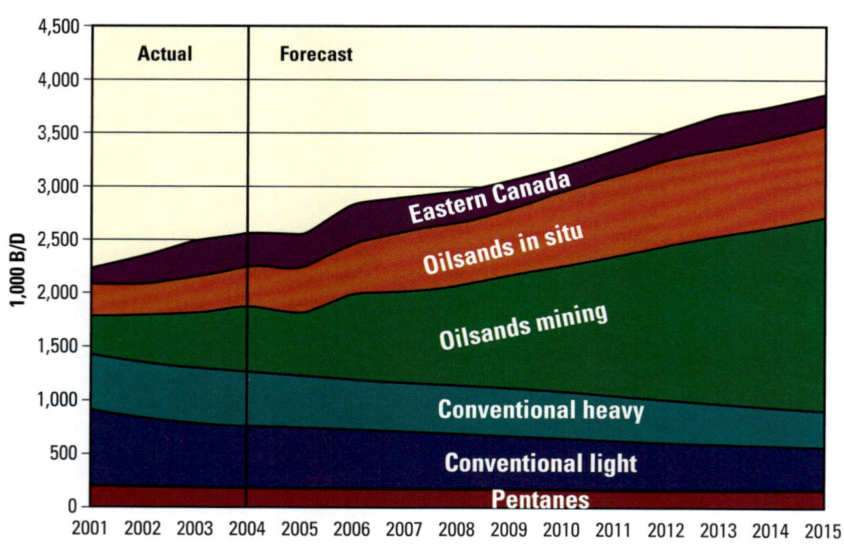

*Moderate case.
Source: Canadian Association of Petroleum Producers

CANADA

Capital: Ottawa
Monetary unit: Canadian Dollar
Refining capacity: 2,041,200 b/cd
Oil production: 2.50 million b/d
Oil reserves: 179.210 billion bbl
Gas reserves: 57.946 tcf

Overview

Conventional oil production in the Western Canada Sedimentary Basin (WCSB)—the backbone of Canada's oil and gas production for 50 years—has been in decline since 1997 as reservoirs there have matured. Alberta Energy Research Institute (AERI; Calgary) said Canada's conventional oil production declined an average of over 6%/year from 2001–2006.

The Canadian Association of Petroleum Producers (CAPP; Calgary) forecasts that conventional oil production in Canada will decline to 726,000 b/d in 2015 from about 1.4 million b/d in 1997 and 1.09 million b/d in 2004. That is an annual decline rate of 4–5% over the coming decade. CAPP projects dramatic growth through 2015 in oil production from the oil sands of Alberta.

With regard to natural gas, Canada produces about 6 tcf/year; 98% of it comes from the WCSB, primarily Alberta. *OGJ* estimated Canadian gas reserves at 57.946 tcf as of January 1, 2007, down by about 13% since 1996. Petroleum Technology Alliance Canada (PTAC; Calgary) estimates that the country's conventional natural gas reserve life index has fallen below nine years.

A study conducted for CAPP indicated that 2.5 times as many wells now must be drilled to yield the same volume of reserves as was the case in the 1990s. At the same time, gas finding and development costs have risen sharply in the past decade, especially since 2002. They averaged $3/Mcf (Can.) in 2005.

Regarding gas from offshore eastern Canada, the Sable Offshore Energy Project off Nova Scotia that began production in 1999, produces over 500 MMcfd from several fields.

Plans are being considered for developing the combined 4 tcf of gas reserves estimated to be in the Hibernia and

Canadian Natural Gas Production Forecast

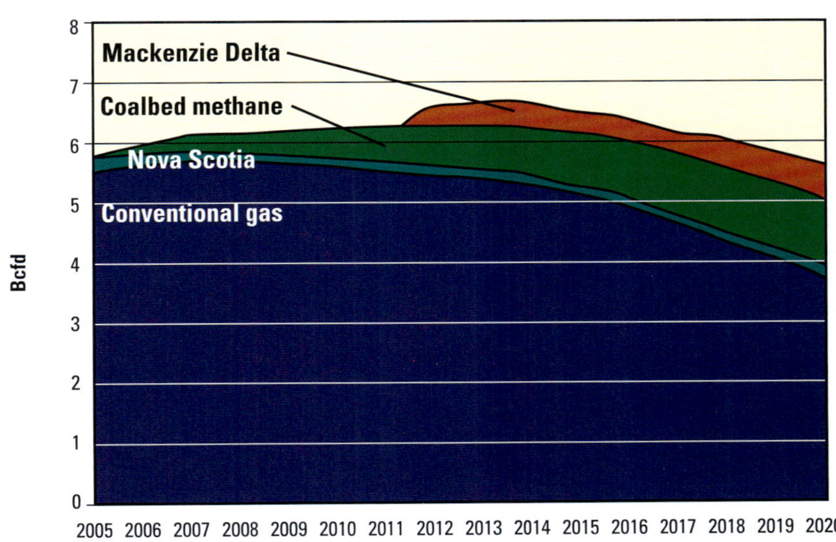

Source: Canadian Association of Petroleum Producers

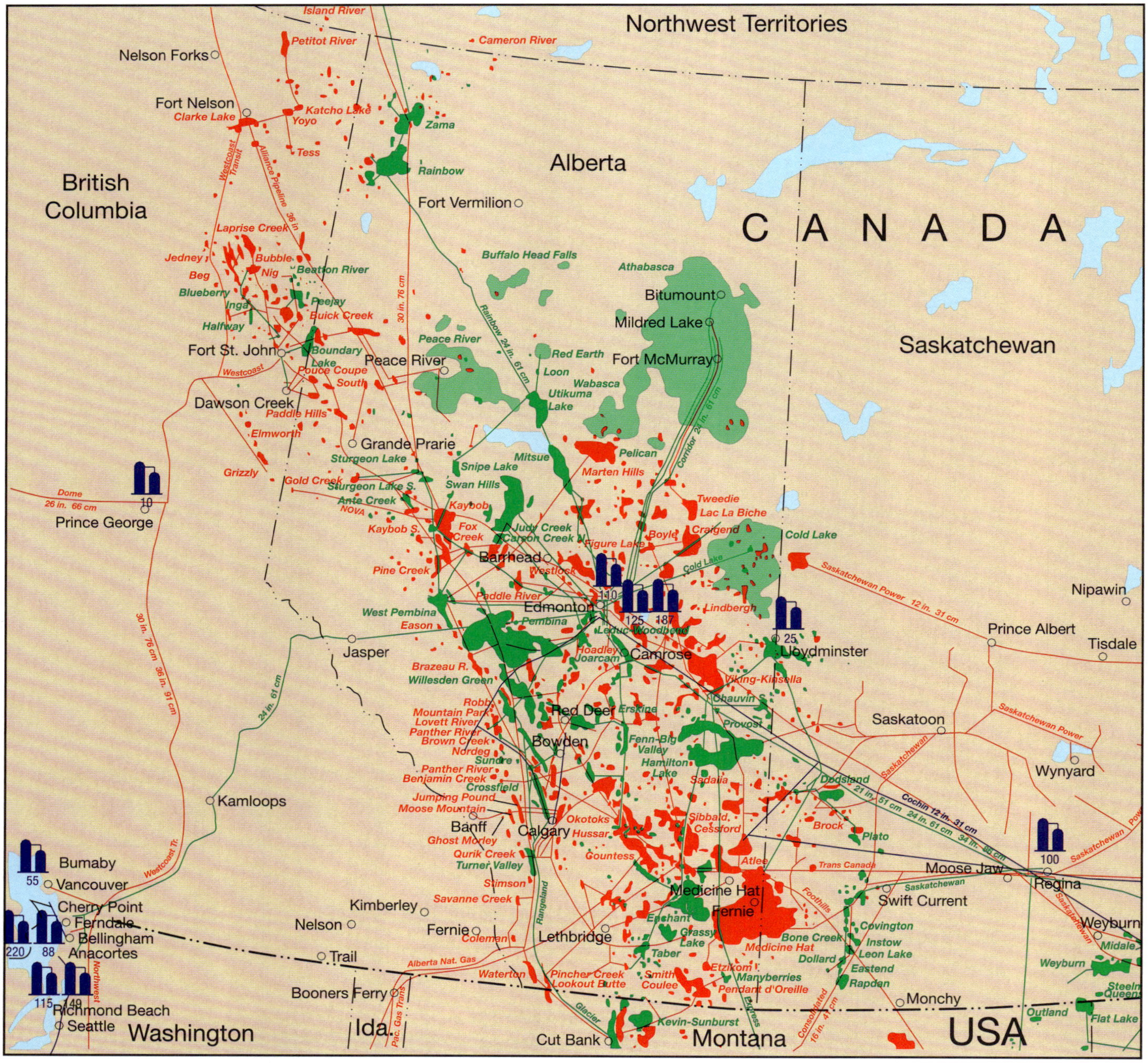

White Rose oil fields offshore Labrador. Significant gas discoveries also have been made off Labrador: 4.6 tcf of marketable gas and 123 million bbl of natural gas liquids, more than half of it in Bjarni field off Goose Bay.

Unconventional natural gas represents another promising resource. Petroleum Technology Alliance Canada (PTAC; Calgary) estimates Canada's unconventional natural gas reserves at approximately 700 tcf for tight sands, 860 tcf for tight shales, and 700 tcf for coalbed methane (CBM).

CBM is of strong current interest. Commercial production—virtually zero before 2003—reached 500 MMcfd in 2006, and CAPP predicted an increase to about 1.5 bcfd by 2010. See the Coalbed Methane section of this chapter for more details.

Technology advances in enhanced recovery represent another growth opportunity for Canadian natural gas. PTAC said new technology could boost recovery rates from conventional gas in place to as much as 72% from a current average of 59%.

The Mackenzie Delta, in the Northwest Territories, holds 9–10 tcf of identified reserves. CAPP also estimates northern Canada overall holds 77 tcf of marketable gas. But no pipelines are currently in place to bring that gas to North American markets.

Liquefied natural gas (LNG) also could augment domestic supply. At least three LNG receiving terminals have been proposed in eastern Canada, representing a total potential supply of 4 bcfd of capacity by 2008. More details are presented in the LNG section of this chapter.

Drilling and Production

Drilling overview

Baker Hughes reported 442 rigs drilling in Canada during the week of December 1, 2006, down 32%, or 211 rigs, from a year earlier (653 rigs). Most rigs (301) were drilling for gas, and the vast majority (438) were land rigs. Alberta had the most land rigs (438).

In early November the Petroleum Services Association of Canada issued its 2007 Canadian drilling activity forecast, predicting "the first slowdown in petroleum industry field activity since 2002" (www.psac.ca).

PSAC expects 21,500 wells to be drilled across Canada in 2007, about 10% fewer than the 23,900 it initially projected 2006. The drop was largely due to reduced gas drilling in Alberta.

In 2007, PSAC expects industry to drill about 1,300 more oil wells than in 2006. Estimates are based on $60/bbl (U.S., WTI) and $6.25/Mcf (Can.). PSAC based its forecast on a crude price of $60.00/bbl (U.S.) for WTI and a gas price of $6.25/Mcf (Can.).

Western Canada

Most of Canada's drilling takes place in the WCSB. Historically, the average Canadian drilling rig utilization for 2004–2006 was 63%. The drilling success rate remains high in western Canada, with the percentage of dry holes dropping to a record low of 5.8% during the first half of 2006.

PROVINCIAL DRILLING[1]

	2007, predicted no. of wells	Change, %	2006, expected no. of wells
British Columbia	1,050	–28	1,450
Alberta	16,010	–12	18,000
Saskatchewan	3,695	flat	3,650
Manitoba	650	+8	600
Ontario	70	+11	63
Eastern Canada[2]	20	–26	27
Northern Canada	5	–50	10
Total	**21,500**	**–10**	**23,900**

[1]Petroleum Services Assoc. of Canada (PSAC). [2]Quebec, New Brunswick, Nova Scotia, Prince Edward Island, Newfoundland & Labrador.

Offshore Eastern Canada

Three oil fields are currently producing in eastern Canadian offshore regions: the Hibernia field since 1997, the Terra Nova field since 2002, and the White Rose field since late 2005. Unless development activity is stepped up off Canada, the nation's offshore oil output is likely to go into decline after 2007, said CAPP.

A group led by Chevron Canada Ltd. (Vancouver) decided in April to suspend work on the Hebron heavy oil field off Newfoundland and Labrador. Hebron, estimated to hold 731 million bbl of oil, 429 bcf of natural gas, and 30 million bbl of NGLs, was to have started up in 2012.

CAPP noted that one or two exploration wells were slated for 2006 and that 4–6 might be drilled in 2007. Rigzone reported seven rigs off eastern Canada in mid-August, five of which were drilling, all off Newfoundland.

The largest operating companies engaged off Nova Scotia (with 200,000 net acres or more) are Canadian Superior, Norsk Hydro, EnCana, Corridor Resources Inc., Marathon Canada Ltd., Petro-Canada, Murphy Oil Corp., and Imperial Oil Ltd.

Petro-Canada (Calgary) announced that oil production from the Terra Nova field resumed in November after 6 months of planned maintenance on the site's FPSO vessel. The company expected output to increase to 110,000–120,000 b/d by year-end. Production had averaged 52,000 b/d before the maintenance work.

Husky Energy Inc. (Calgary) increased production of White Rose oil field off Newfoundland to 110,000 b/d by bringing a fifth well online. Crude from the White Rose SP1 well began flowing mid-year to the SeaRose FPSO vessel. A sixth production well set to come on stream by 2007 would boost White Rose production capacity to 125,000 b/d.

ExxonMobil Canada reported that the Sable Offshore Energy Project reached its 1-tcf production milestone in September. The project encompasses Thebaud, Venture, and North Triumph fields (Tier 1); and Alma, South Venture, and Glenelg fields (Tier 2). Four other fields could be developed for Tier 3: Arcadia, Chebucto, Citnalta, and Onondaga.

EnCana selected the mobile offshore production unit (MOPU) concept for use at the Deep Panuke gas field. The company expects to award the MOPU contract by late 2007 and to lease a jack up capable of processing 300 MMcfd. The company also signed the Deep Panuke offshore strategic energy agreement with the Province of Nova Scotia.

Other Maritimes activity

Husky Energy Inc. (Calgary) completed the White Rose 2006 delineation program, resulting in an increase of the assessment of White Rose oil field's recoverable resources by 190 million bbl of oil. The field is 350 km east of St. John's, NL.

WESTERN CANADA DRILLING ACTIVITY

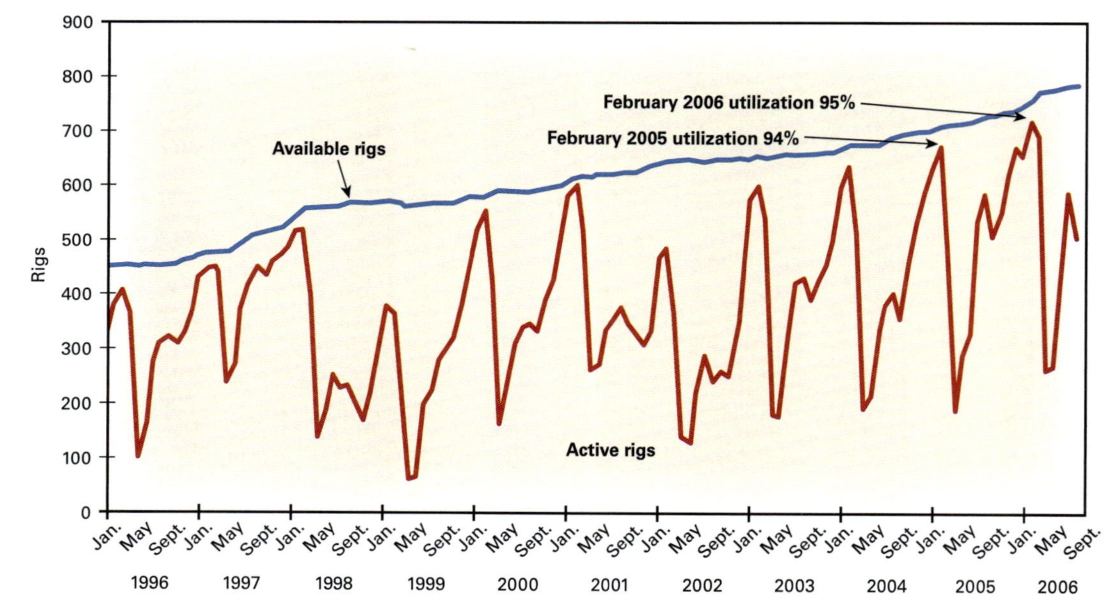

Source: Nickle's Energy Group, Rig Locator

Contributing was Husky's discovery at the North Amethyst K-15 delineation well. This discovery likely contains recoverable resources of 70 million bbl. The total estimate of about 190 million bbl of recoverable resources is incremental to the White Rose proven and probable reserves.

Canadian Imperial Venture Corp. (St. John's, NL) and two partners will soon begin a $4.4 million (Can.) drilling and exploration program on the Port au Port Peninsula in western Newfoundland.

Vulcan Minerals Inc. (St. John's, NL) and a private partner were issued a 5-year license to drill on Parcel 7, covering 395,000 acres in the Anticosti basin in the Gulf of St. Lawrence. Vulcan also announced plans to drill two land wells in the Bay St. George basin in western Newfoundland.

Corridor Resources Inc. (Halifax, NS) planned to start a 2D seismic survey in mid-2006 on two leases in the Elgin subbasin east of McCully gas field in southern New Brunswick. Corridor also drilled the McCully J-66 well to 2,600 m (8,525 ft) MD and cut 46 m (150 ft) of indicated net gas pay in the Hiram Brook formation.

The New Brunswick Dept. of Natural Resources said 3 petroleum companies planned to spend $20 million (Can.) on exploration and more than $80 million on development projects in 2006. They are Contact Exploration Inc. Stoney Creek project, Hillsborough area (oil, gas); Corridor Resources Inc. McCully field, southern New Brunswick (gas); and PetroWorth Resources Inc. southeastern New Brunswick (oil, gas).

WESTERN CANADIAN OIL, GAS WELLS

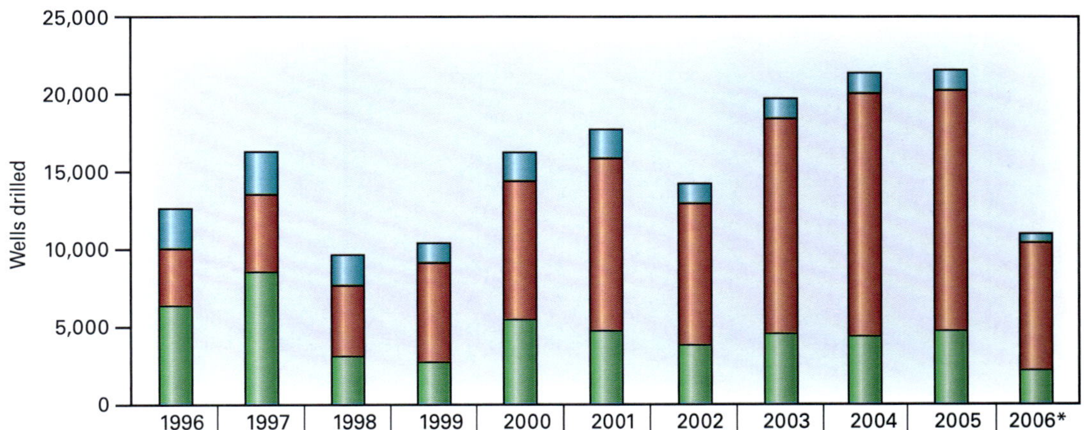

	1996	1997	1998	1999	2000	2001	2002	2003	2004	2005	2006*
Dry, %	19.6	17.2	19.3	13.6	11.7	10	9.1	6.3	5.9	6.5	5.8
Dry	2,459	2,777	1,850	1,427	1,903	1,759	1,289	1,233	1,266	1,414	641
Gas	3,726	4,857	4,585	6,309	8,934	11,177	19,073	13,944	15,645	15,359	8,142
Oil	6,332	8,558	3,142	2,730	5,462	4,689	3,832	4,473	4,427	4,822	2,235

*Through July 2005.
Source: Daily Oil Bulletin, Canadian Association of Oilwell Drilling Contractors

ENERGY TRUSTS MAY END IN 2011

The Canadian government announced plans to end the tax-free status of trusts in 2011, raising the possibility that the E&P industry would be affected by the flight of capital from energy trusts.

The Coalition of Energy Trusts pointed out that such trusts have invested $10 billion (Can.) into the oil patch in the last five years and are responsible for 20% of Canada's crude oil and natural gas production.

Wood Mackenzie Ltd. (Edinburgh) estimated that in the conventional western Canadian oil and gas sector, the removal of tax benefits could result in a transfer of value from the trusts to the government of $640 million (Can.) in 2011 alone, with up to $2 billion lost over the expected life of the trusts' assets.

Deep Panuke Development

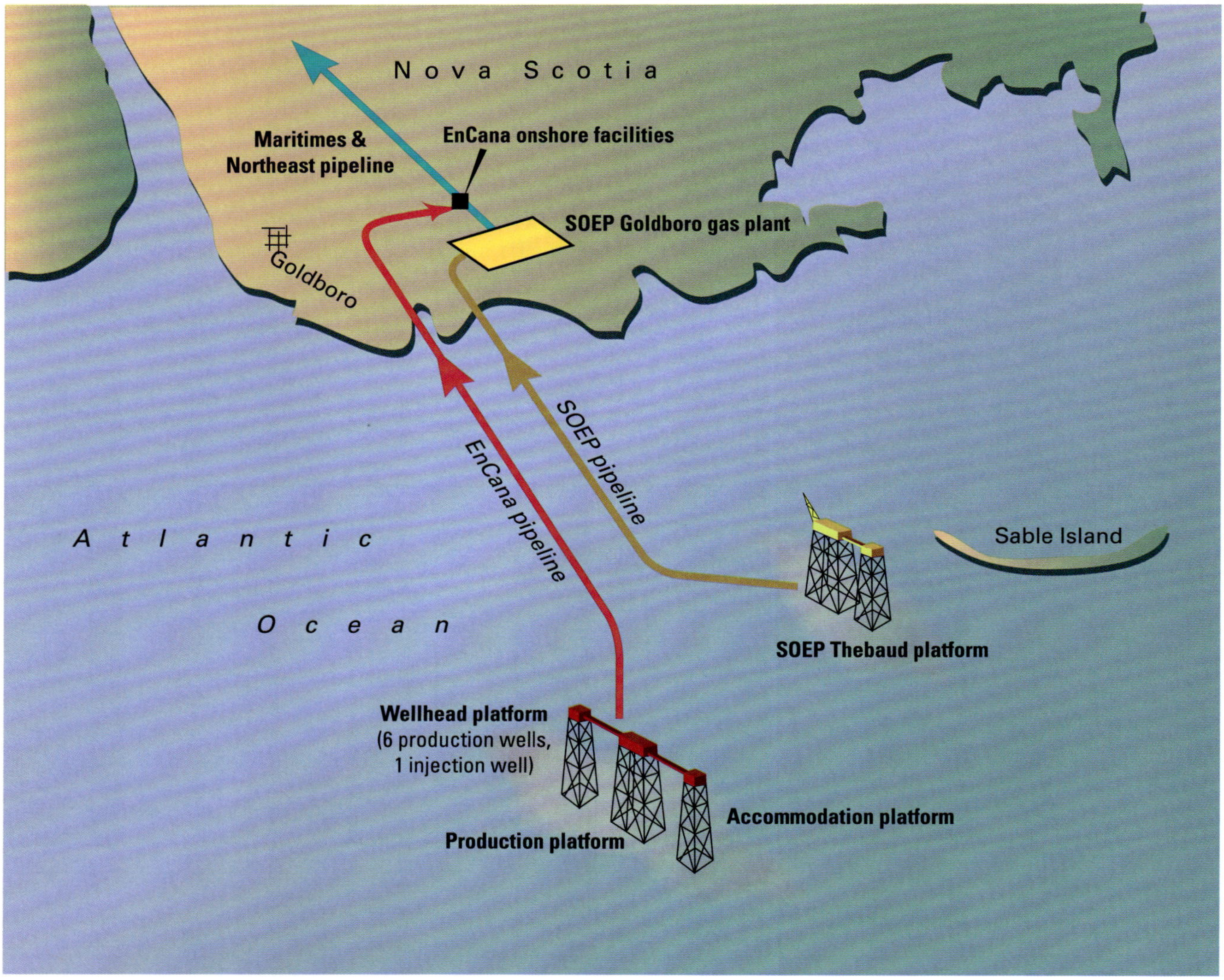

Reserves, Resources Estimates off Newfoundland

	Previous	2006 revised	Increase
Hibernia			
Oil, million bbl	865	1,244	379
NGL, million bbl	160	202	42
Gas, tcf	1.320	1.794	0.474
Hebron complex			
Oil, million bbl	414	731	317
NGL, million bbl	30	30	0
Gas, tcf	0.315	0.429	0.114
Other			
Oil, million bbl	776	776	0
NGL, million bbl	246	246	0
Gas, tcf	8.011	8.011	0
Total			
Oil, million bbl	**2,055**	**2,751**	**696**
NGL, million bbl	**436**	**478**	**42**
Gas, tcf	**9.646**	**10.234**	**0.588**

Remaining oil	Produced, million bbl	Remaining as of Jan. 1, 2006
Hiberina	456	788
Terra Nova and White Rose	166	610

Corridor also gained regulatory approval to build pipelines to connect the McCully gas field with the Maritimes & Northeast pipeline. Corridor plans to begin production in 2007.

Alberta

Cabot Petroleum Canada Corp. discovered gas at its Hinton prospect in northwestern Alberta. The Cabot Hinton 11-16-51-16W5M, drilled to (3,260 m (10,690 ft), encountered three productive zones with 36 m (118 ft) of combined net pay. Initial production was expected to begin at up to 12 MMcfd, then rising to 20 MMcfd following pipeline upgrades.

C-NLOPB RAISES RESOURCE ESTIMATES

The Canada-Newfoundland and Labrador Offshore Petroleum Board (C-NLOPB) increased its estimate of the recoverable oil resources off Newfoundland in 2006 by almost 700 million bbl.

Revisions were made to the producing Hibernia field and to the future development estimates for the Hebron complex (Hebron, Ban Nevis, and West Ben Nevis fields). The revisions increase the total initial resources off Newfoundland and Labrador to 2.751 billion bbl, up by 696 million bbl from the board's previous estimate.

Through year-end 2005, the 3 fields currently producing off Newfoundland—Hibernia, Terra Nova, and White Rose—had produced 622 million bbl of oil.

Saskatchewan

Magnus Energy Inc. (Calgary) announced plans to drill 18 wells near a light oil discovery in the Williston basin near Antler, Saskatchewan. If the field is part of a large regional stratigraphic trap, the seven wells Magnus has drilled could extend the productive Torquay trend by at least 10 miles from existing production at Sinclair, MB.

Northwest Territories

Canada's Northern Oil and Gas Directorate called for bids on two parcels in the Beaufort Sea-McKenzie Delta area. They are 70–160 km (43–100 miles) west of Tuktoyaktuk. One parcel is in the Beaufort Sea, the other off the Yukon.

Oil Sands

Oil sands now account for 39% of Canada's total oil production at approximately 1 million b/d, according to CAPP, and production will grow to 4 million b/d by 2020. The Canadian Energy Research Institute (CERI; Calgary) projects output of 6 million b/d by 2030.

The latest CAPP data show that capital spending on oil sands projects grew from $6.2 billion (Can.) in 2004 to $10.4 billion (Can.) in 2005. The Conference Board of Canada projected that industry will spend more then $100 billion (Can.) on new oil sands development and upgrading projects in the coming decade.

The Athabasca oil sands, along with those at Peace River and Cold Lake (all in Alberta), are the three major bitumen deposits in Canada. The Athabasca formation is the largest and is relatively near the surface, making it more amenable to mining. About 20% of the entire Alberta oil sands deposit can be mined. Deeper deposits at Peace River and Cold Lake require *in situ* recovery methods.

OIL PRODUCTION OFF NEWFOUNDLAND

Source: Canada - Newfoundland and Labrador Offshore Petroleum Board

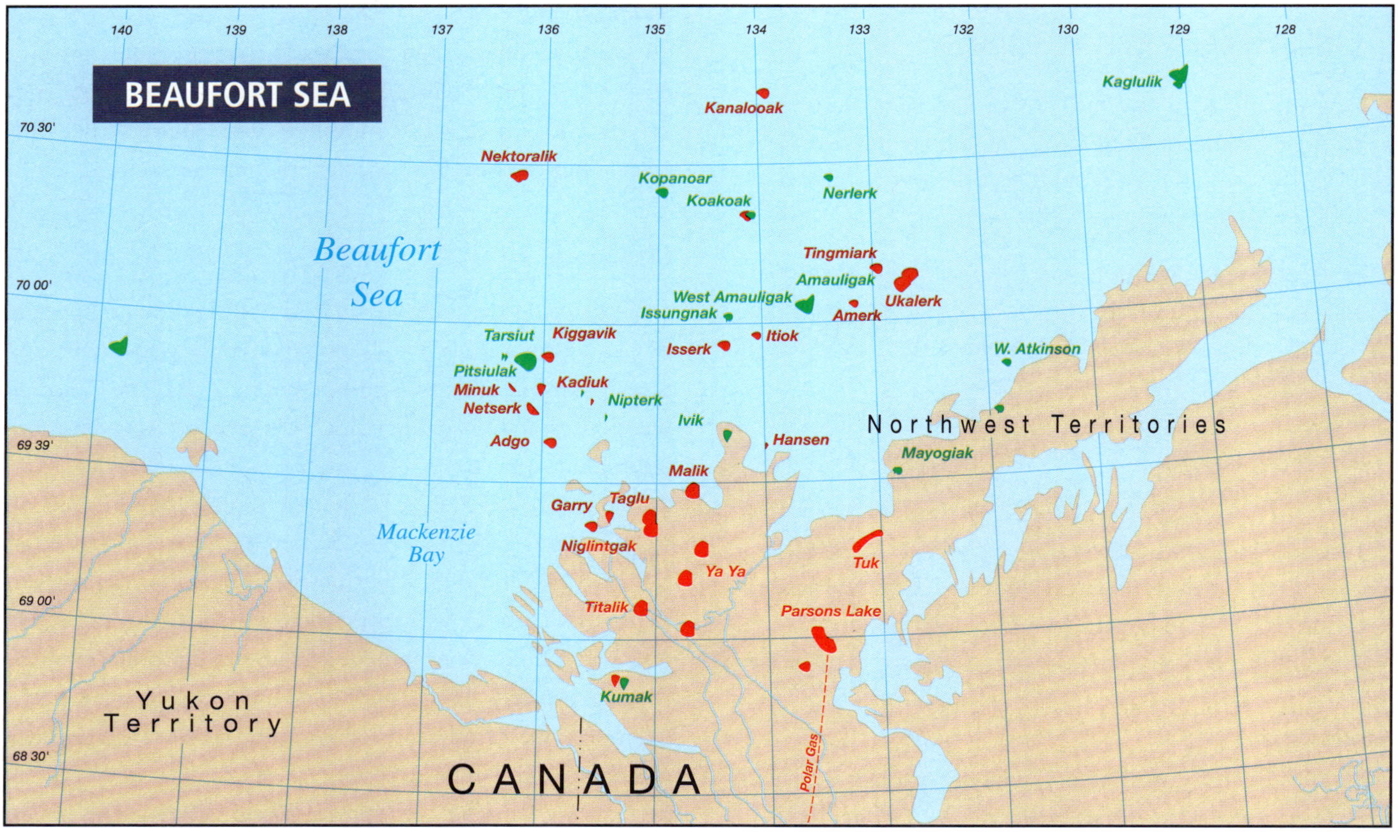

Major Canadian oil sands players include Suncor Energy Inc., Imperial Oil Ltd., Canadian Natural Resources Ltd., Shell Canada Ltd., EnCana Corp., Devon Energy (Devon Canada Corp.; Calgary), Canadian Oil Sands Trust Ltd., and Petro-Canada. All are based in Calgary or have major offices there.

There were four major oil sands projects operating in 2006. Three are in the Athabasca region:

- Suncor, producing 269,000 b/d of synthetic crude (syncrude) and bitumen
- Syncrude, the world's largest integrated mining and upgrading operation, which expected to produce 350,000 b/d in 2006
- Athabasca Oil Sands Project (AOSP), producing about 155,000 b/d of syncrude

The fourth is the Imperial Oil Ltd. project at Cold Lake, producing about 150,000 b/d of bitumen from an in situ project.

Alberta overview

Forecasts show that bitumen production in Alberta from oil sands will almost triple in the next 10 years as many new projects come on stream, said *OGJ*.

Companies continue to lease many potential oil sands areas in Alberta that may hold 173.7 billion bbl of remaining established reserves of bitumen, according to the latest Alberta Energy and Utilities Board (EUB) estimates. The three areas in Alberta with the most potential are Athabasca, Peace River, and Cold Lake. Together, these cover about 140,000 sq km (54,000 sq miles).

Recent entrants include Korea National Oil Corp., South Korea's state-owned oil company, which recently bought the Blackgold Mine (Cold Lake region) from U.S.-based Newmont Mining Corp. and has plans to produce 35,000 b/d starting in 2010.

EUB has estimated that Alberta's oil sands contained 1.69 trillion bbl of bitumen in place and have produced 5 billion bbl through 2005. Of the 173.7 billion remaining established reserves, the agency said 10.2 billion bbl are under active development.

EUB forecasts that total bitumen production for both mined plus in situ bitumen will increase to 460,000 cu m/day (2.9 million b/d) in 2015 from 170,000 cu m/d (1.06 million b/d) in 2005.

The NEB said pipeline capacity could pose a greater constraint on future Canadian oil sands development than water or natural gas availability. Pipeline capacity will be tight by 2007 and more capacity will be needed beyond 2009. Future export pipeline growth will depend on factors ranging from overseas competition to blends required for transportation and by environmental standards.

North American

North American Oil Sands Corp. announced its planned SAGD Kai Kos Dehseh project. It includes an upgrader in Edmonton that would process up to 160,000 b/d of bitumen by 2015 into synthetic oil that can be refined into gasoline, diesel and other fuels.

The company holds interest in oil sands leases in four areas: Leismer, Corner, Hangingstone, and Thornbury. Planned projects include five SAGD facilities and four smaller satellites, and development would start with 10,000 b/d production from the Leismer lease in 2008–2010.

Alberta bitumen production*

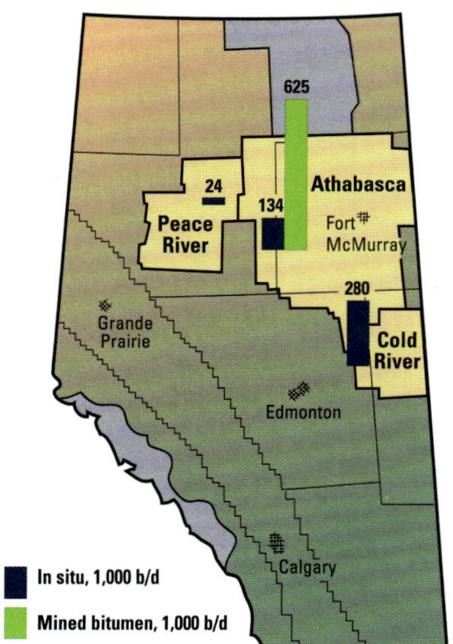

*2005
Source: EUB

Alberta bitumen production

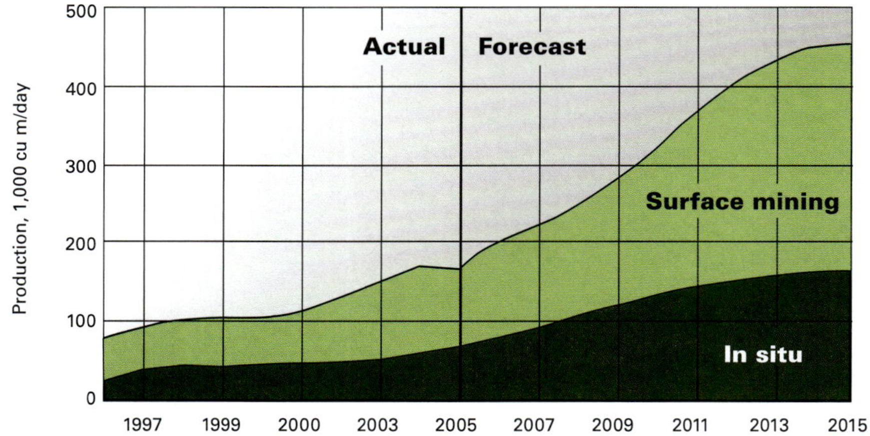

Source: EUB

North American leases

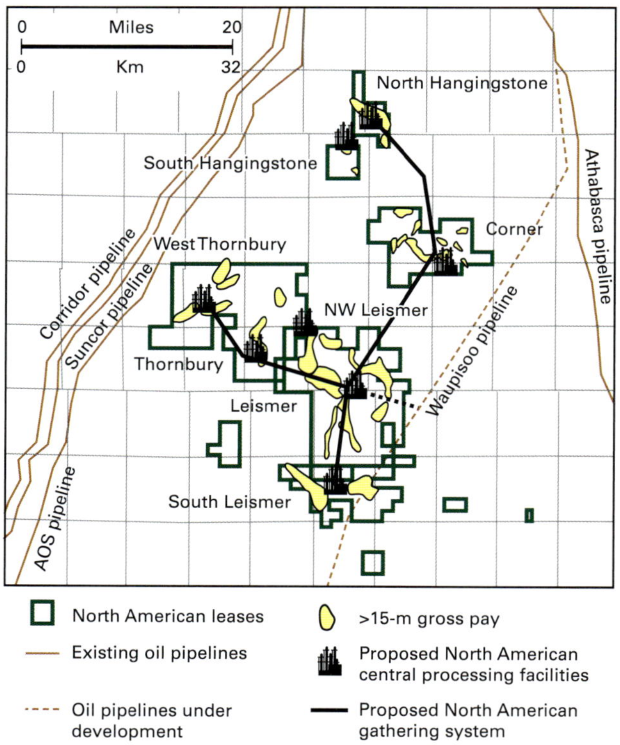

Petro-Canada

Petro-Canada said it plans to produce up to 320,000 b/d of synthetic crude and refined products after 2014 from its oil sands resources. The company estimates that it has about 10 billion bbl of recoverable bitumen in both its minable and in situ resources.

Its plan combines upgrading capacity at Syncrude and Edmonton with a processing agreement at Suncor to produce 135,000 b/d of ultra-low sulfur refined products for its existing Edmonton refinery.

Petro-Canada also is one of the first firms to initiate a commercial SAGD project. Its MacKay River facility has produced for the last three years and the company expected to reach the target production level of 27,000–30,000 b/d by year-end 2006.

Largest growth is expected to come from the Fort Hills project that Petro-Canada and its partners have targeted for first production in 2011. The company said recovery potential is 2.8 billion bbl. Petro-Canada submitted a commercial application for the Fort Hills project to the EUB and expects a regulatory decision by late 2007.

The upgrader will process up to 340,000 b/d of bitumen from the Fort Hills mine north of Fort McMurray and other sources into synthetic crude oil.

Bitumen basics

Oil sands (also called tar sands or bituminous sands) are a mix of clay, sand, water, and bitumen—an organic semisolid such as tar or asphalt that is highly viscous, black, sticky, and soluble in carbon disulfide. The percentage of bitumen in Canadian oil sands ranges from 1–20%. It takes about 1.16 bbl of bitumen to make 1 bbl of syncrude.

Optimism about oil sands is tempered by concerns about availability of natural gas, water, and personnel to support future growth, as well as the prospect of tightening environmental rules on carbon emissions.

When close to the surface, bitumen can be removed using open-pit mining. But when deposits are deeper, bitumen will not flow unless its viscosity is reduced by using in situ techniques that can be complex and expensive.

The simplest in situ treatment is cyclic steam stimulation. After steam injection, a reservoir is allowed to "soak" to achieve uniform heat distribution. Finally, the thinned oil is produced through the same well that was used for steam injection.

In the steam-assisted gravity-drainage (SAGD) process, steam injected into the producing zone creates a high-temperature chamber in the formation. Oil sands viscosity is reduced and the bitumen flows by gravity to a horizontal production well below.

In the newer vapor extraction (Vapex) process, vaporized hydrocarbon solvents are injected into oil sands to increase recovery by viscosity reduction, in situ upgrading, and pressure control.

The toe-to-heel air injection (THAI) process developed by Petrobank Energy & Resources Ltd. (Calgary) combines a vertical air injection well with a horizontal production well. Part of the oil in the reservoir is burned, generating heat that reduces oil viscosity and allows it to flow by gravity to the horizontal production well. e combustion front sweeps through the oil from the toe to the heel of the horizontal producing well.

Shell Canada Ltd.

Shell Canada announced plans to increase production in the Athabasca project in Alberta to 550,000 b/d from 150,000 b/d. Bitumen production is expected in 2009, followed by upgrader operation in 2010.

The company also completed the BlackRock Ventures acquisition and announced plans to raise in situ production by 2008 to nearly 50,000 b/d, primarily from the Peace River area, its Seal and Chipmunk assets, and the first phase of the Orion SAGD project in the Cold Lake region.

Monitoring steam movement

Shell Canada Ltd. explored use of advanced technology to boost steam-based production from its heavy oil leases in the Peace River area. Results showed that time-lapsed seismic, tilt meters, and microseismic can help monitor subsurface steam movement.

Tilt meters can detect changes on the surface of only a few millimeters as injected steam heats the subsurface formations. Microseismic uses downhole geophones to detect the small noises made as rock and sand grains moving in response to heating, indicating the direction of steam advancement through the reservoir.

Shell said these technologies provide a better understanding of how to inject steam efficiently and will lead to expanding production.

North American SAGD Production

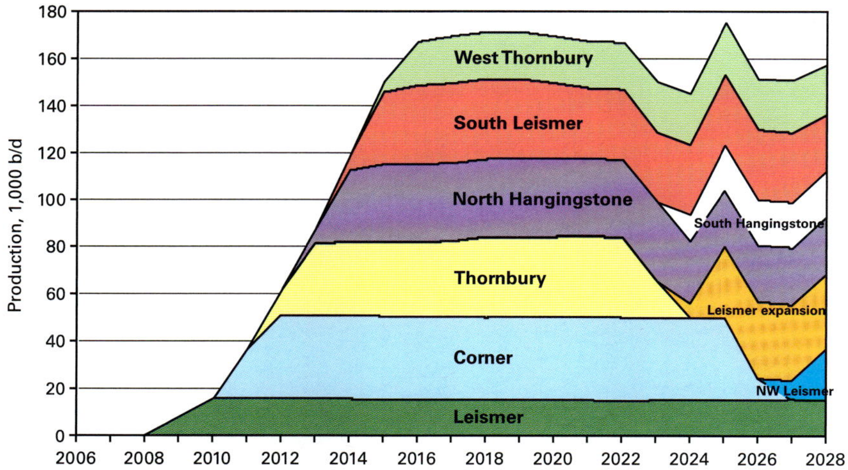

Integrated Oil Sands Strategy

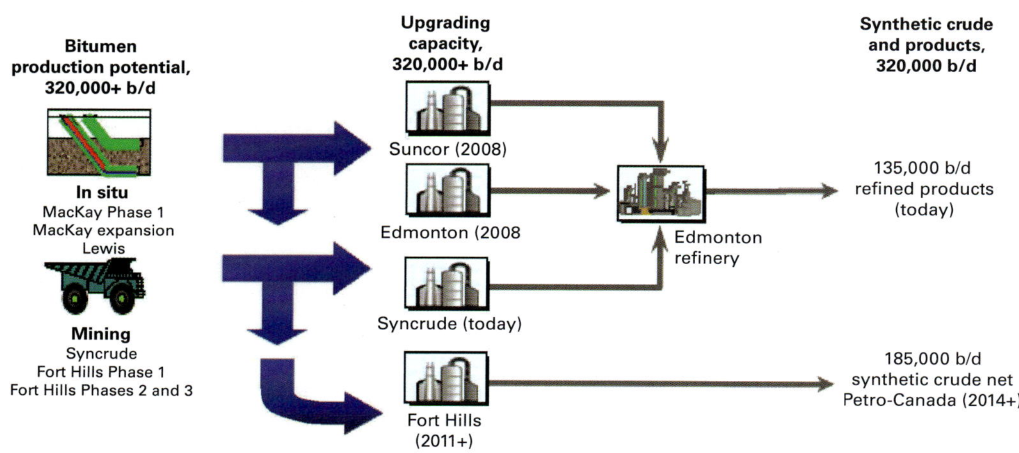

Petro-Canada Resource Potential

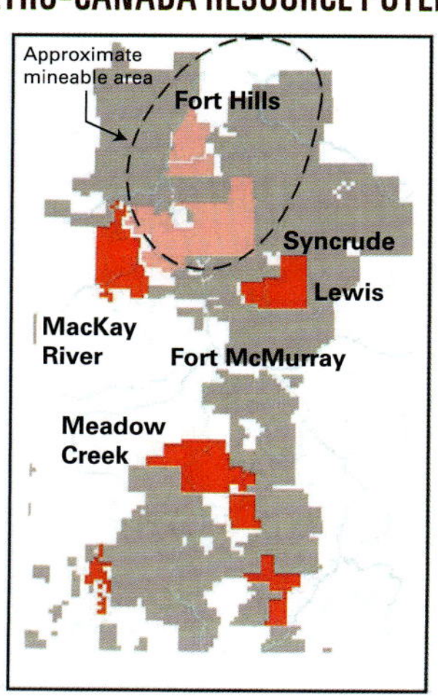

Mining	Petro-Canada interest	Resource, billion bbl
• Syncrude	(12%)	1.1
• Fort Hills	(55%)	1.6
SAGD		
• MacKay River	(100%)	0.3
• MacKay River expansion	(100%)	0.6
• Lewis	(100%)	3
• Meadow Creek	(75%)	1.3
• Other		2.1
Total		**10 billion bbl**

Suncor Energy

Suncor announced plans to spend $4.4 billion (Can.) in 2007 on oil sands development.

Work continued on Suncor's Athabasca oil sands expansion north of Fort McMurray. With commissioning targeted for 2008, that expansion is expected to increase production to 350,000 b/d from 260,000 b/d.

EUB approved Suncor's application to build a third oil sands upgrader, Voyageur, as well as plans to develop the Steepbank mine expansion. The two projects are critical to Suncor's plans to boost production to 500,000 b/d by 2012. Work on both Voyageur and Steepbank should start in 2007.

ConocoPhillips, EnCana joint venture

ConocoPhillips and EnCana Corp. agreed to create an integrated heavy oil joint venture that will involve two 50-50 operating partnerships: one Canadian upstream and one U.S. downstream. The venture, expected to be finalized in January 2007, would spend more than $10 billion over the next 10 years.

EnCana is the largest holder of properties in Canadian oil sands. The upstream partnership involves EnCana's Foster Creek and Christina Lake projects in northeast Alberta, with recoverable bitumen of more than 6.5 billion bbl. The partnership plans to increase production to 400,000 b/d of bitumen by 2015 from the current 50,000 b/d.

The downstream partnership involves ConocoPhillips' 306,000 b/cd Wood River refinery in Roxana, IL, and its 146,000 b/cd refinery in Borger, TX. The partnership plans to expand heavy oil processing capacity at these facilities to 550,000 b/d by 2015 from 60,000 b/d.

Chevron

Chevron Corp. acquired five heavy oil leases with potential for in situ development in the Athabasca region. The leases, covering about 30,350 hectares (75,000 acres), are 39 km (24 miles) southwest of the Athabasca Oil Sands Project.

Deer Creek Energy

Total SA subsidiary Deer Creek Energy Ltd. started the first commercial production from the Joslyn project in the Athabasca region, 60 km (37 miles) northwest of Fort McMurray, Alberta. Cumulative production is estimated at 2 billion bbl of bitumen. Using both SAGD and mining techniques, Deer Creek will bring production to 10,000 b/d of bitumen at plateau.

Husky Energy

Husky Energy (Calgary) completed construction and started steam injection at its 30,000-b/d Tucker oil sands project near Cold Lake. Peak production of 30,000 b/d is expected by 2008. Husky estimates production of 350 million bbl of bitumen over 35 years.

Husky also started a detailed engineering project to increase capacity of its Lloydminster heavy oil upgrader to 150,000 b/d. The $2.3 billion expansion will enable Husky to process additional heavy oil forecast at Lloydminster and at the Tucker oil sands project at Cold Lake. Lloydminster production is expected to average 115,000–120,000 b/d in 2006.

Syncrude Canada

Canadian Oil Sands Trust brought all third-phase expansion units on stream at Syncrude Canada Ltd.'s oil sands plant in Fort McMurray. The project pushes capacity to 350,000 b/d of synthetic crude, gross to Syncrude, and yields an ultralow-sulfur product.

Mackay River SAGD

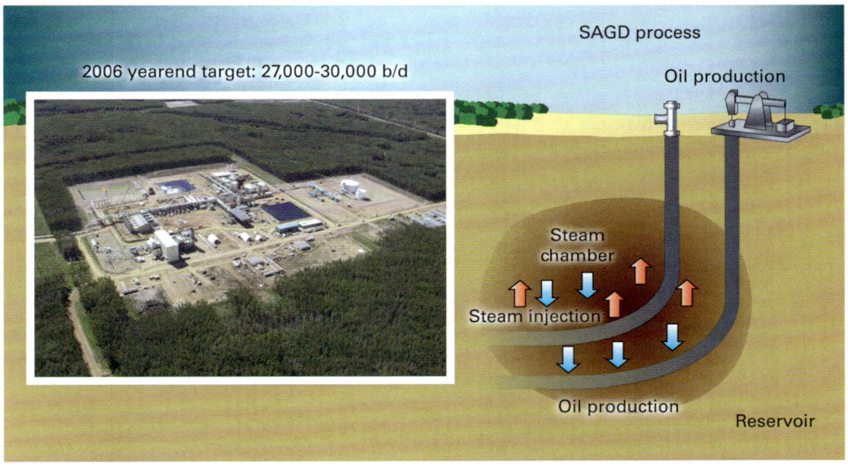

Monitoring Steam Injection

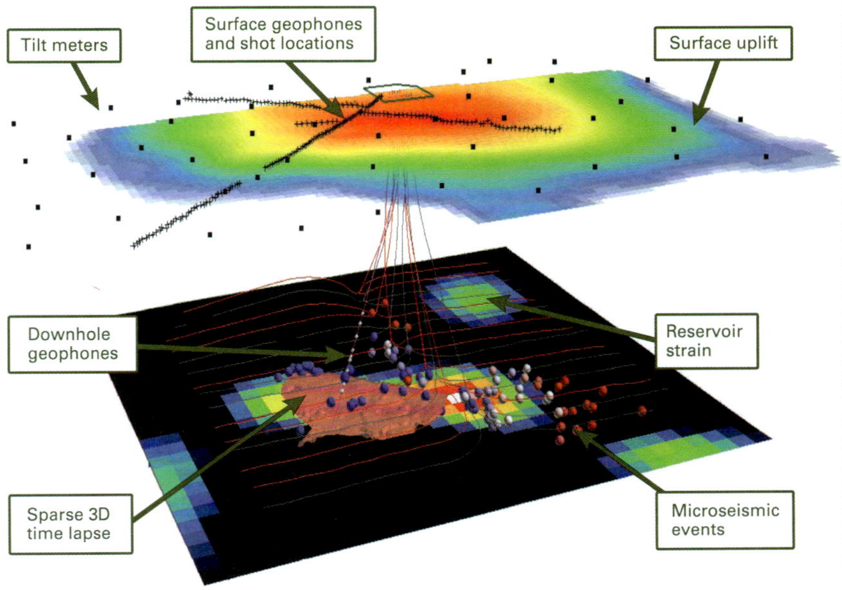

Source: Shell Technology webcast, Sept. 28, 2006

Exploration and Development

Alberta

EnCana Corp. (Calgary) is building gas production at a play in the deep basin of west-central Alberta to which it attributes 2 tcf of unbooked resource potential. The company drilled 38 wells in the first half of 2006, with gas production averaging 95 MMcfd. EnCana estimates original gas in place at 20 to 40 bcf/sq mile and expects ultimate recovery of 2 to 5 bcf.

TUSK Energy Corp. (Calgary) said it plans to develop light oil discoveries in the Mega-Venus focus area of Alberta 300 km (185 miles) northwest of Edmonton. Logs for the most recent well—Gutah 6-17-100-6-w6m—indicated 7.7 m (25 ft) of oil pay in a 13.5-m (44 ft) zone in Devonian Keg River. TD was 2,425 m (7,955 ft). The well flowed 345 b/d and 180 Mcfd of gas.

Galleon Energy Inc. (Calgary) announced plans to develop a light oil discovery near Puskwa, east of Grande Prairie, Alberta. Logs indicated that three other formations including Devonian Granite Wash are gas-bearing. Galleon will drill at least 10 delineation wells. Production could reach 4,000–8,000 boe/d in 2007.

ATLANTIC CANADA BASINS AND NL LAND TENURE MID-2006

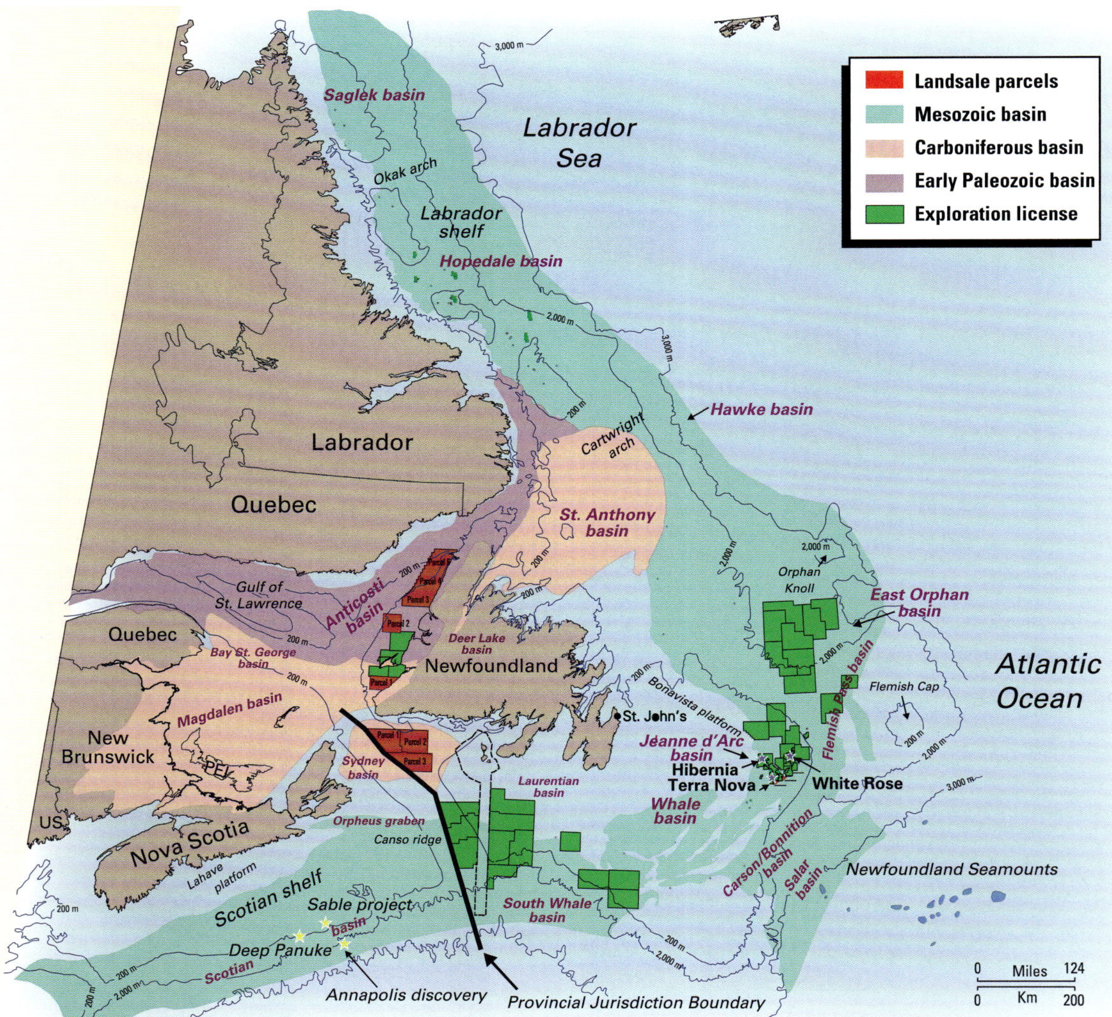

Source: Modified after the GSC, C-NLOPB and Enachescu, 2005

Hopedale Basin Gas Discoveries*

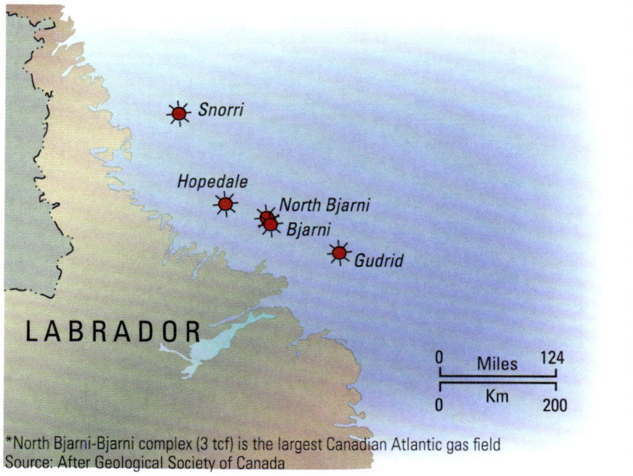

*North Bjarni-Bjarni complex (3 tcf) is the largest Canadian Atlantic gas field
Source: After Geological Society of Canada

Labrador: Hopedale basin

Hopedale is the southernmost rift basin within the Labrador Sea. It is just north of the Orphan basin, an area of current exploration drilling by Chevron, ExxonMobil, Imperial Oil, and Shell. The energy industry now is looking more closely at the Hopedale basin, which covers 175,000 sq km (67,578 sq miles) and extends from the modern-day shelf to the lower slope (200–3,000 m [656–9,843 ft] of water).

Several large gas discoveries were made along the Canadian Labrador shelf during oil exploration the 1970s and early 1980s. Early drilling included 27 wells (21 in the Hopedale basin) that proved the presence of 4.2 tcf of recoverable natural gas. Researchers at Memorial University of Newfoundland also pointed out that about 6 tcf of recoverable gas has been discovered in the Grand Banks.

All five Labrador shelf gas discoveries came from shallow targets in the Hopedale basin. The first, Bjarni H-81 in 1973, was followed by Gudrid H-55 (1974), Snorri J-90 (1975), Hopedale E-33 (1978), and North Bjarni F-06 (1980). These wells tested between 8 and 20 MMcfd.

Computed recoverable resources are 2.2 tcf for North Bjarni, 0.9 tcf for Bjarni, 0.1 tcf for Hopedale, 0.9 tcf for Gudrid, and 0.1 tcf for Snorri. No drilling has occurred in the basin since 1983.

The complex formed by adjacent Bjarni and North Bjarni fields contains more gas (3.1 tcf) than the currently producing Sable Project off Nova Scotia or the large (2.7 tcf) White Rose gas cap in the Jeanne d'Arc basin.

Federal and provincial governments in 2006 introduced discretionary requirements related to flow testing of the first well drilled on a new hydrocarbon prospect that should considerably reduce overall offshore drilling costs.

Newfoundland

Several companies bid a total of $32.4 million (Can.) for exploration licenses off Newfoundland. Of greatest interest was the Jeanne d'Arc basin, followed by western Newfoundland. Bidders included Husky Energy, Norsk Hydro Canada Oil & Gas Inc., NWest Energy Inc., and B.G. Capital Ltd. C-NLOPB expected to issue nine-year exploration licenses in early 2007.

Chevron Canada Resources and partners were drilling what will be the most expensive well ever off eastern Canada, a deepwater wildcat off Newfoundland in the untested East Orphan basin. Orphan is just north of the Jeanne D'Arc basin. Chevron spudded the Great Barasway F-66 well in about 2,350 m (7,700 ft) of water in the EL-1076 license area.

The partners also are testing controlled-source electromagnetic (CSEM) seabed logging technology, previously used in the UK North Sea in 2005.

Electric field receivers are suspended about 50 m (165 ft) above the seabed, and a high-powered electro-magnetic source is towed through the water, emitting a low-frequency

The CSEM Process

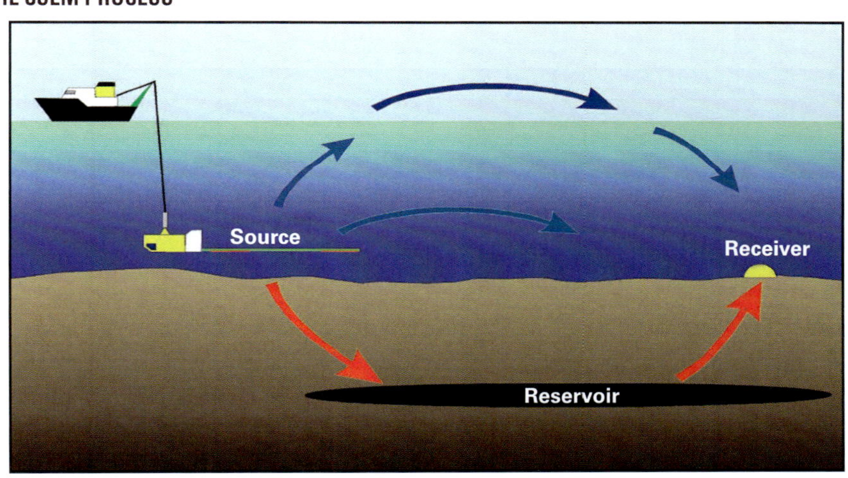

signal. Receivers on the seabed detect the resulting signal at offsets of up to 10–15 km (33–50 miles). Data are interpreted with a combination of geophysical inversion and imaging,

Nova Scotia

In its 2005–06 annual report, the Canada-Nova Scotia Offshore Petroleum Board (C-NSOPB) noted a slight drop in activity over the past three years. C-NSOPB said that 204 wells have been drilled off Nova Scotia The Nova Scotia Department of Energy said that 125 wildcats were drilled offshore, with 20 gas discoveries, and four oil discoveries.

CANADA-NOVA SCOTIA OFFSHORE PETROLEUM BOARD ACTIVITIES[1]

	2005-06	2004-05	2003-04
Drilling program authorizations	1	3	4
Drilling program environmental assessments	1	1	10
Wells approved to drill	[2]4	4	7
Wells drilled	2	2	5

[1]Fiscal year Apr. 1-Mar. 31. [2]Includes 3 wells and 1 sidetrack.

Northwest Territories

Paramount Resources Ltd. (Calgary) said Cameron Hills field could hold 20–30 million bbl. The field is just northeast of the northwest corner of Alberta. Paramount is producing 1,500 b/d of light, sweet crude and 7–8 MMcfd of gas. The company started producing gas in 2002 and oil in 2003 and said it had seen practically no decline in the oil rate. It placed a third well, J-74, on production in winter 2005–2006 at 500 b/d.

Husky Energy Inc. (Calgary) was evaluating a discovery in the Central Mackenzie Valley, made with the Stewart D-57 (formerly Tehwa D-57) wildcat on Tulita District Land Corp. Freehold Block M-38. The well is 35 km (miles) southeast of the Summit Creek B-44 discovery well. On open-hole test, gas flowed at 5 MMcfd. Husky called it "the first successful Cretaceous hydrocarbon discovery in the Central Mackenzie region."

EnCana Corp. (Calgary) applied for a declarations of significant discovery for the Umiak N-05 and N-16 wells at Burnt Lake in the Mackenzie Delta, 47 miles north of Inuvik, on Exploration License 384. EnCana previously reported encouraging hydrocarbon shows at N-16.

WESTERN CANADA CBM WELLS DRILLED

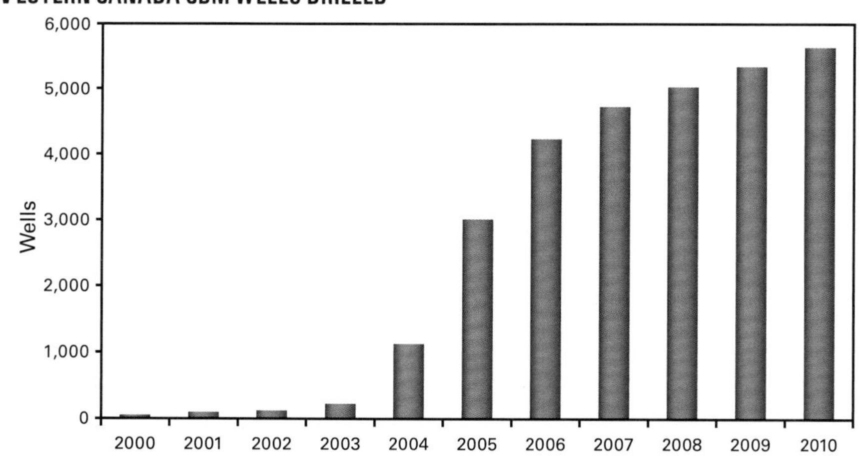

Source: CAPP and First Energy Corp.

Coalbed Methane Distribution

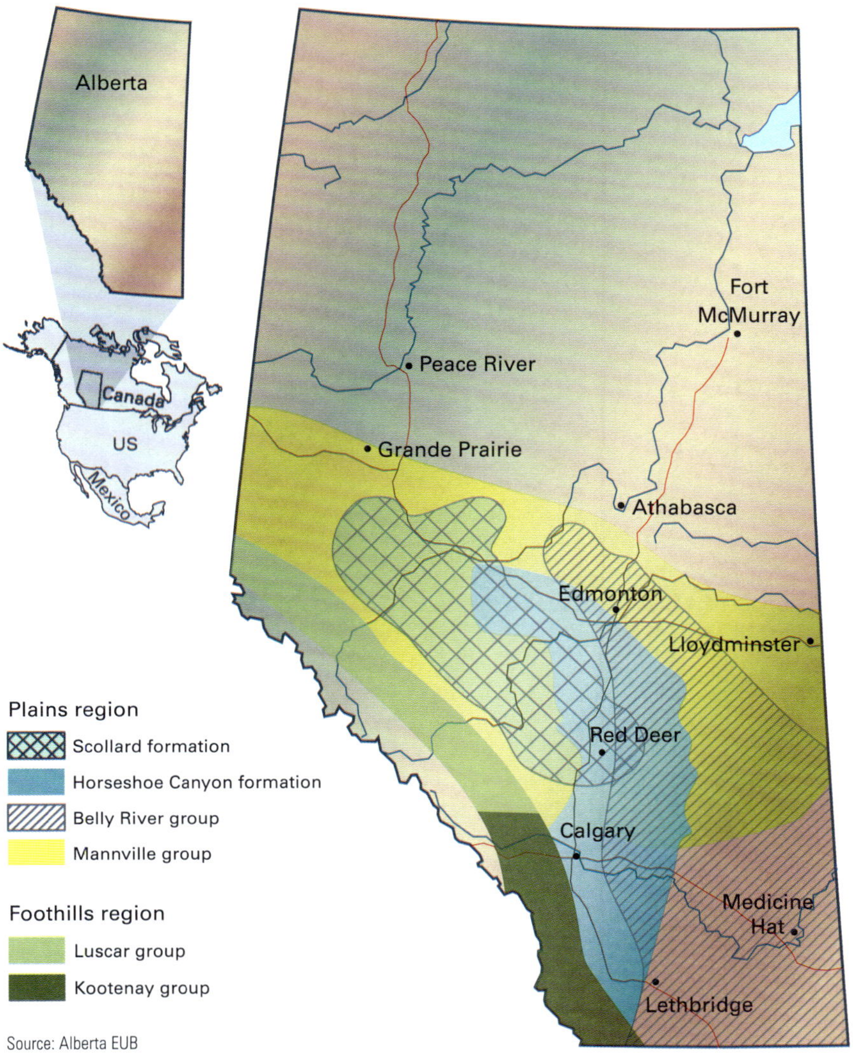

Source: Alberta EUB

Eastern British Columbia and western Alberta have seen limited CBM exploration activity and no commercial production. The coal deposits are structurally complex and the topography presents significant development problems.

Alberta

Canadian Superior Energy Inc. (Calgary) is developing CBM in the Drumheller area, targeting Horseshoe Canyon and Mannville intervals, with first attention to Horseshoe Canyon, where it drilled or participated in 26 wells in 2006. The Mannville coals are deeper and untested; Canadian Superior said it would use horizontal drilling techniques to reach them.

Anderson Energy Ltd. (Calgary) was preparing to participate in 67 Horseshoe Canyon CBM wells in its Tier I area in 2006. It planned to drill two high-impact Horseshoe Canyon CBM projects at Pearce and Blood (between Calgary and the U.S. border).

Nexen Inc. (Calgary) was working to boost coalbed methane production to more than 30 MMcfd by year-end in central Alberta at the first commercial CBM project in water-saturated Lower Cretaceous Upper Mannville coals. The ultimate goal is 150 MMcfd by 2011 from development of about half of the company's CBM lands. Present output is 4 MMcfd.

Saskatchewan

Nordic Oil and Gas Ltd. (Winnipeg, MB) drilled three CBM wells at Joffre and planned to drill up to 15 CBM more, along with Belly River oil wells.

Coalbed Methane

Overview

Commercial production of coalbed methane (CBM) started only recently in Canada, but this unconventional natural gas resource is now a potentially significant contributor to future energy supply. In Alberta alone, the estimated CBM resource is more than 500 tcf with 167 tcf deemed recoverable. CAPP said that 4,000 of the total gas wells drilled in 2006 in the WCSB target CBM.

To date, Alberta has the only commercial CBM production in Canada. At 500 MMcfd, CBM accounted for an estimated 3.6% of Alberta's total gas production in 2006 but is predicted to reach 1.6 bcfd by 2010 and 2.5 bcf/d by 2015.

CBM Blocks, Nova Scotia

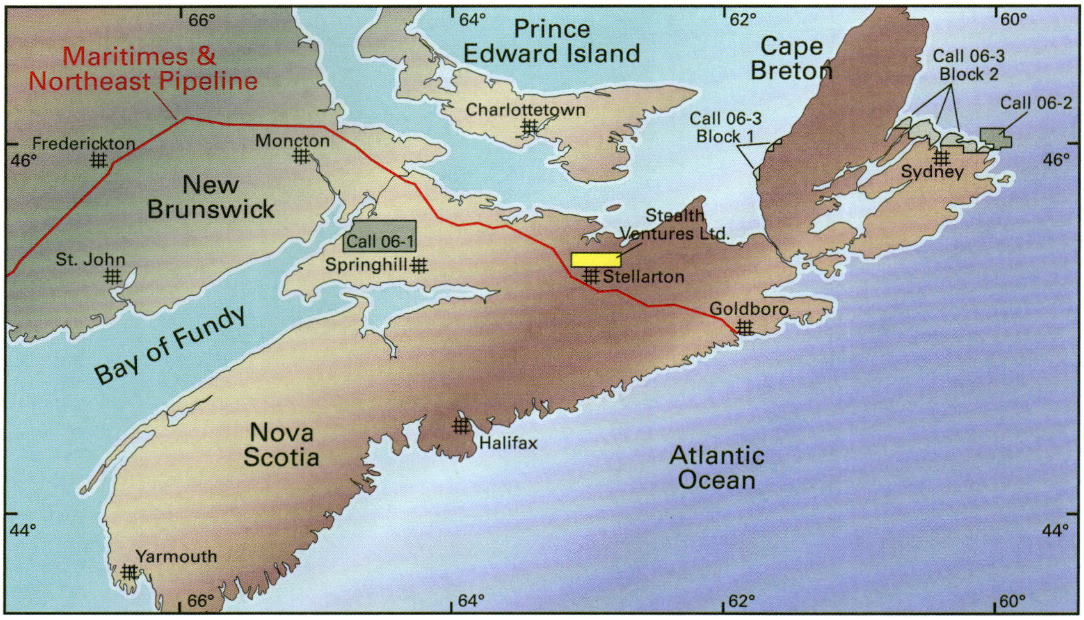

Planned Keystone Pipeline Route

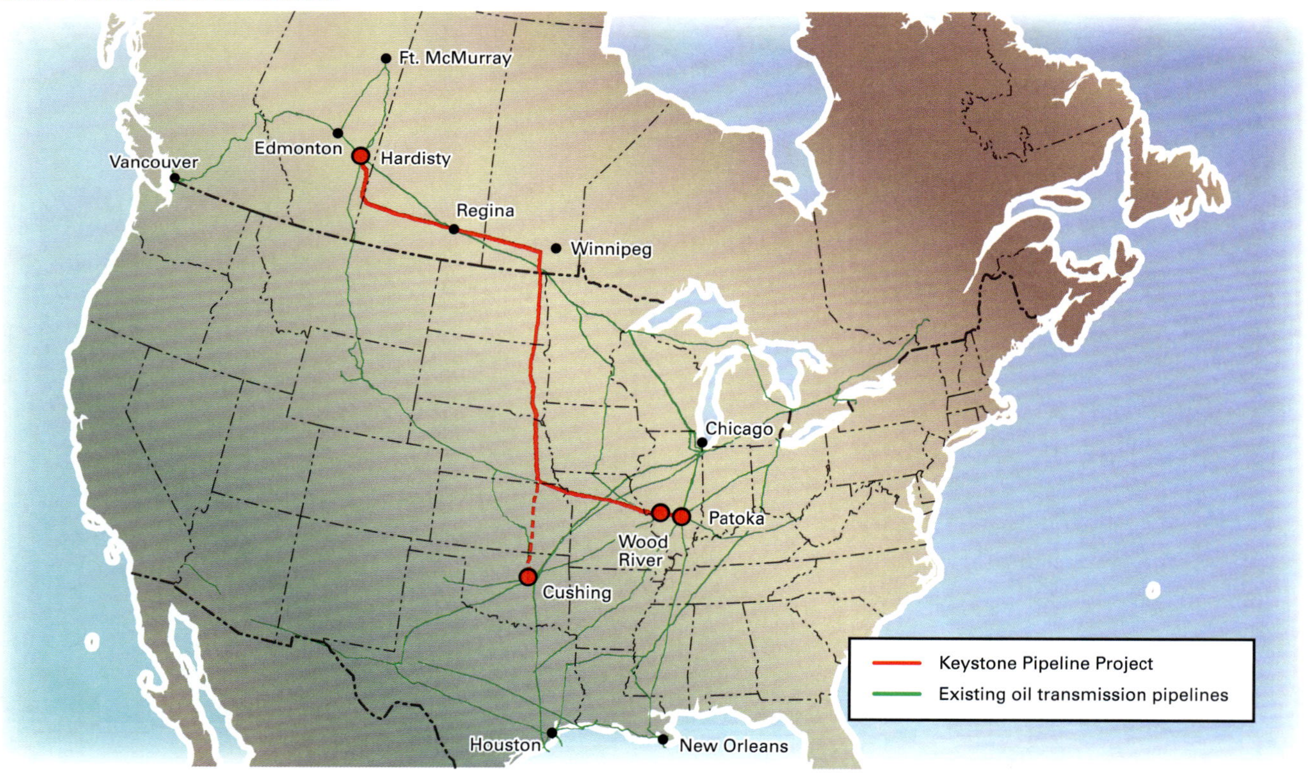

Nova Scotia

The Nova Scotia Dept. of Energy issued two calls for bids on CBM acreage. Call 06-2 offered a 20,267-ha (50,000-acre) block off northeastern Cape Breton. Call 06-3 offered two blocks onshore Cape Breton:

- Block 1, Inverness County, 9,650 ha (23,807 acres), on the west coast near Mabou.
- Block 2, Cape Breton County, 20,175 ha (49,772 acres), on the north coast near Sydney.

Stealth Ventures Ltd. (Calgary) holds acreage containing 1.31 tcf (net) booked reserves as well as the first CBM agreement in the province, for the Stellarton basin. Stealth's first Cumberland CBM well near Springhill intersected more than 20 m (65 ft) of net coal in multiple seams, and the company planned to spud a second well by year-end. Stealth also spudded the Priestville No. 4 CBM well in near Stellarton.

Pipelines

Mackenzie Delta pipeline

The proposed Mackenzie Gas Project would develop an estimated 170 bcm of gas (6 tcf) in the Mackenzie River Delta and build a gas and NGL gathering system as well as a 12–20 bcm/yr (1.2–1.9 bcfd) pipeline to carry gas 1,223 km (760 miles) from the Canadian Arctic to southern markets. Project sponsors are Imperial Oil Ltd. (Calgary); the Aboriginal Pipeline Group (Calgary); ConocoPhillips Canada (Calgary); Shell Canada Ltd. (Calgary); and ExxonMobil Canada Ltd. (Halifax, NS).

However, the regulatory approval process continued to delay work on the project. Numerous issues remain unresolved, including land access arrangements. A Canadian review panel has extended public hearings through April 2007, and the project must then be approved by Canada's National Energy Board before pipeline work can start. Imperial Oil said the delay could increase the project's cost from the current $6 billion (Can.) estimate to $9 billion (Can.).

Keystone Oil Pipeline

Regulatory review continued for the proposed $2.1 billion Keystone Oil Pipeline, which will carry 435,000 b/d of oil-sands crude 2,955 km (1,830 miles) from Hardisty, AB to Wood River and Patoka, IL. The in-service target is 2009.

The Canadian Keystone segment comprises 370 km (230 miles) of new pipeline plus conversion of 853 km (530 miles) of TransCanada's Canadian Mainline gas pipeline to carry oil. A decision on the conversion by Canada's National Energy Board (NEB) is expected by mid-2007.

TransCanada said shippers also have expressed interest in a proposed south extension of the system to Cushing, OK, which would add 468 km (291 miles) of pipeline at an estimated cost of $445 million.

Trans Mountain expansion

Terasen Pipelines Inc. (Calgary)—a unit of Kinder Morgan (Houston)—applied to the NEB for authorization to build and operate loops and oil pumping facilities to increase capacity of the 1,145-km (710-mile) Trans Mountain Pipeline in Alberta and British Columbia by 40,000 b/d.

The $400 million TMX-Anchor Loop project includes 158 km (98 miles) of pipeline from west of Hinton, AB to near Rearguard, BC and two pump stations.

Husky Alberta crude line

Husky Energy Inc. (Calgary) began work to expand its crude oil mainline between Lloydminster, Alberta and its terminal at Hardisty, AB in a $100 million, 80-km (50-mile) project.

The expansion will accommodate increased production from Husky's Tucker oil sands project near Cold Lake and shipments from third parties. It also will allow for increases in future volumes from the company's heavy oil upgrader at Lloydminster. Construction should be completed in late 2007.

Talisman Lynx gas pipeline

Talisman Energy Inc. (Calgary) commissioned the Lynx pipeline, which will gather gas in the Grande Cache area of the northern Alberta Foothills. Parallel pipelines will carry sour gas and fuel gas from the Findley dehydration facility to a new Lynx dehydration facility with raw gas capacity of 130 MMcfd.

Refining and Processing

Second St. John refinery?

Irving Oil Ltd., (St. John, NB) said it is considering building a second refinery at St. John. It would be the first major refinery built in North America in nearly 25 years and the largest private-sector investment in Atlantic Canada. The new refinery could supply as much as 300,000 b/d of products to the U.S Northeast and would cost $5–7 billion. Irving owns the existing 250,000-b/d refinery at St. John, the largest in Canada.

Edmonton hydrogen plant to support oil sands upgrading

Air Products Canada Ltd. began engineering to construct a 105-MMscfd hydrogen plant at the Edmonton refinery of Petro-Canada. Targeted for operation in April 2008, it will be the first Air Products facility to provide hydrogen for processing Canadian oil sands into gasoline and diesel fuel, said the company. Technip (Paris) will design and build hydrogen generation units.

THE COMBINED KINDER MORGAN-TERASEN SYSTEM

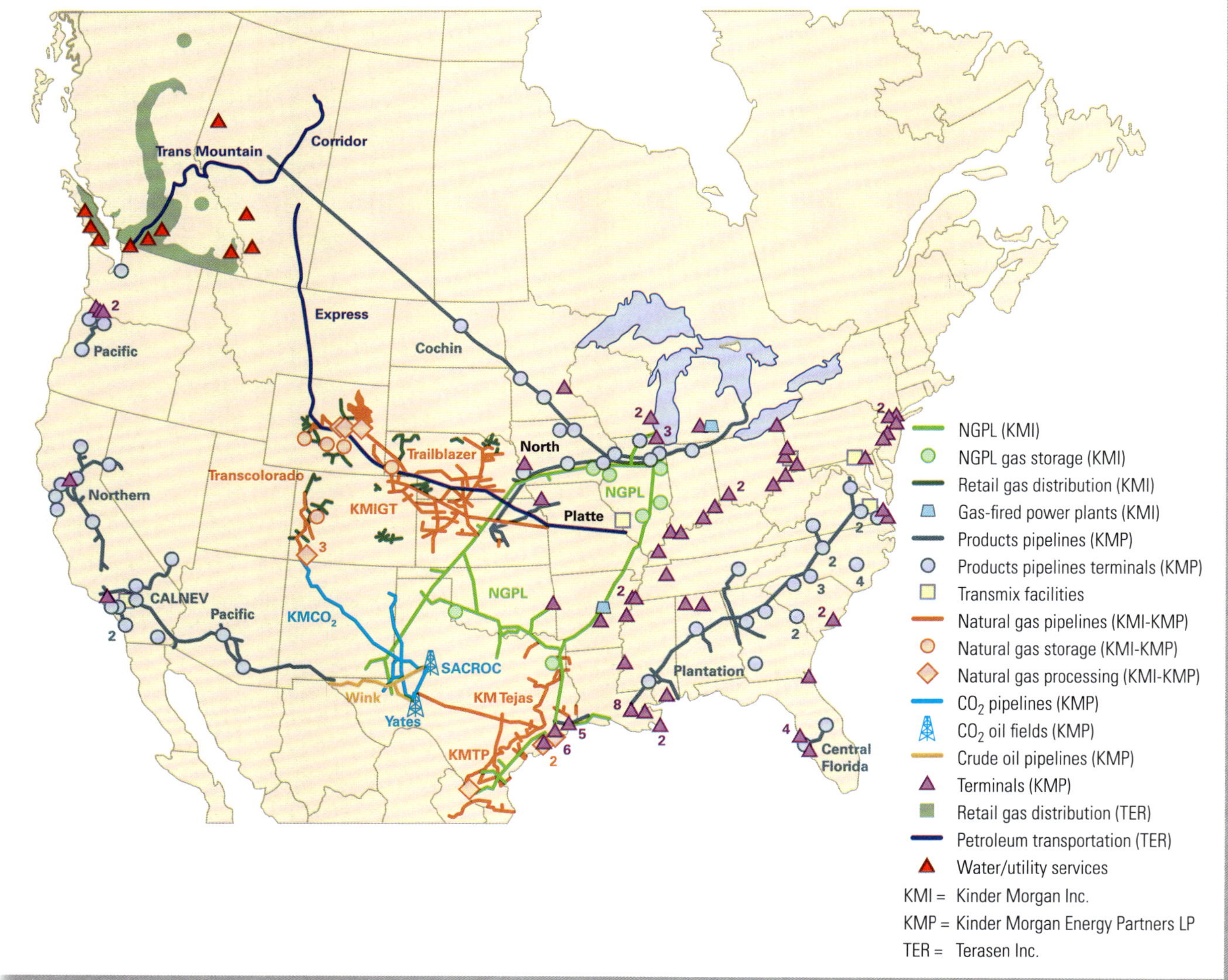

LNG

Kitimat LNG Inc., a unit of Galveston LNG Inc., has proposed a $500 million LNG terminal at Bish Cove, near the Port of Kitimat. Planned sendout capacity is 1 bcfd. It will deliver gas through a 14-km (9-mile) pipeline lateral into the planned Pacific Northern Gas pipeline, which will then move the gas to an interconnection with Duke Energy's Westcoast Energy Main gas transportation system.

Kitimat LNG received provincial environmental approval and a certificate for the terminal and expected federal permits soon. The company planned to start construction in 2006, aiming for operational start-up in 2009.

Pacific Northern Gas Ltd. (Vancouver) and Kitimat LNG announced formation of Pacific Trail Pipelines LP. The new company is to develop 470 km (292 miles) of 30- or 36-in. natural gas pipeline from Kitimat to Summit Lake, BC to serve the proposed LNG terminal. Pacific Trail aims to start construction of the pipeline by 2008.

WestPac LNG Corp. began regulatory review for its proposed $350 million LNG terminal on Ridley Island near Prince Rupert, BC. The terminal could be operational in 2011, with a total receiving capacity of 1 bcfd gas equivalent of LNG. Regasification facilities would deliver 130 MMcfd into existing pipelines.

Canaport LNG, a partnership of Repsol YPF SA and Irving Oil Ltd., announced plans to begin construction of its LNG regasification terminal near Saint John, NB. The site is an existing deepwater marine terminal, Irving Canaport. Construction could be completed by 2009. Baseload sendout will be 1 bcfd; peak-day 1.2 bcfd with expansion capability up to 2 bcfd.

As much as 850 MMcfd of gas will flow from Canaport through New Brunswick through the 145-km (90-mile) Brunswick Pipeline to a connection with the Maritimes & Northeast Pipeline system at the U.S. border near Baileyville, ME.

The Bear Head LNG project was proposed for construction due east of the Canaport site, on Cape Breton Island in Nova Scotia. Anadarko Petroleum Corp. (Houston) said in July 2006 that it had agreed to sell the project to a privately held equity investment firm. As of late September, however, the sale had not been completed. If built, Bear Head would have an initial sendout capacity of 1 bcfd.

Other Company News

Royal Dutch Shell PLC offered $7.7 billion (Can.) in October 2006 to buy the shares that it does not already own in Shell Canada Ltd. in a move to boost the parent company's stake in oil sands. Royal Dutch Shell currently owns 78% interest in Shell Canada. Shell Canada has a substantial position in Canada's oil sands and is pursuing major expansion of production and upgrading capacity.

Burlington Resources, one of the biggest producers in the WCSB, was acquired by ConocoPhillips for about $36 billion. ConocoPhillips Canada became the third largest oil and natural gas production and exploration company in Canada, as well as the second largest gas producer in Canada.

Canadian Natural Resources Ltd. (CNR) announced plans to buy Anadarko Canada Corp. (ACC) for $4.24 billion. Anadarko is keeping its interests in the Mackenzie Delta and other Canadian arctic frontier properties. ACC's land and production base are in Western Canada. Anadarko said it put its Canadian subsidiary up for sale to help reduce debt from its $21 billion purchases of Kerr-McGee Corp. and Western Gas Resources Inc.

Focus Energy Trust (Calgary) struck a $1.2-billion deal to more than double its production by merging with Profico Energy Management, western Canada's largest privately held natural gas company. Focus will gain Profico's Shackleton and southwest Saskatchewan properties, including 46.7 MMboe of proved-plus-probable reserves.

EnCana Corp. (Calgary) and certain affiliates reached an agreement to sell substantially all of their gas storage business interests for $1.5 billion (Can.). The buyer is a private equity fund managed by Riverstone Holdings LLC and The Carlyle Group. EnCana Gas Storage has approximately 174 bcf of working gas capacity at five facilities in Alberta, California, and Oklahoma. EnCana will retain the Hythe (10 bcf) gas storage facility in Alberta for its own use.

Teco Management SA (Norway) announced its acquisition of the Davie Industries shipyard on the Saint Lawrence River in Quebec. Teco intends to build offshore drilling platforms at the shipyard, which had been in bankruptcy since 2001.

Latin America

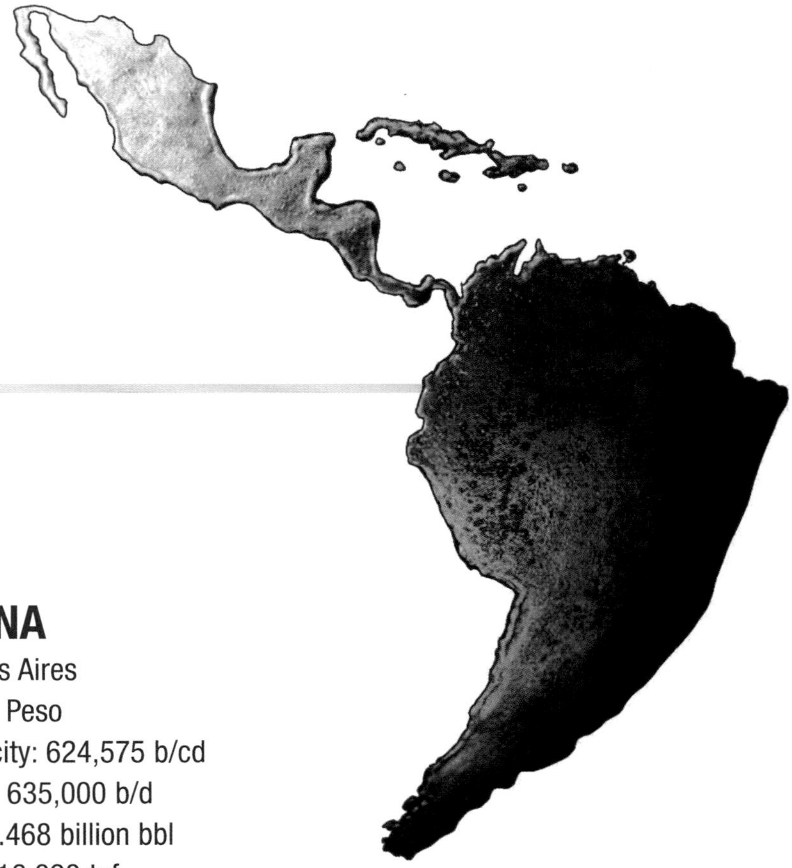

ARGENTINA

Capital: Buenos Aires
Monetary unit: Peso
Refining capacity: 624,575 b/cd
Oil production: 635,000 b/d
Oil reserves: 2.468 billion bbl
Gas reserves: 16.090 tcf

Argentina is one of South America's largest and most important economies. A real growth rate of 5.0% was forecast for 2006 by the U.S. Energy Information Administration (EIA). Although oil production has been declining since 2001, the country remained the third largest producer in South America in 2006. Production of natural gas has grown steadily since 2001.

A major energy development was the signing in October of an accord by Argentine President Nestor Kirchner and Bolivia Pres. Evo Morales that called for $1 billion in investments to increase the flow of Bolivian gas into Argentina. The 20-year accord calls for Argentina to receive 27.7 million cu m/day (978.1 MMcfd) of gas by 2010.

Argentina already held contracts to import 7.7 million cu m/day (271.9 MMcfd), although current shipments average 4.5 million cu m/day (158.9 MMcfd), said Bolivia's state-owned oil company, Yacimientos Petroliferos Fiscales Bolivianos (YPFB).

Exploration and Development

Salta Province blocks attract bids

Salta Province offered seven areas for exploration in a first licensing round. These included the Hickmann, Tartagal Oriental, Chirete, Morillo, Union, and Malvalay areas, most of which fall in the Cretaceous basin. The provincial government received bids totaling $220 million (U.S.).

A consortium of Argentine oil company Tecpetrol and Brazil's federal energy company Petrobras presented an offer for Hickmann.

ARUBA
CURACAO
GRENADA
Bridgetown
BARBADOS
TRINIDAD and TOBAGO
Port of Spain
COSTA RICA
Pto. Limon
Colón
Panama
PANAMA
Santa Marta
Barranquilla
Cartagena
Maracaibo
Tibu
Barrancabermeja
S. Cristobol
S. Silvestre
Cantimplora
Caracas
Lechoso
Jose
Puerto la Cruz
Punzan
Grico
Hato
Sinco
Palmita
Silvestre
Ciudad Bolivar
Penal-Barrackpore-Wilson
Teak
Balata-Bouvallius
Southeast Galeota
Radix
E. Queens Beach
Quarry-Coora-Quinam
Georgetown
Tambaredjo/Jossikreek
VENEZUELA
GUYANA
SURINAME
Karanambo
Volcanera
Cupiagua
Cusiana
Apiay Complex
Chichimene
Bogota
Tulua
Cali
Castilla
Andalucia
Dina
Tello
San Francisco
Roraima
Tumaco
Esmeraldas
COLOMBIA
Quito
ECUADOR
Guayaquil
Amistad
Tumbes
Zorritos
Piedra Redonda
Talara
Bayovar
Iquitos
Concordia
Amazonas
Manaus
Jurua
Southwest
Rio Urucu
Coari
Leste do Urucu
Nova Olinda
Paripi
Para
Maquia
B R A Z I L
Pucallpa
Aguaytia
Aguas Calientes
Acre
Porto Velho
Rondônia
PERU
Camisea
Fields
Lima
La Pampilla
Conchan
Pisco
Paracas
Mato Grosso
BOLIVIA
Trinidad
Lago de Titicaca
Pirin
La Paz
Surubi
Pto. Villarroel
Cochabamba
Sicasica
Colpa
Santa Cruz
Oruro
Tita
Rio Grande
Corumba
Arica
Sucre
Tatarenda
Camiri
Boyuibe
Monteagudo
Vuelta Grande
Villemontes
Caigua
Buena Vista
Sanandita
Tigre
Madre Jones
PARAGUAY
Toro
Bermejo
Campo Duran
Ramos
Lomitas Tranquitas
Tocopilla
Mejillones
CHILE
Rio Pescado
Caimancito
Jujuy
Salta
ARGENTINA
Villa Elisa
Asuncion
Antofagasta
0 100 200 300 400 Miles
0 200 400 600 Km
SOUTH AMERICA

SOUTH AMERICA
0 100 200 300 400 Miles
0 200 400 600 Km
Atlantic Ocean
Georgetown
Tambaredjo/ Jossikreek
Paramaribo
7
Cayenne
GUYANA
SURINAME
FRENCH GUYANA
Belem
Amazon River
44
Manaus
Nova Olinda
Rio Tapajos
Sao Luis
Atum
Curima
1-CES-8
Xareu
Espada
Pescada
Fortaleza
6
Fazenda Belem
Ubarana
Agulha
Canto do
Amaro
Guamare
Serraria Lorena
Faz. Pocinhos
Estreito
B R A Z I L
Rio Teles Pires
Rio Tocantins
Rio Araguaia
Recife
Maceio
Aracaju
269
Salvador
Catu
Rio Sao Francisco
Cuiaba
Brasilia
Sao Mateus
Corumba
PARAGUAY
Aquidauana
Campo Grande
145
Betim
Belo Horizonte
Rio Grande
Parana
Vitoria
Jambarte
Duque de Caxias
232
Campos
RJS-150
Frade
Roncador
10°
5°
0°
5°
10°
15°
20°
55°
45°
35°

SOUTH AMERICA
BRAZIL
PARAGUAY
URUGUAY
ARGENTINA
CHILE
Atlantic Ocean
Falkland Islands (Islas Malvinas)
Golfo San Matias
Santa Cruz
Oruro
Arica
Sucre
Camiri
Tita
Rio Grande
Tatarenda
Boyuibe
Monteagudo
Vuelta Grande
Buena Vista
Caigua
Sanandita
Madre Jones
Tigre
Toro
Campo Duran
Bermejo
Ramos
Lomitas Tranquitas
Rio Pescado
Calmancito
Tocopilla
Mejillones
Antofagasta
Jujuy
Salta
Metan
Corumba
Aquidauana
Campo Grande
Villa Elisa
Asuncion
Formosa
Resistencia
Corrientes
Paso de los Libres
Uruguaiana
Sao Gabriel
Betim
Belo Horizonte
Campos
Duque de Caxias
Rio de Janeiro
Tremembe
Maua
Sao Paulo
Cubatao
Santos
Sao Sebastiao
Apucarana
Irati
Merluza
Curitiba
Sao Francisco do Sul
Perimbo
Florianopolis
Laguna
Canoas
Porto Alegre
Tramandai
Rio Grande do Sul
Cordoba
San Lorenzo
Lumunta
Tupungato
Cacheuta
Mendoza
San Luis
Concon
Valparaiso
Santiago
Refugio
Colorados
Barrencas
El Sosneado
Campana
Buenos Aires
La Plata
Montevideo
Dock Sud
Las Flores
Cnl. Pringles
Concepcion
Colonia Catriel
El Medanito
Guanaco
Plaza Huincul
Challaco
Cerro Bandera
Barda Gonzales
Bahia Blanca
Punta Alta
Puerto Rosales
San Antonio
Las Flores
Comodoro Rivadavia
Rivadavia
Anticlinal Grande
Cerro Dragon
Canadon Grande
Canadon Seco
Cerro Tortuga
Comodoro
Caleta Olivia
Pico Truncado
Santa Cruz
Rio Gallegos
Punta Arenas
Tierra Del Fuego
Rio Grande
Ushuaia
3
20
32
7
145
356
14
181
232
47
163
242
193
181
17
98
38
85
110
37
4
8
189
106
113
25
29
16
20°
25°
30°
35°
40°
45°
50°
80°
55°
70°
60°
50°
40°
0 100 200 300 400 Miles
0 200 400 600 Km

Tecpetrol, Petrobras and Occidental Petroleum Corp. (Los Angeles) each bid for the Tartagal Oriental area. Bidding for the Chirete area were: U.S. oil company Cliveden; a consortium of Occidental and Chinese oil company JHP; and a consortium of the local YPF subsidiary of Spain's Repsol YPF, Argentine oil company Pluspetrol, and Canadian oil company Gran Tierra (Calgary).

Korean oil company Petroterra was the only bidder for the Morillo, La Union and Malvalay areas.

Argentina sedimentary basins

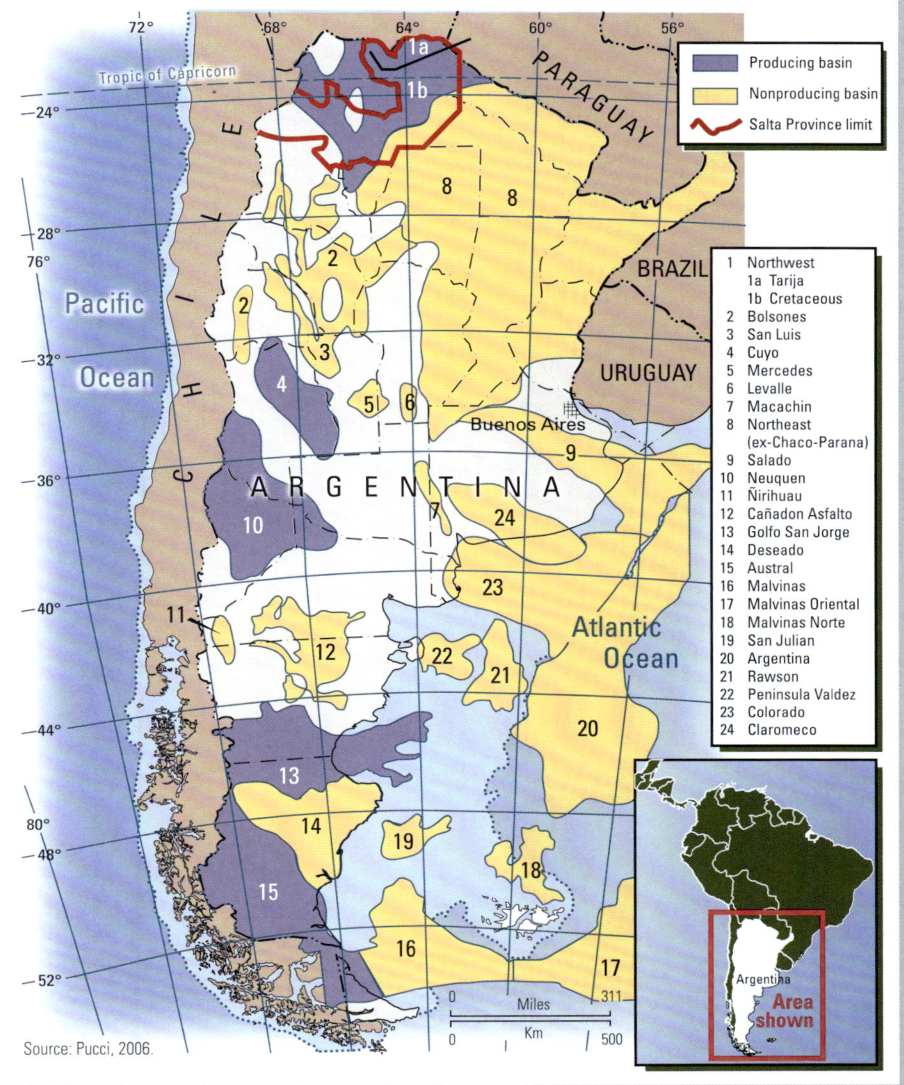

Source: Pucci, 2006.

Seismic study in north

Gold Point Energy Corp. (Vancouver, BC) took farmouts from APCO Argentina Inc.(Tulsa) and Antrim Energy Inc. (Calgary) on the Capricorn License (Yacimiento Norte 1/B Block) in northern Argentina. Gold Point will share the costs of a 3D seismic program (completed in 2006) and an exploration well to earn a 25% working interest in the Martinez del Tineo Oeste Prospect. The well is to be drilled in 2007.

Group to explore deepwater shelf

Petróleo Brasileiro SA subsidiary Petrobras Energía SA (Petrobras), Argentina's Energía Argentina SA (Enarsa), and Madrid-based Repsol YPF SA agreed form a consortium to explore, develop, and commercialize oil and natural gas on Argentina's deepwater continental shelf. The consortium plans to conduct exploration activities in 200–3,000 m (656–9,842 ft) of water about 250 km (155 miles) off the city of Mar del Plata, Buenos Aires.

Areas of Salta Province, Argentina

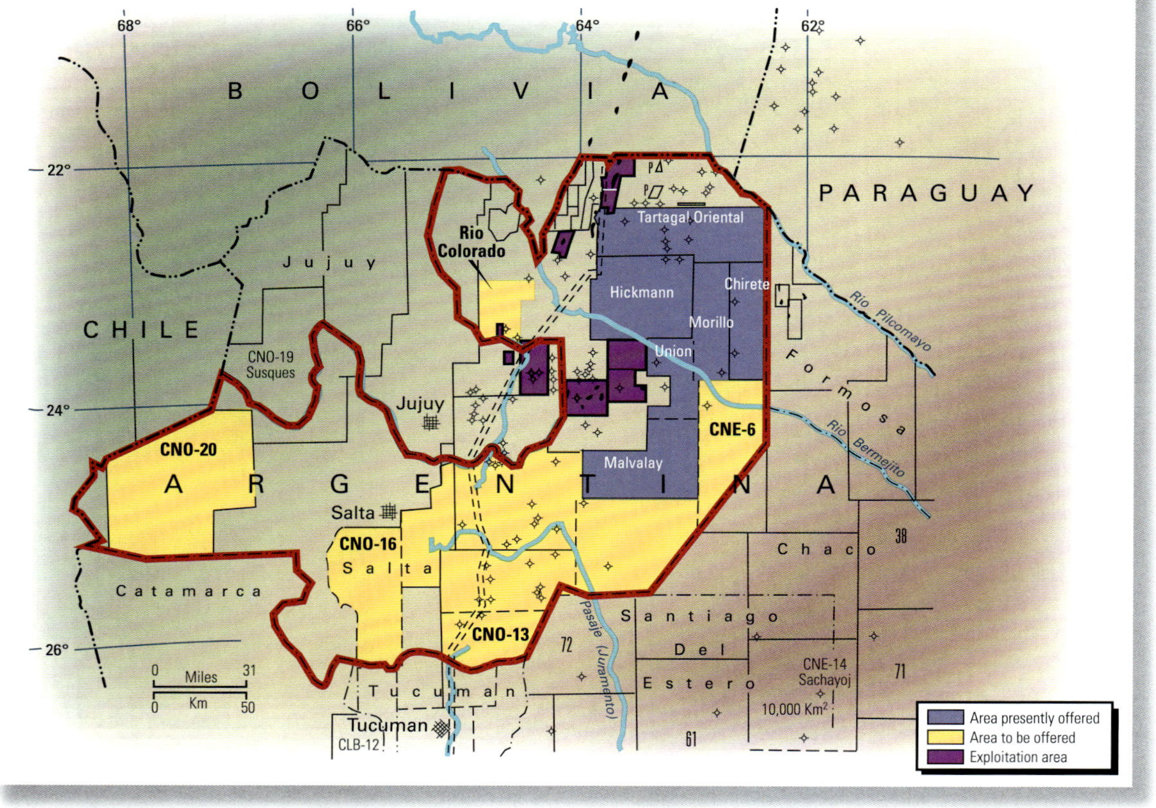

Tierra del Fuego yields two discoveries

A group led by Antrim Energy Inc. (Calgary) and Apco Argentina Inc. holding three licenses in Tierra del Fuego announced two discoveries, in the Las Violatas 105 well as well as in the Los Patos 1005 field.

Technology boosts Falkland prospect

Combining 2D seismic data and controlled source electromagnetic (CSEM) surveys has rendered drillable the Ernest, formerly J1, prospect on North Falkland basin License PL024 in the South Atlantic, said Rockhopper Exploration PLC (London). Ernest is in 160 m (525 ft) of water 100 km (62 miles) from the islands, with an areal extent of 2,880 acres. The surveys also identified encouraging leads on PL023 and PL024, where a 3D seismic survey was to start.

Relook at offshore basins urged

Citing Argentina's declining oil and gas exploration, an experienced consulting geologist based in Buenos Aires urged reexamination of the country's offshore basins in the South Atlantic Ocean.

Juan Carlos Pucci noted that production is not being replaced, the number of wells being drilled has declined, and the number of producing wells has fallen. He also said the number of wildcats also dropped to 182 in 2004, the lowest level in Argentina's history.

The federal government awarded to the state energy company Enarsa all the offshore areas open for bids. And in 2006, Enarsa signed a strategic agreement with Repsol-YPF and other partners to begin exploration of three blocks in the Colorado Basin.

Repsol-YPF was preparing to shoot a 1,890 sq km (730 sq miles) 3D seismic survey to explore the deepwater E-1 block (formerly Blocks CAA-7 and CAA-44). If results of the survey are encouraging, a $30 million wildcat will be drilled.

Between 2005 and 2009, Repsol-YPF said it plans to invest approximately $7 billion (U.S.) in Argentina, $200 million of it for offshore exploration.

Based on a review of geologic data and prior exploration efforts, Pucci suggests that the following offshore areas are worthy of reexamination:

- The continental slope and the Punta del Este subbasin in the Salado Basin
- The continental slope and the southern part of the Colorado basin
- Depths of 4,000–5,000 m (13,125–16,400 ft) in the San Jorge basin
- The fold belt in the southern part of the Austral, Malvinas, East Malvinas, and South Malvinas complex.
- The eastern margin of the Argentina basin bordering the Colorado basin

Argentina's Reserves and Wildcat Drilling

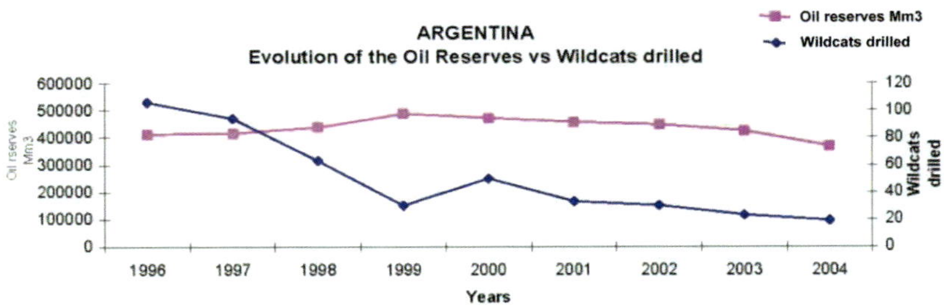

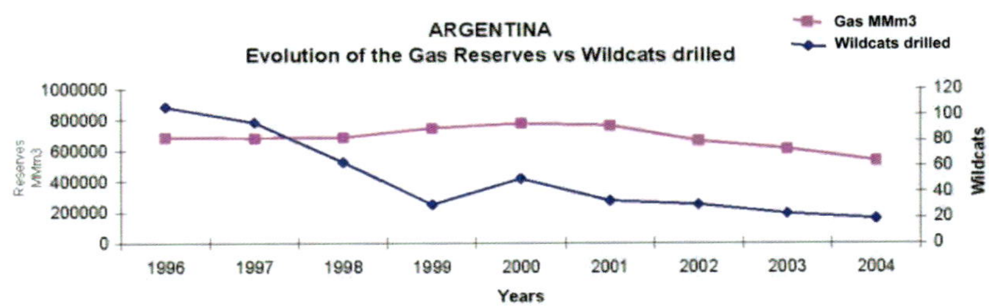

Wildcat Wells Drilled off Argentina

Basin	Wildcats drilled	Operator
Austral (Magallanes)	108	Shell 24, Total Austral 77, Sipetrol 7
Colorado	18	Phillips 9, Hunt 3, YPF 2, Union Texas 3, Shell 1
Gulf of San Jorge	26	Agip 7, Tennessee 6, Sinclair 4, YPF 7, Shell 2
Malvinas Norte (North Falkland)	6	Amerada 2, Lasmo 1, Shell 2, IPC 1
Malvinas	18	YPF 1, Exxon 13, Oxy 3, Total Austral 1
Rawson	1	Exxon 1
Salado	4	Sun 1, Union 2, Amoco 1
San Julian	1	Petrobras 1
Total	**182**	

Drilling and Production

Drilling on the rise

Baker Hughes reported 84 land rigs drilling in Argentina in July, up from 78 rigs drilling a year earlier, and up from 61 rigs three years earlier. In August, Pride International had 27 drilling rigs and 50 workover rigs in the Mendoza, Salta, and Neuquén areas, and another 21 drilling rigs and 54 workover rigs in the Comodoro Rivadavia and Austral areas. Nabors International has five drilling rigs and 11 workover rigs in the country.

ARGENTINA OFFSHORE BASINS AND ACREAGE BLOCKS

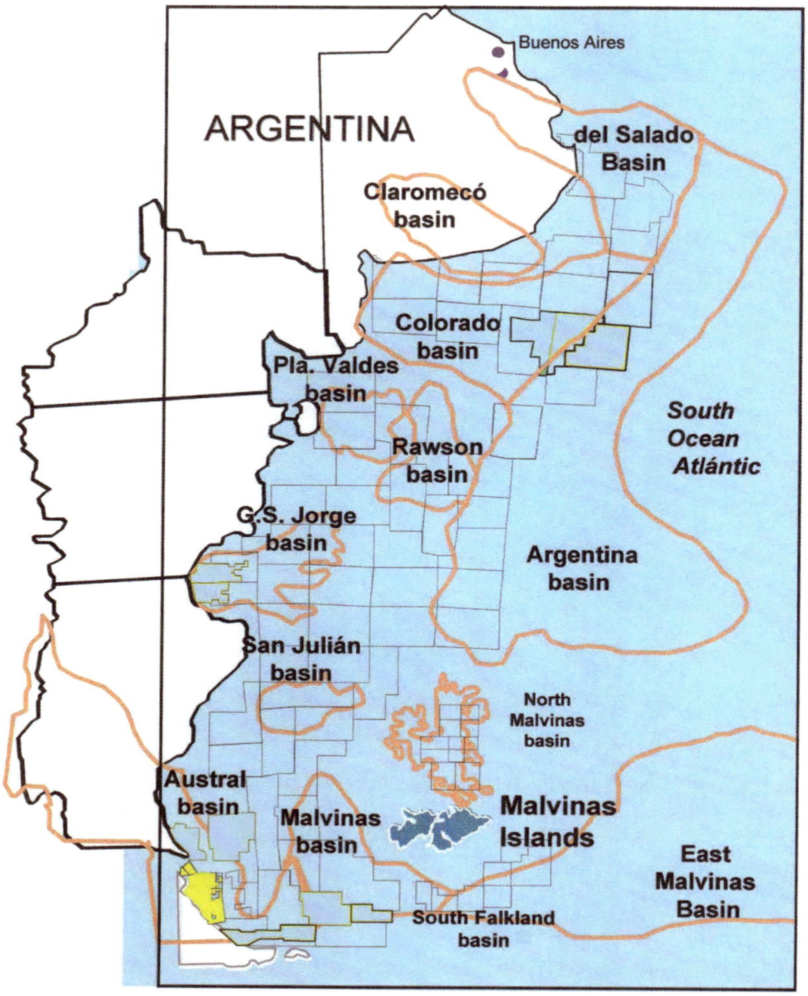

Source: Modified after Turic, 2002

ARGENTINA AUSTRAL BASIN OFFSHORE PRODUCTION, RESERVES

Field	Oil production, cu m/day	Cumulative oil production, cu m	Gas production, cu m/day	Cumulative gas production, cu m	Producing wells
Data as of May 2005					
Argo	295.38	1,487,918	236,322.58	443,081,821	2
Canadon Alfa	978.76	15,886,105	8,271,161.29	84,490,805,361	66
Hidra	268.93	9,594,758	15,700.34	924,744,808	7
Kaus	238.30	1,033,907	23,167.61	75,052,180	1
Magallanes	1,106.00	6,218,837	2,439,709.00	5,654,112,602	32
Poseidon	111.22	87,864	527,410.00	220,705,920	6

Area/field	Proved reserves: Oil, 1,000 cu m	Proved reserves: Gas, million cu m	Operator
Data as of Jan. 1, 2005			
Complejo Hidra (Hidra Sur, Ara Sur, Argo, and Kaus)	1,834	3,435	Total SA
Complejo Alfa (C. Alfa, Ara Sur, and Cullen Norte)	1,776	12,660	''
Vega Pleyade	764	16,889	''
Aries Norte (includes Aries)	684	9,486	''
Carina-Fenix	1,901	32,395	''
Magallanes	1,098	5,183	Sipetrol
Poseidon	309	3,492	''
Offshore total	**8,366**	**83,540**	

At mid-year, Calgary's Antrim Energy Inc. (Calgary) and its partners were moving ahead with their Tierra del Fuego plan, negotiating to extend the drilling contract for a Petroservicios rig.

Other Activity

Petrolifera Petroleum Ltd. (Calgary) was planning a 25-km (16 mile), 6-in., 13,500 b/d pipeline to connect oil discoveries on its Puesto Morales block in the eastern Neuquen basin of northern Argentina. The pipeline was expected to be ready by early 2007.

Five discovery wells on the northern part of the block tested a combined 15,000 b/d from the Upper Jurassic Sierras Blancas and Lower Jurassic Punta Rosada formations. Another well flowed 600 Mcfd of gas without stimulation from zone 9 in Quintuco. Reserves are still to be calculated for all formations.

Antrim Energy Inc. (Calgary) acquired 70% interest in the producing Medianera Block in Rio Negro Province in the Neuquen basin for $1.2 million. The 5,261 ha (13,000-acre) block produces 45–50 b/d from five active wells. The block's 86 wells have recovered 14.8 million bbl of oil from the Tordillo/Quintuco formations at 950–1,150 m. Antrim was planning a block-wide 3D seismic survey to image deeper zones that have never been drilled.

Tecpetrol—part of the energy division of the Argentine Techint Group—was operating several existing and planned polymer projects for enhanced oil recovery (EOR) in Argentina. Tecpetrol operates 12 blocks in the San Jorge, Noroeste, and Neuquina basins. The company's polymer injection activity is in the El Tordillo and Jose Segundo fields in the Gulf of San Jorge basin.

Apache Corp. (Houston) announced that it will negotiate acquisition of the oil and gas operations in Argentina of Pioneer Natural Resources (Dallas) for a total purchase price of $675 million. The transaction was expected to close in 2006. Most of the properties are in the Neuquen Basin, where Apache presently is modestly active. Estimated 2006 production from the assets was 32,500 boe/d. Net proved reserves for the properties were estimated at 101 MMboe.

Planned Projects

Field	El Tordillo	El Tordillo	El Tordillo	Jose Segundo
Location	Chubut	Chubut	Chubut	Chubut
Type of project	Polymer bulk gel	Polymer CDG	Polymer surfactant	Polymer surfactant
Acres	198	82	82	150
Pay zone	CR	CR	CR	MEC-410
Depth, ft	5,500	5,500	5,500	5,500
Oil gravity, °API	21	21	21	21
Planned start	August 2006	August 2006	December 2007	July 2007

Note: CDG = colloidal-dispersion gel.

Existing EOR Projects Not in 2006 Survey

Type project	Polymer	Polymer
Operator	Tecpetrol	Tecpetrol
Field	El Tordillo	Jose Segundo
State/Province	Chubut	Chubut
Start date	November 2005	May 2004
Area, acres	82	155
No. wells, prod.	10	7
No. wells, inj.	2	1
Pay zone, ft	57	23
Formation type	Sandstone	Sandstone
Porosity, %	24	20
Permeability, md	500	80
Depth, ft	5,450	6,680
Gravity, °API	21	25
Oil viscosity at reservoir temperature, cp	5	3
Reservoir temperature, °F.	185	180
Previous production	Waterflood	Waterflood
Oil saturation start, %	50.9	41
Oil saturation end, %	45.2	
Project maturity	JS	JS
Total oil production, b/d	512	240
Enhanced oil production, b/d	99	30
Project evaluation	Promising	Too early to tell
Profitability	Too early to tell	Too early to tell
Project scope	Pilot	Pilot

BELIZE

Capital: Belmopan
Monetary unit: Belizean dollar
Refining capacity: 0
Oil production: N/A
Oil reserves: N/A
Gas reserves: N/A

First commercial oil discovery

Local exploration company Belize Natural Energy Ltd. and its partner, private independent CHX LLC (Denver), completed two wells in late 2005, each capable of producing 500 b/d of 38° gravity sweet crude, according to press reports. The companies were planning more drilling and the acquisition of 3D seismic data.

The Usher-1 and 2 wells are near the town of Spanish Lookout, northwest of the capital, Belmopan, near the border with Guatemala. The well sites are 200 km (125 miles) east of Xan oil field, Guatemala's largest, which produces 17° gravity oil from Cretaceous Coban at 2,315 m (7,600 ft).

The companies have shipped one bargeload of oil to a Gulf Coast refinery, sold some oil locally, and made first royalty and tax payments in late 2005.

In related activity, Spartan Petroleum Corp. (Houston) entered a production-sharing agreement (PSA) in June with the Belize government covering 250,000 acres in the northern portion of the country. Spartan said geophysical work is planned. The PSA is one of five awarded since 2005.

BOLIVIA

Capital: La Paz
Monetary unit: Boliviano
Refining capacity: 41,200 b/cd
Oil production: 45,000 b/d
Oil reserves: 440.0 million bbl
Gas reserves: 24.0 tcf

According to *Oil and Gas Journal (OGJ),* Bolivia had proven crude oil reserves of 440 million barrels in 2006. This was up from only 116 million in 1997, said the country's state-owned Yacimientos Petroliferos Fiscales Bolivianos (YPFB), due to exploration activity and investment in recent years.

Bolivia has the second-largest natural gas reserves in South America (after Venezuela) but remains the continent's poorest country.

Nationalization of hydrocarbon resources

In early 2006, new president Evo Morales embarked on a campaign to re-nationalize all hydrocarbon resources and renegotiate export contracts with Argentina and Brazil.

On May 1, the government nationalized the gas and oil industries. For about a month, more than 3,000 soldiers peacefully occupied 56 facilities. All companies were told to turn production over to YPFB, which would assume supervision of the operations of foreign oil companies much more closely.

Twelve companies are active in Bolivia. The leading operators are Repsol YPF; Petroleo Brasileiro SA (Petrobras), the largest company in Bolivia; BG Group PLC; and Total SA. Repsol YPF and its subsidiary Andina control 35% of Bolivia's known 55 tcf of gas reserves.

The new hydrocarbons law required companies to sign new production-sharing contracts that impose taxes and royalties amounting to 50–80% of the value of production. Companies were told that they would be forced to suspend operations if they did not sign new contracts.

Pres. Morales also decreed that private firms must turn over their Bolivian retail networks to YPFB, and that Brazil's Petrobras, which owns 25% of Bolivia's retail outlets, hand over two refineries in Bolivia. The government later said it would negotiate compensation for the two refineries Gualberto Villarroel (in Cochabamba) and Guillermo Elder Bell (in Santa Cruz de La Sierra). Petrobras bought the plants seven years ago for $100 million.

The Bolivian Hydrocarbon Ministry also launched in July a program to audit multinational oil companies operating in the country to enable the establishment of the new contracts. Prior contracts, which granted ownership of hydrocarbon production and commercialization for the investor companies, violated the Bolivian Constitution, said the ministry.

In late October, key foreign energy companies agreed to remain in Bolivia, operate under state control, and cede larger shares of profits to the government. The contracts signed with 10 foreign oil companies will now be submitted to the Bolivian Congress for ratification.

YPFB estimated that under the new contracts, the state will receive an additional $1.3 billion/year in revenues. It said revenues could be increased to as much as $4 billion within four years, when Bolivia begins exporting 30 million cu m/day of gas to Argentina. YPFB also said the contracts guarantee that the country will receive investments of as much as $2 billion.

Petrobras maintained its rights to explore on the Rio Hondo, Ingre, and Irenda blocks. According to the Hydrocarbon Ministry, Total also is expected to remain in Bolivia and spend about $2 billion exploring for and producing gas from three large fields.

Support from Venezuela

In May, Venezuelan Pres. Hugo Chávez and Bolivian Pres. Evo Morales signed an agreement allowing state-owned Petroleos de Venezuela SA (PdVSA) to participate in exploration and marketing activities in Bolivia as well as the likely creation of a bilateral oil firm, Petroandina.

PdVSA said in July that it had started surveys in Bolivia to identify prospective hydrocarbon areas under control of Bolivian state firm YPFB. Following the survey, PdVSA will determine both the amount of possible future investments and the nature of a joint venture with YPFB.

The Bolivian Hydrocarbon Ministry noted Chávez's commitment to build a petrochemical complex in Villamontes, an asphalt plant in Cochabamba, and a gas liquefaction plant in Rio Grande; and to install service stations in Bolivia. The ministry also said that 200 workers will be sent to Venezuela for training in exploration and production and that Bolivia would import 200,000 bbl of diesel from Venezuela at low prices.

Drilling pace picking up

Baker Hughes reported three land rigs drilling in Bolivia in September 2006, up from two a year earlier. Nabors Industries Ltd. had one drilling rig in the country. Pride International had four land drilling rigs and two land workover rigs; DLS Drilling Logistics and Services Corp. had 51 rigs in Bolivia and Argentina, many working for Pan American Energy (BP PLC and Bridas Group).

In July, the Bolivian government began auditing foreign-controlled gas fields, related to the nationalization of its natural gas reserves. In response, Petrobras cancelled plans for new investments in Bolivia, saying it would invest $22 billion from 2007–2011 to develop natural gas in Brazil, to reduce its dependence on Bolivia.

Natural gas accord with Argentina

Argentine Pres. Nestor Kirchner and Bolivian Pres. Evo Morales signed a natural gas accord calling for $1 billion worth of investments to increase the flow of Bolivian gas into Argentina. The 20-year accord calls for Argentina to receive 27.7 million cu m/day of Bolivian gas by 2010. Argentina already held contracts to import 7.7 million cu m/day, although current shipments average 4.5 million cu m/day, said Bolivia's state-owned YPFB.

Gas exports to Brazil

Before Bolivian nationalization actions in May, Brazil had sought a review of bilateral gas import contracts signed previously by the two countries. The 20-year gas supply agreement also requires investments by Petrobras in Bolivia to increase oil and gas production and to maintain the volume of gas exports from Bolivia to Brazil.

In February, Petrobras presented Bolivia with proposals for cooperation in oil, gas, refining, petrochemicals, and fuel distribution, along with a $5 billion investment plan.

However, Bolivia's Pres. Morales in May expressed a desire to substantially increase (perhaps triple) the price of natural gas that his nation exports to Brazil—about 26 million cu m/day through the 3,150-km (1,957-mile) Bolivia-Brazil pipeline. Brazil has taken a tough negotiating stance on gas purchases.

Bolivia's hydrocarbons ministry had said it was logical that Petrobras would postpone its investment decision while contracts are being negotiated.

BRAZIL

Capital: Brasilia
Monetary unit: Real
Refining capacity: 1,908,278 b/cd
Oil production: 1.710 million b/d
Oil reserves: 11.772 billion bbl
Gas reserves: 10.820 tcf

Overview

According to the U.S. Energy Information Administration (EIA), Brazil is the third-largest energy consumer in the Western Hemisphere, behind the United States and Canada.

Oil

The country has made great strides over the past decade in increasing its domestic oil production, which has been a long-term goal of the Brazilian government, says EIA.

State-owned Petroleo Brasileiro SA (Petrobras) controls more than 95% of crude oil production in Brazil. Rio de Janeiro state contains about 80% of Brazil's total production. Most of the country's crude production is offshore in very deep water and consists of mostly-heavy grades.

Petrobras has aggressively expanded production in recent years. In 2006, it brought the Albacora Leste field online and expanded production from the Golfinho field and the Jubarte field. In 2007, Petrobras expects to continue this expansion, bringing up to three additional projects on-stream: Rocandor P-52; Rocandor P-54; and Espadarte.

Royal Dutch Shell's Bijupira-Salema project in the Campos Basin is the only oil field in Brazil not operated by Petrobras. The project produces 50,000 bbl/d. Shell also hopes to begin production at its BC-10 project by the end of 2008.

Other fields from which operators aiming to bring production onstream in the near future are the Polvo field in mid-2007 (Devon Energy Corp.; Oklahoma City, OK, U.S.A.); the Frade project by the end of 2008; and the Chinook field by 2010 (Norsk Hydro; Oslo, Norway).

Natural gas

The Campos and Santos Basins hold the majority of Brazil's gas reserves. Natural gas production has grown slowly in recent years, however, mainly due to a lack of domestic transportation capacity and partly due to low domestic prices.

Petrobras said that gas use accounted for about 9.3% of overall energy consumption in 2005, up from 7.7% in 2003. It projected that demand would reach 121 million cu m/d by 2011, representing 11% of total energy use. Gas is mostly used as a substitute for fuel oil in industrial and power-generating applications, and domestic prices for natural gas are much lower than international fuel oil prices. Further, the introduction of natural gas imports has lead to a rapid growth in domestic consumption.

Brazil's largest natural gas production occurs in the Campos Basin in Rio de Janeiro state from offshore fields, says EIA. Most onshore production occurs in Amazonas and Bahia states.

In 2006, Petrobras announced an ambitious plan to increase its natural gas production to 520 Bcf by the end of 2008. The project would focus on increasing production from the existing Marlim and Merluza fields and bring two new fields onstream in Espirito Santo state.

Brazil was set to hold its eighth exploration and development licensing round in November 2006, with a deliberate shift toward offerings in gas-bearing basins. A major factor in the shift was Bolivia's decision in May 2006 to nationalize its oil and gas industry; Brazil imports substantial volumes of gas from Bolivia.

The Brazilian Petroleum and Gas Institute (IBP) predicted that oil and gas investments in Brazil during the next five years will total $100 billion and sustain oil self-sufficiency in these fuels, which was achieved in May 2006.

Exploration and Development

Oil, gas production growth seen

In early 2006, Petrobras provided its view of expected Brazilian oil and gas production to the year 2011 and beyond. The company expects total oil production will increase to 2.374 million b/d in 2011 (from 1.684 million b/d produced in 2005) and then to 2.812 million b/d in 2015.

Petrobras also sees total gas production increasing to 70 million cu m/day in 2011 from 27.5 million cu m/day in 2006.

During 2006–2011, fields coming on stream off Brazil include several light crude discoveries in the Espirito Santo basin as well as additional sections of the large fields in the Campos basin. Fig. LA.10 shows the major oil and gas development projects contributing new production.

BRAZIL OIL, NGL PROJECTS

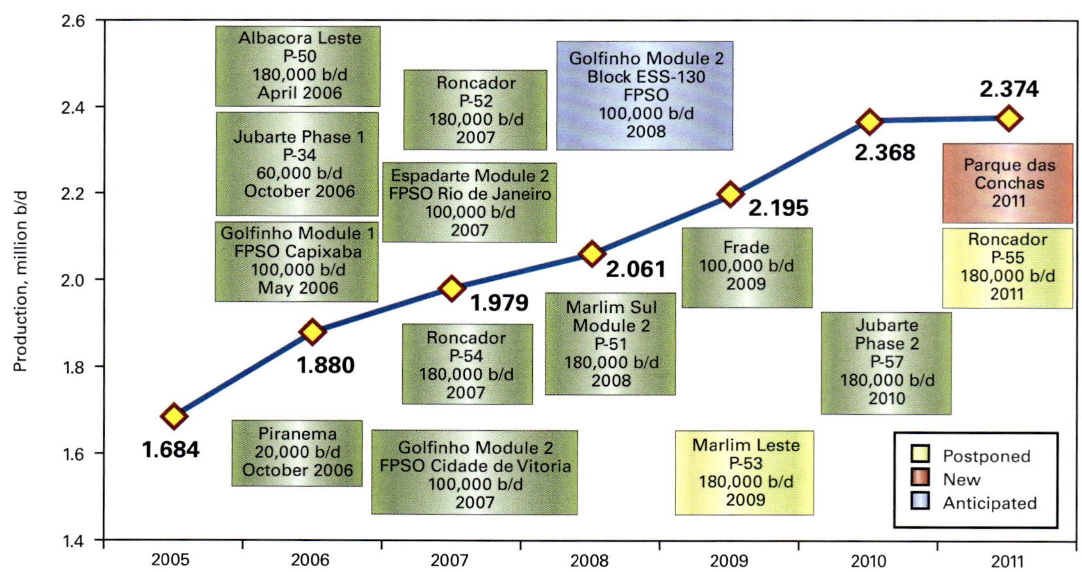

Note: The previous plan scheduled Marlim Leste P-53 in 2008, Roncador P-55 in 2010, and Golfinho Module 3 in 2010. Parque das Conchas includes Abalone, Ostra, Argonauta, and Nautilus fields in Block BC 10. A unit of the Shell Group is the operator for Parque das Conchas

The new oil projects going in from 2006–2011 will add 1,804 million b/d of production, while natural declines lower production to 0.57 million b/d from 1.684 million b/d in 2005. The net increase in 2011 from the 2005 production is 0.69 million b/d.

Petrobras lists several major projects as coming on stream between 2011 and 2015: Marlim Sul P-56; Roncador P-55; Papa-Terra Modules 1 and 2; Marlim Sul Module 4; Roncador Module 4; Cachalote and Baleia Franca; and Baleia Azul.

Papa-Terra in Campos basin Block BC-20 is a mega oil field with estimated reserves of 700 million to 1 billion bbl. The Cachalote, Baleia Franca, and Baleia Azul developments are in the Espirito Santos basin, while the remainder are in the Campos basin.

Gas sector investment

Petrobras estimated that the company and its partners will invest as much as $22 billion in the natural gas chain in Brazil from 2007–2011. Of this amount, about $14.9 billion will be spent on gas exploration and production, with Petrobras investing $11 billion and its partners, $3.9 billion.

Petrobras also estimated it will invest $6.6 billion in gas and energy infrastructure, along with $500 million from its partners. The aim is to build 4,000 km (2,485 miles) of gas pipelines in Brazil and improve gas shipment to the country's north, northeast, and southeast regions by 2008.

Another key focus area will be LNG imports. Petrobras said it is investing $40 million to build one LNG terminal at Pecém, in the state of Ceará, to receive 6 million cu m/day of gas by early 2009, and another $140 million for a terminal at Guanabara Bay in Rio de Janeiro state that would receive 12–14 million cu m/day by the timeframe.

Tracking oil, gas investment

An IBP survey showed that private companies plan to invest $17.5 billion from 2006–10. This includes $7.4 billion in exploration and production, $1.5 billion in refining and transport, $1.5 billion in natural gas, $3.9 billion in petrochemicals, and $3.2 billion in other activities, mainly gas distribution.

Santos basin fields in the spotlight

Petrobras unveiled its Santos Basin Business Unit master plan, which calls for $18 billion in investment in the next 10 years. The production goals for 2010–2011 are 1.05 bcfd

BRAZIL GAS PRODUCTION

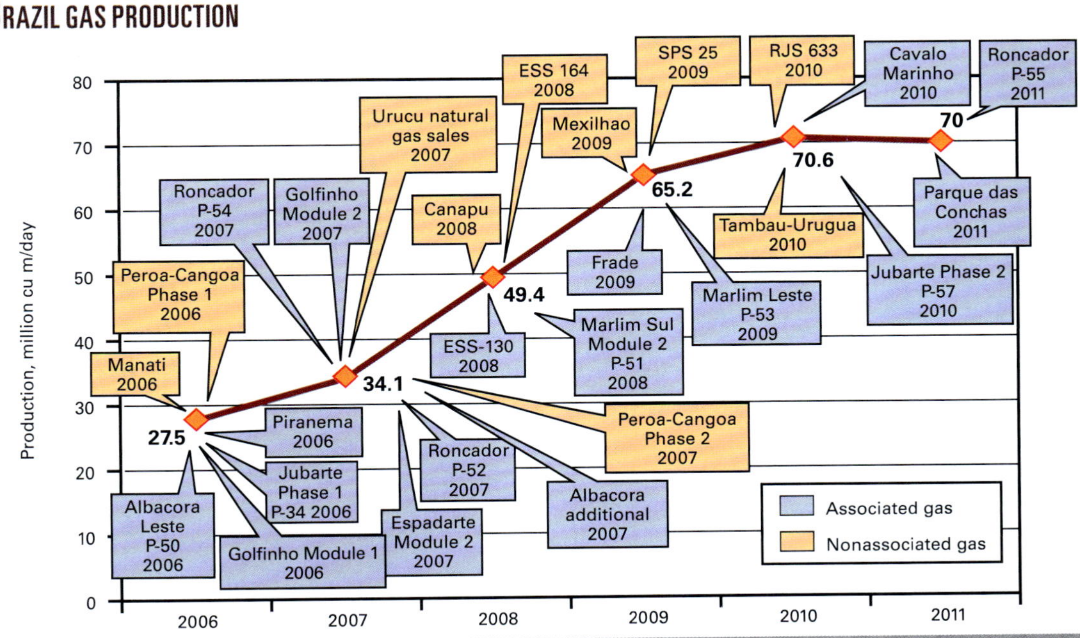

of gas and 100,000 b/d of oil, compared with the current 42 MMcfd and 10,000 b/d. Brazil now imports more than 990 MMcfd of gas from Bolivia.

Petrobras projects that the first large volume of gas from the Santos area should flow in second-half 2008, when Petrobras expects output of 12 million cu m/day from Mexilhão and Merluza fields.

Of the Santos basin's 352,000 sq km (135,907 sq mile), 43% is in water as deep as 1,300 ft and 57% is in 1,300–10,000 ft of water. One-fourth of the concession areas Petrobras holds in Brazil are in the Santos basin. Petrobras and partners hold 11.5% of the basin's acreage.

The most recent discovery in its deepwater sector—the Tupi well—has led to speculation that the so-far gas-prone basin's subsalt realm might rival the Campos basin in terms of ultimate recovery. Petrobras said the well—175 miles off Rio de Janeiro—was the first drilled on Block BM-S-11 and the first to penetrate an evaporatic salt sequence more than 6,500 ft thick.

With true vertical depth of 19,680 ft in 6,975 ft of water, it flowed 4,900 b/d of oil and 5.3 MMscfd of gas from a deep presalt reservoir.

Petrobras, Ecopetrol joint projects

Brazil's Petrobras and Colombia's Ecopetrol announced plans to jointly explore the 170-sq km (65-sq mile) Tucano 156 Block in Bahia. The block is in the Tucano Sul basin, 90 km (60 miles) from the Bahia coast. Ecopetrol will be the operator.

BC-10 development starts

Shell Exploration & Production Co. and partners let the first contracts for development of 3 deepwater heavy-oil fields on Block BC-10 off the coast of Brazil. One contract is for

Petrobras production zones in Santos Basin

- **Cedro and giant Mexilhão fields** 137 km (85 miles) off the Sao Sebastiao Terminal are to start producing early in the next decade, initially at 280–320 MMcfd of gas into facilities with a capacity of 530 MMcfd of gas and 20,000 b/d of condensate.
- **Polo Centro** includes unexplored areas seaward of Merluza and Lagosta fields. If it pans out, production could be tied back to Mexilhão.
- **Merluza field**, discovered in 1979 and producing gas-condensate since 1992, is making 42 MMcfd of gas and 1,600 b/d of condensate. Gas output is to climb to 88 MMcfd in 2008 when the platform begins accepting gas from nearby Lagosta field and from the SPS-25 well area. Installation of a second platform could boost capacity to 350 MMcfd by 2010. Merluza-Lagosta and Mexilhão -Cedro will be connected to a $300 million gas treatment plant at Caraguatatuba. The plant's two units could process 265 MMcfd each and could be expanded to 800 MMcfd.
- **Urugua and Tambau fields** are located at Block BS-500 about 100 miles off Rio de Janeiro. Production there is estimated at 150,000–200,000 b/d of oil and 775 MMcfd of gas in 2011. Urugua's reserves are 250 million bbl of 33° gravity oil and 1.3 tcf of gas, and Tambau's are 1.8 tcf of gas.
- The **Southern Pole**, 125 miles off Sao Paulo, Parana, and Santa Catarina states, produces 9,000 b/d of oil from the SS-11 platform off Parana. Cavalo Marinho field is to go on production in 2008 at 18,000 b/d, and Petrobras believes the area has the potential to deliver as much as 140,000 b/d and 105 MMcfd.

How Santos Basin infrastructure may develop

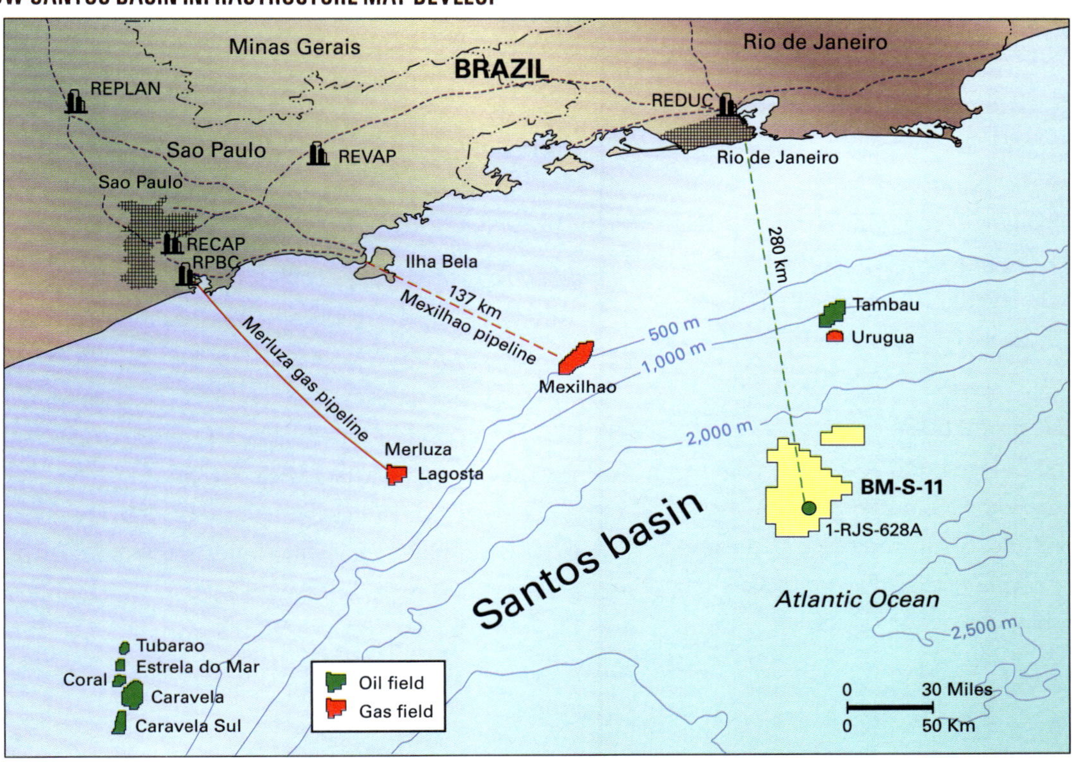

leasing of an FPSO vessel; another is for drilling 10 wells.

The first phase of development will involve Ostra, Abalone, and Argonauta fields. The block is 120 km (75 miles) southeast of Vitoria in Espirito Santo state. Shell partner Petrobras calls the area Parque das Conchas and estimates production will start in 2011.

Shell, the operator, will develop the fields with horizontal subsea wells and manifolds and tie each field back to the FPSO, which will be moored in 1,780 m of water.

Liquids will be separated from gas in large-diameter caissons on the seafloor and then pumped through risers to the FPSO for oil-water separation. Gas will move through a 40-km (25-mile) pipeline to Jubarte. Shell estimates Block BC-10 reserves at 400 million bbl.

Court cancels eighth license round

A federal injunction suspended Brazil's eighth oil and gas licensing round that was set for November 28–29. The court objected to limits in the number of offers a bidder can make, a rule included by the National Petroleum Agency (ANP) for this bid round.

A federal Workers Party congresswoman requested the injunction to "protect" state-owned Petrobras from being limited in the number of offers it could submit. Petrobras dominated previous rounds.

Bidders also were unhappy about ANP's exclusion of Campos and parts of Santos basins from the licensing round. Campos provides some 80% of Brazil's 1.9 million b/d of oil, said the Brazilian Petroleum and Gas Institute.

Two appeals to overturn the injunction failed, but legal experts said the matter probably will end up in a court battle. It is not known whether the winning bids announced during the first days of the round are legally valid or what effect the cancellation may have on future exploration investments.

Shell, Maersk exploration well in Santos Basin

Royal Dutch Shell PLC and Maersk Oil & Gas AS plan to drill in 2007 the first exploration well on the BMS-31 Block in the Santos basin off Sao Paulo. Shell, the block's operator and majority stake holder, is conducting seismic studies on this and its other Brazilian blocks.

Shell also was planning to drill and evaluate an appraisal well on the BS-4 Block in the same basin by yearend 2006. The block has reserves of 300 million bbl of very heavy crude.

Production near for Manati field

Petrobras logged 327 m of gross pay at mid-year in the first of seven development wells it plans to drill in Manati gas-condensate field off Brazil's Bahia state, reported partner Norse Energy Corp. ASA.

The PA29 jack up drilled the well to 1,650 m TD on Block B-CAM-40 in the Camamu and Almada basins. Petrobras estimates Manati reserves at 220 bcf of gas and 1.3 million bbl of condensate. The company expects the field to produce 6 million cu m/day of gas at peak.

Polvo field development

Devon Energy Corp. (Oklahoma City, OK U.S.A.) and SK Corp. (Seoul, South Korea) are accelerating development of Polvo oil field on Block BM-C-8 in the Campos basin. Production from the shallow-water field is to begin in mid-2007 and reach 40,000 b/d.

Devon reported in mid-2006 that it will drill 10 producing wells and three injectors in the first development phase in early 2007. Production will flow without separation to an FPSO vessel designed to handle 90,000 b/d of production and store 1.5 million bbl. Reserves are estimated at 50 million bbl.

Drilling and Production

Fifth-generation semisub to be built

In August, Keppel Offshore & Marine Ltd. announced a $270 million contract with QGP to build a fifth-generation semisubmersible using the new DSS-38 design. This is KFELS's first newbuild for Brazilian waters. The rig will be used for a seven-year contract with Petrobras.

The new rig—scheduled for late 2009 delivery—will be able to drill to 30,000 ft below mud line in water slightly deeper than 9,000 ft. It will have a maximum variable deck-load of 5,550 tonnes, vertical and horizontal riser storage, and a DPS-2 positioning system with eight, 3,000-kW azimuthing thrusters.

New DSS-38 design semisubmersible. (Photo from Keppel Offshore & Marine Ltd.)

Two other rigs, capable of drilling in 2,400 m of water, will be leased from QGP and Schahin Cury for five years, starting in 2009. Petrobras said that some of the rigs could be mobilized to drill in water as deep as 3,050 m.

FPSO for Frade field

A unit of Chevron Corp. (San Ramon, CA U.S.A.) signed a letter of intent with SBM Offshore of the Netherlands for engineering, procurement, construction, and installation of an FPSO vessel in Frade oil field in the Campos basin off Brazil. SBM will convert its Lusan vessel into a unit with capacities of 100,000 b/d of oil production, 106 MMscfd of gas treatment and compression, and 2 million bbl of storage. Installation in 1,080 m of water is scheduled for mid-2008, with first production due in 2009.

Drilling pace up

There were 26 rigs working offshore Brazil in July 2006 according to Rigzone, including eight drillships, 16 semisubmersibles, and two jack ups. BHI reported only 15 rigs off Brazil. Only seven of the 26 are operated by Brazilian companies: Petrobras (4); Schahin Cury (2); and Queiroz Galvão Perfurações SA (1). The other 19 are owned by North American drilling contractors: Diamond Offshore Drilling Inc. (4); Noble Corp. (4); Pride International Inc. (5); Transocean Inc. (5); and one by Norwegian company Petroserv SA.

Petrobras said it will lease four rigs capable of drilling in 2,000 m of water from QGP, Odebrecht, Schahin Cury, and Petroserv. These are being contracted for seven-year periods, beginning in 2010 and renewable for another seven years.

Daily oil production record set in May

Petrobras set a daily oil production record May 25, 2006 when the company's output in Brazil (Table 1) was 1.87 million bbl. Petrobras said the record is "based on the increased recovery of fields located in mature areas in the North and Northeast of the country, as well as the exceptional operating performance of the Campos basin fields."

The new record was 13,000 b/d more than the previous record set on Dec. 19, 2005 and 11% above the 2005 average.

The company plans to bring two other floating production facilities on stream in 2006: the P-34 FPSO, 60,000 b/d from Jubarte field off Espirito Santo, and the Sevan Marine SSP-300 monocolumn FPSO, 20,000 b/d in Piranema field off Sergipe.

Where Petrobras produces oil in Brazil

Production area	Basin	Production on 5-25-06, bbl	% of total*
Amazon	Solimoes	44,337	2.4
Rio Grande do Norte and Ceara	Potiguar and Ceara	72,901	3.9
Sergipe and Alagoas	Sergipe and Alagoas	67,222	3.6
Bahia	Reconcavo	49,900	2.6
Espirito Santo	Espirito Santo and Campos	87,312	4.6
Rio de Janeiro	Campos	1,542,026	82.4
Sao Paulo and Parana	Santos	2,862	0.1
Parana	Xisto (bituminous shale)	4,201	0.2
Total		**1,870,761**	

*Does not total 100% due to rounding.
Source: After Petroleo Brasileiro SA

Oil field start-up makes Brazil self-sufficient

Brazilian Pres. Luiz Inacio Lula da Silva declared his country self-sufficient in oil during the April 2006 start-up of the P-50 FPSO vessel in Albacora Leste oil field in the Campos basin, 120 km (75 miles) off Rio de Janeiro state. Petrobras said the 180,000 b/d FPSO's inauguration would allow Brazil to reach an average production level of 1.91 million b/d later in 2006, a level slightly above that of the country's oil consumption.

Petrobras starts Golfinho test production

Petrobras reported that the Seillean FPSO vessel began operations in early 2006 in Golfinho field in the Espírito Santo basin. Production testing to evaluate reservoir behavior was yielding 21,000 b/d of oil.

Technology key to drive for oil self-sufficiency

Between 2006 and 2010, Petrobras said, it will invest $1.3 billion in research and development (R&D), compared with $398 million in 2005 and $247 million in 2004.

Officials at the company's R&D center (Cenpes) told *OGJ* that among the most important technologies to sustain self-sufficiency and expand offshore oil production in the next 5–10 years are improved subsea separation systems. Cenpes is currently working on two different concepts to improve these systems.

The first is the continuation of the liquid-gas separation and pumping concept to reduce back pressure to the reservoir and thus increase the overall production of wells tied to it. The VASPS (vertical annular separation and pumping system) unit installed in Marimbá-1 well in 2001 is performing well, said Petrobras, enabling 30–40% incremental production. The R&D center is now working to increase VASPS flow capacity and specifications to enable use in 1,500 m water.

The second concept is subsea water separation and reinjection, which frees water-handling capacity on existing platforms to allow new wells to be brought in.

This project—under the Procap-3000 program evaluating operations below 3,000 m water depth—is in the basic engineering-design phase with three vendors. Heavy-oil separation tests were performed successfully in a field-scale laboratory facility. The goal is to apply this concept at Marlim field within by 2008.

This configuration of a subsea water-separation system is installed on the Campos basin. Photo from Cenpes/Petrobras.

The Seillean FPSO will operate in the pilot production phase from the field, where the Capixaba and Cidade de Vitória FPSOs were to be sited in the first half of 2006 and the first half of 2007, respectively. Both units have the capacity to process 100,000 b/d of oil.

Petrobras said that start-up of these operations represents first oil production from deep water outside of the Campos basin.

Refining and Processing

Petrobras and Venezuela's Petróleos de Venezuela SA (PdVSA) announced plans to jointly build a $2 billion refinery in Brazil's northern Pernambuco state. The Abreu e Lima plant, planned to begin operations in 2014, will have a processing capacity of 500,000 b/d and will produce gasoline and diesel having a sulfur content of less than 50 ppm. Construction is scheduled to begin in 2010. The refinery's construction is consistent with Brazil's goal of processing 90% of domestic heavy oil in the country by 2010. The proportion today is 82%.

New fuel aims to cut diesel imports

Petrobras plans to reduce diesel imports with the industrial-scale production of H-Bio, a new form of biodiesel. H-Bio is a mix of vegetable oil and petroleum, developed in the firm's research center (Cenpes).

Petrobras said that H-Bio combines diesel fractions from distillation, cracking, and delayed coking with soy oil and hydrogen as raw materials to produce high-quality diesel. Brazil will blend 2% biodiesel with normal diesel starting in 2008.

Petrobras plans to use 10% soy oil in the two refineries where the technology will be installed in 2007 and 2008. The program could expand to five refineries in following few years.

The company plans to produce 2.56 million cu m of H-Bio fuel from soy oil in 2007, to replace 15% of Petrobras's diesel imports. By 2008, diesel imports will be reduced by 25% when H-Bio production will increase to 4.25 million cu m. Petrobras estimates short-term savings of $145 million/year and medium-term savings of $240 million/year from H-Bio use.

Expansion of Pasqualini refinery

Refap SA and Petrobras worked to expand the 120,000 b/d Alberto Pasqualini refinery in Rio Grande do Sul to 180,000 b/d. The $1 billion project will increase the refinery's ability to process heavy crude and its output of diesel. Petrobras currently imports large quantities of light oil to mix with heavy domestic crudes.

Pipelines

Gas pipelines under construction

Petrobras said it is investing $3.9 billion by 2009 to build 4,191 km of gas pipelines in Brazil.

Bidding halted for Bolivia-Brazil gas line work

Brazilian Gas Transport Co. (BGT) canceled bidding for contracts to expand the 3,150-km (1,957-mile) Bolivia-Brazil pipeline after Bolivia announced in May 2006 the nationalization of its gas sector. BGT said the companies that use the pipeline had all waived their contracts with his company in response to Bolivia's nationalization plan, including Petrobras, Repsol YPF SA, and Total SA.

The line represents a major section of the proposed Southern Gas Pipeline proposed by Venezuelan President Hugo Chávez to link South America from Venezuela to Argentina, with about two-thirds of it crossing Brazil.

BGT said it could restart the bidding process only if there were changes in Bolivia. BGT has paid off all construction costs of the pipeline, which is projected to start making money in 2007.

Petrobras gas pipelines under construction, 2004-07

Project	Pipeline route	Length, km	Diameter, in.	Capacity, MM cu m/day	Installation start-up	Operation start-up
Malha Sudeste	Campinas-Japeri	453	28	5.8	August 2004	December 2005
	Lorena-Poços de Caldas	177	10 and 14	1.0	February 2006	December 2006
	Gasbel II (Loops)	292	16 and 18	6.9	February 2006	January 2007
Malhas Nordeste	Nordestão II Pilar-Mossoró	511	24	10.0	May 2006	August 2007
	Gasfor II Mossoró-Pecém	280	20	12.0	November 2005	November 2006
	Catu-Carmópolis	260	26	10.0	April 2005	November 2006
	Carmópolis-Pilar	175	26	10.0	April 2005	June 2006
	Aratu-Camaçari	43	14	3.0	November 2004	July 2005
	Atalaia-Itaporanga	29	14	3.0	March 2005	June 2006
Gasene	Cabiúnas-Vitória	298	28	20.0	July 2005	January 2007
	Vitória-Cacimbas	131	26 and 16	20.0	August 2004	October 2005
	Cacimbas- Catu	880	28	20.0	July 2005	January 2007
Malha Norte	Urucu-Coari GLP	279	10	10.0	March 2005	July 2006
	Coari-Manaus	383	20	7.5	June 2005	December 2006

Source: Petróleo Brasileiro SA (Petrobras)

Southern Pipeline still on hold

The proposed 10,000-km (6,214-mile) Southern Pipeline (Gasoducto del Sur) gas pipeline would reach from Venezuela to Argentina at an estimated cost of $15–20 billion. Petrobras expressed its view in October that the political will to undertake the project exists, but that it is too soon to estimate a project schedule until feasibility studies by the governments of Brazil, Venezuela and Argentine are completed.

Petrobras plans ethanol pipeline

Petrobras signed an accord with the Goias state government to build a $226 million, 975-km (606-mile) pipeline to transport 4 billion l./year of ethanol, the company said. It will be Brazil's first pipeline to transport ethanol, which currently moves by tanker truck. The pipeline will run from Goias to a refinery in Paulinia, near Sao Paulo.

Brazil is the world's largest producer of ethanol and plans to increase exports, now limited by the need to supply local markets. Current and prospective customers include Venezuela, Nigeria, China, South Korea, India, and the United States.

CHILE

Capital: Santiago
Monetary unit: Peso
Refining capacity: 226,800 b/cd
Oil production: 10,000 b/d
Oil reserves: 150 million bbl
Gas reserves: 3.46 tcf

Chile has one of South America's most robust and open economies. In 2005, its real GDP grew by 6.35%, continuing the strong growth experienced in 2004 (6.2%), according to the U.S. Energy Information Administration (EIA), However, the country has limited domestic energy resources and must import the bulk of the energy it uses.

Oil

Production

Chile has limited oil reserves, and production in 2006 was only 10,000 b/d. In contrast, the country consumed approximately 257,000 b/d. As a result, the country imports significant oil volumes, chiefly from Argentina but also from Brazil, Angola, and Nigeria.

State-owned Empresa Nacional del Petroleo (ENAP) is both a producer and refinery operator, controlling Chile's oil sector. ENAP's international subsidiary, Sipetrol, has pursued investments in places such as Argentina, Colombia, Ecuador, and Egypt. ENAP announced in August that it would invest an additional $36 million in Ecuadorian fields through 2026. Dow Jones reported that officials expressed interest in Ecuador's Block 15, seized from Occidental Petroleum Corp. in May.

The Magallanes basin is the sole oil-producing region in Chile. ENAP has developed 23 fields in the basin, the largest being the Costa Auera.

GeoPark Holdings Ltd. (Hamilton, Bermuda) said in 2006 that it became the first private company to produce oil and gas in southern Chile. The company operates the 440,000-acre Fell Block, in the Austral basin northwest of the Straits of Magellan. Single wells in Molino, Ovejero, and Nika fields produce 5.4 MMcfd of gas and 50 b/d of condensate. GeoPark sells gas to the Methanex methanol plant at Tierra del Fuego and sells condensate to ENAP.

Oil pipelines and refineries

Chile has two crude oil import pipelines. The 270-mile, 115,000-b/d Trasandino links Chile with Argentina, and the Arica-Sica transports gas from Bolivia. ENAP operates all three of the refineries in Chile. The largest is the 113,400-b/d BioBio refinery, north of Santiago. ENAP has made recent investments in both the BioBio facility and the Aconcagua refinery.

Natural gas

Chile imports almost all of its natural gas. ENAP controls the modest domestic production, which occurs mostly in the Magallanes basin. At the same time, increasing energy demand, environmental concerns, and the unreliability of hydropower have prompted the government to encourage gas use. Large-scale imports began 1997, and gas consumption totaled 0.7 bcfd in 2005.

Gas pipelines

Several pipelines connect gas fields in the Magallanes basin to major domestic consumption centers. In addition, seven pipelines built in the late 1990s link Argentina with Chile.

The consortium that operates the Gasoducto del Pacifico line in central Chile proposed to extend the system to serve the Campanario power plant in Cabrero.

Since Argentina's 2004 energy crisis, its gas exports to Chile have fluctuated between 20–50% below contracted volumes. Argentina also more than doubled its gas export tax in 2006. As a result, Chile has been exploring additional gas import pipelines as well as LNG options.

In June 2005, Suez Energy International, a subsidiary of Tractebel (Belgium) began a feasibility study for a pipeline linking Peru's Camisea natural gas project with northern Chile. A 930-mile pipeline system between Pisco, Peru and Tocopilla, Chile, would carry 810 MMcfd. It also would connect to the GasAtacama and NorAndio lines, allowing possible exports to Argentina.

This pipeline is part of the natural gas pipeline "ring" proposed by Peru, Chile, Argentina, Uruguay, and Brazil. The ring would use new and existing pipelines to link natural gas reserves in those countries and reduce the dependence of some of them on Bolivian natural gas.

LNG

ENAP awarded a $400 million tender to BG Group for the construction of an LNG regasification terminal near Quinteros, in central Chile, with a sendout capacity of 330 MMcfd. Construction could begin by 2007, with online operation by 2009.

The Chilean government authorized state-owned Corporacion Nacional del Cobre de Chile (Codelco)—the world's largest copper-producing company—to lead development of an LNG terminal in the Antaofagasta region of northern Chile. It would to supply power plants and industrial consumers. The first phase of development is to be completed by late 2009, delivering natural gas for electricity generation.

Suez has also suggested building an LNG terminal in northern Chile.

COLOMBIA

Capital: Bogota
Monetary unit: Peso
Refining capacity: 285,850 b/cd
Oil production: 530,000 b/d
Oil reserves: 1.453 billion bbl
Gas reserves: 3.996 tcf

Background

Colombia is an important petroleum producer, although political unrest and stagnant reserves have led to decreased exports in recent years. According to the U.S. Energy Information Administration (EIA), the country aims to boost hydrocarbon exploration to preserve its status as a net oil exporter in the longer term.

The government announced in August that it will privatize as much as 20% of state-owned Ecopetrol in a bid to draw much-needed investment to modernize the ailing petroleum sector. The move is Ecopetrol's first opening to private investors, and comes as governments in Venezuela, Bolivia, and Ecuador are taking back control from foreign firms over their natural resources. Economists have estimated the public share offering could be worth $3 billion to $4 billion.

Colombia also continued its efforts to sell its biggest state-owned gas transportation company, Ecogas. A group of local pension and employee funds made a purchase offer in mid-2006, but it was rejected as too low. The government was reviewing other bids in December 2006.

Colombia's longstanding civil conflict—involving two left-wing insurgent groups and a right-wing paramilitary organization—has taken its toll on the country's energy sector. The frequency of attacks against energy infrastructure has declined in recent years, but the country's pipelines and power lines are still regularly damaged by insurgent groups.

Oil

Colombia's oil production has declined steadily since 1999, when it peaked at 830,000 b/d, due to a lack of sizable new reserve discoveries. May 2006 production totaled 538,709 b/d. Much of the country's crude is lighter and sweeter than that of other major Latin American producers.

Colombia's government has taken steps over the past six years to spur investment interest in the upstream sector. In the first five months of 2006, the country's Agencia Nacional de Hidrocarburos (ANH) awarded 15 new exploration licenses to foreign firms.

Countries that signed exploration and production contracts during the last several years with Ecopetrol include Chile's Sipetrol; Colombian-based Argosy Energy International; Petrominerales, a subsidiary of Canada's Petrobank; Brazil's Petrobras; Nexen; Occidental; and Russia's Lukoil.

There has also been renewed interest in Colombia's offshore basins. In 2005, BHP Billiton signed an agreement for exploratory drilling in the Fuerte Block off Colombia's Caribbean coast. ANH has also begun seismic studies of offshore blocks in the Pacific Ocean, which industry analysts believe could hold significant oil reserves.

Oil pipelines and refineries

Brazil's Synergy Group previously announced that it would build a $700 million, 435-mi pipeline linking its Rubiales field to the Ocensa pipeline, aiming for completion of the 80,000-b/d project by 2008.

The Colombian government also has discussed with Venezuela a proposed oil pipeline to export that country's crude oil to Colombia's deepwater Pacific ports. The plan, backed by Chinese oil companies, would include 620 miles of new pipelines and utilize unused capacity on existing ones.

Statement reached on pipeline impact

BP PLC agreed in mid-2006 to pay an undisclosed settlement to 1,000 Colombian farmers who claimed their livelihoods were endangered by construction of the Ocensa oil pipeline. Colombia's largest, the 100,000 b/d line extends from BP's Cusiana and Cupiagua oil fields in the eastern plains to the Caribbean coast. It was laid in the 1990s through part of the state of Antioquia.

Plaintiffs threatened a $28 million lawsuit against BP, alleging that construction of the pipeline ruined farming, livestock raising, fish farming, and gold mining in that area. The "amicable" settlement established a trust fund by the BP Colombia subsidiary to help farmers deal with environmental management, business development, and other issues.

In 2006, four companies (BP, Petrobras, Marubeni, and Glencore) submitted bids to expand the capacity of the Cartegena refinery from 75,000 b/d to 140,000 b/d. Completion of the $800 million project is scheduled for 2010.

Also, Ecopetrol let an initial contract to Technip for an upgrade of the 205,000 b/cd refinery in Barrancabermeja. The project is slated for completion in late 2009.

Natural Gas

The Colombian government's Plan de Masificación de Gas Natural (Natural Gas Mass Consumption Plan) aims to increase domestic natural gas use and establish Colombia as the "gas hub" for the Andean region.

Colombia's gas reserves are spread across 18 basins, seven of which have active production. Most reserves are in the Llanos basin, although the Guajira basin accounts for most of current production. Chevron is the largest natural gas producer in Colombia.

Colombia planned to launch a new natural gas exploration licensing round in late 2006 covering its offshore basins, especially those in the Caribbean Sea. Several companies also have expressed interest in development of coalbed methane resources.

Gas pipelines

BP said it plans to begin construction by 2008 a gas-to-liquids (GTL) plant at the Cuisiana/Cupiagua natural gas field. Initial capacity is expected to be 84,000 b/d of high-quality diesel fuel.

In April 2003, Colombia and Venezuela agreed to build a $320-million, 200-MMcf/d natural gas pipeline linking Colombia's Guajira basin to Venezuela's Maracaibo region. Initially, the pipeline will carry natural gas from Colombia to Venezuela, but there has also been talk of eventually reversing the flow of the pipeline to allow Venezuelan natural gas exports instead.

Colombia has also held discussions with Panama and Ecuador about extending the pipeline into those countries. Construction on the project should begin in July 2006 and be completed in 2008.

Exploration and Development

Petrobras, Ecopetrol joint projects

The national oil companies of Brazil (Petrobras) and Ecopetrol plan jointly to expand production in Tibu oil field in Colombia. The goal is to extend production by as much as 100 million bbl (beyond the 247 million bbl already produced) and to boost output to more than 15,000 b/d from 1,800 b/d. Petrobras will retain 40% of the field's total production.

The first phase of the project will run from January 2007 to mid-2009. Petrobras will invest $40 million for studies to determine the field's potential and for technology.

Lower Magdalena basin

Pacific Stratus Energy Ltd. (Toronto) reported a possible gas discovery in late 2006 on the 97,279-acre La Creciente block in the Lower Magdalena basin in Sucre Department, 100 miles south of Cartagena. The 100% owned La Creciente-1 well went to TD 11,557 ft (3,522 m) and encountered gas shows across a gross 642 ft (195 m) in the Tertiary Cienaga de Oro formation. The company planned to deepen to 11,600 ft (3,536 m), run logs, and conduct tests. The well is on Pacific Stratus's Prospect A, west of Guepaje gas field.

Middle Magdalena basin

Solana Resources Ltd. (Calgary) declared commercial and placed on long-term test the Guariquies-1 well in the Middle Magdalena basin. The well went to TD 10,243 ft (3,122 m). Future appraisal wells will be drilled to 7,000 ft (2,134 m) to target the Oligocene Mugrosa formation.

Guariquies-1 is part of an exploration program that calls for 700 km (435 miles) of seismic and 10 more wells by September 2007. Ramshorn International and Ecopetrol participated in the well.

Rig report

Baker Hughes reported 21 land rigs working in Colombia in July 2006, up from 15 a year earlier and 10 in July 2003. Pride International had 12 land drilling rigs and 8 land workover rigs in Colombia. Nabors Drilling International Ltd. had seven drilling and two workover rigs in Colombia in July. That same month, Petrobank Energy and Resources Ltd. announced that it had contracted two Nabors rigs and resumed drilling on the Orito block.

Putumayo basin

The government awarded Emerald Energy PLC (London) the 365-sq km (141-sq-mile) (Maranta exploration and production contract northeast of producing oil fields in the Putumayo basin. The 18-month first exploration period calls for gathering 30 km (19 miles) of 2D seismic data and reprocessing 40 km (25 miles) of existing 2D data. A second period, at Emerald's option, would drill one well to 11,000 ft (3,353 m).

Fuerte Norte and Sur Caribbean blocks

BHP Billiton Ltd. (Melbourne, Australia) signed contracts with the ANH for exploration and production rights to two blocks in the Caribbean. The Fuerte Norte and Fuerte Sur blocks each covers 1.2 million acres in 50–2,700 m of water. BHP Billiton will operate each block.

Llanos basin

Meta Petroleum Ltd., jointly owned by investment company Elliott Advisors (U.S.) and Synergy Group (Brazil), plans to increase production to 100,000 b/d of upgraded crude oil from 9,000 b/d by further developing Rubiales heavy oil field in the Llanos basin of central Colombia. In mid-2006, Synergy said that the field has 3 billion bbl of heavy crude in place.

Meta has let a contract to Mustang Engineering for FEED and cost estimate services for a heavy oil upgrader complex at Rubiales. The Rubiales field now has 35 wells. Synergy Group said in 2005 that further Rubiales development would require drilling of 600 wells.

Estimated costs include $700 million for infrastructure, including the upgrader and a pipeline to Casanare. Drilling could be completed by 2008, with initial production of 50,000 b/d.

Serafin gas project

PetroLatina Energy PLC began development of the (1991) Serafin gas discovery in Area B of the Tisquirama License. The Serafin well is north of PetroLatina's Los Angeles field. Gross reserves are estimated at 4–8 bcf.

Driller Texas Petroleum Co. encountered a gas-bearing sand in a stratigraphic trap of the Miocene Real Group. Logs indicated 18 ft (6 m) of pay, and the well flowed at up to 16 MMcfd. PetroLatina expects to begin commercial gas delivery to local industries in 2007.

CUBA

Capital: Havana
Monetary unit: Peso
Refining capacity: 301,400 b/cd
Oil production: 40,000 b/d
Oil reserves: 124 million bbl
Gas reserves: 2.5 tcf

Background

Oil & Gas Journal (OGJ) estimates that Cuba has proven oil reserves of 124 million bbl. Domestic crude oil production was expected to average 39,000 b/d in 2006, about equal to its level for 2001–2005. Production was only 15,000 b/d in 1990 but surged following an opening to foreign investment in 1999. However, the country now consumes more than 150,000 b/d and oil imports from Venezuela are particularly important.

Two Canadian companies—Sherritt International Corp. (Toronto) and Pebercan Inc. (Montreal)—have been co-producing about 60% of all Cuban output along with the country's national oil company, Unión Cubapetróleo (Cupet).

Cuba entered into an agreement with Venezuela in April 2005 to buy oil from that country. Venezuela's Pres. Hugo Chávez has played an expanding role in helping Cuba improve its energy situation. Cuba now buys approximately 100,000 b/d of oil from Venezuela at highly discounted prices and is allowed to resell volumes not used.

In 2006, Venezuela agreed to help Cuba re-activate and expand the Soviet-era Cienfuegos refinery (more details below), and the Viet Nam Oil and Gas Corp. (PetroVietnam) and Cupet signed an agreement for oil and gas cooperation.

Since 1999, some 16 international companies have signed contracts to explore on 10 of Cuba's 59 Gulf of Mexico oil and gas exploration blocks. More than 25 oil fields have been discovered and prepared for drilling; the largest being Varadero, with reserves estimated at 3 billion bbl.

There is also an additional projected reserve of 2.8 billion bbl in the southeast section of the Gulf of Mexico, just within Cuba's Exclusive Economic Zone (ZEEC). In 2000, Cuba divided that area into 59 deepwater offshore blocks and, by the end of 2001, investors from a number of nations—including the United Kingdom, Canada, France, Spain and Sweden—had secured deals accounting for some 20 blocks in this region.

Cuba also reached exploration agreements in 2005 with China's Sinopec, ONGC Videsh of India, and Norsk Hydro of Norway. As of November 2006, no major fields had been found. Repsol of Spain pursued drilling in 2004 that only partly met expectations, and Brazil's Petrobras discontinued Cuban exploration after hitting a dry hole in 2001.

By some estimates, Cuba has allocated $1.7 billion to its energy industry since 2004, working with its international partners. The country has the potential to produce 700,000 barrels of oil a day by 2015, according to a study by Ohio Northern University.

Exploration and Development

Repsol gets partners

Repsol YPF Cuba SA, operator of six offshore blocks in Cuba's exclusive economic zone, is conducting 3,000 sq km (1,864 sq miles) of 3D seismic and geological studies and has accepted two partners in the venture. Norsk Hydro ASA (Oslo) and ONGC Videsh Ltd. (OVL; New Delhi) signed an association contract with Repsol and Cupet to explore the blocks. Repsol will remain as the operator.

Activity raises U.S. concerns

Some U.S. lawmakers expressed concern that actions by Cuba could result in development of its oil and gas resources closer to Florida's coast than the U.S. government allows—perhaps as close as 35–40 miles (56–64 km) from the Florida Keys.

The proposed 2007–2011 Outer Continental Shelf leasing plan being developed by the U.S. Minerals Management Service (MMS) would establish a 100-mile (161-km) oil and gas leasing buffer off Florida's coast.

In addition, in December, the Gulf of Mexico Energy Security Act of 2006 became law. One key provision establishes a 125-mile (201-km) no-lease Florida coastal buffer zone to be in effect through 2022 beyond the Sale 181 area.

Oil find on western Block 7

Pebercan Inc. (Montreal, QC, Canada) production-tested an indicated discovery well near the western edge of Block 7 on the North Cuba heavy oil belt east of Havana. The Tarara-100 well, 50 km (31 miles) west of Canasi and Seboruco oil fields, stabilized at 700 b/d through a 12-mm choke. Drilled from shore, the well is drilled to 3,465 m (11,368 ft) measured depth to the Tarara prospect in the Bay of Cardenas.

Pebercan plans to integrate results of the Tarara studies with a new 3D seismic interpretation before it drills Guanabo, the next structure east of Tarara.

U.S. energy execs talk with Cuban officials

U.S. energy executives met with their Cuban counterparts in February in Mexico City in the first private-sector oil summit between the two countries. The Cubans described their country's oil potential and sought support in ending the 45-year-old U.S. trade embargo.

The meeting was sponsored by the U.S.-Cuba Trade Association (U.S.-CTA), Valero Energy Corp., the Louisiana Department of Economic Development, the Texas Port of Corpus Christi, and others. Cuban attendees included the president of Cupet and officials from the Cuban Ministry of Basic Industries and Cuba's ministries of foreign trade, foreign investment, and foreign relations.

U.S.-CTA said it would like to see the United States relax its sanctions for the energy sector as it did in 2000 for food and agricultural product sales to Cuba.

Drilling and Production

Sherritt reports on activity

Sherritt International Corp. (Toronto) reported on a range of activity in early 2006:

- Evaluation of enhanced oil recovery technologies for its fields on the North Cuba fold belt, including carbon dioxide, steam, or gas reinjection
- Planned drilling of an appraisal well at Majaguillar-Corajol, an oil field discovered in 1996 east of Cardenas
- Ongoing development drilling in Santa Cruz field, in Block 7 just off Cuba's central north coast
- Planned exploration drilling at Playa Larga on Block 10 north of Cardenas
- Evaluation of the San Anton exploration well on western Block 9
- Evaluation by a joint venture partner as operator of Block 7 of the Tarara exploration well

The company said it expects its 2006 gross working interest oil production to match 2005 levels, and that it was on track for its planned $140 million in capital outlays in 2006.

Refining

Venezuela to help upgrade refinery

Venezuela's state oil company PdVSA formed a joint venture with Cupet to revamp the unfinished Cienfuegos refinery. Shared costs of the initial investment were estimated at $0.8–1.0 billion. Cupet will hold a 51% stake, PdVSA 49%.

The 76,000 b/d, Cienfuegos refinery was built in the years of Soviet influence over Cuba, but the Soviet Union collapsed before the facility became operational. PdVSA also will supply the refinery with crude and other products. Output could be sold within Cuba as well as abroad.

According to OGJ, Venezuela's Pres. Hugo Chavez is actively seeking to increase refining capacity throughout the region by attempting to buy a small Argentine refinery, doubling capacity at a refinery in Uruguay, and laying the cornerstone for a $2.5 billion refinery in Brazil.

The growing relationship between Cuba and Venezuela is viewed by some U.S. officials as part of a larger political undertaking aimed at diminishing U.S. influence in Latin America.

ECUADOR

Capital: Quito
Monetary unit: U.S. dollar
Refining capacity: 176,000 b/cd
Oil production: 510,000 b/d
Oil reserves: 4.517 billion bbl
Gas reserves: 0

Ecuador has oil reserves of 4.5 billion bbl, the third largest in South America. The country's oil production increased 31% from 2001 to 2005 and was estimated to be 176,000 b/d in 2006. Ecuador is South America's fifth largest oil producer. The state-owned petroleum firm is Petroleos de Ecuador (Petroecuador), and its production arm is Petroproduccion.

Some 13 foreign oil companies had contracts and assets in Ecuador in 2006, including Brazil's Petrobras, Repsol YPF SA, EnCana Corp. (Calgary, AB, Canada), Agip SPA (Milan, Italy), Perenco SA (Paris), and Occidental Petroleum Corp. (Los Angeles, CA, U.S.A.).

Exploration and Development

Development set for Amazon headwaters

Petroecuador's production unit Petroproduccion reported plans to develop Panacocha oil field in the Amazonas province before yearend 2006. Petroproduccion, Panacocha's operator, said the field is thought to contain 42.5 million bbl of recoverable oil and to be capable of producing 6,000–9,000 b/d. Development will cost an estimated $100 million. The field is in an environmentally sensitive area at the headwaters of the Amazon River where there has been little oil exploration and no development.

Drilling and Production

Baker Hughes reported 11 land rigs drilling in Ecuador in July, down from 14 a year earlier. Ecuador's rolling nationalization of oil and natural gas assets, following prosecution of Texaco and protests against Occidental Petroleum Corp.'s work in the Ecuadorian Amazon, apparently dampened exploratory and development drilling.

Although several foreign operators abandoned Ecuador operations in 2006, some 118 wells were drilled through September.

Active drilling contractors included H&P; Nabors International; Drilfor SA, a local company; Changqing Petroleum Exploration Bureau (CPEB); China Petroleum & Chemical Corp. (Sinopec); China Petroleum Corp. (CNPC); Petrex, a subsidiary of Eni SPA; and Perforec Perforaciones Ecuatorianas SA, a wholly owned subsidiary of Saxon Energy Services Inc.

Repsol-YPF was the most active foreign operator, drilling 25 wells to an average MD greater than 10,000 ft (3,048 m).

Perenco drilled 19 wells as operator for the B7-B21 consortium, in the Yuralpa field (Block 21), Oso field, and Lobo field (Block 7).

AEC Ecuador (Alberta Energy Co. Ltd., an EnCana subsidiary), drilled 18 wells. EnCana Ecuador drilled two wells.

Petrobras Energia Ecuador drilled 8 wells in the Palo Azul field in the Oriente foreland basin (Block 18) as operator on behalf of partners Ecuador TLC SA and Teikoku Oil Ecuador, a subsidiary of Japan's Teikoku Oil Co. Ltd.

EnCana Corp. (Calgary, AB Canada) drilled 20 wells early in 2006 before it pulled out of the country. In late February, EnCana announced that it had completed the sale of its Ecuadorian oil and pipeline interests to Andes Petroleum Co., a joint venture of Chinese petroleum companies, for about $1.42 billion.

OMV AG (Vienna, Austria) completed in October the sale of its 5,000-b/d Ecuadorian exploration and production assets to joint-venture partners Burlington Resources Oriente Ltd. and Perenco Ecuador Ltd., a unit of privately held French E&P firm Perenco SA.

ECUADOR DRILLING, JAN.-SEPT. 2006

Operator	Wells, no.	Measured depth, ft
AGIP	2	14,279
Alberta Energy Co. (AEC) Ecuador Ltd.	18	8,555-10,665
Andes Petroleum Ecuador	2	10,200-10,530
EnCana Ecuador	2	11,889-12,350
Occidental	4	7,031-9,200
City Oriente	5	7,725-8,245
Perenco	1	8,400
Perenco, on behalf of Block 7-Block 21 consortium	19	7,825-10,626
Petrobell	5	10,785-11,870
Petrobras, on behalf of EcuadorTLC	8	10,670-11,726
Petroproduccion (PetroEcuador)	22	7,800-10,814
Petrosud-Petroriva	3	10,375-10,467
Repsol-YPF	25	9,158-11,010
SIPEC	1	—
Tecpecuador	2	4,280-4,480

Disputes with Occidental Petroleum Corp.

Occidental Petroleum Corp. (Oxy; Los Angeles, CA U.S.A.) sought arbitration in May 2006 concerning the seizure that month by Ecuador of all of the company's assets in Block 15. The request was filed with the International Centre for Settlement of Investment Disputes (ICSID).

Located in Block 15 are the Eden-Yuturi, Limoncocha, Indillana, Paca Norte, Paca Sur, and Yanaquincha fields. It was Oxy's only producing area in Ecuador. The national government seized the assets in a dispute over tax refunds that Oxy says Ecuador improperly withheld from it. In September, Petroecuador announced that it would establish Rio Napo Operations Co. to operate oil fields formerly owned by Oxy.

Oxy filed its arbitration claim with ICSID in mid-July. The company aimed to restore its rights in Ecuador and pre vent the government from turning operations over to a third party until the claim was settled. In early October, Oxy dropped its arbitration claim against Petroecuador but maintained its claim against the Ecuador government at ICSID. The case was pending in November 2006.

In a second case, Oxy offered Ecuador as much as $1 billion in disputed taxes, investments, and extra revenues to end a legal dispute also concerning Block 15. The issue is whether Oxy in 2000 transferred a 40% interest in Block 15 to Canada's EnCana Corp. without approval from Ecuador authorities.

Oxy proposed giving Ecuador at least $600 million in extra revenues from the disputed area but denies any wrongdoing. As of late November 2006, Ecuador's energy ministry was still studying the case and lawyers for Oxy were evaluating legal options.

Government Actions

New president chosen

In a runoff election November 26, leftist Raphael Correa, U.S.-educated economist, defeated Alvaro Naboa, a businessman from the wealthiest family in Ecuador. President Correa took office January 15, 2007.

According to Prensa Latina news agency, Pres. Correa has said he plans to limit U.S. military activities in Ecuador, renegotiate the country's foreign debt, and oppose any free trade agreement with the United States. Correa is seen as sympathetic to the views of leftist leaders in Bolivia, Cuba and Nicaragua who are allied with Venezuela's President Hugo Chavez.

New energy minister named

Ecuador's new energy and mines minister Alberto Acosta said he aims to diminish dependence on diesel for thermo generation and cut reliance on oil in general through the development of renewable fuels. A further goal is to develop a "regionalized production strategy" for Petroecuador as well as a storage and refining plan.

Also, the ministry will launch Project Black Gold, to combat oil contraband, and a program of online commerce to ensure transparency in oil companies' operations in order to fight corruption.

"Extraordinary profits" tax imposed

Ecuador planned to start collecting a revised "extraordinary profits" oil tax May 1, 2006, the economy ministry told Dow Jones. Ecuador's Congress approved in April the controversial tax on international oil companies operating in that country when oil prices rise above those stipulated in participation contracts.

The tax change calls for the government to receive at least 50% of the revenue when oil prices rise above a benchmark price. The ministry said the tax would help make up for revenue the government should have received from oil production in 2003 and 2005, and could generate as much as $500 million in state revenue in 2006.

In April, Occidental Petroleum said it believed Ecuador's revised tax violates bilateral treaties with the United States and possibly Ecuador's constitution. Oxy produces 100,000

b/d of oil in Ecuador. The economy ministry told Dow Jones that international oil companies might be given the opportunity to renegotiate their operating contracts.

Brazil, Ecuador energy alliance

In July, Brazil's Petrobras signed an agreement with Ecuador's Petroecuador, establishing a 5-year strategic energy alliance covering every aspect of the petroleum industry, including new exploration and production initiatives.

Petrobras has expressed interest in the exploration and development of the Ishpingo-Tiputini-Tambochocha (ITT) field complex in northern Ecuador and several blocks in southeastern Ecuador. The ITT complex is thought to contain about 900 million bbl of proven reserves, which would require $2 billion to produce.

After projects are proposed, Ecuador's president will decide which will be carried out under the alliance and which will be put up for international bids. Petroecuador has been told to prepare for licensing the southeastern blocks sometime in 2007.

However, the November election of Pres. Raphael Correa caused concerns. Petrobras expressed doubts about future investments and voiced concerns over profits reduction in Ecuador. Its unease reflects its experience in Bolivia, where the government of President Evo Morales confiscated Petrobras operations valued at $2.5 billion.

Venezuela, Ecuador oil deals

Venezuela President Hugo Chavez signed in June 2006 refining and technical assistance contracts with Ecuador's outgoing President Alfredo Palacio. Venezuela will refine as much as 100,000 b/d of crude from Ecuador under favorable terms, which officials say will save Ecuador at least $300 million/year. Venezuela also will provide technical advice to Petroecuador. Ecuadorean officials have said the agreements are economic and do not indicate a political alliance with President Chavez.

GUATEMALA

Capital: Guatemala City
Monetary unit: Quetzal
Refining capacity: 0
Oil production: 16,500 b/d
Oil reserves: 83.07 million bbl
Gas reserves: 0

Overview

Oil

Guatemala is one of only two countries in Central America that produces oil. Most of that production is by Perenco (Paris) from Guatemala's northern jungles, along the border with Mexico. The country has virtually no refining capacity, so almost all production is exported to the United States, and Guatemala must import petroleum products.

In September, Guatemala awarded a 25-year oil exploration license (Block A1-2003) to U.S. Oil Guatemala. Additional blocks offered include A1-2005 in the Peten Sur basin and A9-2005, in the center of the country.

The Mexican government in mid-2006 outlined potential plans to build a new Central American refinery, in either Panama or Guatemala (Puerto Quetzal) or Panama. Mexico would supply oil to the refinery, which would have a 250,000 bbl/d crude oil refining capacity.

Natural Gas

In 1999, Guatemala and Mexico signed a protocol on construction of a natural gas pipeline connecting Jaltiplan de Morelos, in southern Mexico, to Puerto Quetzal in Guatemala. The pipeline could be extended to the Honduran and Salvadoran borders (and possibly to Nicaragua and Costa Rica) as part of a wider Central American gas pipeline network.

In 2005, Mexico and countries in Central America signed the Cancún Declaration for implementing regional energy projects. One project would build gas pipelines from various LNG regasification plants in Mexico to the Central American region. However, small Central American markets and fluctuating LNG prices have tended to make importing LNG uneconomic.

Exploration and Development

Peten basin

The Guatemalan part of the Peten basin is underexplored, said independent TrueStar Petroleum Corp. (Calgary, AB, Canada). Peten is one of three basins in the country in which oil has been discovered. The others, Amatique and Pacific, are not currently open to exploration. Guatemala has 137 wells drilled, 76 of them exploration wells. The Guatemalan part of Peten has 10 discoveries, most of which have produced from only one to three wells.

TrueStar holds a 10% interest in oil production sharing contract (PSC) 4-93 and has proposed to acquire the other 90% interest from Cia. General de Combustibles (Buenos Aires), which filed for protection from creditors in late 2000. TrueStar's proposal must be approved by the Ministry of Energy and Mines.

PSC 4-93 governs development of the 554,000-acre oil exploration block A-2-92 in the South Peten basin near the border with Mexico. Successful wells lie east, west, and north of the license area.

All of Guatemala's Peten basin fields produce from the Cretaceous Coban formation. TrueStar has delineated four closed Coban structures. One is Goliath, a large subsurface anticline with surface expression. It is among the shallowest in South Peten. The company has estimated the cost of a 10,000-ft well at about $3 million.

Four fields—Rubelsanto, Chinaja, Caribe, and Tierra Blanca—lie north and west of TrueStar's concession, but none is fully developed. Combined cumulative production is 17 million bbl of oil.

In May, Taghmen Energy PLC (London) reentered the 1XA well on the 6-93 license and kicked off at 5,810 ft. It planned to spud the Huapac 2X well on 6-93 in mid-2006.

Meanwhile, the environment ministry granted technical approval to convert the planned 3D seismic survey on License 7-05 (Tortugas/Atzam) to a 2D project. The first of three well workovers was to start in mid-2006. Both licenses are in the Peten basin.

JAMAICA

Capital: Kingston
Monetary unit: Dollar
Refining capacity: 36,000 b/cd
Oil production: 0
Oil reserves: N/A
Gas reserves: N/A

Jamaica consumed about 23.7 million bbl of petroleum and petroleum products in 2005, according to the Petroleum Corporation of Jamaica (PCJ), the state energy company. The country depends on imported oil to meet over 90% of its energy needs.

Jamaica is among the Caribbean countries that signed the Petrocaribe Initiative in 2005. Under that program, Jamaica can buy crude oil and refined products from Venezuela under favorable financing terms. The similar San Jose Accord enables Jamaica to receive oil and refined products from Mexico.

Exploration and Development

2005 licensing round

In January 2005, PCJ launched a licensing round for 4 onshore and 22 offshore blocks based on a re-evaluation of geological data. Of particular interest is the Walton Basin, a deepwater area between Jamaica and the Pedro Bank. The Pedro Bank itself is also prospective.

In the 2005 round, PCJ granted exploration rights to two companies. Finder Exploration Pty. Ltd. (Perth, Australia) received access to blocks 6, 7, 10, 11, and 12 offshore southern Jamaica. Working with Finder is Gippsland Offshore Petroleum Ltd. (Melbourne, Australia) who began in May a

JAMAICA'S BLOCK OFFERING

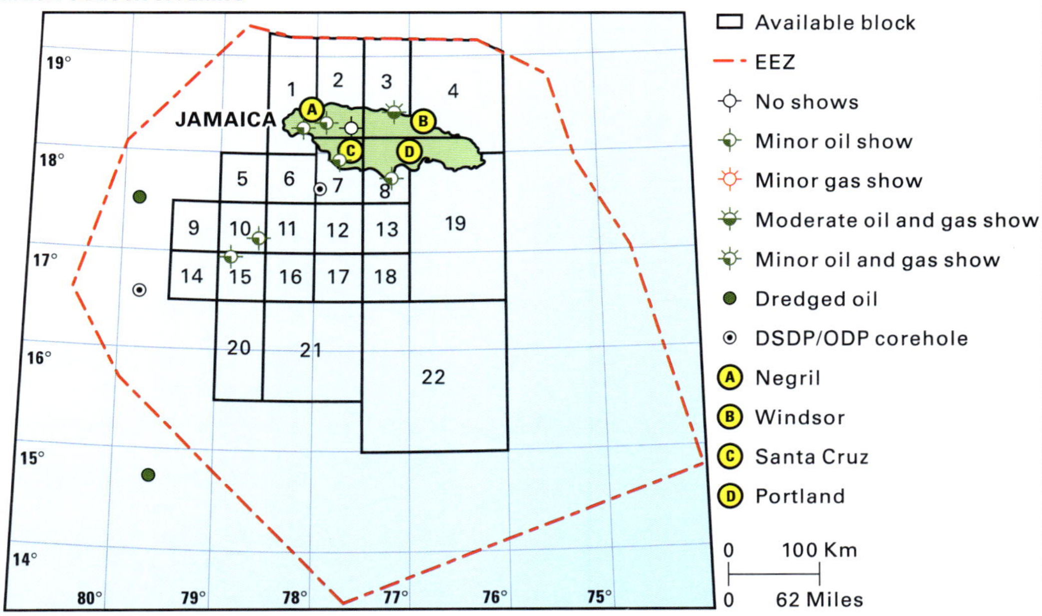

7,180-km (4,460-mile) 2D seismic survey in the Walton basin. Finder expects to identify drillable prospects by year-end 2006 or early 2007.

Rainville Energy (Canada), secured rights to explore offshore blocks 9, 13, and 14. The company anticipates conducting a seismic survey by early 2007.

2007 exploration interest

Toward year-end 2006, PCJ said it was aware of renewed interest in offshore acreage that did not draw bids in the 2005 licensing round. In response, the government presented an information session in January 2007 for prospective developers. The government has given PCJ permission to open negotiations for any open blocks rather than go through the formal bid process.

Malaysian initiatives

In a separate development, the government of Jamaica met in early 2006 with Petronas, Malaysia's state-run energy agency. Following the meeting, Jamaica said it will invite Petronas to conduct offshore exploration in the country. The governments also agreed to consider delivery of additional supplies of Malaysian LNG to Jamaica.

Refining

Petrojam, a subsidiary of PCJ, operates a 36,000-bbl/d facility near Kingston. In 2006, Petrojam began the front-end engineering work to increase the capacity of the facility to 50,000 b/d and improve the plant's ability to produce cleaner products.

Also, a proposed $2 billion, 250,000 bbl/d refinery near St. Elizabeth has entered advanced planning stages. The Jamaican government has tentatively approved the facility, pending an environmental impact statement, said PCJ.

LNG

As of November, PCJ was weighing installation of a floating LNG storage and regasification barge to allow it to meet a 2009 deadline to access LNG for industrial use.

The barge plan could replace a planned $300 million land-based LNG facility that has been under review and is behind schedule. PCJ said that construction could take about 18 months but that negotiations are ongoing and no decision has been made.

The barge facility is proposed for Port Esquivel, the same site planned for the land facility. Capacity would be similar to the 1.15 million tons/year proposed for the land facility. The LNG would come from Trinidad and Tobago, but final supply terms are not final.

A final investment decision could come in early 2007 on the shore terminal, said PCJ, following the late-2006 completion of an engineering study by Mustang Engineering (Houston). However, there is still uncertainty about the configuration, construction schedule, ownership arrangements and final cost of a land-based facility.

MEXICO

Capital: Mexico City
Monetary unit: Peso
Refining capacity: 1,540,000 b/cd
Oil production: 3.273 million b/d
Oil reserves: 12.352 billion bbl
Gas reserves: 14.557 tcf

Background

Oil

According to the U.S. Energy Information Administration (EIA), Mexico's gross domestic product (GDP) increased by 3% in 2005, after growing by 4.4% in 2004. Earlier in decade, Mexico's GDP growth was very small or negative. Recent high global oil prices, coupled with economic recovery in the United States, have driven Mexico's recovery.

The oil sector is a crucial part of Mexico's economy, generating more than 10% of the country's export earnings and approximately 40% of government revenues. Most of Mexico's oil production is from the Cantarell field in the Gulf of Campeche, off the southeastern coast.

The Cantarell field is one of the largest in the world, accounting for 2.0 million b/d of crude oil—60% of total Mexican production in 2005. Since production began to decline in the early 1990s, nitrogen injection has helped to maintain reservoir pressure.

However, Pemex has warned that Cantarell has entered a stage of long-term decline. Production will likely continue to drop by an estimated 14% per year, despite any incremental gains by incorporating additional satellite fields.

The two other major oil production centers in the Gulf of Campeche are Ku- Maloob-Zaap (KMZ) and Abkatun-Pol-Chuc. Production from the KMZ complex, adjacent to Cantarell, has risen by 50% since 1996. Off the coast of Tabasco state, the Abkatun-Pol-Chuc facility produced 299,800 b/d of crude oil in 2005. But production there also has declined steadily, down more than 50% from 1996.

Petroleos Mexicanos (Pemex), Mexico's national oil company, has predicted that the declines in Cantarell production will be offset by new wells coming online in deepwater Gulf of Mexico and in Chicontepec field in Veracruz state. However, many experts say both areas would require at least a decade of exploration and development before significant production begins.

Important onshore production centers in the southern part of the country include Bellota-Jujo and Samaria-Luna. The northern part of the country produced only 83,500 b/d of crude oil in 2005. The largest field in the north is Poza Rica.

Natural gas

According to Pemex, the Southern Region of the country contains the largest share of proven reserves. However, the Northern Region will likely be the center of future reserves growth because it contains almost 10 times as many probable and possible natural gas reserves as the Southern Region.

Mexico's natural gas production has grown in recent years. At the same time, however, gas consumption has grown steadily, driven mostly by the electricity sector, whose share of total gas consumption increased from 16% in 1994 to 33% in 2004. (Pemex itself consumed 43% of domestic consumption in 2004.) As a result of the domestic shortfall in gas production, Mexico imported 766 MMcfd in 2004 and 480 MMcf/d in 2005.

Mexico's natural gas production is relatively spread out across the country. Onshore fields in the north represented 38% of 2005 production, with onshore fields in the south contributing 29% and offshore fields in the Gulf of Campeche represented the remainder. The single largest field is Cantarell, in the Gulf of Campeche. Other major fields include Caan, Culebra, and Muspac.

Calderón Elected President

On September 5, the Federal Electoral Tribunal declared Felipe Calderón of the National Action Party the winner of the presidential election by 0.57 percentage points over Andrés Manuel López of the Democratic Revolution Party. It was the final resolution of balloting conducted on July 2.

Calderón was sworn in December 1. Soon thereafter, he said he favors opening the country's oil industry to international oil corporations for joint production work with Pemex to help increase crude oil exports. Foreign companies currently are allowed only as contractors providing oil field services such as drilling, seismic work, and infrastructure construction.

New tax structure for oil, gas exploration, production

On January 1, 2006, Mexico initiated a new fiscal regime for its oil and gas industry, primarily to ensure the financial stability of Pemex.

The former system taxed total sales rather than profit and required an implicit tax rate of 105% on net cash flow. As a result, Pemex suffered net losses since 1998 and was faltering in its ability to continue exploration and production expansions in promising but costly-to-develop structures, to maintain existing facilities, and to retire its mounting debt.

Under the new regime, the Federal Rights Law will govern Pemex Exploration & Production (E&P) and the Income Law of the Federation will continue to govern the tax regime of other subsidiaries. The new fiscal regime for Pemex E&P consists of five duties:

- Ordinary duty on hydrocarbons
- Duty on hydrocarbons for the oil revenues stabilization fund
- Extraordinary duty on crude oil exports
- Duty on hydrocarbons for the Fund for Scientific and Technological Research on Energy
- Duty on hydrocarbons for fiscal monitoring of oil activities

In summary, the new fiscal regime for Pemex is competitive with tax systems of countries having geological characteristics similar to Mexico's and comparable finding, lifting, and development costs. It allows deduction of taxes and duties payments as well as exploration costs.

It is expected to lead to an increase in exploration and production investment, which should result in higher reserves replacement rates and enhanced hydrocarbon production.

New Fiscal Regime Descriptions

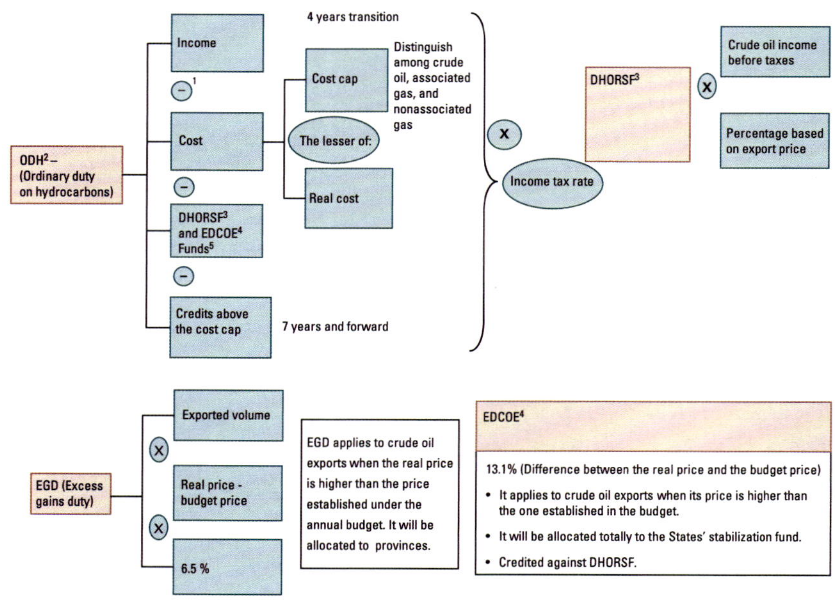

[1]The symbol "-" in the chart means "minus," with the above-mentioned items subtracted from the income. The symbol "X" means "times." [2]To calculate the basis for ODH, it is permitted to deduct costs, the DHORSF and EDCOE funds, and credits above the cost cap. [3]DHORSF: Duty on hydrocarbons for the oil revenues stabilization fund. [4]EDCOE: Extraordinary duty on crude oil exports. [5]Funds are used for fiscal monitoring of oil activities and for scientific and technological energy research.
Source: Pemex

Exploration and Development

Pemex E&P projects

Pemex Exploration & Production (E&P) listed 15 oil and gas E&P projects it said contain combined proved and probable reserves of 21 billion bbl of oil and 18 tcf of gas. Peak production years from those projects range from 2006 to 2029, and seven involve waterflooding, gas lift, or nitrogen injection.

Pemex E&P said that the country's official proved oil reserves dropped slightly to 16.5 billion bbl at yearend 2005 from 17.6 billion bbl a year earlier. Gas reserves slid to 20 tcf from 20.4 tcf. Proved, probable, and possible reserves total roughly three times proved reserves alone.

Tax Rates for Mexico's Oil, Gas Export Duties

Ordinary duty on hydrocarbons

Annual avg. oil export price, $/bbl	Applicable rate over the value of annual oil, gas production minus — the permitted (Fig. 3) deductions, % — 2006	2007	2008	2009
0 - 19.99	87.81	85.61	83.40	81.20
20 - 21.99	87.32	85.24	83.16	81.08
22 - 23.99	83.14	82.10	81.07	80.03
24 - 25.99	82.34	81.50	80.67	79.83
26 - 27.99	81.53	80.90	80.27	79.63
28+	78.68	78.76	78.84	78.92

Duty on hydrocarbons for the stabilization fund Table 1b

Annual avg. oil export price, $/bbl	Applicable rate over the value of annual oil production, %
22.01 - 23	1
23.01 - 24	2
24.01 - 25	3
25.01 - 26	4
26.01 - 27	5
27.01 - 28	6
28.01 - 29	7
29.01 - 30	8
30.01 - 31	9
31+	10

Source: Pemex

Nitrogen injection is temporarily sustaining the output level at the Cantarell complex in the Gulf of Mexico, but Pemex reports sharp production declines in most of its other producing areas.

Mexico's Chamber of Deputies approved a measure that would allow Mexican coal mine operators to produce and use coalbed methane and sell excess gas to Pemex. This would end Pemex's monopoly on gas rights. Senate approval and a presidential signature are still required.

First deepwater gas discovery

Noxal-1, a wildcat drilled by Pemex in April 2006 in the Catemaco fold belt in the Gulf of Mexico, is Mexico's first deepwater discovery. Drilled in 935 m (3,067 ft) of water—a record for Mexico—the well is 102 km (63 miles) northeast of Coatzacoalcos.

At a total depth of 4,000 m (13,125 ft), the fourth interval tested flowed 9 MMcfd of gas from a reserve estimated at 245 bcf, said IHS Energy (Houston). Mexico lacks the technology to develop deepwater fields alone and its laws do not allow outside operators to participate.

Drilling and Production

Rig report

Baker Hughes reported that 82 rigs were drilling in Mexico in October 2006, including 58 land rigs and 24 offshore rigs. This is 11% fewer rigs than a year earlier, when 92 rigs (62 land; 30 offshore) were drilling in Mexico. Most (48) of the rigs were drilling for oil in October 2006; 31 were drilling for gas, and three others had "miscellaneous" objectives.

Pemex lets contract for two light platforms

Fluor Corp. reported that ICA Fluor, its joint venture with Empresas ICA Sociedad Controladora, received a $24 million contract from Pemex E&P to build two lightweight offshore platforms, May D and Sinan NE. ICA Fluor subsidiary Industria del Hierro will perform engineering, procurement, construction, load-out, and seabed fastening. Scheduled completion was November 2006. The platforms will be installed in Pemex's southwest region of the Gulf of Mexico.

Refining

Upgrade for oldest refinery

Pemex plans to upgrade its 194,000 b/cd Gen. Lazaro Cardenas refinery near Minatitlan in Veracruz. It is Mexico's oldest. The planned upgrades will enable the plant to refine additional heavy, high-sulfur oil and increase the production of low-sulfur gasoline and middle distillates.

The upgrades will be made possible through a $216.8 million export sale, backed by a $200 million loan guarantee from the U.S. Export-Import Bank.

PEMEX E&P PROJECTS

Reserves development

Project	Oil reserves[1] million bbl	Gas reserves[1] bcf	Oil peak production[2] 1,000 b/d	Oil peak production[2] Year	Gas peak production[2] MMcfd	Gas peak production[2] Year
1. Amatitlan (Chicontepec)	1,740	2,338	393	2029	439	2029
2. Crudo Ligero Marino (SGP)[3]	766	906	230	2007	641	2007
3. Agua Fria (Chicontepec)	357	637	173	2012	191	2012
4. Ayin-Alux	299	67	74	2014	19	2014
5. San Manuel (SGP)[3]	214	1,084	11	2006	214	2006
6. Ixtal-Manik (SGP)[3]	184	99	55	2008	63	2008
7. Burgos	98	3,006	-	-	1,260	2009
8. Cuenca de Veracruz (SGP)[3]	-	1,043	-	-	825	2007

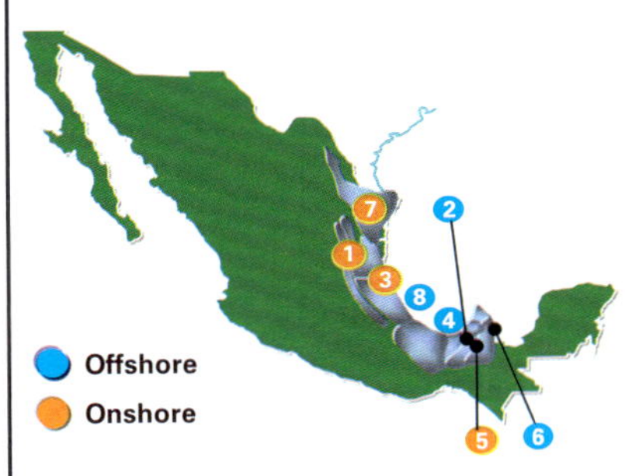

Secondary/enhanced recovery

Project	Oil reserves[1] million bbl	Gas reserves[1] bcf	Injection	Oil peak production[2] 1,000 b/d	Oil peak production[2] Year	Gas peak production[2] MMcfd	Gas peak production[2] Year
9. Cantarell	8,884	2,312	N_2	1,996	2006	727	2006
10. Ku-Maloob-Zaap	4,352	1,048	N_2	794	2010	324	2010
11. Antonio J. Bermudez	2,207	2,495	N_2	161	2008	282	2008
12. Jujo-Tecominoacan	923	1,110	N_2	88	2008	80	2008
13. Poza Rica	670	791	H_2O	51	2011	56	2011
14. Bellota Chinchorro	334	532	GL[4]	53	2008	70	2008
15. Cactus Sitio Grande	308	507	GL[4]	27	2010	73	2010
	21,336	**17,975**					

[1]Proved and probable reserves as of Dec. 31, 2004. [2]Forecast based on portfolio 5.2E. [3]Strategic Gas Program. [4]Gas lift.

MEXICO-EAST
CANTARELL COMPLEX
Gulf of Mexico
Bay of Campeche
Gulf of Tehuantepec
Texas
US
Coahuila
Nuevo Leon
Tamaulipas
Zacatecas
MEXICO
San Luis Potosi
Aguas
Guanajuato
Queretaro
Hidalgo
Tlax.
Veracruz
Michoacan
D.F.
Morelos
Puebla
Guerrero
Oaxaca
Tabasco
Chiapas
Campeche
Yucatan
Quintana Roo
BELIZE
GUATEMALA
HONDURAS
EL SALVADOR
Corpus Christi
Nuevo Laredo
Laredo
McAllen
Reynosa
Brownsville
Matamoros
Cadereyta
Monterrey
San Fernando
Ciudad Victoria
Ciudad Madero
Tampico
Tuxpan
Poza Rica
Salamanca
Queretaro
Tula
Pachuca
Mexico City
Toluca
Cuernavaca
Tlaxcala
Puebla
Jalapa
Veracruz
Azcapotzelco
Minatitlan
Cardenas
Dos Bocas
Ciudad Pemex
Villahermosa
Ciudad del Carmen
Fontera
Campeche
Merida
Morelia
Chilpancingo
Acapulco
Oaxaca
Tehuantepec
Juchitan
Salina Cruz
Tuxtla
Belmopan
Guatemala
San Salvador
Sonsanate
Ulua
Buena Suerte
Monclova
Lampazos
La Presa
Sta. Rosalia
Arcos
Pijitas
Culebra
Canor
Carretas
Brasil
Fortuna
Mision-Trevino
Primavera
Cabeza
Matamoros
S. Bernardo
Benavides
Ramirez
18 de Marzo
Chapul
Huizache
Pobladores
Lerma
Barcodon
Tamaulipas-Constituciones
Ebano Panuco Fields
Arenque
Cabo Nuevo
Arrecife Medio
Isla de Lobos
Tiburon
Esturon
Marsopa
Golden Lane Fields
Bagre
Escuela
Pirache
Miralejos
Remudauera
Capite
M. Pancho
Mecayucan
Angastura
Mirador
Coapo
Novillero
Veinte
Mata Verde
Cacuite
Maculie
Sinán
Citam
Bolontikú
Kix
Yum
May
Xan
Caribe
1 San Diego
W. Chinaja
Rubelsanto
Yalpemech
Tunich
Tabay
Zazil-Ha
Maloob
Zaap
Bacob
Kutz
Lum
Bacab
Ku
Cantarell
Tartunich
Ixtoc
Luch
Batab
Nohoch
Och
Abkutun
Alux
Uech
Pich
Chac
Caan
Kax
Kay
Takin
Chem
235
190
315
185
330
6
21
100
0
100
200 Miles
100
0
100
200
300 Km
25°
20°
15°
100°
95°
90°

US
Texas
Nogales
Nogales
Cananea
Magdalena
El Paso
Ciudad Juarez
Rio Grande
Sonora
Hermosillo
Celulosa de Chuhuahua
Chihuahua
Chihuahua
Delicias
Guaymas
Gulf of California
Baja California
Coahuila
Ulua
Buena Suerte
Navojoa
Cuidad Camango
Jimenez
Topolobanpo
Durango
Gomez Palacio
Sinaloa
Culiacan
La Paz
MEXICO
Durango
Zacatecas
Mazatlan
San Luis Potosi
Zacatecas
Chicura
Canuch
Jayabo
Aguas.
Aguascalientes
Nayarit
Pacific
Ocean
San Blas
Tepic
Lagos
Guanajuato
Guadalajara
Salamanca
Puerto Vallarta
Lago de Chapala
245
Jalisco
Morelia
Colima
Manzanillo
Colima
Michoacan
Texoman
Lázaro Cárdenas
100 0 100 200 Miles
100 0 100 200 300 Km
MEXICO-WEST
30°
25°
20°
15°
115°
110°
115°

Gulf of Mexico
Straits of Florida
Great Bahama Bank
Santa Cruz
Canasi
Puerto Escondido
Yumuri
Seboruco
Varadero
La Luz
Boca de Jaruco
Guanabo
Tarara
Havana
122
Matanzas
Matanzas
Bacuranao
Motembo
2
Cabaiguan
76
Cienfuegos
CUBA
Cristales
Jatibonico
102
Santiago
30°
Yucatan
Cozumel
MEXICO
Quintana Roo
Chetumal
BELIZE
Belmopan
Gulf of Honduras
Caribbean Sea
JAMAICA
34
Kingston
15°
Puerto Cortes
HONDURAS
Tegucigalpa
22
San Salvador
La Union
NICARAGUA
20
Managua
Bluefields
Lake Nicaragua
10°
COSTA RICA
15
Limon
San Jose
Punta Cahuita
Panama City
Golfito
David
PANAMA
La Palma
100 0 100 200 Miles
100 0 100 200 300 KM
CENTRAL AMERICA
85°
80°
COLOMBIA

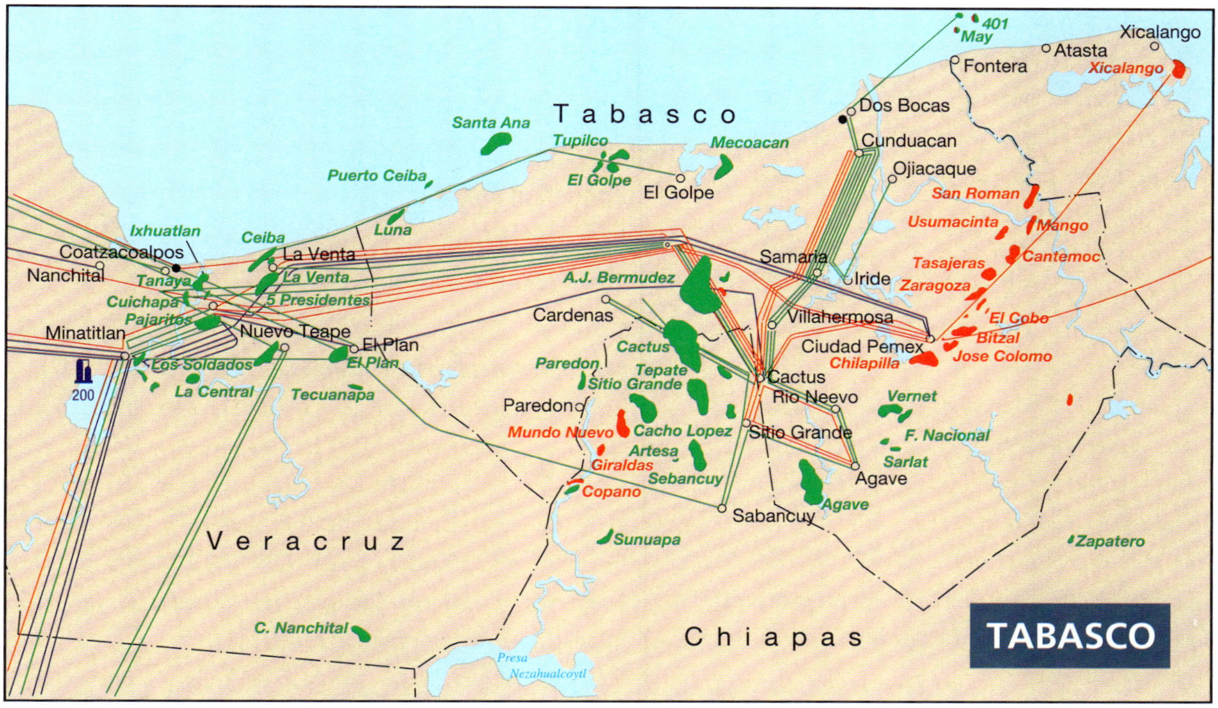

Pipelines

The extensive Pemex oil pipeline network connects major production centers with domestic refineries and export terminals. More than 453 pipelines span 2,900 miles, concentrated chiefly in the south.

Pemex also operates more than 5,700 miles of natural gas pipelines in Mexico. The company has 11 natural gas processing centers that produce NGLs, condensates, and LPG. Pemex also operates most of the country's natural gas distribution network, which includes 12 active connections with the United States.

TransCanada won a contract in 2005 from Mexico's Comision Federal de Electricidad (CFE) to build, own, and operate the 80-mile Tamazunchale Pipeline. The system will extend from the Pemex natural gas processing facility in Naranjos to a gas-fired power plant near Tamazunchale.

The pipeline will have an initial capacity of 170 MMcf/d, but the contract has an option to increase capacity to 430 MMcfd if CFE constructs more gas-fired power plants in the area. TransCanada planned to bring the project online by year-end 2006.

In 2006, CFE awarded a permit to U.S.-based Tidelands for the construction of the 1-bcfd Terranova Oriente bi-direction pipeline, which would connect a proposed storage facility to the U.S. and Mexican grids.

LNG

First terminal commissioned

Commissioning occurred in August on the first LNG terminal in Mexico, soon after arrival of its first LNG cargo. Terminal de LNG de Altamira S de RL de CV is a joint venture of Royal Dutch Shell PLC, Total SA, and Mitsui & Co. Ltd.

Located near Tampico in Mexico's Tamaulipas state, it will have an initial sendout capacity of 500 MMcfd, with possible expansion to 700 MMcfd, and the ability to accommodate ships of up to 200,000 cu m.

Developer Shell plans to import only enough LNG to meet its commitments to Mexican utility Comisión Federal de Electricidad, which has a 15-year contract to receive 5.2 billion cu m/year for power generation. Deliveries were expected to start in October 2006.

Shell, Total, and Mitsui share interests in the terminal. Subsequent LNG cargoes will be delivered by both Shell and Total.

Two Baja terminals

Construction was well under way on Sempra LNG's Energia Costa Azul LNG terminal on the coast of Baja California. The facility was said to be on schedule to begin operation by early

2008. Located 14 miles north of Enseñada, the facility will have an initial sendout capacity of 1 bcf/d. Sempra is evaluating potential interest in expansion to 1.5 bcf/d.

The company has an agreement with BP and Tangguh LNG to provide natural gas from Indonesia. Tankers will dock offshore and unload LNG that will be piped to onshore storage tanks. Gas from the terminal will be delivered to power plants and other industries in the region through a new 45-mile (72-km) pipeline that will be interconnected with existing pipelines.

Also off Baja California, Chevron de México is planning to install a concrete gravity-based LNG import terminal. Front-end engineering work for the 700-MMcfd Terminal GNL Mar Adentro de Baja California was completed some time ago and major Mexican regulatory approvals have been obtained. Chevron said that final design and bid preparation work is being deferred to be line with the timing of securing gas supply agreements.

Gas from the terminal would flow into the existing Baja California pipeline system and be available for distribution to Baja California and the U.S. West Coast. As of November, Chevron de México had not announced any supply contracts.

Possible future plants

Several other LNG plants have been proposed for Mexico. One would be an offshore LNG receiving terminal near the Coronado Islands. Federal officials have approved the project, but there has been objection from local residents and environmental activists.

Both Repsol-YPF and a subsidiary of Suez have described plans to build an LNG terminal at Lazaro Cardenas, in Michoacan state. And other terminals have been proposed for construction at Puerto Libertad, on the Gulf of California, and at the port of Manzanillo.

NICARAGUA

Capital: Managua
Monetary unit: Córdoba
Refining capacity: 20,000 b/cd
Oil production: 14,300 b/d
Oil reserves: N/A
Gas reserves: N/A

U.S. firm to explore offshore in Caribbean

Nicaragua signed two oil and natural gas exploration and production contracts with MKJ Exploraciones, a unit of MKJ Xploration (Louisiana). The six-year concession contracts are for areas covering a combined 8,030 sq km (3,100 sq miles) off the country's Caribbean coast. One is 120 km (75 miles) east-southeast of Puerto Cabezas, the other 169 km (105 miles) northeast of the town of Bluefields.

MKJ will be granted a 30-year production period if it discovers oil or natural gas deposits in commercially marketable quantities. The contract also requires MKJ to present an acceptable environmental impact study to the government before starting operations.

Seismic survey in Pacific Oklanicsa concession

Norwood Resources Ltd. (Vancouver, BC Canada) was acquiring a large 2D seismic survey in early 2006 on the northern part of the Oklanicsa concession along Nicaragua's Pacific Coast, southwest of Managua. The company plans to use the data to confirm the presence of several large structures identified with geological surface mapping.

The concession extends to the border with Costa Rica and covers 845,780 acres on the eastern flank of the Sandino basin. Norwood Resources' initial interest was in potential reservoirs in the Oligo-Miocene Masachapa and Middle Eocene Brito formations in the San Cayetano area. Industrias Oklahoma-Nicaragua is the operator.

Ortega elected president

In November, Daniel Ortega was elected president of Nicaragua. During a civil war in the country in the 1970s, he was a Marxist leader in the Sandinista movement that eventually ousted President Anastasio Somoza Debayle in 1979. He later was elected president, serving from 1985–1990.

Ortega now has taken a more moderate center-left stance and has said he is ready to work with former domestic and international opponents to improve Nicaragua's economic situation. He describes himself as a friend and ally of Venezuela's President Hugo Chavez, who is challenging U.S. influence across Latin America and using petrodollars to support his allies.

PERU

Capital: Lima
Monetary unit: New sol
Refining capacity: 192,950 b/cd
Oil production: 113,000 b/d
Oil reserves: 929.6 million bbl
Gas reserves: 8.723 tcf

Background

Peru has achieved a strong economic performance in recent years, according to the U.S. Energy Information Administration (EIA). GDP grew by 6.7% in 2005, and analysts predicted a 5.1% rate for 2006. The country has controlled inflation and has achieved a positive trade surplus.

Peru is a member of the Andean Community, set up in 1996 by leaders of Bolivia, Colombia, Ecuador, Peru, and Venezuela. One goal of the Community is integration of its members' energy sectors, particularly electricity and natural gas.

Oil

Peru's oil production has increased since 2004 as new projects have come online. The country produced 15% more oil and NGLs in 2005 compared to 2004. However, Peru's oil consumption has also grown over the past 20 years, and the country has been a net importer since 1992, chiefly from Ecuador and other South American countries.

The largest oil producer in Peru is Argentina-based Pluspetrol, which controls over one-half of the country's crude oil production. Other major producers include Occidental Petroleum, Petrobras, and Petro-Tech Peruana.

Peru privatized most of the former state-owned oil company, Petroperu, in 1993. However, Petroperu continues to control the country's only crude oil pipeline, most refineries, and a majority of the retail oil products market.

Peru's crude oil production is concentrated in the northern part of the country. The largest oil blocks are Block 1-AB (Pluspetrol) along the border with Ecuador, Block 8 (Pluspetrol) in the northeastern Amazon region, Block X (Petrobras) in the northwest, and Block Z-2B (Petro-Tech Peruana) off the northwest coast. Blocks 1-AB and 8 account for more than 65% of Peru's total crude oil production. Most of crude oil produced in Peru is heavy and sour, with 1.2% sulfur content.

Natural gas

According to *OGJ,* Peru has proven natural gas reserves of 8.7 tcf, the fifth-largest amount in South America. However, Peru's Ministry of Mines and Energy has indicated that once seismic work is complete on Block 56 (part of the Camisea project, described below), proven reserves could increase to 15–16 tcf.

In 2005, Peru produced approximately 7 bcf of natural gas, up 84% from 2004. In coming the future, Peru likely will become a net exporter of natural gas as the Camisea project comes onstream.

Besides Camisea, the largest concentrations of gas production include the Aguaytia gas field (Maple Gas) in central Peru, Block X (Petrobras) in the northwest region, and Block Z-2B (Petro-Tech) located off the northwest coast.

Exploration and Development

Block XX

In 2005, the Peruvian government signed 15 new exploration and production contracts and aimed to sign 10 more by mid-2006. The first 2006 contract went to Peru's Petrolera Monterrico for Block XX, located in the northern part of the country.

Block Z-34

Independents Gold Oil PLC (Surrey, UK) and Plectrum Petroleum PLC (Aberdeen, Scotland) plan to explore Block Z-34 in the Talara basin. The 3,713 sq-km (1,434 sq-mile) block lies in 100 to 3,000 m (328 to 9,842 ft) of water. The initial work will acquire a modern conventional seismic grid and conduct marine magnetotelluric and electromagnetic studies.

Block Z-2B

The Sechura basin saw its first substantial oil discovery with a well drilled by the Peruvian subsidiary of Petro-Tech International (Houston). The San Pedro 1X well, in the southwestern extremity of Block Z-2B, tested at 1,200 b/d of light oil. The well, on the shelf off Paita, was projected to a TD of 2,255 m (7,400 ft) in 75 m (250 ft) of water.

Block 64

Peru's Ministry of Energy and Mines reported in early 2006 that Occidental Petroleum Corp. and two partners made a light oil discovery in the western Maranon basin of north-central Peru. The discovery is the Situche Central 2X well on Block 64. The discovery flowed 2,400 b/d. Occidental is the operator.

Block 39

The ministry also reported a heavy oil discovery on Maranon Block 39 about 14 km (9 miles) south of the Barrett Resources Pirana find. The Buena Vista 1X well tested at 2,830 b/d. Repsol YPF (Madrid) operates Block 39.

Camisea project

The Camisea project consists of several gas fields in the Ucayali basin of southeastern Peru, principally in Block 88 in the Amazon region, along the Camisea River. Analysts have estimated that Block 88 contains 11 tcf of proven plus probable natural gas reserves and 482 million bbl of NGLs.

An international consortium led by Hunt Oil developed the upstream portion of Camisea, and production began in August 2004. The initial production capacity was 450 MMcf/d of natural gas and 34,000 bbl/d of NGL. However, output capacity was expected to increase once drilling began in May 2006 on Camisea's Block 56, adjacent to Block 88.

Block Z-1

BPZ Energy Inc. (Houston) spudded the first well from the Corvina platform on Block Z-1 in the Pacific Ocean off northwestern Peru. Corvina field is estimated to have 4 tcf of gas in place.

The company also secured all environmental permits required to develop the Corvina gas-to-power project. The permits cover the 16-km (10-mile) gas pipeline to shore, onshore gas processing facilities, and a 160-MW thermal power plant to be built at the coastal village of Nueva Esperanza.

Block 121

Barrett Resources (part of Williams Companies, Tulsa, OK U.S.A.) agreed to sign a contract with Perupetro for the exploration of Block 121 in Loreto department. The seven-year exploration phase of the block will include 2D seismic studies and the drilling of exploratory wells. The block is in the Maranon basin beside Blocks 39 and 67, where other companies have found heavy crude reserves.

Drilling and Production

Block 56, Pagoreni field

The consortium that operates the large Camisea project reported drilling the first of six production wells in the Pagoreni gas-condensate field on Block 56, which holds reserves of 3 tcf. The group expects to invest $600–700 million in the development.

Pluspetrol, operator for the consortium, estimated that the field will produce about 600 MMcfd of natural gas, which will be transported to the Malvinas processing plant and then to the Peru liquefaction facility.

Blocks Z-1 and XXIII

BPZ Energy Inc. (Houston) made plans to spud the first of three wells on Block Z-1 in Corvina gas field off northwestern Peru. The project entails drilling three initial wells, recompleting a shut-in gas well, and constructing a 10-mile (16-km) pipeline to transport gas to a planned 160-MW gas-fired power plant.

BPZ has secured a 2-year rig contract from Petrex SA, a subsidiary of Saipem SPA. BPZ is also negotiating with state-owned Perupetrol for an exploration and development license to work in onshore Block XXIII.

Block 100

Radial Energy Inc. (Houston) sent an Ingersoll Rand drilling rig and drilling equipment to Peru and planned to spud a well in September. The Block 100 prospect is in the Ucayali basin. Ziegler-Peru will be the contract driller and operator. The concession holder, Cia. Consultora de Petroleo, owns the majority interest.

Blocks 122, 128

Gran Tierra Energy Inc. (Calgary) announced plans to explore Blocks 122 and 128 in the Iquitos Arch Trend of the Maranon basin in Peru. The company is operator with 100% working interest. The two blocks total 3.2 million acres.

Pipelines

Petroperu operates the country's sole crude oil pipeline, Norperuano, which links the export terminal at Bayovar to oil fields in Peru's interior. Norperuano has a maximum capacity of 250,000 b/d, but the pipeline currently transports about 80,000 bbl/d.

Transportadora de Gas del Peru (TGP), a consortium led by Techint, built and now operates parallel natural gas and NGL pipelines that carry Camisea production to Lima and to a fractionation plant in Paracas.

In late 2005, talks about a new pipeline linking the Camisea project with northern Chile stalled over a maritime border dispute. If talks resume, the project could be an alternative to using LNG for gas trade between the two countries. Chile gets most of its natural gas imports from Argentina, but gas shortages in Argentina have disrupted supply in recent years. The project would depend on availability of surplus gas from Camisea.

Refining

Peru has six major oil refineries. Repsol-YPF controls the largest, La Pampilla (Lima) with a capacity of 102,000 b/d. The other privately operated refinery is the 3,250-b/d Purcallpa, operated by Maple Gas. Petroperu operates the remaining four refineries. Plans to further privatize downstream facilities have been delayed due to opposition from labor unions and legislators.

LNG

Because gas production from the Camisea project is expected eventually to exceed domestic demand, project sponsors have made plans to export excess production.

Peru LNG Co., a consortium led by Hunt Oil Co. (Dallas, TX U.S.A., signed an agreement with the Peruvian government to export LNG from Peru to Mexico and the western United States. The companies plan to build and operate an LNG plant and export terminal at Pampa Melchorita, about 168 km (105 miles) south of Lima.

The plant will convert 620 MMcfd of gas to LNG and export 4.2 million t/yr. The venture also will lay a 400-km (248-mile) pipeline to transport feed gas from the Ayacucho Mountains to the LNG plant. Work began in 2006, aiming for completion in 2009.

Peru LNG also has talked with ENAP, the state-owned oil company of Chile, about exporting LNG to that country. Although the countries share a land border, the LNG option could be more cost-effective than building a pipeline.

There also has been speculation about a potential LNG partnership with Bolivia, according to EIA. Bolivia has the second-largest natural gas reserves in Latin America but lacks the coastal access needed to pursue LNG exports. One proposal would connect the Margarita gas fields in southern Bolivia to the Peruvian port of Ilo. However, no firm plans had been made as of late 2006.

TRINIDAD & TOBAGO

Capital: Port-of-Spain
Monetary unit: Dollar
Refining capacity: 175,000 b/cd
Oil production: 145,000 b/d
Oil reserves: 990 million bbl
Gas reserves: 25.88 tcf

Exploration and Drilling

Eight blocks draw bids

Nine companies entered bids for eight of the 11 onshore and near-shore blocks offered by the government. The bid round was completed after more than a year's delay caused by the introduction of a new production-sharing contract. Among companies bidding for blocks were Australia's Hardman Oil & Gas Pty. Ltd. and the Indian partnership of Oil & Natural Gas Corp. and Mittal.

Ruby appraisal well

BHP Billiton and its partners drilled an appraisal well on Block 3(a) off the northeast coast. On test, Ruby-1 produced 5,000 b/d of oil through a 7/8-in. choke. The well, drilled to TD 1,750 m (5,750 ft), found 365 m (1,200 ft) of hydrocarbon-bearing sands, including more than 245 m (800 ft) of net pay.

BHP Billiton also confirmed it will spud the Ruby-2 well on Block 3(a) and finish drilling by first-quarter 2007. Ruby-2 is being drilled in 655 m (200 ft) of water.

Blocks 22, 1a, 1b

Petro-Canada signed a contract for Diamond Offshore Drilling Inc.'s Ocean Worker semisubmersible, which will start a six-month, multiwell program on Block 22 in the second half of 2007. Also at that time, the Rowan Gorilla III jack up is scheduled to arrive on the Caribbean island for a 6-month, four-well program on Blocks 1a and 1b.

Cannonball flows gas

BP Trinidad & Tobago said the $250 million Cannonball platform started gas production in March. The unmanned platform is 35 km (22 miles) off Galeota Point, Mayaro, in 70 m (231 ft) of water. Full gas production is expected to reach about 800 MMcfd.

Refining

Refinery upgrades

State-owned Petroleum Company of Trinidad and Tobago Ltd. (Petrotrin) launched a 3-year gasoline optimization program at its 175,000 b/cd Pointe-a-Pierre refinery in South Trinidad. The program will build or revamp eight major plants, plus utilities and supporting units.

Four of the plants will use proprietary technology from UOP LLC (Des Plaines, IL U.S.A.) for reforming naphtha to produce high-octane fuel, producing clean-burning gasoline, and cleaning LPG. It will also revamp a fluid catalytic cracker.

LNG

New rules for future LNG expansions

The government of Trinidad and Tobago in January told oil and gas companies wanting to be part of future LNG expansions that they will have to commit some of their gas for domestic use in that country. The move is driven by increasing concern about the availability of gas for domestic consumption, including for methanol, ammonia, urea, and aluminum projects.

Petrochemicals

Ethylene project studied

Westlake Chemical Corp. signed a memorandum of understanding with the government for construction of a $1.5 billion ethane-based ethylene, polyethylene, and derivatives complex. Construction is to start in late 2007 with operation in 2010.

The facility would produce 570,000 tonnes/year of ethylene from 37,500 b/d of ethane supplied by National Gas Co. of Trinidad & Tobago Ltd. (NGC), which is participating in project evaluation along with National Energy Corp. of Trinidad & Tobago Ltd. The government has expressed an interest in becoming a minority shareholder.

VENEZUELA

Capital: Caracas
Monetary unit: Bolivar
Refining capacity: 1,282,100 b/cd
Oil production: 2.563 million b/d
Oil reserves: 80.012 billion bbl
Gas reserves: 152,380 tcf

Taking Control of the Energy Sector

During 2006, Venezuela's president Hugo Chavez continued actions (begun in 2001) to take greater control of oil-industry assets and operations in the country and to increase the oil-revenue share of Petroleos de Venezuela SA (PdVSA), the state oil company.

By December 31, 2005, all service contracts originally issued to private operators for the production of oil in Venezuela were to be converted to new joint ventures with PdVSA in which that company could take as much as a 90% stake.

By the end of first quarter 2006, PdVSA held an average 60% interest in all of the 32 new joint venture agreements, which represented about 532,000 b/d of oil production out of the country's declared production of 3.2 million b/d.

In addition, PdVSA raised royalties and took back acreage (more than 27,000 sq km in all) where it said companies had failed to invest. PdVSA said companies were compensated a total of $900 million in investment bonds that can be used in the new joint ventures under a program still being finalized.

Regarding taxes, late in 2005 the government had passed a resolution increasing income taxes to 50% from 34% on 4 heavy-oil projects in Venezuela's Orinoco River basin, affecting the operations of ExxonMobil Corp., Chevron, ConocoPhillips, BP PLC, Total SA, and Statoil.

Then in 2006, Venezuela's tax authority Seniat launched a campaign to collect back taxes it said were owed by international operators in that country for the period 2001–04. The Venezuelan government ruled that the tax rate for that period should be 50% rather than 36%. Seniat said the total amount owed by 22 companies it evaluated during the first quarter was $891 million.

In January 2006, the government also passed a resolution limiting deductible expenses on natural gas production to 15% of the wellhead price. This contradicted previous license terms and left various international companies facing new costs.

International Collaboration

Expanding oil exports, building tankers

Venezuela said it plans to spend $56 billion through 2012 to increase the country's current crude production capacity of 3.3 million b/d to 5.18 million b/d.

In parallel, PdVSA wants to spend as much as $2.2 billion on tankers through 2012 that would enable it to send more oil to Asia. Venezuela had 16 tankers as of mid-2006, and PdVSA wants to buy 42 more. At least 14 of the new vessels would be supertankers, said President Hugo Chavez. In August, PdVSA signed a memorandum of understanding with Brazilian construction company Andrade Gutierrez SA to build a shipyard in Venezuela.

Venezuela signed new supply agreements with China, India, Jamaica, Haiti, Paraguay, and Bolivia. Because there is a lack of investment in new production, the country's output is not growing, so supplies to those countries come at the expense of exports to the United States.

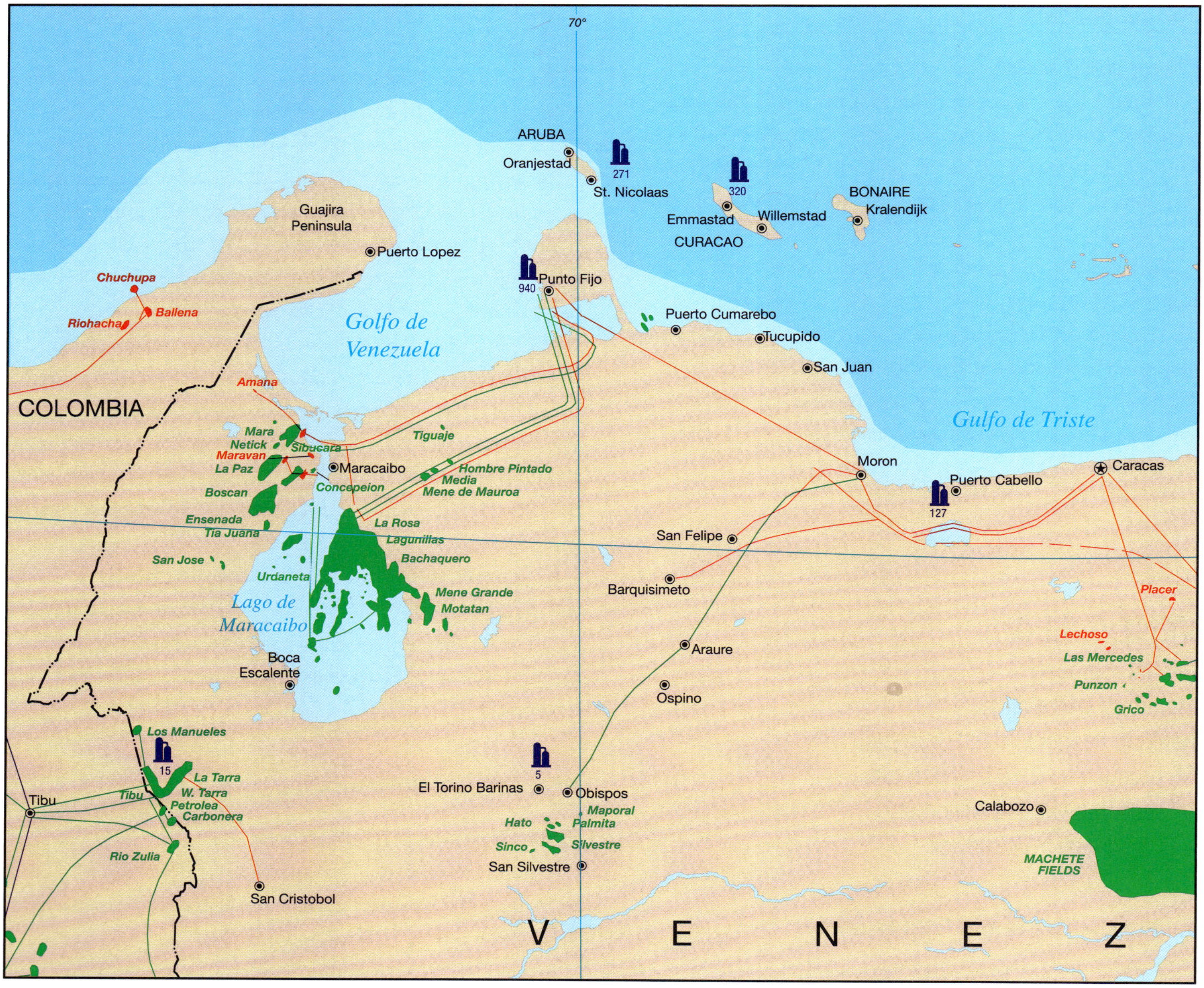

Oil sales to China

Venezuela will cooperate with China in supplying crude oil to help fulfill the Asian giant's energy demands, President Hugo Chavez said during an August 2006 visit to China.

Venezuela shipped only 14,000 b/d to China in 2004. However, in July 2006 exports reached about 155,000 b/d, and President Hugo Chavez said in August that he hopes to double that figure in 2007 and to reach 500,000 b/d in five years.

In 2005 the China National Petroleum Corporation (CNPC) and PdVSA announced plans to work jointly to boost production in eastern Venezuela from the Junin 4 block in the Orinoco belt and from the mature Zumano field.

Cooperation with India

ONGC Videsh Ltd. (OVL), the overseas investment subsidiary of India's state-owned Oil & Natural Gas Corp. (ONGC), and PdVSA met in New Delhi in August to discuss mutual energy cooperation. Indications are that OVL may seek exploration opportunities in Venezuela jointly with other international oil and gas companies. Both sides agreed to establish a group to study energy collaboration.

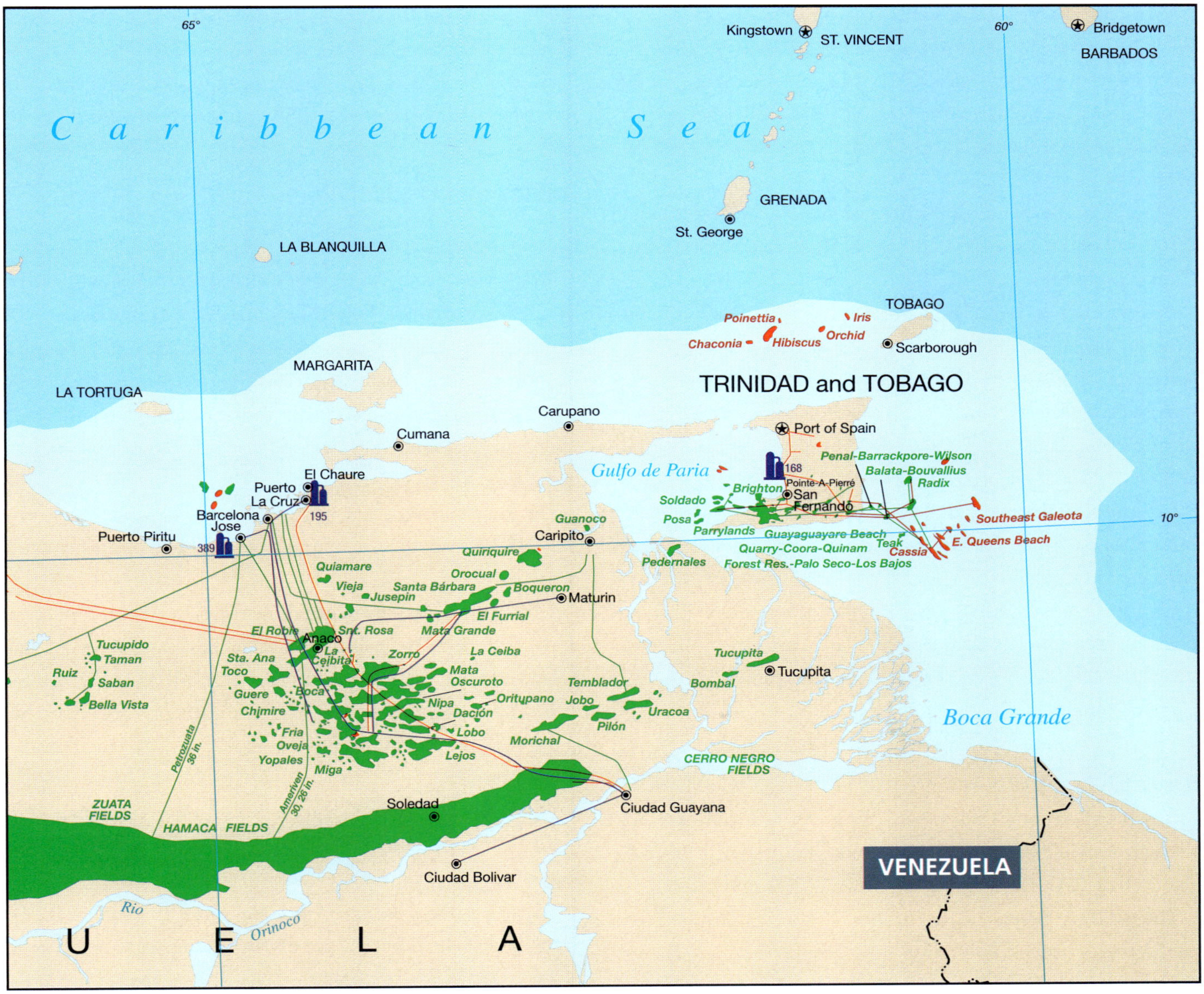

Exploration and Development

Gas discovery off eastern coast

Statoil ASA found dry gas within three intervals drilled in its Cocuina-2X exploration well on Block 4 of Plataforma Deltana, off eastern Venezuela. The well, 240 km from the Orinoco Delta, was drilled to 3,406 m TD. Statoil said it plans to reenter the Ballena 1-X well upon completion of the Cocuina well and to drill at the Orca-1X well in 2007.

The Venezuelan government awarded the Plataforma Deltana Block 4 license to Statoil in 2003, which has partnered with Total SA for development. PdVSA Gas has the option to participate once a commercial discovery has been declared.

Falcon basin: PetroCumarebo

PetroFalcon Corp. (Carpinteria, CA, U.S.A.) said its PetroCumarebo joint venture with PdVSA aims to accelerate activity on the East and West Falcon blocks in northern Venezuela.

PetroFalcon's Venezuela unit, Vinccler Oil & Gas CA, and PdVSA-Corporación Venezuelana del Petróleo (CVP) signed the joint venture conversion agreement in September, giving PetroCumarebo rights to the two blocks for 20 years. Gross production from the two blocks totals 1,200 b/d of oil and 12 MMcfd of gas.

Vinccler was expanding facilities to handle 20 MMcfd and 1,700 b/d at La Vela field and 30 MMcfd and 5,000 b/d at Cumarebo field.

Gas production began in late 2005 from La Vela field and Vinccler began delivering an initial 5 MMcfd into PdVSA's Interconnection Centro Occidente pipeline, which serves the Paraguana Refinery. Vinccler was planning to boost deliveries during 2006. It has reactivated the LV-6X, 7X, and 8X wells and drilled the 9 and 10X wells.

In August, Vinccler and PdVSA started producing 10 MMcfd of gas from Cumarebo field, also introducing it into the Centro Occidente pipeline.

On the West Falcon block, PetroCumarebo said it plans to develop Los Moroches oil and gas field. Vinccler Venezuela also delineated several structural prospects around Cumarebo field.

Three projects: Petrobras-PdVSA

Brazil's Petrobras started drilling for natural gas in Venezuela's Cojedes state under a joint venture with PdVSA. The well is at La Blanca, near the town of Romulo Gallegos.

The well is being drilled under a 2005 agreement to cooperate in three exploration and production projects in Venezuela, including the offshore Mariscal Sucre gas project, a project in the Orinoco Belt, and this project involving drilling in five old oil fields—Adas, La Paz, Lido, Limon, and Nieblas.

Regarding the Orinoco project, Petrobras is to appraise and certify reserves in the Carabobo heavy oil field in preparation for development. Production of 150,000 b/d of oil is expected.

Regarding Mariscal Sucre, the companies said they would form a 35–65% joint venture by yearend 2006 to develop the project. PdVSA will invest $3 billion and Petrobras, $2 billion, to bring on production the four Norte de Paria fields that comprise the Mariscal Sucre project—Dragon, Mejillones, Patao, and Rio Caribe.

The fields, off Venezuela's northeastern coast, are believed to hold 11 tcf in gas reserves. Petrobras said gas from Mariscal Sucre will be sold to Venezuela beginning in 2009, with surplus production exported as LNG, mostly to Brazil.

Drilling and Production

Boost in oil production expected

Production from Venezuela's marginal fields will surge by 2009, predicted Wood Mackenzie Ltd. (Edinburgh) in a report on Venezuela's new *empresas mixtas* (mixed companies, or joint venture companies). The country formed these new joint venture companies with international oil companies after renegotiating contracts under which the companies had been working.

The mixed companies operate the assets, and state company PdVSA takes a majority stake in each, typically 60% or larger.

After two years of reduced investment and production declines during "the painful process of rewriting the marginal-field contracts," a production boost is in prospect, WoodMac said. "By 2009, production from the former marginal fields could rise by 125,000 b/d or more, a 34% increase on today's levels."

These figures exclude production from seven marginal fields relinquished to PdVSA by participants who chose not to convert their contracts.

The new contracts extend license terms by as much as 14 years, which should allow previously untapped oil reserves to be exploited, said WoodMac.

Drilling pace: rapid

In August, PdVSA said it had new, five-year contracts for 27 rigs in a $2 billion drilling program that will create 10,800 new jobs. PdVSA sourced the rigs from 13 companies, predominantly Venezuelan. The government has said it wants to double oil production by 2012.

Regarding land rigs, Baker Hughes reported 71 in Venezuela in July, compared to 56 rigs a year earlier and 32 in July 2003. Regarding offshore rigs, Rigzone reported that 27 (23 drill barges; 3 jack ups; 1 semisub) were drilling off Venezuela in August.

The Venezuelan oil ministry is expected to award new Orinoco exploration and production licenses in early 2007 that likely will lead to additional drilling.

JV for Petrobras, Venezuelan oil firms

Two Venezuelan oil companies signed contracts with Brazil's Petrobras to jointly exploit oil deposits in La Concepcion oil field in Venezuela. The firms are PdVSA unit Companhia Venezuelana de Petroleo (CVP) and privately owned Williams International Oil & Gas Ltd.

PdVSA said the new joint venture, Petroway, would aim to increase oil extraction from La Concepcion field in the Western state of Zulia. The field produces 12,300 b/d of oil.

Refining and Processing

Three new refineries studied

PdVSA is considering spending $10.5 billion to build refineries in Cabruta, Caripito, and Barinas with total capacities of 700,000 b/d. The company said that the Venezuelan and Cuban governments also are working to reactivate Cuba's Cienfuegos refinery, which has a processing capacity of 70,000 b/d.

Venezuela also is considering processing 50,000 b/d from the eastern Franja de Orinoco region at Uruguay's La Teja refinery, as well as building new refineries in Brazil through its alliance with Petrobras.

IEA reclassifies Venezuelan heavy crude

The International Energy Agency (IEA; Paris) has reclassified Venezuelan extra-heavy oil in its reports of worldwide oil production.

Since 2002, IEA had included Venezuela's 17°–32°-gravity synthetic crude—which is upgraded from 8°–9°-gravity Faja crude produced in the Orinoco Belt—in its category for OPEC NGL and nonconventional oil.

Beginning in its March 2006 *Monthly Oil Market Report*, IEA includes Venezuelan heavy crude in its "OPEC crude" category. The change shifts about 600,000 b/d of production into a category deemed subject to Venezuela's OPEC quota.

Venezuelan synthetic crude comes from four upgraders operated by joint ventures involving PdVSA and private partners. Under the new method, IEA will report as production the heavy oil feedstock charged to upgraders and not synthetic crude output, which runs about 10% below feedstock charge.

In the old method, IEA put the synthetic crude in the OPEC NGL and unconventional category and treated heavy oil not upgraded as conventional crude. The change slightly increases total Venezuelan oil production.

Venezuela includes about 34 billion bbl of Faha extra-heavy crude in its reserves estimate of 78 billion bbl.

About 160,000 b/d of Orimulsion, a mixture of heavy oil, water, and a surfactant sold as a boiler fuel in competition with coal, will continue to be counted as OPEC NGL and unconventional oil.

Some U.S. Citgo outlets closing

PdVSA suspended sales to 2,000 U.S. gasoline stations that operate under Citgo's brand. Citgo, a subsidiary of PdVSA, must purchase 130,000 b/d of oil from other refiners to distribute to its 13,000 U.S. retail outlets, which it doesn't own. Citgo said "it is no longer attractive" to maintain distribution of oil products to its outlets in several American states.

Citgo made the decision to interrupt part of its distribution after Venezuelan President Hugo Chavez claimed that the PdVSA subsidiary's businesses indirectly subsidize the U.S. economy.

Citgo said it would withdraw by first quarter 2007 from Iowa, Kansas, Kentucky, Minnesota, Missouri, Nebraska, North Dakota, Ohio, Oklahoma, and South Dakota and would reduce sales in Illinois, Texas, Arkansas, and Iowa.

Pipelines

Caribbean-Pacific oil pipeline plan

PdVSA announced plans for a pipeline through Nicaragua to transport oil from the Caribbean to the Pacific, bypassing the Panama Canal. Officials of the Nicaraguan-Venezuelan oil firm Alba Petroleos de Nicaragua (Albanic) confirmed the plans.

The plans are part of an accord to be signed by Venezuelan Pres. Hugo Chavez and Nicaraguan Pres. Daniel Ortega in 2007. In May 2006, Ortega and Chavez signed a cooperation initiative enabling Venezuela to sell oil to Nicaragua on credit.

The pipeline would be used to boost the export of oil products made in Nicaragua from Venezuelan crude to China and Japan as well as the Pacific Coast of Central America.

Albanic officials also mentioned PdVSA's additional interest in building a refinery in Nicaragua to supply Central America with products. Nicaragua's only refinery is Esso Caribbean & Central America's 20,000 b/d facility in Managua.

The creation of Albanic is widely seen as a sign of Pres. Ortega's growing political alliance with Pres. Chavez. ALBA is the Bolivarian Alternative for the Americas, an initiative started in 2005 by Cuba and Venezuela that aims to counter U.S. efforts to promote hemispheric free-trade integration.

Break shuts Venezuela's ICO gas line

Venezuela's Interconnection Centro Occidente natural gas pipeline experienced a rupture at a joint in early December. Repairs on the shut-in line were expected to take until February 2007. The 36-in line carries gas 70 km from the East Falcon fields to the Paraguana Peninsula refining complex.

Venezuela-Argentina gas line

The presidents of Venezuela, Argentina, and Brazil signed an accord in December 2005 for the construction of an 8,000–10,000 km pipeline to transport natural gas from Venezuela to Argentina through Brazil and Uruguay. The line would cost $10–17 billion and could take five years to construct.

PdVSA, the Andean Development Corp., and Brazil's National Economic and Social Development Bank have signed an agreement to work jointly on financing the project. Other parties also will be invited to participate, said Brazil's Ministry of Mines and Energy.

The Venezuela-Argentina gas line is part of a broad energy integration agreement ratified by Venezuela, Argentina, and Brazil in May 2005.

According to initial estimates, 100–150 million cu m/day of gas would arrive in Brazil at a cost of $2.60/MMbtu, compared with the $3.60/MMbtu that Brazil pays for gas from Bolivia.

Analysts believe that in 20 years, when existing pipelines are interlinked, the network of pipelines would tap into South America's most important gas reserves in Venezuela (about 150 tcf), Bolivia (28.7 tcf), Argentina (23.4 tcf), Brazil (9 tcf), and Peru (6.8 tcf).

The next phase of the project includes a profile of markets in Brazil and in the Southern Cone of Latin America, coupled with additional technical and financial analysis. The largest gas South American reserves are in northern areas, but 70% of the continent's 365 million people live in the south (Brazil, Argentina, Chile, and Uruguay), where the most developed production and pipeline infrastructure are located.

The exact pipeline route has not been defined, but the following estimates have been made.

The first section—from Puerto Ordaz, Venezuela, to Marabá, Brazil—calls for a 66-in. pipeline 2,950 km long. Gas would arrive in Brazil through Rondonia state. The pipeline would extend south to Manaus and continue toward the Amapá state capital, Marabá, where it would branch.

In Manaus, the pipeline could be connected to a system that Petrobras is constructing to transport gas produced at Urucu in the Solimões basin. In Marabá, a 1,387-km segment would branch to Fortaleza.

In Fortaleza the pipeline could be connected to the existing system that extends along the coast to Salvador. From Salvador, Petrobras is building the Gasene gas line to connect Brazil's north and south systems. Petrobras is investing $3.9 billion by 2009 to build 4,191 km of pipelines.

From Marabá the transmission line would continue south in a 1,977 km stretch of 54-in. pipe to São Paulo state. It would distribute 42 million cu m/day in São Paulo.

From São Paulo the pipeline would be routed to the border between Rio Grande do Sul state and Uruguay, then cross Uruguay to Argentina in a 1,875-km section of 38-in. pipeline. This section would transport an estimated 50 million cu m/d.

Heavy Oil

Iran to aid in Orinoco projects

Venezuelan and Iranian officials in September signed a series of accords, including deals to develop oil fields, build factories, and set up a $2 billion investment fund to finance the projects. The agreement is part of Venezuela's 2006–2030 heavy-oil development plan centered within the Orinoco Oil Belt, south of Guárico, Anzoátegui, and Monagas states.

PdVSA estimates that the 27 blocks in the four areas (Carobobo, Ayacucho, Junin, and Boyaca) of the Orinoco Belt area contains 235 billion bbl of crude. PdVSA was producing 620,000 b/d of oil in mid-2006 from the Orinoco, working with ExxonMobil Corp., BP PLC, ConocoPhillips, Total SA, Statoil ASA, and Chevron Corp.

September saw the initial drilling of a certification well (MFK-4E) in an Anzoategui State field on Ayacucho Block 7 by the Iranian company, Petropars Ltd. Petropars was granted a 49% interest in the 540-sq-km heavy oil block, and PdVSA retained 51%. Initial studies indicated 31.2 million bbl of original oil in place, of which Petropars expects to recover at least 20%.

Petropars joins IOCs from Uruguay, Russia, Spain, Brazil, Argentina, China, India, and Belarus that are working with PdVSA.

ORINOCO BLOCKS

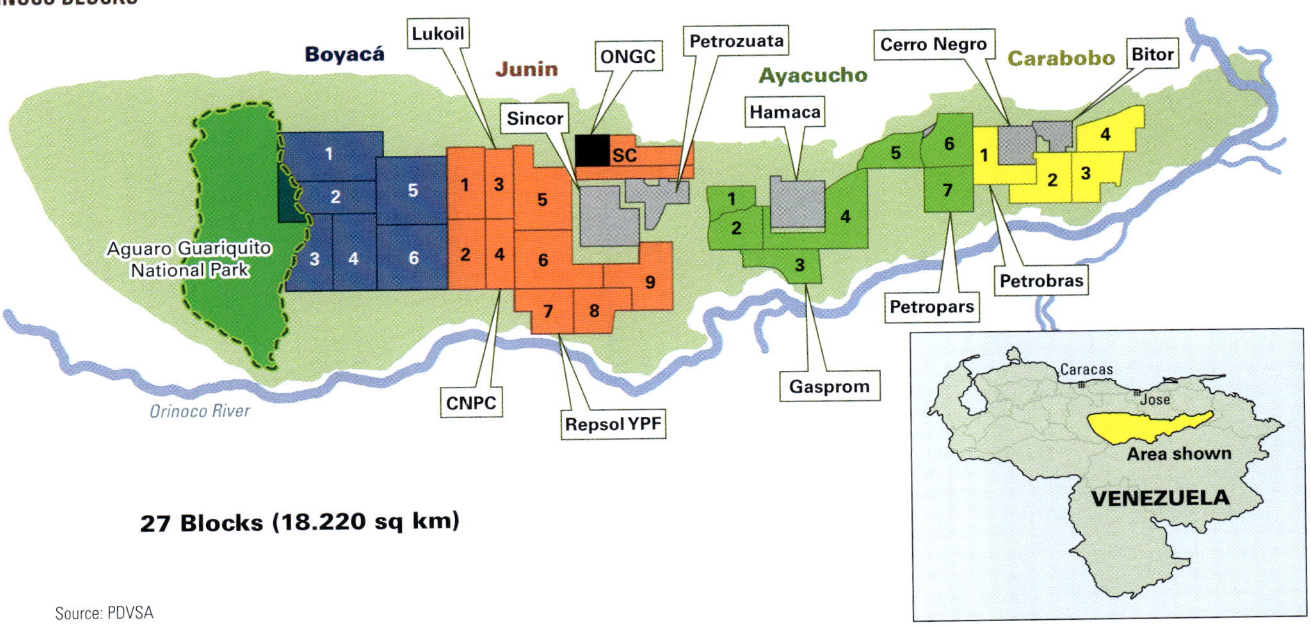

Source: PDVSA

Europe

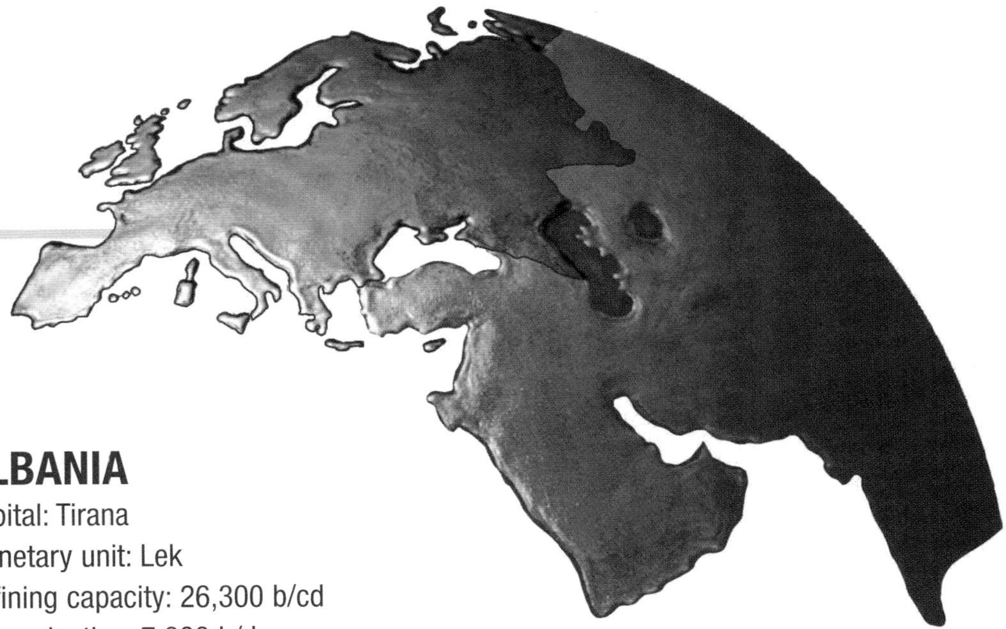

ALBANIA

Capital: Tirana
Monetary unit: Lek
Refining capacity: 26,300 b/cd
Oil production: 7,000 b/d
Oil reserves: 198.13 million bbl
Gas reserves: 30 bcf

Background

The countries of the Balkans region are neither major energy producers nor consumers, according to the U.S. Energy Information Administration (EIA). Although the region holds some important hydrocarbon deposits, these resources are not judged significant on a world scale.

In addition, political and economic instability in the region in recent years has discouraged substantial foreign investment in resource development. Rather, the region has become more important as a transit center for exports of oil from Russia and the Caspian Sea to Western markets.

Oil exploration activity is picking up, especially in Albania and offshore areas in the Adriatic Sea, but oil production in the Balkans remains low. In 2005, the entire region produced only 41,000 b/d, said EIA. As a result, each of the Balkan countries remains a net oil importer.

EIA estimated that Balkan oil consumption in 2005 totaled 237,000 b/d, leaving the region dependent on imports for roughly 80% of its oil needs. The region is well supplied with oil, however, through pipeline links with Russia and Greece as well as port facilities on the Adriatic Sea.

Privatization efforts are continuing in each of the Balkan region's three oil producers, Albania, Croatia, and Serbia and Montenegro. Albania had announced privatization plans in 2004 for the state-owned oil company, ARMO. Several firms expressed interest, but the schedule for the potential sale of shares is still unclear.

EUROPE
0 50 100 150 Miles
0 100 200 Kilometers
North Sea
Irish Sea
Celtic Sea
Bristol Channel
Lyme Bay
Bay of Biscay
Mediterranean Sea
Gulf of Cadiz
IRELAND
N. IRELAND
ENGLAND
WALES
NETHERLANDS
BELGIUM
LUX.
GERMANY
DENMARK
FRANCE
SWITZERLAND
SPAIN
PORTUGAL
CORSICA
SARDINIA
MINORCA
MAJORCA
IBIZA
MOROCCO
ALGERIA
TUNISIA
Corrib
Londonderry
Belfast
Sligo
Ballina
Galway
Dublin
Foynes
Limerick
Tralee
Wexford
Arklow
Cork
Waterford
Whitegate
Bantry
Waterford
Kinsale Head
Moffat
Carlisle
Newcastle
Darlington
Teesside
Middlesbrough
Barrow
Heysham
Lancaster
Morecambe
Formby
Bradford
Eastham
Manchester
Liverpool
Sheffield
Killingholme
Grimsby
Theddlethorpe
Lincoln
Bacton
Ellesmere Port
Aberystwyth
Birmingham
Milford Haven
Pembroke
Neath
Gloucester
Cambridge
Ipswich
Coryton
Oxford
London
Bath
Exeter
Plymouth
Southampton
Fawley
Portsmouth
Weymouth
Dunkerque
Antwerp
Brussels
Groningen
Emden
Leer
Bremen
Lingen
Loningen
Verden
Amsterdam
Rotterdam
Gelsenkirchen
Duisburg
Essen
Dusseldorf
Cologne
Rheinland
Bonn
Verviers
Maubeuge
Charlesville
Grandpuits
Luxembourg
Metz
Frankfurt-am-Main
Karlsruhe
Fredericia
Kaergard
Kalundborg
Skaelskor
Heide
Kiel
Cuxhaven
Brunsbuttel
Wilhemlshaven
Schwerin
Harburg
Hamburg
Beckedorf
Salzbergen
Herringhausen
Buckeburg
Salzgitter
Hamm
Dortmund
Eisenach
Jena
Erfurt
Suhl
Notre Dame de Gravenchon
Port Jerome
Le Havre
Cherbourg
Gargenville
Gonfreville L'Orcher
Plestin Les Greves
Petit Couronne
Brest
St. Malo
Paris
Quimper
Lorient
Laval
Vannes
Donges
Nantes
Angers
Orleans
Blois
Briare
Cholet
Les Sables
Poiters
Chauvigny
Vierzon
La Rochelle
Niort
Roussiness
Royan
Limoges
Le Verdon
Chazelles
Pauillac
Ambes
Bordeaux
Bergerac
Tulle
Les Ancizes
Issoire
St. Etienne
Feyzin
Lyon
Vienne
Le Puy
Villefranche
Bourbonne
Epinal
Dijon
Besancon
Dole
Macon
Bourg
Annemasse
Geneva
Chamberry
Grenoble
Valence
Wallbach
Zurich
Zug
Chur
Bern
Hochst
Vohburg
Ingolstadt
Munich
Bayonne
Toulouse
Castres
Carcassonne
Pamiers
Lourdes
Narbonne
Perpignan
La Junquera
Fos-sur-Mer
Berre l'Etang
Marseille
Toulon
L'Isle
Manosque
Nice
Monaco
Vert-La-Gravelle
Coulommes-Vaucourtois
Villeperdue
Chaunoy
Marolles
Valence en Brie
Villemar
Aufferville
Donnemarie
Flacy
Gisy-les-Nobles
Joigny
Chateaurenard
Trois Fontaines
Revigny
Velye-Germinon
Bussey-Lettree
Dommartin-Lettre
Courdemagnes
Grandville
Forcelles
St. Martin-de-Bossenay
Lavergne
Cazaux
Mothes
Parentis
N. Mimizan
Lugos
Cabeil
Ychoux
Lucats
Lacq sup
Lacq
Pont d'As
Rousse
Meillon
St. Marcet
Proupiary
St. Faust
El Serrablo
Gallician
Gaviota
Ferrol
La Coruna
Aviles
Santander
Vizcaya
Bilbao
Ayolvengo
Otomin
Logrono
Burgos
Vitoria
Huesca
Barbastro
Monzon
Balaguer
Tarrega
Gerona
Palamos
Zamora
Vallodolid
Zargoza
Fayon
Mataro
Barcelona
Tarragona
Chipiron NE
Angula
Salmonete
Dorado
Montanaza
Tarraco
Casablanca
Castellon
Amposta
Vinaroz
Castellon de la Plana
Vallencia
Madrid
Puertollano
Porto
Cabo Ruivo
Lisbon
Setubal
Sines
Almendralejo
Cordoba
Huelva
Sevilla
Rota
Cadiz
Granada
Malaga
Almeria
Cartegena
Murcia
Algeciras
Oran
Arzew
Algiers
Bejaia
Skikda
Tunis
Bastia
Muro
Aleria
Zonza
Porto Torres
Maddalena
Olbia
Alghero
Nuoro
Arbatax
Iglesias
Porto Scuso
Cagliari
Martigny
Aosta
Lugano
Mandello
Brescia
Padua
Biella
Milan
Chivasso
Cremona
Rovigo
Turin
Asti
Aquata
Busalla
Parma
Genoa
Savona
Pistola
Leghorn

Exploration and Development

EOR at Patos Marinza planned

Bankers Petroleum Ltd. (Calgary) reported plans to begin a study for a thermal enhanced oil recovery (EOR) pilot project in the southern part of Albania's Patos Marinza field. The field is estimated to hold resources of 2 billion bbl of original oil in place.

The field is one of the largest onshore heavy oil fields in Europe, located in south-central Albania, about 10 km (16 miles) east of the city of Fier. It has more than 2,500 wells, but cumulative production has been estimated at less than 7%. The two major production horizons are Driza, Marinza and Gorani, at depths of 90–1,800 m (300–5,900 ft).

Under the current development plan, the life of the field is estimated at 25 years. Total field recovery using primary production techniques is estimated to be 10–12%, including the previous production by state-owned Albpetrol.

Bankers said its activities increased production to 4,200 b/d of oil in 2006. The company aims to reach 10,000–15,000 b/d by 2010.

Durresi block work advances

Lundin Petroleum continued technical studies into late 2006 on the Durresi block to determine future exploration drilling activity. Plans calls for drilling a first well on the block in 2007, with the timing subject to rig availability.

The block covers 4,200 sq km (1,600 sq miles) in shallow waters. The Durresi exploration area is located between Cape Rodoni and the Karaburuni Peninsula. Most of the block is located in shallow water less than 100 m (330 ft) deep.

Pipelines

AMBO route a bit more clear

The Albania-Macedonia-Bulgaria (AMBO) Pipeline is one of several trans-Balkan projects proposed to bypass the congested Bosporus and Dardanelles Straits and carry Russian and Caspian crude oil to markets in the West. The AMBO line would begin on the eastern Black Sea coast of Bulgaria at Burgas, cross Bulgaria and Albania, and terminate at the Albanian port of Vlore on the Adriatic.

In October, Albania and Macedonia signed a protocol to define entrance points for the line at Stebleve in Albania and Lakaica in Macedonia. Discussions were under way regarding a similar protocol between Bulgaria and Macedonia.

The 920-km (570-mile) pipeline would carry 750,000 b/d of oil. Project cost is projected to be $1.2 billion, of which $930 million would be provided by several international donors. Construction is expected to take three years once approvals are obtained. The U.S.-based AMBO Pipeline Corp. (New York) has exclusive rights to develop the project.

AUSTRIA

Capital: Vienna
Monetary unit: Euro
Refining capacity: 208,600 b/cd
Oil production: 17,400 b/d
Oil reserves: 50.0 million bbl
Gas reserves: 570 bcf

Pipelines

TAG pipeline loop completed

Construction was completed on the third stage of the Loop II extension to the 380-km (240-mile) Trans Austria Gasleitung (TAG) natural gas pipeline. The loop began operation in January 2007.

The TAG pipeline system runs from the Slovakian/Austrian border near Baumgarten an der March to Arnoldstein on the Austrian/Italian border. It carries Russian gas to Italy, Austria, Croatia, and Slovenia.

The loop addition expands capacity from 37 bcm/yr (3.6 bcfd) to 41 bcm/yr (4.0 bcfd). Addition of a compressor station by 2008 will further boost capacity to 44.2 bcm/yr (4.3 bcfd).

Gas storage

The second-largest Central European natural gas storage facility is slated to begin operation in 2007 at Haidach. Construction began in 2005 on the facility. It will hold up to 2.4 bcm (85 bcf) of natural gas, approximately equal to 30% of annual Austrian demand.

The underground formation is a depleted gas reservoir from which more than 2.9 bcm (102 bcf) has been produced.

The storage project includes a 900-mm (36-inch) pipeline to connect the facility to the German/Austrian gas node at Burghausen.

Refining and Processing

Biofuels plant under discussion

Neste Oil Corp. (Espoo, Finland) and OMV AG (Vienna) signed a memorandum of understanding to jointly build a 200,000 tonne/year biodiesel plant at OMV's 208,600 b/cd Schwechat refinery in Austria.

The plant will be based on Neste Oil technology for producing high-quality diesel from renewable raw materials, such as vegetable oil and animal fat. It is an effort to meet the European Union's challenge for member states to increase biofuel usage to 5.75% of all gasoline and diesel used in traffic by 2010. The facility is expected to start production as early as yearend 2008.

BELGIUM

Capital: Brussels
Monetary unit: Euro
Refining capacity: 790,629 b/cd
Oil production: 0
Oil reserves: 0
Gas reserves: 0

Pipelines

Interconnector gas pipeline

The Interconnector is a subsea gas pipeline running 237 km (147 miles) from Zeebrugge, Belgium, to Bacton, Norfolk, UK. Its flow can be reversed within 12 hours to carry gas in either direction in response to market conditions. Forward-flow capacity from the UK to Belgium is 20 bcm/year (1.9 bcfd).

Effective October 1, 2006, reverse-flow capacity into the UK was increased from 16.5 bcm/year (1.6 bcfd) to 23.5 bcm/year (2.3 bcfd). Work also was begun on a project to raise system operating pressure by October 2007, which would further raise that capacity to 25.5 bcm/year (2.5 bcfd).

Study Sees Uneven Gas Trade in Europe

The wholesale gas market in Europe is developing unevenly and too slowly to suit many traders, according to a study Kingston Energy Consulting (London).

The UK and Benelux countries—Belgium, the Netherlands, and Luxemburg—are judged to have Europe's only "truly liquid trading markets" for gas. However, trading markets are growing rapidly in France, Germany, Italy, and Spain.

The UK accounts for 74% of total European gas trading volumes, followed by the Netherlands, 13%; Belgium, 6%; Spain, 4%, and Italy, Germany, and France, 1% each.

Kingston Energy also noted that the European Union has faulted members for slow progress in liberalizing gas markets and that the European Commission is investigating energy companies for possible violations of competition rules.

Questions about liberalization arose after Russia in January 2006 disrupted gas deliveries to Europe during a dispute with transit country Ukraine and in response to extreme price volatility last winter in the UK.

LNG

Impact of EdF gas plans

Electricité de France (EdF), the partially state-owned French utility, plans to boost its gas presence in northwest Europe by building an LNG regasification terminal in northern France, taking capacity at the Gate LNG project in the Netherlands, and increasing its gas pipeline capacity through Belgium and the Netherlands.

From 2007–2009, EdF will investigate building a 13.14 mtpy LNG regasification terminal at Dunkirk Port in northern France. If the project proves feasible, terminal operations could start in 2011. The terminal's regasified LNG initially would go to France, Belgium, the UK, and Germany.

EdF has also agreed with Dutch Gasunie to take a 10% capital stake in the latter's Gate LNG regasification terminal in Rotterdam, slated to come online in 2010. Gasunie subsidiary GTS and Belgian transporter Fluxys have also agreed to reserve 3 bcm of long-term gas capacity for EdF through the Netherlands and Belgium. This is in addition to the 2 bcm of capacity already acquired by the group on the interconnection between Europe and the UK.

The additional gas supplies will be used in part to fuel an 870-mW combined-cycle gas turbine power plant in southwest Netherlands planned for construction by 2009.

Refining and Processing

Capacity increase at Antwerp refinery

Among the significant increases in worldwide crude refining capacity in 2006 was the boost at the ExxonMobil facility in Antwerp, which went from 275,000 b/cd to 298,000 b/cd.

Lukoil to buy ConocoPhillips's European outlets

OAO Lukoil agreed to buy 376 retail outlets in six European countries from ConocoPhillips, most of them in Belgium. No price was disclosed.

In addition to 156 outlets in Belgium, the agreement includes 83 in Poland, 49 in Finland, 44 in the Czech Republic, 30 in Hungary, and 14 in the Slovak Republic. The deal was expected to close in second quarter 2007.

Astra to sell refinery stake

Astra Oil Trading NV, a division of Belgium's Transcor International NV, announced that it will sell to Brazil's national oil company, Petrobras, a 50% stake in a Pasadena, TX, refinery for $370 million. The move is part of the Petrobras strategy to expand into the United States.

The refinery can process 100,000 bbl of petroleum products daily and is being upgraded to handle more. By 2010, the refinery should have the capacity to process about 70,000 b/d of heavy crude oil, some of which could come from a field off the coast of Rio de Janeiro.

Petrochemicals

Antwerp elastomer production to end

Total Petrochemicals said it will cease elastomer production at its Antwerp facility because of increasingly tough economic and competitive conditions. The facility produces synthetic thermoplastic rubbers, which are used for bitumen modifications, technical compounds, and improving the technical qualities of some plastics.

BULGARIA

Capital: Sofia
Monetary unit: New lev
Refining capacity: 115,240 b/cd
Oil production: 1,000 b/d
Oil reserves: 15 million bbl
Gas reserves: 200 bcf

AMBO pipeline route becomes more clear

The final route for the proposed AMBO (Albania-Macedonia-Bulgaria) oil pipeline has been decided by the first two countries, and codified in a bilateral protocol in October, according to Macedonian press reports.

The anticipated route of the pipeline is from Burgas on the Black Sea to the southern Albanian port of Vlore. Though it has not identified its own exit point yet, Bulgaria stated that it will do so in the near future, so that a second bilateral agreement between it and Macedonia can be signed.

Macedonia's preferred pipeline entry point is said to be in the area of Deve Bair, in the northwest corner of the country. That means it is likely that the Bulgarian exit point will be near Prekolnica or, more likely, Guesevo. As an exit point, the government of Macedonia has chosen the village of Lakavica, from which a connection will be made to Stebleve, Albania.

Progress on the pipeline has been slowed down due to political infighting about the route. However, backers held out for the chosen path, which passes straight towards Skopje and then down through central and southern Macedonia.

Pipeline construction, coupled with potential development of suspected oil deposits in eastern Macedonia, could help restore economic parity to the country. Eastern Macedonia has historically has suffered greater poverty and higher unemployment than Albanian-majority western Macedonia.

Concerns may still arise regarding environmental safeguards and property compensation for those affected by the pipeline. (See related information in the Romania section.)

The European Union accepted Bulgaria as a member country, effective January 2007.

CROATIA

Capital: Zagreb
Monetary unit: Kuna
Refining capacity: 250,317 b/cd
Oil production: 16,800 b/d
Oil reserves: 74.320 million bbl
Gas reserves: 1.1 tcf

LNG terminal under study

A joint venture led by OMV AG (Vienna) has agreed with E.ON Ruhrgas AS (Düsseldorf, Germany) to study the feasibility of building an 8–10 bcm/year LNG regasification terminal in Croatia.

The feasibility and basic engineering studies are expected to be complete by year-end 2008, with the commissioning of the terminal slated for yearend 2011. The terminal will deliver gas to Croatia and other countries in southern and Eastern Europe.

CZECH REPUBLIC

Capital: Prague
Monetary unit: Koruna
Refining capacity: 198,000 b/cd
Oil production: 4,700 b/d
Oil reserves: 15.0 million bbl
Gas reserves: 140 bcf

Gas delivery

The Czech Republic opened its natural gas market to all customers except households in 2006, and will do so for all customers in 2007.

The country's gas transmission system capacity was in the process of being opened to third-party access in 2006, with 33% to be opened in 2008. The gas distribution system is to be unbundled by 2007.

Drilling near Breclav

Moravski naftove doly (MND; Hodonin) began a gas drilling program in Postorna, near Breclav in southern Moravia. MND estimated production at 30,000 cu m/year (1.06 bcf/year) for 8 to 10 years.

The company also reported that it had entered a contract for oil and gas exploration in Yemen, with Oil Search Ltd. (Australia).

Biofuels rule set

The government enacted regulations requiring that motor fuel contain 2% bio-components beginning in 2007. Alcohol will be added to gasoline and rape oil methylester to diesel, according to Access Czech Republic Business Bulletin.

NORTHERN EUROPE
15°
10°
5°
0°
5°
10°
60°
55°
50°
0 50 100 150 Miles
0 100 200 Kilometers
Norwegian Sea
North Sea
Irish Sea
Celtic Sea
Donegal Bay
Galway Bay
Moray Firth
Firth of Forth
Firth of Clyde
FAEROE I.
SHETLAND I.
ORKNEY I.
NORWAY
SCOTLAND
N. IRELAND
IRELAND
WALES
ENGLAND
DENMARK
NETHERLANDS
BELGIUM
LUX.
GERMANY
SWITZERLAND
Midgard
Åsgard
Tyrihans
Ormen Lange
Draugen
Njord
Åsgard Transport
Agat
Troll
Tjeldbergodden
Trondheim
Nyhamna
Mongstad
Kollsnes
Sotra
Bergen
Oslo
Slagentangen
Vall oy Tonsberg
Karsto
Kopervik
Stavanger
Sola
Arendal
Mandal
Goteborg
Skagen
Sullom Voe
Lerwick
Kirkwall
Flotta
Wick
Stornoway
Nigg Bay
Inverness
Mallaig
St. Fergus
Cruden Bay
Aberdeen
Dundee
Finnart
Ardyne Point
Grangemouth
Glasgow
Edinburgh
Ardrossan
Moffat
Londonderry
Belfast
Sligo
Ballina
Corrib
Carlisle
Newcastle
Port Clarence
Teesside
Middlesbrough
Darlington
Barrow
Heysham
Lancaster
Morecambe
Formby
Bradford
Scarborough
Hull
Leeds
Manchester
Liverpool
Eastham
Grimsby
Galway
Ballough
Dublin
Foynes
Limerick
Tralee
Arklow
Wexford
Waterford
Cork
Whitegate
Bantry
Galley Head
Seven Heads
Helvick
Ardmore
Kinsale Head
Ellesmere Port
Lincoln
Theddlethorpe
Bacton
Nottingham
Leicester
Aberystwyth
Birmingham
Coventry
Northamton
Cambridge
Milford Haven
Pembroke
Llandarcy
Neath
Gloucester
Oxford
Ipswich
London
Reading
Bath
Fawley
Bournemouth
Southampton
Portsmouth
Exeter
Plymouth
Weymouth
Dunkerque
Calais
Lille
Brugge
Antwerp
Brussels
Liege
Verviers
Maubeuge
Aarhus
Fredericia
Kaergard
Kalundborg
Esbjerg
Skaelskor
Kiel
Heide
Brunsbuttel
Cuxhaven
Schwerin
Hamburg
Harburg
Wilhelmshaven
Emden
Groningen
Amsterdam
Rotterdam
Salzbergen
Gelsenkirchen
Dortmund
Duisburg
Essen
Dusseldorf
Cologne
Rheinland
Bonn
Eisenach
Suhl
Frankfurt-am-Main
Meerdf
Roxforrde
Notre Dame de Gravenchon
Cherbourg
Le Havre
Port Jerome
Gonfreville L'Orcher
Rouen
Gargenville
Petit Couronne
Grandpuits
Luxembourg
Metz
Karlsruhe
Reims
Vert-La-Gravelle
Coulommes-Vaucourtois
Velye-Germinon
Bussey-Lettree
Paris
Chauroy
Villeperdue
Soudron
Dommartin-Lettre
Courdemagnes
Trois Fontaines
Marolles
Valence en Brie
Donnemarie
Grandville
Forcelles
St. Martin-de-Bossenay
Villemar
Aufferville
Flacy
Gisy-les-Nobles
Joigny
Chateaurenard
Bourbonne
Plestin Les Greves
St. Malo
Brest
Quimper
Rennes
Laval
Lorient
Vannes
Donges
Nantes
Orleans
Blois
Reichstett
Vendenheim
Zurich
Munich
Ingolstadt
Vohburg
Baden-Wurttemberg
210
110
70
106
84
260
78
101
10
91
268
346
227
302
106
262
80
99
331
230
164
223
172
326
188
81
380
406
147
298
357
115
115
21
141
221
221
296
27
106
210
100
12
206
71

DENMARK

Capital: Copenhagen
Monetary unit: Krone
Refining capacity: 176,400 b/cd
Oil production: 342,000 b/d
Oil reserves: 1.277 billion bbl
Gas reserves: 2.542 tcf

Brugdan well yields only data

Operator Statoil AS announced that no commercial oil or gas volumes were proven at the Brugdan 6104/21-1 well in License 006 in the Atlantic, south of the Faroe Islands. The islands are a territory of Denmark, located between Scotland and Iceland.

However, the company said that the well, drilled to TD 4,201 m (13,780 ft) in 480 m (1,575 ft) of water, provided information about drilling in volcanic sub-basalt rock that will be helpful in future exploration off the Faroes.

Brugdan is 70 km (43 miles) northwest of Chevron Corp.-operated Rosebank oil field, a 2004 discovery under development. The Faroe authorities plan a third licensing round for late 2007.

Gazprom, DONG deals

Russia's Joint Stock Company (JSC) Gazprom (Moscow) signed an agreement with Danish utility DONG Energy to deliver natural gas to Denmark. The agreement is the first-ever gas supply deal between the two countries.

Gazprom is to start gas supply to Denmark in 2011 through the North European Gas Pipeline. The agreement commits DONG to buy 1 bcm/year (10.3 bcfd) of gas for 20 years, with an option to increase this quantity.

Under a separate agreement, DONG will begin supplying gas in 2007 to Gazprom's UK unit Gazprom Marketing & Trading. The 15-year contract calls for as much as 600 MMcm/year of gas to be delivered through the Langeled Pipeline, which will connect Ormen Lange gas field offshore Norway with the UK.

Prospects rise in Greenland

Greenland drew strong interest from international oil companies for the island's fourth round of licensing concessions, west of Disko Bay, which covers 92,000 sq km (35,521 sq miles) north of the Arctic Circle.

Offers were filed by affiliates of Exxon, Chevron, Husky, and Denmark's Dansk Olie & Naturgas AS. Eight blocks offered lie in Baffin Bay and the Davis Strait between 67° and 71° N. Lat. Four border Canadian waters.

The Bureau of Minerals & Petroleum (BMP) will process the applications and plans to award licenses in March 2007. BMP plans phase two of the license round from August 2007 to February 2008.

The BMP said the agency believes that the Davis Strait has the potential to become another North Sea, with many oil fields holding 500 million bbl or more.

EnCana Corp (Calgary) became the first and only company to hold a license in the region when it bought rights to the Attamak and Lady Franklin blocks in 2002 and 2005.

Greenland 2004 Blocks

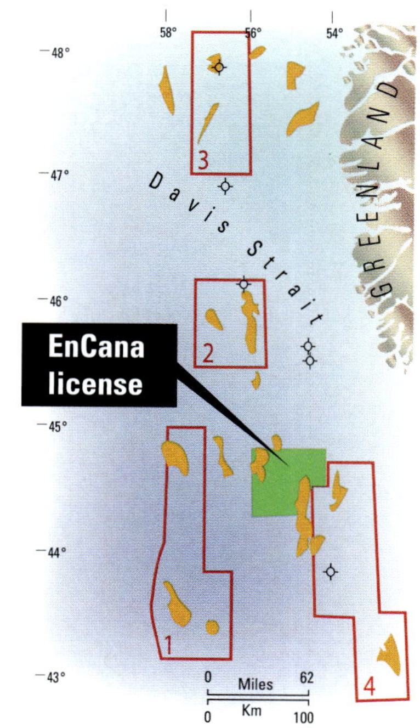

Yellow areas denote giant structural leads of >100 sq km.
Source: Bureau of Minerals & Petroleum

Reports say EnCana's blocks may hold reserves ranging from 400 million bbl to 1.2 billion bbl of oil. Mosbacher Energy Co. (Calgary) said geologists have long wondered if Greenland could match offshore Newfoundland for potential

They pose two related geological questions: At what point did Greenland separate from North America, and is its geological structure similar to that of Newfoundland's Orphan and Jean d'Arc basins?

FINLAND

Capital: Helsinki
Monetary unit: Euro
Refining capacity: 251,800 b/cd
Oil production: 0
Oil reserves: 0
Gas reserves: 0

Biodiesel plant at Porvoo

Neste Oil announced plans to build a second 170,000-tonne/year biodiesel plant at its 200,000 b/cd refinery in Porvoo. Estimated to cost €100 million, the plant will produce biodiesel from animal fats and oil. Production is scheduled for late 2008. A first biodiesel plant under construction is scheduled to start up in 2007.

The plants will use Neste's proprietary process to convert biological triglycerides into a synthetic biodiesel—NExBTL—a nonoxygenate hydrocarbon biodiesel compatible with the existing vehicle fleet, the company said. The company also has biodiesel projects under way in France and Austria.

FRANCE

Capital: Paris
Monetary unit: Euro
Refining capacity: 1,958,794 b/cd
Oil production: 21,500 b/d
Oil reserves: 121.5 million bbl
Gas reserves: 341 bcf

Exploration and Development

French E&P spending growth expected

Investment in 2006 for exploration and production (E&P) in France and French possessions should more than double to €145 million, from €65 million in 2005, said French officials.

Exploration activities

Operators expected to drill 10–15 exploration wells in 2006, spending €67 million compared with only €19 million in each of 2005 and 2004.

Lundin Petroleum AB (Stockholm, Sweden) and Toreador Energy France operated the only successful exploration well of 2005, La Tonnelle-1 on the Nemours Permit in the Paris basin. The well produced 50 b/d of oil during tests in early 2006. Extension work was planned for 2006.

For 2006, 10 requests for onshore exploration permits were submitted. Six are in the Paris basin. The others are in the southern Aquitaine basin, where no new request had been registered since 2001.

Europa Oil & Gas (Holdings) PLC (London) expects to be awarded the Bearn des Gaves exploration permit in the Aquitaine basin in early 2007 following government approval. The permit covers 928 sq km (358 sq miles) west of Lacq and Meillon gas fields. Europa plans to probe a possible western extension of the Lacq play and also sees oil potential on the northern part of the permit.

No well was drilled offshore France in 2005, but the French hydrocarbon ministry reported the highest run of seismic reflection acquisition carried out for years in the waters off continental France and its overseas territories. The coastal French area of interest was the Aquitaine Maritime Permit.

Around the world, seismic work was done on the Guyane Maritime Exclusive Exploration License in northern South America, at St. Pierre et Miquelon off eastern Canada, and on the Juande Nova Maritime Permit in the Mozambique Channel off southeast Africa.

Production activities

For 2006, field development and production outlays in France were expected to increase to €78 million, from €47 million in 2005, driven by the higher cost of developing deeper targets.

Crude production in 2005 fell 5.1% to 1.08 million tons even though operators reworked a number of fields. Gas production saw little investment and plunged 18.6% to 1.81 bcm.

Three producers—Esso Rep, Total E&P France, and Vermilion Rep—each accounted for about 25% of oil 2005 production, while Lundin Petroleum and Madison Energy produce 15% and 5%, respectively. Most of the produced gas is operated by Total E&P France in the Aquitaine basin.

The Paris basin remains the predominant oil producing area with 53%, compared to 47% for the Aquitaine basin. Three fields account for 28% of crude production: Parentis 13% (Aquitaine basin; 141,000 tons); Chaunoy 8% (Paris basin; 84,000 tons); and Itteville 7% (Paris basin; 82,000 tons).

Toreador Resources Corp. (Dallas) reported in early 2006 that the company's 100% owned, multizone 111H horizontal well on test in the Charmottes field, southeast of Paris produced 800 b/d of oil for six days.

Heritage Petroleum PLC (London), in association with European Gas Ltd. (EGL; Alfred Cove, Western Australia; formerly Kimberly Oil NL), is set to develop three permits for coalbed methane production at Bleu Lorraine, near the border with Germany.

EGL aimed to drill three appraisal wells in 2006 at a cost of €5 million. EGL has estimated 576 bcf of gas in place on 36 sq km (14 sq miles) of the Alsting block and 415 bcf of GIP on 32 sq km (12 sq miles) of the Saint Avold block of the 460-sq-km (178-sq-mile) Lorraine Permit, in coals shallower than 1,500 m (4,921 ft).

Gas Delivery

Domestic gas markets opening up

An important step in liberalizing France's gas markets is scheduled for July 1, 2007. After that date, all gas consumers are to have at least nominal freedom to choose a supplier.

The country established four gas-balancing zones, where virtual Gas Exchange Points (PEGs) act as incipient trading hubs, similar to UK's National Balancing Point. The PEGs do not yet have enough activity for reliable price signals, but trade volume is growing.

The government has unbundled the gas transmission system, forcing Gaz de France (GdF) and Total SA, which has the second-largest market share, to create subsidiaries to hold and manage their trunklines.

In the south, the Port-de-Larrau and Euskadour gas pipelines straddle the border, and there are plans to expand the latter. GdF and Total unwound their cross-holdings in joint-venture companies that cover most of southern France and entered into a 3-year gas-release program in 2004 to stimulate competition.

As a result of these and other efforts, there are now 28 registered gas suppliers in France (14 active), in addition to the 22 companies authorized as distributors within a given regional market.

GdF, Suez merger proposed

On a related front, French politicians reacted strongly to new reports in March that Italian electricity company Enel planned to acquire the French utility, Suez. Prime Minister Dominique de Villepin subsequently brokered a competing merger agreement between Suez and GdF.

Despite protests from EU regulators and French unions, the National Assembly approved the privatization in October. However, in November, a Paris court upheld a union plea to delay the merger pending further study. The court also upheld a prior ruling that the merger should not proceed until GdF's European works council had expressed its views. Approval of the merger by the French Senate would still be needed.

The European Commission also has objected to the merger. Concerns focus not only on France, but also on Belgium, where the combined entity would have a near-monopoly on gas and electricity. Other players, including Gazprom, Enel, and Centrica, are awaiting announcement of possible asset-sale requirements to comply with the regulatory demands.

GdF has said that if the merger proceeds, it would consider expanding its Montoire LNG terminal to as much as 16 bcm/year (from the current 10 billion)—with the extra capacity to be offered to third parties—and functionally separating its French LNG business from other operations.

Gazprom marketing directly to French consumers

Russia's gas giant Gazprom launched Gazprom Marketing & Trading France in November in Paris to further develop what it called a total upstream-to-downstream natural gas export strategy. It intends to sell gas directly to medium and large industrial and commercial consumers. The company started operations through UK subsidiary Gazprom Marketing & Trading Ltd.

LNG

Terminal developments

EdF Trading Ltd. (EdFT, London)—a wholly owned subsidiary of the partially state-owned utility Electricité de France (EdF)—is a European wholesale trader of electricity, gas, and coal. EdFT announced in July 2006 that it had reserved one slot/month at GdF's Montoire terminal for 2007. EdFT plans to market some gas to parent company EdF, which has been building a retail portfolio since it got approval to move into the domestic gas business in 2005. EdFT also will sell gas at Belgium's Zeebrugge hub and into the UK through the Interconnector Pipeline.

In August, 4Gas (Rotterdam), an independent developer of LNG terminals in the UK and on the Continent, secured a site option for an LNG terminal at Bordeaux. The joint venture with the Port of Bordeaux would build a €400 million import terminal with initial capacity of 6–9 bcm/year (0.58–0.87 bcfd). Additional funds would be needed for a pipeline connection to the regional distribution system. The specific site is at Le Verdon. Construction could begin in 2008, with start-up in 2011.

4Gas was formed by shareholders of Petroplus International NV (Amsterdam) after its development of the Dragon LNG project in Milford Haven, Wales. 4Gas also is developing terminals in Rotterdam and Nova Scotia, Canada.

The Port of Le Havre Authority tagged a joint venture between Poweo (Paris), an electricity and gas supply company, and Compagnie Industrielle Maritime SNC for a possible new LNG terminal at Antifer, near the Le Havre petroleum port. According to Poweo, the €500 million terminal would have a capacity of 8–10 bcm/year (0.77–0.97 bcfd), with start-up slated for 2011.

In mid-October 2006 EdF won a tender for feasibility studies of an LNG terminal at Dunkirk in northwestern France. EdF would build and operate a terminal with an initial capacity of 6 bcm/year (0.58 bcfd). The studies are expected to take until 2009. The planned start-up is 2011.

Innovative carrier completes trials

Gaz de France (GdF) completed gas trials off the French coast in October for its 154,500-cu-m Provalys LNG carrier, one of the world's largest. Christened in November 2005 at Montoir-de-Bretagne, the vessel is equipped to load LNG in arctic conditions. It is the first conventional-size LNG carrier with a dual-fuel diesel electric propulsion system and the first such ship with the new CS1 membrane containment system. GdF is owner and charterer.

Refining and Processing

Gonfreville refinery upgrade

Total SA has brought on stream a €550 million distillate hydrocracker at its 330,000-b/d Gonfreville refinery in Normandy, France.

GAS TRADING ON FRENCH MARKETS

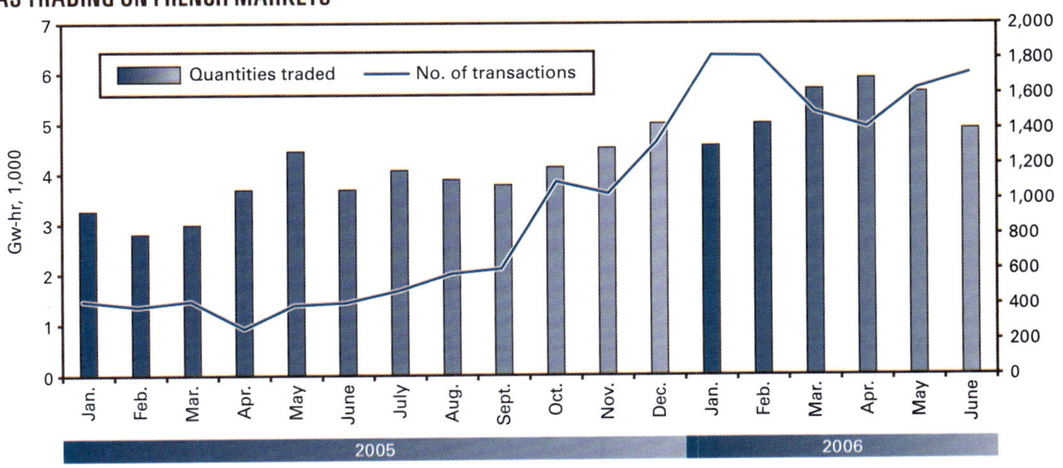

Source: TSO/CRE.

The unit will yield 200,000 tonnes/year (tpy) of jet fuel, 400,000 tpy of naphtha for petrochemicals, 500,000 tpy of high-grade base products for lube oils, and 1.3 million tpy of diesel oil. It will raise the refinery's diesel output capacity to 5.3 million tpy.

The hydrocracker is part of the expansion and modernization of Total's 12 European refineries, spending on which this year will total about €1 billion.

E-85 role in biofuels plan

Members of France's oil industry, independent motor fuels distributors, automobile manufacturers, farmers, and ethanol producers in November signed the "Charter for Ethanol E85" which sets out their commitments over the next 5 years. Esso SAF is the only major oil company that did not sign, but said it would adapt its investment as fuel demand grows.

E85 is a blend of 85% ethanol with 15% unleaded gasoline. Introduction of the fuel at all service stations in France by 2010 as an alternative to unleaded gasoline was deemed feasible in a government study.

The E85 initiative is part of a larger biofuels plan, coupled with a broad energy-savings program in which all gas, heating oil, and electricity suppliers and the auto industry must participate under threat of financial penalty. The aim is to prepare for what French officials call the "post-oil period."

There are also concerns that use of ethanol-based biofuel in a country where most vehicles use diesel could further destabilize France's refining industry, which already must import diesel and export surplus gasoline.

Diesel supply challenges

France needs increased diesel output from existing refineries more than it needs new refineries, said the trade group Union Française des Industries Pétrolières (UFIP). Its members include Total SA, Esso SAF, Shell France, BP PLC, Ineos, and Agip SPA.

Diesel demand has doubled since 1990, reaching 31 million tonnes in 2004, when imports totaled 11 million tonnes. Over the same period, gasoline demand fell by about half, to 16 million tonnes, of which 4 million tonnes was exported in 2004.

Assuming a 2%/year increase of merchandise road transport, an increase in the diesel-fueled share of the vehicle market to 70%, and improvements in vehicle fuel efficiency, UFIP projects that by 2010, refiners will need to produce 6 million tonnes/year more diesel than in 2006. Some supply will come from biodiesel required by the government, but imports of diesel will remain at 11–12 million tonnes/year.

Contracts for biodiesel plants

Diester Industrie let turnkey contracts to Technip for two biodiesel plants in France. The plants, one in the Bassens port zone near Bordeaux and the other doubling the size of the Grand-Couronne plant near Rouen, will use the Axens process.

Capacities each will be 250,000 tonnes/year. The units are to start production by year-end 2007.

Technip also has built biodiesel plants for Diester at Sete and Compiegne, and is building one for the company at Montoir-de-Bretagne.

FRENCH LNG TERMINALS

	Sendout, bcfd	Act./est. start date	Sponsors	Status
Fos-sur-Mer	0.5	1972	Gaz de France (100%)	Operating
Montoire de Bretagne	1.0	1982	Gaz de France (100%)	Operating
Fos Cavaou	0.8	2007	Gdf (66%), total 33%	In construction
Port of Bordeaux	0.6-0.9	2011	4Gags (100%)	Proposed
Antifer	0.8-1.0	2011	Poweo, CIM	Proposed
Dunkirk	0.6	2011	Electricite' de France	Proposed
Montoire Expansion	0.2-0.6	NA	Gaz de France	Contingent

Source: Poten & Partners

Environment

Total summoned for 2001 tanker spill

Total SA was summoned to a French court of summary jurisdiction in early 2006 to be tried for "marine pollution" and "complicity in endangering the life of others" in connection with the December 1999 shipwreck of the tanker Erika in heavy weather and the subsequent heavy fuel-oil spill off the Brittany coast.

The company and all parties involved in the wreck and spill were indicted. The incident polluted 400 km of coastline. Total is blamed for having contravened "its own vetting rules" covering condition of the tanker. Total said the accusations are factually and legally unfounded.

GERMANY

Capital: Berlin
Monetary unit: Euro
Refining capacity: 2,344,192 b/cd
Oil production: 70,000 b/d
Oil reserves: 367.2 million bbl
Gas reserves: 9.076 tcf

Refining and Processing

Coal-to-liquids plant planned

Syntroleum Corp. (Tulsa, OK) and Sustec Industries AG (Switzerland), in a 50-50 joint venture, agreed to develop a 3,000 b/d Syntroleum Fischer-Tropsch (FT) and refining unit as the first phase of a planned 20,000 b/d coal-to-liquid (CTL) plant at Sustec's Schwarze Pumpe industrial facility at Spreetal, 130 km (80 miles) southeast of Berlin.

The project includes expanding Sustec's gasification capacity using its Future Energy process, acquired by the Siemens Power Generation Group (Erlangen, Germany). Siemens will install the 1,000-MW gasification units at Spreetal. The FT unit will produce ultra clean diesel and specialty chemicals from synthesis gas made from gasified lignite coal, municipal waste, and biomass.

FEED work and a feasibility study will shape a decision in 2007 whether to proceed.

Etherification plant planned

PCK Raffinerie GmbH has let a €25 million turnkey contract to Uhde GmbH for a plant for the etherification of light naphtha with bioethanol at its 220,000 b/cd refinery at Schwedt, 110 km (68 miles) northeast of Berlin. Start-up was scheduled for September.

Using a process licensed by Neste Jacobs Oy (Finland), the plant will produce nearly 1 million tonnes/year of light naphtha. The refinery will use the light naphtha as an octane blending component of gasoline.

Petrochemicals

OMV expansion in Bavaria

OMV AG will boost petrochemicals capacity as part of a €1.1 billion investment program in Bavaria through 2010. It plans to spend €640 million to expand ethylene capacity to 450,000 tonnes/year from 340,000 tonnes/year and propylene capacity to 560,000 tonnes/year from 245,000 tonnes/year at its complex in Burghausen.

OMV will build a metathesis plant, enlarge its ethylene plant, and construct a large cracking furnace. It will become sole supplier of propylene to the nearby plastics complex operated by its subsidiary Borealis.

The Borealis complex will be expanded with construction of a €200 million, 330,000-tonne/year polypropylene plant. Total polyolefin capacity at the complex will rise to 740,000 tonnes/year. The new plant will use OMV's Borstar technology.

OMV also plans to invest €150 million to lay a 360-km products pipeline between Munchmunster and Ludwigshafen, giving it access to the Western Europe ethylene network. The initial capacity of 200,000 tonnes/year will be expandable to 400,000 tonnes/year.

Köln ethylene expansion

Ineos Group Holdings Inc. (Grangemouth, Scotland) announced plans to begin in 2007 a €40 million expansion of the ethylene capacity at its olefins complex at Köln. Total ethylene capacity will increase to about 1.2 million tonnes/year from 1.1 million tonnes/year.

Total olefins capacity of the complex is 2.3 million tonnes/year. The new ethylene will feed polyethylene production at Köln or be transported to other Ineos sites through the 495-km (307-mile) Rohrleitungs Gesellschaft pipeline.

Gelsenkirchen-Scholven furnaces

Ruhr Oel GmbH (Gelsenkirchen) hired Linde AG (Wiesbaden) to install 5 new cracking furnaces for an ethylene plant at the company's Gelsenkirchen-Scholven site. The total order value is €130 million, with late 2007 completion planned.

The new furnaces will replace 17 that no longer meet environmental and efficiency requirements. Each new furnace, with a capacity of 100,000 tonnes/year of ethylene, will process heavy hydrocarbons to naphtha and gaseous feedstocks.

Ruhr Oel is a joint venture of BP PLC and Venezuela's PdVSA.

GREECE

Capital: Athens
Monetary unit: Euro
Refining capacity: 413,000 b/cd
Oil production: 1,800 b/d
Oil reserves: 5.0 million bbl
Gas reserves: 35 bcf

Upgrades for Elefsina refinery

Hellenic Petroleum SA has let a FEED contract to a unit of Foster Wheeler Ltd. (Clifton, NJ) for the upgrade of the 100,000 b/cd Elefsina refinery, about 30 km (18 miles) northwest of Athens. Contract terms were not disclosed.

The work includes the selection of technology licensors for a 48,500 b/sd vacuum distillation unit, a 20,000 b/sd coking unit, a 40,000 b/sd hydrocracker, a 95,000 cu m/hr of hydrogen, and a 190 tonne/d sulfur recovery unit.

Foster-Wheeler is to complete the FEED phase in 2007. They also will provide a process design package and prepare the invitation-to-bid package for the engineering, procurement, and construction phase.

Hellenic Petroleum also chose ExxonMobil Research & Engineering Co.'s flexicoking technology for an upgrade project at the Elefsina facility. The 20,000 b/d flexicoking unit will convert vacuum residue to lighter products.

Flexicoking integrates fluid bed coking and air gasification to eliminate use of petroleum coke. It allows refiners to process vacuum resid, atmospheric resid, oil sands bitumen, heavy whole crudes, de-asphalter bottoms, or thermal cracked tar to produce higher-value liquid and gas products.

HUNGARY

Capital: Budapest
Monetary unit: Forint
Refining capacity: 161,000 b/cd
Oil production: 20,200 b/d
Oil reserves: 102.48 million bbl
Gas reserves: 1.21 tcf

Deep gas targeted

Falcon Oil & Gas Ltd. (Vancouver, BC) began drilling operations on its seven-well program in the Pannonian basin of southeastern Hungary, focusing on the Makó Trough. An independent analysis by the Scotia Group (Dallas) estimated theoretical potential recovery from Makó Trough deep-gas exploration at 45.2 tcf at 50% probability and 96.3 tcf at 10% probability.

Falcon operates the deep gas project through its wholly owned subsidiary TXM Energy LLC, based in Budapest. The company acquired two exploration licenses from Makó Energy covering 2,329 sq km (575,263 acres) in south central Hungary, one in the Tizsa area, the other in the Makó area. The Hungarian government has extended both licenses through December 2007.

Falcon/TXM views the play as a basin-centered gas accumulation (BCGA) similar to the Jonah gas field and Pinedale anticline in Wyoming's greater Green River area.

Croatia's Crosco Drilling & Well Services Co. Ltd. is the drilling contractor for TXM, with plans to drill two or three shallow wells in the Tisza area and one to four deep wells in the Makó area. As of late 2006, Falcon/TXM reported promising results from six test wells: Pusztaszer-1, Szukketas-1, Makó-6, Makó -7, and Magyarcsanad-1. The Makó -7 well was expected to reach TD of approximately 6,000 m (19,685 ft) by early 2007.

IRELAND

Capital: Dublin
Monetary unit: Euro
Refining capacity: 71,000 b/cd
Oil production: 0
Oil reserves: 0
Gas reserves: 350 bcf

Corrib protest continues

The Corrib project aims to bring ashore gas from subsea wells in the Corrib field and then deliver it through an onshore pipeline to a gas processing facility at Bellanaboy in County Mayo. The field is approximately 80 km (50 miles) west of the Mullet peninsula in the Slyne Trough in Blocks 18/20 and 18/25. Shell E&P Ltd. is the operator of the Corrib project. Other partners are Statoil AS and Marathon.

In theory, the first gas from the field could be produced by 2009, but that timetable is highly unlikely in view of stiff opposition to the project that flared in 2005 and continued well into 2006.

The Irish administration clamped down on the protests and insisted that work should proceed, but the opposition Labor Party supported the protestors by insisting that facts were being distorted related to planning approval for the complete project.

Some even suggested that the "Shell to Sea" protest campaign had been infiltrated by Sinn Fein, the political wing of the militant Irish Republican Army.

In late October, the Department of Communications, Marine and Natural Resources rejected requests for a new inquiry into alternative sites for the terminal.

Irish press reports in December indicated that protesters were planning to take their campaign to Norway, home of Statoil. At the same time, a reported opinion poll suggested that 70% of people in Mayo wanted the project to go ahead.

Renewable diesel

ConocoPhillips began commercial production of renewable diesel from soybean oil and other vegetable oils at its 71,000 b/d Whitegate refinery in Cork. Whitegate is producing 1,000 b/d of renewable diesel using its existing equipment.

Unlike biodiesel, renewable diesel is blended and transported with petroleum-based diesel. Renewable diesel also is different chemically from biodiesel. Use of the renewable diesel is expected to help reduce carbon emissions.

FALCON'S LICENSE

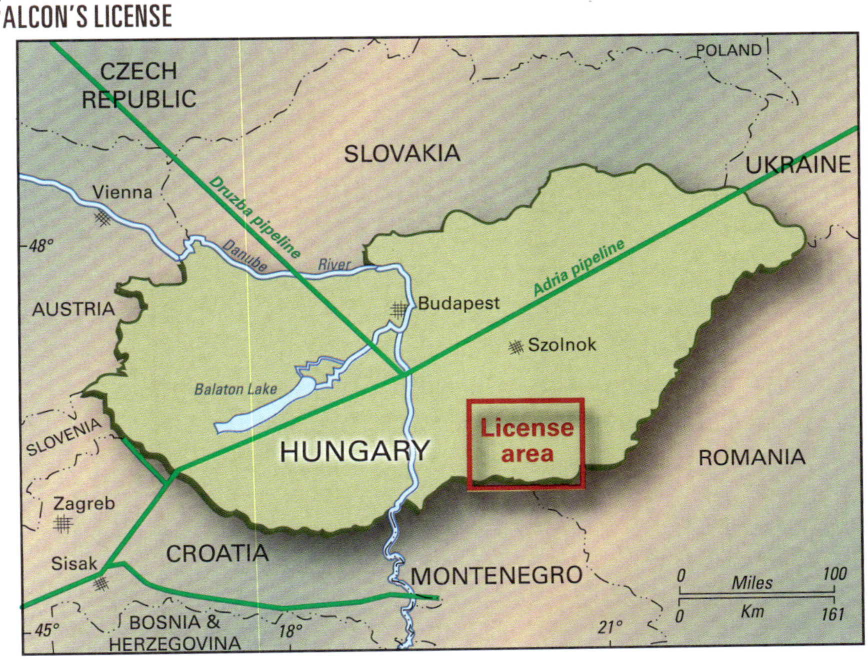

ITALY

Capital: Rome
Monetary unit: Euro
Refining capacity: 2,324,409 b/cd
Oil production: 115,000 b/d
Oil reserves: 621.7 million bbl
Gas reserves: 8.0 tcf

Exploration and Development

Southern Apennines exploration

A joint venture has approved the drilling of the Monte Grosso-2 exploration well on the Serra San Bernardo exploration permit in the southern Apennines. Intergas Piu SRL, a subsidiary of Mediterranean Oil & Gas PLC (London) will operate the well. The other joint venture partners include Eni and Total.

The location is just north of Monte Alpi oil field, one of continental Europe's largest oil fields. Eni operates Monte Alpi. The well is to be drilled to 6,900 m (22,635 ft) in the third quarter of 2007. Monte Alpi is north of Tempa Rossa field, recently cleared for development by Total. The Tempa Rossa project will recover an estimated 200 million bbl of heavy oil.

Po Valley reserves analyzed

An independent geological report has increased proved natural gas reserves held in northern Italy by Po Valley Energy Ltd. (Perth, Australia) to 36.6 bcf from 25.3 bcf.

The review by Italian consultant Ecopetrol takes into consideration results from Po Valley's 2005 exploration program, including the recently drilled Sillaro field near Bologna and the Vitalba field near Milan.

Ecopetrol found that the total proved and probable reserves remained at 104.6 bcf and proved, probable, and possible reserves at 130.4 bcf.

Po Valley is planning to start production from Sillaro, Vitalba, and Santa Maddelana fields in 2007 and to drill the Bezzecca field (formerly Pandino), southeast of Milan.

Ragusa plateau of interest

The Italian government issued preliminary approval for a unit of Cygam Energy Inc. (Calgary) to explore the 83,264-acre Aretusa Permit (d348 C.R.-VG) on the Ragusa plateau in the Mediterranean southeast of Sicily.

Other companies have expressed interest in participating with Cygam's Vega Oil SPA on the project. Targets include Upper Triassic Taormina carbonate and Lower Jurassic Inici carbonates.

Pipelines

Galsi trans-Med gas pipeline

New gas-purchase agreements by five Italian energy companies boosted plans for a third gas pipeline between North Africa and Europe. Enel, Edison, Hera, Ascopiave, and Worldenergy, agreed to import a total of 6 bcm/year (0.58 bcfd) of gas from Algeria through the proposed pipeline, which would connect Algeria with Italy by way of Sardinia.

Sonatrach (Algeria), Enel, and Wintershall (Germany) formed a venture to study feasibility of the pipeline in 2001. Other since joined the venture, now called Galsi. The project envisions 4 pipeline segments:

- 640 km (400 miles) onshore between Hassi R'mel gas field in Algeria and El Kala on the Algerian coast
- 310 km (190 miles) between El Kala and Cagliari on Sardinia in water as deep as 1,950 m (6,400 ft)
- 300 km (185 miles) between Cagliari and Olbia on the northern Sardinian coast
- 220 km (135 miles) between Olbia and Pescaia, southeast of Florence, in water as deep as 900 m (295 ft).

Design capacity of the Galsi line will be 8 bcm/year (0.77 bcfd). Start-up is projected for 2009.

EASTERN EUROPE
0 50 100 150 Miles
0 50 100 150 200 Kilometers
Copenhagen
Malmo
BORNHOLM
RUGEN
Reinkenhagen
Kolberg
Rostock
GERMANY
Stettin
Schwedt
Berlin
Vestrup
Rudersdorf
Roxforde
Frankfut-an-der-Oder
Magdeburg
Staakow
Cottbus
Guben
Grunberg
Leipzig
Dresden
Jena
Zeitz
Litvinov
Kralupy
Pardubice
Kolin
Waidhaus
Prague
CZECH REP.
Trebic
Brno
Znojmo
Munich
Linz
Salzburg
Vienna
Bratislava
AUSTRIA
Graz
Paluzza
Maribor
SLOVENIA
Ljubljana
Zagreb
Belluno
Trieste
Koper
Rijeka
Porto Marghera
Gulf of Venice
Pula
Ferrara
Bologna
Ravenna
Ivana
CROATIA
Gospic
Florence
San Marino
Falconara
Ancona
ITALY
Bellante
Cellino
Rospo
Allano
Chieuti
Lanciano
San Stefano
Cupello
Vasto
Valle Cupo
Portocannone
Monte Stillo
S. Salvo
Marta
Rome
Ripi
Petroliard-Farnesina
Terracina
Candela
Giovinazzo
Bari
Gaeta
Naples
Pietralunga
Taranto
Brindisi
Grottole
Ferrandina
Cerro Falcone
Monte Alpi/Enoc
Tempa Rossa
Pomarico
Pisticci
Rotondella
Nova Siri
Tyrrhenian Sea
Crotone
Luna
San Leonardo
C. Cimiti
Vibo Valentia
Rosarno
Lavinia
Bagnara
SICILY
Palermo
Milazzo
Messina
Gagliano
Mazara del Vallo
Marinella
Bronte
Cisina
San Nicola
Catania
Ardizzone
Marsala
Porta Enpedocle
Gela
Perla
Commareta
Pone
Drillo
Ragusa
Vega
Augusta
MALTA
Mediterranean Sea
Ionian Sea
Adriatic Sea
Gdansk
Kaliningrad
Krasnoborskoye
Ushakovskoye
Kaunas
Vilnius
Grodno
Baranovich
BELARUS
Minsk
Davydovskoye
Vishanskoye
Mozyr
Pinsk
Kobrin
Brest
Plock
Warsaw
POLAND
Poznan
Lodz
Radom
Pulawy
Kielce
Sandomierce
Rozwadow
Trzebina
Grobla
Partinia
Lubaczow
Jaroslaw
Jaslo
Czechowice
Pribor
Zukow
Stablovo
Kovel
Lutsk
Rovno
Dubno
Brody
Ternopol
Lvov
Dashava
Kradobnensk-Kalusz-
Lugskoye
Starynj
Stryy
Bitkov
Kosov
Dolina
Nadvornaya
Kosmachskoye
Chernortsy
Krasnoilsk
SLOVAKIA
Uzhgorod
Miskolc
Budapest
HUNGARY
Pecs
ROMANIA
Bacau
Dranianesti
Ploiesti
Pitesti
Bucharest
Turnu-Severin
Giurgiu
Timisoara
Arad
Szeged
Resita
Belgrade
Pancevo
SERBIA AND MONTENEGRO
BOSNIA & HERZEGOVINA
Banja Luka
Bosanski Brod
Sisak
Sarajevo
Split
Dubrovnik
Aleksinac
Nis
Pristina
Shkoder
ALBANIA
Tirana
Durres
Vlore
Fier
MACEDONIA
Skopje
GREECE
Thessaloniki
Kavala
Prinou
S. Kavala
Komotini
Alexandroupolis
Larissa
Arta
Delfi
Gulf of Corinth
Elefsina
Aspropyrgos
Athens
Argos
Tripoli
Agean Sea
Khania
Rethimnon
Iraklion
CRETE
Rodhos
Soke
Izmir
Burhaniye
TURKEY
Bandirma
Bursa
Izmit
Istanbul
Eregli
S.Eregli
Cst.Eregli
Eregli Deep
Aliphi
S. Akcakoca
Ayazli
Cst. Akcakoca
BULGARIA
Sofia
Vraca
Stara Zagora
Plovdiv
Burgas
Otmanli
Varna
Kavarna
Tjulenovo
Staro Orjachovo
Tolloukhin
Constanta
Navodari
Midia
Lebada
Black Sea
Frasinu-Mislea
Izmail
Odessa
Kherson
Kishinev
MOLDOVA
Ugeny
Lasi
UKRAINE
Kiev
Berdichev
Staro-Konstantinov
Vinnitsa
Kremenchug
Ovruch
Lisichansk
Kharkov
Kursk
Unecha
3 Lines
2 Lines

MALTA

Capital: Valletta
Monetary unit: Lira
Refining capacity: 0
Oil production: N/A
Oil reserves: N/A
Gas reserves: N/A

Central Mediterranean regional structure setting

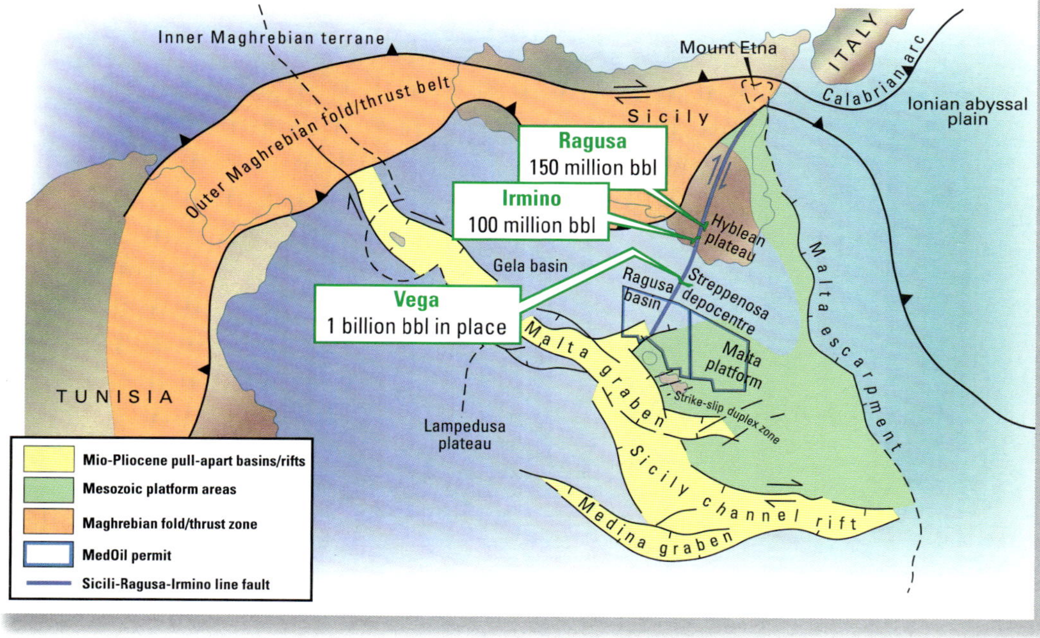

Play types

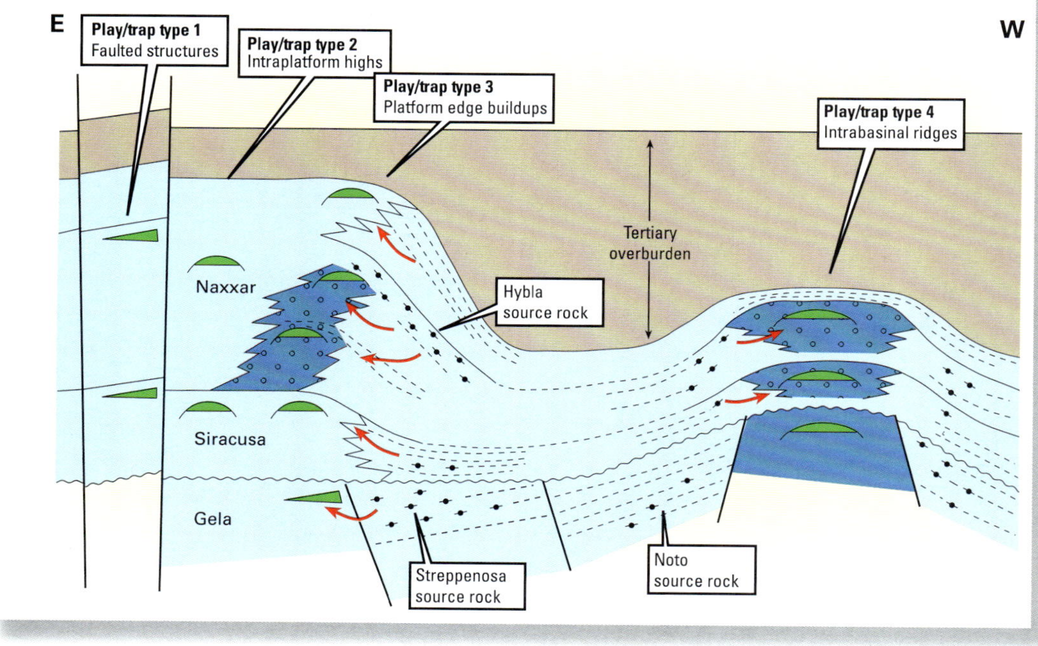

Intrabasin ridge trend promising

MedOil PLC (London) and its technical advisor Exploration Consultants Ltd. (Oxfordshire, UK) identified a new play in the Mediterranean Sea off northwestern Malta. The companies describe the area as having the potential to become a major hydrocarbon province.

They describe the region as the southern extension of the oil-producing Ragusa basin in onshore and offshore southern Sicily.

Evaluating seismic profiles and data from six exploration wells, MedOil identified two geologically distinct provinces: the Malta carbonate platform and the Ragusa basin area. All of the exploration wells were drilled on the carbonate platform.

In the basin are important intrabasin "highs" or ridges where optimal reservoir development and trap formation can be expected. These intrabasin ridges have not been drilled thus far in Maltese territory.

Results from wells drilled in the study area allow interpretation of potential reservoirs in the carbonates of the Lower Coralline and Naxxar formations.

Intrabasin ridges observed on seismic data are thought to be analogous to the intrabasin highs recognized onshore and off Sicily that can control the development of reservoirs and the location of oil fields.

The proximity of the MedOil permits and geologic similarities to the producing basins of Tunisia and Sicily suggest the occurrence of active petroleum systems in the area.

Seismic interpretation indicates that the Ragusa basin may extend into the northwest part of the MedOil ESA, making it likely that analogs of Sicily's producing fields (Vega, Mila, Ragusa, and Irminio) are present, said MedOil.

MedOil developed four potential play/trap combinations, including both established reservoir/trap combinations in producing fields and conceptual situations based on structural and stratigraphic interpretation.

Of these, MedOil described the intrabasin ridge as a proven play/trap combination, noting that the producing fields of Vega, Gela, Ponte Drillo, and Cammarata-Possillo are located on such ridges.

The company said there are other important intrabasin ridges in the Ragusa basin where optimal reservoir development and trap formation can be expected, with potential billion-barrel potential.

NETHERLANDS

Capital: Amsterdam
Monetary unit: Euro
Refining capacity: 1,211,886 b/cd
Oil production: 23,000 b/d
Oil reserves: 100 million bbl
Gas reserves: 50.0 tcf

Drilling and Production

North Sea gas production

Total SA's Dutch unit Total E&P Nederland brought on stream the L4G natural gas field off the Netherlands in the Dutch section of the North Sea, said field partner Lundin Netherlands BV, the Dutch unit of Swedish oil firm Lundin Petroleum AB.

The single L4G subsea development well has a production potential of about 1 MMcmd. The gas will be exported to the Uithuizen terminal in Groningen province through existing facilities.

The field's entire gas output has already been sold to Royal Dutch Shell PLC, ExxonMobil Corp., and Gasunie Trade & Supply BV, a gas-trading venture of the Dutch state,

Also, ATP Oil & Gas Corp. (Houston) said its wholly owned subsidiary ATP Oil & Gas (Netherlands) BV began production from its L06d-S1 well in 35 m (115 ft) of water in the Dutch North Sea. It is expected to produce initially at 15–20 MMcfd net to ATP.

A 41-km (25-mile) tie-back links the subsea well to a Gaz de France platform. ATP Oil & Gas (Netherlands) BV is the operator.

Pipelines

Balgzand-Bacton gas flow to UK starts

The new Balgzand-Bacton Line (BBL) began commercial delivery of natural gas in December from the Netherlands to the UK. The pipeline is operated by BBL Co., a joint venture of E.ON-Ruhrgas (Germany), Fluxys (Belgium), and NV Nederlandse Gasunie.

Most of the 235-km (146-mile) pipeline runs under the North Sea. It achieved its designed operational capacity of 15 bcm/year (1.45 bcfd) when a compressor station entered service in January 2007 in Anna Paulowna, Noord Holland.

The BBL pipeline was the second Netherlands-based gas transport project serving the UK to start up in late 2006 and early 2007. The other was an expansion of the capacity of the Interconnector pipeline in October between Zeebrugge in the Netherlands and the UK by 7 bcm/year (0.68 bcfd) to 23.5 bcm/year (2.27 bcfd).

LNG

Terminal plans proceed

ConocoPhillips and Dutch energy firm Essent Energie BV (Arnhem) began seeking permits for an LNG receiving and regasification terminal to be built at Eemshaven, in Eemsmond, North Groningen.

The companies executed a land option agreement in 2005 with site owner Groningen Seaports, which agreed to deepen the channel in the (Wester) Eems to accommodate LNG vessels. The companies submitted an environmental impact assessment to the province of Groningen. Terminal operation could begin in 2010.

NORWAY

Capital: Oslo
Monetary unit: Krone
Refining capacity: 310,000 b/cd
Oil production: 2.71 million b/d
Oil reserves: 7.705 billion bbl
Gas reserves: 84.26 tcf

Exploration and Development

Tyrihans field: integrated development

Statoil awarded a 440 million kroner ($65.6 million) contract to Norway's Aker Reinertsen (Trondheim)—a joint venture of Aker Maritime ASA and Reinertsen AS—to integrate development of Tyrihans oil and gas field onto the Kristin and Aasgard B platforms in the Norwegian Sea. The contract follows the receipt of government approval to proceed.

Tyrihans South is an oil field with a gas cap and Tyrihans North is a gas and condensate discovery with a thin oil zone. Tyrihans reserves are put at 182 million bbl of oil and condensate and 34.8 billion cu m of rich gas.

Work is to be completed by 2008, with Tyrihans due on stream in 2009 when spare capacity will be available in Kristin's topside facilities.

Gas from Tyrihans will be processed on Kristin and liquids on Aasgard B for transfer to Aasgard C storage and offloading.

The partners will spend 14.5 billion kroner ($2.16 billion) developing Tyrihans, said to be one of the biggest projects on the Norwegian continental shelf.

Deeper oil at Goliat

Eni Norge AS found deeper oil pay during appraisal drilling of its Goliat oil discovery on Production License 229 in the Barents Sea off Norway.

Its 7122/7-4 S well on the South Goliat structure encountered an oil column in Early Triassic Klappmys sandstone. The well also confirmed a deeper oil-water contact in the Late Triassic Realgrunnen subgroup and proved a gas-oil contact in Middle Triassic Kobbe.

The well was drilled to a TD of 2,366 m (7,762 ft) subsea in 372 m (1,220 ft) of water. The license is 50 km (31 miles) southeast of Snøhvit field and 85 km (53 nukes) northwest of Hammerfest. Eni operates PL 229.

Snøhvit oil potential to be studied

Statoil and partners announced plans to conduct feasibility studies on the Snøhvit field to assess commercial viability of Snøhvit oil production and options for development. The field's 2002 development plan envisioned only the exploitation of the field's gas resources. The Snøhvit oil zone is only about 14 m (46 ft) thick, but planning will be undertaken for a possible appraisal well project in 2007 to better identify the oil zone.

EOR at Halten studied

Royal Dutch Shell PLC and Statoil announced a CO_2 feasibility study for enhanced oil recovery on the Halten project. A power station built in Tjeldbergodden will supply the CO_2 for the project.

Shell estimates that the Tjeldbergodden power plant will generate about 2 mtpy of CO_2, which will be injected in the offshore Draugen field to raise oil production. Preliminary studies show that CO_2 injection will increase oil recovery by about 50 million bbl.

Licenses awarded from 19th round

The Norwegian Ministry of Petroleum and Energy, in the country's 19th licensing round, awarded 17 companies interests in March in 13 production licenses covering 33 blocks or parts of blocks on the Norwegian continental shelf.

BG Group is participating in eight of the licenses offered—five in deep waters of the Norwegian Sea and three in the Barents Sea. It will operate five of the licenses.

Statoil gained interests in five production licenses, four in the Barents Sea. It will operate three. In 2007 the company plans to start seismic surveys and to start drilling the first well.

The work program for all the awarded licenses includes four firm wells, seven "drill or drop" decisions, one appraisal well, and at least 9,860 sq km (3,807 sq miles) of 3D seismic survey.

Appraisal tests at Gudrun

Statoil ASA completed an appraisal well on its Gudrun oil, gas, and condensate prospect in the Sleipner area of the Norwegian North Sea. The well, about 190 km offshore in 109 m of water, was drilled to 4,180 m TD and tested successfully from two intervals.

A lower zone was tested at restricted rates of 6,300 b/d of oil and 18 MMcfd of gas, while an upper sand interval was tested at 4,000 b/d and 12 MMcfd, partner Marathon Petroleum Co. (Norway) reported.

Statoil is operator of Gudrun. BP PLC and Gaz de France are also partners.

Victoria discovery

Total announced that its wholly owned subsidiary, Total E&P Norge, signed a sale and purchase agreement with ExxonMobil Exploration and Production Norway to acquire their entire 30% interest in the PL211 license.

This license, in the Norwegian Sea 200 km offshore in water depths of around 400 m, holds the Victoria discovery. The acquisition is subject to the approval of the Norwegian authorities. It will raise Total's participation in the Victoria discovery to 50%.

Total E&P Norge plans to propose to the other partners, Statoil and Eni Norge, that it become operator.

Morvin, Valemon discoveries

Statoil confirmed the Valemon gas and condensate discovery near Kvitebjørn field in the Norwegian North Sea and the Morvin oil discovery on Halten Bank north of Kristin field in the Norwegian Sea. The company spudded the appraisal wells but did not report well tests.

Statoil drilled the Valemon appraisal well, 34/11-5S, to 7,380 m (24,213 ft) TD from its Kvitebjørn platform, calling it the longest high pressure-high temperature well it has drilled. It will complete the well for Kvitebjørn production.

The Smedvig ASA West Alpha semisubmersible drilled the Morvin appraisal, 6506/11-8, which will probably be developed as a satellite to Kristin or Aasgard field.

Tampen area oil strikes

Statoil, operator of production license 050, reported 2 oil discoveries in the Tampen area of the North Sea. Exploration well 34/10-49 and a sidetrack probed five targets—Epidot, Alun, Spinell, Sølvkåpe, and Apollo—in Cretaceous and Jurassic sandstones.

HALTEN CO_2 PROJECT

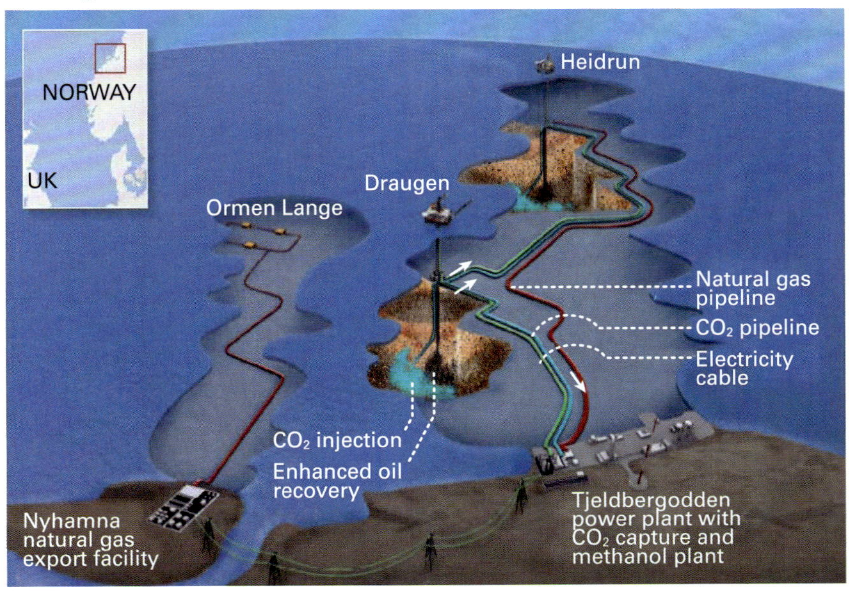

Source: Shell Technology webcast, Sept. 28, 2006

NORWAY'S 19TH LICENSING ROUND RESULTS

Company	Interest, %	Block
Talisman[1]	40	6202/7-9 and [2]12
RWE DEA	30	
DNO	30	
BG Group[1]	70	[2]6202/12
Revus Energy	30	
BG Group[1]	100	6302/5, 7, and 8
Total[1]	100	6406/11
BG Group[1]	70	6505/6 and 9 6506/4
Idemitsu	30	
BG Group[1]	40	6501/1
Idemitsu	20	
Amerada Hess	20	
Noreco	20	
Shell[1]	40	6603/5-8
Statoil	20	
ConocoPhillips	20	
BG Group	20	
Hydro Norsk	30	[2]7124/6 [2]7125/4 and [2]5
Eni	30	
BG Group	20	
SDOE[3]	20	

Company	Interest, %	Block
Statoil[1]	33.5	[2]7123/4
Gaz de France	12	
Total	18.5	
Amerada Hess	3.2	
RWE DEA	2.8	
SDOE	30	
Hydro Norsk[1]	40	7224/6; 7225/4
Statoil	25	
Gaz de France	20	
SDOE	15	
Statoil[1]	50	7226/2 and 3; 7227/1
BG Group	30	
SDOE	20	
BG Group[1]	70	7228/1 and 2
Discover Petroleum	10	
SDOE	20	
Statoil[1]	40	7230/2-6; 7231/4
ChevronTexaco	40	
RWE DEA	20	

[1]Operator. [2]Partial blocks. [3]State's direct financial interest.

The deepest borehole, 34/10-49 S and B, bottomed in Early Cretaceous rocks 5,480 m (17,979 ft) beneath the seabed. The 34/10-49 A sidetrack extended for 3,312 m (10,866 ft) and made the discoveries in Middle Jurassic Epidot and Alun sands.

Statoil estimated combined reserves at 15 million boe. The company said the discoveries could be tied back to Gullfaks field 5 km (3.1 miles) to the northeast.

Farmouts on NCS

PA Resources Norway AS (PAR) will finance the drilling of up to 6 exploration wells at a cumulative cost of 285 million kroner ($42.5 million) through a series of farmouts with DNO ASA subsidiary Det Norske Oljeselskap AS.

PAR obtained interests in 8 licenses on the Norwegian continental shelf, said DNO, which secured 340 rig-days that it expects will allow for drilling of 6–8 wells in 3 years, starting in 2006. The transactions involve the following licenses off Norway: PL 001B; PL 242; PL 028B; PL 305 and 305 B; PL 341; PL 332; and PL334.

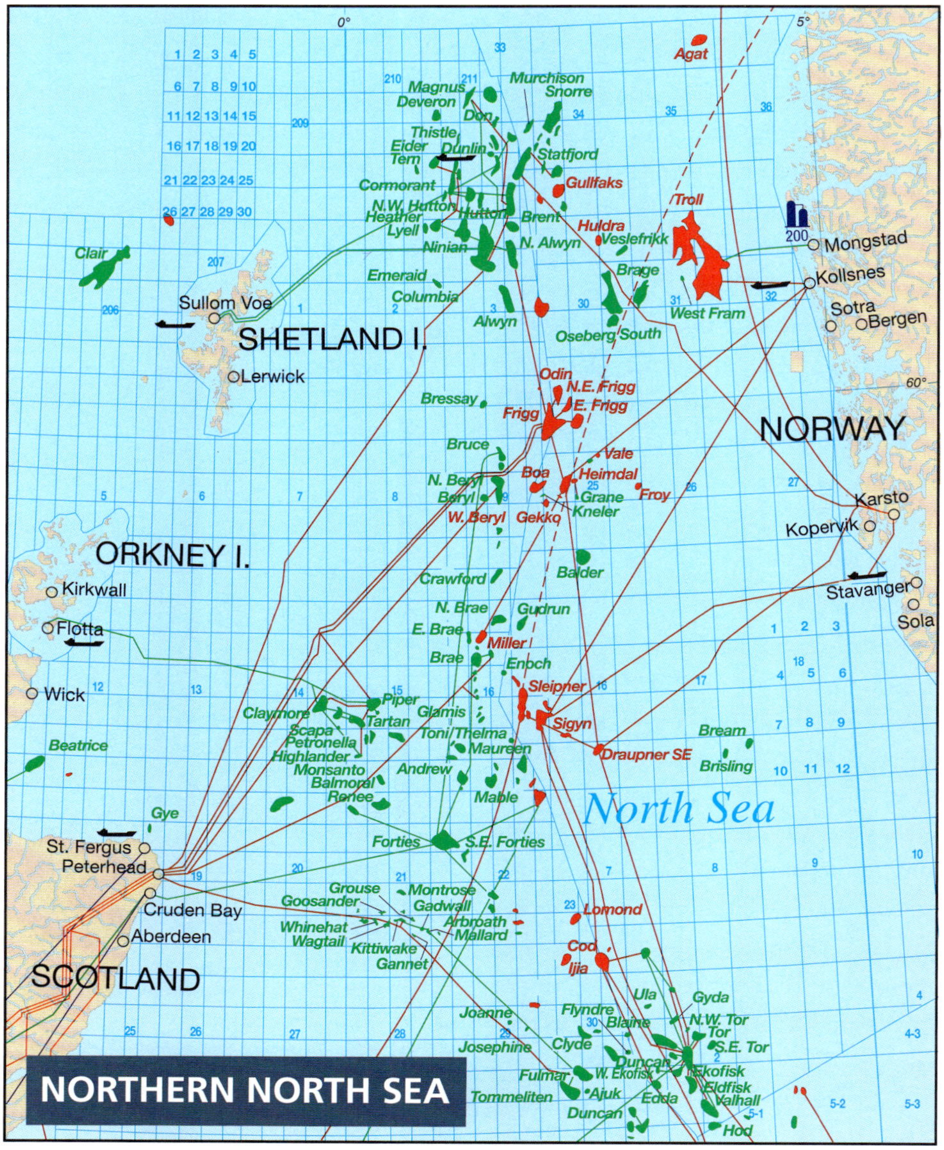

Drilling and Production

Kvitebjørn production reduced in North Sea

Operator Statoil ASA said it will cut natural gas and oil production from Kvitebjørn field in the Norwegian sector of the North Sea by 50% for five months, effective December 23, 2006.

Output will be reduced to 95,000 boe/d in the interest of "sound reservoir management and safe drilling operations" for wells remaining to be drilled, based on reservoir conditions and available methods for drilling in high-pressure, high-temperature reservoirs. That would amount to a net decrease of 15,000 boe/d in 2007 for Statoil, said the company.

Statoil will meet commitments to gas customers by increasing production from other fields.

Production from Kvitebjørn, on stream since 2004, was expected to plateau in late 2005. But in February 2006, estimates of recoverable reserves from the field were increased by 50%, or by 29 bcm of gas and 70 million bbl of condensate.

New Aasgard Blend

Statoil ASA introduced a new light, low-sulfur oil grade, Aasgard Blend, based on condensate and crude from Aasgard, Kristin, and Mikkel fields in the Norwegian Sea.

It consists of Aasgard Crude, Aasgard Condensate, and Kristin Crude, replacing the previous practice of separate crude and condensate liftings, and will be marketed as a crude mix.

Characteristics of the new crude blend are: gravity, 48.9°; specific gravity, 0.7842; sulfur content, 0.08 mass %; pour point, –27° C.; total acid number, <0.01 mg KOH/g; nickel, <0.1 ppm (wt); vanadium, <0.1 ppm (wt); viscosity at 20° C., 1.63 cst.

First Snøhvit gas production well

Statoil tested the first gas production well in May at Snøhvit field in the Barents Sea at the rate of 120 MMcfd. The horizontal well tapped the Jurassic Sto formation. The field is being developed in 250–345 m (820–1,132 ft) of water without sea-surface installations. It is scheduled to start producing gas through a 143-km (89-mile) multiphase pipeline to Melkøya near Hammerfest for trans-shipment as LNG in 2007.

Statoil puts the project's reserves at 193 bcm of natural gas, 113 million bbl of condensate, and 5.1 million tonnes of NGLs.

Deepwater compressor test

Norsk Hydro let a contract to Aker Kvaerner ASA for pilot testing of a new subsea compression technology in the Ormen Lange natural gas field. The project will evaluate a technology that GE Oil & Gas and Aker Kvaerner developed, testing whether a subsea compressor station in water 900 m (2,953 ft) deep is a viable alternative to an offshore platform.

Exploiting gas fields in water deeper than 500 m (1,640 ft) generally has not been profitable because of the cost of conventional offshore platforms.

The pilot subsea compression station would undergo controlled endurance tests during 2009–2011 at a gas treatment facility in Nyhamna, Norway.

Tordis IOR project approved

Statoil ASA got approval from the Norwegian government for a 2 billion kroner ($298 million) improved oil recovery (IOR) project at Tordis oil field in the North Sea.

Scheduled to start in 2006, the project is expected to boost recovery by about 35 million bbl and also reduce discharges of produced water. Statoil said it will be the first project in the world to adopt "full-scale subsea separation." The subsea separation system is due on stream in 2007.

Kongsberg FMC is to build the subsea separation station and Stolt Offshore is to manage marine operations in the field. Tordis is produced through subsea installations tied back to Staoil's Gullfaks C platform.

Find could drive LNG plant expansion

The Statoil-led Snøhvit consortium announced plans to drill additional wells in the Barents Sea in 2007 and 2008 to seek additional gas reserves for a possible future expansion of the Hammerfest gas liquefaction plant at Melkøya in northern Norway.

The plans follow a small but encouraging gas discovery on the Tornerose structure. Exploration well 7122/6-2 was drilled in 408 m (1,339 ft) of water to 3,057 m (10,030 ft) TMD below sea level, about 60 km (37 miles) east of the Statoil-operated Snøhvit field and 100 km (62 miles) northwest of Hammerfest.

Pipelines

Langeled pipeline gas flows

Norsk Hydro ASA announced the opening of the 540-km (340-mile) southern leg of the 44-in. subsea Langeled gas pipeline, from the Sleipner riser platform in the Norwegian North Sea to the east coast of England.

The British and Norwegian prime ministers inaugurated the system and gas delivery began in October from southern Norwegian gas fields. Norway's state-owned Gassco officially became operator of the pipeline and the receiving terminal at Easington.

Statoil ASA designed and laid the southern segment on behalf of Norsk, operator of the Langeled development phase.

When completed, the Langeled pipeline will carry gas from Nyhamna on Norway's west coast through Sleipner and on to Easington, becoming the world's longest subsea export pipeline. Facilities at Nyhamna will treat production from the giant Ormen Lange gas-condensate field, which is about 120 km (75 miles) from Nyhamna in the Norwegian Sea.

Flow from Nyhamna to Sleipner through Langeled's 107-cm (42-in) diameter northern section is to begin when Ormen Lange starts production in October 2007. That reservoir is about 3,000 m (9,842 ft) below the sea floor in 800–1,100 m (2,625–3,608 ft) of water.

Construction is about 90% complete, said Hydro. When Ormen Lange is on stream, Langeled will deliver 70 MMcmd of gas to the UK, supplying an estimated 20% of the country's gas needs for several decades.

Partners in the Langeled joint venture are Hydro, Petoro AS, Statoil, Dansk Olie & Naturgas (Dong), Esso Exploration & Production Norway AS, A/S Norske Shell, and ConocoPhillips.

Ormen Lange, Langeled Pipelines

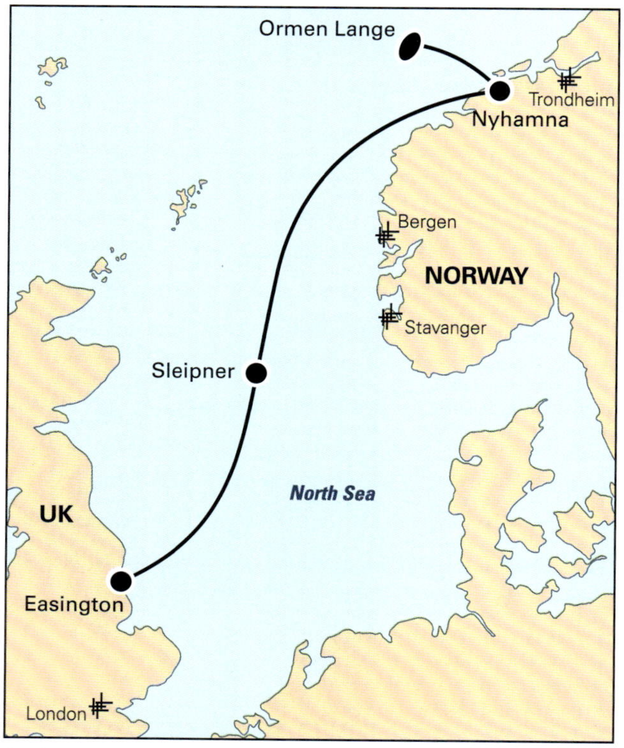

Ormen Lange Location

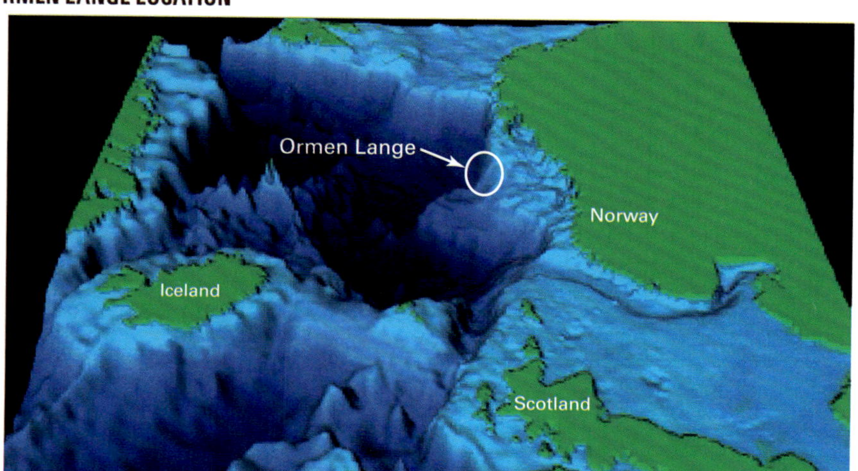

Refining and Processing

Mongstad refinery upgrade

Statoil ASA signed a letter of intent with Aker Kvaerner ASA for engineering plus procurement and construction management assistance services for an upgrade of the 200,000 b/cd Mongstad refinery north of Bergen.

Pending approval from the Norwegian authorities and a Statoil decision by Statoil to proceed with the entire project, engineering work could be completed in 2007 and the refinery upgrade completed in 2008.

The project also includes a new combined heat and power plant and a new gas pipeline from the Kollsnes to Mongstad.

Company News

Norsk Hydro, Statoil to merge oil, gas activities

The Boards of Directors of Hydro and Statoil agreed in late 2006 to recommend to shareholders a merger of Hydro's oil and gas activities with those of Statoil, creating the world's largest offshore operator.

The new company will have a combined production of 1.9 million b/d in 2007 and proven oil and gas reserves of 6.3 billion boe. Hydro will continue as a global aluminum company.

Final closing on the proposed merger, subject to shareholder and regulatory approval, is expected in late 2007. The new firm will be based in Stavanger.

Expanded reach for merged firm

The expected merged Statoil ASA and Norsk Hydro ASA oil and gas businesses will create an upstream player with multiple international operations. According to Wood Mackenzie (Edinburgh), the most important of these will be in the deepwater Gulf of Mexico.

Both companies recently acquired assets in the gulf. Activities in which the merged firm has interests include the Tahiti, Knotty Head, Big Foot, and Jack projects as well as the Frontrunner field, and several development projects in the Independence Hub platform area.

Statoil returned to the deepwater gulf in 2004 through a transaction with Chevron; later the company acquired EnCana Corp.'s deepwater assets for $2 billion and Anadarko Petroleum Corp.'s assets for $900 million. In 2005, Norsk Hydro expanded its deepwater GOM interests with the $2.45 billion acquisition of Spinnaker.

The combined company has commercial interests in 15 countries outside Norway. These include stakes in onshore Algeria, deepwater Nigeria, and Libya; in the Sincor heavy oil project in Venezuela's Orinoco Belt; the deepwater Chinook project in Brazil; and the Azar oil discovery in Iran. Statoil and Hydro also have extensive acreage in the UK North Sea, the Norwegian North Sea, and the Barents Sea.

Hydro to sell gas to Gazprom

Norsk Hydro signed an agreement for sale of gas to Gazprom Marketing & Trading, the UK-based subsidiary of the Russian energy company OAO Gazprom.

Hydro will supply 500 MMcm of gas to Gazprom Marketing & Trading in the British market. The gas agreement duration is from October 2006 through October 2007. The gas will be transported through the southern section of the Langeled gas pipeline.

POLAND

Capital: Warsaw
Monetary unit: Zloty
Refining capacity: 496,500 b/cd
Oil production: 17,000 b/d
Oil reserves: 96.375 million bbl
Gas reserves: 5.82 tcf

Drilling and Production

Wilga-2 well flows gas

FX Energy Inc. (Salt Lake City) started gas flowing from the Wilga-2 well 25 miles southeast of Warsaw. The company expects production of approximately 4 MMcfd of gas and 220 bbl/day of light crude oil.

For several weeks the company plans to vary the producing zones and flow rates. The well has three pay zones of Carboniferous age at 7,732-8,550 ft. FX Energy is operator and has an interest in the well and the 250,000-acre concession. A unit of Polish Oil & Gas Co. also has an interest.

Production also started from the Zaniemysl-3 well; no flow data were available.

FX Energy also began drilling the Winna Gora exploratory well on the Sroda City fault trend, estimating a potential of 25–50 bcf from the prospect. Plans were also in place to drill a second target, the Roskow-1, in the Fences I area in 2006.

Refining and Processing

Gdansk to get HDS unit

Grupa Lotos SA, Poland's second largest refiner, let a contract to ABB of Zurich for the engineering, procurement, and construction of a diesel hydrodesulfurization (HDS) unit at its 90,000-b/cd refinery in Gdansk, Poland. The $130 million award is part of Grupa Lotos's $1.3 billion residue-upgrade project.

The HDS unit will enable Grupa Lotos to meet European Union requirements that diesel contain less than 10 ppm of sulfur, which will come into force in 2009.

ABB Lummus Global's business in Wiesbaden will handle the work. Project start is planned for early 2009. The unit will use Chevron-Lummus Global technology to help boost production of ultra-low sulfur diesel to about 2.3 million tonnes/year.

HDS unit for Plock refinery

Polski Koncern Naftowy (PKN) Orlen SA awarded Technip (Paris) a contract worth €67 million for construction of a diesel oil hydrodesulfurization (HDS) unit at its 376,500 b/cd refinery in Plock.

This project will enable the refinery to increase production of high-purity diesel oil in compliance with the European Norm. Based on the Albemarle ultradeep HDS process, the unit will produce 260 tons/hr of high-purity diesel oil, with a maximum of 10 ppm (wt) of sulfur. Construction is scheduled for completion in 2009.

PORTUGAL

Capital: Lisbon
Monetary unit: Euro
Refining capacity: 304,172 b/cd
Oil production: 0
Oil reserves: 0
Gas reserves: 0

Expansion of Sines petrochem plant

Repsol YPF SA (Madrid) announced plans to nearly double production at the Sines petrochemical complex 160 km (100 miles) south of Lisbon by 2010.

The project will increase the capacity of the facility's naphtha cracker to 570,000 tpy of ethylene from 410,000 tpy, as well as increase polyethylene capacity from 295,000 tpy to 590,000 tpy.

A third initiative will add a power plant, plus new polypropylene capacity of 300,000 tpy.

Blocks 13, 14

Portugal was expected to award oil exploration rights in two blocks offshore Algarve province to a consortium owned by Repsol YPF SA and RWE AG. Blocks 13 and 14 lie in 200 m (650 ft) of water.

ROMANIA

Capital: Bucharest
Monetary unit: New leu
Refining capacity: 516,557 b/cd
Oil production: 98,400 b/d
Oil reserves: 600.0 million bbl
Gas reserves: 2.225 tcf

Drilling and Production

First sales from Bilca project

Europa Oil & Gas (Holdings) PLC (London) and its partners reported in September the first commercial gas sales from the Bilca project in the northern part of the Romanian Carpathians, a prolific oil and gas province.

Flow from three wells developed since 2004—Fratauti-1, Bilca-1, and Bilca-2—was processed in the new Bilca Gasfield Facility and subsequently carried by a 15-km (9 mile) export pipeline to Radauti, to enter the Transgas national pipeline.

The gas is being sold to Romgas SA, the Romanian State Gas Company and partner in the development. Gas will be produced initially at 200,000 cmd (7.06 MMcfd) for the first six months, with an increase to full facility capacity of 300,000 cmd (10.59 MMcfd) if operational conditions warrant.

Europa participated in drilling a deep exploration well on Brates Block (EPI-3), Costisa-1, which encountered gas shows, but it could not be properly tested. Well re-entry is planned for 2007. Exploration wells also are planned for 2007 in the Bilca play and on the Topolita prospect in the Cuejdiu Block (EIII-3).

Production begins at Fauresti field

Toreador Resources began producing natural gas mid-year from its Fauresti field rehabilitation project. Initial flows were 6 MMcfd of gas and 30 b/d of condensate. This production is from five wells the company reentered in late 2005 and early 2006 and recompleted as natural gas producers. Three more wells are to be reentered to finish an 11-well program.

The European Union accepted Romania as a member country, effective January 2007.

Pipelines

Constanta-Trieste project

Continued discussion but little action took place regarding the proposed 1,200-km (760-mile) oil pipeline from the Romania's Black Sea port of Constanta, across Romania, and Serbia, to the Croatian port of Omisalj on the Adriatic. Extension by 160 km (100 miles) to the Trieste oil terminal on Italy's Adriatic coast also has been suggested.

According to analysis in the regional publication, Balkan Insight, the Constanta-Trieste (C-T) line faces stiff competition from two other proposed pipelines that, like the C-T pipeline, also are designed to bypass the congested Bosporus and Dardanelles Straits:

- The 850-km (528-mile) AMBO line, from coastal Burgas, Bulgaria, through Macedonia, to Vlore, Albania, on the Mediterranean
- The Burgas-Alexandroupolis line, a joint project of Bulgaria, Greece, and Russia that would carry Russian oil 350 km (217 miles) from Bulgaria's Black Sea port of Burgas to the Greek port of Alexandroupolis on the Aegean.

Supporters say the C-T line could offer greater capacity than the others, connect at Trieste to the Trans-Alpine Pipeline Network, serve refineries along its route, and enable countries such as Kazakhstan to bypass Russia in delivering their oil to Europe.

However, questions remain about the strength of political and financial commitment to the project, while the others are more advanced in terms of negotiation, organization, and sponsorship.

SLOVAKIA

Capital: Bratislava
Monetary unit: Koruna
Refining capacity: 115,000 b/cd
Oil production: 1,000 b/d
Oil reserves: 9.0 million bbl
Gas reserves: 500 bcf

Slovakia is neither a large producer nor a large consumer of oil and gas. According to the International Energy Agency (IEA; Paris), more than 80% of the country's net primary energy supply is imported, chiefly from Russia.

The IEA has applauded the rapid progress Slovakia has made toward developing an open market economy over the past decade, including relevant changes in the energy sector to meet increasing energy demand.

Challenges ahead include energy-resource diversification, opening energy markets and integrating them into the EU, and increasing energy efficiency to offset rising energy prices and to help control carbon emissions.

Licenses awarded in northeast region

A subsidiary of Aurelian Oil & Gas PLC (London) was awarded three licenses in the Slovak Republic in August. The licenses—Svidnik, Medzilaborce and Snina—form a contiguous area of approximately 2,480 sq km (957 sq miles).

All are in the northeast Carpathian mountain belt, which trends north and west into Poland. The blocks are complimentary to the Bieszczady blocks in southeast Poland, which are just across the border. Previous exploration wells drilled in the late 1970s encountered good shows of oil and gas. Aurelian was planning to begin resource evaluation in late 2006.

Biodiesel enters market

The crude oil refinery group Slovnaft began selling 95-octane gasoline with ETBE (bio-ethyl tertiary butyl ether) in the country. Slovnaft also plans to start adding ETBE to other fuel types as well.

Sale of the ETBE product helps Slovakia meet the EU requirements that 2% of the energy content of transportation fuels be made up of bio-fuels in 2006.

Control of Transpetrol pipeline at issue

The Slovak government continued deliberations with Russian officials regarding control of the Transpetrol oil pipeline system. A 49% stake in Transpetrol was privatized in a tender in 2002 to Yukos Finance, a unit of Russian oil giant, Yukos.

However, Yukos entered bankruptcy in August 2006, and a complicated sequence of transactions left control of the 49% stake unresolved as of year-end 2006.

Complicating the situation is the fact that Slovakia's right of veto on the disposition of the stake expires in April 2007. After that time, the likelihood of its sale to a Russian firm increases.

Three companies have expressed interest in Transpetrol: the Russia's Gazprom and Russneft, and Slovakia's largest private equity group, Penta Investments.

Energy professionals fear that if the stake remains under Russian control, the Družba pipeline that Transpetrol uses will not be employed to transport Caspian Sea oil west, as the EU wants. Instead, the company would continue to transport exclusively Russian oil, preventing the desired diversification of supplies and increased energy security for Slovakia and Europe.

SPAIN

Capital: Madrid
Monetary unit: Euro
Refining capacity: 1,271,500 b/cd
Oil production: 3,000 b/d
Oil reserves: 150.000 million bbl
Gas reserves: 90 bcf

Refining and Processing

La Rabida refinery

Compania Espanola de Petroleos SA (CEPSA) let two contracts related to the expansion and upgrade of its 100,000 b/cd La Rabida refinery at Huelva. The project will increase output of middle distillates.

One is a lump-sum contract to Foster Wheeler Iberia SA for FEED and early procurement services for new crude, vacuum, and gas concentration units. The new units will be based on a design package by UOP (Des Plaines, IL, U.S.A.).

The second contract is to UOP itself for eight process units employing proprietary technologies for hydrocracking and sulfur removal.

The new complex will include a 90,000 b/sd crude distillation unit, a 30,500 b/sd vacuum distillation unit, a 32,000 b/sd gas concentration unit, a 4,200 b/sd LPG Merox unit, a 27,000 b/sd naphtha Merox unit, a 31,500 b/sd distillate Unionfining unit, a 41,200 b/sd atmospheric and vacuum gas oils Unicracking unit, and an amine regeneration and sour water stripper units.

Cartagena refinery

Repsol YPF SA (Madrid) chose the technologies of IFP NA subsidiary Axens NA for the planned expansion of its 100,000 b/cd Cartagena refinery in Murcia.

The new refining train will include a 50,000 b/d hydrocracker that will convert a blend of heavy vacuum and heavy coker gas-oils into high-quality, ultra-low-sulfur middle distillates for the European market.

Axens also is committed to provide a C_5-C_6 alkane isomerization unit to boost gasoline pool octane, two hydrodesulfurization units to produce ultralow-sulfur diesel from straight run and light coker gas-oil, and a coker naphtha HDS unit. Startup of the new facilities is slated for 2010.

Castellon refinery

BP Oil Refineria de Castellon SA awarded a detailed engineering, procurement, and construction supervision contract to Foster Wheeler Iberia SA for a delayed coker at BP's 104,500 b/cd refinery in Castellon de la Plana. The project aims to reduce residual fuel oil production at the facility.

Contract terms and reconfiguration details were not disclosed. The planned 20,000 b/sd coker will use Foster Wheeler's proprietary delayed coking technology. The new coker is scheduled for completion during 2008.

SWEDEN

Capital: Stockholm
Monetary unit: Krona
Refining capacity: 434,000 b/cd
Oil production: 0
Oil reserves: 0
Gas reserves: 0

Government aim: Break oil dependence

Sweden's Minister for Sustainable Development said that the country had the chance to be an international model as the first government to break its dependency on fossil fuels.

A series of incentives and penalties would boost its renewables sector and reach a state of energy self-sufficiency by 2020 to yield environmental and economic benefits. Proposed government measures include:

- Tax relief for conversion from oil to renewable fuels for homes, public buildings, and the transport sector
- Greater research into renewable resource development and use
- Continued investment to further expand use of district heating
- Financial incentives to increase use of use biofuels in district heating

Alternative motor fuels

Statoil ASA (Stavanger, Norway) said sales of its E85 biofuel grew by 270% in Sweden in 2006 compared with 2005. Svenska Statoil's petrol stations sold 19.5 million l (5.2 million gal) of E85 in 2006 at 170 of its sites in Sweden. The company plans to add 90 outlets in 2007.

Statoil said that 36,711 "environmental cars" were purchased in Sweden during 2006, up 156% over 2005. They represented 13.5% of all newly registered cars, compared to 5.2% in 2005.

The company also began adding 5% rapeseed oil (rape methyl ester) to diesel sold at its retail outlets in Sweden. The company said it expects to cut carbon emissions by 115,000 tonnes/year by replacing 46 million l/year (12 million gal/year) of diesel with Biodiesel 5, the blend that contains the rapeseed oil component. Statoil said all diesel-powered vehicles could use it with no engine adjustments.

Security concerns about Russian pipeline

Plans by Russia's Gazprom to build a natural gas pipeline across the Baltic Sea generated worry across the political spectrum in Sweden. A range of leaders expressed concerns not only about potential environmental impacts but also about potential use of the pipeline infrastructure as a platform for Russian intelligence-gathering.

The pipeline, to be managed by the Russo-German company Nord Stream, would span 1,200 km (745 miles) between Vyborg, Russia and Greifswald, Germany. It would not enter Swedish territorial waters but would cross Sweden's economic zone in the Baltic.

The Nord Stream plan is being studied by the Swedish Environmental Protection Agency, which will then share its views with Finland, Russia, Denmark, and Germany.

LNG terminal planned for Nynäshamn

Nynas Oil Group (Stockholm) said it intends to partner with AGA AS (Oslo, Norway) and Fortum Oyj (Espoo, Finland) to apply for a license to build an LNG receiving terminal near the Nynas refinery in Nynäshamn. Initial plans call for a storage volume of 20,000 cu m (706,293 cf). The proposed completion date is 2009.

TURKEY

Capital: Ankara
Monetary unit: Lira
Refining capacity: 714,275 b/cd
Oil production: 42,000 b/d
Oil reserves: 300 million bbl
Gas reserves: 300 bcf

Exploration and Development

Akcakoca subbasin project moves ahead

Toreador Resources Corp. (Dallas) entered the final stages of construction of the South Akcakoca subbasin project in the Black Sea off Turkey. Toreador and its joint venture partners are assessing the reserve potential along the Akcakoca trend, in water too deep for jack-up rigs.

Partner Stratic Energy Corp. (Calgary) said construction of the onshore production facility was progressing well as major equipment underwent testing in December.

An 18-km (11-mile) onshore pipeline, already constructed, links the production facility to the national grid. Toreador expected to finalize gas plant installation and precommissioning in December. Completion of a related offshore pipeline was scheduled for early 2007.

Stratic said the Akkaya and Dogu Ayazli production platforms would be commissioned first, followed by the Ayazli platform. First-phase production of 50 MMcfd is expected in January 2007.

Bismil, Cizre licenses awarded

TransAtlantic Petroleum Corp. (Calgary) was awarded three exploration licenses in southeastern Turkey near the border with Iraq. The 4-year licenses cover a combined 162,762 acres.

Two licenses are near Bismil on the Tigris River adjacent to the Molla and Karakilise producing oil fields. TransAtlantic will target an underexplored Paleozoic play at 3,000 m (9,845 ft).

The third license is near Cizre 60 km (37 miles) from the Iraq border and targets a deep Zagros-type subthrust play. Combined work commitments will total $300,000 for two years.

Pipelines

BTC pipeline deliveries start

The Baku-Tbilisi-Ceyhan (BTC) crude oil pipeline went into operation in mid-2006. The 760-km (1,087-mile) pipeline is the first to allow crude oil to be exported from the Caspian region without relying on the Russian pipeline system, crude oil swaps with Iran, or passage through the Bosporus and Turkish Straits.

It carries oil from the Azeri-Chirag-Guneshi oil field in the Caspian Sea to the port of Ceyhan on the Mediterranean Sea in southern Turkey.

BTC Pipeline Company (BTC Co.) announced that the first cargo of 600,000 bbl of oil from the pipeline was loaded onto the tanker British Hawthorn at Ceyhan on June 4. The official inauguration ceremony was held July 13.

The pipeline's throughput capacity is 1 million b/d, expected to be reached in 2008.

BTC Co. was created in 2001 to build own and operate the pipeline. BTC Co. shareholders are BP, State Oil Company of Azerbaijan (SOCAR), Unocal, Statoil, Turkiye Petrolleri Anonim Ortakligi (TPAO), Eni, Total, Itochu, Inpex, ConocoPhillips, and Amerada Hess. BP is the pipeline operator.

Cyprus ponders offshore exploration

Cyprus and Egypt signed several cooperation agreements in May after a summit meeting between leaders of both countries. Topics included potential establishment of joint private-sector projects for oil and natural gas exploration offshore Cyprus, said the Middle East News Agency (MENA).

Previous studies have suggested the seabed around Cyprus could contain unproven reserves of 6 billion to 8 billion bbl of crude oil. Preliminary data appear to give encouraging" signs of hydrocarbons. The Greek Cypriot press reported that oil exploration studies could start in February 2007.

In January 2007, Cyprus officials signed a similar agreement with Lebanon regarding oil and gas exploration in the Mediterranean between the two countries.

All of the agreements drew warnings, however, from Turkish leaders as well as Turkish Cypriot officials. They expressed concerns that offshore oil and gas deposits must be accessed jointly by both Turkish and Greek Cypriot communities and that revenues must be shared.

UNITED KINGDOM

Capital: London
Monetary unit: Pound
Refining capacity: 1,887,068 b/cd
Oil production: 1.480 million b/d
Oil reserves: 3.875 billion bbl
Gas reserves: 17.0 tcf

Exploration and Development

Shoei prospect

A group led by ConocoPhillips reported a large gas-condensate discovery in the Central UK North Sea. Reserves have been estimated at 100–275 million bbl of condensate.

The prospect—formerly known as Shoei but to be renamed—lies on Blocks P011 and P0032 30/7a, in 81 m (265 ft) of water. It is 9 km (6 miles) west of the producing Judy oil and gas field.

Production is slated to begin in 2010, with possible tie-back to the ConocoPhillips Judy platform. Partners in the project with ConocoPhillips are BG Group and Eni UK Ltd. The group will seek development approval in 2007.

Saxon, Guillemot fields

Petro-Canada secured the GlobalSantaFe Glomar Arctic III semisubmersible for use in its 2006-2007 UK work program. This includes development of the Saxon oil field (Tay formation, Central North Sea), in-fill drilling in the Guillemot oil field (Blocks 021/24 and 021/25, Central North Sea), and three exploration wells. Production from the Saxon field is expected in 2007.

Babbage well

Dana Petroleum PLC (Aberdeen) reported the successful drilling and flow testing of Babbage gas field in the UK Southern Gas basin. The appraisal well, on North Sea Block 48/2a, targeted a crestal area of the field with a view to proving gas reserves and potential productivity.

The vertical well, drilled to more than 3,350 m (11,000 ft) TD, flow-tested at 10.7 MMcfd of gas without stimulation. The company called Babbage one of the largest undeveloped gas fields in the UK North Sea, with as much as 390 bcf in place in Permian Rotliegend.

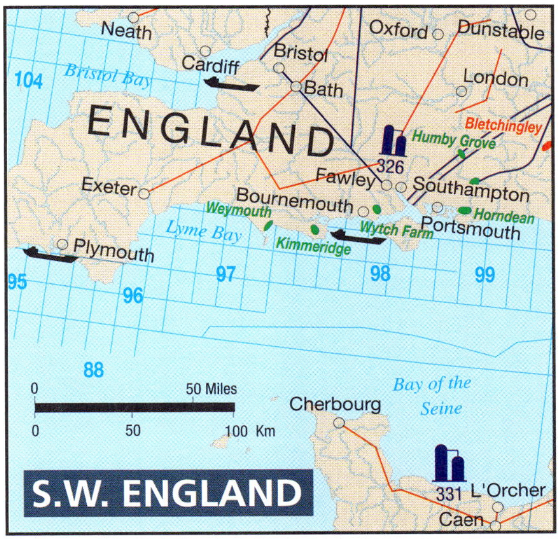

E.ON Ruhrgas UK North Sea is operator. Partners include Dana and Centrica Resources Ltd.

Vulcan East well

Silverstone Energy Ltd. (Aberdeen, Scotland) struck natural gas with the 49/21-10A Vulcan East exploration well, drilled to 2,475 m (8,125 ft) TD in the southern UK North Sea.

The Vulcan East well discovered an accumulation separate from adjacent Vulcan field, said Silverstone, the operator. The discovery is estimated to contain 150-200 bcf of gas in place.

Silverstone and partners, ConocoPhillips and BP PLC, are reviewing the well results to determine the viability of bringing the field onto production. Silverstone is a 50-50 joint venture between Lime Rock Partners (Houston) and independent Storm Ventures International Inc. (Calgary).

Cygnus well

Endeavour International Corp. (Houston) announced its first discovery in the Southern Gas Basin in the UK of the North Sea when the Cygnus 44/12-2 well tested as a gas discovery.

The company also expects to close on the acquisition of interests in seven producing fields in the Central North Sea section of the United Kingdom Continental Shelf and to begin exploratory drilling.

Drilling and Production

Rhum field

BP PLC and Iranian Oil Co. UK Ltd. began natural gas production and export from what they called the UK's largest undeveloped gas discovery, Rhum gas field, 386 km (240 miles) northeast of Aberdeen.

The high-pressure, high-temperature (HPHT) reservoir flowed through the 3/29a-5 well at an initial rate of 130 MMscfd of gas. Production is expected to plateau at 300 MMscfd of gas when all three development wells are on stream.

Rhum has estimated reserves of 800 bcf of gas. It is a subsea tieback to BP-operated Bruce field. Gas will be exported from Bruce via the Frigg pipeline system to St. Fergus and associated condensate will be piped via Bruce into the Forties Pipeline System.

The development of a HPHT gas reservoir using a long-distance subsea tie-back is a world first, BP said.

High activity expected on UKCS

Oil and gas exploration and production firms operating in the UK Continental Shelf (UKCS) are planning to drill about 180 wells in 2006–2008, said UK-based North Sea consultancy Hannon Westwood. Around $2 billion/year of investment will be required to boost production from the mature province.

Companies will compete intensely for drilling rigs, said the consultancy, noting that there is a rig utilization rate of 93-100% of Europe's 72-rig offshore fleet. There are 143 companies with UKCS interests, many of which are small independent North Sea companies. The wells are expected to add 500 million boe/year of reserves.

The strong competition for exploration opportunities has been demonstrated by the record number of applications for acreage under the UK's 24th licensing round. The results were expected in first quarter 2007. Approximately 140 companies have applied for licenses.

Causeway wells

Antrim Energy Inc. (Calgary) secured a semisubmersible rig to drill three Causeway appraisal and development wells in 2007 on UK North Sea Block 211/22a. AGR Peak Well Management will conduct the drilling program. Peak earlier had managed the drilling on East Causeway.

Antrim will spud the first appraisal-development well in 2007 about 2 miles southwest of the East Causeway discovery. The company expects to drill to 3,500 m (11,500 ft) and to appraise the Middle Jurassic Brent Group of sandstones.

The drilling program follows Antrim's East Causeway discovery 211/23d-17z, which flowed on final test up to 7,500 b/d of light, sweet oil.

Goosander field

Production began from Goosander oil field on Block 21/12 in the Greater Kittiwake Area of the UK North Sea. Dana Petroleum PLC (Aberdeen, Scotland) said it and 50-50 partner Venture Production PLC (Aberdeen), the operator, expect the subsea waterflood development to reach a peak annual average production rate of 9,500 b/d.

The partners reentered and completed the Goosander discovery well, 21/12-3, for tieback to the Kittiwake platform. They plan to drill a water-injection well in 2007. The field is expected to produce 15.2 million bbl of oil and 2.6 bcf of natural gas over 7–10 years.

Clair field

Chevron Corp. conducted an assay of crude from the Clair oil field off the UK west of Shetland, from which BP PLC, the operator, began production in 2005. Plateau production will be 60,000 b/d of oil and 15 MMcfd of natural gas.

Clair lies 75 km (47 miles) west of Shetland in 140 m (460 ft) of water. The reservoir, which covers five license blocks—206/7a, 206/12, 206/8, 206/13a, and 206/9—has an estimated 5 billion bbl of oil in place. Initial development will access 250 million bbl of oil reserves.

Oil from the field will move to the Sullom Voe terminal in the Shetland Islands through a 105-km (65-mile), 22-in. pipeline. Gas will flow through a 10 km (6-mile), 6-in spur to the Magnus enhanced oil recovery trunkline.

Chevron's development partners included BP, ConocoPhillips, Royal Dutch Shell PLC, and Amerada Hess Corp. BP is the operator.

Clair TBP Distillation

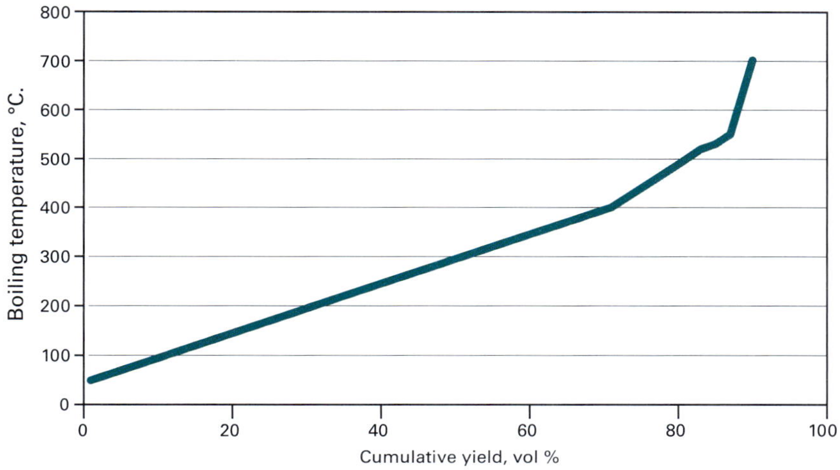

Glenelg field

Total SA has started production from Glenelg gas and condensate field in 100 m of water in the UK North Sea, 240 km (150 miles) east of Aberdeen.

Production is from a highly deviated well drilled a total length of 7,300 m (23,950 ft) from the Elgin platform into a high-temperature, high-pressure reservoir 5,600 m (18,375 ft) below sea level. Total said the well has potential to produce 30,000 boe/d of gas and condensate.

Total is the operator of Glenelg. Other interests are E.ON Ruhrgas, BG International, Gaz de France, and Eni SPA.

Chestnut field

Venture Production PLC (Aberdeen) and partners Bow Valley Petroleum (UK) Ltd. and Atlantic Petroleum UK Ltd. were proceeding to develop the Chestnut oil field, in Block 22/2a. Total recoverable reserves are estimated at 6–9 million bbl. First production is scheduled for 2007.

Two subsea wells will be tied back to the Sevan Marine SSP300 floating production vessel. The SSP300, newly constructed, is anticipated to be installed in the field during 2007.

Kilmar field

ATP Oil & Gas Corp. (Houston) began gas production at an undisclosed rate from the Tors development in Kilmar field on UK North Sea Block 43/22. ATP operates the Southern Gas basin block.

Production is from a multilevel well drilled to 3,360 m (11,025 ft) TVD. The well is the first of at least four scheduled in the first phase of development. When fully developed, Tors will produce 50–75 MMcfd net to ATP. Tors has 73 bcf of gas equivalent net proved reserves and 42 bcf of net probable reserves.

Forvie North field

Total SA reported that production started from Forvie North gas and condensate field on UK North Sea Block 3/15, 440 km (273 miles) northeast of Aberdeen, in 120 m (393 ft) of water. Total owns the field, developed as a satellite to Alwyn North, 33 km (20 miles) away. A high-pressure, gas-condensate pipeline was built from Forvie North to Alwyn North.

The Forvie North subsea system has expansion capacity to allow the tie-in of additional wells. At plateau, Forvie North is expected to produce 20,000 boe/d.

North West Hutton platform removal

BP PLC let a contract to Heerema Marine Contractors for the main decommissioning work on its North West Hutton platform 130 km northeast of the Shetland Islands in the UK North Sea. The platform sits in 140 m (460 ft) of water. Offshore removal is to be completed by 2009.

The 20,000-tonne topsides and the 17,000-tonne steel jacket down to the tops of the footings will be moved to the Able yard on Teesside for recycling and disposal. BP said no company has tried to dismantle a steel jacket structure of this weight and in this water depth.

Pipelines

Three projects boost gas delivery

Three major gas transport projects serving the UK started up in late 2006.

The 540-km (340-mile) southern leg of the Langeled pipeline opened, bringing gas from the Statoil ASA-operated Sleipner and Troll fields in the Norwegian North Sea to Easington, Hull in the eastern UK.

When the northern leg is completed and Norway's giant offshore Ormen Lange field starts production in 2007, Langeled will deliver 70 MMcmd, supplying an estimated 20% of UK gas needs. (See "Norway" section for more details.)

UK GAS INFRASTRUCTURE, PROPOSED IMPROVEMENTS

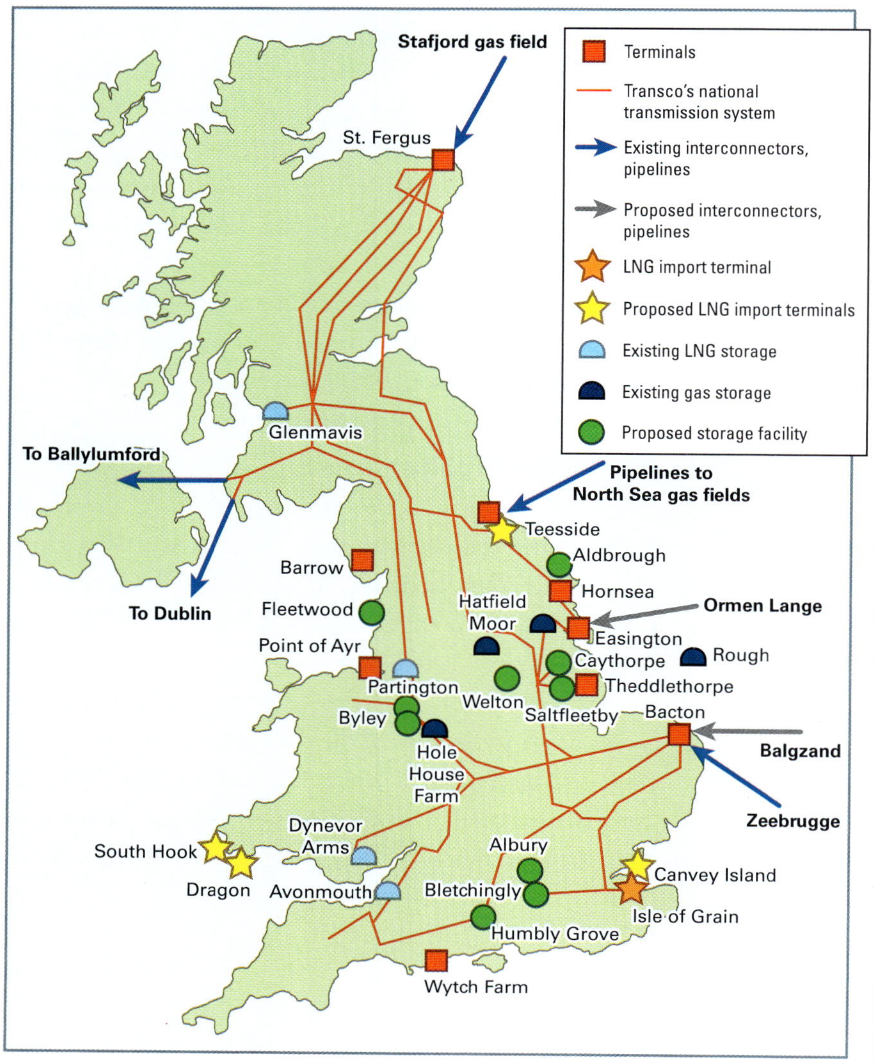

The new Balgzand-Bacton pipeline began delivering natural gas from the Netherlands. The 235-km (146-mile), 36-in. pipeline will have a capacity of 15 bcm/year when a compressor station enters service in 2007 in Noord-Holland province. The pipeline is a joint venture of E.ON Ruhrgas (Germany), Fluxys (Belgium), and NV Nederlandse Gasunie.

Also, the capacity of the Zeebrugge-Bacton Interconnector pipeline between the Netherlands and the UK was increased by 7 bcm/year (0.68 bcfd) to 23.5 bcm/year (2.27 bcfd).

Policy

Tax warning

In December 2005, the Chancellor of the Exchequer doubled the supplementary corporation tax rate on oil and gas production to 20% in a move that pushed the total corporate tax on production begun after 1993 to 50% and on older production to 75%.

BP PLC, Europe's biggest oil company, estimated that it would face a $600 million, one-time deferred tax charge if the tax increase were approved.

In response, a variety of UK labor, community, and business groups insisted that the government must change the fiscal regime to ensure production of the 16-27 billion bbl of oil and gas equivalent estimated to be recoverable from the UK Continental Shelf (UKCS).

The UK Offshore Operators' Association (UKOOA) said that the UK needs a "radical review" of the extraction of oil and gas if it is to remain competitive in attracting investment.

The association said that increasing UK production—requiring investment of £300 billion ($570 billion)—must become a priority, stressing that this investment might not be made unless the regulatory and tax regime accommodates resource maturation and high production costs of the UKCS.

Oil and gas currently supply 75% of the nation's energy demand, and this is expected to rise to 80% by 2020. Gas is expected to provide over half of the UK's electricity needs by 2020.

Cabinet-level energy post

The UK Offshore Operators Association (UKOOA) recommended the appointment of a secretary of state for energy to develop and provide leadership on energy policy. UKOOA made the recommendation in its response to the government's energy review.

The association also called for creation of an energy policy that reflects the fact that the UK will increasingly depend on oil and gas for its energy supply.

Subsea salt cavern storage

UK lawmakers should consider allowing natural gas to be stored in subsea salt caverns, said the secretary of state for Trade and Industry during a parliamentary debate on supply security.

Current UK storage laws do not address undersea storage facilities. A report by the British Geological Survey identified the southern North Sea and a small area in the Irish Sea is suitable for gas storage in salt caverns.

'Looming energy gap' seen

The UK will require supply from all energy sources to sustain its standard of living through 2050, said 150 scientific, technical, economic, and sociological experts. They contributed to a November 2005 white paper aimed at helping the UK government address a "looming energy gap." Changes in the way energy is used also will be necessary.

The Geological Society of London (GSL) and five sister institutions sponsored the white paper. "How to Plug the Energy Gap" stressed a continuing need for fossil fuels, renewables, and nuclear fission, as well as energy conservation. The document is on the GSL Web site: www.geolsoc.org.uk.

LNG

South Hook

CB&I John Brown Ltd. (The Woodlands, TX) awarded Cape PLC (Wakefield, UK) a contract on the construction of the new South Hook LNG terminal at Milford Haven, Wales. Cape said it will provide the "common user access service" for the main process unit on the terminal, which is to be completed by 2008.

CB&I won contracts in 2004 and 2005 from South Hook LNG for Phases 1 and 2 of the project. The terminal will be owned and operated by South Hook LNG Terminal Co., a joint venture of Qatar Terminal Co. Ltd. and ExxonMobil Qatargas (II) Terminal Co. Ltd.

LNG for South Hook will be supplied from the Qatar Liquefied Gas Co. Ltd. (Qatargas II) LNG plant being built at Ras Laffan Industrial City in Qatar. Qatargas II will supply 15.6 mtpy of LNG to the UK for 25 years.

Teesside

ConocoPhillips Petroleum UK Ltd. said it will submit an application on behalf of Norsea Pipeline Ltd. for planning permission for an LNG regasification facility and combined heat and power plant on the site of Norsea's Teesside, UK, oil terminal.

ConocoPhillips is a Norsea Pipeline shareholder. Others are Total SA, Statoil ASA, Eni SPA, and Norsk Hydro AS. ConocoPhillips expects receipt of planning permission in 2007. Subject to official approvals, the facilities could be in operation in 2011.

Middle East

BAHRAIN

Capital: Manama
Monetary unit: Dinar
Refining capacity: 248,900 b/cd
Oil production: 172,000 b/d
Oil reserves: 124.56 million bbl
Gas reserves: 3.25 tcf

Bahrain has become a major regional financial center, but its economy still depends heavily on oil. Petroleum revenues make up about two-thirds of government revenue and export earnings, according to the U.S. Energy Information Administration (EIA).

Bahrain's two major industries are refining and aluminum smelting, so oil prices dramatically affect its economy. The relatively high oil prices of recent years have helped Bahrain's real GDP to grow at over 5% for 2004 and 2005, said EIA. A rate of 5.5% is projected for 2006.

Sitra refinery upgrades

A key facility in Bahrain's hydrocarbon sector is the Sitra refinery of Bahrain Petroleum Company (Bapco), with a capacity of approximately 250,000 b/cd. Work continued in 2006 on a major project to modernize diesel production at the refinery.

JGC Corporation (Yokohama, Japan) is managing the building or upgrading of several refinery units. This includes building a new hydrocracker unit with a capacity of 60,000 bpd to produce low sulfur diesel. Construction began in 2005 and completion is expected in early 2007.

This new hydrocracker will allow increased production of low-sulfur diesel (concentration <10 ppm) that meets the strict European and US environmental regulations.

In a separate project, Foster Wheeler Corp. (Clinton, NJ, U.S.A.) continued work on a new gas treatment plant and sulfur recovery unit at the refinery to reduce sulfur dioxide emissions. The work is to be completed in 2008.

The completion date for a unit to produce high-quality lubricant base oil was postponed from 2008 to 2009, said Neste Oil (Porvoo, Finland), which will market the oil. The unit will produce 400,000 tonnes per year (tpy) of sulfur-free, very high viscosity index base oil.

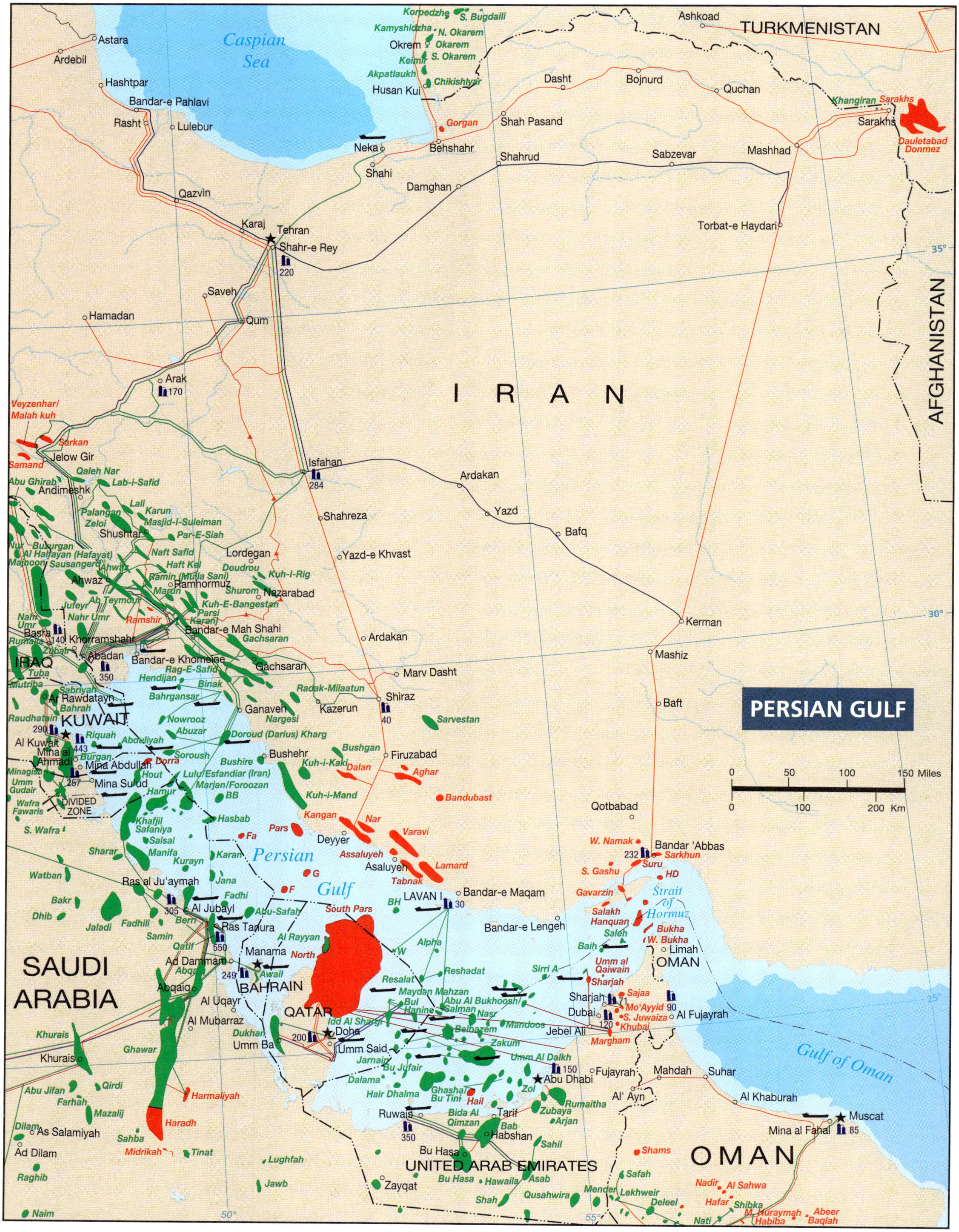
PERSIAN GULF
TURKMENISTAN
IRAN
AFGHANISTAN
IRAQ
KUWAIT
SAUDI ARABIA
BAHRAIN
QATAR
UNITED ARAB EMIRATES
OMAN
Caspian Sea
Persian Gulf
Strait of Hormuz
Gulf of Oman
DIVIDED ZONE
0 50 100 150 Miles
0 100 200 Km
35°
30°
25°
50°
55°
Astara
Ardebil
Hashtpar
Bandar-e Pahlavi
Rasht
Lulebur
Qazvin
Karaj
Tehran
Shahr-e Rey
220
Saveh
Qum
Hamadan
Arak
170
Neka
Shahi
Behshahr
Gorgan
Damghan
Shahrud
Shah Pasand
Dasht
Bojnurd
Quchan
Ashkoad
Sabzevar
Mashhad
Sarakhs
Khangiran
Dauletabad Donmez
Torbat-e Haydari
Okrem
Husan Kui
Korpedzhe
S. Bugdaili
Kamyshldzha
N. Okarem
Okarem
S. Okarem
Keimir
Akpatlaukh
Chikishlyar
Isfahan
284
Ardakan
Yazd
Bafq
Kerman
Mashiz
Baft
Shahreza
Yazd-e Khvast
Lordegan
Nazarabad
Ramhormuz
Ahwaz
Shushtar
Andimeshk
Jelow Gir
Veyzenhar/ Malah kuh
Sarkan
Samand
Abu Ghirab
Qaleh Nar
Lab-i-Safid
Lali
Karun
Masjid-i-Suleiman
Palangan
Zeloi
Par-E-Siah
Naft Safid
Haft Kel
Ramin (Mulla Sani)
Doudrou
Kuh-I-Rig
Marun
Shurom
Kuh-E-Bangestan
Ab Teymour
Ramshir
Parsi
Karanj
Gachsaran
Bandar-e Mah Shahi
Bandar-e Khomeine
Rag-E-Safid
Binak
Hendijan
Bahrgansar
Nur
Buzurgan
Al Halfayan (Hafayat)
Majnoon
Sausangerd
Jufeyr
Nahr Umr
Basra
140
Khorramshahr
Abadan
350
Zubair
Rumaila
Tuba
Mutriba
Sabriyah
Ar Rawdatayn
Bahrah
Raudhatain
290
Al Kuwait
443
Riquah
Mina al Ahmadi
Mina Abdullah
257
Mina Su'ud
Minagish
Umm Gudair
Wafra
Fawaris
S. Wafra
Nowrooz
Abuzar
Abdullyah
Burgan
Soroush
Dorra
Hout
Hamur
Khafji
Safaniya
Lulu/Esfandiar (Iran)
Marjan/Foroozan
BB
Hasbab
Doroud (Darius) Khargh
Bushire
Bushehr
Ganaveh
Kazerun
Nargesi
Radak-Milaatun
Shiraz
40
Sarvestan
Marv Dasht
Ardakan
Firuzabad
Bushgan
Kuh-i-Kaki
Dalan
Aghar
Bandubast
Kuh-i-Mand
Kangan
Nar
Varavi
Pars
Fa
Deyyer
Assaluyeh
Asaluyeh
Lamard
Tabnak
G
F
Qotbabad
W. Namak
Bandar 'Abbas
232
Sarkhun
Suru
HD
S. Gashu
Gavarzin
Salakh
Hanquan
Saleh
Bukha
W. Bukha
Limah
Baih
Umm al Qaiwain
Sharjah
Sajaa
Mo'Ayyid
S. Juwaiza
Khubai
Margham
Sharjah
71
Dubai
120
90
Al Fujayrah
Jebel Ali
Fujayrah
Mahdah
Suhar
Al Khaburah
Muscat
Mina al Fahal
85
Al' Ayn
Shams
Nadir
Al Sahwa
Hafar
Shibka
M. Huraymah
Habiba
Abeer
Baqlah
Nati
Deleel
Lekhweir
Safah
Mender
Bandar-e Maqam
Bandar-e Lengeh
LAVAN I.
30
BH
W
Alpha
Sirri A
Reshadat
Resalat
Maydan Mahzan
Bul Hanine
Abu Al Bukhoosh
Salman
Nasr
Mandoos
Belbazem
Zakum
Umm Al Dalkh
150
Abu Dhabi
Zubaya
Rumaitha
Arjan
Zol
Ghasha
Bu Tini
Hail
Bida Al Qimzan
Tarif
Bab
Habshan
Sahil
Ruwais
350
Bu Hasa
Hawaila
Asab
Shah
Qusahwira
Dalama
Hair Dhalma
Jarnain
Bu Jufair
Umm Said
Doha
200
Idd Al Shargi
Dukhan
Umm Bab
Zayqat
Lughfah
Jawb
South Pars
North
Awali
Manama
249
Al Rayyan
Abu-Safah
Berri
Ras Tanura
550
Qatif
Ad Dammam
Abqaiq
Al Uqayr
Al Mubarraz
Al Jubayl
305
Ras al Ju'aymah
Fadhli
Jana
Kurayn
Karan
Manifa
Salsal
Sharar
Watban
Bakr
Dhib
Jaladi
Fadhili
Samin
Khurais
Ghawar
Abu Jifan
Qirdi
Farhah
Mazalij
Harmaliyah
Haradh
Dilam
As Salamiyah
Ad Dilam
Sahba
Midrikah
Tinat
Raghib
Naim

IRAN

Capital: Tehran
Monetary unit: Rial
Refining capacity: 1,451,000 b/cd
Oil production: 3.85 million b/d
Oil reserves: 136.27 billion bbl
Gas reserves: 974.00 tcf

Exploration and Development

Azadegan field

In early 2004, Japan's biggest oil and gas explorer, Inpex Holdings Inc. (Tokyo) signed a $2 billion deal to develop the Azadegan onshore oil and gas field. Located in southwestern Iran, it has an estimated 26 billion bbl of oil in place and reserves of 6 billion bbl.

Inpex, partly owned by the Japanese government, had planned to begin production from Azadegan in 2007, aiming to extract 260,000 b/d of crude oil by 2012. However, the company missed several consecutive deadlines for project start-up.

Inpex said one major reason for the delays was the failure of Iran to remove all land mines in critical areas. The company also was under pressure from the United States, which opposed the development deal. U.S. officials were trying to keep international pressure on Iran to give up its nuclear ambitions.

Under pressure from Iran, Inpex agreed in October 2006 to reduce its concession to 10% from the original 75%. The other 65% stake went to the National Iranian Oil Company (NIOC). In turn, NIOC suggested it might transfer the 65% share to other companies, perhaps even Chinese or Russian firms.

However, in December, according to Iran Daily, NIOC launched the first domestic buyback arrangement, aiming to build roads and other infrastructure at the field. A contract for initial drilling was expected to be finalized by mid-2007.

South Pars gas field

NIOC talked in October with Total SA (Paris) about a project to develop South Pars gas field and the Pars LNG project. NIOC expressed hope that South Pars Phases 6–8 would be commissioned by March 2007.

Development and Production

Yadavaran oil field

Iran invited China Petrochemical Corp. (Sinopec) to finalize an agreement initiated in 2004 for development of Iran's Yadavaran oil field. Final approval is expected in 2007. The Petroenergy Information Network, operated by Iran's oil ministry, said Sinopec also would secure oil and natural gas supplies.

Under the 2004 informal agreement, Sinopec would pay Iran up to $100 billion over 25 years for oil and gas and for a 51% stake in Yadavaran field. China would be able to buy 250 million tonnes of LNG and 150,000 b/d of Iranian crude oil at market rates over that period.

Natural Gas Infrastructure

National Iranian Gas Co. (NIGC) plans steady expansion of transmission and processing infrastructure in a program that envisions increases in imports and exports of natural gas. The expansion will take gas service to parts of the country not yet reached by distribution systems and boost exports by pipeline and as LNG.

Much of the gas will come from the giant offshore South Pars gas field, which is being developed in stages. Supplementing that supply will be imports already under way from Turkmenistan and starting from Azerbaijan.

Market expansion

Domestic consumption is on the rise. Gas use in 2005 reached an estimated 105.1 bcm, compared with 65 bcm in 2000. NIGC expects a rise to 156.2 bcm in 2009. The number of Iranian cities using gas reached 550 in 2005 and will reach 690 in 2009.

NIGC believes pipeline exports can reach 70 bcm/year in 2009 and 110 bcm/year in 2020. Export targets are Turkey, Europe, Azerbaijan's Nakhichevan enclave along the Iranian-Armenian border, Armenia, Pakistan, India, the UAE, Kuwait, and Oman.

NIGC is targeting LNG exports from three planned liquefaction projects to China, Thailand, and India of 35 MMcmd in 2009 and 180 MMcmd in 2020. It said Iran's gas imports from Turkmenistan soon will peak at 8 bcm/year.

To accommodate market growth, NIGC expects Iran by 2009 to have 29,400 km (18,268 miles) of gas transmission pipeline in place and a distribution network of 125,000 km (77,670 miles). Domestic gas consumption that year will represent 69% of the Iranian energy market and 54 million customers will have gas service—80% of the population.

NIGC plans to invest about $18 billion through 2009 in high-pressure gas pipelines, compressor stations, gas processing plants, underground storage, distribution networks, and maintenance.

Pipeline projects

Many of the pipeline projects planned by NIGC connect to the Iranian Gas Transmission (IGAT) system.

The IGAT IV pipeline will carry 110 MMcmd of gas from South Pars and the Parsian gas plants to consumption areas. The project includes 1,030 km (640 miles) of pipe. The main part of the pipeline was connected with the Pol Kaleh compressor station in Isfahan in 2004, and a 351-km (218-mile) section to Fars Province became operational in 2004.

A second stage of IGAT IV will include a spur line to Kerman, a line to a Fars petrochemical plant, a second line to Yazd, and a line from Isfahan to Mobarekeh.

The IGAT V trunkline will carry 75 MMcmd of sour gas from South Pars Phases 6–8 to Khoozestan oil fields for injection. It will run 504 km (313 miles) from Assaluye to Agha Jari.

The IGAT VI pipeline generally will parallel IGAT V to serve Bushehr and Khoozestan provinces, including oil field injection. It will extend 492 km (305 miles) and carry up to 90 MMcmd.

IGAT VII, 860 km (535 miles) long, will carry gas from South Pars Phases 9–10 for use in Sistan and Baluchestan province in southern Iran and possibly for export to the UAE, Pakistan, and India.

IGAT VIII, a 1,050-km (652-mile) line, will carry South Pars gas to the Parsian gas plant and north to a connection with a line serving Tehran.

To serve the northern and eastern provinces of Semnan, Khorasan, Golestan, and Mazandaran, NIGC plans a second pipeline between Parchin and Sangbast [790 km (490 miles)].

A LOOK AT IRAN'S GAS PLANTS, TRANSMISSION PIPELINES

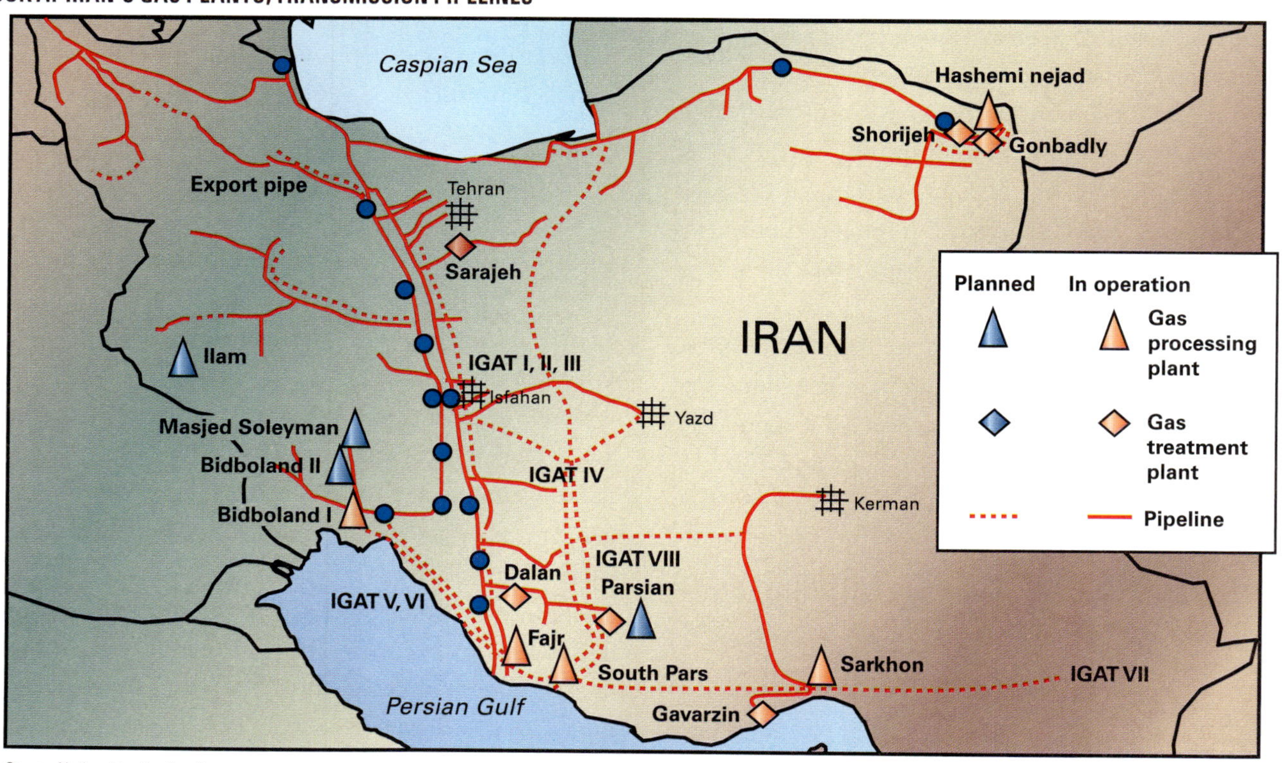

Source: National Iranian Gas Co.

NIGC'S PROJECTIONS FOR GROWTH IN IRANIAN...

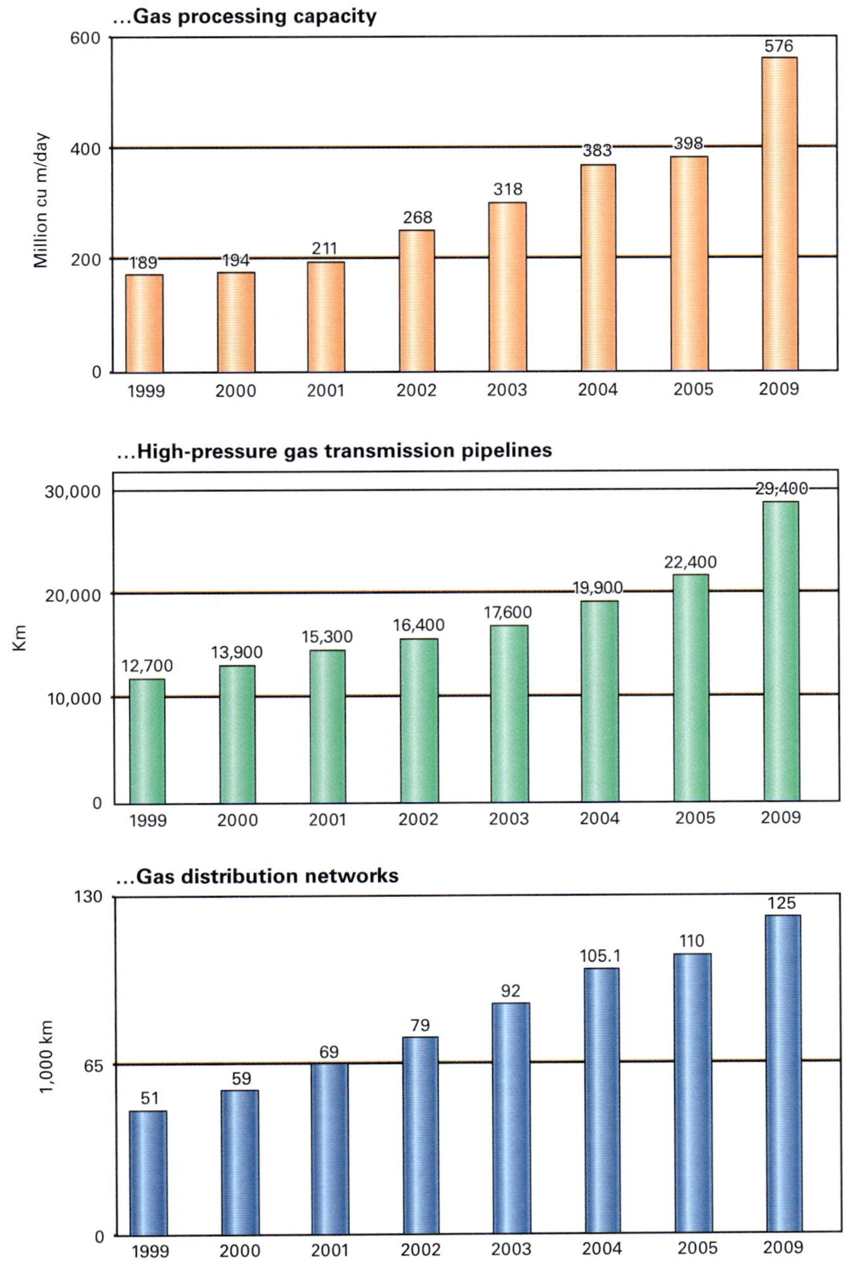

Source: National Iranian Gas Co.

Also planned is a 110-km (68-mile) segment from Miami to Jajarm. The system will handle South Pars gas delivered through IGAT VIII.

To serve the western and northern provinces of Hamadan, Kordestan, Zanjan, and Western and Eastern Azerbaijan provinces, NIGC plans a 280-km (174-mile) pipeline between Saveh and Bijar, and a 192-km (120-mile) pipeline between Bijar and Miandoab. Other segments will boost pipeline lengths planned for this region to 950 km (590 miles).

Gas processing

Seven gas processing plants recently have been completed or are planned and under construction.

The Parsian plant began treatment operations in 2003, dehydrating 20 MMcmd of gas from Tabnak field and stabilizing 12,000 b/d of condensates. Construction of processing facilities at this plant will lead to yields of ethane, pentanes, butane, and propane.

The Bidboland II plant will sweeten and process 57 MMcmd of gas at facilities to be built about 14 km (9 miles) southeast of the existing Bidboland plant. It will process gas from Pazanan, Gachsaran, and Bibi Hakimeh fields and yield 15 bcm/year of sweet gas as well as ethane, propane, butane, and natural gasoline.

In a separate project, NIGC plans a gas processing plant 25 km (15 miles) northwest of the city of Ilam in western Iran. With an inlet capacity of 10 MMcmd, it will process gas from Tange Bijar and Kamankooh fields.

NIGC also plans a small processing plant at Masjed Soleiman with inlet capacity of 1 MMcmd and is studying a plant able to process 14 MMcmd at South Gesho sour gas field in Hormuzgan Province. The latter would remove 600 tpd of sulfur and 9,000 b/d of condensate, sending sweet gas to southeastern markets.

Storage projects

NIGC has identified several reservoirs that might be converted for use as underground gas storage. One is Sarajeh gas and condensate field, 40 km (25 miles) east of Qom. NIGC estimates that it would take about two years to work over old wells and drill new wells to deplete the field and convert it to storage.

Other prospects are the Yortsha Dome saltwater reservoir, 25 km (15 miles) south of Varamin, and the Talkeh saltwater Dome in central Iran. NIGC also thinks there could be reservoirs amenable for storage use in Abardejno, Siahkooh, Marehtapeh, Prandak, and Eastern Azerbaijan provinces.

Gas Imports

More gas from Turkmenistan

Iran wants to increase its natural gas imports from Turkmenistan and said it is willing to pay a higher price, according to the office of the Iranian foreign minister. Preliminary negotiations about the suggested increase of the Turkmen gas price from the current $42–$60/1,000 cu m were held in Ashgabat in February 2006.

According to the Turkmen Oil and Gas Ministry, Turkmenistan delivered 5.8 bcm of gas to Iran in 2005. Contracts signed in 2005 called for the supply of 8 bcm in 2006.

LNG

Fast-track plan for Qeshm Island

Liquefied Natural Gas Ltd. (Perth) announced plans to accelerate development of its 3.45-mtpy Qeshm Island LNG project, off Iran. The first phase of the liquefaction plant is expected to be operational in 2010, delivering 1.15 mtpy. The LNG would be sold in India or East Africa.

Developers were working with Iranian authorities to select a plant site. National Iranian Oil Co. (NIOC) has signed a deal to provide as much as 530 MMcfd of gas to the Qeshm Island facility.

Nuclear Program Issues

Iran impact on IOC decisions

Frontier Strategy Group (Cambridge, MA, U.S.A) said the standoff between Iran and the West regarding its nuclear development program has substantial implications for international oil and gas companies. No matter what the outcome of the diplomatic crisis, Frontier concluded that the following scenarios are likely:

- China, India, and other developing countries will invest more aggressively in energy assets in Africa and Latin America, perceived as "safer" regions than the Middle East.
- Nuclear energy may become a more tempting option for Europe as concerns about gas supply grow.
- Latin American governments will gain new leverage over oil and gas firms, but this region still may be attractive.

Nuclear Program Standoff

Interactions continued throughout 2006 among Iran, international agencies, and various other countries regarding Iran's pursuit of a nuclear program. Actions and reactions had the not-unexpected effect of causing spikes in oil prices and increasing concerns about energy security.

For example, following the July 12 UN Security Council Resolution, the declaration by Iranian Pres. Mahmoud Ahmadinejad that Iran would not back down from its nuclear program sent October-dated U.S. light crude futures up 37¢ to $70.37/bbl.

Apart from the direct effect of sanctions on supply, analysts also feared that Iran could retaliate by disrupting oil exports traversing the Strait of Hormuz, through which an estimated 20% of the world's supply passes every day.

Others also feared that threats by Iran to pull out of the Non-Proliferation Treaty and expel UN nuclear inspectors could further inflate the risk premium in oil prices.

Barclays Capital Inc. (London) said, "Iran is rapidly becoming the most important non-economic influence on the oil market today." Analysts at Raymond James & Associates (Houston) said that the odds had greatly increased that there will be an Iranian-related oil supply disruption in the next two years.

IRAQ

Capital: Baghdad
Monetary unit: Dinar
Refining capacity: 597,500 b/cd
Oil production: 1.915 million b/d
Oil reserves: 115.00 billion bbl
Gas reserves: 112.00 tcf

Particularly in central and southern areas, sectarian violence continued to ravage the country in 2006 in the aftermath of the overthrow of former president, Saddam Hussein. Nouri al-Maliki became prime minister in May, taking on the challenges of establishing centralized government operations, rebuilding infrastructure, and reestablishing social and commercial activities.

Oil Production

Iraq's oil ministry said oil production averaged about 2 million b/d in April, rising to 2.3 million b/d in October. Northern oil fields near Kirkuk were reportedly producing 250,000 b/d in March 2006, with production from southern fields at approximately 1.5 million to 1.75 million b/d. The country expected to produce 3 million b/d of crude oil by yearend 2006.

Ministry officials said that the national production target for 2010 is 4.5 million b/d, to be achieved in part through increased foreign oil-sector investment.

Oil Exports

Both the U.S. Energy Information Administration (EIA) and Iraqi officials estimated total net exports at 1.6–1.7 million b/d in the June to October period. However, by year-end, that level had dropped to 1.1 million b/d—the lowest since 2003—because of continued pipeline sabotage in the north and bad weather and logistics problems in the south.

Oil from northern fields is exported primarily through a pipeline system from Kirkuk to the port of Ceyhan, Turkey. Prior to the U.S.-led war in 2003, this pipeline system carried 700,000 to 800,000 b/d, but recurring sabotage by militants has made reliable operation very difficult.

Oil from southern fields is piped to export terminals at Basra and Khor al-Amaya on the Persian Gulf. Iraq has relied for its export revenues on light crude from southern fields.

Northern pipeline system

Militant attacks effectively shut down the northern pipeline system from mid-2004 to early 2006, according to the Iraqi oil ministry.

Ministry officials said Iraq is targeting sustainable exports from the north of 50,000 to 100,000 b/d and has plans to increase that to 1 million b/d by 2010. Reaching that target will depend on repair schedules and increased security for a second, 26-in (66-cm) pipeline.

In June, Iraq resumed crude oil pipeline exports from northern fields for the first time in four months.

Southern export terminals

Oil ministry officials estimated that March exports through the country's two main Persian Gulf terminals would reach 1.5 million b/d. They also said, however, that exports could not exceed 1.6 million b/d from the south for a variety of reasons, including a lack of investment in southern oil fields.

Following an Asian tour in late 2006, the Iraqi oil ministry said that the government of Japan had promised a loan of $3.5 billion to help build a refinery and a floating loading platform off Basra.

Exploration and Development

Kurdistan tract promising

A unit of Western Oil Sands (Calgary) began exploring a 914,000-acre block in Kurdistan, southeast of Kirkuk, under an agreement with the Kurdistan Regional Government Sulaymaniyah Administration.

WesternZagros Ltd., seeking conventional oil and gas in the Zagros fold belt, has identified three large structures—Banwanoor, Kalar, and Shakal—from satellite imagery and active oil seeps. The block lies 60 km (37 miles) southeast of and on trend with the southeastern edge of supergiant Kirkuk oil and gas field. Drilling could begin as soon as late 2007.

Seismic equipment donated

Japan Petroleum Exploration Co. (Japex; Tokyo) is donating $15–20 million worth of seismic survey equipment to Iraq under a technical cooperation agreement made in 2005 and renewed in March 2006 for a year. The Iraqi oil ministry said the equipment is intended to help the government update seismic data from several oil fields last studied in the 1970s and 1980s.

The agreement calls for joint studies of development of an oil field near Baghdad, evaluation of an oil field and adjacent unexplored prospects in northeastern Iraq, and evaluation of Paleozoic reservoirs on a block in the Western Desert of Iraq and of deeper formations in the south.

New field study due in Kurdish area

The government of the Autonomous Region of Kurdistan entered into a second memorandum of understanding (MOU) with K Petroleum Co. (KPC) for integrated reservoir field studies in northeastern Iraq, reported Heritage Oil Corp. Heritage and Kurdish firm Eagle Group are joint venture partners of Heritage Erbil Oil Ltd., the parent of KPC.

The MOUs cover a combined 1,300 sq km (502 sq miles). More than 100 prospects have been identified. Field work in the first MOU area began in January.

Large oil discovery for Tawke project

DNO ASA (Oslo, Norway) said its Tawke-1 exploration well in the northern Kurdish region of Iraq discovered 100 million bbl of recoverable oil in a single zone using a 30% recovery factor. The well is located within the Dihok PSA area.

The unspecified shallow discovery zone flowed 5,000 b/d of good quality oil, a rate restricted by test equipment. Of six intervals tested, two deep and two shallow intervals tested oil. The company planned to begin production in early 2007 and to drill 18 development wells later in the year, including oil producers and water injectors.

To deliver Tawke oil to the main northern pipeline, DNO has installed 8 km (5 miles) of a 42-km (26-mile) pipeline. Central processing facilities with 50,000 b/d capacity will be transported to the Tawke area for installation and commissioning.

Al-Ahdab field

The China National Petroleum Corp. agreed to terms for new exploration rights for al-Ahdab oil field in southern Iraq, with initial planning by a joint Iraqi-Chinese committee set to begin in November 2006.

Oil revenues: $16 billion lost

Iraq lost an estimated $16 billion of potential oil export revenues from January 2004 to March 2006 because of insurgent attacks, criminal activity, and aging and poorly maintained infrastructure, a September audit report by the U.S. special inspector general for Iraq reconstruction said.

The country also is paying billions of dollars to import refined petroleum products to support the consumption needs of its citizens.

The report also noted that Iraq's oil exports remained below established targets and that electric power generation was far below demand when there were few insurgent attacks from late April to early June of 2006.

Improvement has been slow, in part because of the lack of a permanent government and in part because of the limited initiative of some Iraqi ministries, said the report.

TURKEY
SYRIA
IRAN
IRAQ
LEBANON
ISRAEL
JORDAN
KUWAIT
WEST BANK
GAZA STRIP
SAUDI ARABIA
MIDDLE EAST
Red Sea
Lake Van
Lake Urumiyeh
Aksaray
Mersin
Ceyhan
Dortyol
Batman
Tabriz
Bostanabad
Shanpur
Uromiyeh
Naqadeh
Mosul
Irbil
Al Qawayr
Qayyarah
Kirkuk
Al Hadr (Hatra)
Ash-Sharqat
Halabja
Bayji
Kifri
Sar-e Pol-e Zahab
Kermanshah
Khanaqin
Naft-e Shah
Sumar
Samarra
Ilam
Tibni
Baniyas
Tartus
Homs
Tripoli
Tadmur
Mayadin
Al-Haditha
Hit
Beirut
Sidon
Damascus
Baghdad
Mehran
Jelow Gir
Haifa
Nazareth
Ar Rutbah
As Suwayrah
Karbala
Al Hashimiyah
Al Kut
Al Hayy
Ad Diwaniyah
Tel Aviv-Yafo
Amman
Jerusalem
Hebron
Abu
Sukhayr
Ar Rifai
Ash Shatrah
As Samawah
Suq ash Shuyukh
Badanah
Jalibar
Az Zubair
Safwan
Ar Rawdatayn
Elat
Al Jawf
Sakakah
Rafha
Al Bad
Al Qaysumah
Ash Sharmah
Duba
Al Waih
Marsa Alam
Umm Cajj
Medina
Riyadh
Dumah
As Salamiyah
Ad Dilam
Al Hillah
Yanbu al Bahr
Badr Hunayn
0 100 200 Miles
0 100 200 300 Km
35°
30°
25°
36°
40°
44°

In 1997, China signed a $700 million contract to explore al-Ahdab over 23 years. Plans had called for 90,000 b/d of production, but field development was suspended because of UN sanctions on Iraq and security problems since the start of the U.S.-led war in 2003.

Al-Ahdab field, in the mid-south region of Iraq, has estimated reserves of as much as 1 billion bbl. It is considered commercially viable because of its proximity to existing pipelines and refineries.

ISRAEL

Capital: Jerusalem
Monetary unit: Shekel
Refining capacity: 220,000 b/cd
Oil production: 100 b/d
Oil reserves: 1.960 million bbl
Gas reserves: 1.275 tcf

Background

In the Eastern Mediterranean region, Israel has the largest and most developed economy. According to the U.S. Energy Information Administration (EIA), growth of real GDP in 2005 was 5.2%, and was forecast to be 4.7% for 2006.

Oil

In 2005, Israel produced only minimal quantities of oil. The country imported approximately 249,000 b/d of crude oil and products to meet domestic needs, primarily from Egypt, the North Sea, West Africa, and Mexico.

However, in recent years, Israel has increased imports from Russia and the Caspian, including Azeri oil transported through the new Baku-Tbilisi-Ceyhan (BTC) pipeline.

More than 20 small independent firms are active in hydrocarbon exploration throughout the country. Israel's Petroleum Commission has estimated that the country could contain 5 billion bbl of oil reserves.

Natural Gas

Gas production in 2004 was estimated at 2.8 MMcf. However, Israel plans to dramatically increase the share of natural gas in its fuel mix, especially for electricity generation; demand is expected to reach 282 bcf by 2010. Supply options being developed include imports from Egypt's Nile Delta and production from domestic offshore reserves.

Anticipating future needs and the potential for disruption of imports from within the region, Israel also is considering building an LNG regasification facility or an underwater pipeline to bring gas from Turkey.

Over the past several years, several energy companies have discovered small but commercially viable quantities of natural gas off the coast of Israel and the Gaza Strip.

Israel's and the Palestinian Authority's offshore gas reserves are under concession to two main groups:

- Yam Thetis group, comprising the U.S.-based Nobel Energy, Israel's Avner Drilling, and Delek Exploration
- A British Gas (BG) partnership with U.S.-based Isramco and others

Yam Thetis operates Israel's two largest gas fields, Mari and Noa (Border), with estimated combined reserves of 1.5–1.7 tcf. In July 2006, YamThetis signed an agreement that augments an 11-year deal to supply natural gas to Israel Electric Corporation (IEC) and the Ashdod refinery.

The new agreement will increase quantities delivered significantly above the current 64 bcf/year and will also support a second gas-fired power plant near Tel Aviv. Excess gas will supply the Hagit and Gezer power plants, scheduled to convert to natural gas in 2007.

BG first discovered gas in 1999 in the Palestinian Authority's territorial waters with its Gaza Marine-1 well, which indicated possible reserves of 1.4 tcf. The company's 25-year contract includes gas exploration and set-up of a distribution network.

Israeli, British, and Palestinian Authority officials as well as private-sector parties have been involved in on-again off-again negotiations regarding a BG plan to import natural gas from the Gaza fields. Those negotiations resumed in mid-2006 after earlier disagreements about gas price and customer base.

BG also is considering the option to export gas to LNG plants in Egypt for sale to Europe and North America.

Isramco announced in early 2005 its independent plans to drill in the Med Ashdod lease (Gad 1 well) with U.S. and Israeli partners. Isramco has a stake in the Med Yavne license, operated by BG. There are no development plans yet for Med Yavne, which contains the productive Or-1 well.

Exploration and Development

Heletz lease

Israel's petroleum commissioner offered an exploration/exploitation license until March 29, 2007, for part of the former Heletz lease, 55 km (34 miles) south of Tel Aviv, which was terminated in mid-2006. The three-year lease will have a possible four-year extension.

Heletz was discovered in 1955 as the eastern Mediterranean's first oil field. Some 17 million bbl have been recovered of an estimated 19 million bbl of 29° gravity oil in place. The field has 88 wells, of which six wells are producing 70 b/d.

Exploration off Haifa

Israel's MNI approved a permit enabling PetroMed PLC (Toronto) to begin exclusive exploration on 3,500 sq km (1,350 sq miles) of prospects offshore Haifa in northern Israel.

PetroMed said it would be ready in early 2008 to drill "in a world-class offshore turbidite formation."

The blocks of interest are 50–70 km (30–45 miles) from the coast in 1,000–2,000 m (3,280–6,560 ft) of water. The company plans to explore for deeper targets similar to discoveries made in the Nile Delta.

Completion of Ma'anit-1 well

Zion Oil & Gas Inc. (Dallas) filed an initial public offering in early 2006, seeking to raise $14 million to complete the exploratory Ma'anit-1 well, TD 4,830 m (15,842 ft), and to drill an appraisal well.

The Ma'anit-1 wildcat, which Zion believes may be a multizone oil and gas discovery, is on the 95,800-acre Ma'anit-Joseph license, 60 km (37 miles) north-northeast of Tel Aviv. Water flow into the wellbore from interbedded volcanic zones interfered with earlier completion attempts.

Oil shale pondered

Israel has sizeable deposits of oil shale—an estimated 14–15 billion tonnes with 600 million tonnes recoverable—mostly in the Rotem basin of the northern Negev desert, near the Dead Sea.

Under consideration by the government was a proposal from Haifa-based AFSK Hov Tom (Haifa) to build a large plant in Mishor Rotem that could produce 60,000 b/d of oil from 6 million tonnes of shale and 2 million tonnes of bitumen (from the Ashdod refinery).

Israeli-Hezbollah conflict

On July 12, armed conflict broke out in the south of Lebanon between the Shi'a Islamic militia called Hezbollah and Israel. The month-long fighting, triggered by Hezbollah attacks on Israeli soldiers, included rocket and artillery exchanges, air strikes, movement into Lebanon by Israeli forces, and a naval blockade of Lebanon by Israel.

New Israeli Prime Minister Ehud Olmert came under considerable criticism for Israeli actions during the conflict. He had assumed the office in April, three months after his predecessor, Ariel Sharon, suffered a severe stroke.

The conflict caused billions of dollars worth of damage to Lebanon's infrastructure and massive population displacement. A ceasefire brokered by the United Nations ended the fighting in mid-August.

The month-long battle—coupled with pipeline explosions in Nigeria and Iranian insistence on pursuing a nuclear development program—helped push oil prices to new highs. Front-month crude futures hit a record high of $75.89/bbl in early trading July 13.

Refining

Israel's two major refineries historically have been owned and operated by Oil Refineries Ltd. The facilities are re at Haifa (130,000 b/d) and Ashdod (90,000 b/d).

In July, the Ashdod refinery was sold to Israel's largest fuel retailer Paz Oil Company, Ltd. The majority of shares in the Haifa refinery were to be offered for sale in late 2006.

Following the war with Lebanon, Israel's National Infrastructure Ministry (NIM) was reportedly considering moving the refinery and associated petrochemical facilities out of the port city due to environmental and security concerns.

Pipelines

Israel's main operational oil pipeline is the 245-km (152-mile) Trans-Israel Pipeline (or Tipline). The 42-inch (106-cm) line from the Red Sea port of Eilat to the northern Mediterranean port of Ashkelon can carry an estimated 1.0–1.2 million b/d.

In July 2006, Israel signed an agreement with the State Oil Company of Azerbaijan (SOCAR) to import and transport Azeri Light Crude through the pipeline.

New line could carry Russian gas

Officials from Israel's Ministry of National Infrastructure met in early 2007 with OAO Gazprom executives to discuss possible delivery of Russian natural gas to Israel.

Local reports said Israel wants to secure 2–5 bcm/year, to be delivered through a new pipeline from Turkey beginning in 2011.

Israel and Turkey were reported to be interested in constructing a 610-km (380-mile) pipeline across the Mediterranean from the Turkish port of Ceyhan to Ashkelon, Israel. Project cost was estimated at more than $2 billion.

Increased electric power generation projects in Israel are driving gas demand, which is expected to average 8 bcm by 2010. Israel currently relies primarily on costly imported oil.

El Arish-to-Ashkelon pipeline

Rather than buying gas from politically sensitive Gaza, Israel elected to buy gas from Egypt. In 2005, Israel Electric Corp. (IEC) signed a $2.5 billion gas supply contract with East Mediterranean Gas Pipeline Co. (EMG) of Egypt. IEC is to receive 1.7 bcm/year from EMG for 15 years.

The EMG-IEC agreement also calls for an 80-mile subsea pipeline to be built from Egypt's port of El Arish to Israel's port of Ashkelon. The pipeline is to be operational in 2007.

If BG (as described above) decides to export gas from off Gaza to Egypt, El Arish could also be the landing point for that pipeline. In 2005, the Palestinian Authority and Egypt signed a protocol to arrange sales of offshore Gaza gas to Egypt, with plans to export it as LNG to world markets.

National gas transmission system

Israel's Ministry of National Infrastructure (MNI) revived plans for a national gas transmission system, excluding the Palestinian Authority. Completion was scheduled for 2010.

The MNI had granted state-owned Israel National Gas Lines a license in 2004 to build and operate the system. In mid-2006, the MNI awarded contracts for the first several stages of construction, to start by year-end.

KUWAIT

Capital: Kuwait City
Monetary unit: Dinar
Refining capacity: 889,200 b/cd
Oil production: 2.20 million b/d
Oil reserves: 99.00 billion bbl
Gas reserves: 54.50 tcf

Gas production plans

Kuwait announced the discovery of 35 tcf of free natural gas and a large quantity of light oil in what it called its northern oil fields. Commercial gas production is planned to start by 2008, said state-owned Kuwait Oil Co. (KOC).

The firm expects an initial output of 180 MMcfd, which it would increase to 600 MMcfd by 2011 and to 1 bcfd by 2014–15.

The company is searching for more gas in five exploration sites in the same region, which it did not name. The gas is needed for its desalination plants and for a planned large refinery.

Kuwait said it also planned to start exploration for natural gas in what it called undisputed parts of offshore Dorra field, which Iran also claims. The energy ministry said Kuwait would invite bids for seismic surveys of part of Dorra field and other Kuwaiti offshore fields.

KOC officials also have pledged to increase Kuwait's oil production capacity to 3 million b/d by 2010, to 3.5 million b/d by 2015, and 4 million b/d by 2020 from the current 2.6 million b/d.

Oil reserves reviewed

Kuwait officials were working to clarify the volume of Kuwaiti estimated oil reserves, according to the Energy Ministry. The Council of Ministers and the National Assembly as well as Kuwait Petroleum Corp. are taking part in the evaluation.

Kuwait has estimated its oil reserves at 100 billion bbl, but it has been reported in the media that internal Kuwaiti records indicate reserves of 48 billion bbl.

Project Kuwait still on hold

Project Kuwait is a $7 billion, 25-year plan, first formulated in 1997 by the Supreme Petroleum Council, to increase the country's oil production—and to help compensate for declines at the mature Burgan field—with help from international oil companies.

In particular, the project would boost output at five northern oil fields—Abdali, Bahra, Ratqa, Raudhatain, and Sabriya—from their current rate of around 650,000 b/d to 900,000 b/d within three years. The government also hopes to boost total production capacity to 4 million b/d from 2.5 million b/d.

Project Kuwait has support from key members of the ruling family and Kuwait Petroleum Corp. The Kuwait energy minister has said the project is essential to achievement of the country's production targets.

OPEC ACTIONS ON OIL OUTPUT

From July 2005 until late 2006, the official output ceiling for members of the Organization of Petroleum Exporting Countries (OPEC), excluding Iraq, stood at 28 million b/d. The International Energy Association (IEA; Paris) estimated, however, that the group's crude output has exceeded that total.

Average 2005 crude oil output by all 11 OPEC members was 29.76 million b/d. Iraqi crude production in 2005 averaged 1.8 million b/d. IEA estimated that total OPEC crude production averaged 29.87 million b/d in the first quarter of 2006.

At mid-year, OGJ projected that OPEC crude output would average just less than 30 million b/d for the remainder of 2006. Coupled with the group's NGL supply, this would put average 2006 worldwide oil supply at 85.5 million b/d.

OPEC announced on October 20 it would cut output by 1.2 million b/d starting November 1 by 1.2 million b/d (to 26.3 million b/d). However, there was little evidence in mid-November of any serious rollbacks.

At its December 14 meeting in Nigeria, OPEC voted to cut crude output by 500,000 b/d—to 25.8 million b/d—effective February 1, 2007, "in order to balance supply and demand."

Late in 2006, the new president of Ecuador, Rafael Correa, announced his country's intention to rejoin OPEC; Sudan and Angola also expressed interest in joining the cartel. The combined production of these three countries is 2.4 million b/d, about 8% of current OPEC production.

OPEC OIL PRODUCTION

Country	First quarter 2006	First quarter 2005	Change	Change, %	OPEC quota, 1,000 b/d
	1,000 b/d				
Algeria	1,360	1,333	27	2.0	894
Indonesia	923	950	–27	–2.8	1,451
Iran	3,847	3,883	–36	–0.9	4,110
Iraq	1,717	1,797	–80	–4.5	—
Kuwait[1]	2,515	2,398	117	4.9	2,247
Libya	1,667	1,613	54	3.3	1,500
Nigeria	2,233	2,357	–124	–5.3	2,306
Qatar	820	777	43	5.5	726
Saudi Arabia[2]	9,325	9,079	246	2.7	9,099
United Arab Emirates	2,613	2,383	230	9.7	2,444
Venezuela	2,630	2,130	500	23.5	3,223
Total OPEC	**29,650**	**28,700**	**950**	**3.3**	**28,000**

[1]Not included in 2006 quota. [2]Kuwait and Saudi Arabia production each include half of Neutral Zone.
Source: Oil & Gas Journal

But the project has been repeatedly delayed because of political opposition and resistance from nationalists and Islamists in parliament to the idea of allowing foreign companies into the country's oil sector.

Legislation that would facilitate Project Kuwait was introduced again in the Kuwaiti parliament in early 2005. The bill was approved by the Finance and Economic Committee in June 2005, but with amendments limiting its scope to four of the five fields, excluding Bahra.

In October 2005, new members of that committee tried but failed to take the item off its agenda. The government later requested that the full parliament prepare to consider the issue in January 2006.

However, no action occurred in January. Parliamentary elections on June 29 resulted in a new 50-seat assembly, which was expected to lead to yet more confrontational politics as well as delays to Project Kuwait.

Refinery at Al-Zour

Kuwait National Petroleum Co. (KNPC) pre-qualified 11 companies for construction of a 600,000 b/d refinery 100 km (62 miles) south of Kuwait City in the Al-Zour area.

The companies include Technip, Foster Wheeler Italiana, GS Engineering & Construction Corp., SK Engineering & Construction Co., Hyundai Engineering & Construction Co., Stone & Webster International Inc., and Petrofac International Ltd.

There will be two contracts for manufacturing units in the refinery, one for utilities and services, and another for storage tanks and a pier. Target date for completion is 2010.

Kuwait has three refineries—at Al-Ahmadi, Mina Abdullah, and Shuaiba—with combined capacity of about 915,000 b/d. KNPC said it plans to upgrade Al-Ahmadi and Mina Abdullah by early 2011, but close Shuaiba once Al-Zour and the upgrades are completed, giving the country a total refining capacity of around 1.4 million b/d.

OMAN

Capital: Muscat
Monetary unit: Riyal
Refining capacity: 85,000 b/cd
Oil production: 738,000 b/d
Oil reserves: 5.500 billion bbl
Gas reserves: 30.00 tcf

Exploration and Development

Block 49 data

Circle Oil PLC (London) was encouraged by preliminary data from a 4,800-km (2,983-mile) airborne gravity and magnetic survey on Block 49. The 15,438 sq km (5,960-sq-mile) block is on the southern edge of the Rub al Khali basin and the eastern flank of the Ghudun Salt basin. A 2D seismic is planned for early 2007.

The block's two main play concepts are the Silurian Sahmah play, which may extend into Block 49 from eastern Saudi Arabia, and the Lower Haima-Huqf play. Several billion barrels of oil have been discovered in Proterozoic-Cambrian reservoirs associated with the Ara salt in the South Oman basin southeast of Block 49. There is evidence of a similar salt basin, named Ghudun, beneath the western part of the block.

SW OMAN EXPLORATION AREA

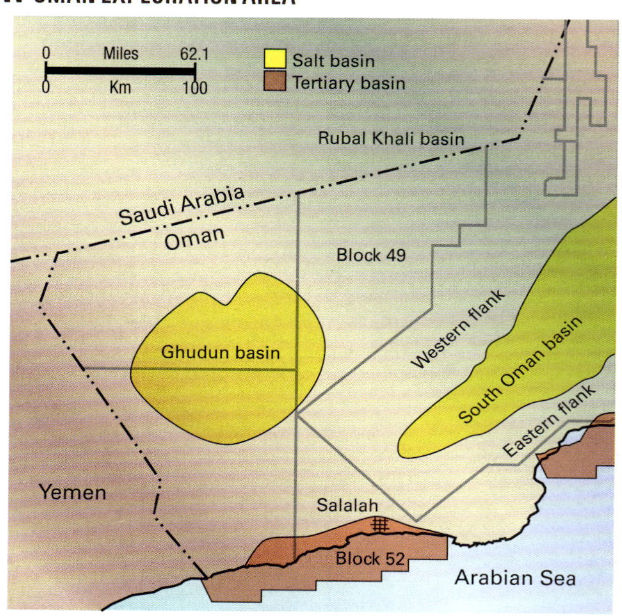

Source: Circle Oil PLC

Partners with Oxy on Block 54 are Mitsui E&P Middle East BV and Liwa Energy Ltd. Liwa Energy is the oil and gas subsidiary of Mubadala Development Co., which is owned by the Abu Dhabi government.

Block 58 award

PTTEP (Thailand) signed an exploration and production sharing agreement with the Ministry of Oil and Gas for 2,264 sq-km (874 sq-mile) Block 58 in southwestern Oman. PTTEP will own and operate the block, which is PTTEP's second venture in Oman. Terms were not disclosed, but the Oman government reserved the right to a 20% interest if a commercial discovery is made.

The company plans to start production at mid-year from Shams-4 field on Block 44, a 1,162-sq km (449 sq-mile) tract 300 km (186 miles) west of Muscat.

Block 44 find

Thailand's PTT Exploration & Production PLC (PTTEP) discovered oil, condensate, and natural gas with an appraisal well in the northern part of the onshore Block 44. Two reservoir formations, with a total thickness of 96 m (315 ft), were discovered with the Shams-4 well, which was drilled to 3,750 m TD (12,300 ft).

Natural gas test flowrates ranged from 6.7 MMcfd to 29 MMcfd. PTTEP plans to put Shams-4 on production by 2007. Shams-4 is about 4 km (2.5 miles) from the Shams-1 gas field, which was to start producing 50 MMcfd of gas and 4,000 b/d of condensate in mid-2006.

Block 54 award

Oman awarded a production-sharing contract to a group led by Occidental Petroleum Corp. for Block 54, a 5,620 sq-km (2,170 sq-miles) area in southeastern Oman.

The PSC area is adjacent to Block 53, about 500 km (310 miles) southwest of Muscat, where Oxy operates Mukhaizna oil field, Oman's sixth largest field. Block 53 is judged to have 2.4 billion bbl of original oil in place, from which Oxy hopes to recover 1 billion bbl with a large steamflood.

Nimr-Karim wells

Petroleum Development Oman (PDO) let a contract to PT MedcoEnergi Internasional Tbk. (Jakarta) to develop 18 small oil fields in the Nimr-Karim area of southern Oman. The work program will include exploratory appraisal, subsurface studies, well engineering, facilities construction, operations, and maintenance. The contract area had been producing 18,000 b/d of crude oil.

Drilling and Production

EOR in Qarn Alam, Marmul fields

Royal Dutch Shell PLC was working with Petroleum Development Oman (PDO) to start a steamflood in the fractured carbonate in Qarn Alam field. Most previous steamfloods have been in sandstones.

The viscous 220-cp oil in the field is mostly in the matrix that is surrounded by fractures. The field currently produces under gravity drainage that is a slow process. PDO fills the fractures with gas and the oil gradually flows from the matrix into the fractures.

QARN ALAM STEAM INJECTION

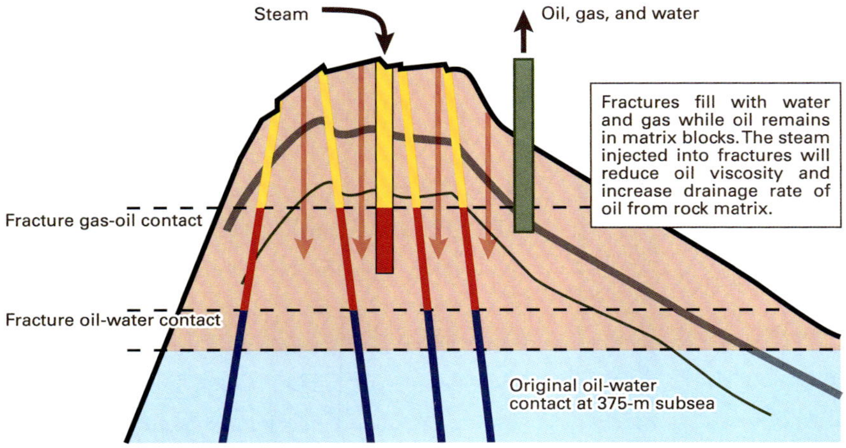

Source: Shell Technology webcast, Sept. 28, 2006

By injecting steam in the fractures, PDO expects to lower the oil viscosity by a factor of about 100 to accelerate oil flow. The company expects the field's oil recovery factor to increase to 20–25% from the current 2–3%.

Another project will inject polymer and surfactant to enhance oil recovery in Marmul field—a technique piloted effectively by Shell in the 1980s. PDO was planning a 27-well development. Marmul has slightly less viscous oil than Qarn Alam and its Al-Qalata reservoir has had water injected for several years.

PDO plans to start polymer injections on the field's flanks and then as soon as possible start injecting surfactants with polymer to reduce oil saturation. PDO expects to recover about 10% of the trapped oil.

Well tests on Block 8

Indago Petroleum Ltd. subsidiary Indago Oman Ltd. began testing the West Bukha-2A (WB2A) well on Block 8 off Oman, pursuing a production target of first-quarter 2008.

WB2A, a sidetrack of the original West Bukha-2 wellbore, penetrated the Shuaiba and Kharaib reservoirs of the Lower Cretaceous Thamama formation to a depth of 4,529 m (14,860 ft) and was then plugged back to 4,383 m (14,400 ft) after electric wireline logs were obtained.

On the first of two 4-hr flow tests of the uppermost Shuaiba section of the Thamama reservoir, the well achieved an average stable flow rate of 3,524 b/d of 42° gravity oil and 1 MMscfd of gas on a 36/64-in. choke at 2,512 psi.

On the second test, the well's flow rate increased to an average 4,392 b/d of oil and 1.4 MMscfd of gas.

Indago said it planned to conduct a second flow test in the Mishrif-Mauddud reservoir following the Thamama testing period. Previously, flow rates of as much as 13.6 MMscfd and 2,910 b/d of condensate have been obtained in the Mishrif formation, which lies above the Thamama.

Mukhaizna steamflood

Occidental Petroleum and partner Liwa Energy (a subsidiary of Mubadala Development Company, Abu Dhabi) were making plans to start an ambitious steamfllood project at the Mukhaizna field.

Located halfway between Haima and Rima, the field holds an estimated 1 billion bbl of heavy-oil resources. Production in 2006 averaged 10,000 b/d.

Some 85 wells were to be drilled in 2006 and 70 in 2007, then more than 1,800 between 2008 and 2012. The production target for year-end 2008 is 50,000 b/d, increasing to a plateau of 150,000 b/d in 2012.

QATAR

Capital: Doha
Monetary unit: Riyal
Refining capacity: 200,000 b/cd
Oil production: 820,000 b/d
Oil reserves: 15.207 billion bbl
Gas reserves: 910.520 tcf

Drilling and Production

Plan to boost Al Shaheen oil output

Maersk Oil Qatar Co. and state-owned Qatar Petroleum agreed to more fully develop Al Shaheen field on Block 5 in the Persian Gulf, boosting production gradually to 525,000 b/d by late 2009 from 240,000 b/d in the first quarter.

The $5 billion 2005 field development plan calls for drilling more than 160 producing and water injection wells in six years, clustering 19 production and accommodation platforms in three new locations, laying subsea pipelines, and adding facilities to gather associated gas. Work is to start immediately.

Previous outlays at Al Shaheen, which underlies the northern part of supergiant North gas-condensate field 110 miles north of Doha, are about $2 billion.

Dolphin Gas Project

The $3.5 billion Dolphin Gas Project, a major initiative of Dolphin Energy Ltd. (DEL), approached completion in 2006.

The upstream segment of the project will develop and produce wet gas from Qatar's North Field to be piped to Ras Laffan for processing. As of mid-2006, a total of 24 wells had been drilled in North Field, and twin 80-km (50-mile) subsea lines to Ras Laffan had been installed.

After processing, 2 bcfd of dry gas will move through a 370-km (230-mile) subsea pipeline from Ras Laffan to Taweelah, Abu Dhabi. The line was completed in mid-2006.

Starting in 2007, gas will be distributed within the UAE and Oman for power generation, water desalination, and industrial processing. Deliveries are planned to continue for 25 years.

DEL is owned by Mubadala Development Company (Abu Dhabi), developer of the Dolphin Gas Project with partners Total SA (Paris) and Occidental Petroleum (Los Angeles, CA, U.S.A).

Al Khaleej gas project: second phase

Qatar Petroleum Co. and ExxonMobil Middle East Gas Marketing Ltd. began the second phase of the Al Khaleej Gas (AKG-2) project in giant offshore North field. AKG-2, estimated to cost more than $3 billion, is scheduled to be operational in 2009.

It is expected to produce 1.58 bcfd of wet gas and recover associated condensate and natural gas liquids for local sale. Development will involve building offshore and onshore facilities, including two wellhead platforms, gas treating and liquids recovery facilities, and fractionation units.

Energy cooperation pact with Japan

Japan and Qatar signed a cooperation agreement aimed in part at securing long-term energy supplies for the import-dependent Asian nation. Details were not announced. Qatar's Ministry of Energy and Industry said the pact would not result immediately in projects or increased energy sales.

Refining and Processing

First commercial GTL plant opened

The $950-million Oryx gas-to-liquids (GTL) plant was inaugurated at Ras Laffan Industrial city in June. Plant designer Foster Wheeler Ltd. (UK) calls it the first plant to convert gas to liquid fuel on a commercial scale.

The joint venture of state-owned Qatar Petroleum Co. and Sasol Ltd. (Johannesburg) uses the proprietary low-temperature Sasol Slurry Phase Distillate process, based on Fischer-Tropsch technology.

The plant will convert 330 MMcfd of lean gas from Qatar's North gas field into 34,000 b/d of ultra-low-sulfur diesel, 24,000 b/d of diesel, 9,000 b/d of naphtha, and 1,000 b/d of LPG.

Major new Ras Laffan complex

Qatar Petroleum and ExxonMobil Chemical Qatar Ltd. have signed a heads of agreement to progress studies for a proposed $3 billion, world-scale petrochemical complex in Ras Laffan Industrial City, Qatar. Start-up is planned for 2012.

The complex will include a 1.3 mtpy steam cracker and associated derivative units, including polyethylene and ethylene glycol, which will employ ExxonMobil's proprietary steam cracking furnace and polyethylene technologies.

The facility will use feedstock from Qatar's North field and will serve Asia and Europe.

Pearl GTL project launched

Qatar Petroleum Co. and Royal Dutch Shell PLC launched the Pearl gas-to-liquids (GTL) project in Qatar's Ras Laffan Industrial City. Shell will fund 100% of the project, estimated to cost $18 billion. Initial contract awards began in 2006, said Shell.

The complex will have two 70,000 b/d GTL trains and associated facilities. Products will include naphtha, diesel-like GTL fuel, normal paraffins, kerosine, and lubricant base oils.

Gas input to the GTL plant will come from a processing plant that will extract about 120,000 boe/d of condensate, LPG, and ethane from 1.6 bcfd of gas from offshore North field.

Production from the first train is expected to begin by 2010, with start-up of the second train due a year later.

LNG

India shipping contract

A consortium of Japanese shipping lines—Mitsui OSK Lines, NYK Lines, and K Line—won a 25-year contract from India's Petronet LNG Ltd. to deliver LNG from Qatar to India. The consortium will carry 2.5 mtpy of LNG from Qatar's RasGas liquefaction plant to Petronet's Dahej regasification terminal at Gujarat.

Qatargas Trains 6 and 7

A joint venture of France's Technip SA and Japan's Chiyoda Corp. won a $4 billion contract to build Trains 6 and 7 at the Qatargas plant in Ras Laffan.

Qatargas III, a joint venture of Qatar Petroleum, ConocoPhillips, and Mitsui & Co. Ltd. owns Train 6, which is to begin production in 2009.

Qatargas IV, a joint venture of QP and Royal Dutch Shell PLC, owns Train 7, production from which is expected to start up at yearend 2010.

The trains each will have a production capacity of 7.8 mtpy, with the output largely to be sold to the United States.

New LNG carriers ordered

Qatar Gas Transport Co. (Nakilat) ordered six 265,000-cu-m LNG carriers from South Korean firms Daewoo Shipbuilding & Marine, and Samsung Heavy Industries. Each yard will build three vessels. Qatar Liquefied Gas Co. will lease the carriers to transport LNG from the Qatargas II project to Europe and the United States.

Qatargas II, a joint venture of Qatar Petroleum and ExxonMobil Corp., comprises Trains 4 and 5, each with a planned production capacity of 7.8 mtpy. Train 4 is scheduled to start production in 2008 and Train 5 in 2009. Nakilat said it intends to acquire a fleet of 61 carriers by 2010.

SAUDI ARABIA

Capital: Riyadh
Monetary unit: Riyal
Refining capacity: 2,095,000 b/cd
Oil production: 8.990 million b/d
Oil reserves: 259.800 billion bbl
Gas reserves: 239.500 tcf

Exploration and Development

Manifa field project

The Manifa Arabian Heavy Crude project includes development of the giant offshore Manifa oil and gas field, construction of gas-oil separation plants and crude stabilization units and separators, as well as the expansion of the Khursaniyah Gas Plant to process the associated sour gas.

The production target is 900,000 b/d of Arab Heavy Crude when the project is completed in mid-2011.

Saudi Aramco let a FEED and project management services contract in October 2006 to two units of Foster Wheeler Ltd. for onshore portions of the project. Completion of the FEED is expected in late 2007. Lump-sum contract awards to perform offshore work are expected in early 2007.

Gas discovered with Kassab-1 well

Saudi Aramco reported a gas discovery 50 km (31 miles) south of its giant Ghawar field and 220 km (137 miles) southeast of Riyadh. The Kassab-1 well flowed on test at 16.2 MMscfd on a 44/64-in. choke at a wellhead pressure of 1,520 psi. The well was tested at 4,800 m (15,750 ft) in the Lower Devonian Jauf sandstone reservoir underlying the prolific Upper Permian Khuff reservoir.

Offshore deep gas targeted

Saudi Aramco planned to test at least three more reservoirs in a Persian Gulf gas discovery 160 km (99 miles) north of Dhahran between Jurayd and Harqus oil fields.

The Karan-6 well—the first of 11 deep gas prospects—flowed 40 MMcfd of gas through a 38/64-in. choke with 5,033 psi wellhead flowing pressure on an open-hole test of the Permian Khuff A reservoir at 3,320 m (10,888 ft). Flow was limited by the test string. Aramco said the well might produce 80 MMcfd with a normal completion.

The company plans to test the Khuff B, C, and D reservoirs, which logs indicate are capable of producing at rates comparable to the initial test.

Aramco also plans to test some targets off Saudi Arabia from 2006 to 2010 at depths exceeding 6,100 m (20,000 ft). This is Aramco's first deep offshore gas exploration since the 1980s.

Drilling and Production

Production target: 12 million b/d

Saudi Arabia's oil and gas investments will total some $70 billion by 2011, said an official with the Ministry of Petroleum and Mineral Resources at a September OPEC seminar in Vienna.

And at the Asia Oil & Gas Conference in Kuala Lumpur that same month, the head of Saudi Aramco said that the Saudi production capacity is estimated to reach 12.5 million b/d by 2009, compared with 11 million b/d in early 2006.

The Aramco chief said that the country would maintain its surplus production capacity of 1.5 million to 2 million b/d as production grows. Major projects planned by Aramco include:

- Expansion of refining capabilities in South Korea S-Oil, in the Philippines through Petron, in the United States through the Motiva joint venture with Royal Dutch Shell PLC, and in China's Fujian and Shandong provinces with Sinopec
- Development of two export refineries at Jubail and Yanbu' in Saudi Arabia
- Construction of a world-class, integrated refining and petrochemical facility at Aramco's Rabigh refinery with Sumitomo Chemical
- Expansion of the capacity of the country's Master Gas System to enable the use of natural gas as the primary fuel and feedstock for domestic industries and utilities

Oil, gas field developments outlined

In response to rising global oil demand and limited spare production capacity, Saudi Aramco reported at the May Offshore Technology Conference (Houston) that it is developing several fields and increasing drilling. The company listed the following developments that target increased production capacity:

- Haradh: 300,000 bo/d in 2006
- Khursaniyah: 500,000 bo/d in 2007
- Nuayyim expansion: 100,000 bo/d in 2008
- Shaybah expansion: 250,000 bo/d in 2008
- Kurais: 1.2 million bo/d in 2009

The Kurais development includes the 2.6 billion-bbl Abu Jifan; 22.8 billion-bbl Khurais; and 1.4 billion-bbl Mazalij oil fields. Khurais also is expected to produce 350 MMcfd of gas and 70,000 b/d of NGL.

Saudi Arabia has 85 oil fields, with 320 reservoirs, and about 260 billion bbl of oil reserves. Reserves calculations do not include enhanced oil recovery, producing oil of less than 12° gravity, or stringers that were once thought noncommercial but now can be produced with modern technology.

Aramco in 2005 had 83 rigs drilling and expected to have about 120 rigs drilling by year-end 2006 before reaching a plateau of 125 rigs in 2007. This compares to 52 rigs active in 2004.

About 20 rigs are drilling offshore, compared to five or six the past. All are drilling in the Persian Gulf but exploration drilling is anticipated to start in the Red Sea in the future.

By 2011, Aramco expects to drill about 200 oil and gas exploration wells, while joint ventures will drill an additional 25 gas exploration wells. Gas is a target for exploration because of rising demand. The country uses all of the 8 bcfd it produces domestically and is expected to need 9 bcfd soon.

Khurais contracts awarded

Saudi Aramco signed two contracts for the development of its 1.2 million b/d Khurais Increment Program. The crude contract went to Italy's Snamprogetti SPA, and the gas contract was let to a consortium of Hyundai Engineering & Construction Co. Ltd. and Foster Wheeler Energy Ltd.

Aramco said the program is the largest crude increment undertaken in its history. It is slated for completion by mid-2009.

The program covers three oil fields: Khurais, Abu Jifan, and Mazlij. The Khurais gas facility will process a total of 563 MMscfd of sour associated gas and 70,000 b/d of condensate from the fields.

Snamprogetti will build four identical crude processing and gas compression trains as well as inlet and outlet facilities to receive wet and sour crude through upstream trunklines. The Hyundai-Foster Wheeler contract includes construction of two gas conditioning trains, crude and NGL storage facilities, product shipping facilities, flares, and all related support facilities.

Gas drilling in Empty Quarter

A joint venture of Royal Dutch Shell PLC, Total SA, and Saudi Aramco began exploratory drilling for nonassociated gas and condensate in the southern part of Saudi Arabia's Empty Quarter. Drilling was scheduled to end in early 2007.

The venture, South Rub al-Khali Co. Ltd. (SRAK), spudded Isharat 1, the first of seven wells planned to be drilled by 2009. Seismic survey work was under way, aiming to acquire at least 16,000 km (10,000 miles) of 2D data by 2011.

Rub al-Khali basin wildcat

LUKOIL Overseas spudded its first wildcat aimed at deep gas-condensate in Block A in the eastern Rub al-Khali basin. The well is projected to 5,000 m (16,400 ft) on the dome of the Tukhman structure. The drillsite is south of supergiant Ghawar oil and gas field in the 29,928 sq-km (11,555 sq-mile) block held 80% by LUKOIL Saudi Arabia and 20% by Saudi Aramco.

Shaybah expansion contracts

Saudi Aramco signed five major contracts to expand production from Shaybah oil field in the Rub' al Khali region on the eastern edge of Saudi Arabia's Empty Quarter. Currently, 141 wells produce more than 500,000 b/d. The expansion will boost production by 250,000 b/d by yearend 2008.

The field produces 42° gravity Arab Extra Light crude oil from Shuaiba pay at an average depth of 1,417 m (4,650 ft). The crude is virtually sulfur-free and high in its gasoline fraction.

Saudi Aramco hired SNC-Lavalin (Montreal, QC, Canada) for engineering, procurement, and construction of the central producing facilities and Hyundai Heavy Industries for power generation. Three Saudi Arabia-based contractors received contracts for power transmission and site development.

Zuluf, Central Safaniya fields

Saudi Aramco signed two lump-sum turnkey contracts with subsidiaries of J. Ray McDermott SA (Houston) for detail design, procurement, fabrication, transportation, and installation of drilling, production, and pipeline facilities off Saudi Arabia.

One contract involves drilling support structures and wellhead platforms in Zuluf oil field, plus wellhead production platforms in Central Safaniya oil field to support 2007 production.

The second contract is associated with the 22-km (14-mile) subsea portion of the new 66-km (41-mile) BKTG-1 pipeline to carry 220 MMscfd of gas from the Abu Ali Plant to the Khursaniyah Gas Plant. The subsea portion is to be installed by mid-2007.

Refining and Processing

New refineries at Yanbu' and Jubail

Saudi Aramco signed two contracts for construction of two grassroots joint-venture export refineries. The cost of each project was estimated at $6 billion. Start-up is targeted for 2011.

The plant at Jubail on the Persian Gulf will be designed and built by Total SA. A twin plant is to be constructed in Yanbu' on the Red Sea coast, under a contract with ConocoPhillips. Each plant will have a capacity of 400,000 b/d, converting Arabian heavy crude to ultra low-sulfur refined products.

Also under study was expansion of the Ras Tanura refinery, possibly transforming it into an integrated refining and petrochemical complex.

Rabigh complex

Rabigh Refining & Petrochemical Co. (PetroRabigh) let an engineering, procurement, and construction contract to two Foster Wheeler Ltd. subsidiaries for work related to its refining and petrochemical complex under development in Rabigh. PetroRabigh is a joint venture of Saudi Aramco and Sumitomo Chemical Co. Ltd. (Tokyo).

Foster Wheeler will perform detailed design, procurement, and management of the construction contractors for the utilities and offsite facilities.

The project will upgrade Saudi Aramco's 400,000 b/cd Rabigh refinery to raise output of transportation fuel and make it the basis of a complex producing a total of 2.4 million tonnes/year of petrochemical solids and liquids. The project is scheduled for completion in 2008.

Haradh GOSP-3 start-up

The third of three gas-oil separation plants in the Haradh area of Saudi Arabia's Ghawar field came on stream in early 2006, according to *Saudi Aramco Week*. GOSP-3 lies 170 km (105 miles) south of al-Hasa and about 6.4 km (4 miles) east of the town of Haradh.

Oil and gas flow from several of 32 new wells and then through more than 160 km (100 miles) of pipeline to the Abqaiq stabilization plant and 'Uthmaniyah gas plant for further processing.

The facility produces 300,000 b/d of Arabian Light crude and 140 MMscfd of associated natural gas.

CO plant

Lurgi AG (Frankfurt) won a contract worth €137 million to build what it called the world's largest carbon monoxide plant in Saudi Arabia. The contract from Saudi International Petrochemicals Co. (Sipchem) commits Lurgi, through a joint venture with Air Liquide (Paris), to build a 345,000 tpy CO plant in Jubail Industrial City. The total value of the project is €157 million. Start-up by 2009 is planned.

Petrochemicals

Jubail complex

Saudi Kayan, a joint venture of Saudi Basic Industries Corp. and Al-Kayan Petrochemical Co., awarded Flour Corp. an engineering, procurement, and construction management services contract for the utilities and offsite (U&O) facilities of the joint venture's 4 mtpy petrochemical complex to be built in Al-Jubail.

The facility will produce both specialty amine derivatives and polycarbonates for the first time in Saudi Arabia, said Fluor. Engineering on the U&O facilities began in July and will continue through 2008. Construction is slated to begin in February 2007, and completion is targeted for December 2009.

Saudi Kayan also awarded an engineering, procurement, and construction contract to KBR (Houston) for a 1.35 mtpy olefins plant in Jubail Industrial City.

Saudi European Petrochemical Co., (SEPC), a unit of Saudi Basic Industries Corp., awarded Samsung Engineering Co. Ltd. (Seoul) a contract to construct a 500,000 tpy polypropylene plant at SEPC's Ibn Zahr complex in Al-Jubail. The addition will nearly double the plant's current capacity. The additional capacity is expected to come on stream by third quarter 2008.

International Activities

Landmark accord with China

Following a summit meeting in Beijing, Saudi Arabia's King Abdullah bin Abdulaziz al-Saud and Chinese Pres. Hu Jintao agreed to a landmark agreement in early 2006 on oil, natural gas, and minerals as part of five main areas of cooperation between the two nations.

The new agreement on hydrocarbons is the first between the two governments on overall cooperation in the field of energy. No details were revealed, but a Saudi official said, "China is one of the most important markets for oil and Saudi oil is one of the most important sources of energy for China."

The energy deal was said to set a framework for specific energy investments between the two sides, and that specific agreements would soon be signed by the two countries' respective national oil companies.

This point apparently was underscored by a report in the *Beijing Morning Post* on January 24, 2006, that said the two countries planned to build a large crude oil storage facility holding 25 to 30 million tonnes in China's Hainan province. The facility was said to be part of a joint project that also includes a refinery and a natural gas storage facility.

China's demand for oil is increasing by some 15% per year. In 2005, China imported 130 million tonnes of crude oil, up 3.3% over 2004. China's imports from Saudi Arabia more than doubled in recent years, from 8.8 million tonnes in 2001 to more than 20 million tonnes in 2005.

The agreement also caps substantial energy cooperation between China and Saudi Arabia since 2003, which includes a number of projects already under way. For example, China's Sinopec is drilling for gas in the Saudi desert and building a refinery with Saudi Aramco in China's Fujian province And Saudi Aramco has started engineering work with Sinopec on a second refinery in Qingdao city.

Energy pact with India

India and Saudi Arabia signed in January 2006 a strategic energy partnership, the Delhi Declaration, to work cooperatively in the upstream and downstream sectors. The signing was announced near the end of the visit by Saudi Arabia King Abdullah bin Abdulaziz al-Saud to India.

The energy partnership envisions increased volumes of crude oil supplies from Saudi to India through long-term contracts that would be renewed automatically. Currently, India imports 25 mtpy of Saudi crude oil. The declaration also is expected to promote Saudi investments in refining, marketing, and storage in India, as well as India-Saudi ventures in gas-based fertilizer plants in Saudi Arabia.

SYRIA

Capital: Damascus
Monetary unit: Syrian pound
Refining capacity: 239,865 b/cd
Oil production: 422,500 b/d
Oil reserves: 2.50 billion bbl
Gas reserves: 8.50 tcf

Block 26

Gulfsands Petroleum PLC (Houston) announced plans to spud two wells in 2006 on 11,000-sq km (4,250-sq mile) Block 26, which it operates in northeast Syria. Both are in the northeast section of the block.

The first is the Souedieh North well. A vertical well will be drilled to 2,200 m (7,215 ft) TD to test Cretaceous reservoirs similar to those producing in the adjacent Souedieh and Karachok oil fields. This prospect has estimated recoverable reserves in excess of 100 million bbl of oil.

The second is Tigris-1 on the Tigris prospect, which underlies the Souedieh oil field. Souedieh is the largest known oil field in Syria, from which medium-gravity oil is produced from mid-Cretaceous sands.

Tigris-1will be drilled to 4,500 m (14,760 ft) TD, targeting Carboniferous and Devonian sandstone reservoirs. The prospect has estimated recoverable reserves of more than 500 million boe.

Based on study of other existing well and seismic data, Gulfsands said it had identified 31 prospects and leads and plans to drill a total of at least four wells on Block 26 by mid-2007.

Block 39

The ministry of petroleum and mineral resources awarded Loon Energy Inc. (Calgary) the right to negotiate a production sharing contract on Block 9. The block covers 3,900 sq miles in the Homs subbasin in northwestern Syria, locale of Cherrife, Ash Shaer, Palmyra, and Al Rasem, some of the country's largest light oil and gas fields. Six wells have been drilled on the underexplored block.

Block XVII

Dual Exploration Inc. took a farmout from Stratic Energy Corp. (both in Calgary) of a 31.67% participating interest in Block XVII in southwestern Syria, subject to government approval.

The block covers 5,076 sq km (1,960 sq miles) in the Palmyride fold belt south of Palmyra, Cherrife, and Abu Rabah gas fields. It has three previous wells.

The companies plan to spend $5.2 million in three to four years on seismic data and one exploratory well.

Petro-Canada to sell stake

Petro-Canada agreed to sell its stake in a joint venture—involving 220 wells in 36 fields in eastern Syria—to India's Oil & Natural Gas Corp. and China National Petroleum Corp. for $676 million (Can.). The sale is subject to Syrian government approval.

Royal Dutch Shell PLC operated Al Furat for Petro-Canada, whose share represents 58,000 boe/d before royalties.

Petro-Canada said it plans to continue exploration elsewhere in Syria. The company signed several drilling rig contracts in 2006 for future work, including expected drilling on Block II in early 2008.

UNITED ARAB EMIRATES

Capital: Abu Dhabi
Monetary unit: Dirham
Refining capacity: 781,250 b/cd
Oil production: 2.591 million b/d
Oil reserves: 97.8 billion bbl
Gas reserves: 214.4 tcf

Drilling and Production

Boosting oil output from Abu Dhabi fields

Abu Dhabi Co. for Onshore Oil Operations (ADCO) let a FEED contract to Foster Wheeler International Corp. to boost production from three mature oil fields: Sahil, Asab, and Shah (SAS). Contract value and current production figures were not available.

ADCO plans to increase total production from the fields to more than 1.8 million b/d by 2010.

Development of the SAS fields—180 km (112 miles) south of Abu Dhabi City—also is intended to enhance the environmental aspects of the fields and to provide facilities to enhance operation of the fields through 2040. The contract is scheduled for completion in mid-2007, with commissioning in 2010.

Government to run offshore Dubai fields

The international group that operates Dubai's mature offshore oil fields said it will relinquish its concession to the emirate's government in April 2007. Dubai Petroleum Establishment (DPE), a new government entity, will become operator of the fields.

A consortium of Dubai Petroleum Co. (DPC), wholly owned by ConocoPhillips, and Dubai Marine Areas Ltd. (Duma), of which DPC is part, has worked in Dubai since 1961.

DPC/Duma operates Fateh, Southwest Fateh, Falah, and Rashid fields. Others with interests are ConocoPhillips, the operator; Total SA; Repsol YPF SA; RWE Dea AG; and Wintershall AG.

Although Dubai is part of the United Arab Emirates—which is a member of the Organization of Petroleum Exporting Countries—its production is exempt from OPEC quotas and not widely reported. It is believed to have fallen recently to just above 100,000 b/d. UAE production is about 2.5 million b/d, almost all from Abu Dhabi.

Refining and Processing

Ruwais ethylene plant

Borouge, a joint venture between Abu Dhabi National Oil Co. (ADNOC) and European plastics provider Borealis, let a $1.3 billion, lump-sum turnkey contract to Linde Group and Consolidated Contractors Co. for construction of an ethylene cracker in the UAE.

The steam cracker, having an ethylene capacity of 1.5 mtpy, is planned as an expansion of a petrochemical complex in Ruwais, Abu Dhabi. From ethane feedstock, the cracker will produce polyethylene and polypropylene for export.

The project, scheduled to start in early December, is expected to be on stream by 2010.

Fujairah refinery considered

ConocoPhillips and International Petroleum Investment Co. (IPIC) of Abu Dhabi agreed to conduct a feasibility study regarding construction of a 500,000 b/d refinery in Fujairah, UAE.

If the firms decide to proceed, it is expected that ConocoPhillips and IPIC would form a joint venture to own and operate the refinery. The two firms also signed a memorandum of understanding to identify upstream and downstream opportunities for joint investment.

Petrochemicals

New units planned for Ruwais

Abu Dhabi Melamine Industry Co. was planning an 80,000-tpy melamine project in Ruwais. The project's estimated cost is $260 million, with completion scheduled for first-quarter 2009. Also, Abu Dhabi Polymers Co. Ltd. was planning a polyethylene and a polypropylene unit. Both are expected to be completed in 2010.

LNG

$1 billion storage site planed

Two leading Dubai government entities—the Dubai Multi Commodities Centre (DMCC) and Techno Park—formed a partnership with LNG Impel, a wholly owned unit of Galveston LNG Inc., to develop a $1 billion LNG storage facility at Techno Park in Dubai.

The partners said Dubai LNG Storage Hub, with an expected total storage capacity of 40–65 bcf, will be the first of its kind in the world. The open season bid process for storage capacity began in August.

Pipelines

Dolphin assumes operation of Al Ain-Fujairah line

Dolphin Energy Ltd. (DEL; Abu Dhabi) assumed operation and maintenance (O&M) responsibilities for its 182-km (113-mile) natural gas pipeline from Al Ain, Abu Dhabi, to Fujairah. The pipeline is the first initiative under its Dolphin Project.

According to a joint agreement, Emirates General Petroleum Corp. (Emarat) managed O&M for the 24-in (60.96-cm) pipeline from its commissioning in January 2004 until December 2005. When DEL's main export pipeline from Qatar is completed in early 2007, DEL will manage it as well.

DEL is owned by Mubadala Development Co. (on behalf of the Abu Dhabi government), Total SA, and Occidental Petroleum Corp.

The $3.5 billion Dolphin Project will produce and process raw gas from Qatar's North field to a processing plant in Ras Laffan, Qatar, and transport dry gas from there by marine pipeline to the UAE, beginning in 2007.

An onshore segment currently transports 135 MMcfd of gas from Oman to Union Water & Electricity's 656-MW electric power station and desalination plant on the UAE East Coast.

The offshore segment will transport gas on a long-term basis from North field to Fujairah in early 2007 and to Oman via a reversal of the onshore segment in 2008. DEL says the pipeline carries the first-ever flow of gas from one Gulf Cooperation Council member country to another.

Company News

Dana Gas PJSC to acquire Canadian firm

Private gas company Dana Gas PJSC (Sharjah, UAE) announced plans to acquire Centurion Energy International Inc. (Calgary) for $1.25 billion (Can.) net. Centurion has oil and gas exploration and production with properties in Egypt, Tunisia, and off West Africa. Shareholders are expected to approve the offer at a special meeting set for January 2007.

YEMEN

Capital: San'a
Monetary unit: Rial
Refining capacity: 140,000 b/cd
Oil production: 415,000 b/d
Oil reserves: 3.0 billion bbl
Gas reserves: 16.9 tcf

First oil delivery from Kharwah-1

Operator OMV AG (Austria) and its partners began production of 1,000 b/d of oil from its Kharwah-1 well on Block S2 (Al Uqlah) in central Yemen. Plans call for delivery of 11,000 b/d by 2008.

The group then aims to expand oil production to 32,000 b/d by 2010 under a second development phase. The field spans 1,000 sq km (621 sq miles) and has 50 million bbl of proved oil. OMV expects it to have a lifetime of at least 20 years.

OMV also won the right to operate Block 29 in the Jeza-Qamar basin in eastern Yemen under the country's third licensing round. It will work in a joint venture with Pakistan Petroleum Ltd. Block 29 covers 9,237 sq km (3,566 sq miles). Exclusive negotiations for a production-sharing agreement were under way.

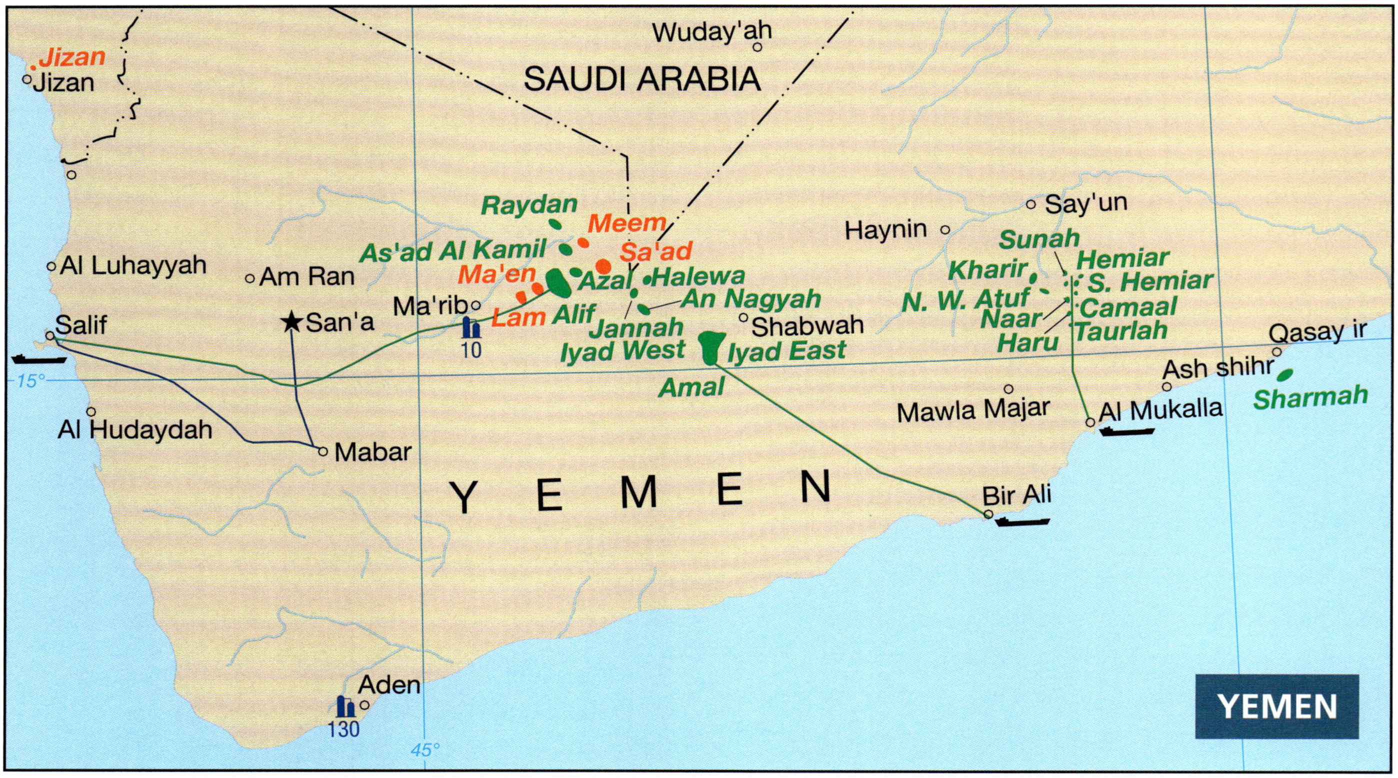

Block 22

Yemen Mayfair Petroleum Corp. and partner Oracle Energy Corp. (Vancouver, BC) announced plans to spud the Fatima-1 exploration well on Block 22 in western Yemen in mid-October.

Projected TD is 2,250 m (7,382 ft) to target the Middle Miocene that tested 42° gravity paraffinic crude oil and dry gas in Jaizan basin wells to the north (in southwestern Saudi Arabia) in the early 1990s.

Block 22 covers 8,151 sq km (3,150 sq miles) in the Tihama basin.

Block 10

Operator Total SA reported an oil discovery and successful production test of the Jathma-1 well in the East Shabwa Development Area (ESDA) on Block 10 in Yemen.

Jathma-1 is the first of a three-well exploration program targeting a new area in the northern part of Block 10. The well reached TD of 3,175 m (10,415 ft) and tested 1,900 b/d of 35° gravity oil.

Total also reported the KHA 1-16 oil well in the ESDA on Block 10 produced more than 8,400 b/d when tested through a 70/64-in. choke.

Test results extend Khairir oil field to the east beyond current 3D seismic survey data. The well was drilled on the eastern-most 3D seismic parameter as part of a continuing appraisal of the field's basement reservoir. KHA 1-16 reached a MD of 3,298 m (10,820 ft).

Blocks 3, 15, 35

MND Exploration & Production Ltd., a subsidiary of the Czech oil and gas producer MND, took a farmout under which it will participate with Oil Search Ltd. (Port Moresby, Papua New Guinea) in three exploration blocks in Yemen.

MND acquired interests in Gardan Block 3, Mukalla Block 15, and Al Hood Block 35. A land rig was under contract to start drilling three wells in fourth quarter 2006, starting with Prospect B on Block 35.

Spud date for Riyan, Oil Search's first well on Block 15 in the Gulf of Aden, is January 2007.

Tests in Masila basin

DNO ASA (Oslo) tested oil in the Godah-1 sidetrack on Block 32 in the Masila basin in Yemen.

The well, drilled to a TD of 1,841 m (6,040 ft) downdip of the original location, encountered a full oil column and flowed at 1,839 b/d gross with no water.

The original Godah-1 exploration well was drilled on a separate structure 14 km (9 miles) east of Tasour field, primarily to target the Qishn S1A sandstone.

DNO and partners have agreed to drill the Godah-2 appraisal well farther downflank of the Godah-1 sidetrack to evaluate the areal extent of the oil accumulation and possibly prove the oil-water contact.

Africa

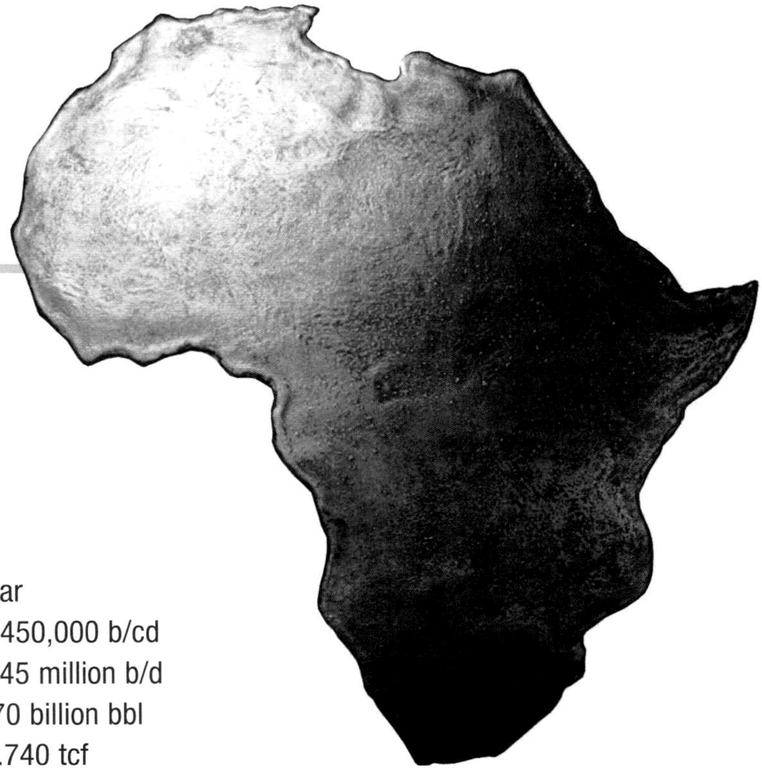

ALGERIA

Capital: Algiers
Monetary unit: Dinar
Refining capacity: 450,000 b/cd
Oil production: 1.345 million b/d
Oil reserves: 12.270 billion bbl
Gas reserves: 161.740 tcf

Exploration and Development

Block 405b shows promise

First Calgary Petroleums Ltd. (Calgary) was awaiting an exploitation license in late 2006 from Sonatrach for the giant MLE gas condensate field, in the central part of Menzel Ledjmet Block 405b in eastern Algeria's Berkine basin. The field covers more than 85 sq km (33 sq miles) on the block's east side.

MLE is one of the largest gas discoveries in Algeria in recent years. It had gross probable and possible reserves of 6.6 tcf and 903 million bbl in Lower Devonian to Triassic pays at the end of 2005. The start of production is expected in 2009.

First Calgary has drilled five exploration wells and an appraisal well. All were successful in multiple geological horizons in what the company calls its Central Area Field Complex. Further appraisal will start in 2007, aiming to secure the next commercialization stage following MLE.

The company had drilled and tested the Zemlet er Rakeb-1 exploration well in the northwest part of the block earlier in 2006. It gauged hydrocarbons at rates sufficient to support drilling ZER-2 about 4 km (2.5 miles) to the southwest.

WEST AFRICA
Atlantic Ocean
Mediterranean Sea
Ionian Sea
Gulf of Guinea
Bight of Biafra
MOROCCO
ALGERIA
TUNISIA
LIBYA
WESTERN SAHARA
MAURITANIA
MALI
NIGER
CHAD
SENEGAL
GAMBIA
GUINEA-BISSAU
GUINEA
SIERRA LEONE
LIBERIA
IVORY COAST
BURKINA FASO
GHANA
TOGO
BENIN
NIGERIA
CAMEROON
CENTRAL AFRICAN REPUBLIC
EQUATORIAL GUINEA
GABON
CONGO (FORMER ZAIRE)
CABINDA
ANGOLA
NAMIBIA
BOTSWANA
REPUBLIC OF SOUTH AFRICA
SICILY
GREECE
CRETE
Rabat
Mohammedia
Casablanca
Kechoula
Essaouira
Djebel-Jeer
Safi
Sidi Rhalem
Sidi-Kacem
Sidi Ifni
Las Palmas
Tarfaya
Oran
Algiers
Arzew
Skikda
Bejaia
Bizerte
Tunis
Gela
Athens
Tripoli
Tobruk
Sider
El Golea
Hassi Messaoud
Bel Rhazi
Touat
Ilatou
SBAA
Azzene
Reggane
In-Salah
Wafa
In Amenas
Farrud
Umm Farud
Amal
Nafoora
Augila
Defa
Messla
Sarir
Murzuk
Amasralad
Le Camp
2 Lines
3 Lines
Chinguetti
Nouakchott
Bamako
Niamey
Ouagadougou
Conakry
Freetown
Bissau
Monrovia
Abidjan
Espoir
Panthere
Bellier
Foxtrot
Saltpond
Accra
Lome
Porto Novo
Lagos
Abuja
Kano
Kaduna
Makurdi
Lake Chad
N'Djamena
Komé
Bolobo
Miandoum
Doba
Bangui
Yaounde
Kribi
Ceiba
Sao Tome
Libreville
Cape Lopez
Port Gentil
Tschengue
Rembo-Kotto
Asswe
Sette-Cama
Gamba-Ivinga
Mossaka
Brazzaville
Kinshasa
Pointe Noire
N'Kossa
Emeraude
GCO
Guntala
Essungo
Mavanga
Malева North
Sulele
Benfica
Tobias
Luanda
Lobito
Windhoek
Walvis Bay
Gaborone
Luderitz
Johannesburg
Kudu
Ibhubesi
Cape Town
Cape of Good Hope
Mossel Bay
Sable
F-A-
Oribi
0 300 600 Mi.
0 200 400 600 Km.

EAST AFRICA
Mediterranean Sea
Indian Ocean
Red Sea
Gulf of Aden
Persian Gulf
Mozambique Channel
Ionian Sea
Aegean Sea
SICILY
GREECE
CRETE
TURKEY
CYPRUS
SYRIA
LEBANON
ISRAEL
JORDAN
IRAQ
IRAN
PAKISTAN
SAUDI ARABIA
OMAN
YEMEN REPUBLIC
TUNISIA
LIBYA
EGYPT
NIGER
CHAD
SUDAN
ERITREA
DJIBOUTI
ETHIOPIA
SOMALIA
NIGERIA
CAMEROON
CENTRAL AFRICAN REPUBLIC
CONGO (FORMER ZAIRE)
UGANDA
KENYA
GABON
CONGO
RWANDA
BURUNDI
TANZANIA
CABINDA
ANGOLA
ZAMBIA
MALAWI
MOZAMBIQUE
ZIMBABWE
NAMIBIA
BOTSWANA
SWAZILAND
LESOTHO
REPUBLIC OF SOUTH AFRICA
MADAGASCAR
Tunis
Tripoli
Cairo
Khartoum
Asmera
Djibouti
Addis Ababa
Mogadishu
Nairobi
Kampala
Kigali
Bujumbura
Dodoma
Lilongwe
Lusaka
Harare
Maputo
Mbabane
Maseru
Pretoria
Gaborone
Windhoek
Luanda
Kinshasa
Brazzaville
Libreville
Yaounde
Bangui
N'Djamena
Antananarivo
Riyadh
Sanaa
Baghdad
Damascus
Amman
Jerusalem
Beirut
Nicosia
Athens
Kuwait
Manama
Doha
Abu Dhabi
Port Sudan
Mombasa
Dar es Salaam
Beira
Durban
Cape Town
Johannesburg
Lake Victoria
Lake Tanganyika
Lake Malawi
Lake Chad
Lake Kariba
Songo-Songo
Pande Temane
Kudu
Ibhubesi
0 300 600 Miles
0 200 400 600 Km

SPAIN
SARDINIA
MOROCCO
TUNISIA
ALGERIA
Madrid
Puertollano
135
Murcia
120
Cartagena
Cordoba
Malaga
Almeria
Algeciras
Cagliari
Bizert
34
Tabarka
Annaba
300
Skikda
Bejaia
Carree
Algiers
60
Blida
Setif
Constantine
Batna
Biskra
Qairouan
Kasserine
Douleb
Arzew
60
Oran
Beni Saf
Tiaret
Sidi Bel Abbes
Saida
Oujda
Fes
3 Lines
3 Lines
2 Lines
2 Lines
Oued Gueterini
Djabel Onk
Guergult Kihal
Ras Toumb
Sidi Behara
La Skhirra
Gafsa
Laghouat
Touggourt
Hassi R'Mel
Hassi R'Mel
DJ Bissa
Hassi R'Mel South
3 Lines
Djorf
Torba
Draa Tamra
Boukhezana
Guenafide
Sabria
El Borma
Keskessa
Ait Kheir
Ghardaia
Mokta
N. Goussa
Guellala NE
Oued Noumer
Guellala
Hassi Messaoud
Rhourde el Baguel
Hassi Messaoud
Ouargla
Haoud Berkaoui
Nassim
Hawa
Hassi Bahamou
Messdar
El Gassi
30
Zotti
El Agreb
Bir Berkine
Chouech es Saida
Hassi Berkine
Fedjet Ezzine
El Golea
Hasel Touareg
Bridea
Wadi El Tec
Our Ghourd
Hassi Chargul N.
Menzel Lejmat
Rhourde El Adra
El Merk
Hassi Chargul
In Amedjane
Rhourde Nousa
El Quar
Tin Zemane
Ohanet
Ektaia
Sedoukhane
Zerafa West
Zerafa
Krechba
Timimoun
Bel Rhazi
Ablad
Bouhadid West
Bouhadid
Qued Zine
Teguentour
Tin Fouye Tabankort
Achab
Stah
Amassak
Alrar
Dechelra West
Dechelra
Sbaa
Azzene
Dechelra
Reg
Garet El Guafoul
Garet el-Befinat
Gour Mahmoud
Hassi Mezoula
Edayen
Irtalene
In Am
Issaouane
Tim
Gara
Irtalene West
Tigfermas
Hassi M' Sari
In Salah
Hassi Moumene
W. Ihansatone
Hassi Tabtab
Qued Talha
Hassi Hassine
In Salah
El Adeb Larache
Assakaifaf
Reggane
Tit
Djebel Zinl
Oubarakat
Makarrane North
Berga
En Bazzene
Reggane
DJ Belda
Bahar el Hammar
Qued Djaret
Tirechoumine
Krebbed
Azzei Matti
DJ Thara
Gara Azzel Matti
Tibaradine
Krib Doura
Adrar Morrat

ITALY
GREECE
SICILY
Ionian Sea
Sea of Crete
CRETE
Mediterranean Sea
Gulf of Sidra
Gulf of Gabes
Gulf of Hammamat
Cape Bon
LIBYA
NORTH AFRICA
Cagliari
Palermo
Gagliano
Reggio
Augusta
Siracusa
Gela
Athens
Bizerte
34
Tunis
Sousse
Qairouan
Sfax
Gabes
Zarzis
Remada
Mellitah
Tripoli
120
Misratah
Sidra
Es Sider
Ras Lanuf
220
8
Benghazi
Zueitina
Marsa el Brega
Tobruk
20
Sarir
10
Waha
Adri
Sabha
Awbari
Murzuk
Ghadames
In Amenas
Qued Zar
Birsa
Tazerka
Oudna
Sidi el Kilani
Lalla Saida
Isis
Ashtart
Miskar
Tama
Bouri
Bahr es Salaam
Al Jurf
Makhrouga
Tigi
Nassim
Adam
Oued Zar
Hawa
Qued Zar
Wafa
El Merk
Ouen Essar
Mereksen
Ain Antar
Stah
Alrar
Zarzatine
Tihalatine
Edjeleh
Daja
Amasralad
Oued Amasralad
Le Camp
Atshan
Bir Tlacsin
Z-66
Kebir
Emgayet Fields
FF1
FF2
FF3
LL
El Hamra
Gazeil
Oued-Tahara
C-NC115
El Sharara
H-NC115
A-NC115
Elephant
Antelat
Mabruk N.
Mabruk E.
Mabruk W.
Bahi
Bazuzi
Almas
Arbad
Tabet
Hateiba N.
Hateiba S.
Attahadi
Dahra-Hofra
Tagrifet
Daba
Fidaa
Zella
Aswad
Karim
Ed Dib
Mellugh
Hakim
Agoco
Raguba
Harem
Khuff
Samah
Sabah
Khalifa
Barrut T.
En Naga North
Barrut
En Naga
Veba Discoveries
Beda
Dor
Marada
Meghil
Tmed
Waha S
Defa
Belhedan
Assumud
Sahl
El Meheiriga
Sorra
Wadi
Raleh
Harash
Intisar Fields
Augila
Amal
Libba
Fatah
Jakhira
Nafoora
Bu Attifel
Latif
Rimal
Katib
EE-80
B-82
K-80
HH-80
Masrab
Gialo
Argub
Messla
Chadar
Farigh EE
Farigh UUU
Sarir L.
Sarir N.
Sarir S.

Africa's role in liquids output, LNG growth

An evaluation by IHS Energy (Englewood, CO, U.S.A.) provided the following highlights:

- In 2006, Africa supplied 12% of world liquids production and 29% of active LNG shipping capacity.
- Africa will supply 30% of an estimated 13.65 million b/d growth in global liquids production capacity by 2010, excluding the U.S. onshore and Canada.
- African finds totaled nearly 25% of world liquids reserves discovered from 2000–04 and 12% of world gas reserves discovered in the same period, excluding the U.S. onshore and Canada.
- Companies discovered 300 billion boe, two-thirds of it liquids, in Africa through 2004. Of that, 85% was found in 10 basins—22% in Libya's Sirte basin alone.
- Africa's remaining liquids resources at the end of 2004 were estimated at 105 billion bbl. That includes Nigeria 35.6 billion bbl, Libya 26.8 billion bbl, Angola 13.6 billion bbl, Algeria 14 billion bbl, and Egypt 3.4 billion bbl. A further 11.2 billion bbl was in sub-Saharan and Saharan countries.
- A substantial part of Africa's oil production growth will come from new giant fields in Nigeria, Angola, and Algeria. Giant deepwater discoveries off Angola and Nigeria are expected to add more than 2.2 million b/d.
- National oil companies, facing increasing competition from international oil companies for access to hydrocarbon resources, hold 216 licenses in Africa, compared with 95 tracts in 1995.
- Africa will contribute more than 25% of the expected growth in global LNG capacity by 2010. Africa had 50 mtpy of LNG capacity at the end of 2005, most of it in Algeria and Nigeria.

Zotti block drilling planned

Petro-Canada signed several drilling rig contracts and announced that it expected exploratory work in the Zotti Block in Algeria to start by year-end 2006.

Timimoun basin

Statoil ASA secured two drill rigs and was planning for exploratory drilling in 2006 in the Hassi Mouina block. It signed separate contracts with Nabors Drilling International Ltd. and Algeria's Entreprise Nationale de Forage (Enafor). The contracts provide the rigs through March 2008 with extension options.

The Hassi Mouina block covers 22,993 sq km (8,878 sq miles) in the Timimoun basin. Sonatrach previously drilled one well on the block and discovered gas, which Statoil will develop.

Separately, a group led by Total SA also reported a gas discovery in the Timimoun basin. The MJB-3 exploration well flowed 8.3 MMcfd of gas from an undisclosed formation. The group was evaluating the discovery and continuing its exploration and appraisal program.

Reggane gas search

A group led by Repsol YPF SA, which made two natural gas discoveries in the northern Reggane basin of central Algeria in early 2006, announced plans to drill at least eight more wells by the end of 2007. Further seismic evaluation was under way.

The group said that the Reggane-5 and Sali-1 wells on the Reggane North Permit 1,328 km (825 miles) southwest of Algiers tested gas from the Lower Devonian formation.

Reggane-5 flowed gas at a maximum 24 MMcfd on a 32⁄64-in. choke with 4,238 psi wellhead pressure from two levels. Sali-1 flowed as much as 3.8 MMcfd on a 1-in. choke with 122 psi wellhead pressure.

The Reggane North permit consists of Blocks 351c–352c, which cover 12,012 sq km (4,638 sq miles). Other interests are Sonatrach, RWE Dea AG, and Edison Gas.

Illizi basin gas find

Petroceltic International PLC (Dublin) said two wells confirmed the commercial potential for production from shallow reservoirs on part of the 10,872 sq-km (4,199-sq mile) Isarene permit in the Illizi basin of eastern Algeria.

ISAS-1 tested 360 Mcfd of wet gas in two zones between 851 and 856 m (2,972 and 2,808 ft) and 850 Mcfd of wet gas from a second zone at 558–564 m (1,831–1,850 ft).

About 40 km (25 miles) west, the Hassi Tab Tab-2 well flowed 12.56 MMcfd of wet gas from two zones between 907 and 943 m and 2.44 MMcfd of wet from a zone at 436–442 m (1,430–1,450 ft).

Isarene is 120 km (75 miles) south of In Amenas, Algeria's largest wet gas project.

Drilling and Production

First gas from In Amenas

The In Amenas natural gas project in southeastern Algeria started production in May. Output was expected to build to 870 MMcf/day of gas, plus 55,000 b/d of liquids, by July.

In Amenas is a joint development of Statoil (Oslo), BP (London) and Algeria's state-owned Sonatrach. It is some 850 km (528 miles) south of Hassi Messaoud.

Investment levels for oil, gas development

Sonatrach said investments by foreign oil companies in Algeria came to $1.3 billion in 2005. Looking ahead, the company said some $8.6 billion will be invested in the development of Algerian fields between 2006 and 2010—about $2.5 billion (29%) by foreign partners and $6.1 billion by Sonatrach.

Sonatrach made eight discoveries in 2005—six by itself—and oil production grew 4% over 2004, reaching 232.3 million tonnes. Production in partnership grew by 18% in 2005. Oil exports amounted to 141.3 million tonnes.

The company said it was aiming to increase 2006 oil production by 1.5% over 2005 to 235.6 million tonnes.

Pipelines

Galsi gas purchases announced

New gas-purchase agreements by five Italian energy companies gave a push to plans for a third gas pipeline between North Africa and Europe.

The companies—Enel, Edison, Hera, Ascopiave, and Worldenergy—agreed to import a total of 6 bcm/year of gas from Algeria through the proposed pipeline connecting Algeria with Italy by way of Sardinia.

Algeria's Sonatrach, Enel, and Wintershall (Kassel, Germany) formed a venture to study the feasibility of the pipeline in 2001. Other companies joined the venture, now called Galsi.

The project envisions four pipeline segments:

- 640 km (397 miles) onshore between Hassi R'Mel gas field in Algeria and El Kala on the Algerian coast
- 310 km (192 miles) between El Kala and Cagliari on Sardinia in water as deep as 1,950 m (6,398 ft)
- 300 km (186 miles) between Cagliari and Olbia on the northern Sardinian coast
- 220 km (137 miles) between Olbia and Pescaia, southeast of Florence, in water as deep as 900 m (2,953 ft)

The pipeline will have a capacity of 8 bcm/year and is projected to start up in late 2009.

NK-1 oil pipeline progress

ABB Group (Zurich) won a $215 million contract from Sonatrach to construct two pumping stations and automation systems on the 665-km (413-mile) NK-1 oil pipeline linking oil fields in Algeria at Haoud El Hamra with the Mediterranean port of Skikda.

The NK-1 project, scheduled for completion in late 2008, will add 18 mtpy to Algeria's oil transport capacity.

Medgaz pipeline completes permit stage

Medgaz SA (Madrid) in December made a firm investment decision to proceed with the design, construction and operation of the Algeria-to-Spain gas pipeline that will serve European markets.

That decision marked the end of the permit phase and the beginning of the construction phase. The firm said that all government approvals have been granted, and engineering, procurement and construction contracts were awarded.

Detailed engineering and material fabrication will take place in 2007, and actual construction will start in early 2008. First gas delivery is slated for mid-2009. Initial capacity will be 8 bcm/year.

Shareholders in the pipeline project are Algeria's state-owned Sonatrach, CEPSA (Madrid), Iberdrola (Bilbao), Endesa (Madrid), and Gaz de France (Paris).

Spanish gas system capacity

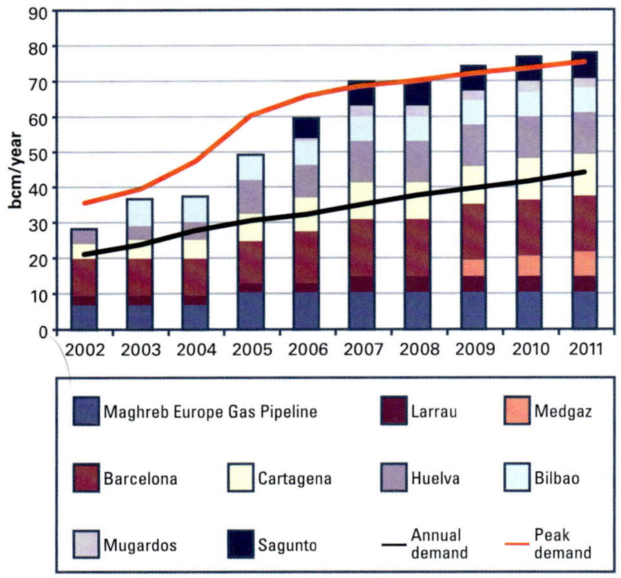

Source: CNE (2004).

Route, bathy-morphological features

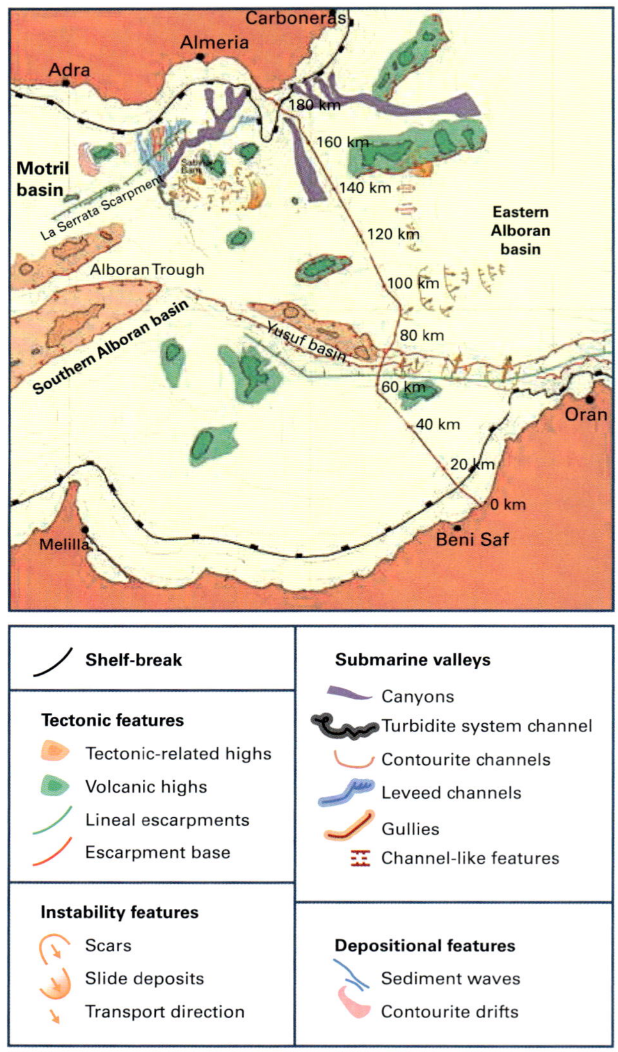

In Algeria, the Medgaz line will connect with the Hassi R'Mel-Beni Saf gas pipeline operated by Sonatrach. Through a reception terminal at Almeria, Spain, the line will hook up with the Almería-Albacete gas pipeline of Enagas (Madrid), facilitating its connection to the Spanish and European gas grid.

The total offshore distance of the pipeline is 210 km (130 miles) across the Mediterranean seabed. Maximum water depth for the central section will be 2,160 m (7,087 ft).

LNG

Group plans LNG plant

Repsol YPF SA (Madrid), Gas Natural SDG SA (Barcelona), and Sonatrach agreed to form a new company, Soc. de Licuefaccion, to build and operate a liquefaction plant based on natural gas from Gassi Touil field in eastern Algeria. Plans call for the 4 mtpy LNG plant to come on stream in 2009.

Policy

Oil windfall profits tax

The Algerian Energy Ministry was awaiting comment as of October from international oil companies regarding a windfall profits oil tax slated to take effect in 2007. The tax was part of a law enacted in September. The same law provides state firm Sonatrach to take a 51% interest in contracts involving international oil companies.

Companies from outside Algeria will have to pay the oil tax in any month when the price of Brent crude averages more than $30/bbl. The tax rate will vary from 5 to 50%, depending on a company's total production.

Anadarko Petroleum Corp., Royal Dutch Shell PLC, BHP Billiton Ltd., ENI SPA, and Hess Corp. had oil operations in Algeria in 2006. Analysts said the companies were awaiting an explanation of the new tax structure.

ANGOLA

Capital: Luanda
Monetary unit: Kwanza
Refining capacity: 39,000 b/cd
Oil production: 1.385 million b/d
Oil reserves: 8.00 billion bbl
Gas reserves: 2.00 tcf

Exploration and Development

Block 17 potential raised

State-owned Sonangol and operator Total E&P Angola reported that the Orquidea-2 appraisal well confirmed and expanded the potential of the Orquidea oil discovery on Block 17.

Drilled in 1,165 m (3,822 ft) of water 2 km (1.2 miles) from the discovery well, Orquidea-2 encountered the same Miocene objectives as did Orquidea-1 and also identified deeper Oligocene reservoir levels. Both the Tertiary Miocene and Oligocene objectives are oil-bearing.

The two-well success on the Orquidea structure, near the Lirio, Cravo, and Violeta finds, confirms the potential for development of a fourth production zone on Block 17, for which preliminary design was under way.

The production zone is in the northwestern area of the block and will complete the Girassol and Dalia zones, soon to be followed by the Pazflor production zone.

Block 31: Strikes 11 and 12

Operator BP Exploration (Angola) PLC and state-owned Sonangol announced two more oil discoveries—their eleventh and twelfth—in ultra-deepwater Block 31 off Angola.

The Titania well was drilled in 2,152 m (7,060 ft) of water, 23 km (14 miles) northwest of the Hebe oil discovery announced in late 2005. Titania reached a TD of 5,339 m (17,516 ft) below sea level, penetrating Oligo-Miocene sands. The well, in the southeast part of the block, flowed on test at 2,045 b/d through a 20/64-in. choke.

The Urano well flowed on test a maximum 1,970 b/d of oil through an 86/64-in. choke. The well reached a TD of 4,578 m (15,020 ft) below sea level. The well was drilled in 1,938 m (6,358 ft) of water, 345 km (214 miles) northwest of Luanda.

Titania and Urano are the first two wells drilled through salt in Block 31, into the sandstone reservoir beneath.

Plans were being made in late 2006 for development of all 12 discoveries—the first four, made earlier in the northeast part of the block, and the eight more recent discoveries including Titania and Urano. Block 31 covers 5,349 sq km and lies in 1,500–2,500 m (4,920-8,200 ft) of water.

ANGOLA BLOCK 17

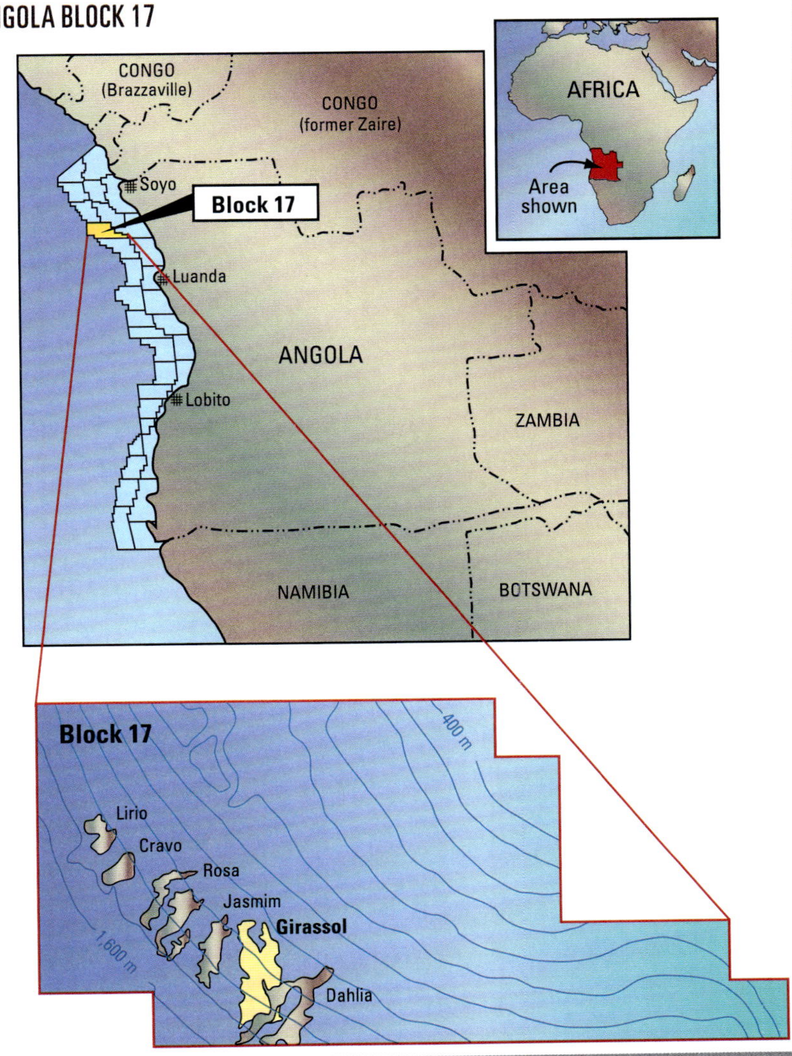

Block 32: Three oil discoveries

Total E&P Angola, in participation with state-owned Sonangol EP, reported three oil finds on Block 32 in ultradeep water off Angola.

One was with the Salsa-1 well, the sixth exploration well drilled on the block. It is in the southeastern portion of the block in 1,806 m (5,925 ft) of water. It tested 3,686 b/d of oil from a Miocene reservoir. Salsa-1 is 15 km (9 miles) southwest of the Mostarda-1 discovery.

Also, the Gengibre-2 delineation well confirmed the commercial potential of the adjacent 2005 Gengibre-1 oil discovery. Gengibre-2—161 km (100 miles) off the Angolan coast in 1,695 m (5,564 ft) of water—was drilled to a TD of 4,342 m (14,245 ft). On test, the well flowed oil at 4,540 b/d from a Miocene reservoir and 5,100 b/d from an Oligocene reservoir.

Also, the Mostarda-1 well, drilled in 1,758 m (5,768 ft) of water, tested 5,347 b/d of 30° gravity oil from one interval. The discovery is in the eastern section, 15 km (9 miles) southeast of the Gengibre-1 strike.

Technical studies were under way to evaluate development options for the three discoveries.

ANGOLA'S DEEPWATER FIELDS

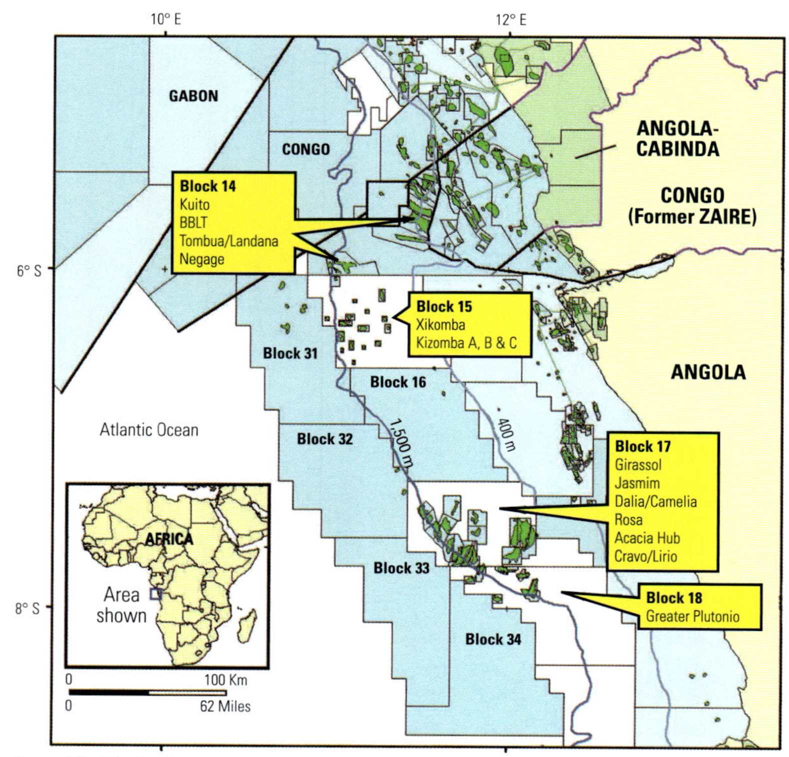

Source: Wood MacKenzie

Drilling and Production

Dalia field production

Oil production from the offshore Dalia field was expected to reach 250,000 b/d by mid-2006, said operator Total SA. The company described Dalia as the largest deepwater development to be brought on stream in 2006 and among the largest projects of its kind in the world.

Located on deepwater Block 17, 135 km (84 miles) offshore in 1,200–1,500 m (3,935–4,920 ft) of water, the field holds 1 billion bbl of recoverable oil. Production is from 37 subsea wells.

Block 14: Landana North reservoir

Chevron Corp. subsidiary Cabinda Gulf Oil Co. Ltd. began oil production from the Landana North reservoir in the Tombua-Landana development area. It is Chevron's third operated deepwater development off Angola.

The Landana North-1 well on deepwater Block 14 was the first of 46 wells planned in the project. The well is tied back to the Benguela Belize-Lobito Tomboco compliant piled tower, which allows for early production as well as the gathering of reservoir data.

The project, 80 km (50 miles) offshore in more than 365 m (1,200 ft) of water, will employ a compliant piled tower with one subsea center. Projected peak production from the completed development is 100,000 b/d of oil by 2010.

Super-large FPSO readied

BP PLC was making arrangements in late 2006 to install what is being called a super-large FPSO vessel on deepwater Block 18 in its Greater Plutonio fields off Angola.

The 360,000-dwt vessel, named after the fields, is 310 m long, 58 m wide, and 32 m high (1,017 ft long, 190 ft wide, 105 ft high). Built by Hyundai Heavy Industries, the Greater Plutonio vessel can hold 2 million bbl of crude oil and can handle production of about 200,000 b/d. Its living quarters accommodate 120 crew members.

Kissanje blend crude assayed

ExxonMobil Corp. conducted an assay of the Kissanje Blend crude from the Kizomba B deepwater project offshore Angola. Esso Exploration Angola (Block 15) Ltd. started Kizomba B production in 2005.

The $3.5 billion project involves development of 1 billion bbl of oil in Kissanje and Dikanza oil fields, which lie in 1,000 m (3,280 ft) of water about 346 km (215 miles) northwest of Luanda. Production from the fields is expected to peak at 250,000 b/d.

Kizomba B includes what Esso Angola parent ExxonMobil Corp. calls the world's largest FPSO vessel, able to store 2.2 million bbl.

Esso Angola completed the Kizomba B project in 31 months. Total production from Kizomba A, Kizomba B, and the Xikomba project, which went on stream in 2003, was expected to peak at more than 550,000 b/d by yearend 2005.

Kizomba C subsea systems

ExxonMobil Corp. let an engineering, procurement, and construction contract to Vetco Gray for subsea systems in Mondo oil field in the first phase of the Kizomba C deepwater development project. Kizomba C is on Block 15 in 740 m (2,428 ft) of water about 370 km (230 miles) west of Luanda.

In addition to Mondo, it will develop the Saxi and Batuque discoveries, all with subsea wells tied back to FPSO vessels used in the Kizomba A and B developments.

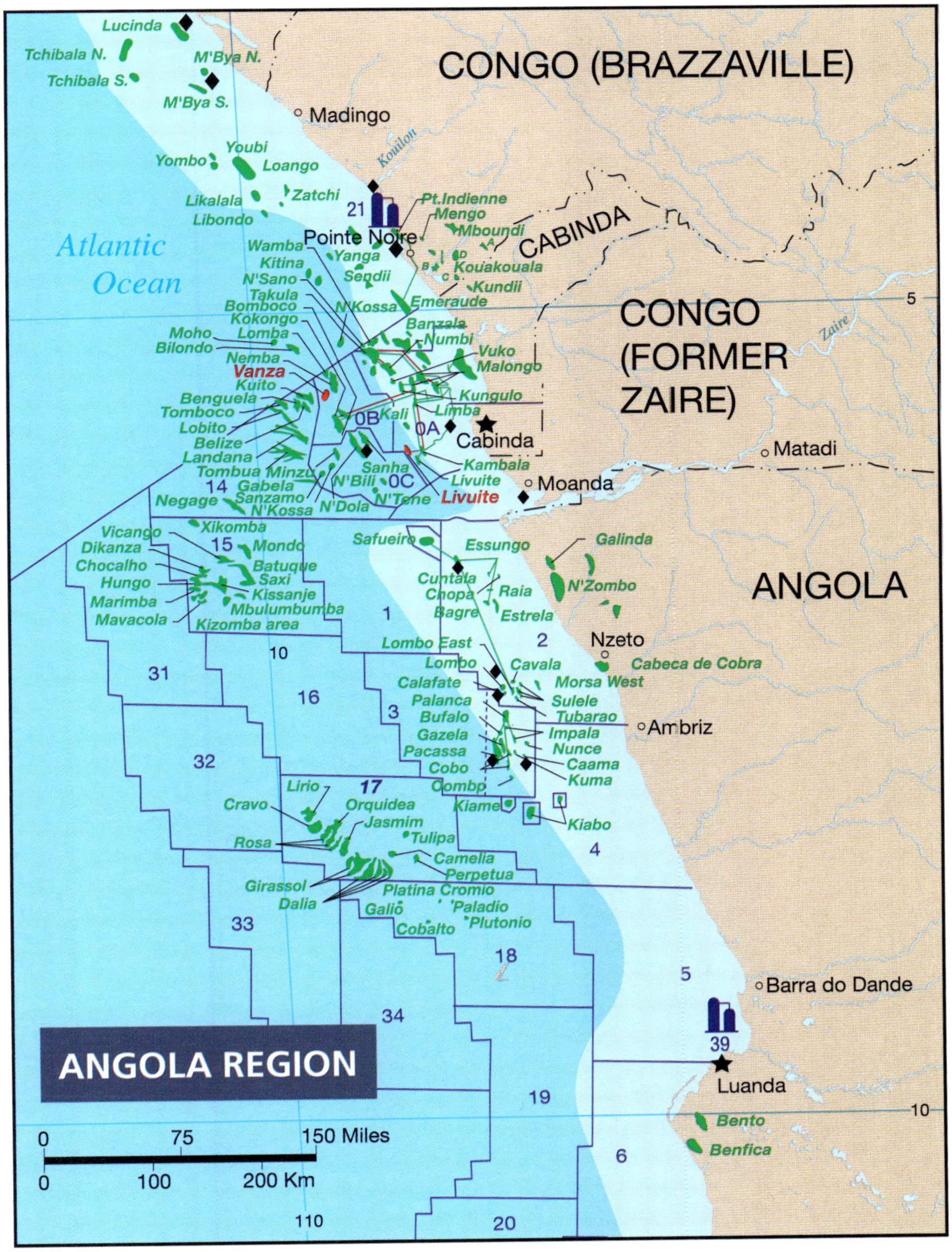

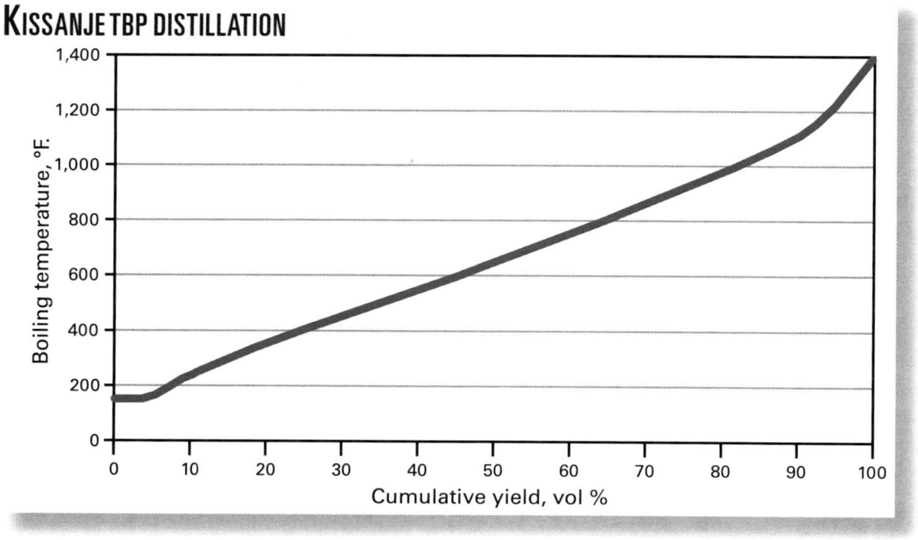

Belize oil field on stream

Belize oil field on deepwater Block 14 off Angola came on stream, reported Cabinda Gulf Oil Co. Ltd., the operator and a subsidiary of Chevron Corp. The Benguela, Belize, Lobito, and Tomboco fields form the BBLT development, one of Chevron's biggest projects.

Block 14 lies in the Lower Congo basin in 390 m (1,280 ft) of water covering 602 sq km (1,560 sq miles). The block is the deepwater extension of the Chevron-operated Cabinda concession, which contains seven commercial discoveries.

BBLT is being developed in two phases. Phase 1, called Benguela Belize, combines an integrated drilling and production platform hub facility, supported by a compliant piled tower. Use of the tower marks the first time this technology has been used outside the Gulf of Mexico, Chevron said. Phase 2, called Lobito Tomboco, will produce through subsea wells tied into the central production hub.

CAMEROON

Capital: Yaounde
Monetary unit: CFA franc
Refining capacity: 42,000 b/cd
Oil production: 87,500 b/d
Oil reserves: 400 million bbl
Gas reserves: 3.9 tcf

Chad-Cameroon oil project

See the section on Chad for information about this project.

Dissoni block oil strike

Total E&P Cameroon struck oil with the first exploration well drilled on the Dissoni Block in the Rio del Rey basin off Cameroon. The well encountered 50 m (165 ft) of sandstone oil pay. Total, the operator, was considering development.

In 2005, Cameroon's Societe Nationale des Hydrocarbures (SNH) awarded the Dissoni Block to Total and Pecten Cameroon Co. in the country's first-ever production sharing contract.

Operator change for PH-77 license

Noble Energy Cameroon Ltd. announced it will become operator of the PH-77 license off Cameroon by acquiring ConocoPhillips's 50% participating interest. State-owned SNH may assume a 50% participating interest if there is a commercial discovery.

The license covers 1.125 million acres (1,760 sq miles) between the coast and the country's border with Equatorial Guinea. It is contiguous with Equatorial Guinea Blocks O and I, off Bioko Island, for which Noble Energy is technical operator.

In 2005, the company tested 24 MMcfd of gas and 1,225 b/d of condensate in its O-1 well on a Block O prospect that it designated Belinda.

Kombe-N'sepe block change

A unit of private independent Kosmos Energy (Dallas), acquired a 35% working interest in the 748,000-acre (1,170-sq mile) Kombe-N'sepe block, south of Douala in the Douala basin coastal strip.

Kosmos and its partners—Perenco (London), the operator; and state-owned SNH—aimed to acquire 2D seismic data in late 2006, planning to later drill an exploration well under a multiyear agreement. The work will focus on the under-explored northern extension of the late Cretaceous turbidite play that extends from Gabon to Cameroon.

Perenco also was planning to appraise the former Elf Aquitaine's 1982 M'Via oil and gas discovery on adjacent acreage.

License round plans

According to the U.S. EIA, Cameroon was expected to open licensing rounds for blocks in the Rio del Rey region and on the Bakassi Peninsula. Nigeria withdrew from the latter area in mid-2006. Industry experts believe Bakassi acreage could contain significant amounts of oil because it borders areas in the Gulf of Guinea that, in the past, have yielded numerous oil discoveries.

Natural gas development

Perenco (London) signed a 25-year contract with Cameroon's SNH to develop the offshore Sanaga Sud natural gas fields, in the Douala/Kribi-Campo basins. Gas from the fields will generate power at the Kribi plant. The majority of Cameroon's natural gas reserves are located in the Rio del Rey, Douala, and Kribi-Campo basins.

CHAD

Capital: N'Djamena
Monetary unit: CFA franc
Refining capacity: 0
Oil production: 180,000 b/d
Oil reserves: 1.5 billion bbl
Gas reserves: 0

Oil revenues dispute with World Bank

Early in the year, the Chadian government violated a 1999 agreement with the World Bank regarding use of oil revenues. The government voted to use more than the agreed-to percentage of oil revenues for general-budget, near-term use, saying it needed more rapid access to the funds for security and other societal priorities.

After several months of back-and-forth threats to shut down production and freeze loan accounts, the parties compromised in July. Chad will be allowed to use up to 30% of oil revenues for the general budget rather than the original 20%. The other 70% is to go to longer-term development programs.

Tax dispute with producers

In October, the government resolved a two-month dispute with Petronas and Chevron. The president of Chad had accused the producers of not paying nearly $500 million in oil-profit taxes and ordered them to leave the country. The two firms agreed to pay outstanding taxes for 2005 and 2006, totaling $289 million. The settlement allowed the two companies to remain in Chad.

Working to add new fields

The project operating group spent $232 million in 2005, bringing total capital outlays on oil production efforts to $4.7 billion. Some $137 million of 2005 outlays went to develop new fields beyond the initial three: Kome, Bolobo, and Miandoum. The remainder was for running seismic surveys and drilling exploration wells.

Chad-Cameroon oil project overview

The Chad-Cameroon Petroleum Development and Pipeline Project aims to boost production of oil from fields in landlocked Chad and transport it by pipeline across Cameroon to the coastal city of Kribi for export. Oil revenues are targeted to improving economic and social conditions as well as upgrading infrastructure in the two nations, both of which are quite poor.

The foundations of the project were established in the late 1980s, followed by oil-resource evaluation in the early and mid 1990s, and the start of pipeline construction in 2000. Oil production from three fields near Doba—Kome, Bolobo, and Miandoum—began in 2003, and the 1,075-km (670-mile) pipeline was completed in 2004. Capacity is estimated at 225,000 b/d.

Production from a smaller field, Nya, began in 2005, and first flow from the Moundouli field is expected in 2007. Exploration of other fields was under way in 2006.

Current participants in the project include ExxonMobil, Chevron Corp., Malaysia's Petronas, and the World Bank—which loaned $93 million for pipeline construction—as well as the governments of Chad and Cameroon. Esso Exploration & Production Chad Inc. operates the Doba oil fields.

In 2006, the government created a national oil company, Soc. Des Hydrocarbures du Tchad (SHT), to become the fourth member of the production group.

The project's annual report showed that Chad's oil production averaged just above 170,000 b/d in 2005 as output ran somewhat below the planned initial 230,000 b/d. As of year-end 2005, the project had shipped more than 133 million bbl of 20.5° gravity crude from the terminal at Kribi, Cameroon.

The U.S. Dept. of Energy's Energy Information Administration (EIA) estimated average crude oil production at 160,000 b/d for the first half of 2006.

Chad-Cameroon Pipeline

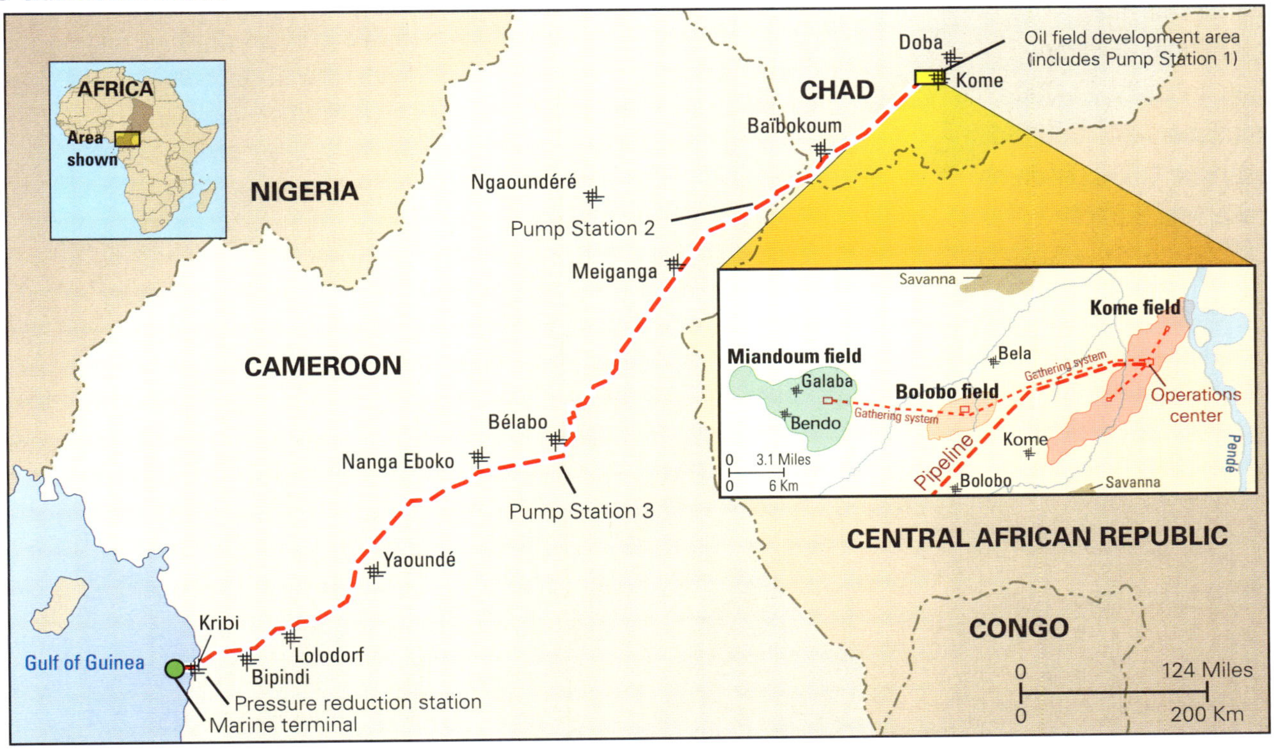

Source: Chad Esso Inc.

Exploratory drilling in the Maikeri-Poutouguem area proved the commercial feasibility of a potential new oil field that could add as many as 17 new producing wells. The project submitted a concession application to the government.

Engineers were assessing the feasibility of a potential new development based on test well results at Timbre. Also, major exploration efforts were under way east of the town of Sahr in the East Doseo basin, where a large area is now accessible for test drilling after road and infrastructure improvements.

CPC gets exploration rights

Taiwan's state-run Chinese Petroleum Corp. (CPC) and Chad signed an agreement under which CPC will invest $30 million over four years to explore for oil and gas in Chad. CPC obtains exploration rights to 26,250 sq km (10,135 sq miles) in the area of the Lake Chad and Doba basins.

Growing Number of Oil Fields in Doba Basin, Chad

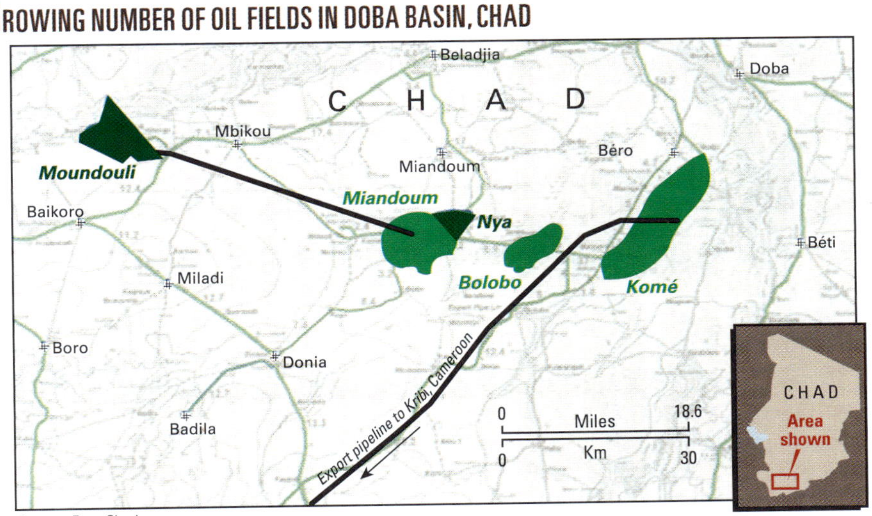

Source: Esso Chad

CONGO (BRAZZAVILLE)

Capital: Brazzaville
Monetary unit: CFA franc
Refining capacity: 21,000 b/cd
Oil production: 240,000 b/d
Oil reserves: 1.600 billion bbl
Gas reserves: 3.2 tcf

MTPS Block oil discovery

Total SA reported an oil discovery on the Mer Très Profonde Sud (MTPS) Block about 180 km (112 miles) offshore Pointe Noire. Drilled in 1,970 m (6,463 ft)of water to 3,370 m TD (11,056 ft), the Aurige Nord Marine 1 exploration well tested 4,970 b/d of oil. Subsidiary Total E&P Congo is operator.

The MTPS Block covers more than 5,000 sq km (1,930 sq miles) and lies in water as deep as 3,000 m (9,850 ft).

Aurige Nord Marine 1 is the third oil find on the block, following Andromède Marine 1 in 2000 and Pégase Nord Marine 1 in 2004. Total's equity production in 2005 from the Congo (Brazzaville) averaged 95,000 b/d of oil.

Oil discovery on Moho/Bilondo permit

Total announced an oil find in the Moho/Bilondo permit, approximately 80 km (50 miles) offshore, in water 600 to 900 m deep (1,968 to 2,950 ft).

Drilled to a total vertical depth of 2,269 m (7,444 ft), the Mobi Marine 2 well identified two new oil-bearing structures at the base of the upper Miocene.

CMS Nomeco legal case

Oil producer CMS Nomeco continued to battle with three hedge funds that sued the company in an attempt to collect bad debts owed by the government of Congo (Brazzaville). In August 2006, the U.S. Court of Appeals for the Fifth Circuit ruled in favor of the oil company.

CMS Nomeco has operated offshore under a contract with the Congo (Brazzaville) government since 1979 to produce oil. The company pays royalties to the government in the form of crude oil from the offshore production.

The hedge funds acquired—at deep discount—a $35 million defaulted loan from 1982 that was incurred by the Congo (Brazzaville) for highway construction, then secured a judgment for payment against the government. The hedge funds were seeking to collect the debt through garnishment of the royalties that CMS Nomeco pays the government in order to operate in the country's offshore waters.

Previously, a federal district judge had ruled in favor of the hedge funds, saying that CMS Nomeco should turn over to the hedge funds the oil that it provides to the Congo (Brazzaville) as royalty payments. However, courts in the Congo (Brazzaville) have ruled that the oil producers cannot divert the royalties in that way.

According to the law firm Thompson & Knight LLC (Dallas) that represents CMS Nomeco, the case has critical implications for all U.S. firms doing business in countries that have defaulted on loans. More than 70 countries are either in or near default.

The firm maintains that if U.S. companies are required to become guarantors of defaulted sovereign debt in foreign countries, they will be at a competitive disadvantage in the international arena.

The Fifth Circuit Court of Appeals ruled in August that the lower court committed several reversible errors and decided in favor of CMS Nomeco and co-defendant, Nuevo Congo, both U.S. affiliates of Perenco Holdings (London).

To head off further action by the hedge funds, CMS Nomeco and Nuevo Congo began a process to seek action by Congress and the Bush administration in the case. One approach may be to revisit how the 1978 Foreign Sovereign Immunities Act —which historically has protected the assets of foreign governments from U.S. judgments—affects a foreign country's resources.

A lawyer for one of the hedge funds said trying to overturn the court's decision through legislation would be a mistake, making it extremely difficult for private banks to lend money to the Third World.

The Moho/Bilondo permit is on the Haute Mer block and is the first subsea development on the continental shelf of Congo (Brazzaville). Total said this well will add further reserves to the current Phase 1 development of Moho/Bilondo, launched in August 2005.

That first-phase development plan envisions 12 subsea wells tied back to a floating production unit. Output of 90,000 b/d at peak will flow through a 30-km (19-mile) line to Djeno. The field is expected to come onstream in 2008.

Work was expected to begin on Blocks I and II by year-end 2006, following a presidential decree ratifying technical studies preparatory to the acquisition of 400 km (248 miles) of 2D seismic data on Lake Albert in 2007. The seismic campaign will target potential extensions to geological plays already identified in the region.

Heritage's exploration work program on the Ugandan side of the basin was under way. The company spudded the vertical Kingfisher-1 well on Block 3-A from the shore of Lake Albert. It set 10 ¾-in. intermediate casing to 1,060 m (3,478 ft) and was preparing to drill out with a 9 7/8-in. hole. The exploration well is targeting multiple objectives and is expected to reach 3,000–4,000 m TD (9,842–13,125 ft).

CONGO (FORMER ZAIRE)

Capital: Kinshasa
Monetary unit: Zaire
Refining capacity: 0
Oil production: 20,000 b/d
Oil reserves: 180.00 million bbl
Gas reserves: 35 bcf

Albert graben basin, Blocks I and II PSA

Heritage Oil Corp., Calgary, signed a production-sharing agreement with Congo (former Zaire) for a 39.6% working interest in the 6,000 sq-km (2,316 sq-mile) Blocks I and II in the Albert graben basin onshore and off Congo.

The PSA includes the portion of Lake Albert that lies in Congo adjacent to the company's Block 3A and Hardman Resources Ltd.'s Block 2 licenses in Uganda.

Heritage said the recent success at Hardman's Block 2 license, where that company production-tested its Waraga-1 well at a cumulative maximum flow rate of over 12,000 b/d, proves that there is a potentially very significant new hydrocarbon system in the basin.

WHERE UGANDA'S FIRST DISCOVERIES HAVE BEEN MADE

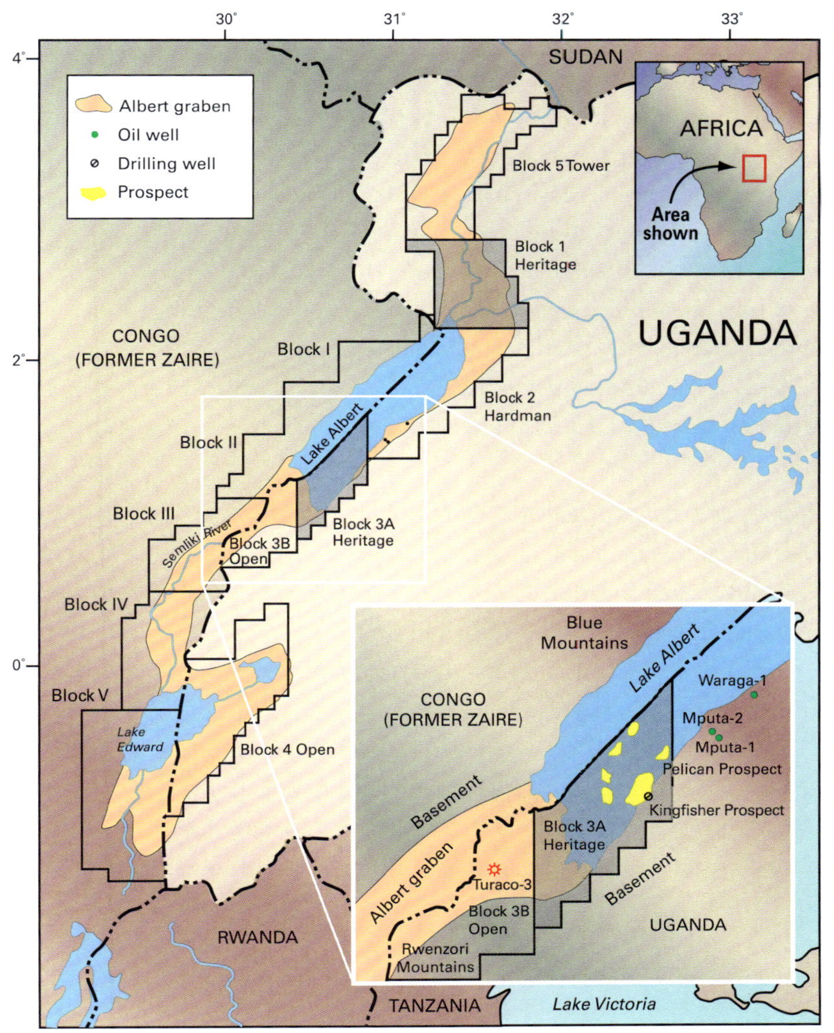

Source: After Heritage Oil Corp.

Tullow gets Congo blocks

The government has awarded Tullow Oil PLC (London) an interest in Blocks I and II in the northeastern part of the country. The blocks lie along Lake Albert, where Tullow holds an interest in Uganda blocks 1, 2, and 3A across the lake.

Several large oil discoveries have been made on the Uganda side of the Albert graben.

The two Congo blocks total 20,000 sq km of acreage on land and in Lake Albert.

Tullow planned to start technical studies on the blocks in 2006 and a seismic program in 2007 to target potential extensions to the geological play types already identified in the region.

Nganzi block

A subsidiary of SOCO International PLC (London) acquired an 85% working interest in a production-sharing contract that covers the 800 sq-km (308 sq-mile) Nganzi Block onshore along the Atlantic coast. SOCO DRC Ltd. is operator, and state La Congolaise des Hydrocarbures (Cohydro) holds 15%.

A 2005 aeromagnetic and gravity survey over the onshore extension of the coastal basin indicated the presence of a deep presalt source graben in the northern part of the basin in the Nganzi Block. Regional mapping shows the graben to be on trend with the source basin for M'Boundi oil and gas field in southern Congo (Brazzaville).

Several leads, interpreted as large horst blocks, are to be further evaluated by 2D seismic in 2007.

National petroleum assessment

Fusion Petroleum Technologies Inc. (The Woodlands, TX, U.S.A.) launched a national petroleum assessment of Congo (former Zaire), under an amended protocol that gives assessment sponsors preferential rights to all onshore acreage in the country for the duration of the project.

The addendum will protect the rights of companies that support the evaluation of the country's interior frontier basins, the company said. Several major companies have joined the project, and Fusion said that 10 to 13 are expected to participate.

EGYPT

Capital: Cairo
Monetary unit: Pound
Refining capacity: 726,250 b/cd
Oil production: 667,500 b/d
Oil reserves: 3.7 billion bbl
Gas reserves: 58.5 tcf

Overview

The Gulf of Suez accounts for about half of Egypt's oil production; the rest is from its three other key areas: The Western Desert, the Eastern Desert, and the Sinai Peninsula. The Gulf of Suez is likely to continue to be the main oil producing region, according to the Ministry of Petroleum.

But aggressive exploration in the deep water of the Mediterranean, and onshore in the Nile Delta and Western Desert, has resulted in several recent discoveries.

Oil discoveries also have been made in recent years in the Mediterranean and Nile Delta, areas traditionally considered gas-prone.

Due to major recent discoveries, natural gas is likely to be the primary growth engine of Egypt's energy sector for the foreseeable future, according to the U.S. Dept. of Energy's Energy Information Administration (EIA). Gas production increased by more than 220% from 1999 to 2004. Estimated 2006 production of 5.5 Bcf/d, if correct, would represent a further 150% increase since 2004.

To support its goals of doubling natural gas exports by 2011, the government aims to add 30 tcf to its proven gas reserves by 2010. Most of this increase would come from new natural gas discoveries offshore from the Nile Delta, and from finds in the Western Desert.

In the Nile Delta, recent offshore field developments include Port Fuad, South Temsah, and Wakah. In the Western Desert, the Obeiyed Field is an important natural gas area under development.

Two areas in the Western Desert have shown strong potential. Obeiyed, with probable natural gas reserves estimated at 5 tcf, was producing 300 MMcf/d in 2006. Production in the Khalda concession was approximately 275 MMcf/d.

Exploration and Development

Egas license round

The Egyptian Natural Gas Holding Company (Egas) was established in 2001 by the Petroleum Ministry to guide natural gas-related development, separately from the Egyptian General Petroleum Corporation (Egpc).

In December 2006 and January 2007, Egas reported the award of 11 blocks in its 2006 licensing round.

- Onshore Block 1 (Tanta Block, in the north-central region) to RWE Dea (Hamburg, Germany)
- Offshore Obaiyed Block 11 to OMV (Austria).
- Offshore Block 6 (N-Hapy Block) and onshore Block 8 (S-Diyur Block) to Gujarat State Petroleum Corp. (GSPC; India)
- Offshore Block 9 to a partnership of Statoil (Norway) and Sonatrach International Petroleum Exploration & Production Corp. (Sipex, a subsidiary of Algeria's Sonatrach)
- Onshore Block 2 (Rommana Block, in the northeast) to a consortium of Sipetrol International (Santiago, Chile), PTTEP (Thailand), and Centrica Energy (UK)
- Offshore Block 8 (Sidi Adb El Rahman Block, in the northwest) to a consortium of PTTEP, Edison International (U.S.A.), and Sipetrol
- Offshore Nile Delta Block 4 (El Arish) to Tharwa Petroleum [part of Egpc and Egas
- North Maghara Block on the northern Sinai Peninsula to Petzed (Cairo)
- South Siwa Block (Western Desert) and East Magawish Block in the offshore Gulf of Suez to Al Thani Corp. (Qatar)

Ganope license round

The government established the Ganoub El Wadi Petroleum Holding Company (Ganope) in 2003 to enhance exploration in Upper Egypt. In November 2006, Ganope announced the award of exploration and development rights to Block Mesaha under its 2006 Bid Round 1.

The block is in Upper Egypt, in the southwest corner of the Western Desert. It was awarded to a joint venture consisting of Melrose Resources PLC (Edinburgh, Scotland), Oil Search (Port Moresby, Papua New Guinea), and Hellenic Petroleum (Greece).

Ganope aimed to have 15 concessions in place by the end of 2006 and 25 by 2008. All Egas and Ganope awards are subject to completion of documentation and approval by the Egyptian parliament.

Deepwater oil and gas

A 2004 global study of deepwater petroleum potential by Wood Mackenzie Group concluded that about 50 billion bbl of crude and 28 billion boe of gas have been discovered to date, and that an additional 114 billion bbl of oil and 68 boe of gas reserves are yet to be found. The study analyzed about 110 deepwater reservoirs (below 400 m, or 1,300ft) in 30 sedimentary basins and river deltas in 18 countries.

Wood Mackenzie estimated that oil and gas production from deep water could grow to provide 9% of the world's total by 2010, compared to about 5% in 2004.

The study also estimated that future exploration in the deep water of the Mediterranean will add 70 tcf of gas reserves and 3 billion bbl of oil reserves to Egypt's hydrocarbon storehouse. The study ranked Egypt's deepwater gas potential second in the world, after the Gulf of Mexico.

Most of the production now coming from the Mediterranean is from shallow structures. But recent discoveries by BP in north Alexandria and Italy's ENI in Port Fouad indicate deeper structures are present. Companies are particularly interested in offshore Mediterranean and onshore in the Nile Delta and northern parts of the Sinai Peninsula, said the Ministry of Petroleum.

Apache progress

Apache Corp. (Houston) said its Syrah 5X appraisal well in Egypt's Western Desert tested 47.6 MMcfd of gas from the Jurassic Lower Safa sand, extending Syrah field to the northwest. Syrah field, on the Khalda Concession, is slated to begin production in third quarter 2008.

The Alexandrite 1X well in the Matruh concession flowed 19.8 MMcfd of gas and 4,045 b/d of condensate, indicating that it is commercial. Initial production was slated for February 2007 from a recompletion in the Alam El Bueib 6 (AEB 6) formation. Within the new interval, Apache tested 11 m (37 ft) of net pay in two AEB 6 segments just below 3,650 m (12,000 ft).

Several wells in the Western Desert also had encouraging results. The Qasr 34 appraisal well tested 18.4 MMcfd of gas and 725 b/d of condensate after reaching a TD of 4,267 m (14,000 ft) in Jurassic Lower Safa rock. The well extends the Qasr field, 2.5 km to the northwest, adding 2,200 acres to the field. Logs showed 30 m (96 ft) of net pay in the upper sand of the Lower Safa at 4,062–4,097 m (13,328–13,442 ft).

The Zasr 36 well flowed 2,945 b/d of oil and 2.1 MMcfd of gas. The well was a new Alam El Bueib (AEB) discovery in the field. Apache drilled the well to 3,787 m (12,424 ft) in AEB 3G sands and targeted AEB 3D and 3E sands. Log analysis showed 17 m (58 ft) of net pay in the AEB 3D and 3E sands.

The Hathor Deep 1X well on the Khalda Offset concession tested 12 MMcfd of gas in the AEB 6 formation and 1,237 b/d of oil from the AEB 3D formation. Apache received approval from state Egpc to proceed and has installed surface equipment; approval from the petroleum minister was pending.

The Qasr 40 well logged 18 m (60 ft) of net oil pay in the AEB 3E sands and in secondary pay intervals in the AEB 3A and 3C. The new well is a step-out northwest from the Qasr 31.

The Kenz 35 well was drilled 1 km (0.6 miles) northwest of the nearest producing well in the Kenz field on the Khalda Ridge. Kenz 35 logged 43 m (140 ft) of AEB net pay as well as 5 m (18 ft) of net pay in the Upper Bahariya. It will be completed as a gas and condensate producer.

Giza North-1 well

BP and its partners found an estimated 1 tcf of natural gas in early 2007 with its Giza North-1 exploratory well in the North Alexandria concession off Egypt in the Mediterranean Sea. The well, 56 km (35 miles) north of Alexandria City, was drilled to 2,040 m (6,692 ft) in 668 m (2,192 ft) of water.

The partners plan an appraisal well in second quarter 2007 to evaluate the complex. BP has made three previous discoveries in the same Pliocene formation: Taurus, Libra, and Fayoum.

Discussions with Greece

In early 2006, Egypt and Greece signed a memorandum of understanding for oil and gas exploration, including discussion of the best means to export gas to Greece (as LNG, compressed natural gas, or by pipeline). Greece would become a transit point in moving gas on to Europe. Greece has begun to study a plan to transfer CNG to the Greek islands of Crete and Rhodes in cooperation with Egypt.

Drilling and Production

About 60 firms are currently operating under 102 concession agreements in Egypt, including BP, BG Group, Eni, Shell, Apache, and Petronas. In 2005–2006, some 63 agreements and amendments were signed that involved $2.188 billion and commitments to drill a total of 258 wells, according to the Ministry of Petroleum.

Saqqara field close to production

Operator BP said that the large offshore Saqqara field was expected to begin commercial production in 2007. When discovered in 2003, it was Egypt's biggest oil find since 1989, adding 80 million bbl to reserves. The Ministry of Petroleum estimates peak production of 40,000 to 50,000 b/d. The field is next to the existing El-Morgan field.

Pipelines

Gas exports to Israel

Construction of a subsea pipeline to carry Egyptian natural gas to Israel was scheduled to begin in early 2007, according to Egyptian press reports. The export arrangement was agreed to in mid-2005.

The East Mediterranean Gas Company (EMG) partnership is to supply $3 billion worth of natural gas over 20 years to Israel for power generation. Egyptian company Petrojet is scheduled to start building the pipeline in early 2007.

Export line extension

An existing pipeline carrying Egyptian natural gas exports was extended 370 km (230 miles) in early 2006, from the Jordanian seaport of Aqaba to Rehab in the north of that country.

The first part of the line was opened in July 2003, carrying Egypt's first gas exports from El-Arish to Aqaba. The extended line has a capacity of 1 bcfd.

Discussion was also under way to expand the line further, from Rehab to the Syrian/Turkish border, making it possible to market gas to Europe. Longer-term expansion of the network could extend it in stages by another 738 km (460 miles), from Beni-Suef in the south to Aswan in the north.

LNG

Damietta and Idku status

Egypt has two LNG plants onstream. The facility at Damietta, which shipped its first cargo in early 2005, has one liquefaction train and a design capacity of 7.56 bcm/yr (0.73 bcfd). A second facility at Idku has two trains, each with a design capacity of 3.6 mtpy. It came onstream in 2005.

LNG from Damietta is sent to the Spanish market and to Mediterranean and U.S. markets. In June 2006, partners Eni, BP and Union Fenosa signed a framework agreement for the expansion of the plant from the current 770 MMcfd to 1.4 bcfd by 2009.

The second LNG export terminal is the two-train Egyptian LNG (ELNG) facility at Idku, near Alexandria. It was built by BG in partnership with Petronas. Output from the first train is taken by Gaz de France. The second train supplies BG Group with LNG, for export to the Lake Charles, LA, terminal in the United States, and to Italy.

Each train has a capacity of 500 MMcfd. In early 2006, BG announced plans to increase production at Train 1 and Train 2 by 5–10% by 2010.

Gas from BG's Simian/Sienna offshore fields flows to the Idku facility. The site has space for four more trains, and gas from fields in the West Delta Deep Marine area of the Mediterranean are viewed as a likely source of additional supply.

Nile Delta gas for LNG

Two Egyptian units of Royal Dutch Shell PLC took a farmout from Centurion Energy International Inc. (Calgary) to acquire 50% interest in the West El Manzala and West El Qantara concessions, onshore in the Nile Delta. The concessions cover a combined 800,000 acres (1,250 sq miles).

Shell and Centurion will develop LNG opportunities if threshold quantities of gas are discovered on the concessions, subject to government approvals. Operator Centurion spudded the first of the five wells in February 2006.

Centurion said several prospects and leads have been confirmed on the two blocks and that 15 exploration and appraisal wells were planned for 2006.

Refining and Processing

More output from 9 plants

Work continued on an Egpc program to increase the capacity of Egypt's nine existing refineries to produce additional jet fuel, kerosene, gasoline, and high-value solvents.

One key effort would be the expansion of the Middle East Oil Refinery (Midor). With a design capacity of 100,000 b/sd, it produces LPG, 95-octane gasoline, and jet fuel and diesel that meet year-2005 European specifications. An 20% increase in throughput was under evaluation.

The Ministry of Petroleum also was considering a new $9.5 billion grassroots refinery that would be co-located with a major petrochemical complex and have a capacity of 200,000 to 300,000 b/d. Products would be clean-burning transport fuels and petrochemical intermediates (such as ethylene and propylene) meeting European specifications.

Fourth Khalda gas plant train

Khalda Petroleum Co. (KPC) hired Petrofac (London) to design and construct a fourth gas-conditioning train at Salam on the Khalda Concession in the Western Desert. KPC is a joint venture of Apache and Egyptian General Petroleum Corp.

The fourth train, combined with the third, will increase total conditioning capacity for production from Apache's Jurassic gas reserves to 710 MMcfd of gas and 66,000 b/d of condensate. The expansions are scheduled for completion by late 2008.

Petrochemicals

The Egyptian government has created a Master Plan to guide development of the nation's petrochemical manufacturing industry. Implementation will cost an estimated $10 billion over 20 years, creating 14 petrochemical complexes producing 15 million tons of product annually.

The plan is being implemented through the Egyptian Petrochemicals Holding Company (Echem), created in 2002.

Seven areas have been identified for possible petrochemical projects: Alexandria, Beheira, Kafr El Sheikh, Dahkalia, Damietta, Ismailia, and Suez.

Another Echem goal is to upgrade the performance of currently operating petrochemical companies, such as Egypt Petroleum Company (EPC), Sidi Kerir Petrochemicals (Sidpec), and the linear alkyd benzene unit of Ameryia Petroleum Refining Company (APRC).

A number of petrochemical projects were in various stages of development, with planned completion dates stretching to 2009. For example, BG Group is said to be considering the following:

- Two plants at Damietta to produce 1.26 mtpy of methanol and 1.2 mtpy of ammonia/urea, respectively (late 2009)
- A Port Said facility to produce 350,000 tpy of propylene and polypropylene (mid-2008)
- A polyvinyl chloride production plant (150,000 tpy) at Alexandria (2008)
- A polystyrene production plant (200,000 tpy) at Alexandria (2008)
- A polyethylene plant (750,000 to 1 million tpy) at Coastal Road (late 2009).

EQUATORIAL GUINEA

Capital: Malabo
Monetary unit: CFA franc
Refining capacity: 0
Oil production: 320,000 b/d
Oil reserves: 12.0 million bbl
Gas reserves: 1.3 tcf

Oil royalty raised

Equatorial Guinea's Pres. Teodoro Obiang ratified a new hydrocarbon law that raises the minimum royalties that oil and gas companies must pay to 13% from 10%.

In a November research note, Raymond James & Associates Inc. cautioned that the new law also gives Equatorial Guinea the right to a 20% share in contracts with foreign operators, and mandates payment by producers of "any windfall tax that may be imposed by the state."

Licensing

The government extended the closing date of its 2006 licensing round to March 31, 2007, from January 31, 2007, to allow companies to fully evaluate available acreage.

LNG at Bioko Island

Equatorial Guinea LNG Co. Ltd. (EG LNG) awarded a FEED contract to Bechtel Corp. for initial work aimed at possible construction of a second LNG train—capacity 4.4 mtpy—at Bioko Island. Completion of the FEED work by mid-2007 will guide a decision on whether to proceed.

EG LNG also was in discussions with gas resource holders in Equatorial Guinea, Nigeria, and Cameroon to secure gas supply to support construction of Train 2.

WEST AFRICA DEEPWATER OIL DISCOVERIES THROUGH 2005

	Exploratory wells	No. of discoveries	Discovery volume — Million bbl	Avg. discovery size — Million bbl
Nigeria	120	28	7.3	261
Angola	145	53	11.0	208
Eq. Guinea	22	8	1.4	175
Congo (Brazz.)	23	5	0.7	140
Ivory Coast	4	1	0.2	200
Mauritania	18	5	0.5	100
Total	**332**	**100**	**21.0**	**211**

Sources: IHS Inc., OPEC

The company reported that it was ahead of schedule in building Train 1. It has a capacity of 3.4 mtpy and is slated to begin first shipments in mid-2007.

Petrobras stake in Block L

Brazil's Petrobras reported that the government of Equatorial Guinea approved the company's acquisition of a 50% participating interest in a production-sharing contract covering Block L in the deepwater portion of the Rio (River) Muni basin. The block covers 4,250 sq km (1,640 sq miles) in water 500 to 2,200 m deep (1,640 to 7,215 ft).

Participants in Block L are Chevron Equatorial Guinea Ltd., Amerada Hess Equatorial Guinea Resources Inc., Energy Africa Equatorial Guinea Ltd., and Sasol Petroleum International (Pty.) Ltd.

Chevron will as remain operator of the block, but Petrobras has the option to become operator in the event of a commercial discovery.

Block L is near the prolific Block G, operated by Amerada Hess, in which eight oil fields have been discovered, including the Ceiba field that is already in production. Drilling of the first exploratory well on Block L was planned by year-end 2006. If successful, first oil may occur by 2010.

West Africa's Oil Reserves

	Total discovered			Remaining, split		
	Produced	Remaining	Total	Onshore	Offshore	of which deep water
	Billion bbl					
Angola	6.0	14.4	20.4	0.011	14.40	10.5
Eq. Guinea	1.0	1.6	2.6	0	1.60	1.3
Congo (Brazz.)	2.0	2.0	4.0	0.320	1.65	0.7
Gabon	3.0	1.4	4.4	0.600	0.80	na
Ivory Coast	0.1	0.4	0.5	0	0.40	0.3
Mauritania	0	0.5	0.5	0	0.50	0.5
Subtotal	**12.1**	**20.3**	**32.4**	**0.9**	**19.35**	**13.3**
Rest of West Africa						
Cameroon	1.0	0.3	1.3	0	0.30	0
Guinea Bissau	0	0	0	0	0	0
Sierra Leone	0	0	0	0	0	0
Subtotal	**1.0**	**0.3**	**1.3**	**0**	**0.30**	**0**
Nigeria	**26.0**	**35.2**	**61.2**	**16.4**	**18.80**	**7.3**
Total West Africa	**39.1**	**55.8**	**95.0**	**17.3**	**38.50**	**20.6**

Sources: IHS Inc., OPEC

West Africa's Undiscovered Oil Volume*

	USGS undiscovered oil at 2000	Discovered oil, 2000-05	Undiscovered oil at 2005
Angola	14.00	6.00	8.00
Eq. Guinea	2.50	0.50	2.00
Congo (Brazz.)	5.80	0.70	5.10
Gabon	8.10	0.15	7.95
Ivory Coast	0.60	0.30	0.30
Mauritania	0.05	0.50	0
Subtotal	**31.0**	**8.20**	**23.40**
Rest of West Africa			
Cameroon	1.50	0.10	1.4
Guinea Bissau	0.04	—	0.04
Senegal	0.05	—	0.05
Gambia	0	—	0
Benin	0	—	0
Ghana	0.20	—	0.20
Subtotal	**1.79**	**0.10**	**1.69**
Nigeria	**37.00**	**4.20**	**32.80**
Total	**69.80**	**12.50**	**57.80**

*US Geological Survey estimate.
Sources: USGS, IHS Inc., OPEC

West Africa in the Spotlight

An evaluation by the Secretariat of the Organization of the Petroleum Exporting Countries (OPEC) highlighted the strong exploration record and the start of many new fields in recent years in West Africa, particularly in the deep offshore.

The fact that more oil is to be found indicates that the region will become even more important in the future, according to the study.

West Africa includes Nigeria and several non-OPEC countries. The latter group includes seven important producers and three small producers.

The important producers are Angola (at the southern limit of West Africa), Cameroon, Congo (Brazzaville), Gabon, Equatorial Guinea, Ivory Coast, and Mauritania (at the northern limit of West Africa). The small producers are Benin, Ghana, and Senegal.

OPEC's medium-term world oil supply forecast shows that West Africa's oil production excluding Nigeria is expected to increase to 3.8 million b/d by 2012, driven by more than 26 new deepwater developments. Nigeria's production capacity will also rise to 4 million to 4.2 million b/d by 2012, driven by more than 15 developments in deep and shallow water.

WEST AFRICA OIL RESERVES BY BASIN

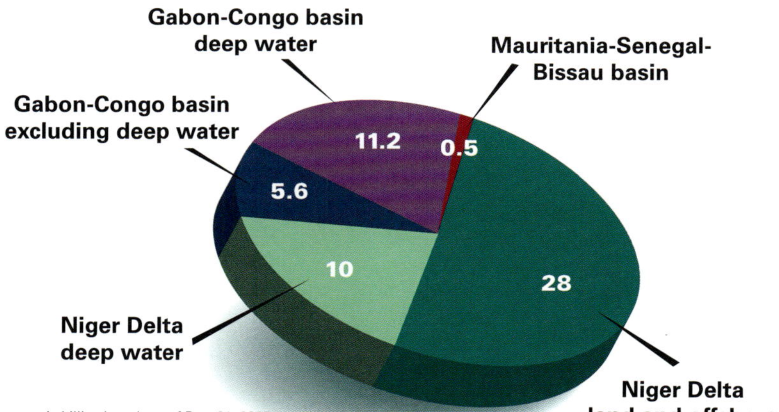

Reserves in billion barrels as of Dec. 31, 2005.
Sources: OPEC, US Geological Survey

NET PRODUCTION BY INDUSTRY GROUP IN 2005

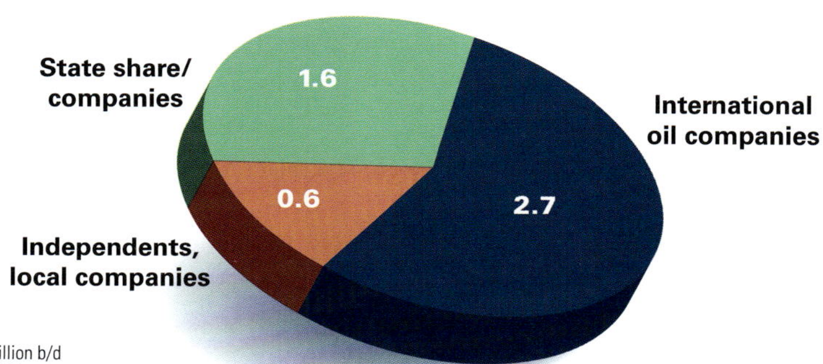

Values in million b/d
Sources: OPEC, company reports

ESTIMATED 2005 OIL PRODUCTION BY TYPE OF ENVIRONMENT

	Total	Onshore (Split)	Offshore (Split)	Deep water (Split)	Deep-water production in 2000
	Million b/d				
Angola	1.23	0.01	0.50	0.77	0
Eq. Guinea	0.36	0	0.06	0.27	0.07
Congo (Brazz.)	0.24	0.08	0.16	0	na
Gabon	0.25	0.14	0.11	0	na
Ivory Coast	0.09	0	0.03	0.06	0
Cameroon	0.08	0.02	0.06	0	na
Subtotal	**2.25**	**0.30**	**0.90**	**1.13**	**0.07**
Nigeria	**2.60**	**1.20**	**1.30**	**0.13**	**0**
Total	**4.90**	**1.50**	**2.20**	**1.30**	**0.10**

Sources: OPEC, IHS Inc.

GABON

Capital: Libreville
Monetary unit: CFA franc
Refining capacity: 17,300 b/cd
Oil production: 232,000 b/d
Oil reserves: 2.00 billion bbl
Gas reserves: 1.00 tcf

Avouma platform in place

The production platform for Avouma oil field off Gabon arrived at the field, Pan-Ocean Energy Corp. Ltd. said. Production was slated to start by year-end 2006 from two horizontal wells at 10,000 b/d (3,136 b/d net to Pan-Ocean). Drilling was planned to begin in October.

Avouma will be the second producing field on the Etame Marin Permit. Etame field produces 18,000 b/d. A 16-km (10-mile) pipeline connecting the Avouma platform with an FPSO facility on Etame field was close to completion.

Pan-Ocean said Etame co-venturers received government approval for an exclusive exploitation area around the Ebouri discovery. A development plan was to be filed with the government in late 2006. Ebouri oil field is expected to come on stream in late 2007 at 5,000 b/d.

IVORY COAST

Capital: Abidjan
Monetary unit: CFA franc
Refining capacity: 65,200 b/cd
Oil production: 30,000 b/d
Oil reserves: 100.0 million bbl
Gas reserves: 1.0 tcf

West Espoir oil well on stream

The first of five West Espoir offshore oil wells came on stream in July, said Tullow Oil PLC. Canadian Natural Resources Ltd. (Calgary) is the operator. Drilled on Block CI-26 Special Area E, the well flowed at more than 5,000 b/d of oil equivalent. The block covers 139 sq km (54 sq miles).

Development plans call for four additional production wells and two water injection wells. The West Espoir project is expected to reach peak production of 10,000 b/d in late 2007.

The project includes a new wellhead tower tied back via a 5.5-km (3.5-mile) pipeline to the Espoir FPSO vessel, which processes production from both East and West Espoir fields. The two fields together produce more than 35,000 b/d of oil.

Rigless installation

BJ Tubular Services completed what it called the region's first rigless conductor installation, for CNR International (Cote d'Ivoire) SARL. BJ Tubular had a contract to preinstall 12 conductors and supply cold cutting services for West Espoir oil field.

The conductors were installed in more than 125 m (400 ft) of water with 135 m (440 ft) of mudline penetration.

Block CI-27 wildcat tests

Foxtrot International LDC (Abidjan) tested natural gas from a 20-m (65-ft) perforated interval on the 930 sq km (360 sq-mile) Block CI-27 in 400 m (1,312 ft) of water.

The Mahi-1 exploration well encountered a 135-m (443-ft) hydrocarbon column. It flowed 32 MMcfd of natural gas and 250 b/d of condensate through a 1-in. choke.

Mahi-1 is 5 km (3 miles) south of the offshore Foxtrot gas field, which produces about 80 MMcfd. Block CI-27 also contains the Manta gas field, for which an application for development was has been submitted. The block, operated by Foxtrot International, has estimated gas reserves of 300 bcf.

LIBYA

Capital: Tripoli
Monetary unit: Dinar
Refining capacity: 378,000 b/cd
Oil production: 1.70 million b/d
Oil reserves: 41.464 billion bbl
Gas reserves: 52.650 tcf

Libya has the largest oil reserves in Africa (39 billion bbl) and is the eighth-largest oil producer among the members of OPEC. Libya wants to increase oil production to 2 million b/d by 2008–2010 and to 3 million b/d by 2015, from 1.6 million b/d today.

Since 1961, Libya has produced high quality, low-sulfur, sweet crude oil, primarily sold into the European market. The country is now the second-largest oil supplier to Europe. Libya's National Oil Corp. (NOC) set up subsidiaries with offices in Europe: Umm Al-Jawaby Oil Service Co. Ltd. (London) in 1983, and Mediterranean Oil Services GmbH (Düsseldorf) in 1988.

Since the United States lifted economic sanctions against Libya in 2004, NOC has reinstituted lease sales, which are again attracting the attention of many international players as well as national oil companies looking to diversify assets abroad.

Leasing

Third lease round

Libya's NOC solicited bids in mid-2006 for its third bidding round. The acreage on offer covered 41 blocks in 14 exploration areas across the Murzuq, Sirte, Ghadames, Kufra, and Cyrenaica basins, as well as the offshore sector. Most of these areas are relatively untested, with only 9 onshore wells and 1 offshore well in the areas on offer.

Winning bids were announced in December. Barrows Co. Inc. (New York) characterized Russian firms OAO Tatneft and OAO Gazprom as the "big winners." A summary on the Libyan NOC website shows that Tatneft won licenses for three contract areas (Ghadames and Sirte basins) and Gazprom won a license to explore offshore Area 19 (total $40 million bonus).

Chinese Petroleum Corp. (CPC; Taiwan), Germany's Wintershall AG (Germany), and a partnership of Petro-Canada and Repsol YPF took one license each (total $18 million bonus).

The total work commitment arising form the third bidding round is 41,100 km (2D) and 10,000 km (3D) seismic and 35 exploration wells. E&P-sharing agreements (EPSAs) were scheduled to be signed in January 2007.

Following the opening of bids in December, NOC announced that a fourth bid round would be held in first-quarter 2007 with "the focus on gas."

Exploration and Development

Gas discovery in Murzuq basin

Woodside Energy (NA) Ltd. (Tripoli)—a subsidiary of Woodside Petroleum Ltd. (Perth) and operator of the NC210 Block in the Murzuq basin in Libya—made a gas discovery in late 2006 with its C1-NC210 exploration well on the block, said joint venture partner Repsol YPF.

The well, 1,000 km (620 miles) south of Tripoli and 150 km (93 miles) south of producing Al Wafa gas field, was drilled to 808 m (2,650 ft) TD. Wireline logs indicate the potential for several hydrocarbon-bearing zones.

An initial production test of the Devonian Awaynat Wanin formation flowed at 5.7 MMscfd through a 72/64-in choke. The Carboniferous Mrar M7 reservoir flowed 5.8 MMscfd of gas on test through a 72/64-in. choke. Calculated absolute open flow is 13 MMscfd.

Sirte basin oil find

RWE Dea AG (Hamburg, Germany) discovered oil in late 2006 with its A1-NC 193 well in the Sirte basin, 500 km (310 miles) southeast of Tripoli. Drilled to 1,285 m (4,214 ft), the well intersected the Paleocene Dahra formation and on test flowed 35° gravity oil at 410 b/d through a 32/64-in choke.

LIBYAN BASINS

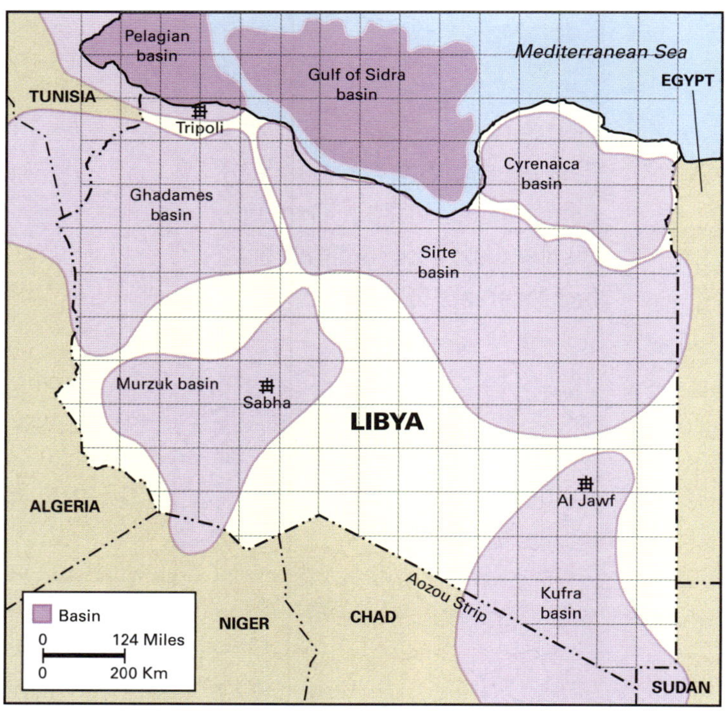

Sources: OGJ archives, National Oil Co. of Libya

Libya, third bid round results*

Basin	Contract area (block)	Existing wells, no.	Winning operator	Operator commitment: 2D seismic, km	3D seismic, sq km	New wells, no.	Production share, %	Bonus, $ million
Offshore	19 (1, 2, 3, 4)	0	OAO Gazprom	4,000	2,000	6	10	10.1
	20 (1, 2, 3, 4)	*0*	*ExxonMobil Libya Ltd.*	*20,000*	*1,000*	*4*	*22.3*	*10*
	43 (1, 2, 3, 4)	*1*	*ONGC Videsh Ltd.*	*1,000*	*4,000*	*1*	*28*	*10*
Ghadames	82 (I)	2	OAO Tatneft	2,000	750	5	10.4	10
	98 (2, 4)	2	Tatneft	2,500	500	5	10.4	10
Sirte	69 (1, 2, 3, 4)	4	Tatneft	5,000	750	6	12	10
	137 (3, 4)	0	Petro-Canada, Repsol YPF	1,000	500	1	18	10
Murzuq	162 (1, 2)	0	Chinese Petroleum Corp.	1,600	0	3	7.8	5
	113 (3, 4)	*1*	*Inpex Corp.*	*1,000*	*500*	*3*	*12.9*	*10*
Kufra	201 (1, 2, 3, 4)	0	Wintershall AG	3,000	0	1	13.5	3
Total		**10**		**41,100**	**10,000**	**35**	**—**	**88.1**

*Winning bid list from Libya's National Oil Corp., Dec. 20, 2006; Barrows Co. Inc., New York. Italicized rows represent the three single bids approved Dec. 24, 2006.

A1-NC-193 is the first in a 10-well minimum drilling campaign in six RWE Dea licenses: NC 193, 194, 195, 196, 197, and 198. A comprehensive seismic acquisition campaign had gathered 3,200 line-km (1,988 miles) of 2D and 2,400 sq km (926 sq miles) of 3D data in the identified licenses.

More oil found on NC-115

A group led by Repsol YPF SA (Madrid) reported its seventh discovery on Block NC-115 in the Murzuk basin, 800 km (497 miles) south of Tripoli. The well, which produced 2,300 b/d of light crude, is near late-2005 discoveries that produced 2,060 boe/d and 4,650 boe/d, respectively. Current production from NC-115 is 200,000 b/d.

Exploration in Cyrenaica Basin

Exxon Mobil Corp.'s subsidiary, ExxonMobil Libya Ltd., signed an EPSA with Libya's NOC early in the year to begin exploration activity offshore Libya.

The agreement covers the large Cyrenaica Basin Contract Area 44, which was awarded in the second round of EPSA IV licensing in October 2005. The contract area comprises 2.5 million acres (3,900 sq miles) and is located offshore in water depths ranging from 3 m to more than 3,050 m (10 feet to more than 10,000 feet).

U.S. operators return

Libya's plan to nearly double its oil production to 3 million b/d by 2010–11 was spurred by successful licensing rounds and the return of U.S. operators that were exiled in 1986.

But operators evidently were proceeding slowly to begin off drilling programs on newly won or reoccupied acreage. Only eight to nine rotary rigs have operated in Libya for the past several years, and no change was apparent in late 2005. The land rig market in North Africa was reported to be tight at the start of 2006.

Libyan officials express the hope that operators will drill 50 wildcats/year by 2010 and undertake large seismic programs. A tightness of land rigs could retard drilling schedules.

Three onshore basin—Sirte, Ghadames, and Murzuk—produce most of Libya's oil. If oil prices remain high, winners of the license blocks could carry out significant exploration in several nonproducing and low-producing basins.

Occidental Petroleum Corp. and the Oasis group of companies were resuming operation of their assets in the Sirte basin, operated by Libya's NOC in the interim. Oxy, which took over its historical Libyan assets in July 2005, has a net working interest in about 130,000 sq km including those assets and the acreage it won in licensing rounds. The properties have significant potential for future growth from enhanced oil recovery projects, the company said.

In late 2005 Oxy said the properties would contribute 22,000 b/d to the company's year-end 2005 exit rate. Oxy lifted its first Libyan crude oil in late September. The company said its work program is proceeding on nine exploration blocks it was awarded in 2005. Oxy is the country's largest land holder.

OXY'S PORTFOLIO IN LIBYA

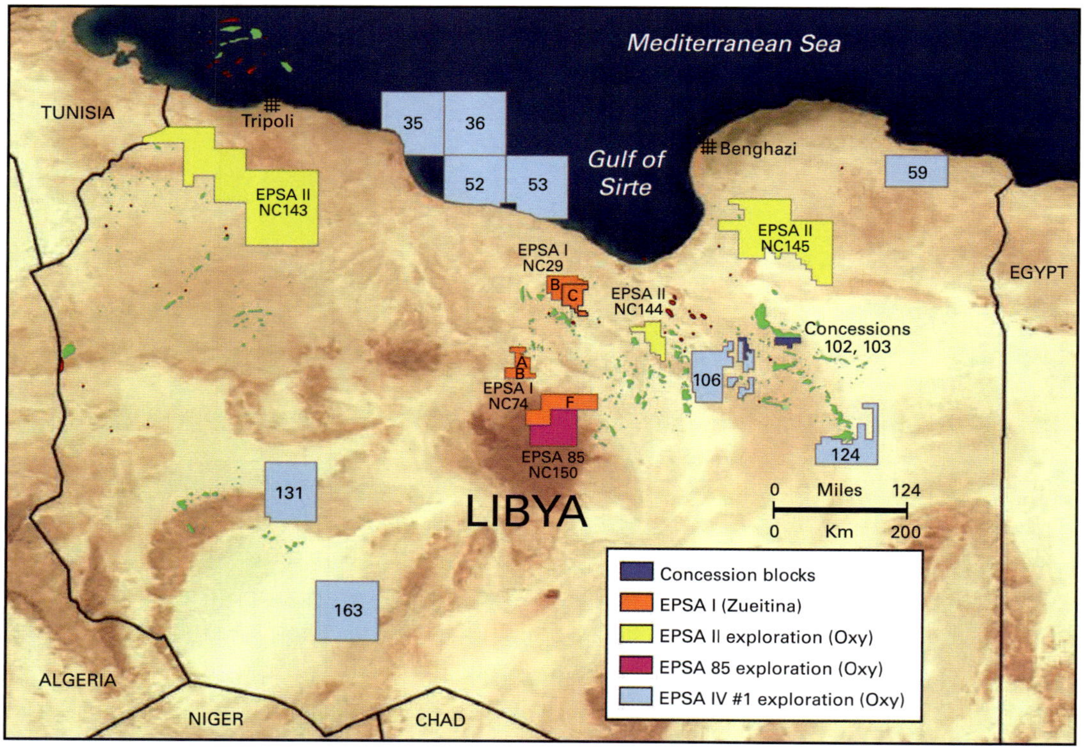

Source: After Occidental Petroleum Corp.

The Oasis group, which signed the agreement with Libya in late December 2005, operates 13 million acres that produce 350,000 b/d of oil. Interests are ConocoPhillips, Marathon Oil Corp., Amerada Hess Corp., and LNOC. ConocoPhillips said the lands contain sizable undeveloped oil and gas resources.

ConocoPhillips to resume Libyan operations

ConocoPhillips reported that, after reaching an agreement with Libya's NOC in late 2005, it will return to its former oil and gas production operations in Libya along with co-venture partners Marathon Oil Co. and Amerada Hess Corp. The group suspended operations in Libya in 1986.

Under the new agreement, ConocoPhillips and its partners will return to their former exploration and production interests in the Waha concessions in Libya. The concessions, which produce about 350,000 b/d of oil, cover nearly 13 million acres (20,300 sq miles) in the Sirte basin.

ConocoPhillips said it would expect to add about 45,000 net b/d of oil to its production profile, based on the current gross production figure from the concessions.

Pertamina exploring two blocks

Indonesia's state oil and gas firm PT Pertamina signed contracts with Libya to explore and develop two blocks, one onshore and the other offshore.

Pertamina is expected to spend $3.6 billion to develop the Sirte Block in the Sahara Desert and the Sabrata Block in the Mediterranean Sea. The two blocks cover a total of 3,671 sq km (1,415 sq miles).

Work will take place over a 30-year contract period under an agreement with Germany's Commerzbank AG. Pertamina said it would conduct seismic tests and drill two wildcat wells on the Sirte Block and another two wells on the Sabrata Block over a five-year exploration period.

The company was starting exploration in both blocks in 2006.

MADAGASCAR

Capital: Antananarivo
Monetary unit: Malagasy franc
Refining capacity: 0
Oil production: 0
Oil reserves: N/A
Gas reserves: N/A

Block 1101 PSC

Candax Energy Inc. (Calgary) was awaiting presidential ratification of a production-sharing contract on the 15,000 sq-km (5,792 sq-mile) Block 1101 in Madagascar. The work program calls for geological and geochemical studies, shooting 2D seismic, and drilling one well in the 2-year first phase.

Majunga Offshore Profond block

BG Group acquired an interest in the Majunga Offshore Profond exploration block in Madagascar under an agreement with Vanco Madagascar Ltd., marking BG's entry into that country.

The area work program includes the drilling of an exploration well, planned for early 2007. The block, which covers about 15,840 sq km (6,115 sq miles), lies in 200 to 3,000 m of water (656 to 9,842 ft) off northwestern Madagascar, about 100 km (62 miles) from Mahajanga.

Block 3109

Madagascar Oil Ltd. (London) announced that it raised $60 million for developing a heavy-oil accumulation at Tsimiroro in western Madagascar and for exploration programs. Madagascar Oil and Tullow Oil PLC (London) formed a 50-50 joint venture to develop conventional light oil on Block 3109. The block covers 11,050 sq km (4,266 sq miles) onshore southwest of Antananarivo in the Morondava basin.

MALI

Capital: Bamako
Monetary unit: CFA franc
Refining capacity: 0
Oil production: 0
Oil reserves: N/A
Gas reserves: N/A

Taoudeni basin evaluated

A first-time comprehensive review of existing exploration data on the Taoudeni basin in Mali and Mauritania showed large to regional scale structures, said Baraka Petroleum Ltd. (Perth).

The company was discussing farmouts with interested parties.

Baraka said the review "confirmed the expected stratigraphy and favorable structural style, as well as delineating the extent of the Ordovician and Infra-Cambrian play fairways in the company's 260,000 sq-km (100,411 sq-mile) gross acreage holding in the Taoudeni basin."

Ordovician and Infra-Cambrian formations produce 1.23 million b/d of oil and 445,000 b/d of condensate in the nearby and related Ahnet and Illizi basins in Algeria.

STATUS OF MALI OIL AND GAS BLOCKS

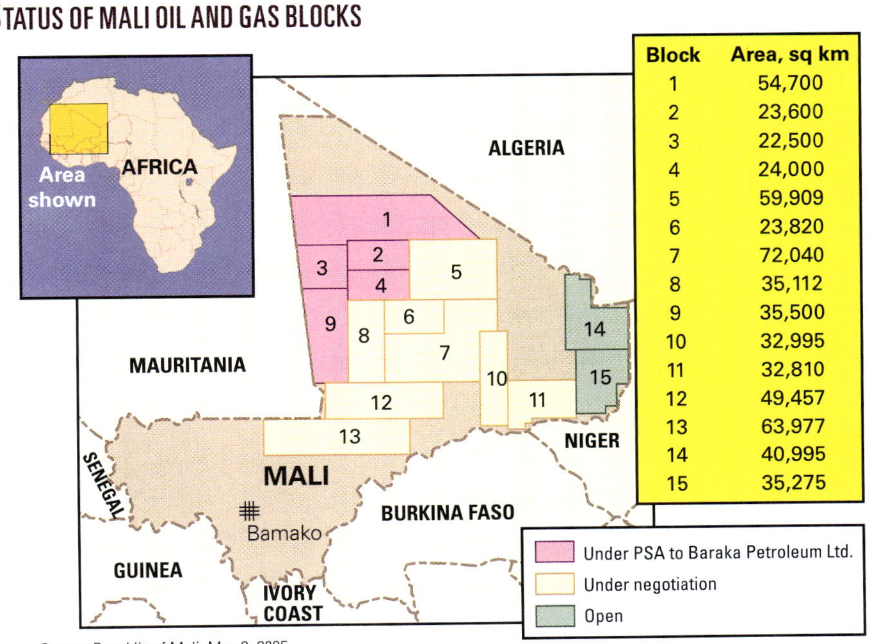

Block	Area, sq km
1	54,700
2	23,600
3	22,500
4	24,000
5	59,909
6	23,820
7	72,040
8	35,112
9	35,500
10	32,995
11	32,810
12	49,457
13	63,977
14	40,995
15	35,275

Source: Republic of Mali, May 2, 2005

MALI SEDIMENTARY BASINS

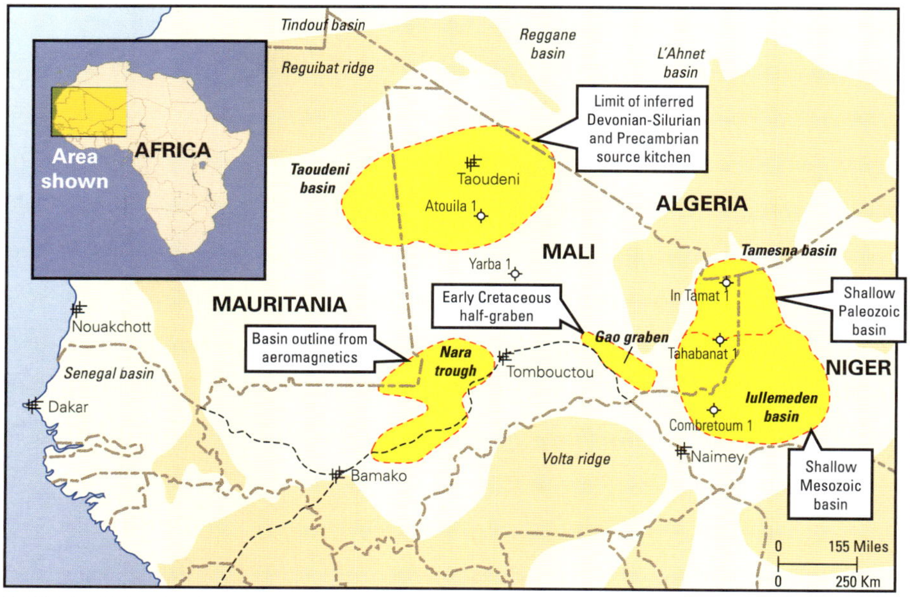

Source: Modified from Isis Petroleum Consultants

Baraka identified 40 oil and gas leads along the defined structural trends. The company noted that Taoudeni is practically unexplored, with four wells and 12,500 line-km of seismic in an area more than twice the size of Texas.

A comprehensive regional airborne gravity and magnetic survey of Baraka's five Mali blocks was scheduled to be finished by September.

MAURITANIA

Capital: Nouakchott
Monetary unit: Ouguiya
Refining capacity: 0
Oil production: 32,000 b/d
Oil reserves: 100.0 million bbl
Gas reserves: 1,000 bcf

Chinguetti oil field on stream

Woodside Energy Ltd. (Perth) reported that the Chinguetti oil field off Mauritania started production. Output averaged 30,000 b/d in October according to participant Roc Oil Co. Ltd. (Sydney) and is expected to reach 75,000 b/d at peak. A ten-year productive life has been projected.

Because of somewhat lower than expected production rate, the 123 million bbl reserve estimate was being reviewed and a time-lapse (4D) seismic survey was proposed for 2007.

The $270 million project—Mauritania's first offshore development—comprises six production wells, five water injectors, and one gas injector. In 800 m (2,625 ft) of water 90 km (56 miles) off capital city Nouakchott, the field is producing into an FPSO vessel. The Chinguetti gas handling/reinjection facilities can handle 80 MMcfd.

A dispute between the Woodside joint venture and the Mauritanian government over changes to four other Woodside-operated production-sharing contracts remains unresolved, but the dispute does not affect Chinguetti. Revenue shares from nearby discoveries Tiof, Tevet, and Banda are being disputed.

Gas discoveries multiply

In mid-2006, Gaz de France said it was starting a feasibility study of an LNG project in Mauritania based on deepwater gas discoveries. Industry observers estimate that at least 4 tcf of recoverable gas is needed to support a commercial LNG project.

Because of the barely explored nature of the country's offshore, it is unclear what total recoverable volume has been identified so far. The development time frame appears to be no earlier than 2012–2014.

In an agreement finalized with Dana Petroleum PLC (London) in 2006, Gaz de France gained interests in lightly drilled offshore blocks 1, 7, and 8. At least one of these, Block 7, contains a major gas discovery, Pelican. And at least four of the discovered oil fields—Chinguetti, Tevet, Tiof, and Banda—operated by other companies are known to contain varying volumes of gas.

Meanwhile, the operators exploring the area's nine blocks were planning to continue wildcatting the area and appraising the discoveries made so far. Dana Petroleum operates blocks 1, 7, and 8; Woodside Petroleum Ltd., Perth, operates blocks 2 to 6 and has an interest in Block 7; and Al Thani Corp. of Sudan operates Block 19. The nine blocks total more than 22,650 sq miles (14.5 million acres).

The other major gas players with interests off Mauritania are BG PLC, Malaysia's Petronas, and Wintershall AG (Germany).

Northern area

In the northern offshore area, the largest single known non-associated gas deposit is Pelican, discovered in 2003 as a 93-mile stepout from the oil discoveries in the Chinguetti area. Pelican's multiple Upper Cretaceous sands are distinct from the Miocene-aged sandstone systems that hosted the Chinguetti, Tiof, Tevet, and Banda discoveries.

After the Pelican-1 gas discovery, Dana Petroleum and others recorded 1,540 sq km (595 sq miles) of 3D seismic data between Pelican and shore where geochemical results said oil may have been displaced to the east.

Dana Petroleum plugged the Flamant-1 wildcat on western Block 8 in September but expressed encouragement. It said the shallower part of the well had gas-bearing intervals, indicating mature hydrocarbon source rock in the area. TD is 10,742 ft.

The Flamant-1 drillsite is 193 km (120 miles) northwest of the Pelican discovery and about 50 km (31 miles) south of the border with Western Sahara. Block 8 covers 11,797 sq km (4,555 sq miles). Pre-drilling estimates put the Flamant structure size at more than 6 tcf of gas.

Discoveries and Blocks off Mauritania

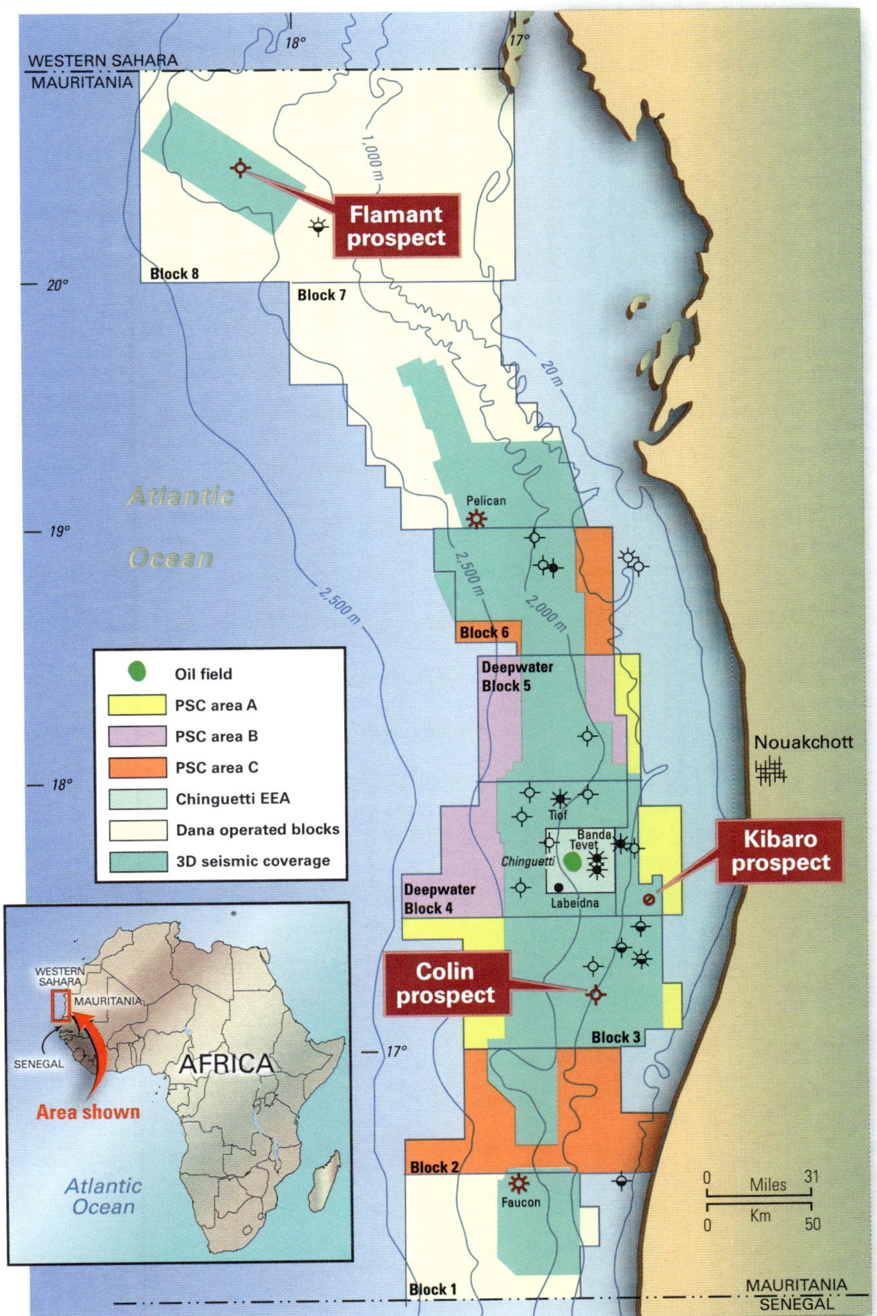

Source: After Hardman Resources Ltd.

In December, Dana reported an oil discovery in Cretaceous sandstones in its Aigrette-1 well on Block 7, which covers 10,090 sq km (3,895 sq miles). The well was drilled to TD 5,152 m (16,900 ft) in 1,358 m (4,455 ft) of water. It is about 43 km (27 miles) northwest of and on trend with Pelican.

The company said tests confirmed a 20-m (65-ft) oil column in good quality Cretaceous age reservoir sands.

Southern area

Dana Petroleum drilled Faucon-1 to TD 3,675 m (12,054 ft) and found 14 m (46 ft) of net hydrocarbon-bearing Cretaceous sandstone in two intervals above a further 270 ft of potential sandstone reservoir. The 200 to 400 bcf estimated to be in place was not enough for stand-alone development but proved a working petroleum system.

The findings provided encouragement for other prospects identified on Block 1, including Petrel, and for the St. Louis exploration license off northernmost Senegal.

Tullow Oil PLC's Energy Africa Ltd. unit acquired the St. Louis license with Senegal's state Petrosen in late 2003. Tullow acquired a 1,200-km 2D seismic survey in 2004 and identified several prospects.

THE DISCOVERED RESOURCE BASE OFF MAURITANIA

Field/ discovery name	Block	Discovery year	Water depth, ft	Remarks
Chinguetti field	4	2001, oil	2,624	2P reserves 123 million bbl; started production Feb. 25, 2006
Tiof field	4	2003, oil	3,540	Potential phased development; contingent resource volume 287 million bbl
Banda discovery	4	2002, gas oil leg	1,000	110 m gas/24 m oil column; contingent 1.2 tcf gas recoverable. Mean oil 344 million bbl STOIIP
Tevet field (Mio.)	4	2004, oil	1,600	41 million bbl contingent resource pending concept selection
Tevet field (Cret.)	4	2005, oil	1,525	8 m gross oil-bearing sands
Pelican discovery (Cret.)	7	2003, gas	5,575	364 m gross gas column with numerous sands
Labeidna discovery	4	2005, oil	4,150	Thin sands over 116 m gross pay
Faucon discovery (Cret.)	1	2005, gas	3,800	10 m gas-bearing sands; 200 bcf gas in place (per operator)

Source: Modified from Hardman Resources Ltd.

The Faucon-1 well results led to a conclusion that it will be beneficial to gather 3D seismic data on St. Louis before drilling an exploration well off Senegal. This survey was planned to start in late 2006 and could be combined with an incremental survey in the southern part of Block 1.

The 2006 drilling program off Mauritania began with Colin-1 on Block 3. Hardman Resources found no commercial hydrocarbons but concluded that the Miocene reservoir B sand quality was much better than at Chinguetti, where the B sands are known to be gas-bearing.

MOROCCO

Capital: Rabat
Monetary unit: Dirham
Refining capacity: 154,901 b/cd
Oil production: 300 b/d
Oil reserves: 1.069 million bbl
Gas reserves: 60 bcf

Wildcat on Moulay Bousselham permit

A group led by the Heyco Maroc affiliate of Harvey E. Yates Co. (Roswell, NM, U.S.A.) planned to spud the Menadra-1 wildcat on the Moulay Bousselham permit onshore in northern Morocco.

Projected to 1,620 m (5,315 ft) , the well will test a thrusted anticlinal structure tentatively identified as Cretaceous. Regional oil-source correlations along trend indicate a mixed Jurassic and Silurian source for hydrocarbons.

Sebou concession

Circle Oil subsidiary Circle Oil Maroc Ltd. (COML) signed an agreement with Morocco's Office National des Hydrocarbures et des Mines (ONHYM) for the 269 sq-km (104 sq-mile) Sebou Concession in the Rharb basin. COML will gather 3D seismic data before starting a six-well drilling program in 2007.

MOZAMBIQUE

Capital: Maputo
Monetary unit: Metical
Refining capacity: 0
Oil production: 0
Oil reserves: N/A
Gas reserves: 4.5 tcf

Mozambique has three gas discoveries—Temane, Pande, and Buzi—all near the coast in the southern part of the country. Gas deliveries began in February 2004 from the Temane fields through a $1.2 billion, 865-km (537-mile) pipeline to Secunda, South Africa. This is Mozambique's only hydrocarbon production.

Searching the Rovuma basin

Recent high oil and gas prices have led operators into exploration in more of the world's remote, nonproducing, and underexplored areas. One of these is the Rovuma basin in coastal and offshore sections of Tanzania and northeastern Mozambique.

In March, in Mozambique's second licensing round, Anadarko Petroleum Corp. (Houston) won exploration and production rights to the 2.64 million-acre (4,125 sq-mile) Offshore Area 1. And in early 2007, Anadarko signed a contract with Mozambique to proceed with evaluation.

The block includes about 90,000 onshore acres (1,406 sq miles) and extends offshore 56 km (35 miles) eastward in water as deep as 1,825 m (6,000 ft). It borders Tanzania to the north and extends southward about 160 km (100 miles). Only two wells have been drilled in the area, said Anadarko.

The contract commits Anadarko to a five-year initial exploration term, with options to extend that phase for three years, and permits a 30-year production term following any commercial discoveries. The company also agreed to acquire 2D and 3D seismic and drill seven wells during the initial exploration term.

The company said it had already identified multiple leads across an area "equivalent in size to 460 typical Gulf of Mexico lease blocks."

Meanwhile, Eni SPA won exploration rights to Offshore Area 4. This unexplored 17,646 sq-km block (6,813 sq miles) is seaward of Anadarko's block, some 2,000 km (1,242 miles) northeast of Maputo in as much as 2,600 m of water (8,530 ft).

To the north, in southernmost Tanzania, Artumas Group (London) is developing Mnazi Bay gas field and was awarded the 15,000 sq-km (5,790 sq-mile) Rovuma block in northern Mozambique onshore southwest from Mnazi Bay.

Agip SPA's 1982 Mnazi Bay-1 discovery well went to TD 3,489 m (11,447 ft) in late Cretaceous. It flowed 14 MMcfd of biogenic gas from deltaic sandstones of Oligocene age. Artumas was planning to drill three wells in 2006 to develop Oligocene and Miocene formations.

Produced gas will be processed in a plant shipped from Houston and transported through a 27-km (17-mile) pipeline to Mtwara, Tanzania.

Artumas has identified eight other prospective structural leads in the concession that include deeper potential hydrocarbon bearing structures in Eocene and Cretaceous formations. Mean reserves were estimated at 670 bcf of gas in 2005, based on the discovery well alone.

WHO WON IN MOZAMBIQUE'S ROVUMA BASIN

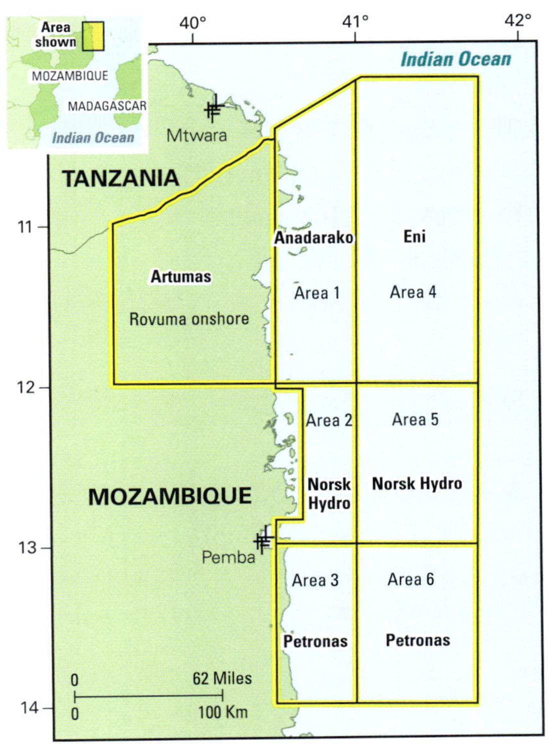

On the Rovuma block, Artumas also planned to follow up on Esso's 1986 Mocimboa-1 well. That well encountered hydrocarbon shows in several intervals, and subsequent log studies identified bypassed gas in Cretaceous. Potential exists for increased pay thickness updip on the structure, said Artumas.

Sasol gets license in Mozambique

Sasol (Johannesburg, South Africa) received conditional approval from the government of Mozambique to explore for hydrocarbons in a 10,000 sq-km (3,861 sq-mile) area off the Mozambican coast.

Sasol began work to satisfy conditions set out for its environmental impact assessment by Mozambican authorities. The company expected this to enable the start of exploration activities in early 2007.

The exploration area includes Blocks 16 and 19, directly offshore the gas field that currently supplies natural gas to Sasol's petrochemical operations.

NIGERIA

Capital: Abuja
Monetary unit: Naira
Refining capacity: 438,750 b/cd
Oil production: 2.20 million b/d
Oil reserves: 36.220 billion bbl
Gas reserves: 181.900 tcf

Nigeria is the largest oil exporter in Africa. The U.S. Dept. of Energy's Energy Information Administration (EIA) estimated average net exports of 2.1 million b/d during the first half of 2006.

Nigeria has several LNG projects on tap for monetizing primarily associated gas. Development of deepwater resources also continues and is expected to add more than 1 million bo/d of production in the next few years.

Violence disrupts production

Nigeria is a member of OPEC and until recently has been Africa's leading oil producer and the world's sixth biggest crude exporter with a normal output of 2.5 million b/d. But militants—chiefly the Movement for the Emancipation of the Niger Delta (MEND)—threatened that production through attacks on oil production facilities in the delta.

In December, Raymond James analysts estimated that the country lost 600,000 to 800,000 b/d in hydrocarbon production during the year. Oil & Gas Journal reported in early October at least five Nigerian oil workers had been killed.

The International Energy Agency (IEA; Paris) said sabotage and accidents in traditional producing areas of the Niger Delta were offsetting production gains from deepwater fields.

Its August Oil Market Report said that shuttered Nigerian production peaked at 785,000 b/d in July before receding to 750,000 b/d by the beginning of August. The country's average production in July was 2.26 million b/d, compared to 2.5 million b/d as recently as late 2005.

At mid-year, deepwater production had not been hurt by the attacks plaguing onshore output. IEA said that new deepwater projects were coming on stream in a trend that "may well continue, with rising deepwater supply providing an offset to sporadic attacks on facilities nearer the coast and onshore."

But it added: "Such a scenario likely undermines plans for production capacity to reach 2.9–3 million b/d by 2007."

Fear of potential disruptions of crude supplies from Nigeria was a major cause of 2006 fluctuations in world oil process. EIA estimated that Nigeria had net crude oil exports of 2.2 million b/d in June.

"In a world that currently has perhaps just 1.5 million b/d of spare upstream capacity, credible threats [against] 2.5 million b/d Nigeria will almost inevitably have a significant impact on prices," said Barclays Capital Inc. (London).

Exploration and Development

OPL 286-DO license

BG Group, through a partnership with Sahara Energy Exploration & Production Ltd., was awarded a deepwater license at mid-year off the western Niger Delta of Nigeria. OPL 286-DO, previously OPL 213, is in 200–1,000 m of water (655–3,280 ft) about 250 km (155 miles) southeast of Lagos. BG Exploration & Production Nigeria Ltd. will be block operator. Its work program includes drilling an exploration well in the first 5 years.

Deepwater block yields oil strike

Esso Exploration & Production Nigeria Deepwater West Ltd. and partners made the first oil discovery on deepwater Oil Prospecting License 214 about 112 km (70 miles) off Nigeria. The Uge-1 well, drilled to 5,130 m (16,831 ft) TD in 1,263 m (4,144 ft) of water, encountered more than 90 net m (300 net ft) of oil pay.

Uge-1 is about 145 km (90 miles) south-southeast of Esso's Erha deepwater oil and gas field, which recently began production and was expected to ramp up to 150,000 b/d by late 2006.

Block operator Esso, Chevron Nigeria Deepwater B Ltd., Phillips Deepwater Exploration (Nigeria) Ltd., and Oxy Nigeria Exploration & Production Ltd. each have an interest in the block, as do Nigerian Petroleum Development Co. and Sasol E&P Nigeria Ltd.

Tubu, Madu oil fields

Chevron Nigeria Ltd. (CNL), operator of National Engineering & Technical Co. Ltd. (NETCO)—a wholly owned subsidiary of Nigerian National Petroleum Corp. (NNPC)—authorized two units of Foster Wheeler Ltd. to carry out further design engineering for the Tubu and Madu oil field developments in permit areas OML 52 and OML 85, respectively.

Foster Wheeler Energy Ltd. and Foster Wheeler (Nigeria) Ltd., through a joint venture agreement with NETCO, will evaluate the economics of the planned development and investigate possible configurations for surface facilities.

Foster Wheeler (Nigeria) Ltd. was already working on the front-end design for the non-associated gas wellhead platforms and pipelines portion of CNL's Olokola Gas Supply Project in Nigeria.

Block OML 122 gas discovery

Equator Exploration and Peak Petroleum announced in May the discovery of gas in the exploratory Owanare 1 well, on OML 122. They said Owanare 1, the second well drilled on OML 122 by the Peak-Equator combine, encountered high pressures and temperatures at 4,100 m (13,450 ft) and was suspended for possible production. The well cut pay at 1,683–1,732 m and 2,183–2,196 m (5,520–5,682 ft and 7,162–7,205 ft). Gas in place is estimated at 185 bcf.

Block 332 transaction

BG Group acquired a 45% interest in and become operator of Block 332 off Nigeria in the western Niger Delta under a farm-in agreement with Sahara Energy Exploration & Production Ltd. The agreement comes as BG, along with partners Sahara and Seven Energy Nigeria Ltd., signed a production-sharing contract in January 2006 for Block 332 with Nigerian National Petroleum Corp.

The work program will be carried out in two phases, the first involving the acquisition of 3D seismic data and the drilling of one exploration well. Block 332 is in 100–1,000 m (328–3,280 ft) of water, 100 km (62 miles) southeast of Lagos.

OML 90 transaction

Syntroleum Corp. (Tulsa, OK, U.S.A.) signed a Heads of Agreement (HOA) with Britania-U Nigeria Ltd. to acquire a 40% participating interest in an oil discovery in OML 90 offshore in the Ajapa field. The field lies 234 km (145 miles) south of Lagos.

The Ajapa #1 well was drilled in about 10 m (30 ft) of water and it flowed at stabilized rates of 4,210 bo/d from about 30 m (100 ft) of net pay. The tested crude oil was high quality 40 API gravity.

Syntroleum estimated the Ajapa Field could contain from 15 million to more than 70 million barrels of gross oil. The company was planning an appraisal drilling program to confirm the reserve and production potential of the Ajapa field. The first appraisal well was planned for 2006.

Drilling and Production

East Area oil-gas project starts

An ExxonMobil Corp. subsidiary started gas reinjection at the East Area Additional Oil Recovery Project off Nigeria in late July. This action stabilized oil production at 120,000 b/d from six fields that once produced more than 200,000 b/d, the company said. The six fields are Iyak, Ubit, Oso, Usari, Idoho, and Mfem.

The $1.3 billion project is designed to recover more than 530 million bbl of oil from the OML 67 and 70 blocks and reduce routine gas flaring. The fields are in shallow water 27 km (17 miles) off southeastern Nigeria. OML 67 is near the border with Cameroon and Equatorial Guinea.

Erha oil and gas field on stream

ExxonMobil Corp. unit Esso Exploration & Production Nigeria Ltd. started production from deepwater Erha oil and gas field off Nigeria. The field lies in 1,200 m (3,937 ft) of water on Block 209, about 160 km (100 miles) southeast of Lagos. The development includes Erha and Erha North, a satellite development expected to come on stream in late 2006.

GULF OF GUINEA GOLDEN RECTANGLE OIL, GAS FIELDS*

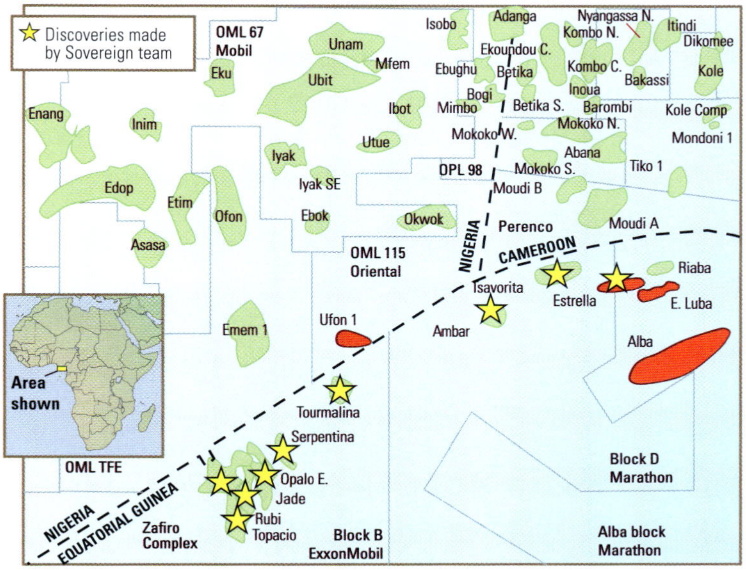

*Rectangle dimensions 80 km E-W by 60 km N-S. Area contains more than 5 billion bbl of proved reserves and is producing more than 1 million b/d

By year-end, Erha production was expected to ramp up to 150,000 b/d and Erha North to produce 40,000 b/d. Associated gas production will be about 300 MMcfd, which will be reinjected. The Erha and Erha North developments consist of 32 subsea wells tied to an FPSO vessel. Total cost is estimated at $3.5 billion. ExxonMobil has estimated that Erha field may contain 1 billion bbl of reserves.

Nigeria-Sao Tome JDZ

Chevron Corp. spudded the Obo-1 wildcat on 704 sq-km (272 sq-mile) Block 1 about 306 km (190 miles) north of Sao Tome and 200 km (125 miles) south of Port Harcourt. Water depth is approximately 230 m (750 ft). The nearest existing wells to Block 1 are Akpo and Egina, less than 20 km (12 miles) north in OPL 246.

LNG

Supply from Brass LNG

A subsidiary of Suez Energy International (Brussels), Suez LNG Trading SA, signed a memorandum of understanding with Brass LNG to buy 2 mtpy of Nigerian LNG for 20 years. First delivery of LNG is expected in 2010.

The LNG is destined primarily for North America. The LNG supply will come from Brass LNG's new 10 mtpy two-train liquefaction plant.

In a second transaction, BG Group PLC subsidiary BG Gas Marketing Ltd. signed a memorandum of understanding to buy 2 mtpy of LNG from Brass LNG for 20 years. BG expects deliveries to start during 2010 to Lake Charles, LA, and Elba Island, GA (U.S.A.) A liquefaction plant is to be built near Nigeria's Brass oil export terminal.

Brass LNG includes Nigerian National Petroleum Corp. and partners Eni SPA, ConocoPhillips, and Total SA. Production from the first two Brass LNG trains (5 mtpy capacity each) is scheduled to start in 2011. Most of the LNG is intended for export to Europe and the United States. Feed gas will come from the partners' production.

Bonny Train 5 comes online

Nigeria LNG Ltd. (NLNG) began production from the fifth train at the Bonny LNG plant. Also, the first LNG cargo from Train 4, which started up in November 2005, was shipped to the Lake Charles, LA, terminal in the United States.

Trains 4 and 5 increase the company's LNG production capacity to more than 17 mtpy.

NLNG will ship more than 14 mtpy of LNG to Europe and more than 8 mtpy of LNG to the United States after Train 6 comes on stream in 2007. It also is planning a further expansion to add more than 8 mtpy of LNG capacity, also targeted for North America.

Olokola project

LNG Japan Corp. (Tokyo), a joint venture of Sojitz Corp. and Sumitomo Corp., announced plans to participate in Nigeria's Olokola LNG project. LNG Japan will join state-owned Nigerian National Petroleum Corp., Chevron Corp., Shell Gas & Power Developments BV, and BG Group.

The first stage of the $7 billion project is to build facilities capable of producing 11 mtpy of LNG by 2010. LNG Japan said it plans to sell its 330,000-tpy share of LNG in the United States and Europe.

In April, NNPC awarded a $14.5 million FEED contract for Olokola LNG to Delta Afrik Engineering Ltd. (Lagos).

KWALE FLARE GAS WILL GENERATE POWER

A project at the Kwale oil-gas process plant in Nigeria aims to recover associated gas that otherwise would be flared, putting it to use generating electricity.

Officials of the International Finance Corp. and World Bank announced in November that the Kwale project would be the first effort supported by Global Gas Flaring Reduction (GGFR), a public-private partnership to be registered as a Clean Development Mechanism (CDM) under the Kyoto global warming reduction accords.

Fewer than 20 countries account for more than 85% of flared and vented gas worldwide, according to GGFR estimates. Nigeria currently leads the world, with more than 24 billion cu m (850 bcf) flared or vented in 2004, the most recent year for which figures are available, said the GGFR.

The capturing of flared gas at Kwale will eliminate a projected 1.5 million tons of carbon dioxide annually, said the GGFR, making it the tenth largest of 2,160 projects in the CDM pipeline under the Kyoto accords.

Elsewhere, a project in Angola will use flare gas as input to an LNG liquefaction plant. In a project backed by Chevron Corp., production from 20 locations will be consolidated for export.

GGFR members include multinational oil companies BP PLC, Chevron, Eni, ExxonMobil Corp., Marathon Oil Corp, Norsk Hydro AS, Shell, Statoil ASA, and Total SA.

Member countries include Algeria, Angola, Cameroon, Chad, Ecuador, Equatorial Guinea, Indonesia, Kazakhstan, Nigeria, Qatar, and the Russian province of Khanty-Mansijsysk. Canada, Norway, the UK, the United States, and members of the European Union are listed as donor countries.

GGFR also receives support from the Organization of Petroleum Exporting Countries' secretariat and the World Bank Group.

International Initiatives

Rail, oil projects

South Korean and Nigerian officials envision upgrading the African country's rail system in exchange for oil interests.

In a memorandum of understanding signed in November, South Korea will provide Nigeria with long-term, low-interest commercial loans and have its companies take part in the second phase of Nigeria's railway modernization program.

In exchange, Nigeria will transfer shares in an operational oil field to South Korea.

The $10 billion railroad project calls for 1,500 km (930 miles) of track linking oil capital Port Harcourt on the Gulf of Guinea to Maldugun in Nigeria's northeastern region.

Nigeria made a similar pact with China in October in an $8.3 billion contract to construct a railway line from the nation's economic capital Lagos to Kano, the largest commercial city in the north.

The Chinese, hoping to secure oil interests in Nigeria, recently granted the African country a $2.5 billion loan facility, a substantial amount of which would be used on the rail project.

Proposed refinery

Indian Oil Corp. (IOC) said it hoped to build a 6 mtpy refinery in Nigeria's Edo state. IOC officials visited Nigeria in August to discuss the $2 billion proposal.

IOC hopes to secure a 12- to 15-year crude supply contract in conjunction with the refinery proposal and said it will negotiate for equity stakes in Nigerian production. IOC now imports 2 mtpy of crude from Nigeria National Petroleum Corp. under a term contract and hopes to double or triple the rate.

China interest in OML 130

CNOOC Ltd. (China National Offshore Oil Corp.) closed on its acquisition of a 45% interest in offshore OML 130 for $2.268 billion from South Atlantic Petroleum Ltd. (Sapetro) of Lagos.

The transaction was approved by both the Nigerian National Petroleum Corp. and the Chinese government, which is CNOOC's majority owner. This was the first investment offer made by CNOOC since its unsuccessful bid for Unocal Corp. in mid-2005.

The license contains the Akpo gas and condensate field, which covers 1,295 sq km (500 sq miles) in 1,100–1,800 m (3,600–5,900 ft) of water. It is located 200 km (124 miles) off Port Harcourt. OML 130 contains the Egina, Egina South, and Preowei discoveries.

Scheduled on stream in late 2008, Akpo is expected to reach peak production of 225,000 boe/d.

SOUTH AFRICA

Capital: Cape Town
Monetary unit: Rand
Refining capacity: 488,297 b/cd
Oil production: 20,000 b/d
Oil reserves: 15.00 million bbl
Gas reserves: 0

Exploration and Development

South Coast Gas Project

Pioneer Natural Resources Company began development activities on its South Coast Gas Project in Block 9, offshore Mossel Bay. Pioneer is partnering with PetroSA on the project.

Once on-stream, the South Coast gas fields will provide feedstock for PetroSA's onshore gas-to-liquids plant in Mossel Bay. Engineering plans were being finalized for subsea systems, pipelines, umbilicals and surface equipment modification. Development drilling began in early 2006, with a total of six wells likely to be drilled.

The project will include the subsea tie-back of gas from the Sable field, from which Pioneer and PetroSA have been jointly producing oil since 2003. Six additional gas accumulations will also tie back to the project, which will utilize existing production facilities on the F-A platform for transportation via existing pipelines to the GTL plant.

Production is expected to begin in the second half of 2007 and increase to an average of 100 MMcf/d of gas and 3,000 b/d of condensate through 2012.

West coast drilling proposed

BHP Billiton was in discussion with the South African government regarding its proposal to begin exploratory drilling offshore South Africa's west coast, in the Orange Basin. The company said operations on Block 3B/4B Hondeklip Bay and Saldanha could begin as early as 2007 once project terms are confirmed.

Several others are operating in the west coast region, including PetroSA, Pioneer Natural Resources, Energy Africa, Anschutz, Forest Exploration, and Canadian Natural Resources.

EM fields development

PetroSA announced that it would start development drilling in offshore Block 9 EM fields in early 2007. Without intervention, PetroSA said, gas flow from the South Coast Project is expected to end in 2013. The company wants the EM fields to remain viable until the FA Field ceases to flow around 2020.

The company also said it would end production from both the Oribi and Oryx Fields in 2007 and from the Sable Field in 2008

Sable Field milestone

PetroSA reported in February 2007 that its Sable oilfield has achieved a production milestone of 20 million bbl of crude oil production in the month of January. Sable is 150 km (93 miles) southwest of Mossel Bay.

Ibhubesi field

Forest Oil International applied in late 2006 for production rights to develop the giant Ibhubesi gas field in the Orange River Basin, located off the west coast in Block 2A, north of Saldanha Bay. The field contains an estimated 13 tcf in gas reserves.

The company's plan for the 5,000 sq-km (1,930 sq-mile) concession area envisions development in four phases, each taking three to five years. Existing exploration wells would be completed for use in the development phase, with new wells drilled as required. As many as 100 wells could be drilled in Block 2A during the project.

PetroSA and Anschutz Overseas are partners with Forest Oil. The company has drilled eight appraisal wells to confirm reserves and was conducting an environmental impact study.

2007 license round plans

Petroleum Agency SA announced plans for a 2007 offshore license round, with a bid deadline of July 31. Seminars have been scheduled for March in Johannesburg, London, and Houston. The acreage offered is in the northern Orange basin shallow water, southern Orange basin deepwater and shelf, proximal Bredasdorp basin, and Algoa offshore basin.

Refining and Processing

New refinery at Richards Bay

Planning was under way for a $4 billion, 300,000 b/d oil refinery to be built in Richards Bay, 200 km (124 miles) north of Durban in KwaZulu Natal province by Drako Oil and Energy (Johannesburg). The company said that it had already secured a 100% loan and purchased a 606-ha (245-acre) site. Target completion date is 2010.

Drako aims to build two other refineries of the same size. The second site is at Coega, at the southern tip of South Africa; the third site is still to be determined.

Latest GTL technology tested

PetroSA, Statoil, and Lurgi reported successful demonstration of a low-temperature synthesis technology said to extend the commercial application of gas-to-liquid (GTL) conversion systems. The partners formed GTL.F1 in 2005, aiming to scale up and commercialize GTL technology.

Since 2004, the low-temperature (200–250°C) Fisher Tröpsch technology was tested at PetroSA's production complex at Mossel Bay. Statoil said that the goal of demonstrating and readying the technology for full-scale production had been achieved.

The GTL complex was of semi-commercial size (1,000 b/d) and the reactor was described as the world's largest in its class. The complex operated continuously in 2006 and produced as planned for much of that time, said Statoil.

Statoil has developed patent-protected GTL technology since 1986. The foundation of the technology is the Fisher-Tröpsch process, a chemical reaction that converts synthesis gas (carbon monoxide and hydrogen) to liquid hydrocarbons, mainly diesel and naphtha.

Upgrade planned for Sapref

Sapref refinery is South Africa's largest, providing 35% of total refining capacity. Processing 180,000 b/d of crude oil (8.5 mtpy), it is a joint venture of Shell SA Energy and BP Southern Africa.

The U.S. Dept. of Energy's Energy Information Administration (EIA) reported that Shell and BP plan to invest $100 million in the facility by 2011, primarily to reduce emissions. The refinery is 16 km (10 miles) south of Durban.

SUDAN

Capital: Khartoum
Monetary unit: Pound
Refining capacity: 121,700 b/cd
Oil production: 295,000 b/d
Oil reserves: 5.000 billion bbl
Gas reserves: 3.0 tcf

Overview

The UN Security Council asked the International Criminal Court (ICC) in March 2005 to investigate the violence in the western region of Darfur. The United States has called the violence genocide, though Khartoum denies the charge.

The ICC in February 2007 named the first two suspects accused of committing war crimes. The ICC also said it was monitoring the spillover of violence into Chad and the Central African Republic.

Some experts estimate that 200,000 people have been killed and 2.5 million others driven from their homes in Darfur since 2003, when rebels took up arms against the government, charging it with neglect.

Khartoum disputes those figures, has argued that the ICC has no jurisdiction over its nationals, and has said that it would not allow any of its citizens, including rebels, to be tried outside Sudan. Sudanese media reported that Khartoum was planning to launch its own war crimes trial in March 2007.

Civil war. January 2005 brought a settlement of Sudan's 20-year civil war between the Arab and Muslim north and the largely black African Christian south. The peace accord calls for a referendum in 2011 on independence for the south.

The war dramatically changed the makeup of the country's oil sector. Western oil companies pulled out of the country because of human rights abuses during the conflict. After the peace settlement, Chinese, Malaysian and Indian companies moved into Sudan's oil sector and now dominate it.

While its two largest oil fields are in the south, Sudan has begun to sell exploration concessions in the Darfur and northern areas.

Exploration and Development

First White Nile well on Block Ba

White Nile Ltd. said it expected to spud the first exploration well on the Block Ba concession in South Sudan early in the second quarter of 2007. The firm's contracted drill rig was in Africa and was being transported to the Block Ba concession area in Southern Sudan in February.

White Nile said that seismic work had identified numerous drill targets and that it had prioritized four structures where it believes the productive Muglad Basin extends into the concession area, including one large structure of over 50 sq km (19 sq miles).

Legal dispute: White Nile, Total

A legal dispute continued in 2006 between Total SA (Paris) and small independent, White Nile Ltd. (Guernsey, UK) over the right to drill in a promising area of Southern Sudan called Block Ba.

Total was awarded the block by the central government in Khartoum in 1980 and performed seismic work until the country's civil war broke out in 1985.

Total suspended its work during the war but paid a fee to the Khartoum government to keep its rights to the block. The war ended in 2002 and Total aimed to restart exploration on the block. It began talks with Khartoum to amend the existing contract.

But in 2005, White Nile made an arrangement with the Sudanese People's Liberation Movement, a former southern rebel group that now shares power with the Khartoum government. White Nile received a 60% stake in Block Ba. The remaining 40% went to Nile Petroleum Corp., the southern government's own national oil company.

Total has brought the case to court in the United Kingdom, arguing that its 1980 contract is still valid under the peace agreement signed by the Khartoum government and the southern leadership in 2005.

As of early 2007, the companies were battling over court-ordered disclosure by White Nile of certain documents pertaining to the case.

Total said in February that it might consider giving the Southern Sudan government a stake in the exploration permit the company is claiming for Block Ba, in order to oust White Nile.

New field in Block 4

Sudan has discovered a new oilfield in the southern Kordofan region. The Neem field began producing 24,000 b/d in mid-2006, with expected peak production of 40,000 b/d.

Neem is on Block 4, operated by the Greater Nile Petroleum Operating Company. GNPOC is a consortium of China's state-owned CNPC, Malaysia's Petronas, India's ONGC, and Sudapet. GNOPC also operates Sudan's main oil pipelines.

Lundin returns to Sudan

Lundin Petroleum AB (Sweden) presented a plan in mid-2006 to Khartoum for development of the Sarehat oil square 5/1. The company holds a 24.5% interest in Block 5B that covers the southern part of the Muglad basin. The block is on trend with the discoveries made by Chevron and later GNPOC in Blocks 1, 2 and 4 and recently by the Petronas/ONGC/Sudapet group in Block 5A.

The program includes the acquisition of seismic data, building of infrastructure, and drilling of up to three exploration wells beginning in late 2006.

Block 5B is operated by WNPOC, a joint operating company between Petronas (Malaysia) and Sudapet on behalf of a consortium. The block, in the Muglad Basin, covers 20,119 sq km (7,768 sq miles). Muglad is the main oil-producing basin in Sudan and has seen the development of several large oilfields in the last few years.

Oil production outlook

Sudan's total hydrocarbon output is expected to reach 650,000 boe/d in 2007 and 800,000 boe/d in 2008. The minister of energy and minerals told an energy conference in January 2007 that Sudan aims to reach 1 million bpd by 2010.

Sudan's two main oilfields, producing heavy-sweet Nile Blend and Dar Blend, are located in the south. Industry experts estimate the region has hundreds of millions of barrels of recoverable reserves.

Most Western oil companies pulled out of Sudan during the north-south war because of human rights abuses during the conflict. Companies from Malaysia, China, and India are leading oil development efforts in Sudan.

One example is the Thar Jath field, in Block 5A, which came onstream in mid-2006. India's Oil and Natural Gas Corp. (ONGC), with a 24.125% interest in the field, said the production target for year-end 2007 is 50,000 b/d.

Block 5A crude flows

White Nile (5B) Petroleum Operating Company Ltd. (WNPOC), the operator of Block 5A, announced in August the first commercial lifting of crude oil produced from the Block's Thar Jath field.

About 400,000 barrels of crude oil was loaded onto an oil tanker at the Bashayer Marine Terminal for delivery to a long-term contract buyer.

WNPOC is a joint company of Petronas of Malaysia and Sudapet of Sudan. It acts operator for Block 5A , Block 5B and Block 8. Partners (in various combinations) for the three blocks include ONGC Videsh Ltd., Hi Tech Petroleum, and Lundin Petroleum AB, as well as Petronas and Sudapet.

The Thar Jath field in the Muglad Basin began producing in late June 2006. Initial production of 30,000 b/d was anticipated to rise to 50,000 bpd by year-end 2006 and to70,000 b/d by mid-2007.

The crude oil from the field is transported through a 172-km (107-mile) pipeline that links at Heglig to the existing 1,542-km (958-mile) pipeline system that goes to Port Sudan. The Thar Jath crude is categorized as sweet heavy oil (API 22°) with low gas oil ratio.

WNPOC also operates Block 5B and Block 8.

OPEC INVITATION

Sudan was invited in 2006 to join the Organization of the Petroleum Exporting Countries (OPEC) and was reviewing the invitation as of early 2007. The country has had observer status since 1999. Angola became OPEC's twelfth member in January 2007.

Pipelines

Melut-Port Sudan oil pipeline

The long-delayed 1,350-km (840-mile) pipeline from the oil fields of the Melut basin to Port Sudan on the Red Sea coast officially opened in early 2006. The new 28-in pipeline will allow Sudan to increase its oil production to about 500,000 b/d, from 300,000 b/d.

The terminal at Port Sudan was expected to begin exporting oil by August 2006.

Petrodar is the pipeline operator, a partnership that includes China's state-owned CNPC, Malaysia's Petronas, Chinese company SINOPEC, Sudanese Sudapet, and Dubai-based al-Thani.

Refining

Khartoum refinery upgrade

With help from China National Petroleum Corporation (CNPC), Sudan doubled the capacity of its main refinery at Khartoum up to 100,000 b/d. CNPC owns 50% of the facility; the Sudanese government owns the other 50%. The $341 million expansion of the refinery was completed in mid-2006.

The refinery produces mostly gasoline, and about half of that output will be exported. Production of diesel fuel is also expected to increase significantly in coming years.

TANZANIA

Capital: Dodoma
Monetary unit: Shilling
Refining capacity: 14,900 b/cd
Oil production: 0
Oil reserves: 0
Gas reserves: 230 bcf

PSAs signed for three east-coast licenses

Dominion Petroleum Ltd. (Bermuda) entered into three production-sharing agreements (PSAs) with Tanzanian Petroleum Development Corp. covering more than 10 million acres (15,625 sq miles) in the onshore Mandawa, Kisingire, Lukuliro, and Selous licenses along the eastern coast of Tanzania.

Dominion identified 16 prospects and planned to drill four wells beginning in late 2006. The company said that the Jurassic Mandawa sedimentary basin contains potential reservoir sandstones and limestone and a number of potential traps.

The Kisingire PSA comprises two large, adjoining areas: Kisingire and Lukuliro. The Songo Songo gas fields lie to the east of this area. The geology comprises an early sedimentary basin underlying a Jurassic rift. Royal Dutch Shell PLC's shallow drilling in the 1980s identified oil source rocks in this basin.

Dominion will seek additional seismic data to identify prospects in the Selous PSA in southern Tanzania, a relatively underexplored area dominated by the Permian Selous basin.

Reserves raised for Songo Songo gas field

Wells in Songo Songo gas field off eastern Tanzania showed a pressure decline of less than 2% in 2005, and analysis in 2006 permitted a reserves increase despite no new drilling, said operator EastCoast Energy Corp. (Tortola, British Virgin Islands).

Proved and probable reserves were raised 14% to 560 bcf. This amounted to a 41% rise in proved reserves to 241 bcf and a 25% hike in proved and probable reserves to 320 bcf.

Songo Songo's five wells produced 14.7 bcf in 2005, bringing cumulative production to 19.3 bcf since the start of commercial operation in 2004. Production peaked at 72.8 MMcfd in August 2005.

EastCoast plans to drill one or two wells on the northern part of the Songo Songo West prospect, just west of the gas field in 2007 in 20 m (65 ft) of water. Gas in place is estimated at a most likely 600 bcf. The company is interpreting seismic data shot on the Nyuni license under a farmout from Aminex PLC (London).

EastCoast also let a contract to Petrofac Engineering Ltd. (Surrey, UK) to find the most efficient way to boost capacity to 120 MMcfd by late 2007. If a third processing train were installed, capacity would then be limited by the 232-km (145-mile), 105–110 MMcfd pipeline from the Songo Songo field to Dar es Salaam.

TUNISIA

Capital: Tunis
Monetary unit: Dinar
Refining capacity: 34,000 b/cd
Oil production: 65,000 b/d
Oil reserves: 400.00 million bbl
Gas reserves: 2.30 tcf

Exploration and Development

El Louza block

TGS-NOPEC Geophysical Co. (Naersnes, Norway) mobilized a vessel to acquire 600 sq km (231 sq miles) of multiclient 3D seismic data on the 4,100 sq-km (1,583 sq-mile) El Louza block in the Mediterranean.

MedOil PLC (London) said the survey is designed to accurately map the oil volume potential of the M'Sela-1 oil discovery.

It will also cover four separate exploration prospects and a potentially large stratigraphic trap in the "Isis formation" that tested oil in the discovery well and was the producing stratum in nearby Isis oil field.

WHERE INTEREST IS GROWING IN THE CENTRAL MEDITERRANEAN

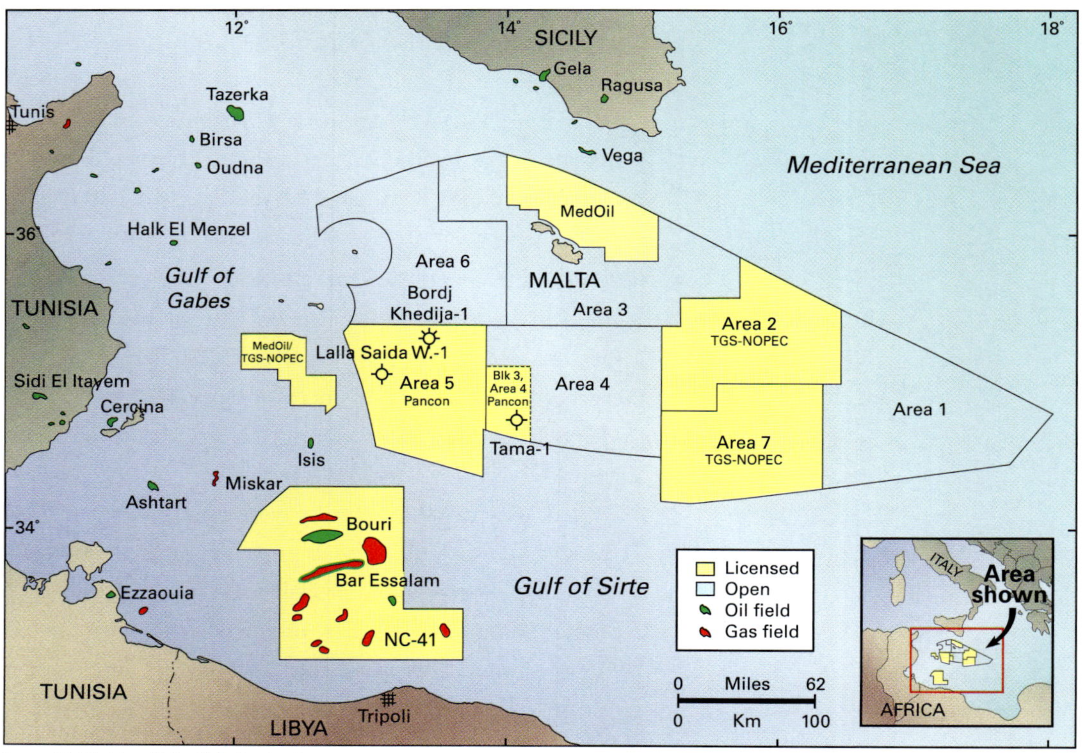

Source: Modified from Malta Oil Exploration Department

M'Sela tested 1,200 b/d of 34° gravity oil from one reservoir and 118 b/d of 16° gravity oil from a deeper reservoir. Permit interests are MedOil 95% and TGS-NOPEC 5%. MedOil is also negotiating with Maltese authorities for a 1-year extension of its exploration study area.

Bazma Permit

Cygam Energy Inc. (Calgary) said its Rigo Oil Co. Ltd. subsidiary acquired the Bazma Permit west of the 400,000-acre Jorf Permit where Rigo's interest is 37.5%. Several companies have expressed interest in participating in Bazma. Oil and gas pipelines cross both permits. Bazma is prospective in Triassic Tagi sandstones and Permian carbonates.

Kerkouane, Chorbane blocks

Grove Energy Ltd. (Vancouver, BC, Canada) acquired two large exploration blocks from Anschutz Overseas Tunisia Corp. (Denver), a private independent.

The Kerkouane offshore-onshore block covers 6,720 sq km (2,595 sq miles), and the Chorbane onshore block covers 2,428 sq km (938 sq miles). Collectively, they contain 12 prospects and leads.

Kerkouane, which contains the Dougga gas-condensate discovery made by Shell in the early 1980s, is adjacent to Grove's Pantelleria permits. Chorbane has two leads adjacent to Sidi el Kilani oil field and on trend with Miskar and Hasdrubal gas fields. The Grove work programs call for drilling one well on Kerkouane and acquiring 250 line-km of 2D seismic data on Chorbane.

Drilling and Production

Oudna oil field

Lundin Petroleum and Atlantis Holding Norway AS planned to start production from Oudna oil field off Tunisia in September. Initial gross production is expected to be 20,000 b/d of 41.5° gravity oil from Miocene Lower Birsa sands. The field has estimated proved and probable reserves of 11.5 million bbl.

Oudna has a single subsea production well tied back to an FPSO vessel formerly on nearby Isis field.

Refining

Skhira refinery

Tunisia announced a list of prequalified bidders to build a 120,000 b/d refinery in the southern coastal city of Skhira. Officials said the government plans in mid-2007 to select a contractor to build, own, and manage the refinery for at least 30 years. Construction could begin in December 2007.

Tunisia has a hydroskimming facility at Bizerte with nameplate distillation capacity of 34,000 b/d. Snamprogetti SPA built the Bizerte refinery in 1962–1963. Société Tunisienne des Industries de Raffinage operates it.

UGANDA

Capital: Kampala
Monetary unit: Ugandan shilling
Refining capacity: 0
Oil production: 0
Oil reserves: 200 million bbl
Gas reserves: 0

President announces first oil discovery

Uganda's Pres. Yoweri Museveni announced in early October the first discovery oil in the country, after years of exploration. Hardman Resources Ltd. (Melbourne) discovered the oil in June, at three western fields called Waraga 1, Waraga 2, and Mputa.

The president said that he expected production to begin in 2009, initially at 6,000 to 10,000 b/d. Construction of an oil refinery was also being planned. A Ministry of Energy official said that the three fields have reserves of 100 million to 300 million bbl, with 30 million bbl ready for extraction.

The Associated Press said the announcement was the first acknowledgment by a government official that Uganda has oil reserves.

Discoveries draw interest

Discoveries near Lake Albert in landlocked Uganda are awakening exploration interest all along the Albert graben, but *Oil & Gas Journal* observed that it might be a considerable time before crude oil begins flowing. It could take one to three years to determine commercial viability.

To justify construction of a 1,125-km (700-mile) pipeline from the Albert graben to the Indian Ocean coast at Mombasa, Kenya, would take certification of 300–500 million bbl of recoverable oil.

Smaller reserves might support construction of Uganda's first refinery. Even modest discoveries could be economic because the local costs of refined petroleum products are very high, particularly in western Uganda.

On the exploration side, four blocks are in force in Uganda, starting near the border with Sudan. Operators have applied for blocks on the Congo (former Zaire) side of the graben, and awards could be forthcoming soon.

Block 3A oil found

Heritage Oil Corp. (Calgary) found 40 m (131 ft) of oil-bearing zones in its Kingfisher-1A sidetrack well on Uganda's Block 3A in Lake Albert, which spans 1,991 sq km (769 sq miles). The well reached 2,962 m (9,717 ft) and was cased to 2,502 m (8,208 ft). Plans were being made to secure a rig to drill to 3,000–4,000 m TD (9,842–13,123 ft).

In a November 2006 evaluation, test data suggested that the well could flow at stable rates of about 5,600 b/d and also indicated an extremely high permeability of more than 2,000 md.

The Kingfisher prospect is a large structural high expressed at surface on the bed of Lake Albert. Seismic data indicate an area extent of up to 70 sq km (27 sq miles). Heritage plans to gather 3D seismic over the Kingfisher structure during 2007.

WHERE UGANDA'S FIRST DISCOVERIES HAVE BEEN MADE

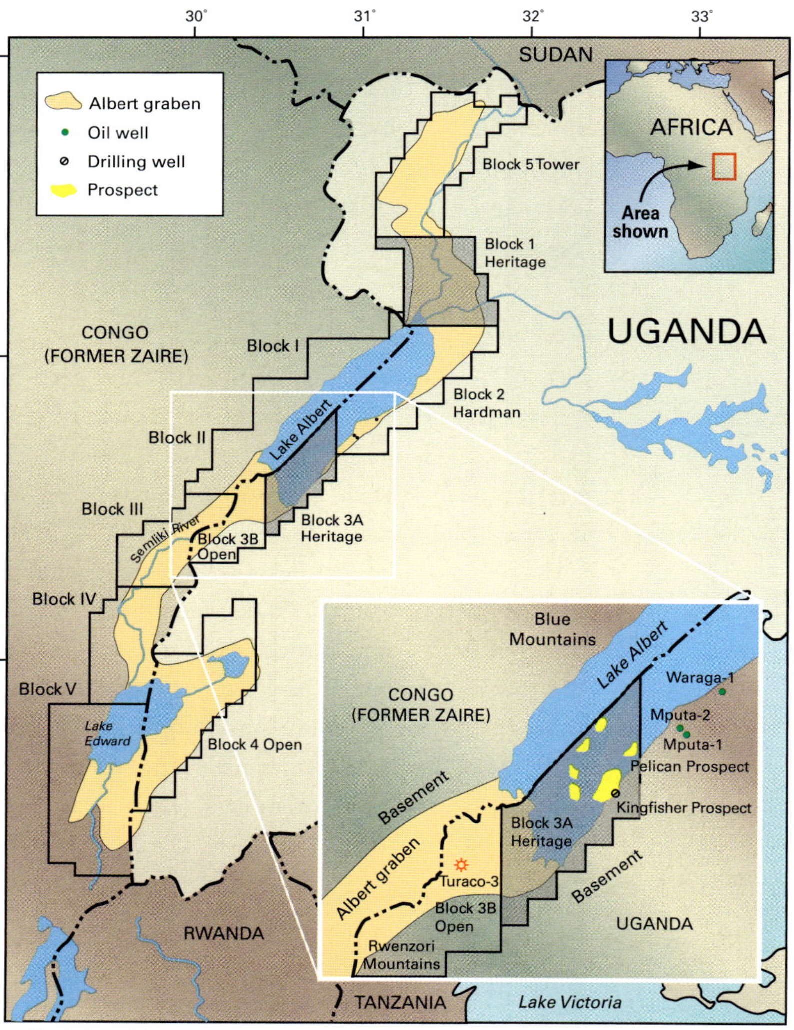

Source: After Heritage Oil Corp.

Block 2 development

The Ugandan government, Hardman Resources Ltd. (Melbourne), and Tullow Oil PLC (London) agreed in October to accelerate appraisal and development of Block 2 on the Ugandan side of the Albert graben. (Tullow was in the process of buying Hardman in late 2006.)

The partners planned to spud the Nzizi-1 well, an updip appraisal well southwest of the Mputa oil discovery, by early November.

They also planned further 3D and 2D seismic surveys to better define the volumes of existing discoveries at Waraga and Mputa fields, develop infill locations, and explore the northern area of the block to identify drilling prospects in 2007–08. Drilling of offshore prospects in Lake Albert could occur by year-end 2007.

Oil discovered could be used to fuel a potential 50-MW local electric power station and supply a small refinery.

Block 5 development

Tower Resources PLC (London) was starting farmout work on 1.5 million-acre Block 5. The block is an unexplored graben known as Rhino Camp near the northern end of the Albert graben.

It is highly probable that reservoir, structure, and seal are present at Rhino Camp. Tower considers the main exploration risk to be the thermal maturity of source rocks.

Further geophysical interpretation is being undertaken and new geophysical activity is planned to determine how deep the basin is and how analogous Block 5 may be to Block 2.

Other action in Africa

Several other African nations saw growing initial interest in exploration and development in 2006.

Benin

Capital: Porto Novo

The Ministry of Mines, Energy and Water launched a late-2006 bid round by re-offering two land and two deepwater blocks. Coastal blocks A and B and deepwater blocks 5 and 6 are in the Benin coastal sedimentary basin.

Gambia

Capital: Banjul

Buried Hill Energy (Cyprus) PLC (London) said its Gambia subsidiary entered into 30-year licenses on the A1 Alhamdulilah and A4 blocks in the Atlantic off Gambia. The acreage is separated from Mauritania by the Senegalese coast.

Ghana

Capital: Accra

The government awarded Tullow Oil PLC (London) interests in the Shallow-water and Deepwater Tano blocks along the marine boundary with Ivory Coast. Tullow also took a farmout to acquire an interest in the adjacent West Cape Three Points offshore license.

All three licenses are in the Greater Ivoirian basin on trend with Espoir oil and gas field off Ivory Coast.

Ethiopia

Capital: Addis Ababa

Swedish independent Lundin Petroleum AB signed a production-sharing contract for Blocks 2 and 6 in the southeastern Ogaden basin, spanning more than 24,000 sq km (9,266 sq miles). Indications of light oil, gas, and condensate have been documented.

Malaysia's state-run Petronas let a contract to China's Zhongyuan Petroleum Exploration Bureau (ZPEB) to begin seismic surveys of the Ogaden basin. ZPEB will shoot seismic surveys in the areas of Fer Fer, Warder, and Genale.

In 2004–2005, the government granted five oil exploration licenses to two Malaysian companies. Petroliam Nasional Bhd. holds four licenses in the Gambela and Ogaden basins. Pexco holds a license in Ogaden.

Kenya

Capital: Nairobi

A group led by Woodside Petroleum Ltd. (Perth) spudded an exploration well on the Pomboo prospect on Block L-5 in the Indian Ocean.

Aminex PLC (London) signed a participation agreement with Upstream Petroleum Services Ltd. (Dubai) regarding Blocks L9 and L10 in the Lamu basin off southern Kenya. UPSL was acquiring seismic and geochemical data under a technical evaluation agreement with the government.

Former Soviet Union

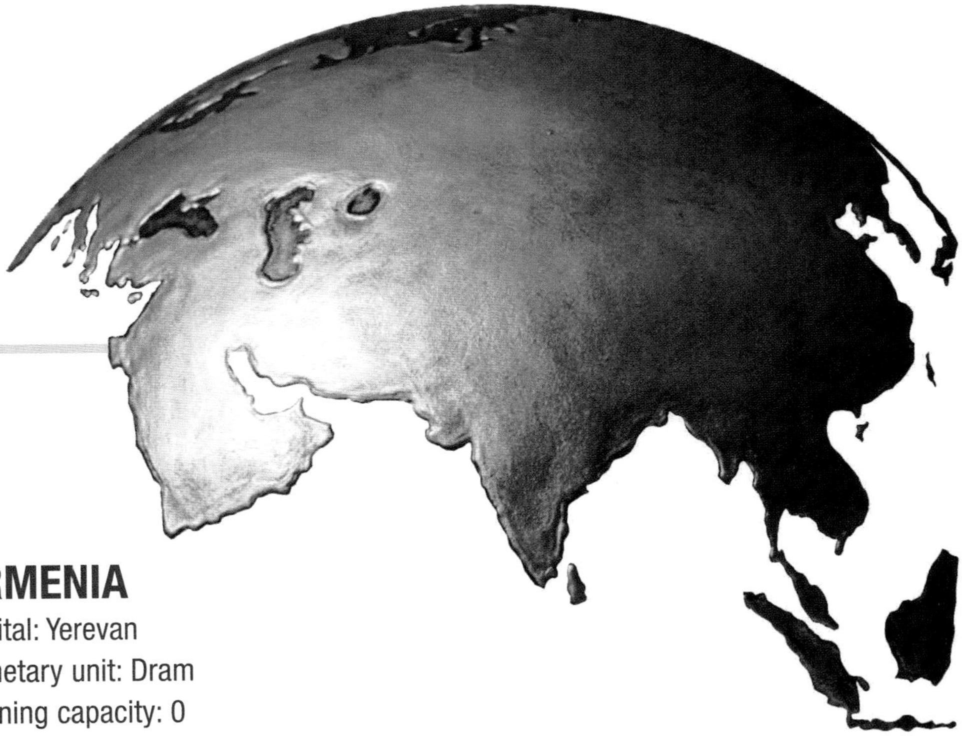

ARMENIA

Capital: Yerevan
Monetary unit: Dram
Refining capacity: 0
Oil production: 0
Oil reserves: 0
Gas reserves: 0

Exploration and Development

Block 2 drilling planned

Transeuro Energy Corp. (Vancouver, BC, Canada) began acquiring seismic data in its Block 2 license area. The company contracted with AESP (Armenia) to provide 131 km (81 miles) of new seismic in the Oktemberian region, 85 km (53 miles) west of Yerevan. The company said this is the first seismic to be acquired in Armenia in more than eight years.

Initial seismic focus will be on the Karmir gas prospect where two old Soviet wells have produced non commercial gas to surface. Survey work then will be extended west, over the "SWOKT" structure where six old Soviet wells suggest a large shallow gas field may exist.

Survey results plus reprocessed historic seismic data will help finalize locations for three planned wells. The first wellsite is on the Karmir prospect. Drilling to approximately 2,500 m (8,200 ft) is planned, to test the Lower Sand Clay (LSC) and Lower Multi-colored Suite (LMS).

The second well will test the Southwest Oktemberian prospect located updip of the Karmir structure. Well logs suggested numerous possible gas reservoirs in the LSC from 320–860 m (1,050–2,820 ft).

Likely locations for a third well are in the Shorakhpur area or in the Oktemberian region.

Former Soviet Union

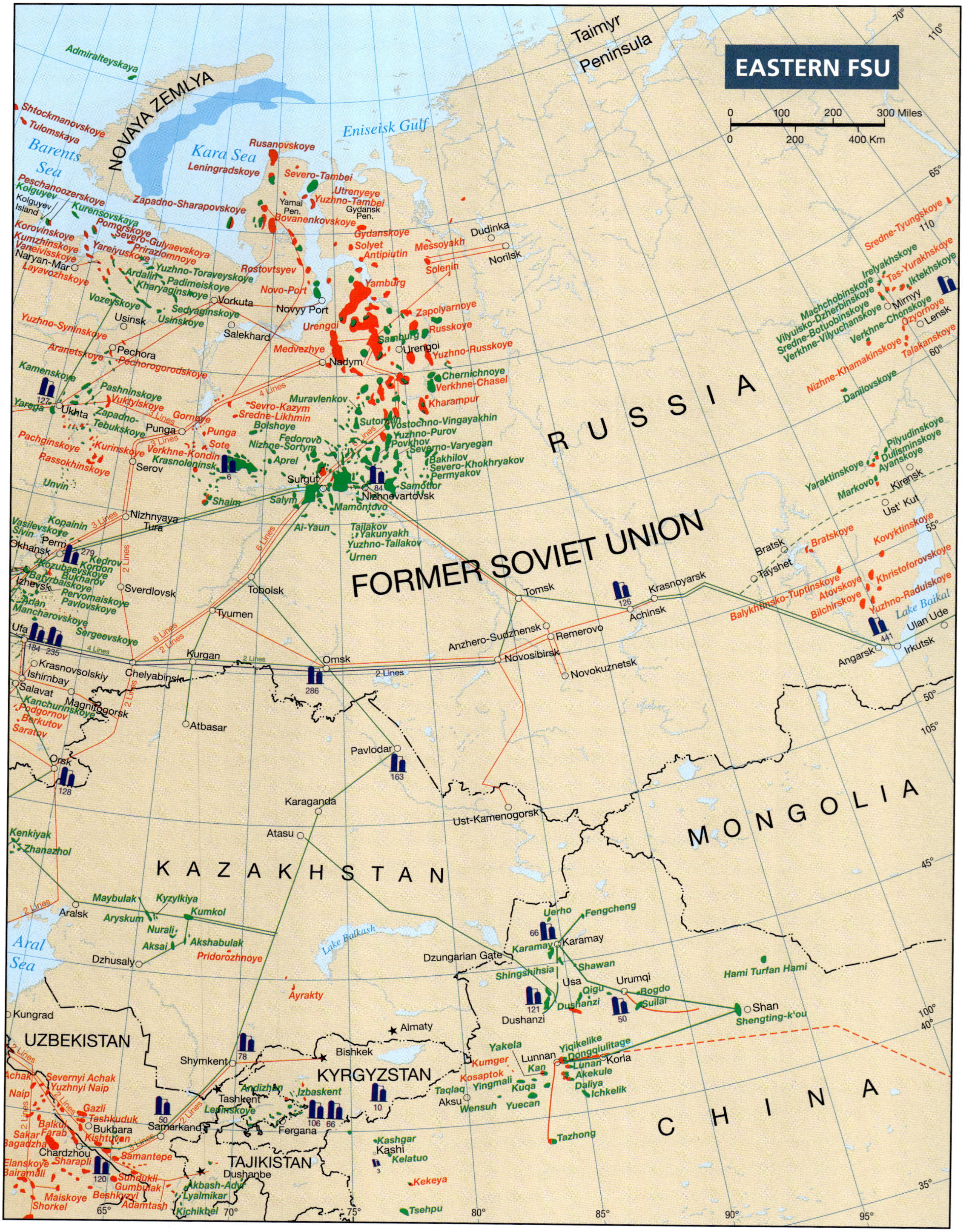

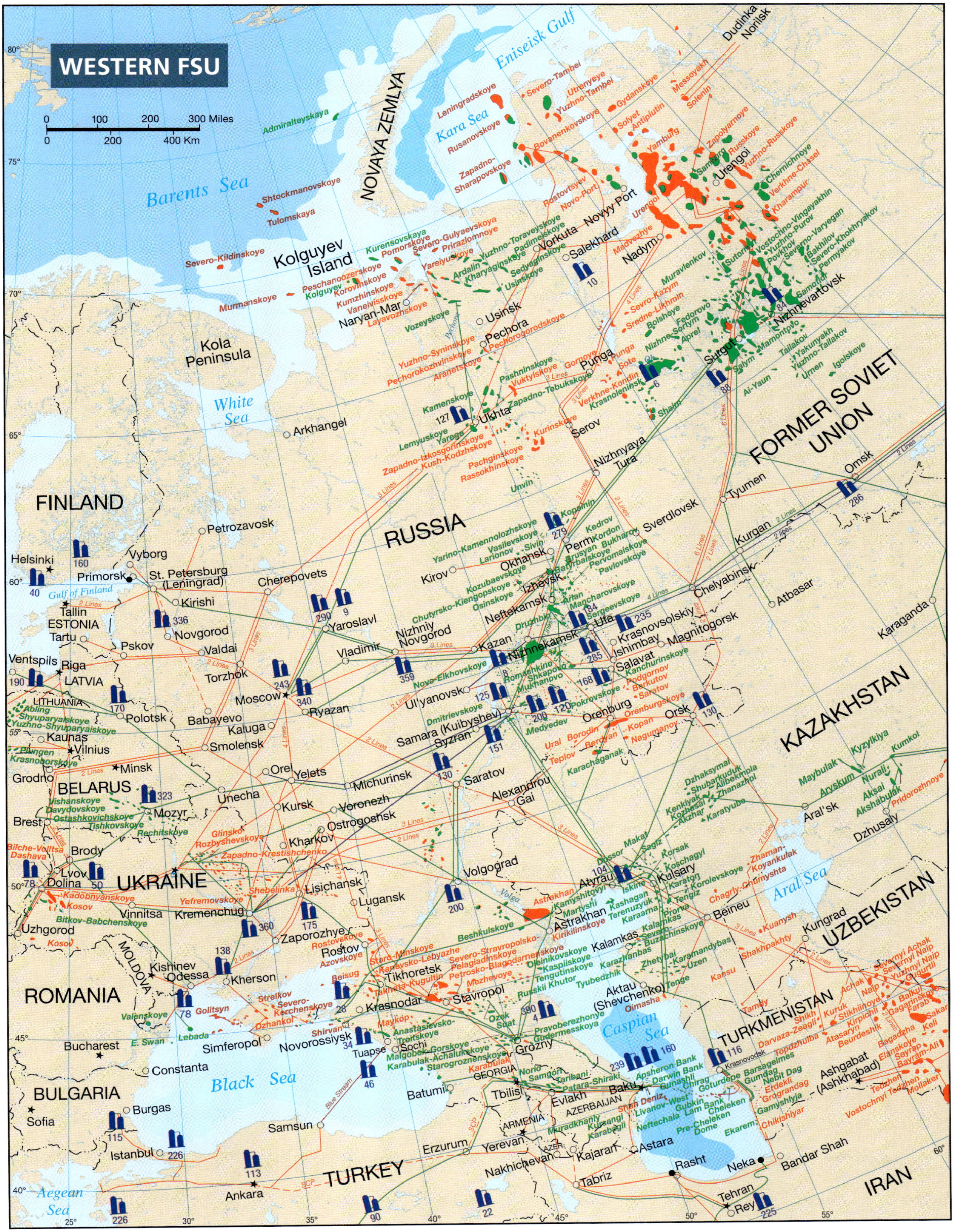
WESTERN FSU
0 100 200 300 Miles
0 200 400 Km
Barents Sea
Kara Sea
Eniseisk Gulf
White Sea
Gulf of Finland
Black Sea
Caspian Sea
Aral Sea
Aegean Sea
NOVAYA ZEMLYA
Kolguyev Island
Kola Peninsula
FINLAND
RUSSIA
FORMER SOVIET UNION
KAZAKHSTAN
UZBEKISTAN
TURKMENISTAN
UKRAINE
BELARUS
ESTONIA
LATVIA
LITHUANIA
MOLDOVA
ROMANIA
BULGARIA
GEORGIA
ARMENIA
AZERBAIJAN
TURKEY
IRAN
Helsinki
Vyborg
Primorsk
St. Petersburg (Leningrad)
Tallin
Tartu
Pskov
Novgorod
Kirishi
Petrozavosk
Cherepovets
Arkhangel
Naryan-Mar
Usinsk
Pechora
Ukhta
Vorkuta
Salekhard
Novyy Port
Nadym
Urengoi
Surgut
Nizhnevartovsk
Serov
Nizhnyaya Tura
Sverdlovsk
Tyumen
Omsk
Kurgan
Chelyabinsk
Atbasar
Karaganda
Perm
Okhansk
Kirov
Izhevsk
Neftekamsk
Ufa
Krasnousolskiy
Ishimbay
Magnitogorsk
Salavat
Orsk
Orenburg
Kazan
Nizhnekamsk
Nizhniy Novgorod
Yaroslavl
Vladimir
Moscow
Ryazan
Kaluga
Valdai
Torzhok
Babayevo
Smolensk
Ul'yanovsk
Samara (Kuibyshev)
Syzran
Saratov
Michurinsk
Orel
Yelets
Kursk
Voronezh
Ostrogoshsk
Kharkov
Unecha
Minsk
Mozyr
Grodno
Brest
Vilnius
Kaunas
Polotsk
Riga
Ventspils
Brody
Lvov
Dolina
Uzhgorod
Vinnitsa
Kremenchug
Lisichansk
Lugansk
Zaporozhye
Rostov
Kishinev
Odessa
Kherson
Simferopol
Novorossiysk
Tuapse
Sochi
Krasnodar
Tikhoretsk
Stavropol
Grozny
Volgograd
Alexandrov Gai
Astrakhan
Atyrau
Kulsary
Beineu
Kalamkas
Aktau (Shevchenko)
Aral'sk
Dzhusaly
Kungrad
Krasnovodsk
Ashgabat (Ashkhabad)
Baku
Tbilisi
Batumi
Yerevan
Nakhichevan
Evlakh
Kajaran
Astara
Tabriz
Rasht
Neka
Bandar Shah
Tehran
Rey
Erzurum
Samsun
Ankara
Istanbul
Burgas
Sofia
Constanta
Bucharest
Galati

Pipelines

Gas pipeline from Iran

A 140-km (87-mile), $220 million pipeline to carry gas from Iran to Armenia was in the final stages of construction in at year-end and due to start up by April 2007. The pipeline covers 100 km (62 miles) in Iran and 40 km (25 miles) in Armenia.

Iran is to deliver 36 bcm (1.27 tcf) of natural gas over 20 years. Terms could be revised in the future to add five additional years and increase gas deliveries to 47 bcm (1.66 tcf). Armenia plans to use the gas to produce electricity. In fact, the country is expected to compensate Iran in part for the costs of the pipeline with deliveries of electric power.

Russia controls Armenia's gas distribution system, and uncertainty remains regarding the degree of control that Russia will exercise over the new pipeline.

Iran's parliamentary speaker told Armenian officials that his country was interested in building a second natural gas pipeline to Armenia, according to media reports.

AZERBAIJAN

Capital: Baku
Monetary unit: Manat
Refining capacity: 398,978 b/cd
Oil production: 625,000 b/d
Oil reserves: 7.0 billion bbl
Gas reserves: 30.0 tcf

Exploration and Development

Alyat-Deniz field

Eurasia Energy Ltd. (Aberdeenshire, Scotland) planned negotiations with State Oil Co. of the Azerbaijan Republic in November to develop a 600 sq-km (232 sq-mile) shallow-water block in the Caspian Sea 70 km (43 miles) south of Baku.

SOCAR completed the Alyat-87 well in Alyat-Deniz field on the block in April. It went to TD 3,610 m (11,844 ft) and flowed 300 b/d of oil and 3.35 MMcfd of gas. SOCAR began drilling a second well.

AIOC CONTRACT AREA, PIPELINE CORRIDORS

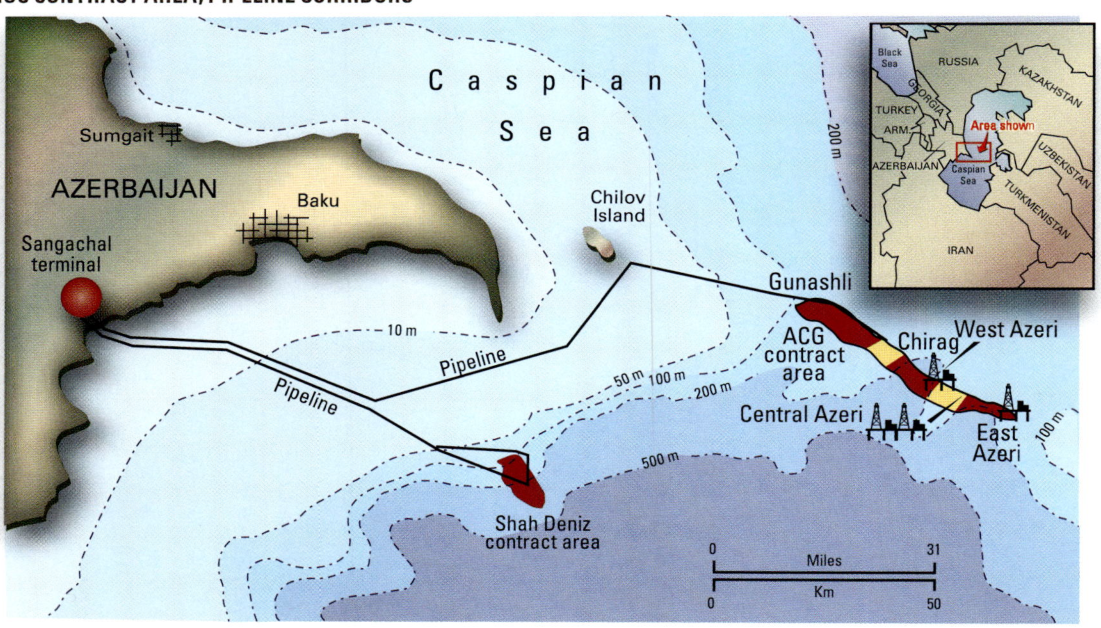

AIOC = Azerbaijan International Operating Co., a consortium of foreign oil companies

Also on the block are seven prospective exploration structures that trend southeasterly from the coast to 70 km (43 miles) offshore in 20–50 m (66–164 ft) of water. The structure names are Hamamdag-Deniz, Garasu, Sangi-Mugan, Ulfat, Aran-Deniz, Dashly, and Sabayil.

SOCAR placed Alyat-Deniz cumulative production at 12.5 million bbl of oil and 1 bcm of gas, and estimated remaining reserves at 15 million bbl and 1 bcm.

Drilling and Production

Shah Deniz field comes on stream

Commercial gas production began in December from Shah Deniz gas-condensate field in the Caspian Sea off Azerbaijan. Technical problems arose in January 2007, but production resumed in mid-February.

Consortium partner Total SA said the field's State 1 production plateau is expected to be 300 MMcfd. Shah Deniz covers about 860 sq km (332 sq miles) about 70 km (43 miles) south of Baku, in 50–600 m (164–1,969 ft) of water.

Shareholders include BP PLC (technical operator), Statoil ASA (commercial operator), State Oil Co. of the Azerbaijan Republic, Total, Naftiran Intertrade Co. Ltd., LukAgip NV, and Turkiye Petrolleri Anonim Ortakligi.

Gas reserves are estimated at 25–35 tcf. The field is expected to produce 8.6 bcm/year (0.83 bcfd) of gas in Stage 1 and 37,000 b/d of condensate, which will be shipped to Ceyhan, Turkey, for processing.

Gas is being exported to Azerbaijan, Georgia, and Turkey through the $1.3 billion, 700-MMcfd South Caucasian Pipeline. Operator BP PLC began line-fill of the SCGP in June. The pipeline extends 692 km (430 miles) from Baku to Tbilisi, Georgia, and Erzurum in eastern Turkey, paralleling the Baku-Tbilisi-Ceyhan crude oil pipeline.

Azeri crude assayed

BP Oil International Ltd. assayed the Azeri Light crude first produced in the Caspian Sea off Azerbaijan in December 2005. The oil comes from the West Azeri (WA) field, part of the Azeri-Chirag-Gunashli (ACG) complex.

Giant ACG, 100 km (62 miles) east of Baku, has reserves of 5.4 billion bbl of oil and is planned as a 30-year development and production project.

Plateau production from WA is expected to be 300,000 b/d in addition to the expected plateau production of 340,000 b/d from Central Azeri.

ACG is being developed in several phases. Chirag has been producing since 1997, followed by Phase I of Azeri in early 2005, and the late-2005 start-up of WA. Production from East Azeri is scheduled to come on stream in 2007.

ACG Phase III involves development of deepwater Gunashli field, already sanctioned and expected to begin production in 2008. Plateau production in 2009 is expected to surpass 1.1 million b/d.

Pipelines

BTC pipeline deliveries start

The Baku-Tbilisi-Ceyhan (BTC) crude oil pipeline went into operation in mid-2006. The 760-km (1,087-mile) pipeline is the first to allow crude oil to be exported from the Caspian region without relying on the Russian pipeline system, crude oil swaps with Iran, or passage through the Bosporus and Turkish Straits.

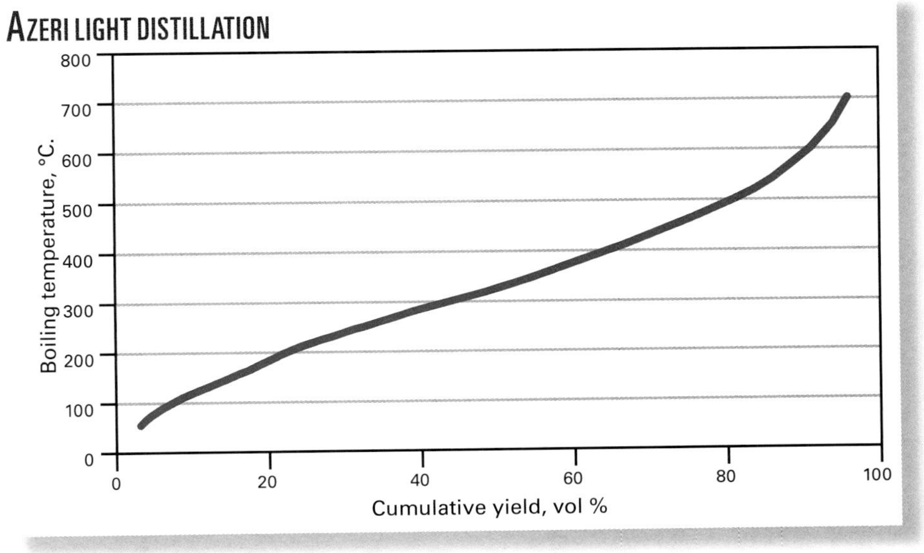

It carries oil from the Azeri-Chirag-Guneshi oil field in the Caspian Sea to the port of Ceyhan on the Mediterranean Sea in southern Turkey.

BTC Pipeline Company (BTC Co.) announced that the first cargo of 600,000 bbl of oil from the pipeline was loaded onto the tanker British Hawthorn at Ceyhan on June 4. The official inauguration ceremony was held July 13.

The pipeline's throughput capacity is 1 million b/d, expected to be reached in 2008.

BTC Co. was created in 2001 to build own and operate the pipeline. BTC Co. shareholders are BP, State Oil Company of Azerbaijan (SOCAR), Unocal, Statoil, Turkiye Petrolleri Anonim Ortakligi (TPAO), Eni, Total, Itochu, Inpex, ConocoPhillips, and Amerada Hess. BP is the pipeline operator.

GEORGIA

Capital: Tbilisi
Monetary unit: Lari
Refining capacity: 0
Oil production: 2,000 b/d
Oil reserves: 35.0 million bbl
Gas reserves: 300 bcf

Exploration and Development

Block 12, Taribani field

Frontera Resources (Houston) drilled three re-entry wells—Dino-2, Well #23, and Niko-1—in Block XII in the Taribani Field with promising results. Located in eastern Georgia, this large, undeveloped oil field covers approximately 80 sq km (31 sq miles) with productive horizons situated in Miocene and Pliocene age reservoirs.

Reservoirs are situated at depths of between 2,200 and 3,500 m (7,218 and 11,483 ft). Independent estimates indicate 118 million bbl of P3 reserves from four of 12 identified horizons within the field.

Frontera also was planning for re-entry of the Dino-2 well in mid-2007, to be followed by new wells drilled into the Zone 9 horizon.

BTC AND SCP PIPELINES

In January 2007, the company also reported that its 3D seismic survey shot over the "C" prospect in the Basin Edge Play Unit, Block 12, confirmed a structure larger than previously mapped. It could hold as much as 500 million bbl of recoverable oil, said Frontera. Cretaceous reservoir objectives were identified at depths from 2,700–3,700 m (8,858–12,140 ft). Exploratory drilling was planned to start in the second quarter.

Kumisi-1 and Manavi-12 wells

CanArgo Energy (Tbilisi) began drilling in February 2007 on its Kumisi-1 exploration/appraisal well in the southern part of Georgia. It will evaluate the West Rustavi-16 Cretaceous gas-condensate discovery made in Soviet times. The well is expected to reach TD of 3,700 m (12,140 ft) in mid-2007.

The prospect is just south of Tbilisi, near the route of the new South Caucasus gas trunkline. The government has committed to purchase any gas discovered at Kumisi.

At year-end, the company reported that it has reached total depth of 5,109 m (16,762 ft) with its Manavi-12 appraisal well in Ninotsminda/Manavi Block. The well drilled into the Cretaceous interval to appraise the oil discovery made by the Manavi M11 exploration well in 2003.

Preliminary log analyses indicate a potentially significant hydrocarbon column. A short initial production test at year-end was to be followed by a more comprehensive production test in early 2007.

Ninotsminda field

CanArgo Energy Corp. said its Ninotsminda Oil Co. Ltd. unit signed a contract to supply gas from Ninotsminda field to state-owned Georgian Gas Transportation Co. Horizontal well N100H2 at the field has tested at 13 MMcfd and 301 b/d of condensate.

Georgian Gas will deliver the gas to the state-owned Gardabani gas-fired thermal power plant and will repair the 40-km (25-mile) pipeline from the field to the plant.

Deliveries of 7 MMscfd initially are to grow as Georgian Oil and CanArgo rework wells in the field.

The ability to sell gas in Georgia gives confidence to move ahead with its program to appraise the potentially large Kumisi gas prospect, CanArgo said.

Licensing

Five blocks offered

The State Agency for Regulation of Oil and Gas Resources accepted applications to participate in the licensing of Blocks V, VI-A, VI-B, VII-B, and VIII. They encompass a total area of some 18,605 sq km (7,183 sq miles). Four of them—V, V, VI-A, VI-B, and VII-B—are in western Georgia; Block VIII is in south-central Georgia along the Turkish border.

Pipelines

BTC pipeline begins operation

The Baku-Tbilisi-Ceyhan (BTC) crude oil pipeline went into operation in mid-2006. The 760-km (1,087-mile) pipeline is the first to allow crude oil to be exported from the Caspian region without relying on the Russian pipeline system, crude oil swaps with Iran, or passage through the Bosporus and Turkish Straits.

It carries oil from the Azeri-Chirag-Guneshi oil field in the Caspian Sea to the port of Ceyhan on the Mediterranean Sea in southern Turkey.

See the Azerbaijan section for additional details.

Azerbaijan gas flows to Georgia

Azerbaijan's Azarigaz began exporting gas in January 2007 along the Haciqabul-Qazax-Tbilisi trunkline to Georgia, which is seeking to reduce its dependence on increasingly expensive Russian supplies.

The State Oil Co. of the Azerbaijani Republic will initially supply 3 million cu m/day of gas (105.9 MMcfd). By the end of January, gas is to come to Georgia from Azerbaijan's Shah Deniz field through the Baku-Tbilisi-Erzurum pipeline.

Georgia is expected to require 1.8 bcm of gas (63.6 bcf) in 2007 and had an agreement with Russia's OAO Gazprom to supply 1.46 bcm (51.5 bcf) of that amount. But in November 2006, Gazprom announced that prices for the Georgian market could more than double in 2007. As Georgia balked at paying the new price, Gazprom threatened to cut off supplies.

Pipeline, storage upgrade

To bolster its energy security, Georgia was reportedly considering plans to repair its main gas lines and to build a gas storage facility in the southern Ninotsminda district initially capable of storing as much as 500 million cu m (17.67 bcf) of gas.

The U.S.-financed Millennium Challenge Corp. will study the feasibility study of the project, which is expected to attract foreign investors.

In November 2006, the local Millennium Challenge Georgia Fund (MCGF) launched its Energy Infrastructure Rehabilitation Project after signing contracts with Georgia's Geoengineering and Azerbaijan's Khazardenizneftegaztikint.

The Azeri firm will work on the northern section of the gas main to make urgent repairs at four sites. The Georgian company will work on two sites on the southern section of the pipeline.

Blasts damage two lines

Supplies of natural gas from Russia to Georgia and Armenia were disrupted in January 2006 when explosions in North Ossetia damaged the main North Caucasus-South Caucasus (NC-SC) pipeline and the Mozdok-Tbilisi (M-T) auxiliary pipeline.

Officials in Georgia blamed the incidents on Russia. The Russian foreign ministry at first labeled the blasts accidental but later announced a criminal investigation, suggesting sabotage by insurgents.

KAZAKHSTAN

Capital: Astana
Monetary unit: Tenge
Refining capacity: 345,093 b/cd
Oil production: 1.060 million b/d
Oil reserves: 30.00 billion bbl
Gas reserves: 100.00 tcf

Background

Kazakhstan has significant oil and gas resources and is expected to become one of the world's top 10 oil producers within the next decade. Oil reserve totals are comparable to Nigeria and Libya, and no other country in Europe or Eurasia, except Russia, has more gas reserves. Significant additional oil and gas discoveries are likely if government policies stabilize and investor interest remain high.

However, researchers at Louisiana State University concluded that the high cost of doing business in the country means that Kazakhstan will need to improve its institutional framework to successfully compete for foreign investment.

High levels of risk and uncertainty continue to plague the oil and gas sector as the government makes significant changes to the petroleum tax legislation and takes an aggressive approach in "rebalancing" contractual arrangements with industry.

High levels of bureaucracy, regulatory burden, and corruption persist, and economic factors appear to be increasingly subordinated to geopolitical objectives aimed to strengthen relationships with China and Russia. The rapid pace of change and the high degree of uncertainty present significant challenges and risk to foreign investment.

Oil and gas in the economy

The oil sector currently accounts for nearly 30% of gross domestic product and 57% of total annual export revenues. Foreign direct investment over the past decade amounted to nearly $30 billion, of which $20 billion was from the United States.

Ownership structure

Because of mergers and acquisitions, the ownership structures of Kazakhstan's oil enterprises have changed several times since the country gained independence from the Soviet Union in 1991.

Seven enterprises account for more than 90% of oil production. Led by the Tengizchevroil Consortium (TCO), the list of foreign firms includes most of the major international companies, including BP, BG, Chevron, ConocoPhillips, Eni-Agip, ExxonMobil, Royal Dutch/Shell, Total; regional and national companies, such as CNPC, Lukoil, Rosneft, Romanian national oil company Petrom, and Turkish National Oil Co.; and a host of smaller companies, from Canada, China, Denmark, Hungary, the United States, the UK, and South Korea.

Oil production, consumption

After the country gained independence, oil production declined, reaching a low of 414,000 b/d in 1995. However, from 1999 to 2004, it grew by 15% per year. In the first half of 2005, production grew at 10%. Government forecasts predict levels of 3.0–3.4 million b/d by 2010.

KAZAKH OIL AND GAS SECTOR, 2001-04

	2001	2002	2003	2004
	%			
GDP growth rate	13.5	9.8	9.2	9.4
Share of oil & gas sector in GDP	24	26	27	29
Share of oil & gas sector in industrial output	39	42	43	43
Share of oil & gas sector in gross exports	—	50	53	57
Share of oil & gas sector FDI in gross FDI	67	51	46	63

Sources: Deloitte, Ernst and Young, Agency of Statistics of the Republic of Kazakhstan

KAZAKHSTAN'S MAJOR OIL PRODUCERS IN 2004

Company (consortium)	Production, million tons	Share, %
Tengizchevroil LLP (TCO)	13.6	23
PetroKazakhstan (PKK)	9.9	17
Kazmunaygaz JSC (KMG)	8.9	15
Karachanganak Petroleum Operating BV (KIO)	8.5	14
Mangistaumunaygas (MMG)	5.3	9
CNPC-Aktobemunaygas JSC (AMG)	4.8	8
Karazhambasmunay (KM)	2.3	4
Other	5.9	10

Kazakhstan Oil Production and Consumption

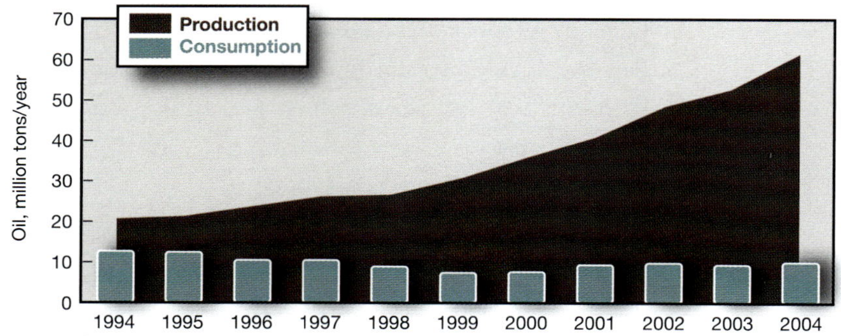

Source: BP PLC

Through the 1990s, demand for oil in Kazakhstan declined before showing a modest gain in 2000. At current consumption levels, significant quantities of oil are available for export. During 2004, Kazakhstan exported 942,000 b/d of oil and condensate, and for the first six months of 2005, exports averaged 1.1 million b/d.

Over the next 15 years, crude oil and liquid production could increase more than threefold, to 3.0–3.5 million b/d, according to an International Monetary Fund study. At that level, Kazakhstan is expected to be one of the world's top 10 producers.

Kazakhstan's Main Oil and Gas Fields

Field	Contract type	Proved + probable reserves, billion bbl	2004 production, million bbl/year
Aktobe	R/T[1]	1.0	34
Emba	R/T	0.7	19
Karachaganak	PSC[2]	2.4	49
Kashagan	PSC	15.0	0
Kumkol	R/T	0.6	54
Tengiz	R/T	9.0	101
Uzen	R/T	1.0	36
Kurmangazy	PSC	[3]5.5-7.7	0
Tub-Karagan	PSC	[3]2.2	0

[1]Royalty/tax system. [2]Production-sharing contract. [3]Kurmangazy and Tub-Karagan estimates represent potential reserves. The first exploratory well drilled at Tub-Karagan was a dry hole.

Gas production, consumption

In 2004, Kazakhstan became a net exporter of gas, helped in part by legislation requiring companies to stop gas flaring. A supply-demand imbalance remains in the country; however, because most gas is located in the west while future demand is expected in the south from Shymkent to Almaty.

Production of natural gas has recently averaged around 14 bcm/year (1.35 bcfd), and the government development plan projects a quadrupling of gas production to 60 bcm (2.11 tcf) by 2010.

Kazakhstan Medium Term Petroleum Outlook*

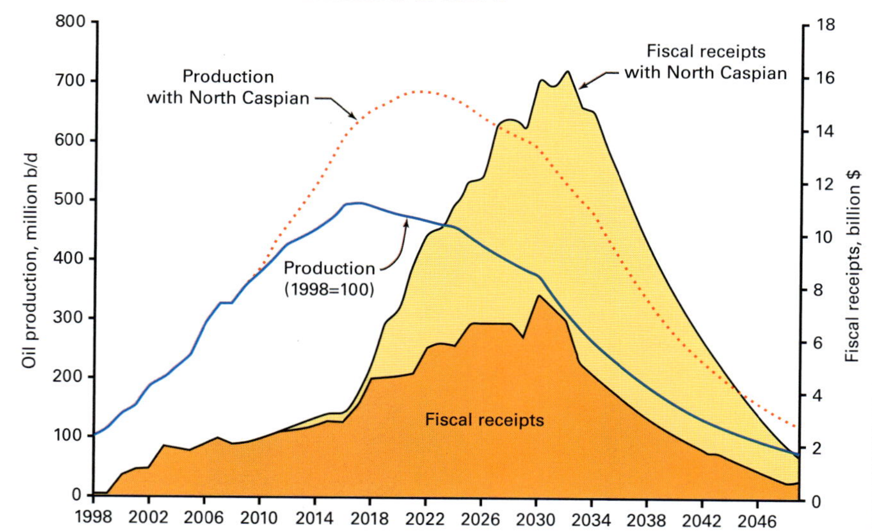

*Solid lines are based on proven crude oil reserves. Production and revenues are likely to be much higher because of expected additional North Caspian fields other than Kashagan (dotted line). Enhanced recovery and field development are likely to extend the curves outward from their peaks.

Source: International Monetary Fund staff projections.

Drilling and Production

Kashagan field production delayed

Production from the Kashagan oil field in the North Caspian Sea has been delayed by environmental problems and technical challenges. Production, planned to start in 2005, has been postponed to 2009. Development is being led by Eni SPA (Rome).

Kazakh state-owned energy company KazMunaiGaz is investigating the project to determine why higher costs and delays have occurred.

Kazakh gas production and consumption

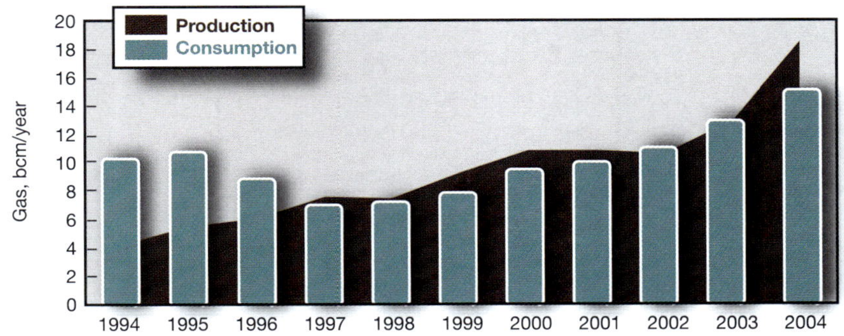

Source: BP PLC

Kashagan has recoverable reserves of 13 billion bbl. Initial production is expected to be 75,000 b/d, climbing to a plateau of 1.2 million b/d in several stages.

Kyzyloi field, Akkulka block

CanArgo (Guernsey, UK) continued development on both the development program for the Kyzyloi gasfield, and on further exploration in the surrounding Akkulka Block.

The company's latest exploration well, AKK02, reached total depth with gas indications, and casing was being run in early 2006.

Further testing also was carried out on the AKK04 exploration discovery, with stabilized production rates in excess of 8.83 MMcfd. Testing of the AKK01 and AKK03 exploration wells was also scheduled to begin in 2006.

On the Kyzyloi field itself, testing was also being carried out on the KYZ109 well, with gas flow to surface. Its location on the extreme western edge of the field may lead to revised estimates of the extent of the field area. Initial Kyzyloi development wells have tested gas at a combined rate of 25 MMcfd.

E Block production

Max Petroleum PLC (London) completed and begun production from the ZMA-2X well, the second of five wells planned on the Zhana Makat A structure on the E Block in the Pre-Caspian basin of Kazakhstan. ZMA-2X encountered an oil column about 18 m (59 ft) thick.

The operator also began drilling the ZMA-4X well, which will be followed by the ZMA-3X and ZMA-5X, all updip on the Zhana Makat A structure. Max said it intends to drill up to 30 shallow wells, six intermediates, and three deep wells by the end of 2007.

Morskoe Block production

Big Sky Energy Corp. (Calgary) began production from its Morskoe Block on the Caspian Sea northeast shore. The company was trucking 5,400 b/d of clean oil from a tank battery that serves several wells to an offloading point in Prorva oil field, 30 km (19 miles) away. The company planned to rework existing wells and drill new wells to exploit deeper reservoirs on the block.

Refining and Processing

Refinery network

Kazakhstan has refineries at Atyrau, Shymkent, and Pavlodar. Oil fields in West Kazakhstan deliver crude to refineries in Russia through the Atyrau-Samara pipeline; refineries in Shymkent and Pavlodar use West Siberian crude shipped through the Omsk-Pavlodar-Shymkent-Chardzhou pipeline.

Kazakhstan's refineries

Source: US Energy Information Administration

After 1991, refinery utilization levels declined significantly and remain at depressed levels. Equipment and processes are outdated and inefficient, encumbered with disproportionate tax, payroll, and social burdens.

Fourth Karachaganak train

Karachaganak Petroleum Operating BV (KPO) let an engineering, procurement, construction, and commissioning contract to Petrofac (Aberdeen, Scotland) to add a fourth stabilization and sweetening train at the Karachaganak Processing Complex in northwest Kazakhstan.

The train will process gas and condensate from Karachaganak oil and gas-condensate field. Liquids capacity at the complex will increase to 10.3 million tonnes from 7.7 million tonnes. Installation is scheduled to be completed in 2009.

Oil Refining in Kazakhstan

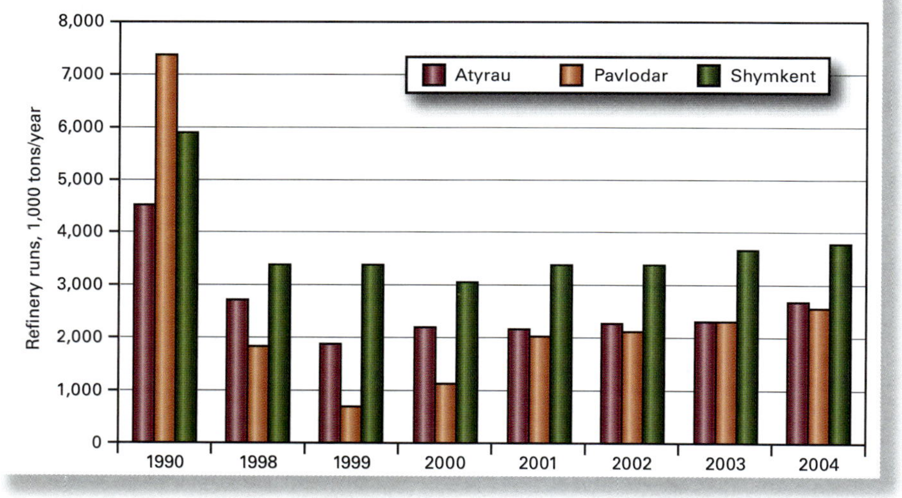

Kazakhstan Refinery Capacity Utilization

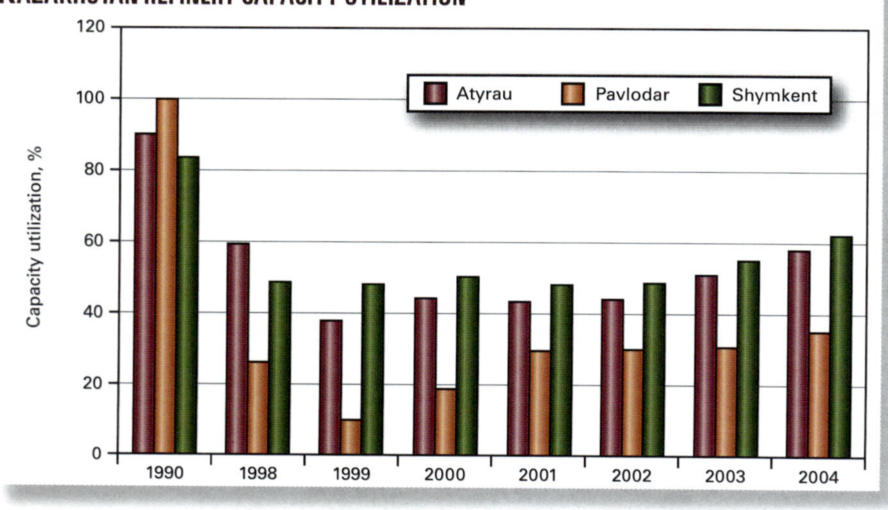

Pipelines

Kazakh oil through BTC line

Kazakhstan signed an agreement with Azerbaijan that will allow transport of Kazakh oil through the Baku-Tbilisi-Ceyhan pipeline. The pact gives Kazakhstan an export route to the Atlantic Basin that does not depend on Russia.

Kazakhstan will supply as much as 25 mtpy oil for the 1,800-km (1118-mile) line, which passes through Azerbaijan, Georgia, and Turkey. Kazakh oil will be shipped by tanker from the northern Caspian port of Aktau to Baku. The project envisions eventual construction of subsea pipelines between Kazakhstan and Azerbaijan.

Kazakhstan oil output is expected to reach 3 million b/d by 2015, largely on the strength of Kashagan offshore field, which is projected to start commercial production by 2009.

CPC line capacity to be doubled

Kazakh Pres. Nursultan Nazarbayev said in April that his country and Russia will increase the capacity of the Caspian Pipeline Consortium (CPC) pipeline from 28 mtpy of crude oil to 67 mtpy, but did not specify when the increase would take place.

CPC's 1,500-km (932-mile) Tengiz-Novorossiisk pipeline carries crude oil from fields in western Kazakhstan to Russia's Black Sea coast for onward transport by ship and pipeline. In 2005, the CPC increased oil exports by 35.5% to more than 30 million tonnes.

Kazakh pipeline infrastructure

Current infrastructure delivers Kazakhstan oil westward to the Black Sea by pipelines and rail, northward to Russia by pipeline and rail, southward through the Persian Gulf through swaps with Iran, and eastward through China by rail.

Pipeline to China

An oil pipeline to China, linking the Kumkol fields in Central Kazakhstan to Western China, is currently near completion. The first section, completed in 2003, crosses western Kazakhstan from the oil fields of the Aktobe region to the oil hub of Atyrau.

In December 2005, Kazakhstan inaugurated the second stage, the 988-km (614-mile) Atasu-Alaskhankou line. The first delivery at Atataw Pass was made in April 2006. The section initially is transporting up to 210,000 b/d of crude oil.

The third stage of work will complete the Kenkiyak-Kumkol pipeline in central Kazakhstan. Once the Kenkiyak-Kumkol-Atasu pipeline is in operation, officials said, the plan is to add crude produced in Kazakhstan's far western regions, including the offshore Kashagan field.

Other export lines

Kazakhstan has three major export pipelines—Tengiz-Novorossiisk (Caspian Pipeline Consortium), Atyrau-Samara, and Kenkiyak-Orsk—with a total capacity (including railways) estimated at 1.5 million b/d. The CPC pipeline is Kazakhstan's principal export route to world oil markets.

The Atyrau-Samara pipeline has recently been upgraded to increase capacity to 30 million metric tons/year. Kazakhstan also signed an agreement with Russia guaranteeing transport quotas for the next 15 years. The relative significance of the link is likely to diminish as CPC realizes full capacity.

Kazakhstan also recently signed an agreement with Azerbaijan regarding use of the Baku-Tbilisi-Ceyhan (BTC) pipeline. The pact will allow transport of Kazakh oil that comes by tanker from the northern Caspian port of Aktau to Baku to be transported through the BTC.

The CPC pipeline links supergiant Tengiz oil field with the Novorossiisk terminal on the Russian Black Sea coast. Its capacity is 28 million metric tons/year. From early 2005 to mid-2006, CPC transported over 2 million metric tons a month of Kazakh oil—around 530,000 b/d—representing nearly half of the total current oil exports and virtually all of the light oil and condensate production in Kazakhstan.

The CPC pipeline is the first and only privately operated pipeline to be built in the former Soviet Union. The shareholder structure is complex and includes the governments of Russia, Kazakhstan and Oman as well as eight corporations. CPC owns and operates the pipeline.

The corporate shareholders have initiated the engineering design on a $1.6 billion build-out program to achieve CPC's full capacity of 67 million tons a year. This would be realized in stages—10 million tons capacity addition by 2007, another 10 million tons in 2008, and the final build-out in 2009.

Caspian region oil pipelines

Source: Updated from US Energy Information Administration

North Caspian oil and gas infrastructure

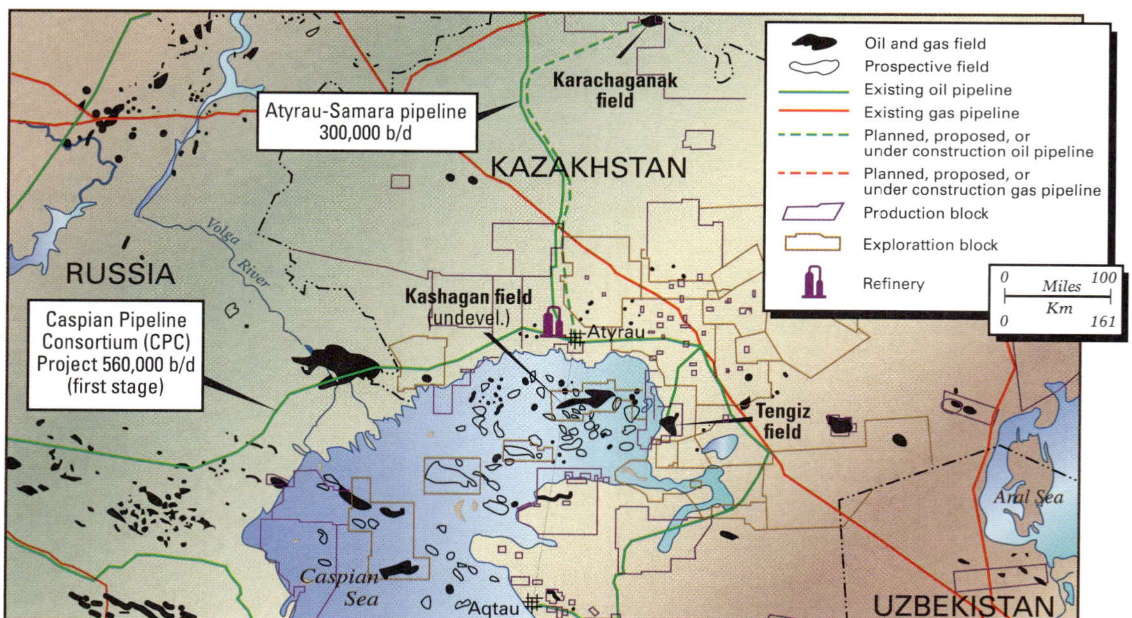

Source: US Energy Information Administration

Kazakh oil fields: Three giants

The majority of oil production growth in Kazakhstan is expected to come from four fields: Tengiz, Karachaganak, Kashagan, and Kurmangazy.

Karachaganak, Tengiz, and Kashagan account for the majority of foreign direct investment in the country. Tengiz and Karachaganak are producing; Kashagan is expected to start up in 2008.

Kurmangazy is considered highly prospective but is still in the exploration phase. Kurmangazy is reported to have potential reserves of 5.5–7.7 billion bbl, but this figure is considered speculative.

Tengiz field

The Tengiz field, located along the northeast shore of the Caspian Sea and is one of the largest and deepest oil fields in the world. It lies three miles deep underneath a half-mile thick salt dome. Its recoverable reserves are estimated at 6–9 billion bbl.

For the first half of 2005, the field produced 267,000 b/d oil and condensate. Tengiz production is transported westward to the Black Sea port of Novorosiisk through the CPC pipeline.

A three-year, $4 billion investment program for gas reinjection is expected to double production to 500,000 b/d by 2007, and according to company estimates, Tengiz could potentially produce 700,000 b/d by 2010. Total investment in Tengiz over 40 years is expected to reach $20 billion.

Karachaganak field

Karachaganak oil and gas-condensate field is located in northern Kazakhstan near the border with Russia's Orenburg field. Considered one of the world's largest gas-condensate fields, its production averaged 230,000 b/d during the first half of 2005. Peak production is expected to increase to 500,000 b/d by 2010 under a significant investment program.

After a link with CPC was completed in 2003, Karachaganak's bargaining position and ability to reach world markets improved significantly.

Kashagan field

Kashagan field is located off the northern shore of the Caspian Sea. Comparable in size to Prudhoe Bay, Alaska, with geologic and hydrocarbon characteristics similar to Tengiz, the field is estimated to have recoverable reserves of 7–9 billion boe, with potentially 10–13 billion using secondary recovery methods.

Kashagan's first production is expected in 2008 at 75,000 b/d, increasing to 450,000 b/d by 2012 and 900,000 b/d by 2013. A plateau of 1.2 million b/d is expected by 2016.

Kazakhstan's three largest fields

Ownership	Tengiz	Karachaganak	Kashagan
		%	
BG		32.5	
Chevron	[1]50	20	
ConocoPhillips			9.26
Eni/Agip		[1]32.5	[1]18.52
ExxonMobil	25		18.52
Inpex			8.33
LukArco	5		
Lukoil		15	
KMG	20		8.33
Royal/Dutch Shell			18.52
Total			**18.52**
2005 production			
Oil + condensate, b/d	267,000	230,000	[2]75,000
Gas, MMcfd	—	530	—
Recoverable reserves			
Oil + condensate, billion bbl	6-9	2.4-6	[3]7-9
Gas, tcf	>10	16-46	—
Investment to date, $ billion	**10**	**4**	**5**
Expected investment, $ billion	**23**	**8**	**29**

[1]Denotes operator or co-operator. [2]First production is expected to begin in 2008. [3]With secondary recovery techniques, the top of the range of recoverable reserves is estimated to be 13 billion bbl.

KYRGYZSTAN

Capital: Bishkek
Monetary unit: Som
Refining capacity: 10,000 b/cd
Oil production: 1,000 b/d
Oil reserves: 40.0 million bbl
Gas reserves: 200 bcf

Blocks 1 and 2

Groundstar Resources Ltd. (Calgary) signed a memorandum of understanding to negotiate exploration and production sharing and joint venture agreements with Kyrgyzneftegaz. The agreements relate to operation of an unspecified producing oil field and exploration Blocks 1 and 2 in the Fergana basin.

Block 1 covers 120 sq km (46 sq miles) with Jurassic objectives at 2,000 m (6,560 ft). Block 2 covers 370 sq km (143 sq miles) in the Kugart depression and has objectives in multiple zones to 4,000 m (13,125 ft).

Seismic survey on NG-72-00 license

Cambrian Oil & Gas PLC (London) was acquiring 100 line-km of 2D seismic data on the Tash Kumyr NG-72-00 license. The data will be used to highgrade areas that have positive geochemical anomalies.

Cambrian let a contract to JSC Kyrgyzgeophysica to conduct the survey. The company was aiming to define up to three locations for drilling in mid-2006.

Cambrian was pursuing three Kyrgyz initiatives: the Beshkent-Togap field water injection project; the Tash Kumyr exploration project; and an exploration license area with five prospects and a recently acquired intermountain basin exploration license area adjacent to Lake Toktogul.

Bashkent-Togap oil field has more than 70 producing wells. It has produced 4.1 million bbl of 65–76 million bbl in place, and potential exists to recover 6 million bbl, said Cambrian.

Gazprom to invest in oil, gas exploration

Gazprom executives met with Kyrgyzstan officials in early 2006, and the Russian company agreed to invest hundreds of millions of dollars in the exploration and development of Kyrgyzstan's oil and gas reserves. Plans were made to set up a joint Gazprom-Kyrgyz group to undertake projects of mutual interest.

LITHUANIA

Capital: Vilnius
Monetary unit: Litas
Refining capacity: 190,000 b/cd
Oil production: 3,600 b/d
Oil reserves: 12.0 million bbl
Gas reserves: 0

Exploration in Baltic Sea

Grupa Lotos S.A. (Warsaw, Poland) applied for licenses to explore and develop oil reserves offshore Lithuania and Latvia.

The exploration activity on the Baltic shelf in Lithuanian and Latvian waters is part of the company's 2006–12 development strategy. Officials said that the oil consortium plans to invest more than $150 million (U.S.) to increase annual oil production from 1.6 million bbl in 2005, to approximately 7.1 million bbl in 2012.

Antanavas-1 wildcat

Geonafta (Gargdzdai, Lithuania) began to drill its Antanavas-1 wildcat exploration well offshore in the Plunge Block. The company chose the Antanavas-1 drillsite, located in the Baltic Depression Priekule-Telsiai Terrace, on the basis of its 150 km 2D seismic survey, shot over the most prospective section of the 1,440 sq km (556 sq-mile) block.

Antanavas-1 is 18 km (11 miles) east of the Plunge Field. It is being drilled to a total measured depth of 1,880 meters (6,168 ft) or 1,700 meters (5,678 ft) total vertical depth, targeting Middle Cambrian Deimena sandstones.

Mazeikiai refinery upgrades

AB Mazeikiu Nafta (Mazeikiai District, Lithuania) let a contract to Foster Wheeler Italiana SPA (Milan) for modernization of the Mazeikiai refinery in the northwest part of the country. The only refinery in the Baltic states, it is being upgraded to meet the European Union's "Auto-Oils V" specifications for the sulfur content of gasoline required by 2009.

Foster Wheeler will perform basic design, front-end engineering, detailed design, and procurement services. The project will boost capacity of the fluid catalytic cracking unit to 60,000 b/sd from 45,000 b/sd and revamp several other facilities. Completion was expected by year-end.

In a separate program at the same refinery, Mazeikiu Nafta contracted with Haldor Topsoe to revamp a 41,000 b/sd hydrotreating unit. The hydrotreater will be converted by early 2007 into an ultra-low sulfur diesel unit, which will produce diesel containing less than 8 ppm sulfur.

RUSSIA

Capital: Moscow
Monetary unit: Ruble
Refining capacity: 5,340,961 b/cd
Oil production: 9.475 million b/d
Oil reserves: 60.0 billion bbl
Gas reserves: 1,680 tcf

Exploration and Development

Oil giant discovery in Caspian

OAO Lukoil announced in early 2006 its discovery of a multipay oil and gas-condensate field on the Severny license in the Caspian Sea. The company named it the Filanovsky field and described it as Russia's largest discovery in a decade.

The discovery well on the Yuzhno-Rakushechnaya structure flowed at the rate of 5,840 b/d of light, low-sulfur oil. The field, 220 km (137 miles) south-southwest of Astrakhan, is said to contain probable and possible recoverable volumes of 600 million bbl of oil and 1.2 tcf of gas.

Lukoil said 75% of the field's reserves are oil, whereas most other fields it has discovered in the Caspian region are predominantly gas. Early data indicate that the new field should produce 100,000 b/d.

Filanovsky is just northwest of Lukoil's Yu. Korchagin field, which is under development.

Gazprom spending plans

OAO Gazprom said in early 2007 that its board had approved the company expending $19.95 billion (U.S.) in 2007 as capital investments and long-term financial investments.

Investment is targeted to several gas production projects. These include building infrastructure for the Kharvutinskaya area of the Yamburgskoye field, the Bovanenkovskoye and Kharasaveyskoye fields; developing the Yuzhno-Russkoye field; and developing the Shtokmanovskoye and Prirazlomnoye fields

Facility construction will continue on the Yen-Yakhinskoye, Urengoyskoye, Zapolyarnoye, and other fields.

Major gas transmission projects will include the Northern Tyumen Oblast-Torzhok gas pipeline, expanding the Urengoy gas transmission hub and the northwestern Russia gas transmission network.

Capital investments will be made for geological exploration and production drilling.

Long-term financial investments for 2007 include financing upstream projects in the Tajikistan and Kyrgyzstan, Block 112 of offshore Vietnam, India's Bengal Bay, and the Ustyurt region of the Uzbekistan. Long-term loans also are planned for Gazprom subsidiaries operating on the Yuzhno-Russkoye field and in the Krasnoyarsk Krai, as well as in Venezuela, and the Caspian Sea offshore.

Lukoil's Severny record

Reporting on exploratory drilling through 2004 in the Northern Caspian region, Lukoil described finding commercial hydrocarbon flow from multiple pays at five structures.

The five are Khvalynskaya, '170-km', Shirotnaya, Sarmatskaya, and Rakushechnaya. Eight wells had been drilled on the five structures, finding oil and gas, gas, and gas-condensate.

The company gauged the largest gas flows, in excess of 35 MMcfd, at Khvalynskoye and Sarmatskoye fields. It tested the largest oil flows, 2,750 b/d, at '170-km' field and Yu. Korchagin.

Lukoil hopes to begin production as early as 2008 from the region.

Offshore Lagansky block

Lundin Petroleum AB (Stockholm) delayed work at the Morskoye prospect on the Lagansky block off Russia in the Caspian Sea. A two-well drilling program is now scheduled to start during second quarter 2007, targeting three main Cretaceous and Jurassic reservoirs at 800–1,600 m (500–1,000 ft).

Lundin bought Valkyries Petroleum Corp. in 2006, a company with exploration and producing interests in Russia and a 70% stake in Lagansky block. The 2,000 sq-km (772 sq-mile) block lies between shore and Lukoil's Filanovsky discovery.

North Irael block

The LLC NK Recher-Komi joint venture reported a modest oil discovery on the North Irael block in the Pechora basin in Russia's Komi region.

The No. 61 exploration well recovered 59.3 bbl of clean oil from 5 m of net pay in the Devonian interval at 1,625 m subsea, said Arawak Energy Corp. (Anguilla, British West Indies). Arawak and Lundin Petroleum AB (Stockholm) hold interests in Recher-Komi.

The block is producing 1,600 b/d of oil from five of six wells drilled earlier.

Snezhnaya field

Imperial Energy Corp. PLC (London) tested a Lower Cretaceous new pool light oil discovery in Snezhnaya field on Block 77 in the Tomsk Region of Western Siberia.

Snezhnaya 139, a planned Jurassic appraisal/production well, flowed 41° gravity oil to surface with no water and little gas from 5 m of net pay in the Lower Cretaceous at 2,713–2,715 m.

Imperial earlier reported a Lower Cretaceous discovery at Aikagalskaya field on Block 69 about 180 km (112 miles) from Snezhnaya field.

Production from the Snezhnaya Jurassic was expected to be more than 2,500 b/d by mid-2006.

Eastern Siberia oil field development

Urals Energy Public Co. Ltd. (Nicosia, Cypress) assumed financial and operating control of Dulisminskoye oil and gas-condensate field in the East Siberian basin in April and described a $26 million development budget for 2006.

Estimates of proved and probable reserves total 109 million bbl of 40° gravity oil and 1.9 tcf of gas in Lower Cambrian Yaraktin reservoirs at 1,950 ft and 2,000 ft. The goal is to boost output to 12,000 b/d of oil by the end of 2008 from less than 1,000 b/d in mid-2006.

Full development is believed to require 44 more wells. Intended completion date is 2009.

Gabyshevskoye field

Lukoil-Perm, the operating arm of Lukoil Overseas Holding Ltd., began developing Gabyshevskoye oil field in the Oktyabrsky district in the south of the Perm region of Russia. It didn't disclose drilling plans.

A 1997–1998 seismic survey identified three structural highs designated Gabyshevskoe, North-Gabyshevskoe, and Malo-Tartinskoe. The field's recoverable oil reserves are estimated at 187,000 tonnes. The field was producing 12.3 tonnes/day per well in mid-2006 from three wells. Peak production of 20,000 tpy is expected in 2008.

Oil found in Perm region

Lukoil-Perm discovered oil with an exploration well drilled to 1,800 m (1,120 ft) in the Oktyabrsky and Chernushinsky districts in the southern Perm region. It encountered oil in Verean, Lower and Mid-Visean, and Upper Devonian-Tournaisian age sands of a field it has designated Sharonovskoe.

Lukoil-Perm plans to bring the field on line in 2009. Peak production of 400 b/d is expected in 2012.

Ust-Vakh, Kovykta work

Halliburton International Inc. (Russia) won a $33 million contract from TNK-BP to provide drilling and formation-evaluation services for the Ust-Vakh field development in Western Siberia. Halliburton began work in the Ust-Vakh field in 2005, providing services on eight rigs and drilling a total of 85 wells.

Two TNK-BP subsidiaries let contracts worth a total of more than $60 million to Petrofac Ltd. (Woking, UK) for management services in development of giant Kovykta gas-condensate field and related pipeline construction in eastern Siberia.

At Kovykta, Petrofac will provide project management services for construction of gas production facilities and a condensate-loading system with a capacity of 2.5 bcm/year (0.24 bcfd). Petrofac also will provide project management services for building a 645-km (400-mile) pipeline from Kovykta processing facilities to Irkutsk. Pipeline gas deliveries were expected to start by year-end 2006.

Kovykta, one of Russia's largest gas fields, is 450 km 9 miles) north of Irkutsk, in the Zhigalovo and Kazachinsko-Lensk Districts. Reserves are 2 trillion cu m of gas and more than 83 million tons of condensate.

Sakhalin Island developments

Sakhalin-1

Export system commissioned

An international consortium led by Exxon Neftegas Ltd., a subsidiary of ExxonMobil Corp., announced in September the commissioning of the oil export system for the Sakhalin-1 project.

The multiphase project on Russia's northeast Pacific coast includes a series of complex extended-reach wells, including a land-based rig that can reach more than five miles offshore.

Interest holders in Sakhalin-1 are ExxonMobil Corp., Japan's Sakhalin Oil & Gas Development Co., OAO Rosneft, and India's state-owned Oil & Natural Gas Corporation OVL.

The first oil for Sakhalin-1 began flowing into the export system in August, and the first tanker was due to begin loading at the newly constructed DeKastri terminal in Khabarovsk Territory late in September.

However, Russia's Federal Service for Ecological, Technological, and Nuclear Monitoring announced that oil shipment for export would likely be delayed pending resolution of safety-related objections regarding construction of the terminal.

ExxonMobil estimated that production would ramp up to an estimated peak rate of 250,000 b/d by year-end 2006, following completion of the offshore processing facility.

A 225-km (140-mile) pipeline transports crude—produced from wells in the Sea of Okhotsk and processed onshore at Chayvo—west across Sakhalin Island and the Tatar Strait to the DeKastri terminal in the Khabarovsk Krai in the Russian Far East.

The Sakhalin-1 project began operations in October 2005 and has been producing as much as 50,000 b/d of oil, along with natural gas production of some 60 MMcfd.

The consortium has been selling its oil in the Russian Far East and marketing its gas to two domestic customers in the Khabarovsk Krai. Sakhalin-1's proximity to Japan, the second-largest net oil importer in the world, means that it is likely to be a primary customer.

The ExxonMobil unit serves as project operator. Partners are Japan's Sakhalin Oil & Gas Development Co. Ltd., Russia's RN-Astra and Sakhalinmorneftegas-Shelf, and India's ONGC Videsh Ltd.

Pre-accord signed for gas

ExxonMobil Corp. signed a preliminary agreement on principles with China National Petroleum Corp. (CNPC) for the supply of natural gas from the Sakhalin-1 project. Delivery will be to northeastern China, but quantities, pricing, and schedule were not stated.

Earlier, ExxonMobil said it is ready to deliver 8 billion cu m/year of gas to China. Currently, gas is delivered in amounts of up to 1 bcm to Russia's Khabarovsk territory.

The Orlan platform, operated by Parker Drilling crews, was used to drill extended-reach offshore targets for the Sakhalin-1 Consortium. (Photo courtesy of Exxon Neftegas Ltd.)

Sakhalin-2

Phase 1 status

Development of hydrocarbon resources offshore Sakhalin Island is proceeding is a series of follow-on phases. In Phase 1 of the Sakhalin-2 project, a consortium has been locating and producing oil and gas since 1999 on the northeastern part of Sakhalin Island and immediately offshore, in the Sea of Okhotsk, from the Piltun-Astokhskoye oil field and the Lunskoye Piltun-Astokhskoye gas field.

Original consortium members were Royal Dutch Shell PLC, Mitsui & Co. Ltd., and Mitsubishi Corp.

The project marks Russia's first production sharing agreement (PSA), signed in 1994 with Sakhalin Energy Investment Company Ltd. (SEIC). Phase 1 has been producing about 80,000 b/d of oil during summer, moving the product to Japan in tanker ships.

Phase 2 status

Phase 2 envisions construction of two offshore platforms, a gas liquefaction plant to convert the gas to 9.6 million tonnes/year of LNG, a gas condensate refinery, two lines of an 800-km (500-mile) pipeline, and an oil offloading terminal.

The LNG plant would be built at Prigorodnoye on Aniva Bay in the south. It would be Russia's first. Each of two trains would have a capacity of 4.8 mtpy. First shipments were scheduled to start in 2008 to Japanese utility companies.

However, during 2006, several issues seriously hindered progress. Most notable were a doubling in estimated project cost to $22 billion, plus a series of actions by Russia that threatened to shut down the project because of alleged environmental violations related to pipeline construction.

Negotiations led the consortium to sell to Gazprom a 50% stake plus one share interest in the project for $7.45 billion, reducing the interests of the other three participants. The four parties signed a protocol to formalize the offer in December. Final resolution was expected by first quarter 2007.

Many observers called Russia's action a violation of the rule of law and the sanctity of contracts that could cloud other joint projects in Russia, such as development of the Kovykta gas field.

In early 2007, the European Bank for Reconstruction & Development (EBRD) announced that it would not offer financial support for Sakhalin-2 because of the pending change in majority control of SEIC. EBRD was considering offering a $300 million syndicated loan to SEIC.

Sakhalin-3

Sakhalin-3 will develop the Ayashsky, Kirinsky and Odoptinsky blocks on the shelf of Sakhalin, with combined estimated reserves 4.6 billion barrels (620 million tons) of oil and 770 bcm (27.2 tcf) of gas. Russia expects to auction the Sakhalin-3 oil and natural gas project in 2007 after new rules that limit foreign investment are approved.

The winner of the tender, expected in 2007 after the new rules are implemented, would have a minority stake in a deal most likely controlled by Gazprom or Rosneft.

In late 2006, ONGC Videsh Ltd. (OVL; India)—overseas subsidiary of state-owned Oil & Natural Gas Corporation—began discussions with Gazprom and Rosneft for compiling a joint bid to buy into Sakhalin-3. OVL holds an interest in Sakhalin-1.

Sakhalin-4 and Sakhalin-5

License blocks

Phases 4 and 5 of Sakhalin development involve two offshore blocks: East-Schmidt (11,800 sq km; 4,555 sq miles) and West-Schmidt (9,400 sq km; 3,630 sq miles). Both are at the north end of the island, with harsh climate, complex ice conditions, and water depths of 100–200 m (62–124 ft).

More than 20 prospective structures have been identified and considerable exploratory seismic work has been conducted. Joint investment to date exceeds $80 million.

Rosneft and BP signed shareholder and operating agreements in late 2006 for joint activity on the two blocks. Planned exploration activities include drilling of six wells. Spending through 2011 is projected to total $700 million.

Drilling and Production

Offshore development trends

Offshore production in Russia is expected to grow significantly in the future, according to analysts with PFC Energy (Houston). By 2020, offshore fields could account for 20% of the country's total oil and gas output. And two-thirds of future offshore production is expected to come from the Barents Sea and Kara Sea.

Russia is moving to revise regulations and tax policy that will affect offshore development. The once-expected shift to holding auctions instead of tenders now appears unlikely. Production sharing agreements (PSAs) like those for Sakhalin Island will become rare.

The involvement of international oil companies and service companies will be essential for development off Russia because Russian companies lack extensive offshore experience. However, foreign companies operating in Russia likely will be prevented from taking more than 49% interest in projects.

Tomsk oil production begins

Imperial Energy Corp. PLC (London) began production from wells Snezhnaya 135 and Dvoinoye 2 on Block 77 in the Tomsk region of western Siberia. Tomsk's main reservoir is in the deep Jurassic.

The company also expected to bring four new wells in Snezhnaya field on stream, increasing Block 77 production to 1,500–2,000 b/d of oil.

YUKOS DECLARED BANKRUPT

In August, a Moscow arbitration court declared OAO Yukos bankrupt, the latest development in the former oil giant's three-year court battle with the Russia government over back taxes.

Russian tax authorities and state-owned Rosneft are the biggest creditors of Yukos. Liquidation proceedings are expected to result in an auction of its remaining assets. Rosneft already bought Yukos unit Yuganskneftegas. Gazprom has said it is interested in acquiring some assets, particularly the Tomskneft production subsidiary.

Former Yukos chief executive Mikhail Khodorkovsky, was jailed in May 2005 after being convicted of fraud and tax evasion. He remains in a prison camp at Krasnokamensk, Chita, near the border with China.

In early April, the Yukos vice-president tasked with saving the company from bankruptcy was arrested on charges of misappropriation of funds and money-laundering. Vasily Aleksanian was held on charges of embezzlement and money laundering.

Pipelines

East Siberia Pacific Oil Pipeline

Overview

The 4,188-km (2,600-mile) East Siberia Pacific Oil Pipeline (ESPO) is designed to deliver as much as 80 mtpy of oil (1.6 million b/d) from Russia to eastern markets. Estimates of construction cost range from $6 billion to $12 billion.

In the first construction phase, now under way, a 2,400-km oil pipeline will be built from Taishet in Eastern Siberia to Skovorodino—the midway point—near the Chinese border.

This phase also includes construction of a rail oil terminal at the line's eastern end—either Perevoznaya Bay or Nakhodka, on Russia's Pacific coast.

The second phase, depending on the development of Eastern Siberian oil fields, would build the further link between Skovorodino and either Perevoznaya Bay or Nakhodka on the Russian Pacific coast. Russia has not yet decided when construction of the eastern half of ESPO would begin.

China is also pushing for a pipeline spur from Skovorodino to Daqing that would carry as much as 30 mtpy of crude oil to that country.

First section completed

OAO Transneft built about 530 km (330 miles) of the planned western segment of the ESPO in 2006 and plans to lay about 1,250 km (775 miles) of the pipeline in 2007.

Plans call for completion of the Taishet-Skovorodino section by 2008. From Skovorodino, oil would be sent on to China by rail or a branch line of the pipeline.

In a related development, Transneft authorized Urals Energy Public Co. Ltd.'s Dulisma oil field to be connected to the ESPO line. The field, 350 km (217 miles) northeast of Ust-Kut, is expected to produce 12,000 b/d by year-end 2008, with output rising to 30,000 b/d by 2011.

Japanese participation

Japan and Russia continued to discuss a proposal under which Japanese firms would join Russian development of an East Siberian oil field in connection with the ESPO project. No specific Siberian oil field was selected, but several Japanese firms expressed interest in investing in the project.

The Japanese view an agreement on field development as a way to speed construction of the eastern half of ESPO so that Siberian oil could be shipped more quickly to Japan. Russia had not yet decided when to start building the eastern half of the line.

Route changes

Under Transneft's original plan, approved by regulatory authorities, ESPO would have run along the shoreline of Lake Baikal for about 100 km.

However, in April, Russian Pres. Vladimir Putin directed that the route be shifted more than 40 km (25 miles) to the north to avoid potential ecological damage to the lake in the event of pipeline damage. Trasneft said the shift could raise costs by as much as $1 billion.

Several entities along the proposed route continued to push for inclusion of pipeline branches. For example, Khabarovsk Territory officials say that branch lines to refineries in Khabarovsk and Komsomolsk-on-Amur would cut dependence on railroads for oil shipments and would triple output.

Nord Stream Pipeline

Construction began in late 2006 on an onshore portion of the proposed 1,200-km (745-mile) Nord Stream pipeline, formerly called the North European Gas Pipeline. The line will run under the Baltic Sea from Russia to Germany.

The 917-km (570-mile) onshore section will run from Babayevo, which is 800 km (500 miles) east of St. Petersburg, to the coast. It will connect existing pipelines from Siberian gas fields to Nord Stream. Original plans also called for an LNG liquefaction plant near St. Petersburg.

The Nord Stream consortium (Zug, Switzerland) is comprised of Gazprom and subsidiaries of Germany's BASF AG and E.ON AG.

Pipeline operation is scheduled to begin in 2010 with an initial capacity of 27.5 bcm/year (2.66 bcfd). Capacity could be doubled if a parallel second line is built in Phase 2. Total capital spending for a two-line offshore section would exceed €4 billion, to be financed in full by Gazprom.

Druzhba spur

A leak in July closed one of the two mid-sized spurs of Russia's Druzhba oil pipeline, Russia's main oil export line to Europe. There were conflicting reports on pipeline damage and on the size of the spill. The pipeline normally pumps 1.2 MM b/d of crude to Europe. The damaged spur had supplied 250,000 b/d to Lithuania's Mazeikiu refinery.

Operator Transneft said long-term closure of the branch might be needed to conduct much-needed repairs. The company said the Druzhba-1 line is 42 years old and made of metals that are now outlawed.

LNG

First LNG delivery to UK

Gazprom delivered its first shipment of LNG in April to the UK—140,000 cu m purchased from Gaz de France and delivered to BP PLC. It was the largest LNG shipment to reach the UK's Isle of Grain terminal in north Kent, which can receive and process up to 3.3 mtpy of LNG.

LNG from the Isle of Grain terminal is expected to provide 5% of UK supply needs.

Yamal Peninsula project

Repsol YPF SA announced plans to form a joint venture with Tambeineftegaz and Anadarko Petroleum Corp. for an integrated LNG project on Russia's Yamal Peninsula.

The project would develop South Tambey gas field and construct an LNG plant. The LNG would be transported to Repsol YPF's Canaport facility (St. John, NB, Canada) for regasification and eventual sale to the U.S. East Coast markets.

Gazprom plans

Gazprom plans to become a major LNG supplier and is building its LNG expertise, particularly in North American markets, said an official with Gazprom Marketing & Trading Ltd. (London).

The company intends to participate not just in gas production but in all elements of the value chain, including liquefaction, shipping, and regasification.

Gazprom is expected to acquire in 2007 a controlling interest in the Sakhalin-2 project, which involves liquefaction of produced gas at a new facility.

Another project of interest would liquefy gas from the Shtokman gas and condensate field in the Barents Sea. Plans call for startup of the first phase in 2012, producing 15 mtpy of LNG.

A third project is Baltic LNG. A plant near St. Petersburg would liquefy 700 MMcfd of gas that Gazprom transports through the Unified Gas System. LNG output would be 3–5 mtpy, forecast to start in 2010. The primary market would be North America.

Policy Issues

Russian-Ukraine dispute on gas prices

In early January 2006, Ukraine and Gazprom locked horns in a dispute about natural gas prices and transit fees that culminated in Gazprom cutting supplies to Ukraine for several days and led to natural gas shortfalls in many European countries. Details are presented in the Ukraine section of this document.

Energy demand security

Pres. Putin raised doubts that he will ratify the European Energy Charter on cooperation with Europe, which Russia entered in 1996. The charter would broaden access to Russian energy supplies and transportation.

In November he called for changes to the charter involving access to gas and oil pipelines. Pres. Putin said he wants Russia to become integral to the European gas market and advocated interdependence between his country and the EU to secure "demand security." However, his request aggravated concerns that Russia is reverting to Soviet-style use of energy to influence international politics.

Gas-sector reform urged

Massive volumes of natural gas are wasted annually in Russia's gas sector, and there is a lack of investment in developing new fields to offset depletion of key producing fields.

These factors "will begin to affect Russia's position as a secure and reliable supplier" unless structural and regulatory reforms are introduced swiftly, said a study by the International Energy Agency (Paris).

The 200-page study, released at the close of the G8 Summit in July, is available on the IEA Website: http://www.iea.org. The title is "Optimizing Russian Natural Gas: Reform and Climate Policy."

Oil exports to Asia-Pacific

Pres. Putin said that at least 30% of Russia's oil exports will be heading to Asia and the Pacific in the next decade as part of a strategy of diversified marketing aimed at creating energy security for producing countries. Current export levels to those regions total about 3%.

Limiting foreign investment

Russia has been consolidating state control over what it terms strategic resources. A pending bill would increase the field size considered "strategic" to 70 million tons of crude (about 513 million barrels) and 50 bcm (1.77 tcf) of gas. Foreign investors would be limited in how much of a stake they could hold in such deposits.

Shtokman Field

Overview

The Russian part of the Barents Sea is estimated to contain about 22 billion boe. PFC Energy (Houston) said that by 2020, approximately 20% of Russia's hydrocarbons will come from its offshore fields, and two thirds of the future offshore production is expected to come from the Barents Sea and Kara Sea.

The Shtokmanovskoye (Shtokman) gas-condensate field, one of the largest in the world, lies 555 km (345 miles) east of Murmansk in the Barents Sea, in 350 m (217 ft) of water.

The Russian government has estimated reserves at 3.7 tcm of natural gas and more than 31 million tonnes of condensate. The hydrocarbons are in four main reservoirs at depths of 1,900–2,300 m (1,180–1,430 ft).

The massive project poses many technical and logistical challenges and has an estimated cost of $10–20 billion. Gazprom has said it expects to begin production from Shtokman in 2010.

In 2000, the Russian government signed a production-sharing contract with ZAO Sevmorneftegaz, controlled by Gazprom and Rosneft. In 2004–2005, Gazprom signed preliminary contracts with Hydro, Statoil, Total, ConocoPhillips, and Chevron Corp.

No foreign partners

In early October, Gazprom stunned the international energy community when it announced that it would develop Shtokman field without foreign partners and would pipe the gas to Europe rather than convert it to LNG for North American markets as had been planned.

Five companies had competed for what were to have been two minority interests in the project: ConocoPhillips, Total SA, Chevron Corp., Statoil ASA, and Norsk Hydro Oil & Energy. Many of them have been working in the field with Gazprom under short-term contracts for more than 15 years.

Gazprom attributed the decision to economics, saying foreign companies were "unable to provide the capital required." Also cited was the uncertainty of LNG sales in comparison with pipeline deliveries secured by long-term contract.

Some observers said the move also may reflect exasperation over Washington's resistance to Moscow's entry into the World Trade Organization and over the strong U.S. support given to the BTC oil pipeline between Azerbaijan and Turkey by way of Georgia, avoiding Iran and Russia.

SHTOKMAN FIELD

Refining

Tatarstan refinery

CJSC Nizhnekamsk Refinery let a contract to Foster Wheeler Ltd.'s Paris-based subsidiary Foster Wheeler France SA to design a process package involving its proprietary delayed coking technology.

The coker will be part of an integrated refining and petrochemical grassroots complex to be constructed at Nizhnekamsk, in Tatarstan. The upgrade will allow conversion of heavy residue feed into transport fuels.

The planned complex, expected to cost $3 billion, will consist of a refinery with 140,000 b/d distillation capacity, a deep conversion refinery, and a petrochemical plant. Construction in three phases is slated for 2008–10.

Oil Exports

According to an analysis published by the Center for Petroleum Business Studies (Moscow), Russia's ability to export crude oil is limited by available export facilities. The most important oil export projects include further development of the Druzhba and Baltic pipeline systems, as well as new construction of the Murmansk oil hub and Angarsk (East Siberian) pipeline.

Russia's total crude oil exports outside the FSU are forecast to reach 5.9–6.1 million b/d by 2007 and 8.7–8.9 million b/d by 2010.

Russian gas to Europe

Gazprom has been providing gas to Europe for more than 30 years, and European gas demand is expected to grow about 2%/year, according to the International Energy Agency (IEA; Paris).

In 2005, Gazprom sold 156.1 bcm (5.51 tcf) of gas in Europe, more than a quarter of European gas supply, according to the Gazprom 2005 annual report. That is about 2.9 bcm (0.10 tcf) more than it provided in 2004.

With European production declining, members of the EU will import as much as 75% of their natural gas in 2015, compared with 57% in 2005, according to IEA.

Controlling Energy Resources

Russia is taking a pragmatic new approach to the control of its energy resources, according to a study by the U.S. Naval War College. It is establishing new hybrid national energy companies, with state-owned PetroKremlin becoming an entirely new model for a national energy company.

Russia's crude oil exports, 1991-2005

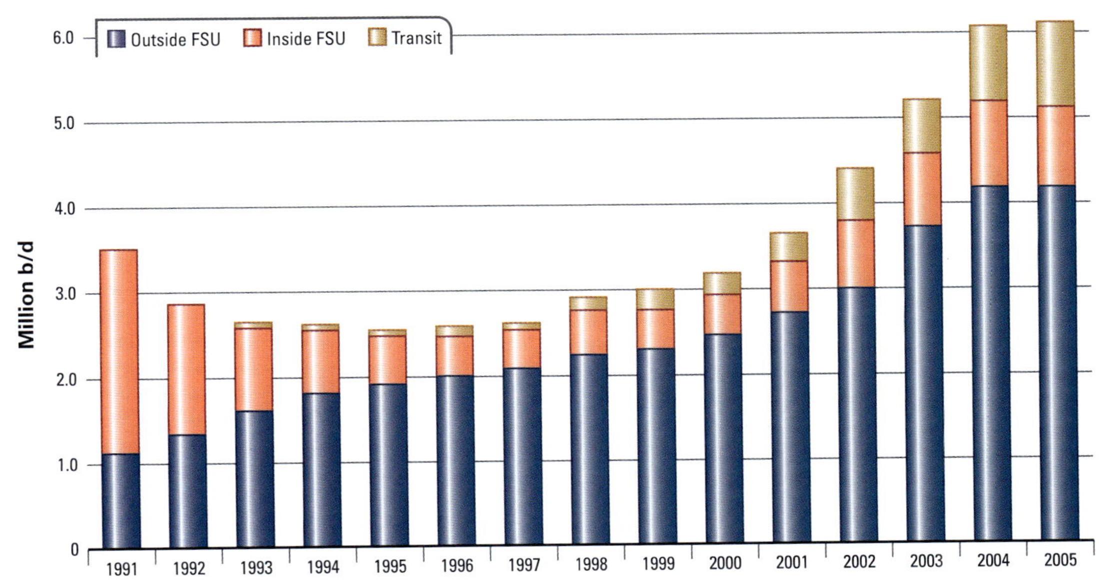

Source: World Energy Analysis & Forecasting Group, PetroMarket Research Group (February 2006)

Major Russian Oil Export Outlets

	2004		2007 (projected)		2012 (projected)	
	Capacity	Use	Capacity	Use	Capacity	Use
	Million b/d					
Eastern Europe (pipelines)	1.38	1.22	1.78	1.37	1.98	1.56
Druzhba	1.38	1.22	1.68	1.28	1.68	1.32
Northern branch	0.98	0.88	1.28	1.08	1.28	1.10
Southern branch*	0.40	0.33	0.40	0.20	0.40	0.22
Adria (Croatia)	—	—	0.10	0.09	0.30	0.24
Black Sea (ports-terminals)	2.23	1.02	2.43	1.31	2.95	1.66
Novorossiysk	0.98	0.85	0.98	0.90	1.20	1.10
Tuapse	0.18	0.10	0.18	0.16	0.18	0.16
South Ozereyevka (CPC)	0.63	—	0.83	0.08	1.13	0.16
Odessa (Ukraine)	0.26	0.03	0.26	0.08	0.26	0.08
Pivdenny (Ukraine)	0.18	0.04	0.18	0.09	0.18	0.16
Baltic Sea (ports)	1.71	1.00	1.90	1.58	2.06	1.78
Primorsk	1.10	0.84	1.20	1.05	1.20	1.10
Ventspils (Latvia)	0.32	—	0.32	0.20	0.32	0.20
Butinge (Lithuania)	0.24	0.14	0.28	0.24	0.28	0.25
Vysotsk	0.05	0.02	0.10	0.09	0.10	0.09
Porvoo (Finland)	—	—	—	—	0.16	0.14
Barents Sea (Murmansk)	0.07	0.01	0.20	0.18	0.60	0.54
Far East	0.08	0.04	0.32	0.27	2.04	1.84
De Kastri (port)	0.04	0.01	0.12	0.11	0.24	0.22
Sakhalin-2 (terminal)	0.04	0.03	0.20	0.16	0.20	0.18
Taishet-China (pipeline)	—	—	—	—	0.60	0.54
Taishet-Nakhodka (pipeline)	—	—	—	—	1.00	0.90
Central Asia (Kazakhstan-China line)	—	—	0.40	0.20	0.60	0.30
Total	**5.47**	**3.29**	**7.03**	**4.91**	**10.23**	**7.68**

*Druzhba southern branch figures exclude capacity used for additional deliveries via the Adria and Odessa-Brody pipelines.

Source: Center for Petroleum Business Studies, January 2006

Russian vs. Traditional National Energy Companies

Russian national energy company (NEC)	Traditional export oriented NEC
51% state owned, 49% investor owned	100% state owned
Can sell equity to foreign "strategic partners" to cement state-to-state energy ties	May form joint-ventures with foreign partners
Can use capital raised from stock sales to finance mergers and acquisitions and develop reserves independently	Often too capital-starved to independently develop the huge reserves they own (i.e., Pemex in Mexico, NIOC in Iran)
Russia not an OPEC member	Typically found in OPEC member states
Several large state companies compete internally	One exclusive national producer
Russian oil sector partially nationalized	Most exporter NECs result from complete oil and gas sector nationalization
Rosneft and Gazprom have global ambitions	With the exception of Norway, NECs of the world's largest oil exporters generally restrict operations to their home soil
The Russian oil and gas sector has always been primarily operated by Russians	Until nationalization in the late 1960s and early 1970s, most OPEC producers' oil and gas was produced by Western firms
Russian state energy companies have aggressively moved to acquire assets in neighboring countries to maintain economic and political control	Most other NECs focus primarily on supplying oil, gas, and sometimes petrochemicals and refined products
Both are based on the premise that energy is a key foreign policy lever	
Both operate from a position of power because they control either all, or at least most of their countries' oil and gas reserves	
Both represent the supply-side face of resurgent resource nationalism	
Both types of NEC generally operate less efficiently than fully privatized energy producers	

Russia's privatization, reversal, 1992-2005

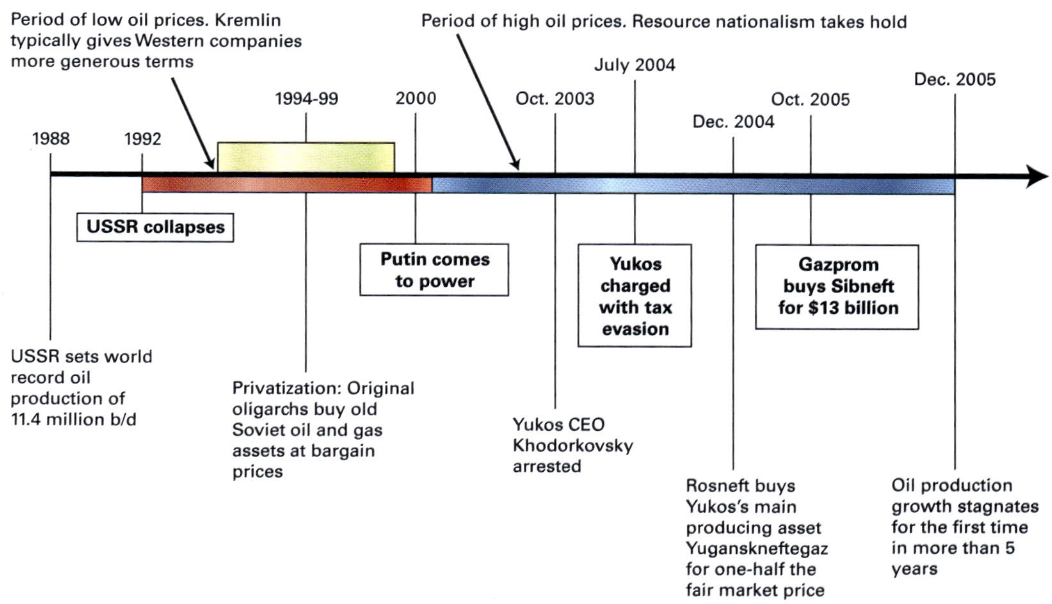

The growing government share of Russia's oil production, 1998-2005

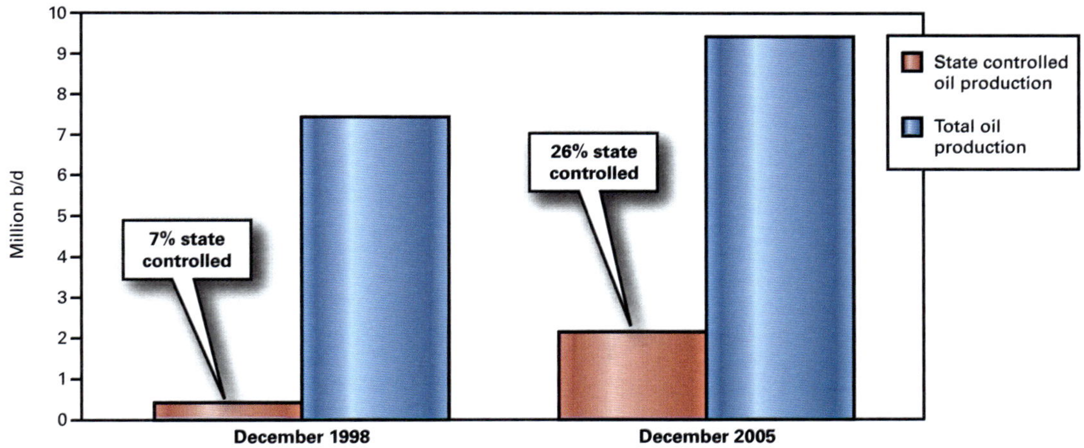

Sources: US Energy Information Administration, OAO Rosneft, Gazprom, and Russian business daily Vedomosti

Control plus capital

The country is not expected to follow the OPEC path of 100% state-controlled companies. Instead, it will permit foreign investors to own up to 49% of the state companies.

In this way, the Kremlin can use foreign capital to maintain and possibly boost production and acquire competitors while preserving control over what may be the world's largest combined oil and gas reserves.

Most governments with strategic holdings in a national champion energy producer tend to reduce their position over time as part of a gradual privatization. Russia has done the exact opposite.

Pres. Putin has worked to restore state control over natural resource industries that had been largely privatized after the Soviet Union fell. He views the country's natural resources as one of Russia's few remaining sources of international power.

PetroKremlin's Growth

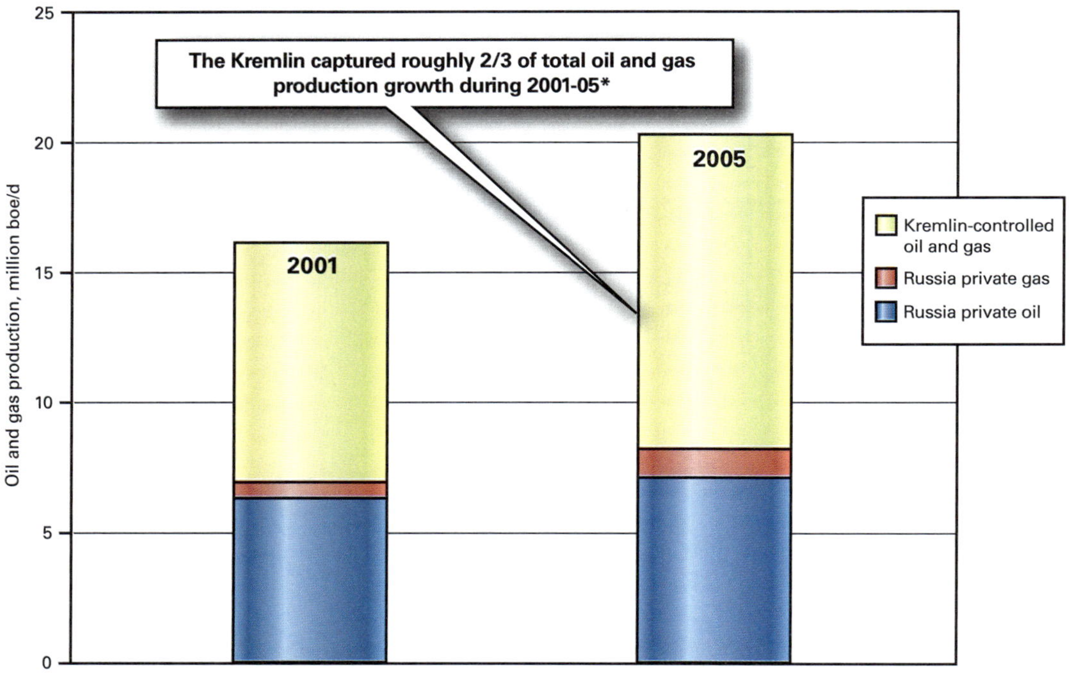

*Total Russian oil and gas production, 2001: 16:36 million boe; 2005: 20.30 million boe.
Sources: US Energy Information Administration, Oil and Capital

Currently, the Russian state companies produce the equivalent of 12 million b/d of oil, making PetroKremlin by far the world's largest state energy operation. If the nationalization trend continues, PetroKremlin could be the world's third largest national oil producer after Saudi Aramco and the National Iranian Oil Company (NIOC).

Implications

To replace reserves and maintain production, Russian companies need large investments in new exploration, drilling, and transportation infrastructure. Today's investments will determine Russia's production during 2010–15.

Russia also must develop clearer rules to guide Western energy firms pondering investments in the country. If the Kremlin does not quickly welcome new investment, its ability to deploy energy effectively as a foreign policy tool may suffer.

More broadly, other major energy exporters may be tempted to follow the new Russian approach. National energy companies worldwide may see value in relying on capital markets and Western contractors rather than Western producing companies, which often want equity ownership in oil and gas. This action could trigger further consolidation of Western oil and gas producers.

TAJIKISTAN

Capital: Dushanbe
Monetary unit: Ruble
Refining capacity: 0
Oil production: 0
Oil reserves: 12.0 million bbl
Gas reserves: 200 bcf

Gazprom licenses granted

The government granted two licenses to Russia's OAO Gazprom for evaluating natural gas and oil fields in Tajikistan. The company's affiliate, Zarubezhneftegaz, will conduct geological surveys.

Gazprom officials said in November that the company allocated $7 million (U.S.) for prospecting in the republic in 2006, and another $12 million for 2007.

The Rengan gas field, 20 km (12 miles) south of Dushanbe, has probable natural gas reserves of 35 bcm (1.24 tcf). The Sargazon field is in the Dangarin district of the Khatlon region, 150 km (90 miles) southeast of Dushanbe, with probable gas reserves of 30 bcm (1.06 tcf).

Gazprom said it may also receive rights to explore and develop two other fields: the Sarikamysh Block, 35 km (22 miles) west of Dushanbe, said to contain 113 bcm (4 tcf) in gas reserves; and the Shaambary Block, 20 km (12 miles) west of Dushanbe with about 30 bcm (1.05 tcf).

About a dozen oil and gas fields have been discovered in Tajikistan, of which two gas and five oil deposits are now producing. The country consumes more than 1.2 bcm (42 bcf) of natural gas per year. In 2005, Tajikistan produced 29.4 million cu m (1.04 bcf) of gas.

Gas from Uzbekistan to increase

In recent years, gas imports to Tajikistan from Uzbekistan were declining, from 720 million cu m in 2004 to a projected 610 million cu m in 2006. However, Tajik energy officials project that imports for 2007 will increase by about 6.5% compared to 2006, to a level of 650 million cu m.

Uztransgaz announced in 2006 that it would almost double its price for gas sold to Tajikistan's state gas company, Tajikgaz, effective January 1, 2007. Many Tajik domestic consumers struggle to pay their gas bills. Uztransgaz—a subsidiary of Uzbekistan's state oil and gas company—receives only about 40% of payments for gas in cash and writes off the remainder for the right to transport gas through Tajikistan.

TURKMENISTAN

Capital: Ashgabat
Monetary unit: Manat
Refining capacity: 236,970 b/cd
Oil production: 165,000 b/d
Oil reserves: 600.00 million bbl
Gas reserves: 100.0 tcf

Exploration and Development

Supergiant gas find claimed

The reported discovery of a supergiant gas field near the town of Iolotan in the Amu Daria basin of Turkmenistan would, if confirmed, help justify construction of a gas pipeline to China.

The field is said to contain 650 bcm (23 tcf) of gas at 5,000 m (16,400 ft), but it was not clear whether this is the recoverable volume or the amount of gas in place. Turkmenistan, which already pipes gas to several countries and intends to serve several more, invited China National Petroleum Corp. to develop the field, according to press reports.

Iolotan is east-northeast of supergiant Dauletabad-Donmez gas field, discovered in 1976 with estimated ultimate recovery of 1,388 bcm (49 tcf) from a reservoir at about 2,987 m (9,800 ft).

Tests at Lam field

Dragon Oil (Turkmenistan) Ltd, a unit of Dragon Oil PLC (Dublin), reported success in the Cheleken Contract Area in the Turkmenistan section of the Caspian Sea. Its Lam field well, LAM 10/112, drilled from the refurbished LAM 10 platform, tested oil from Zones 4, 5, and 6 at a combined rate of 2,950 b/d.

Well LAM10/112 was drilled to 3,629 m TD (11,906 ft) in reservoir Zone 7. A dual completion method allowed two reservoir intervals, at different reservoir pressures, to be produced at the same time.

Plans call for the fourth well, LAM 10/113, to be drilled to 3,700 m TD (12,140 ft) from the LAM 10 platform. Dragon also was making plans to drill a series of development wells in the Cheleken contract in the first half of 2007.

Drilling and Production

Dyarbekir field

Turkmenistan Pres. Saparmurat Niyazov authorized his country's ministry of oil and gas to make an agreement with Petronas Carigali (Turkmenistan) Sdn. Bhd. (Malaysia) for production from Dyarbekir oil field, 50 km (31 miles) off Turkmenistan.

The action provided the legal framework for the start of commercial oil production in 2006 from the field, which lies on Block 1. Petronas has been exploring Block 1 since 1996 and has made two other discoveries, Makhtumkuli and Ovez.

Pipelines

Turkmen gas sales to Pakistan

Turkmenistan agreed to supply 90 MMcmd (3.2 bcfd) of natural gas to Pakistan over 30 years through the proposed $3.3 billion Turkmenistan-Afghanistan-Pakistan (TAP) pipeline.

Energy ministers from the three countries signed the gas supply memorandum of understanding in February. India also has expressed its willingness to participate in TAP.

Pakistan will face a gas shortfall by 2010 and is working to import LNG and pipeline gas from Turkmenistan, Iran, and Qatar to meet energy demand linked to its strong economic growth.

More gas imports

Iran wants to increase its natural gas imports from Turkmenistan and said it will accept the fact that it may have to pay a higher price, according to the office of the Iranian foreign minister.

Preliminary negotiations about the suggested increase of the Turkmen gas price from the current $42–60/1,000 cu m were held in Ashgabat in February 2006. According to the Turkmen Oil and Gas Ministry, Turkmenistan delivered 5.8 bcm (204.8 bcf) of gas to Iran in 2005. Contracts signed that year called for the supply of 8 bcm (282.5 bcf) in 2006.

Turkmenistan produced 7.03 bcm (248.2 bcf) of gas in January 2006, up 24% over 2005, according to the country's National Statistics and Information Institute. Gas exports grew 28%. In addition to Iran, contracts signed late last year stipulate the delivery of 30 bcm (1.06 tcf) of Turkmen gas to Russia and 40 bcm (1.41 tcf) to Ukraine in 2006.

GAIL, EIL seek participation

GAIL India Ltd. joined with Engineers India Ltd. (EIL) to explore business opportunities in Turkmenistan.

GAIL talked with TurkMenGaz about participation in exploration and production, pipeline, and city gas distribution projects, while EIL's interest was in upgrading the Chardzou refinery at Seidi and building a new refinery there.

While EIL has prepared an internal plan to tap opportunities in Turkmenistan, GAIL is interested in participating in the extended Turkmenistan-Afghanistan-Pakistan (TAP) pipeline project and its resources implementation. The project was redesignated TAPI after India's inclusion.

UKRAINE

Capital: Kiev
Monetary unit: Hryvnia
Refining capacity: 879,759 b/cd
Oil production: 97,000 b/d
Oil reserves: 395.0 million bbl
Gas reserves: 39.0 tcf

Exploration and Development

Shell, UGV sign major exploration agreement

Ukrgazvydobuvannya (UGV) and Shell Exploration and Production signed a wide-ranging oil and gas exploration Joint Activity Agreement (JAA) covering licenses, agreed work program levels and the terms of joint activities.

UGV is a subsidiary of national joint stock company Naftogaz Ukrainy (NAK). NAK and Shell signed an agreement in 2005 to conduct joint studies in the Dniepr Donets basin, in central-eastern Ukraine. The agreement covers an area of more than 31,000 sq km (11,970 sq miles) and is thought to contain potentially significant resources of natural gas.

Under the terms of the JAA, Shell has farmed-into eight UGV held licenses in the Dniepr Donets basin with access to deep potential reservoirs, which partly lie beneath large scale shallower fields already in production. The project comprises seismic evaluation and drilling of deep exploration wells over three years. Work was planned to start in 2006.

Azov Sea concession

Shelton Canada Corp entered into an agreement with Chernomornaftogas to acquire a concession known as the North Kerchenskaya field in the southern part of the Azov Sea. This concession holds the North Kerchenskaya structure, which was discovered initially by, seismic and subsequently found to be gas bearing through drilling.

Two gas wells have defined an estimated resource of up to 4.5 bcm (160 bcf) of recoverable natural gas. Gas was tested from both wells at rates up to 2.5 MMcfd (70,800 cmd) from Miocene age reservoirs at drill depths of approximately 1,300 m (4,265 ft).

North Kerchenskaya field is 25 km (16 miles) offshore with water depths of 12 m (39 ft). It is adjacent to properties on North Bulganakskoye and East Kazantipskoye, which produce natural gas from sediments of similar age. The concession will be jointly developed and operated with Ukraine's Chernomornaftogas.

Russian-Ukraine gas dispute

In early January 2006, Ukraine and OAO Gazprom locked horns in a dispute about natural gas prices and transit fees that culminated in Gazprom cutting supplies to Ukraine for several days and led to natural gas shortfalls in many European countries.

According to FACTS Inc. (Honolulu), the dispute will likely have a long-term influence on supply strategies, going beyond European frontiers and boosting LNG receiving terminal projects, likely providing an opportunity to increase LNG flows from Asia to Europe.

The dispute also provides a preview of what Russian price expectations might be in negotiations for natural gas exports from Sakhalin I to China.

Russia and Ukraine settled their dispute on January 4, three days after Gazprom had shut off gas delivery. Under the two 5-year agreements signed, Russia will sell natural gas to Ukraine at a price four times as expensive as the previous contract.

Russia will add its volumes to the larger and much cheaper supplies from Turkmenistan, and sold by RosUkrEnergo (half-owned by Gazprom) to national joint stock company Naftogaz Ukrainy (NAK) at a common price of $2.60/MMBtu. In return, Russia will increase the transit fee for gas deliveries through Ukraine by 47% and pay in cash instead of gas.

Coming just one year after a similar crisis between Russia and Belarus, the Russo-Ukranian crisis triggered serious discussion in Europe about two aspects of security of supply.

The first concerns the region's infrastructure. Most of the Russian gas bound for Europe moves through Ukraine. Only two other, more limited-volume, transit routes exist: the Yamal-Europe pipeline through Belarus and the recently inaugurated Blue Stream, crossing the Caspian Sea to link Russia and Turkey.

A lack of interconnecting infrastructure, however, prevents natural gas delivered to Turkey from moving to northern Europe.

The other concern lies with dependence on a single supplier. Many observers considered the dispute as much a political argument as a business disagreement, tarnishing Gazprom's reputation as a reliable supplier.

PFC Energy (Houston) said a host of European countries face Gazprom price pressure. Some have not agreed to new contract terms with higher prices, and some may be unable to pay. Confrontation over payment could create another EU energy security scare.

Prykerchenska area

Vanco Energy Co. (Houston) won the right to conclude a production sharing agreement with Ukraine for the Prykerchenska Tender Area in the Black Sea. The company described this as Ukraine's first deepwater license award and Ukraine's first production sharing.

Covering 3.2 million acres (5,000 sq miles), the Tender Area is located offshore the Crimea near the Kerch region.

Vanco also awarded a deepwater drilling contract to a unit of Atwood Oceanics, Inc. for one firm well and one optional well in the Prykerchenska block. The company expects to drill in late 2007 off the shelf in the northeastern corner of the block in water depths greater than 300 m (985 ft).

UZBEKISTAN

Capital: Tashkent
Monetary unit: Soum
Refining capacity: 222,271 b/cd
Oil production: 105,000 b/d
Oil reserves: 594.00 million bbl
Gas reserves: 65.00 tcf

Khauzak field development

Lukoil Overseas began development drilling in the Khauzak contract area of Dengizkul gas field in southwestern Uzbekistan. Khauzak is the first field to be developed under the Kandym-Khauzak-Shady-Kungrad production-sharing agreement (PSA) signed in 2004 by a group that includes Lukoil Overseas, and Uzbekneftegaz.

Original gas in place is estimated at 329 bcm (11.6 tcf). Commercial production is expected to start by year-end 2007. Peak gas production is projected to exceed 10 bcm/year (0.97 bcfd), while cumulative production could reach 207 bcm (9.5 tcf), said Lukoil.

The partners plan to drill 37 development wells. The Kandym-Khauzak-Shady-Kungrad project involves drilling more than 180 development wells, building more than 1,500 km (932 miles) of pipeline, and constructing a gas processing plant with capacity of 8 bcm/year (0.77 bcfd) in the Kandym area.

Uzbekistan, South Korea agreement

South Korean Pres. Roh Moo-hyun signed an agreement with Uzbek Pres. Islam Karimov to allow joint exploration and development of oil and gas fields in the central Asian country.

Under a memorandum of understanding signed among South Korea's National Oil Corp. (KNOC), Korea Gas Corp. (Kogas), and Uzbekistan's state-run Uzbeknefgaz, the South Korean firms have the exclusive right to explore and develop two oil fields and two gas fields.

KNOC will carry out preliminary exploration for six months in Chust-Pap and Namangen-Terachi in the eastern part of the country with Uzbeknefgaz. Kogas will cooperate in developing Surgil and Uzunkui gas fields.

Gas plant construction starts

State-owned Uzbekneftegaz began construction of a $63.6 million LPG unit at its gas processing subsidiary, Shurtanneftegaz, in the Kashkadarya region.

Shurtanneftegaz is one of three gas processing companies in Uzbekneftegaz and processes 11 bcm/year (1.06 bcfd) of gas.

The first stage of construction is to be completed by year-end, yielding a capacity of 45,000 tpy. After three stages of construction, the plant will produce 175,000 tpy of a propane-butane mix by 2010.

Uzbekneftegaz also plans several projects to increase LPG production at the Mubarek gas processing plant and the Shurtan gas production complex.

China

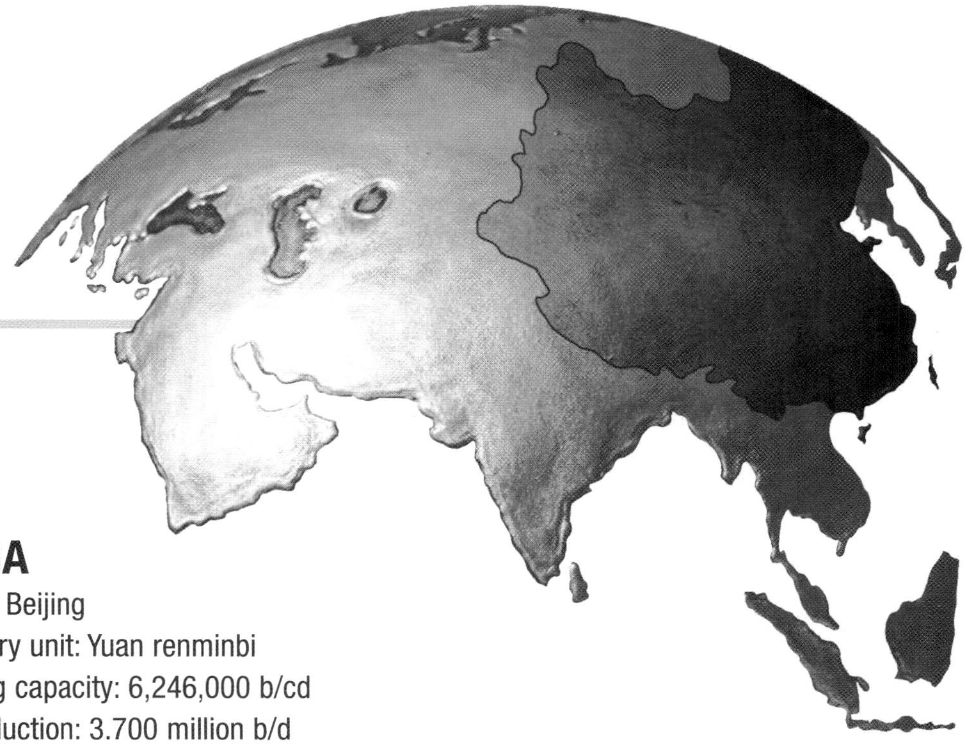

CHINA

Capital: Beijing
Monetary unit: Yuan renminbi
Refining capacity: 6,246,000 b/cd
Oil production: 3.700 million b/d
Oil reserves: 16.00 billion bbl
Gas reserves: 80.00 tcf

Background

A 2006 report by the U.S. Dept. of Energy (DOE) reviewed China's growing demand for oil and gas and the implications of that growth for other countries.

DOE's Energy Information Administration (EIA) has estimated that China will consume 7.4 million b/d of oil in 2006, an increase of nearly 0.5 million b/d from 2005. China's 2006 increase in oil demand will represent 38% of the world total increase.

China's demand for oil is increasing by some 15%/year. In 2005, China imported 130 million tonnes of crude oil, up 3.3% over 2004. Its crude imports from Saudi Arabia more than doubled, from 8.8 million tonnes in 2001 to more than 20 million tonnes in 2005.

Oil production

Production of oil from China's two major onshore fields, Daqing and Liaohe, is on the decline. Daqing oil field—the largest, and actively productive since 1967—produced less than 50 million tonnes of oil in 2003. That was the first time in 27 years that annual production was below that level.

Production is expected from new finds made in 2004 near Shengli field in the northeast, and far-western reserves in Xinjiang provinces have been opened. The Tarim, Junggar, and Turpan-Hami basins have combined reserves of 20.9 billion tonnes of oil and 10 tcf of gas.

However, said DOE, those new discoveries will not offset declining production.

Only about 15% of China's oil production capacity is from offshore resources, but that production has been growing at an average of 15.3% per year. It reached 28.4 million tonnes in 2004, and the government is aiming for 67 million tonnes by 2010.

WESTERN CHINA
0 50 100 150 200 250 Miles
0 100 200 300 400 Kilometers
RUSSIA
KAZAKHSTAN
MONGOLIAN REPUBLIC
KIRGIZSTAN
PAKISTAN
INDIA
PEOPLE'S REPUBLIC OF CHINA
NEPAL
BHUTAN
BANGLADESH
MYANMAR (BURMA)
THAILAND
LAOS
VIETNAM
Xinjiang
Qinghai
Xizang
Gansu
Ningxia
Shaanxi
Sichuan
Guizhou
Yunnan
Guangxi
Pavlodar
Karaganda
Atasu
Ust-Kamenogorsk
Lake Balkhash
Dzungarian Gate
Karamay
Uerho
Fengcheng
Shingshihsia
Shawan
Usa
Qigu
Dushanzi
Urumqi
Bogdo
Suilai
Shanshan
Hami Turfan Hami
Hami
Shan
Shengting-k'ou
Almaty
Bishkek
Yakela
Kumger
Kosaptok
Lunnan
Kan
Yiqikelike
Dongqiulitage
Taqiaq
Yingmaii
Aksu
Kuqa
Lunan
Korla
Akekule
Wensuh
Yuecan
Bostan
Daliya
Ichkelik
Wug Chat
Kashgar
Kashi
Kelatuo
Kekeya
Tsehpu
Tazhong
Ulan Bator
Zunnbayan
Tsagaan Els
Pai-yang-ho
Shih-you-kou
Ya 'exia
Laojunmiao
Danbei
Baiyanghe
Shiyougou
Yumen
Gaotai
Xianshuiquan
Shizigou
Qigequan
Huatugo
Youshashan
Wunan
Lenghu
Jiandinshan
Nanyishan
Youquanzi
Kaitemilike
Lenghusanhao
Lenghusihao
Mahai
Lenghuwuhao
Yuka
Toufengshan
Sebeiyihao
Sebeiehao
Yanhu
Golmud
Tianjum
Koko Nor
Xining (Sining)
Lanzhou
Yinchuan
Shenglijin
Liujiazhuang
Lizhuangzi
Majiatan
Jingbian
Dashuikeng
Baiyanjie
Yuancheng
Mafang
Huachi
Maling
Chenghab
Zhilou
Yongping
Yan'an
Qingyang
Ganguye
Xiashiwan
Shih-tan-miao
Houma
Tianshui
Wushan
Xi'an (Sian)
Huang He
Manali
Hoshiarpur
Nagqu
Lhasa
Brahmaputra
Katmandu
Thimphu
Ganges
Shillong
Chhatak
Sylhet
Haripur
Kailastila
Patharia
Rashiapur
Kishorganj
Beanibazar
Kamta
Fenchuganj
Titas
Habiganj
Dhaka
Bakhrabad
Begamganj
Semutang
Feni
Chittagong
Kutubdia
Patiya
Bay of Bengal
Jiangyou
Deyang
Zhongba
Yingshan
Wubaiti
Penglaizhen
Gong
Nanchong
Chengdu
Leshan
Weiyuan
Chongqing
Wolonghe
Shiyougou
Yibin
Xishui
Luzhou
Guiyang
Tengcong
Kunming
Liuzhou
Tiandong
Nanning
Beihai
Mandalay
Irrawaddy
Mekong
Ayadaw
Yenangyaung
Chauk/Lanywa
Mann
Htaukshabin
Paduakpin
Pyaye
Pyalo
Prome
Myanaung
Shwepyitha
Mae Fang
Mae Suhn
Mae Sot
Utiaradit
Louanphrabang
Vientiane
Hanoi
Haiphong
Tien Hai
Weizhou 11-1
Weizhou 10-3
Weizhou 6-1
Gulf of Tonkin
Dongfang
Yangtze

RUSSIA
MONGOLIAN REPUBLIC
Ulan Bator
Komosamolsk
Khabarousk
Heilongjiang
Qing 'an
Daqing
Harbin
Qianguo
Jilin
Huadian
Dongfeng
Fusong
Chongjin
Liaoning
Tieling
Fushun
Shenyang
Liaoyang
Jinzhou
Panshan
Anshan
Yingkou
Dandong
NORTH KOREA
Pyongyang
Sea of Japan
Aershan
Zunnbayan
Tsagaan Els
Saihantala
Chifeng
Beijing
Tianjin
Huai'an
Cangzhou
Hohot
Chin-hsi
Qilu
Changli
Renqiu
Huanghua
Bohai Gulf
Jin Xian
Luda
Lushun (Dalian)
Seoul
Onsan
SOUTH KOREA
Ulsan
Pusan
Yosu
Korean Channel
JAPAN
Hiroshima
Sasebo
Kashima
Nagasaki
Oita
Osumi Strait
Shih-you-kou
Nei Mongol
Ningxia
Danbei
Yumen
Baiyanghe
Gaotai
Shandan
Gansu
Tianjum
Ulan
Koko Nor
Xining (Sining)
Lanzhou
Tianshui
Yinchuan
Shenglijin
Liujiazhuang
Lizhuangzi
Majiatan
Donghongzhuang
Hongjinzi
Dashuikeng
Youfangzhuang
Baiyanjin
Mafang
Yuancheng
Huachi
Maling
Chengbao
Anzai
Wugi
Zhilou
Pingliang
Shih-lian-miao
Yulin
Suide
Jingbian
Zichang
Yongping
Quinghuazhi
Yan'an
Qingyang
Ganguye
Xiashiwan
Xianyang
Xi'an (Sian)
Shaanxi
Wushan
Shanxi
Shijiazhuang
Hebei
Yanshan
Zibo
Kenli
Jinan
Zhaoyuan
Gaoqing
Weifang
Qingdao
Shandong
Yellow Sea
Weicheng
Anyang
Wenmingzhai
Wennan
Wenzhong
Pucheng
Wenliu
Huzhuangji
Baimiao
Machang
Luoyang
Zhengzhou
Huang He
Henan
Fangcheng
Lianyungang
Jiangsu
Baoying
Liuzhuang
Gumin
Caozhuang
Lianmengzhuang
Anhui
Huangque
Hefei
Zhenwu
Nanjing
Shanghai
Shuanghe
Hubei
Yingshan
Wuhu
Tuoshi
Qianjiang
Wuhan
Yichang
Mianyang
Anqing
Hangzhou
Zhenhai
Ning-Bo
Donghai
Bo Yuan Ting
Ping Hu
East China Sea
Zhongba
Jiao
Jiangyou
Yingshan
Gong
Wubaiti
Bajiaochang
Penglaizhen
Chengdu
Nanchong
Yangtze
Garze
Sichuan
Suining
Wusheng
Dazu
Wolonghe
Chongqing
Shiyougou
Weiyuan
Leshan
Batang
Ceshui
Yueyang
Jiujiang
Jinxi
Zhejiang
Hunan
PEOPLE'S REPUBLIC OF CHINA
Jiangxi
Hemgyang
OKINAWA
Guizhou
Anshun
Quanzhou
Fuzhou
Fujian
Taipei
Chuhuangkeng
Jinmen
Formosa Strait
Chiting
Cingcaohu
Jinshuei
Paoshan
Yongheshan
Taoyuan
Tiejhanshan
Sinying
Paonanghsi
TAIWAN
Kaohsiung
Pingtung
Kunming
Yunnan
Liuzhou
Bose
Guangxi
Tiandong
Mashan
Nanning
Guangdong
Guangzhou
Hong Kong
Lianjiang
Maoming
Beihai
Zhanjiang
Xijiang 24-3
Xijiang 30-2
Lufeng 13-1
Huizhou 26-1
Lufeng 15-1
Huizhou 21-1
Liuhua 4-1
Liuhua 11-1
VIETNAM
Irrawaddy
Mekong
Red
Weizhou 11-1
Weizhou 10-3
Weizhou 6-1
Hanoi
Tien Hai
Haikou
HAINAN DAO
Jinfeng
Wenchang 8-3
Wenchang 9-2
Wenchang 14-3
Wenchang 13-1
Wenchang 19-1
Wenchang 13-2
Gulf of Tonkin
Dongfang
Yinggehai
Yacheng
Yacheng 13-1
South China Sea
Mae Fang
Mae Suhn
LAOS
THAILAND
Vientiane
PHILIPPINES
0 100 200 300 Miles
0 200 400 Km
EASTERN CHINA

Interest in offshore exploration has centered on the Bohai Sea area, east of Tianjin—believed to hold more than 1.5 billion bbl of reserves—and the Pearl River Mouth basin.

Natural gas, LNG

Though natural gas has not been a major source of fuel in China, the country has embarked on a major expansion of its gas infrastructure. Most notable is construction of the West-East pipeline connecting the Tarim and Ordos basins to the southern coast. China also is considering importing gas by pipeline from Russia and Kazakhstan.

China is building its first LNG import terminals. The $846 million facility at Shenzhen, near Guangdong, will import 3 mtpy of LNG from Australia. A second is under construction in Fujian province and is to be completed in 2007. Jiangsu, Zhejiang, Liaoning, Hebei, Shandong, and Shanghai provinces are also considering LNG receiving terminals.

Refining and SPR

China's major oil companies are increasing refining capacity.

- China National Offshore Oil Corp. (CNOOC) is building a 240,000 b/d refinery in Huizhou, Guangdong province, expected to be operational in 2008.
- A $3.5 billion expansion of the Quanzhou refinery in Fujian will triple its capacity to 240,000 b/d.
- Saudi Aramco and China Petroleum & Chemical Corp. (Sinopec) have agreed to build a refinery in Quanzhou to process heavy Saudi crude.
- CNPC is planning a major expansion of the Dushanzi refinery in Xinjiang, which will be partially supplied by the new pipeline from Kazakhstan.

After a decade of consideration, China has unveiled plans for a strategic petroleum reserve (SPR). It will hold state-controlled crude stocks—90 days of imports—at four sites: Dalian, Huangdao (Qingdao), Zhenhai, and Zhoushan (Ningbo).

Energy-market impacts

As Chinese oil companies invest in energy assets—and in foreign companies that own such assets—concern has grown that such action will 'remove' energy resources from the market, constricting supply and raising world prices.

Because China can be expected to consume most of the oil produced from its purchased assets, the view of DOE is that the effects of the purchases should be economically neutral. They will simply displace what the Chinese otherwise would have bought on the open market.

China's oil consumption

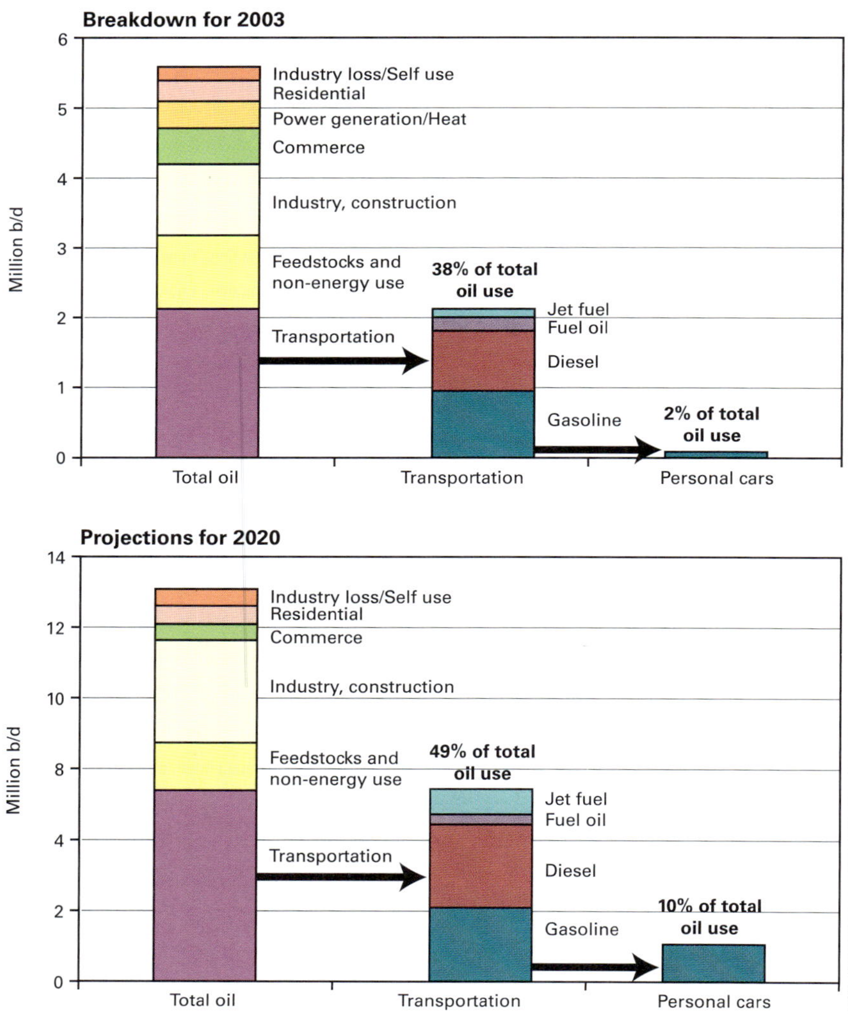

Source: US Department of Energy

And if China develops resources in areas where private commercial interests are unwilling to invest, such action might enlarge total global oil supply.

Exploration and Development

Block 29/26

Husky Oil China Ltd., subsidiary of Husky Energy Inc. (Calgary, AB, Canada), reported a "significant" discovery with its Liwan 3-1-1 well on offshore Block 29/26 in the South China Sea. The discovery could contain a potential recoverable resource of 113-170 bcm (4-6 tcf) of natural gas, making Liwan 3-1-1 one of the largest gas discoveries off China.

Liwan 3-1-1—called by Husky the first deepwater discovery and the deepest well off China—was drilled to 3,843 m TD (12,608 ft) in 1,500 m (4,920 ft) of water in the Pearl River Mouth basin. Block 29/26 is 250 km (155 miles) south of Hong Kong.

Further drilling on the 3,965 sq km block will follow a 3D evaluation. Additional appraisal to refine reserves estimates is unlikely until 2008–2009.

Sampaguita

The 1976 Sampaguita gas-condensate discovery in the South China Sea northwest of Palawan Island could be an accumulation of as much as 566 bcm (20 tcf) of gas, said Forum Energy PLC (London).

Interpretation of 3D seismic data indicates a potential of 368 bcm (13 tcf) in an area of 290 sq km (112 sq miles). Previous estimates indicated a likely resource of 142 bcm (5 tcf). The license is 250 km (155 miles) west of Malampaya gas and oil field.

Block 25/34

Primeline Energy Holdings Inc. (London) completed a 549 sq-km (212 sq-mile) 3D seismic survey over the Lishui 36-1 discovery and adjacent areas of Block 25/34 in the East China Sea. Interpretation by year-end 2006 will guide selection of 2007 drilling locations on the 7,006 sq-km (2,705 sq-mile) block.

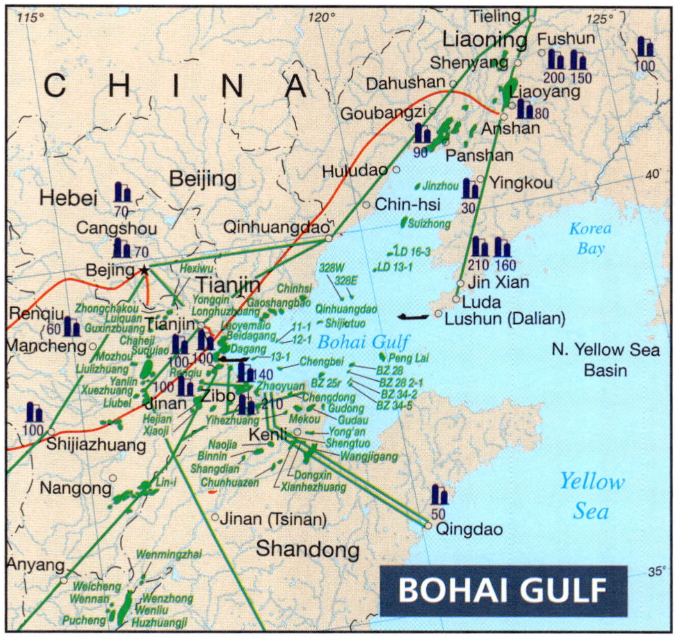

Bohai Bay fields

Anadarko Petroleum Corp. started oil production from the CFD 11-6/CFD 12-1S development project in Bohai Bay. The development project straddles Blocks 04/36 and 05/36 in 23 m (75 ft) of water.

Anadarko expected 10 wells to be producing a total 15,000 b/d by year-end. Gross production of 22,000 b/d from 22 wells is expected by mid-2007.

CNOOC Ltd. made an oil discovery on Structure BZ28-2S in the Yellow River Mouth Sag of central northern Bohai Bay. CNOOC drilled the well to 2,125 m TD (6,972 ft) in 22 m (72 ft) of water.

The BZ 28-2S-1 well penetrated 85 m (279 ft) of oil pay and 8 m (26 ft) of gas pay. Drillstem tests conducted over two intervals yielded a combined 1,900 b/d of oil and 2 MMcfd of gas.

Sichuan basin

A unit of China's Sinopec let a contract to GX Technology Corp. (Houston) for a comprehensive full-wave imaging program at the largest gas field in western Sichuan Province.

GXT seeks to identify promising prospects based on seismic data previously acquired. Initial imaging was under way, with final results to be delivered in first half 2007.

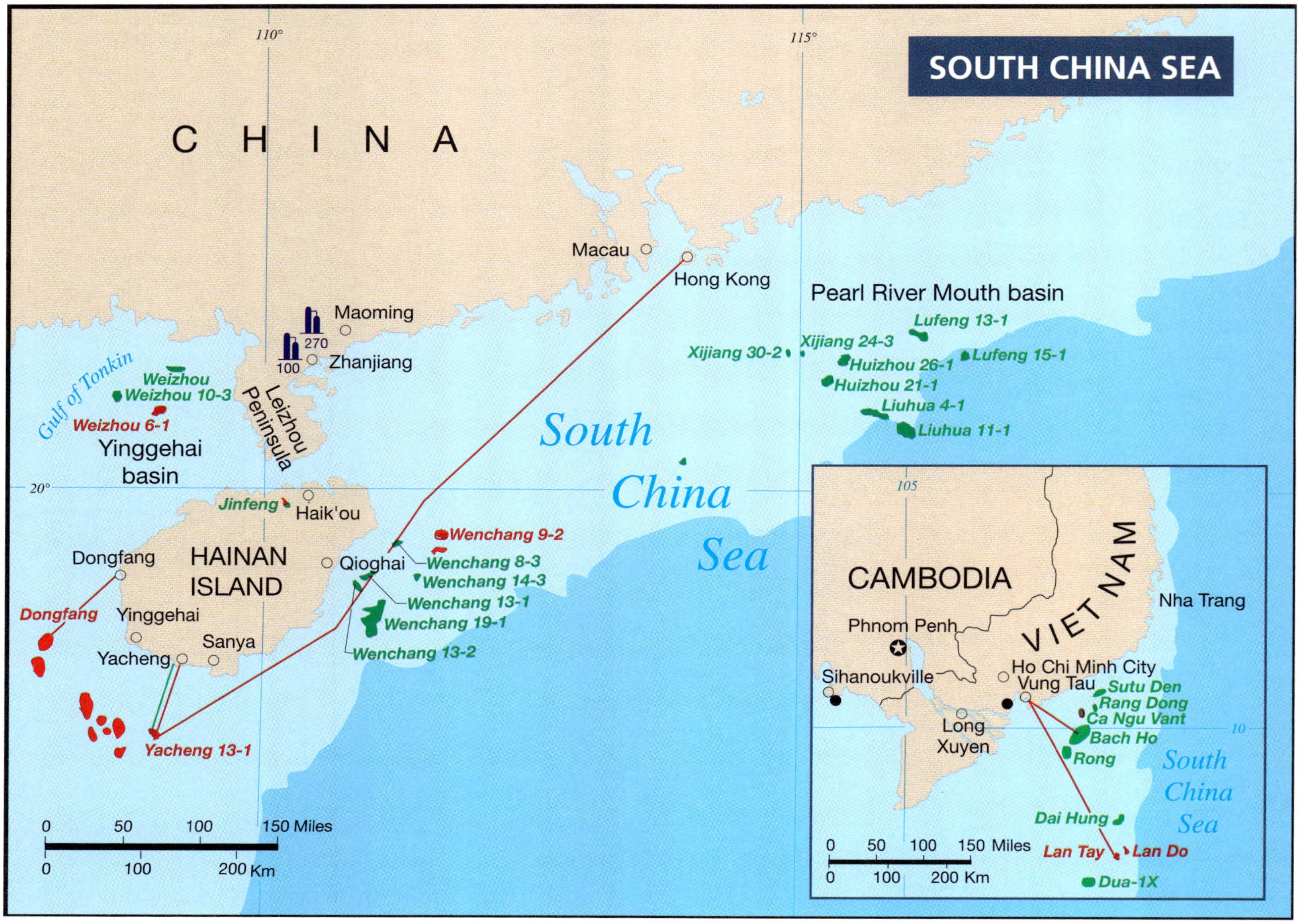

Block 22/12

Roc Oil Co. Ltd. (Sydney) and partners began appraisal of the Wei-6-12S-1 exploration well on Block 22/12 in 30 m (98 ft) of water in the Beibu Gulf off China.

On test, the well, drilled to 2,535 m TD, flowed at a combined, stabilized 5,750 b/d of oil from three separate hydrocarbon columns.

Pressure gradient information suggests that the oil column extends downdip from the discovery well.

Qian 112 oil field

China North East Petroleum Holdings Ltd. planned to start fracing three wells in Qian 112 oil field southeast of Qian'an in Jilin Province. The Qian 5-2, 3-5, and 12-16 wells follow the recently completed Qian 12-14 well.

CNEH said Qian 112 field has proved reserves of 1.2 million bbl in the Cretaceous Fuyu and Gaotaizi formations. It said the Gaotaizi is the main target in the production wells.

CNEH's 2006 drilling program aimed for completion of 40 wells by the end of third quarter. By yearend 2007, the company plans to complete 81 more wells, bringing total producing wells to 102.

Block 11/34

Devon Energy Corp. (Oklahoma City, OK, U.S.A.) signed a production-sharing contract with CNOOC for Block 11/34 in the South Yellow Sea off eastern China. The block covers 10,878 sq km (4,200 sq miles) in 15–38 m (50–125 ft) of water, 400 km (250 miles) northeast of Shanghai.

CNOOC can take up to a 51% working interest if there's a commercial discovery. Devon will conduct a seismic survey and drill exploratory wells during the exploration period.

Block 16/19

CNOOC Ltd. reported a discovery in the ENI China BV Huizhou (HZ) 25-4-1 well in the eastern South China Sea. The well is on Block 16/19 in the Pearl River Mouth basin, about 180 km (112 miles) southeast of Hong Kong.

The well was drilled to 3,900 m TD (12,795 ft) in 102 m (335 ft) of water. On test, HZ25-4-1 flowed at 5,000 b/d. CNOOC has the right to acquire up to 51% working interest in any commercial discoveries on the block.

The discovery is 60 km from the producing HZ 19-3/2/1 oil fields on Blocks 16/08 and 16/19. The fields have two platforms, 14 wells, and a subsea pipeline to an FPSO vessel.

Drilling and Production

BZ 25-1/25-1S oil field

CNOOC Ltd. brought on stream Platform A, part of second-phase development of Bo Zhong (BZ) 25-1/25-1S field in 20 m (66 ft) of water in southeast Bohai Bay, 127 km (79 miles) southeast of Long Kou City.

Platform A is producing more than 2,200 b/d of light crude oil from two wells. Five other platforms started production in 2004 and 2005. BZ 25-1/25-1S field oil production totaled now 30,000 b/d.

DF 1-1 gas field

CNOOC Ltd. started second-phase production from Dong Fang (DF) 1-1 natural gas field in Bei Bu Gulf in Yinggehai basin in the western South China Sea. Initial production of 35 MMcfd of gas came from three wells, boosting field production to 187 MMcfd.

Design production capacity of the second phase is 111 MMcfd. Total DF 1-1 field production is expected to reach 2.8 bcm/year (98 bcf/year) of gas.

Most gas from second-phase development will supply a methanol project in Hainan Province. The field lies in 63–70 m (207–230 ft) of water about 110 km (68 miles) off Dong Fang City

HZ 21-1 gas field

Huizhou (HZ) 21-1 gas field in the eastern South China Sea came on stream and is producing 54 MMcfd through four wells, partner CNOOC said.

The field is in 115 m (377 ft) of water on Block 16/08 in the Pearl River Mouth basin. The processed gas will go to a Zhuhai gas terminal through subsea pipelines.

HZ 21-1 operator CACT Operators Group plans joint development of HZ 21-1 and Pan Yu 30-1 gas fields. The latter is expected to come on stream in 2007. HZ 21-1 field is expected to produce 0.56 bcm/year (19.7 bcf/year). CACT includes Agip SPA, Chevron Corp., and CNOOC.

Sakhalin-1 gas pre-accord

ExxonMobil Corp. signed a preliminary agreement on principles with CNPC for the supply of natural gas from the Russian Sakhalin-1 project.

The agreement calls for delivery to northeastern China. Quantities, pricing, and delivery schedules were not disclosed. Earlier, ExxonMobil said it is ready to deliver 8 bcm/year (0.77 bcfd) to China.

Refining and Processing

Ningbo refinery complex

China agreed in principal to the construction of a $10 billion refinery and ethylene complex in Ningbo by a joint venture of Formosa Plastics Group (Taipei) and China Petroleum & Chemical Corp.

Described as the largest-ever foreign investment project on the Chinese mainland, the proposed facility, subject to approval by China's State Council, will have a capacity of 1.2 mtpy of ethylene and 10 mtpy of oil. It will target sales in Taiwan.

Pipelines

First oil through China-Kazakhstan line

PetroChina successfully transported its first shipment of crude oil in July through the 963-km (598-mile) China-Kazakhstan pipeline to its oil tank farm and refinery in northwestern Xinjiang Uygur Autonomous Region. The arrival of the oil marked the full commercial operation for China's first cross-country crude pipeline.

The pipeline originates at Atasu in northern Kazakhstan, enters China at Alashankou port on the Sino-Kazakhstan border, and terminates at PetroChina's Dushanzi Petrochemical Co. site.

The system will have an initial capacity to pump 200,000 b/d of crude oil, with plans to reach 400,000 b/d by 2010.

Mekong River oil transport

China has begun receiving oil shipments from Thailand by a Mekong River route in early 2007 under an experimental program. Two ships, each carrying 150 tonnes of oil, arrived in Yunnan Province. The river route is viewed a possible alternative to the Strait of Malacca route.

China in March signed an agreement with Laos, Myanmar, and Thailand on shipping oil on the waterway. Yunnan officials said China would ship about 70,000 tonnes/year of oil from Thailand alone by river and that possible shipment of even larger volumes was under study.

LNG

Zhong Ximentang Island

Shanghai LNG Co. Ltd. was building an LNG terminal at Zhong Ximentang Island in China's Zhejiang province, with plans for the facility to become operational in the first half of 2009.

The complex, which includes docking facilities, a regasification plant, and an undersea pipeline, will have an initial capacity of about 3 mtpy—enough to process newly contracted supplies.

In October, Petronas subsidiary Malaysia LNG signed deals with Shanghai LNG to supply up to 3.03 mtpy of LNG for 25 years. Shanghai LNG is a joint venture of the Shenergy Group Ltd. and CNOOC Gas & Power, a wholly owned unit of CNOOC.

Rizhao City

China and EurOrient Financial Group signed an agreement to build an LNG receiving terminal and related facilities in Rizhao City in Shandong Province. EurOrient said it was considering the Rizhao Lanshan Port as a possible location.

With an initial capacity of 1.5 mtpy, the $965 million facility will supply gas to Rizhao City and the greater Qingdao territory.

The project will be financed and developed in phases. The first phase will handle as much as 500,000 tonnes of LNG. Potential suppliers are Australia, Indonesia, Malaysia, Qatar, Oman, and Russia.

If a construction permit is received by December 2007, construction could start first quarter 2008.

Petrochemicals

Lurgi-based propylene plants

Datang International Power Generation Co. Ltd. and Shenhua Ningxia Coal Industry Group awarded Lurgi AG (Frankfurt, Germany) two contracts for initial work on two commercial-scale propylene plants that will use Lurgi technology to produce plastics from coal.

Total capital investment for both projects amounts to more than €2 billion. Lurgi says the Chinese plants, expected to be the largest in the world, could begin operation by 2009, producing 500,000 tpy of polypropylene.

Expanding Verbund capacity

BASF-YPC Co. Ltd.—a 50-50 joint venture of BASF AG (Ludwigshafen, Germany) and Sinopec—announced plans to expand the capacity of a 600,000 tpy steam cracker and invest in additional downstream plants at the chemical Verbund site in Nanjing.

The $500 million project will expand the steam cracker's capacity to 750,000 tpy of ethylene. Also, plans call for expansion of the ethylene oxide (EO) plant and development of EO derivatives, development of the C4 value chain, including C4-specialties, and extension of the acrylics value chain to produce super-absorbent polymers.

Tianjin ethylene plant

Sinopec launched a 1.0 mtpy ethylene project in the Binghai New Zone, in north China's Tianjin municipality. The project, scheduled for completion in 2009, will serve as the core of the petrochemical industry in the Chinese port city.

The new facility will raise ethylene production capacity at the company's Tianjin branch to 1.2 mtpy and its refining capacity to 12.5 mtpy.

Sinopec also plans to build two crackers in Shanghai and Wuhan of Central China's Hubei Province. The Shanghai plant is designed with a capacity of 1.0 mtpy, while the Wuhan facility would produce 800,000 tpy. Approval from Chinese authorities was pending.

Daya Bay complex

All production units of the $4.3 billion Daya Bay petrochemical complex in Huizhou Guangdong Province became fully operational in early 2006.

The facility is a joint venture of CNOOC, Shell Petrochemicals Co. Ltd., and Guangdong Guangye Investment Group Co. Ltd. Its world-scale cracker unit produces 800,000 tpy of ethylene and 430,000 tpy of propylene. The complex is expected to produce 2.3 mtpy of total products.

China: Petrochemicals competitor?

China can become a major competitor in petrochemicals, said SRI Consulting (Houston) in a 2006 report. The study compares costs for petrochemical plant construction and operation in China compared to the United States, Japan, and Germany.

China's main competitive advantage is low-cost labor. The cost of local sourcing, especially labor cost, is significantly lower in China than in the developed countries. A chemical plant built in China can cost one-third less than the same plant on the U.S. Gulf Coast, said SRI.

Chengdu ethylene plant

China's National Development and Reform Commission granted approval to PetroChina to build a $2.5 billion ethylene plant at Chengdu, the capital of Sichuan Province. The new ethylene cracker, scheduled for completion by 2010, has a design capacity of 800,000 tpy.

Coalbed Methane

CBM outlook

According to China's major coalbed methane enterprise, in 2005, China invested $300 million in CBM exploration, completed drilling 607 wells, and put four blocks into full operation.

Looking ahead, one industry aim is to achieve 10 bcm (0.35 tcf) of production by 2010, and to increase CBM proved reserves by 300 bcm. Another is to establish an industrial system for CBM development and use. This would include nine pipelines extending 1,390 km (864 miles), carrying 5.73 bcm/year (0.55 bcfd).

Liulin block

A unit of Fortune Oil PLC (London) began drilling its first pilot well on the Liulin Block in the eastern Ordos basin of Shanxi Province. The basin is one of the world's largest CBM resource areas.

The FL-EP2 vertical production well, spudded in November, will be drilled to a TD of 1,155 m (3,790 ft).

A second vertical pilot production well targeting 610 m TD (2000 ft) was to be drilled by year-end. Further pilot production and slim-hole data wells are planned for 2007.

Shilou North block

Verona Development Corp. (Vancouver, BC, Canada) was drilling two wells to assess CBM potential south of Liulin, on the 1,015 sq-km (392 sq-mile) Shilou North Block in the eastern Ordos basin in Shanxi Province. The wells target seams of the Late Carboniferous Taiyuan and Early Permian Shanxi formations.

Coal to Liquids

Shaanxi project

China granted approval to state-owned Yankuang Group, part owner of Yanzhou Coal Mining, to develop a coal-to-oil project in the northwestern province of Shaanxi.

Initial capacity of 1 mtpy of oil products eventually will rise to 5 mtpy.

Japan partnership

Japan's New Energy and Industrial Technology Development Organization (NEITDO) said it will work with China to convert coal into liquid fuels. The agency said it is more economical to try the technology in China than in Japan, which has fewer available mines.

Liquefaction experiments were to start in Beijing by year-end. Chinese companies will eventually build plants near mines to turn coal into liquid fuels.

Under the planned program, a Chinese power company will start operating a liquefaction plant in 2010 to process 3,000 tpd of coal mined in the Inner Mongolia Autonomous Region.

Also in 2010, a Chinese coal company will develop a mine in the Uighur Autonomous Region of Xinjiang and start testing a plant to process 2,500–3,000 tpd of coal.

Sasol technology

In June 2006, China signed an agreement with South Africa's Sasol Ltd. regarding construction of two 80,000 b/d coal-to-liquids facilities using that company's CTL technology.

International Activities

China is cooperating more with its neighbors to secure supplies of oil and natural gas from regional sources instead of the Middle East, said *Oil & Gas Journal*.

Saudi, Chinese cooperation

In early 2006, Saudi Arabia and China agreed to a landmark accord on oil, natural gas, and minerals as part of five main areas of cooperation. The new agreement on hydrocarbons is the first between the two governments on overall cooperation in the field of energy. Specific agreements between the countries' respective national oil companies are to be negotiated.

Aramco, Sinopec agreement

Saudi Aramco and Sinopec signed a memorandum of understanding that commits Aramco to supplying Sinopec and its affiliates with 1 million b/d of Arabian crude by 2010.

The memorandum focuses on collaboration in the areas of crude processing and commercial storage, natural gas exploration and production, petroleum products, and petrochemicals derivatives.

The MOU also covers the countries' joint developments in China—the Fujian Refining and Ethylene Project and the Qingdao Refinery Project.

Aramco and Sinopec agreed to establish the Fujian Ethylene Joint Venture and the Fujian Marketing Joint Venture in 2006. The Refining and Ethylene Integrated Project would begin production by 2009.

Further negotiations were under way regarding the terms of Aramco's participation in the Qingdao Refinery Project, which has an on-stream date of 2008.

Central Asia activity

China has significantly stepped up its involvement and investment since 2005 in energy projects in Central Asia, said Global Insights (Boston, MA, U.S.A.), driven by the need to secure oil and gas resources to support the country's growing economy.

A series of state-to-state energy-sector "cooperation" deals in the region have led to successful investment by Chinese state-owned companies.

Several major upstream hydrocarbon projects are moving forward, especially in Kazakhstan, but increasingly in Uzbekistan. It also appears that China and Turkmenistan are poised to establish a relationship that could see Turkmen oil and gas eventually delivered east to China.

Depending on the degree of cooperation achieved by Russia and China, the combined weight of their investments could further marginalize U.S. and European political influence in Central Asia, which is already secondary to the neighboring powers.

Cooperation with Asian neighbors

In April 2006, China signed an agreement with Turkmenistan. One provision involves sale of gas by Turkmenistan to China and construction of a pipeline to deliver it. China will purchase 30 bcm/year (2.9 bcfd) of gas over 30 years, starting in 2009, when the proposed Turkmenistan-China gas pipeline is commissioned.

Another provision focuses on joint exploration and development of all fields on the right bank of Amu Darya River under a production-sharing agreement.

The agreement with Turkmenistan added to a multibillion-dollar series of deals China made recently with other Asian countries to import oil and gas:

- Kazakhstan opened a pipeline to deliver oil to China in December 2005.
- Laos, Myanmar, and Thailand agreed to ship oil to China along the Mekong River, as an alternative to shipment through the dangerous Straits of Malacca.

CNPC in Africa

China's CNPC also has been strengthening its oil and gas investment in Africa:

- Niger: The company started drilling the wildcat Saha-1 well at the Tenere block, and will carry out similar operations at the Bilma block.
- Algeria: CNPC started exploration and development projects and signed in November a cooperation agreement with Sonatrach to look for development opportunities. The company previously expressed interest in building a 15 mtpy greenfield refinery at Tiaret.

M&As abroad

The Chinese central government has been urging state-owed oil companies to intensify merger and acquisition activity abroad to help meet the country's growing oil and gas demand:

- CNOOC announced plans to acquire a 45% interest in Oil Mining License (OML) 130 off Nigeria.
- Chinese companies have shown an interest in Canadian oil sands. In 2005, Sinopec Group paid $105 million for a 40% interest in Synenco Energy Inc.'s proposed $4.5 billion Northern Lights project. And CNOOC paid $122 million in 2005 for a 16.69% share of private MEG Energy Corp., which plans a steam-assisted gravity drainage project near Fort McMurray, AB, Canada.
- Gas Authority of India Ltd. (GAIL) signed a memorandum of understanding in early 2006 with Sinopec and CNOOC to cooperate in exploration, development, and/or production efforts.

Asia Pacific

AFGHANISTAN

Capital: Kabul
Monetary unit: Afghani
Refining capacity: 0
Oil production: 0
Oil reserves: 0
Gas reserves: 1.75 tcf

Resource base reassessed

Afghanistan's undiscovered oil and gas resource base is much larger than previously understood, said the U.S. Geological Survey after a multiyear study.

The agency estimated mean undiscovered resources at 15.7 tcf of gas, 1.6 billion bbl of oil, and 562 million bbl of natural gas liquids. Most of the gas is indicated to be in the Amu Darya basin and most of the oil in the Afghan-Tajik basin.

The two basins cover a combined 518,000 sq km (200,000 sq miles) in northern Afghanistan, and the USGS assessed 85,500 sq km (33,000 sq miles) of that. It subdivided four petroleum systems into eight assessment units.

Formations with the greatest potential for gas are Upper Jurassic carbonate and reef reservoirs beneath an impermeable salt layer in relatively unexplored parts of northern Afghanistan. Greatest potential for oil is in Cretaceous to Paleogene carbonate reservoir rocks associated with thrust faulting and folding in the Afghan-Tajik basin.

Afghanistan has sought to mount a licensing round, but no exploration is reported in the country in recent years.

The U.S. Trade and Development Agency funded the study, which USGS conducted jointly with the Afghanistan Ministry of Mines and Industry.

Assessment results are available at the Web site of the USGS Central Energy Team: http://energy.cr.usgs.gov/oilgas.

Afghanistan's Amu Darya and Afghan-Tajik Basins

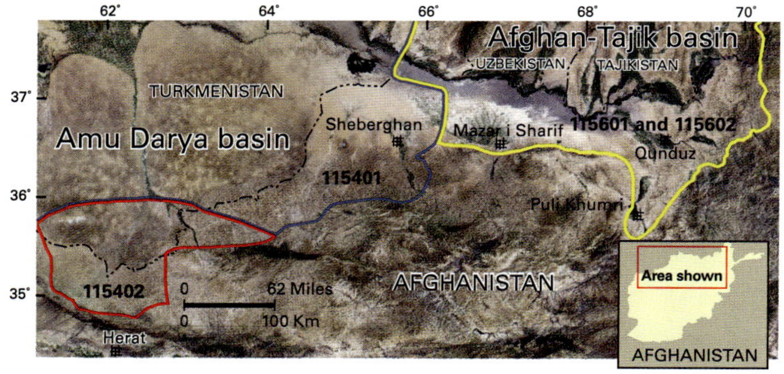

Amu Darya Jurassic-Cretaceous Total Petroleum System (115401, within blue line); Kalaimor-Kaisar Jurassic Total Petroleum System (115401, within red line); Afghan-Tajik Jurassic Total Petroleum System (115601, within yellow line); Afghan-Tajik Paleogene Total Petroleum System (115602, within yellow line). Some system boundaries extend west and north behond the view of the image.

Source: Image from National Geospatial Intelligence

Afghan Assessment Units and Undiscovered Volumes*

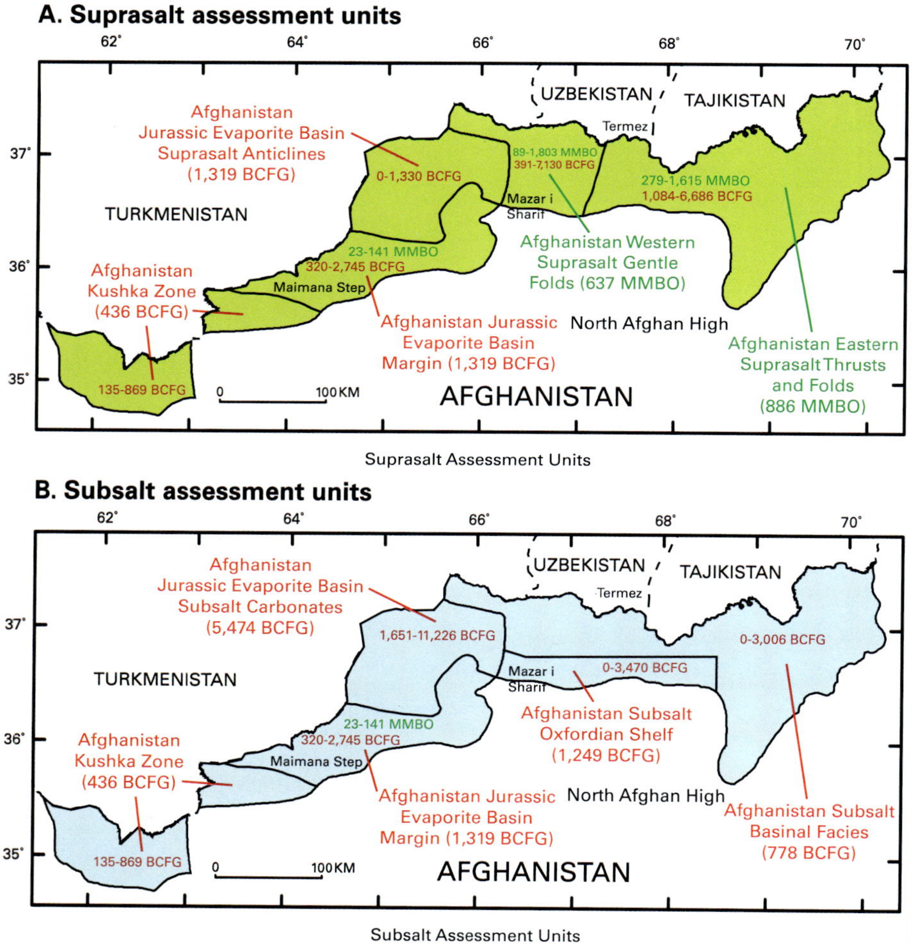

*The Afghanistan Jurassic Evaporite Basin Margin Assessment Unit (1150107) and the Afghanistan Kushka Zone Assessment Unit (11540201) encompass both subsalt and suprasalt intervals.

Topping plant completed

Privately held Oxiana Energy Inc. (Princeton, NJ, U.S.A.; Tashkent, Uzbekistan) disclosed the completion in 2005 of a 3,000 b/d topping plant at Angot oil field, in the Sar-i-Pol province of northern Afghanistan. Oxiana owns the facility with local partner Shamal Inter Neft.

Oxiana was awaiting permission under a new Afghan hydrocarbon law to participate in the rehabilitation and development of nearby oil fields, which will provide feedstock for the plant.

AUSTRALIA

Capital: Canberra
Monetary unit: Dollar
Refining capacity: 704,659 b/cd
Oil production: 450,000 b/d
Oil reserves: 1.592 billion bbl
Gas reserves: 30.37 tcf

Background

Oil

Australia is slipping in its attempts to maintain self-sufficiency in oil, an *Oil & Gas Journal* assessment said. Despite a rally during 2005–2006 as a result of some new production in the Carnarvon basin off Western Australia, the outlook is sobering.

The country's total expected crude oil and condensate production rate for 2006 was about 450,000 b/d, well behind the peak production of over 800,000 b/d achieved in 2000.

There were no significant oil discoveries from third quarter 2004 to first quarter 2006, and if success in exploration does not improve,

Northern Afghanistan

the nation's self-sufficiency in liquid petroleum could fall to about 50% by 2010—a marked contrast to the 95% level in 2000.

Natural gas

On the other hand, natural gas discoveries and production are increasing. Three main finds in 2005—Pluto and Hurricane in Carnarvon basin, and Caldita in the Timor Sea—recorded substantial gas accumulations. Most of the development projects brought on stream in 2006 and those under construction will provide gas for both domestic consumption and for export as LNG.

Total gas production expected for 2006 is 45.5 bcm, up from 41.2 bcm in 2005. LNG production of 10.6 mtpy in 2005 (about 40% of total gas production) was expected to rise to over 13 mtpy in 2006.

Australia's gas scene is also enhanced by an increasing production of methane from coalbeds, mostly in southeast Queensland. CBM now supplies 30% of Queensland's gas demand.

Addressing flagging oil output

The Australian government, petroleum industry, and researchers launched a plan aimed at increasing oil production, making Australia one of the world's top five LNG exporters, and doubling domestic use of natural gas.

The action was driven by increasing concern about falling oil and condensate production. The plan establishes an alliance between the upstream oil and gas industry, represented by the Australian Petroleum Production and Exploration Association (APPEA), and the federal, state, and Northern Territory governments.

It aims to increase economic growth and resource security. Strategic goals are to ensure that by 2015:

- Oil self-sufficiency levels will remain at least at 2006 levels;
- LNG production will exceed 50 mtpy;
- Natural gas as a competitive feedstock for processing will be doubled;
- Gas will provide primary energy for 70% of all new electric power generation capacity.

An issues paper was being circulated for public comment, and a final strategy document was to be published by early 2007.

Exploration and Development

Pluto field

Based on further appraisal drilling, Woodside Petroleum (Perth) said the best estimate of dry gas in its Pluto field, 190 km (118 miles) off the Burrup Peninsula, is 4.1 Tcf, or about 14% more than the last estimate at 3.6 Tcf.

Also, Woodside said the Burrup Peninsula will be the site for an LNG plant to process production from the Pluto field. Site work was to begin by year-end 2006. A final investment decision is expected by mid-2007, with first LNG production anticipated for 2010 or 2011.

Woollybutt field

A joint venture led by ENI Australia approved a budget to develop the South Lobe section of Woollybutt oil field in the Carnarvon basin off Western Australia.

The new project, in production license WA-25-L, will connect two horizontal wells (Woollybutt 4 and 6) to an FPSO vessel through subsea pipelines.

Well drilling was slated for midyear 2007, with first production of 10,000 b/d expected in 2008.

WA-205-P permit

Chevron Corp. discovered natural gas with its Clio-1 wildcat in 915 m (3,000 ft) of water on permit WA-205-P, about 145 km (90 miles) off northwestern Australia. It was completed in September, 20 km (12 miles) northwest of Gorgon gas field.

Drilled to a TVD of 4,725 m (15,500 ft), the well encountered 190 m (623 ft) of net gas sands in the Triassic Mungaroo formation. Chevron planned to start a follow-on 3D seismic survey by year-end.

Georgina basin

Independent Texalta Petroleum Ltd. (Calgary) and its Australian partners were granted Exploration Permits EP 103 and EP 104, covering approximately 5.5 million acres (8,595 sq miles) in the unexplored Georgina basin northeast of Alice Springs. Reprocessing of existing seismic data was initiated.

All of the southern Georgina formations are thought to lie no deeper than 1,000 m. Texalta said the geology is similar to that of the Williston basin in North America, and government researchers said that individual trapping situations could contain oil and gas reserves in excess of 100 million bbl.

Basker field

Anzon Australia Ltd. (North Sydney) moved toward production start-up from the lower interval of Basker-2 well in Basker oil field, in the Gippsland basin, 75 km (47 miles) off Victoria. Basker-Manta Joint Venture developed the field.

The Basker-2 well started production at 9,500 b/d of oil from the upper of two intervals; average flow then dropped to 8,000 b/d. Commingled flow from the two intervals was expected by mid-2006.

The well, in 155 m (508 ft) of water in Bass Strait, is a subsea completion, linked to the first FPSO ever used in the Gippsland basin.

Angel field

Woodside Energy Ltd. (Perth) and its partners in the North West Shelf Venture were launching a $1.6 billion (Aus.) project to develop the Angel gas and condensate field off Western Australia. Three production wells were planned for 2006–2007, with production due on stream in 2008.

A remotely operated processing platform will be located in 80 m (262 ft) of water about 49 km (30 miles) east of the venture's North Rankin production facility, to which it will be connected by a new pipeline.

Block W05-16

Chevron Corp. subsidiary Chevron Australia Pty. Ltd. and partners received exploration rights to Block W05-16 in the Carnarvon basin off northwest Australia. The permit covers 2,645sq km (1,020 sq miles) adjacent to the gas fields of the Gorgon project. Seismic work was to begin in 2006.

Puffin program

Peak Group (Aberdeen, Scotland) began its first multiwell, multioperator well management program off Australia. The program covers eight wells and includes three other operators: Santos Ltd. (Adelaide), Eni SPA (Rome), and ConocoPhillips (Houston).

The Puffin 9 exploratory well was to be drilled to 3,000 m (9,842 ft) below the seabed in the Puffin oil field of AED Oil Ltd. (Melbourne), on permit AC/P22 in the Timor Sea.

Two development wells will follow—Puffin 7 and Puffin 8—to be completed subsea and tied back to an FPSO vessel.

VIC/P46 permit

Beach Petroleum Ltd. said it will expedite exploration and possible development of the VIC/P46 exploration permit in the Otway basin. Beach was to become operator of the permit under a farmout agreement with Essential Petroleum Resources Ltd. (South Melbourne).

Venture partners said numerous prospects have been identified and were planning a $20 million 3D seismic and drilling program. Beach said drilling would begin in 2007.

South Australia

Adelaide Energy Pty. Ltd. announced plans to drill as many as four onshore wells over the next five years, targeting oil in southeast South Australia. The first appraisal well is to spud by mid-2007.

The program follows the company's recent award of PEL 255, which includes Block OT2006-A in the Otway basin.

Reindeer-Caribou field

Apache Corp. was evaluating options for natural gas and condensate development based on test data from its Gnu-1 exploration well in the Reindeer-Caribou field, on the WA 209P permit in the Carnarvon basin.

The Gnu-1 well, in 61 m (200 ft) of water, was drilled into the Early Jurassic North Rankin formation and on test produced 26 MMcfd of gas and 61 b/d of condensate. Production could start in 2008.

Blacktip gas field

Eni Australia Ltd. began development of Blacktip natural gas field, 330 km (205 miles) southwest of Darwin, in the southern Bonaparte Gulf. Plans include a 108-km (67-mile) subsea pipeline to an onshore treatment plant to be built at Wadeye in the Northern Territory, capable of treating 1.3 bcm/year of gas (4.59 bcf/year).

First gas from the Blacktip project is expected on stream in early 2009. The field has a reserves estimate of 1.2 tcf of gas plus 150 million bbl of condensate.

Perseus-over-Goodwyn project

The Perseus-over-Goodwyn project headed by Woodside Energy (Perth) is located 130 km (81 miles) offshore Western Australia in Block WA-1-LR, in water 130 m (426 ft) deep.

Phase-one plans call for drilling three subsea wells on Perseus that will be tied into the Goodwyn platform production facilities by a 22-km (14-mile) subsea pipeline. Gas and condensate then will be piped to the North Rankin platform and from there to the onshore LNG facility on the Burrup Peninsula.

The new wells are expected to go onstream during the first half of 2007.

Stybarrow field

Located 65 km (40 miles) northwest of Exmouth in Block WA-225-P(2), the Stybarrow field is one of the deepest development areas in Australia, in 800 m (2,625 ft) of water. BHP Billiton (Melbourne) and Woodside Energy (Perth) continued development work, aiming for production start-up in 2008 at a level of about 80,000 b/d.

Greater Sunrise fields

This complex gas development project straddles the joint development zone between Australia and East Timor in the eastern Timor Sea. Details are presented in the East Timor section of this document.

Thylacine-Geographe fields

Woodside Energy and its partners continued development of the Thylacine-Geographe gas fields in the offshore Otway basin, which are expected to come on stream in 2007.

Gas will flow by subsea pipeline to onshore treatment facilities near Port Campbell. The two fields combined are expected to produce 950 bcf of raw gas, 12.2 million barrels of condensate, and 1.7 million tonnes of liquid petroleum gas.

Kipper field

A consortium led by ExxonMobil Corp. began design work to develop the Kipper gas-condensate field in the Gippsland basin of southeast Victoria.

The field is in 100 m (328 ft) of water about 45 km (28 miles) off the coast. Holding about 17.5 bcm (620 bcf) of gas and 30 million bbl of condensate, it is the largest undeveloped gas field in the region.

Production would come ashore for treatment in the ExxonMobil-BHP Bass Strait production facilities at Longford in Victoria. Gas would enter the system through the West Tuna platform.

If a final investment decision is positive in 2007, production could begin by 2010.

Drilling and Production

Cliff Head field on stream

First commercial offshore production in the Perth basin off Western Australia began at the Cliff Head, said operator ROC Oil Co. Ltd. (Sydney) and its partners. The first export of crude was delivered to BP Australia Ltd.'s Kwinana Unloading Facility in May.

The field is located on the WA-31-L production license in 18 m (59 ft) of water, 11 km (7 miles) off Geraldton in the Indian Ocean.

Initial production in July was 12,500 bpd. Cumulative production reached 1 million bbl in September and 2 million bbl in December. Proved and probable reserves are estimated at 14 million bbl.

Bayu-Undan production

Development of the Bayu-Undan fields in the Timor Sea reached a milestone with first transport of gas in early 2006 to the new single-train LNG facility at Wickham Point in Darwin. LNG then was loaded into a Japanese carrier.

Northern Australia's first LNG production project, the $2 billion (Aus.) ConocoPhillips Darwin facility can process 3.24 mtpy. Gas flows to the plant through a 500-km (310-mile) subsea pipeline.

Plans called for initial delivery of three collective spot LNG cargoes to Tokyo Electric and Tokyo Gas. Together, TE and TG will buy 3 mtpy over 17 years.

ConocoPhillips has approval to produce as much as 10 mtpy and was seeking additional gas supplies and markets to justify a second and perhaps a third LNG train.

Enfield field

Production began in July from five subsea wells in the Enfield field in Carnarvon basin to an FPSO vessel, with a rate of 74,000 b/d achieved in September. The field is off Western Australia, 50 km (30 miles) northwest of Exmouth.

However, flow had dropped by year-end and partners Woodside Energy (Perth) and partner Mitsui E&P Australia announced plans for additional wells and a workover of existing wells in 2007.

Yolla field

Gas production began in May from the offshore Yolla field in the Tasmanian Bass basin. Development was conducted by a consortium headed by Origin Energy (Sydney).

A 147-km subsea line carries gas to the Victorian coast, and a 32-km (20-mile) onshore section then takes it to a gas treatment plant at Lang Lang about 80 km (50 miles) southeast of Melbourne.

The field complex is estimated to hold 307 bcf of gas, 14 million bbl of condensate, and 800,000 bbl of oil in recoverable reserves.

Casino gas field on stream

Production began from Casino gas field operated by Santos Ltd. (Adelaide) in the offshore Otway basin of western Victoria. Casino is on permit Vic/P44 about 30 km (19 miles) offshore in 70 m (230 ft) of water. The field has been developed using subsea wellheads and a subsea pipeline.

First gas went to TRUenergy's Iona onshore processing plant near Port Campbell.

Callawonga-1 production

Oil production began in November from Callawonga-1, on PEL 92 about 7 km (4 miles) north of Christies oil field and 90 km (56 miles) northwest of Moomba. Operator Beach Petroleum NL (Adelaide) describes it as the most westerly output achieved in the Cooper/Eromanga region.

The Callawonga-1 well cut 8 m (26 ft) of net oil pay in Jurassic Namur sandstone. Initial free-flow production was 2,100 b/d. Beach estimates P50 reserves at Callawonga field at 1.5 million bbl.

OTWAY BASIN FIELDS

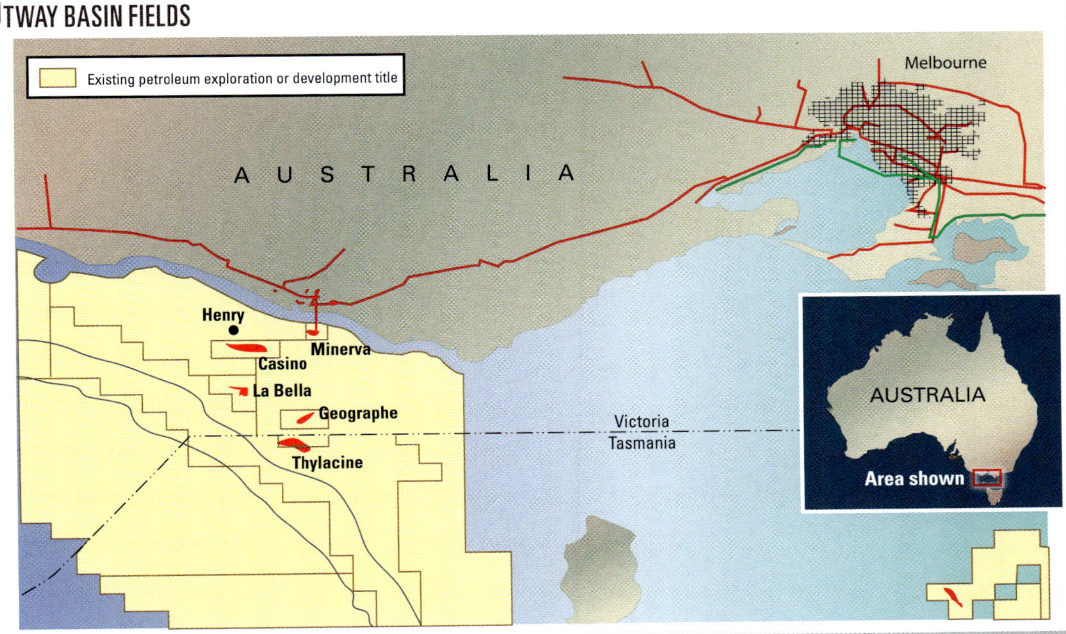

Greater Gorgon Project

The Greater Gorgon Project is the largest among several potential LNG export projects in Western Australia. It is being pursued by the Australian subsidiaries of Chevron Corp., ExxonMobil Corp., and Shell. The $15 billion (Aus.) project, 130 km (81 miles) off Western Australia, involves several elements:

- Development of the Greater Gorgon gas fields, including subsea pipelines to Barrow Island
- Two 5-mtpy LNG trains on Barrow Island for liquefying produced gas
- An LNG shipping facility
- Assessment of a domestic gas plant and pipeline to deliver gas to the mainland
- Greenhouse gas management by injecting carbon dioxide extracted from the natural gas into deep formations beneath Barrow Island

The plan was restructured in early 2005 to include nearby ExxonMobil-operated Jansz field, and the framework agreement gave Chevron 50% of the project, while ExxonMobil and Shell have 25% each.

Gorgon joint venture partners have awarded contracts for upstream and downstream elements of the project. Progress slowed in early 2006, however, due to escalating costs, concerns about environmental impacts on Barrow Island (a nature reserve), and uncertainty about sales agreements for some of the produced LNG volumes.

Environmental approval given

In December 2006, Western Australia's environmental minister cleared the way for environmental approval to be granted for the Gorgon Project. He overruled a mid-2006 decision by the Western Australia Environmental Protection Authority that had cast doubt on the project.

However, the project partners now must commit an additional $60 million (Aus.) to comply with certain environmental directives.

The group also will be required to contribute $40 million (Aus.) toward resolving environmental issues in the Pilbara and West Kimberley regions.

The project also needs final approval from the federal environmental minister.

Government grant for CO_2 demo

The project received an Australian government grant of $60 million (Aus.) towards the proposed carbon capture storage project on Barrow Island. The $850 million (Aus.) carbon dioxide geo-sequestration project is expected to be the largest of its kind in the world.

The plan is to inject 125 million tonnes of excess CO_2 produced during the life of the LNG project. Gas from the Gorgon field has about 12% CO_2 content.

A commercial-scale demonstration project will liquefy the CO_2 and inject it into reservoirs 2.5 km (8,200 ft) below the island. Government funding is conditional on meeting environmental approvals.

Sale of LNG output

Chevron signed letters of intent to sell 1.5 mtpy of LNG to Osaka Gas Corp. and a total of 2.7 mtpy to Tokyo Gas and Chubu Electric Power Co. Inc.

Shell was reported to be negotiating the sale of 14 mtpy over 20 years with India's Gujarat State Petroleum Corp.(Gandhinagar). ExxonMobil had not announced any sales at year-end.

Exploration activity

Chevron Australia Pty. Ltd. won the rights to deepwater exploration permit area WO5-13. The area is 160 km (100 miles) northwest of the Western Australian coast in the Exmouth Plateau region. The plateau is the deepwater frontier of the Carnarvon basin, which includes the North West Shelf and the Greater Gorgon Area.

The company also reported an indicated gas discovery at its Chandon-1 wildcat on permit WA-268-P in the Greater Gorgon region. It did not run tests but called the discovery "sizable."

Cooper basin activity

Santos Ltd. (Adelaide) announced plans to drill 100 wells in the Cooper and Eromanga basins in a new oil exploitation program. The initial program targets 6–7 million bbl of proven plus probable net reserves. Contingent on success, Santos might drill an additional 70 wells, targeting an additional 4–5 million bbl of proven plus probable reserves.

Beach Petroleum Ltd. (Adelaide) was drilling three wells in the Cooper basin in 2006. The company announced a new pool oil discovery in the Poolowanna formation in the Sellicks field. Beach holds interests in nine exploration areas and four oil production joint ventures in the Cooper/Eromanga basins.

Stuart Petroleum Ltd. (Adelaide) holds 5,051 sq km (1,950 sq miles) in PEL90, PEL93, PEL102, and PEL113 in the Cooper basin. The company drilled seven Cooper wells in the first half of 2006, following four earlier discoveries.

Minor shareholder Australian-Canadian Oil Royalties Ltd. (ACOR) said the Tintaburra oil field may contain 84 million bbl of proved plus probable oil in place.

LNG

Kwinana terminal

Industrial conglomerate Wesfarmers Ltd. (Perth) announced plans to build a $138 million (Aus.), 175-tpd LNG terminal and regasification plant in Kwinana, south of Perth. Construction was to start in late 2006, with the plant coming online in 2008.

The plant will supply natural gas for heavy-duty trucks as well as domestic industrial and electric power generation markets. During the first 10 years of plant operation, Santos Ltd. (Adelaide) will supply up to 37 PJ (34.4 bcf) of gas feedstock from its share of production in the offshore John Brookes field.

Browse project

Woodside Energy Ltd. (Perth) hired Foster Wheeler Energy Ltd. to evaluate technical issues related to a proposed offshore LNG development to process gas from the Browse basin off Western Australia. Contract value was not disclosed.

The Browse LNG project comprises the East and West Browse joint ventures operated by Woodside. Fields involved would be Torosa, Brecknock, and Calliance, in 35–700 m (115–2,300 ft) of water, all approximately 400 km (250 miles) offshore.

Two successful appraisal wells were drilled in 2005, and four more were planned for 2006. Total gas reserves are estimated at 20 tcf with 300 million bbl of condensate.

Projects under study

Billiton BHP proposes to bring gas ashore from the 1979 Scarborough gas discovery on the Exmouth Plateau to an

GORGON, OTHER WESTERN AUSTRALIA LNG

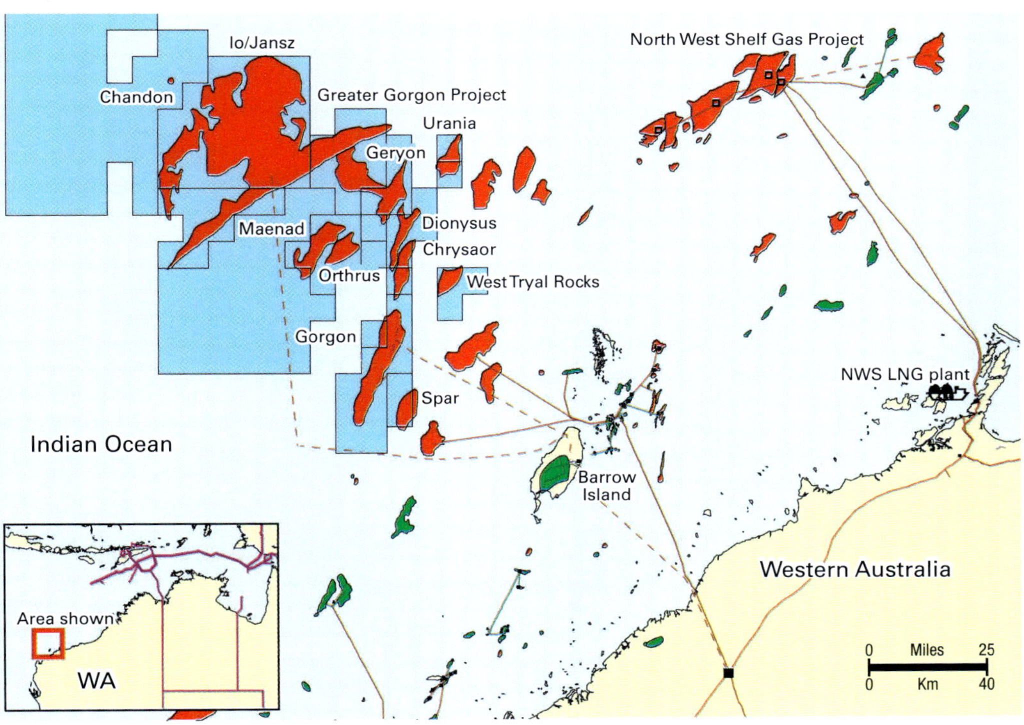

Source: Wood Mackenzie PathFinder.

LNG plant to be built at Onslow, Western Australia. The LNG would go to its proposed Cabrillo Port regasification plant off California.

Methanol Australia Ltd. (Melbourne) has proposed the Tassie Shoals artificial island LNG and methanol project in the Timor Sea. Initial plans call for gas from the nearby Evans Shoal gas field to supply the plants, which will be built on concrete islands set on the seafloor in about 70 m (230 ft) of water.

Coalbed Methane

GAIL, Arrow joint work

Gas Authority of India Ltd. and Arrow Energy NL (Brisbane) signed a memorandum of understanding outlining areas for cooperation and investment in CBM projects in Australia and India. The joint venture may also involve other parties.

The alliance will seek to export gas to Asia as GTL products and compressed natural gas.

GAIL will invest in as many as ten of Arrow's CBM development projects in Queensland. Top priorities are sites in the Clarence-Moreton basin, the Styx River basin, and the Nagoorin Graben basin.

Stratford prospect

Renewed testing at the LMG03 CBM well in the Gloucester basin resulted in a peak flow rate of 537 Mcfd, said Molopo Australia Ltd (Melbourne).

The lowermost target seam is at 405 m (1,330 ft) at the well, on the Stratford prospect, 100 km (62 miles) north of Newcastle. Lucas Coal Seam Gas (Pty.) Ltd. (Sydney) is the operator.

The initial flow rate was more than 400 Mcfd after reduced pumping to dewater.

The companies also were drilling LMGC1 at 258 m (846 ft). It has intersected several anticipated seams and intervals marking the Craven Subgroup in Gloucester Coal Measures.

Surat basin field

Arrow Energy NL (Brisbane) made its first natural gas deliveries from Kogan North coalbed methane field in Queensland. The project is the first to produce CBM from this part of the Surat basin.

Gas was flowing into the Roma-Brisbane pipeline at 2 TJ/day (1.86 MMcfd) to CS Energy Ltd., which is Arrow's 50-50 partner and the Queensland government's explorer and gas utility.

The venture will supply 4 PJ/year (3.72 bcf/year) of gas to CS Energy over 15 years.

Refining and Processing

From 2004 to 2006, Australia's four oil refiners—Caltex, Shell, BP, and ExxonMobil—focused considerable effort on clean fuel development. The companies spent an average $400 million each in upgrading plants and equipment at the country's seven refineries to comply with federal specifications for diesel with less sulfur content and gasoline with less benzene content.

One recent proposal envisions a GTL plant in central Australia based on Amadeus basin gas discoveries. A second proposes a $450-million condensate processing plant for Port Darwin, using feedstock from condensate associated with gas production in the Timor Sea. Construction of the GTL plant could begin in 2007, with an operational date of 2009.

Pipelines

Dampier-Bunbury expansion

Industrial growth in Western Australia prompted an increase in the size of the proposed Stage 5 expansion of the Dampier-Bunbury natural gas trunkline to handle an additional 375 TJ/day (349 MMcfd) of natural gas to come online around 2008.

Dampier Bunbury Pipeline Co. was working to secure government approvals and to finalize renegotiated customer contracts by mid-2006.

The project will lay 1,150 km (715 miles) of pipeline alongside the existing pipe. The current Stage 4 expansion of 200 km (125 miles) of looping line will supply an extra 100 TJ/day (93 MMcfd).

PNG-Australia gas pipeline dead

Australian Gas Light Co. and Malaysia's Petronas, partners in the proposed 3,000-km (1,864-mile) gas pipeline from the Southern Highlands of Papua New Guinea (PNG) to

Australia, decided in August to stop work on its Australian leg. AGL cited a lack of critical foundation customers and escalating costs as the reasons for the decision.

Gas for the pipeline was to have been produced under the banner of the PNG Gas Project. In early 2007, project partners—Oil Search, ExxonMobil, AGL, Nippon Oil, and the PNG government—abandoned plans to proceed with the pipeline to Australia. Oil Search said the partners would instead develop the gas fields for LNG and petrochemical production.

Petrochemicals

Ammonium plants

Indian-owned Burrup Fertilisers Pty. Ltd. started production from its $800 million (Aus.) ammonia plant on the Burrup Peninsula of northwest Western Australia.

The new facility can produce as much as 760,000 tpy of liquid ammonia, using natural gas from fields on the North West Shelf. The first shipment was scheduled for mid-2006.

Plenex Ltd. (Melbourne) withdrew from a proposal to build second large ammonium plant on the peninsula, citing cost and feedstock-supply issues. Plenex said its partner, explosives manufacturer Dyno Nobel Ltd. (Sydney,) would continue to study the feasibility of the plant, but the project is in doubt.

Environmental Issues

Otway CO_2 trial

The Australian Cooperative Research Centre for Greenhouse Gas Technologies (Canberra) was awarded production licenses PPL11 and PPL13 in the onshore Otway basin of western Victoria in which to conduct its first carbon dioxide sequestration trial. The licenses cover Buttress CO_2 field and nearby depleted Naylor gas field.

The pilot gas injection, storage, and monitoring program was to begin by year-end 2006. Under the program, CO_2 produced from Buttress (3 MMcfd) will be piped to Naylor and injected through a well into the Cretaceous Waarre reservoir. Injection will continue for two years.

Monitoring was initiated to establish baseline data and will continue for 4–5 years to gauge the movement of CO_2 in the reservoir.

Funding will come from the Victorian government, petroleum companies, and overseas research groups. Results will be particularly relevant to work in the Gorgon field, the gas from which contains 12% CO_2.

BANGLADESH

Capital: Dhaka
Monetary unit: Taka
Refining capacity: 33,000 b/cd
Oil production: 4,000 b/d
Oil reserves: 28.0 million bbl
Gas reserves: 5.0 tcf

Block 10, Begumganj field

Three state energy companies announced in November plans to develop Begumganj gas field onshore on Block 10, 177 km (110 miles) southeast of Dhaka. The field has proved reserves of 32 bcf but is thought to potentially hold as much as 400 bcf.

Participants are Bangladesh Petroleum Exploration & Production Co. Ltd. (Bapex), Bangladesh Gas Fields Co. Ltd., and Sylhet Gas Fields. The joint development is a first for the country, said the energy ministry. Production of approximately 10 MMcfd is expected.

Block 9, Narsingdi field

Bapex began drilling a second well in October on the Narsingdi gas field in onshore Block 9. The field is owned by Bangladesh Gas Fields Limited (BGFL).

The field is already producing approximately 20 MMcfd from the first well. This production rate is expected to double when the second well comes onstream in early 2007.

Petrobangla, the national oil company, estimates total recoverable gas reserves in the Narsingdi Field at approximately 215 bcf.

Exploration activity in Bangladesh

Sangu, Hatia drilling

Cairn Energy PLC (Edinburgh, Scotland) was planning to drill three new wells offshore Bangladesh in the Bay of Bengal by year-end.

Sangu-10 is a production well on the Sangu Field. Sangu is Bangladesh's only offshore gas field to date. The others will be an appraisal well on the South Sangu structure and an exploratory well at Hatia Island. If time allows, an appraisal well might also be drilled at Meghama Island.

Cairn said that Meghama and Hatia may contain reserves of more than 4 tcf.

Gas pipeline to India

According to press reports in August, the Indian government decided to alter the route of a proposed major gas pipeline so that it will not pass through Bangladesh. The line would carry gas from the giant Myanmar gas fields in Blocks A-1 and A-3 in the Bay of Bengal to Calcutta.

Instead, the pipeline will reportedly be routed through India's northeastern states. The change is expected to triple the cost of the pipeline but avoid the costs of transit fees and right-of-way acquisition.

BRUNEI

Capital: Bandar Seri Bagawan
Monetary unit: Dollar
Refining capacity: 8,600 b/cd
Oil production: 197,000 b/d
Oil reserves: 1.10 billion bbl
Gas reserves: 13.8 tcf

Oil flow at Champion West field

Brunei Shell Petroleum Co. Sdn. Bhd. (BSP), a 50-50 joint venture of Royal Dutch Shell PLC and the Brunei government, reported in early 2006 the start of oil production from Phase III of the Champion West field, 90 km (56 miles) offshore. The well flowed at an initial rate of 16,700 b/d, and also was producing natural gas.

Seria North Flank onstream

BSP initiated oil production from its Seria North Flank wells in late 2006.

The Seria Field is the oldest producing field in Brunei Darussalam, and the Seria North Flank project is one of several projects aimed at extending production from mature fields.

Blocks L and M awarded

Brunei National Petroleum Company (PetroleumBrunei) awarded exploration rights to Block L to Loon Energy (Calgary) and its partner QAF Brunei (Bandar Seri Begawan, Brunei), and Block M to a group comprising ChinaOil (a subsidiary of Petrochina International), Valiant International Petroleum (London), and Jana Corporation.

Block L, largely unexplored, covers an area of 2,253 sq km (870 sq miles) that includes Brunei-Muara District, the Jerudong area, and part of the Tutong District. The giant Champion and Seria oilfields are located within 25 km (16 miles) of the block.

Block M covers 3,011 sq km (1,163 sq miles). Belait Field, a small part of Block M, produced the first oil found in Brunei.

Under the production sharing agreement. Loon and QAF are to invest a minimum of $20 million (U.S.) during the first three years of the license to reprocess existing seismic, acquire new seismic, and drill at least two exploration wells.

EAST TIMOR

Capital: Dili
Monetary unit: U.S. dollar
Refining capacity: 0
Oil production: N/A
Oil reserves: N/A
Gas reserves: N/A

Offshore permits awarded

East Timor offered six offshore petroleum exploration permits at mid-year in its inaugural bidding round. Nine bids were made for the six blocks. Eni SPA (Rome) won five and Reliance Industries Ltd. (Mumbai, India) received one.

All permits are off East Timor's southern coast. None are in the Joint Development Area administered with Australia. Water depths vary from shallow to as deep as 3,000 m (9,840 ft) in the Timor Trench.

Reliance received Block K in the far east of the offshore region. Eni received Blocks A, B, C, E, and H. Blocks A and B are in the north, abutting the southern East Timorese coast.

Barossa-1 well test

A joint venture of ConocoPhillips (Houston) and Santos Ltd. (Adelaide, Australia) recorded strong gas flow from its Barossa-1 well on Permit NT/P69 in the eastern Timor Sea, 295 km (183 miles) northeast of Darwin, Australia. The main test flowed gas at 30 MMcfd and 7–9 bbl/MMcf of condensate.

Development of the field is part of a program to provide reserves to feed a second LNG train at the venture's Darwin liquefaction plant. The reservoir lies about 4,250 m subsea in 233 m of water.

Shell drilled two earlier wells on the structure, most recent in 1998, but abandoned the prospect as uneconomic at the time.

AC/P34 find

Coogee Resources Ltd. (Perth) reported a second oil discovery at its Swallow-1 well in Ashmore Cartier Permit AC/P34 in the Timor Sea. The wildcat intersected a 25-m (82-ft) oil column in the primary Plover formation objective.

The company said Swallow could hold as much as 2 million bbl of recoverable oil and produce at rates in excess of 5,000 b/d. The find follows the earlier discovery at Coogee's Swift North-1 well.

Combined total reserves around Montara are estimated at 38 million bbl of recoverable oil. All accumulations are within tieback range of the planned Montara FPSO facility. Production should start in mid-2008.

Puffin-9 well discovery

AED Oil Ltd. (Melbourne) confirmed an oil discovery in two zones in its Puffin-9 exploration well in the Timor Sea. The well encountered 6 m (20 ft) of upper oil sand and 3.8 m (12.5 ft) of lower sand. Both sands are known to be oil-bearing in other parts of the complicated Puffin structure.

Plans were being made to drill the Puffin-7 development well to tap reserves at the original Puffin-1 wellsite drilled by ARCO in the 1970s and the subsequent Puffin-5 well. Puffin-7 was to be brought on stream by yearend.

Greater Sunrise fields

The governments of East Timor and Australia finalized an agreement in January 2006 to share royalties on a 50-50 basis from the Greater Sunrise gas fields of Woodside Petroleum Group (Perth) in the Timor Sea. Parliaments of both nations must ratify the agreement.

The field straddles the offshore border between Australian waters and the joint development zone, the Timor Gap, administered by both countries.

The accord also sets aside the pursuit of conflicting maritime boundary claims for a period of 50 years as long as the field development goes ahead within ten years of the treaty's effective date.

Woodside plans to pipe the gas to Darwin for treatment, probably involving an expansion of the existing ConocoPhillips treatment plant at Wickham Point.

Woodside stopped work on the $3.7 billion project in 2005 because of concerns about the sovereign risks involved. Also, because Woodside is working to bringing onstream its Pluto and Browse basin projects off Western Australia, a start-up date for the Greater Sunrise development is uncertain.

JPDA agreement

Australia and East Timor finalized an agreement in March 2006 to begin joint oil and natural gas exploration in the Timor Sea. The agreement covers the controversial Timor Sea Joint Petroleum Development Area (JPDA). The area was set up when East Timor became independent in 2002.

The agreement stipulates that 90% of hydrocarbon revenues will go to East Timor. According to private estimates, the Timor Sea could contain 500 million bbl of recoverable hydrocarbon liquids and 5 tcf of natural gas.

INDIA

Capital: New Delhi
Monetary unit: Rupee
Refining capacity: 2,255,540 b/cd
Oil production: 682,000 b/d
Oil reserves: 5.6246 billion bbl
Gas reserves: 37.96 tcf

Background

India's total hydrocarbon resources, including those in deepwater, are estimated at around 28 billion tonnes of oil equivalent. Analysts believe that more than 81% of the country's hydrocarbon potential remains to be explored.

Since 2000, four discoveries per year of oil and gas have been reported on average. However, 15 finds were reported in 2005, and India's total gas reserves have increased by over 60% since 2002.

Gas supply

According to FACTS, Inc. (Singapore), India's gas industry is at a crossroads because demand is outpacing domestic production. Although passage of the New Exploration Licensing Policy (NELP) in 1999 raised domestic gas prices and encouraged new exploration and development, domestic production alone will remain insufficient to meet rising domestic demand.

To improve security of supply, India has been seeking gas from a variety of sources, including increased domestic production, investments in oil and gas production in other countries, and gas imports. Among India's import options, liquefied natural gas (LNG) is expected to increase in importance.

Domestic gas production increased from 1.637 bcfd in 1995 to 2.6 bcfd in 2000. By 2008, some 550 MMcfd of additional gas is expected to be available from the Krishna-Godavari (KG) basin, increasing to 1 bcfd by 2010.

Until 2003, indigenous production—through state-owned companies Oil & Natural Gas Corp. (ONGC) and Oil India Ltd. (OIL)—provided the bulk of natural gas needs. With the opening of the hydrocarbon sector in the mid-1990s, private companies formed joint ventures (JVs) with multinationals or India's NOCs to explore, develop, and produce gas.

India's proven gas reserves are estimated at 38.8 tcf. Its reserves-to-production ratio is 31 years. Most reserves lie off northwest Mumbai in the Arabian Sea and in the state of Gujarat.

India's NOC[1] JV-private sector domestic gas production

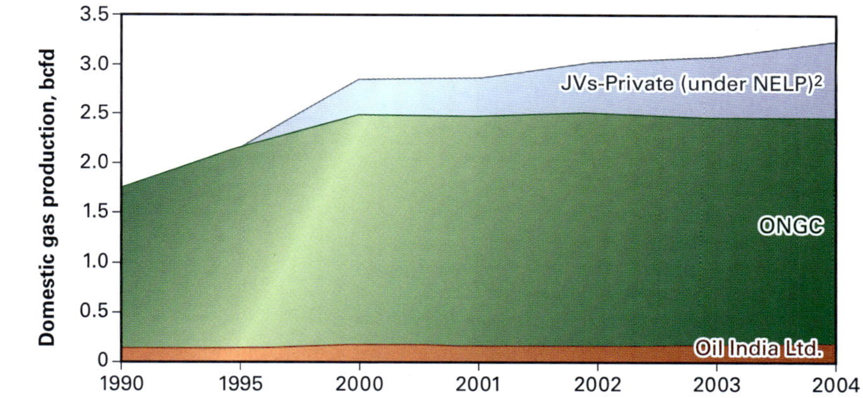

[1]National oil companies. [2]New Exploration Licensing Policy.
Source: SAIC analysis based on data from 2004-05 annual report of India's Ministry of Petroleum and Natural Gas, http://www.petroleum.nic.in

Gas supplies

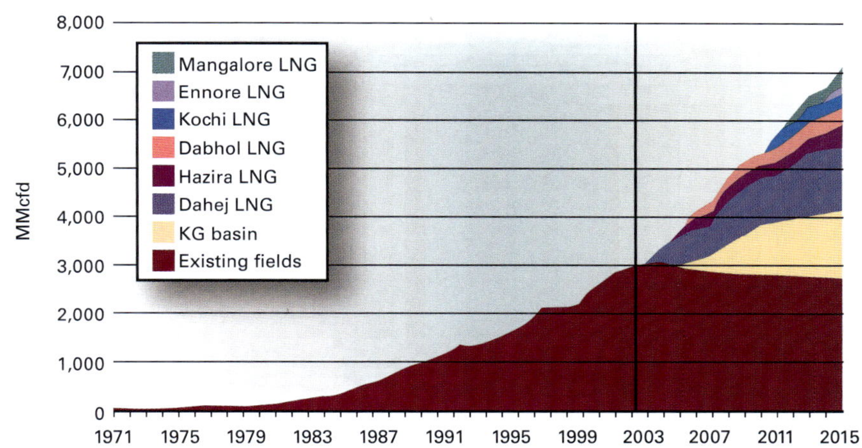

Key ONGC overseas investment in equity oil, gas

Country	Equity oil and gas participation	Participation start	Hydrocarbon involved
Russia	A 20% stake in Sakhlin-1 production-sharing contract. At $2.8 billion, it is India's largest foreign investment. Sakhlin-1 contains 2.3 billion bbl of oil and 17.1 tcf of natural gas reserves.	1996	Natural gas and oil
Vietnam	A 45% stake in a production-sharing contract agreement for a natural gas field with reserves of 2.04 tcf	1998	Natural gas
Myanmar	A 20% participating interest in "Shwe" natural gas field, with estimated recoverable reserves of 4-6 tcf	2000	Natural gas
Sudan	A 25% stake in Greater Nile Petroleum Operating Co. (purchased from Talisman Energy Inc., Calgary)	2003	Oil
Qatar	An appraisal, development, and production-sharing agreement for appraisal of the Najwat Najem oil structure in Qatar	2005	Oil
Syria	A 60% participating interest in exploration Block-XXIV	2004	Oil
Iran	A consortium of ONGC, Indian Oil Corp., and Oil India Ltd. has an exploration service contract for the Farsi block in the Persian Gulf	2002	Oil
Iraq	An exploration and development contract for Block 8 in Iraq's Western Desert	2001	Oil
Cuba	A 30% stake for deepwater exploration blocks in Cuba with a potential of over 4 billion bbl of oil (Cuban government approval pending)	2005	Oil and natural gas
Egypt	ONGC has entered into a concession agreement for petroleum exploration and exploitation in the North Ramadan area, Gulf of Suez	2005	Oil

Source: SAIC analysis based on information from Wall Street Journal

SOUTHERN ASIA
KAZAKHSTAN
UZBEKISTAN
TURKMENISTAN
KYRGYZSTAN
TADZHIKISTAN
IRAN
AFGHANISTAN
PAKISTAN
INDIA
NEPAL
BHUTAN
BANGLADESH
SRI LANKA
PEOPLE'S REPUBLIC OF CHINA
Arabian Sea
Bay of Bengal
Andaman Sea
Kumkoi
Pridorozhnoye
Ayrakty
Darvaza-Zeagli
Shikh
Topdzhulba
Kuruk
Achak
Gazil
Chardzhou
Ashkhabad
Samarkand
Tashkent
Chemkent
Fergana Fields
Izbaskent
Mali-Su
Fergana
Dzhalalabad
Bishkek
Alma-Ala
Dushanbe
Shatiyak
Tedzhen
Batram-Ali
Shurtan
Sundukli
Sheberghan
Kichekbei
Dauletabat-Donmez
Karabil
Karachop
Mazar e Sharif
Yetym-Tag
Djar-Kuduk
Islim
Angot
Hodja-Gugerdag
Hodja-Bulan
Djumin
Ak-Darya
Herat
Kabul
Powtar Fields
Tacheng Basin
Karamay
Karama
Usa
Uerho
Fengcheng
Shingshisia
Shewan
Algo
Junggar Basin
Bogdo
Dashanc
Suilai
Hami Turfan Hami
Yigikelike
Dongqiulitage
Lunan
Akekum
Daliye Akekule
Shengting-k'ou
Turpan
Turpan Basin
Ichkelik
Kumger
Kosaptok
Kan
Taqlaq
Yingmali
Bostan
Wensuh
Yuecan
Wug Chat
Kashgar
Kashi
Kelatuo
Kekeye
Tsehpu
Tazhoung
Lenghu
Xianshuiquan
Shizigou
Jiandinshan
Lenghusihao
Olgequah
Nanyishan
Huatugo
Youquanzi
Kelfemlike
Lenghuwuhao
Youshashan
Wunan
Toufengshan
Yanhu
Sebeiyhao
Sebefehao
Mayai
Toot
Dakhni
JoaMair
Peshawar
Dhulian
Balkassar
Islamabad
Rawalpindi
Manali
Quetta
Dhodhak
Panjpir
Rodho
Lahore
Hoshiarpur
Pirkoh
Nandpur
Multan
Loti
Kandkot
Qadirpur
Sui
Indus Basin
Punjab Basin
Sukkur
Dadu
Mari
Zamzama
Bhit
Miano
Kadanwari
Sawan
Panipat
Ambala
New Delhi
Budaun
Tando Alam (OGDC)
Karachi
Shakti
Bhagyam
Sonro
Dabhi S.
Mangala
Aishwariya
Vijaya/Vandana
Mathura
Saraswati
Raageshwari
Guda
GR-F
Dhabi
Nari
Mazari
Matli
Liari
Tajedi
Makhdumpur
Turk
S. Mazari
Laghari
Kaskheli
Jabo
Kameshwari
Kaloj
Sanand
Jamnagar
Bereja
Dholka
Ahmadabad
Cambay
Kanpur
Ankleshvar
Anklesvar
Kosamba
Olpad
Surat
Mumbai High
R-12
Panna
BS-13
BS-12
South Bassein
Heera
Neelam
Mumbai (Bombay)
R-15A-1
Ratna
Nagpur
Kathmandu
Thimphu
Digboi
Sibsagar
Jorhat
Nowgong
Bihar
Sylhet
Chhatak
Haripur
Kailistilla
Dhaka
Beani Bazar
Patharia
Fenchugani
Rashidpur
Habiganj
Fenchugani
Bengal Basin
Bakhrabad
Feni
Semutang
Shalbazpur
Begumganj
Kutubdia
Sangu
Patiya
Calcutta
Haldia
Chittagong
Ayadaw
Chauk
Yenang
Mann
Htauks
Paduakpin
Pyato
Pyaye
Prome
Myana
Shwepyit
Visakhapatnam
Narsapur
Narsa
D6-K2
D6-K1
D6-A1
Krishna-Godavari
Hyderabad
Yangon (Rangoon)
Yadana
Bangalore
Madras
Mangalore
Bhuvanagiri, Kamara
Karaikai
Narimanam
Kovilkallappal
Kerala
Cochin
Sapugaskanda
Colombo
Lhasa
Lunpola Basin
Ya'exi
Laojun
116
120
160
162
104
66
50
10
20
95
36
120
156
62
46
30
660
185
120
132
12
60
27
20
66
61
33
6
164
131
180
10
152
50
26

Ulan Bator
MONGOLIA
Erlian Basin
Eerdous Basin
Qinghai Hu Basin
Jiuqan (Juixi) Basin
Lanzou
Lhasa Basin
Dian-Qian-Gui Basin
Bose Basin
Pai-yang-ho
Ya'exia
Shih-you-kou
Danbei
Baiyanghe
Laojunmiao
Shiyougou
Shenglinjin
Liujiazhuang
Majiatan
Lizhuangzi
Hongjinzi
Donghongchuang
Dashuikeng
Zichang
Baiyanjin
Wuji
Yongping
Maling
Quinghazhi
Danbei
Gangua
Chinkanshou
Changchun
Kongchuling
Tangluku
Unggi
N. China Basin
Fahniu
Shuguang
Chin-hsi
Huanxing
Yuoy
Jinzhou
GanSuizhong 36-1
Bohai Basin
N. KOREA
Pyongyang
Seoul
S. KOREA
Beijing
Tiamin
Layemajo
Chinshsi
Gaoshangbao
Pohi Bay
Takang
Bozhong 28-1
Bozhong 34-2/4E
Yanlin
Shengil
Dawangzhuang
Qingdao
Lin-4
Subei Basin
Zhongyuan
Wecheng
Wenmingzhai
Wennan
Wenzhong
Pucheng
Wenliu
Liuzhuang
Machang
Nanyang Basin
Ahhui
Fumim
Lianmengzhuang
Huanggue
Wangil
Sha-Yen
Zhangdian
Xiamen
Welgang
Shuanghe
Dongzhuang
Qianjiang
Shanghai
Donghai
Bo Yuan Ting
E. China Basin
Ping Hu
Hangzhou
Yellow Sea
Guanghau
Wuhan
Wanchang
Zhangang
Xijiayuan
Tuoshi
Huayuan
Gaochang
Xingqou
Quingjiang Basin
Jianghan Basin
Zhongba
Zhongtaishan
Jinhuazhen
Leiyinpu
Guanan
Wubaiti
Gihua
Wolongha
Wanshunchang
Penglaizhen
Longnushi
Daxinxi
Weiyuan
Shiyoug
Shilongxnou
Fuzhou
Chuwei
Taipei
Paoshan
Shoal Basin
Yungoshan
TAIWAN
Kaohsiung
Sanshui Basin
Santou
Guiyang
Wanglongchang
Taihehums
Myitkyina
Shiwan-Dashan Basin
Tiandong
Guangzhou
Hong Kong
K'unming
Macao
Maoming
Lufeng 13-1
Lufeng 15-1
Hizhou 26-1
Liuhua 11-1
Liuhua 4-1
Xijiang 30-2
Weizhou 11-1
Fuizhou 21-1
Beihai
Weizhou 10-3
Weishou 6-1
Tien Hai
Haik'ou
Jinfeng
Zhujiuang Basin
HAINAN
Dongfang
Wenchang 9-2
Wenchang 8-3
Wenchang 14-3
Wenchang 13-1
Wenchang 19-1
Wenchang 13-2
Yacheng 13-1
PHILIPPINES
Manila
Limay
Batangas
ANMAR
URMA)
Chauk/Lanywa
LAOS
Mae Fang
Mae Suiti
Vietiane
Fang
VIETNAM
Pyaye
Prome
Myanaung
THAILAND
Phu Horm
Nam Phong
Chonnabot
Mae Sot
Sirikit
KDN-1
Pyaye
Pyato
Khon Kaen
Shwepyitha
Syriam
Martaban
Bung Ya
Sawng
Neung
Bangkok
CAMBODIA
Phnom Penh
Nha Trang
Yetagun
Sriracha
Ho Chi Minh City
South China Sea
Guloc
Malampaya
Camago
Nido
West Linapacan
Matinlok
Cadlao
Plamuk
Kapong
Platong
Surat
Nang Nuan
Platong S.
Pladang
Tantawan
Chaba
Dara
Trat
Erawan
Sabun
Baanpot
Jakrawan
Pailin
Punan
Arthit
Bongkot
Songkhla
17-B-1
Bua Ban
Pilong
Sutu Den
Rang Dong
Bach Ho
Rong
Dai Hung
Dua-1X
Bursa
MALAYSIA
Damar
Jemeh
Bintang
Sepat
Noring
Inas
Bedong
Laho
Natuna
S.W. Emerald
Barton
Tembungo
S. Furious
Erb W.
St. Joseph
Kinarut
Erb S.
Samarang
Glayze
SABAH
Jintan
Bijan
Brunei
B12
B11
Lutong
E11
F28
E8
SARAWAK
Bunya Nibung
Tarakan
17
100
70
30
172
170
42
50
100
110
29
270
520
817
70
74
140
310
650
10
60
146
134
160
206
24
100
50
50
56
200
450
300
270
104
170
180
26
62
301
160
185
9
45

The country imported 254 MMcfd of LNG in 2004. This will likely increase to 1.5 bcfd by 2010 and to 2.9 bcfd by 2015.

Four major pipeline projects to import gas from Bangladesh, Myanmar, Iran, and Turkmenistan have been discussed, but all face formidable geopolitical and commercial hurdles. It appears that pipeline imports in 2015 will be negligible.

Overseas resources

The quest for energy security has led ONGC to team with international oil and gas companies in exploration and development in recent years.

India recently has won exploration rights in Egypt, Qatar, and Cuba. India's equity participation in Vietnam, Russia, and Myanmar is primarily for natural gas. The majority of other investments are for oil.

Energy demand

Analysis by the U.S. Energy Information Administration (EIA) and the International Energy Agency (IEA; Paris) indicate a robust increase in economic, energy, and natural gas growth in India.

According to EIA, total energy demand in India will grow to 29.3 quadrillion btu (quads) in 2025 from 14 quads in 2002.

Natural gas represents 6.5% of total primary energy consumption, but EIA expects this share to grow to 10% by 2025. Overall, India's total gas demand is projected to reach 7.7 bcfd in 2025, compared to 2.5 bcfd in 2002.

Coal will continue to be the primary energy source for electric power generation, but gas use in the power sector is projected to account for 70% of the total increase in gas consumption.

Natural gas is expected to account for 12% (1.7 quads) of fuel use for power generation in 2025, compared to just 5% (0.3 quads) in 2002.

GAS DEMAND

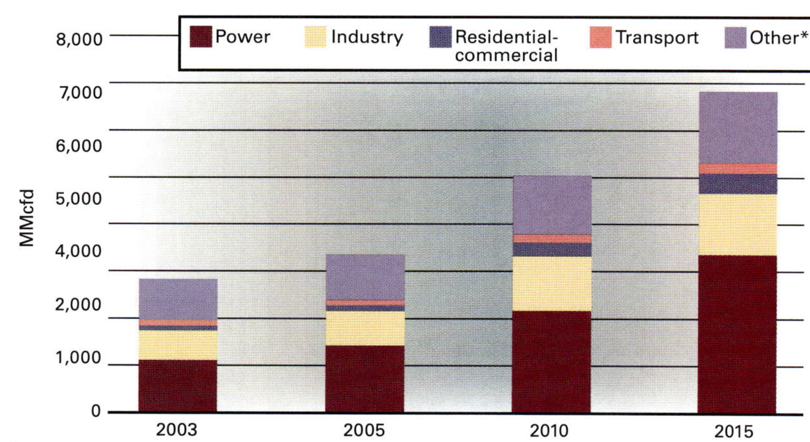

*Includes fertilizer production.

NATURAL GAS DEMAND, SHARE OF TOTAL ENERGY IN INDIA

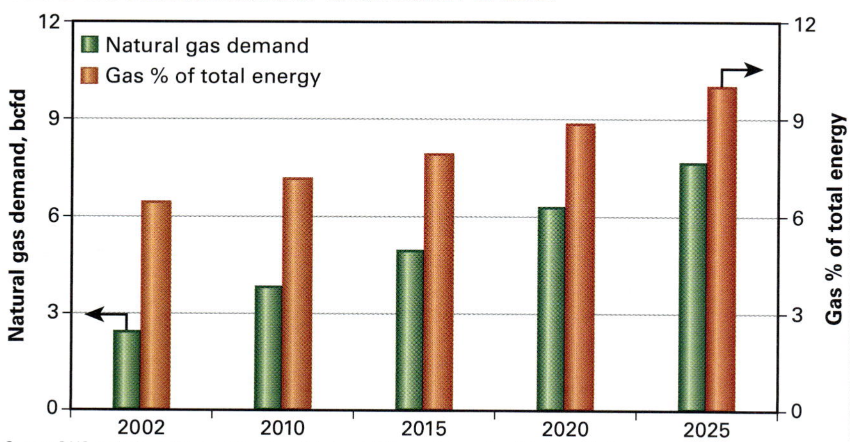

Source: SAIC analysis and forecast based on data from US Energy Information Administration's 2005 "International Energy Outlook" (IEO 2005)

Exploration and Development

ONGC recap

Oil & Natural Gas Corp. Ltd. reported in July on 10 discoveries it made in 2005–06, including five natural gas finds in the deep waters off the Andhra coast in the KG basin. Other discoveries included:

- RV-1 in the Mumbai offshore
- B-9-1 in the Saurashtra offshore
- GS-15-E in the KG offshore
- Turputallu-I in the KG onshore
- Mekeypore-3 in the Assam shelf

ONGC said the five KG basin discoveries on Block KG-DWN-98/2—D-1, A-1, U-1, W-1, and E-1—together with two earlier finds, have reserves of up to 3 tcf of gas.

The company acquired Block KG-DWN-98/2 from Cairn Energy PLC (Edinburgh, Scotland), which earlier had made two discoveries with reserves of just under 2 tcf.

Block D6 oil find

The first deepwater oil discovery off India's east coast is a Cretaceous find that opens a new petroleum province in the D6 block, KG basin. That was the report from Reliance Industries Ltd. (Mumbai), operator of Block D6, and partner Niko Resources Ltd. (Calgary).

MA-1, the combine's first Cretaceous test on the block, went to TD 3,783 m (12,410 ft) and penetrated 26 m (85 ft) of net oil pay and 72 m (236 ft) of net gas pay.

The oil zone tested at more than 6,700 b/d of oil and 10.96 MMscfd of gas. The upper rich-gas zone tested at more than 32 MMscfd of gas and 3,370 b/d of condensate.

Western Rajasthan gas discovery

Focus Energy Ltd. (New Delhi), reported a gas discovery on Block RJ-ON/6 in westernmost Rajasthan, near the border with Pakistan.

The SGL-1 well, in the Shahgarh area of the Jaisalmer district, flowed 15 MMscfd of gas from Cretaceous sandstones at 3,100 m (10,170 ft). The amount of producible gas is yet to be determined.

Mangala project

A unit of Cairn Energy PLC (Edinburgh, Scotland) let a contract to Mustang Engineering, a subsidiary of John Wood Group PLC (Aberdeen, Scotland).

Mustang will provide front-end engineering design verification, detail engineering, procurement, and construction management services for the 150,000 b/d Mangala production facility to be located in the Rajasthan Desert in western India.

Shell, ONGC cooperation agreement

Royal Dutch Shell PLC and ONGC of India signed a memorandum of understanding (MOU) covering upstream and downstream projects, mostly in India.

The agreement covers exploration and production of hydrocarbons and distribution of natural gas and oil products. It also will look into refining, petrochemicals, and coal gasification.

While giving Shell greater access to Indian projects, the MOU will allow ONGC more opportunities outside the country.

One focus will be on raising production from existing fields and jointly bidding in future exploration rounds. Downstream, a joint steering committee could recommend a long-term oil products supply arrangement and joint operation of terminals and depots.

Drilling and Production

Strong local demand for natural gas, a string of moderate development successes, and improvements in the regulatory environment are driving drilling operations in India, *Oil & Gas Journal* said.

The government's recent efforts to make licensing more flexible and the possibility of partial divestment of state-owned operating companies suggest wider drilling and development activity in the future.

Two of the largest finds in India in the last few years are Reliance Industries Ltd.'s deepwater gas discovery in the KG basin in October 2002 and Cairn Energy Ltd.'s gas and oil discoveries in Rajasthan in 2004–2005.

Oil & Natural Gas Corp. Ltd. (ONGC)

Oil and Natural Gas Corp. is the largest and most active operator in India. In mid-2005, the Indian government approved investment of about $4.75 billion in an LNG, power, and petrochemicals project in Mangalore to be implemented by ONGC subsidiary, Mangalore Refinery & Petrochemicals Ltd.

India's Petroleum Activity

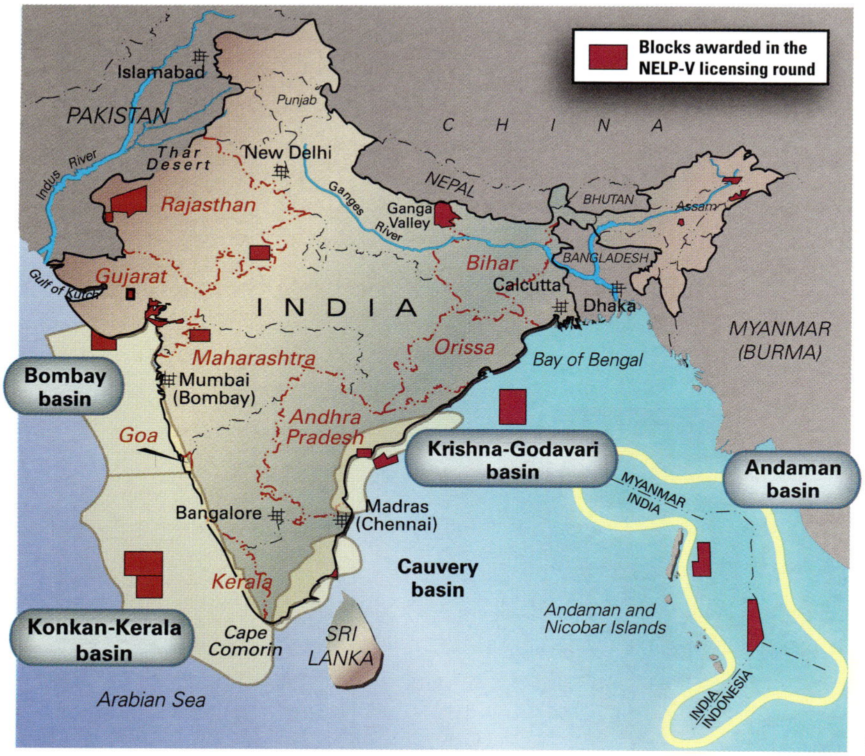

Beyond India's borders

Through ONGC Videsh Ltd. (OVL), ONGC has 18 exploration and production properties in 13 countries. OVL operates six properties overseas, holding 100% interests in four of them.

OVL made a $1.7 billion investment in Sakhalin several years ago; that remains its largest single investment outside of India.

In July 2005, OVL signed an agreement with the Mittal Group, a large steel maker, to form a joint venture called ONGC-Mittal Energy Ltd.

In late 2005, ONGC became a joint venture member with Norsk Hydro and Repsol YPF (Madrid) for deepwater exploration on Blocks 25, 26, 27, 28, 29, and 36 off northern Cuba.

ONGC is also looking at activities in Vietnam, Indonesia, Egypt, Canada, Myanmar, and Nepal. The company sold its Gulf of Mexico interests in favor of opportunities in Sudan.

ONGC also will assume 30% of the costs of developing Cairn Energy's Mangala oil discovery in Rajasthan.

During 2004–05, ONGC made four oil and gas discoveries: Vashista in the eastern offshore, D-33 in the western offshore, Wamaj in Gujarat, and Tiphuk on the Assam Shelf. Total reserves amount to 49.4 million tonnes.

The company's Sagar Samriddhi deepwater drilling campaign was using three deepwater drillships in 2006 under long-term contracts, with 47 deepwater wells budgeted.

Brownfield development is scheduled for the Bombay High area with infill drilling that will include extended-reach drilling and multilateral completions.

In the KG basin, ONGC completed its first smart "e-well" offshore, at the G-1-GS15 field.

ONGC also was looking at in situ combustion and EOR for Gujarat's heavy onshore oil and coal gasification projects as well as possible participation in coalbed methane development.

Oil India Ltd. (OIL)

Oil India is the country's second largest domestic E&P company. Starting in 1981, OIL was 100% government owned with a mandate to pursue domestic exploration and development. In 1996, 1.87% of the company was distributed to employees and others. A new 10% equity issue was under consideration in 2006.

In 2004–05, the company made three discoveries in Assam: Samdang (Eocene), West Zaloni, and North Tanali. From these projects in the northeast, and from gas and heavy oil projects in Rajasthan, OIL produces 3.4 million tonnes of crude oil and 2 bcm of natural gas per year.

Gujarat State Petroleum Corp. (GSPC)

The company operates 11 oil and gas fields in India with partners Niko Resources Ltd., Hindustan Oil Exploration Co. (HOEC), and Heramec India Pty Ltd.

GSPC is involved in oil, gas, and coalbed methane exploration projects with Gas Authority of India Ltd. (GAIL); Delhi-based Jubilant Enpro Ltd., part of the Jubiliant Group; GeoGlobal Resources Inc.; HOEC; India Oil Co. (IOC); ONGC; and Prize Petroleum Corp. Ltd.

Niko Resources

Canada's Niko Resources Ltd. operates the Hazira gas and oil field offshore India and the onshore Surat gas field. In fiscal 2006, Niko planned to drill two gas wells and 6–8 oil wells at Hazira and install oil-handling facilities.

In the D6 block, the company expected to drill six exploration wells and two development wells, continue evaluating seismic data, construct facilities, and build infrastructure.

In the NEC 25 block, the company will drill four exploration wells, work on seismic, and advance FEED and development plans.

Refining and Processing

Vizag plant expansion

State-run refiner Hindustan Petroleum Corp. Ltd. (HPCL) was seeking partners for investments, oil procurement, and product marketing as part of its proposed Visakhapatnam (Vizag) refinery expansion.

HPCL plans to expand capacity to 16.5 mtpy from 7.5 mtpy.

After completing the $2 billion expansion, HPCL would dismantle an uneconomical 1.5-mtpy plant, making the net effective capacity at Vizag 15 mtpy.

New delayed coker

Reliance Petroleum Ltd. (Mumbai) awarded a process and detailed engineering contract to Foster Wheeler for installation of an 8-drum delayed coking unit as part of the Jamnager Export Refinery Project in Gujarat. Foster Wheeler said it will be the largest delayed coker in the world. Startup is scheduled for 2008.

Barmer refinery plans on hold

ONGC placed on hold its plans to build a 7.5-mtpy refinery near Barmer to handle crude oil produced in Rajasthan state, citing uncertain economics. First production of 150,000 b/d of oil from the Rajasthan site is expected in late 2008.

The company said it could instead transport the heavy crude from Rajasthan to its subsidiary, Mangalore Refinery & Petrochemicals Ltd. (MRPL), for processing in southern Karnataka state.

ONGC proposed a special pipeline to Mundra port in Gujarat for further shipment to the MRPL refinery.

Upgrades at Manali complex

Chennai Petroleum Corp. Ltd. (CPCL) committed to invest more than $1 billion to upgrade its 130,660 (total) b/d refinery complex at Manali. Planned investments include a $647 million residue upgrade and actions to increase the capacity of Refineries I and III. Completion is scheduled for 2010.

Other work will add diesel and naphtha hydrotreating units (so products can meet European fuel standards) and enhance the polypropylene plant to produce other olefins.

Jamnagar refinery upgrades

Reliance Petroleum Ltd.—a subsidiary of Reliance Industries Ltd. (Mumbai)—plans to add an 83,000 b/d alkylation unit at its 580,000 b/d export refinery under construction in Jamnagar.

ExxonMobil Research & Engineering Co. will provide technology for producing alkylate having high octane and low vapor pressure for gasoline blending. The new unit should start up by year-end 2008.

Reliance also let a contract to Linde AG (Wiesbaden, Germany) for five hydrogen units that will increase hydrogen production capacity to 525 MMscfd at Jamnagar.

NGL plant at Dahej

ONGC awarded Toyo Engineering Corp. a ¥20 billion order to build an NGL extraction plant at Dahej in Gujarat state. On start-up in mid-2008, the plant will extract liquids for use as petrochemical feedstock using LNG from Petronet LNG Ltd.'s nearby 5-mtpy terminal.

Petrochemicals

Assam complex

A consortium led by GAIL India Ltd. promoted a $1.2 billion integrated petrochemical complex in India to be built in Dibrugarh, Assam. The complex will contain a gas separation plant, hydrocracker unit, and downstream polymer plant. It will produce 220,000 tpy of ethylene and 60,000 tpy of propylene.

GAIL's partners are OIL, Numaligarh Refinery Ltd. (NRL), and the Assam state government.

The project will be completed in 60 months from the date of financial closure. The site has been selected and environmental clearance has been obtained.

OIL and ONGC will provide 6 MMcmd of gas as feedstock, and NRL will provide 160,000 tpy of petrochemical-grade naphtha.

Gujurat complex

The government approved a $1.05 billion aromatics and olefins complex that ONGC plans to build next to its subsidiary Mangalore Refinery & Petrochemicals Ltd. and its Dahej petrochemical facility in Gujarat. The facility will have a dual-feed cracker with capacity of 1.1 mtpy of ethylene plus units for producing polyethylene, polypropylene, and styrene butadiene rubber.

Licensing

NELP-VI bid round

Awards for India's sixth round of bidding for oil and gas exploration blocks under the New Exploration Licensing Policy (NELP-VI) were to be announced in early 2007. More than 75 companies participated.

The government offered a total of 55 blocks—25 onshore, 6 in shallow water, and 24 in deep water. Acreage totaled 355,000 sq km (137,065 sq miles). Both the number of blocks and the acreage were the largest ever offered by the Indian government.

Ten coalbed methane blocks also were offered in the NELP-VI leasing round. More than 50 bids were expected, from 18 Indian and eight foreign companies.

India's Planned, Operating LNG Regasification Terminals

Project name	Dahej	Hazira	Kochi	Ratnagiri (Dabhol)	Ennore
Location	Gujarat	Gujarat	Kerala	Maharashtra	Chennai
Operator	Petronet LNG Ltd. (PLL)	Royal Dutch Shell PLC and Total Gaz Electricité	PLL	Ratnagiri Gas & Power Pvt. Ltd.[1]	Gas Authority of India Ltd. (GAIL)
Capacity, million tons/year	5 (to be expanded to 10 by 2008)	2.5 (to be expanded to 5)	2.5	5	7.5
Status	Operating since January 2004	Operating since April 2005. Lost its sole customer, GSPC, in December 2005.	Planned for 2009	80% complete. Suspended in 2001.[2] To commence operation in Sept. 2006	Planned for 2009
Cost	$550 million	$600 million	$440 million	Unknown	Unknown
Supply source	Qatar-Ras Laffan LNG Co. Ltd. (RasGas), (25 year take-or-pay for as much as 7.5 million tonnes/year)	Australia, Qatar, Abu Dhabi, Oman (To be operated on spot basis)	Qatar. RasGas	Discussions with Australia, Malaysia, Oman, Qatar, and Abu Dhabi	Possibly Iran (South Pars field)
Contract price	$2.53/MMbtu for first 5 years	Unknown. Flexible contract duration offered	Unknown	NA (analysts estimated maximum ceiling LNG supply price of $3.50/MMbtu	Unknown
Regasified product transmission, distribution	Existing pipelines	Existing pipelines. GSPC[3] gas supply agreement	Existing pipelines	Existing pipelines	New pipelines

[1]Partners are GAIL and National Thermal Power Corp. [2]Original operator Enron Corp. filed bankruptcy. Project halted. [3]Gujrat State Petroleum Corp.
Sources: Platts, Petronet LNG, Royal Dutch Shell PLC, SAIC analysis of most favored projects, SAIC's WLNG model. All amounts listed are estimates and may not reflect actual capacities or costs.

Foreign Direct Investment policy

The Ministry of Commerce and Industry's foreign direct investment (FDI) policy for oil and gas was updated in April 2006.

Up to 100% foreign participation is now permitted in all activities other than refining, although those pertaining to trading and marketing of petroleum products must divest 26% equity to the public or an Indian partner within five years.

Public-sector involvement in refining projects is limited to 26%; private companies are permitted up to 100%.

About 140 production-sharing contracts have already been signed, representing an estimated investment of more than $4 billion.

India, Saudi energy pact

India and Saudi Arabia signed in January 2006 a strategic energy partnership, the Delhi Declaration, to work cooperatively in the upstream and downstream sectors.

Previously, Saudi Arabia signed a similar accord with China regarding oil, gas, and minerals.

The India-Saudi partnership envisions increased volumes of crude oil supplies from Saudi to India through long-term contracts that would be renewed automatically. Currently, India imports 25 mtpy of Saudi crude oil.

The declaration also is expected to promote Saudi investments in refining, marketing, and storage in India, as well as India-Saudi ventures in gas-based fertilizer plants in Saudi Arabia.

Coalbed Methane

Joint CBM work in India, Australia

India's GAIL and Arrow Energy NL (Brisbane) signed a memorandum of understanding outlining areas for joint cooperation and investment in coalbed methane (CBM) projects in Australia and India. The proposed joint venture may also involve other parties.

The alliance will seek prospects for exporting gas to Asia as gas-to-liquids and compressed natural gas.

GAIL will invest in and accelerate as many as 10 of Arrow's CBM development projects in Queensland that are awaiting capital.

LNG

Background

India imported 0.25 bcfd of LNG in 2004. This will likely increase to 1.5 bcfd in 2010 and to 2.9 bcfd in 2015. Two receiving and regasification terminals are operational, at

LNG imports

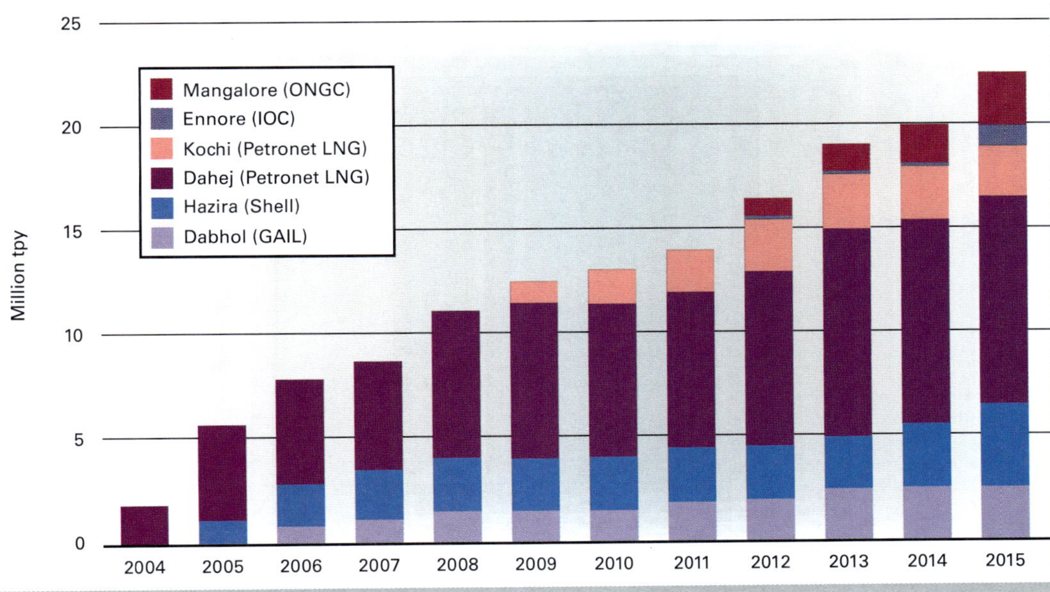

India's LNG Terminals

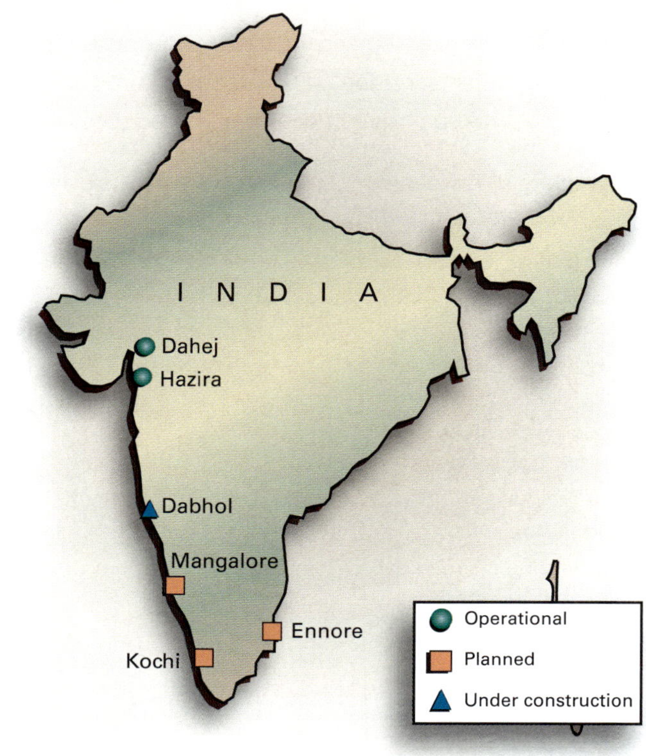

Dahej and Hazira. Three others are in various stages of planning or construction, at Dabhol in Maharashtra, at Kochi, and at Ennore.

Existing terminals

Dahej became operational in 2004. Capacity is 5 mtpy. Petronet LNG—a private joint venture of state-owned ONGC, GAIL, IOC, Gaz de France, Asian Development Bank, and other private investors—owns the terminal, which receives LNG from Qatar. Petronet plans to double its capacity by 2008.

Hazira has been operated by Royal Dutch Shell PLC since 2005. Shell lost its sole dedicated customer, Gujarat State Petroleum Corp. (GSPC), in December 2005, making continued operation uncertain.

Shell was seeking other customers in 2006, including owners of a power plant proposed for construction nearby. The company also was considering combining the 2.5-mtpy regasification facility with a container terminal to improve economics.

Terminals planned or being built

Dabhol, a 5-mtpy terminal, is largely complete but unused in the wake of the 2001 collapse of Enron Corp. Various parties were evaluating options for the terminal (and the adjacent power plant), which likely could not be operational before 2009.

Kochi, backed by Petronet, will have a capacity of 2.5 mtpy. It is expected to be operational by 2009, processing LNG received from Iran. GAIL signed an agreement midyear to supply the Kayamkulam power plant with gas from the Kochi facility.

India's Proximity to Major LNG Suppliers

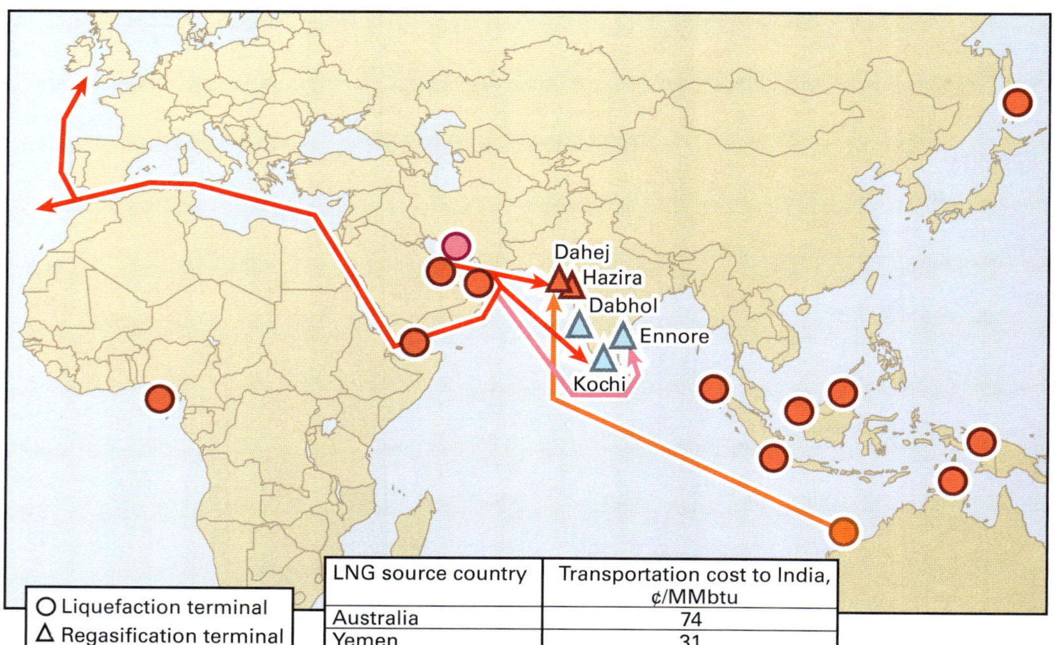

LNG source country	Transportation cost to India, ¢/MMbtu
Australia	74
Yemen	31
Iran	28
Qatar	23
Oman	18
Malaysia	39

Source: SAIC's WLNG model

Ennore is an IOC-backed project that includes an 1,800-MW power plant in addition to the 2.5-mtpy regasification terminal. It could be operational by 2015.

Sources of supply

India is close to almost 50% of the world's LNG supply, situated between rapidly growing LNG importers to its east and big OPEC exporters to its west.

India's LNG terminals receive gas from Qatar, UAE, or Oman. Longer term, Iran also is poised to provide significant LNG supplies, if geopolitical risk and price issues can be resolved. Future LNG supplies also might come from Brunei, Indonesia, Malaysia, or Australia.

Cairn projects in South Asia*

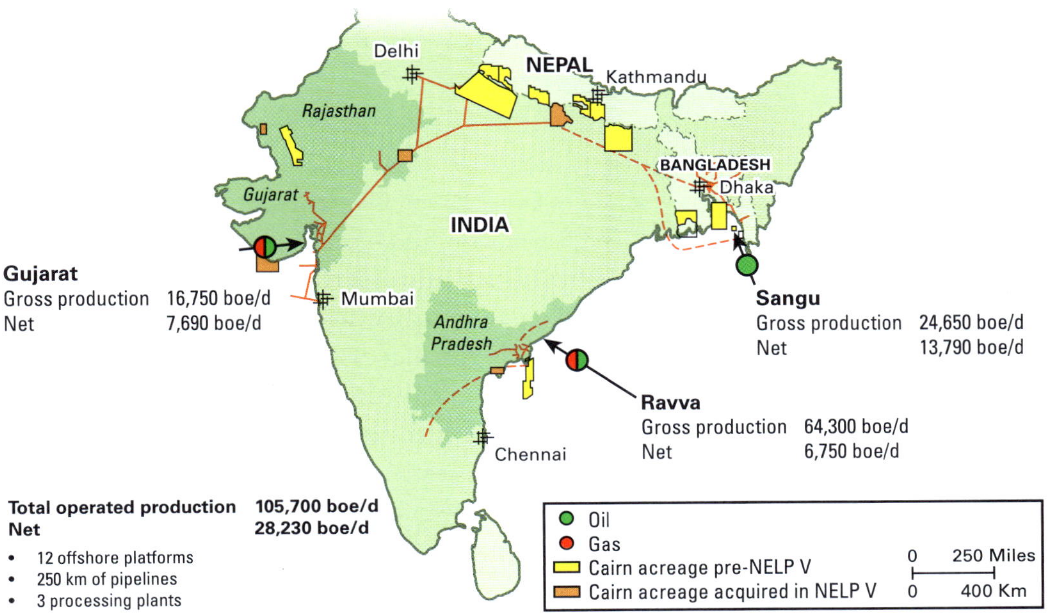

*All figures interim 2005, rounded.

Cairn Energy role

Scotland's Cairn Energy PLC is a major participant in India's hydrocarbon sector. Production and cash flow from the company's Ravva field off eastern India and Sangu field off Bangladesh helped to fund subsequent exploration elsewhere in India.

Cairn began drilling in western India in 2000, discovering the Lakshmi and Gauri gas fields in Gujarat, followed by drilling in Rajasthan in 2002.

Several years of wildcat drilling led to successes in western Rajasthan, including the 1 billion-bbl Mangala field, discovered in 2004. Neighboring Bhagyam and Aishwariya fields also show promise.

Target production from the three (combined) is 120–150,000 b/d, with Mangala production expected to begin in late 2007. Estimates indicate that 380–700 million bbl should be recoverable from the three.

Cairn fields in Rajasthan

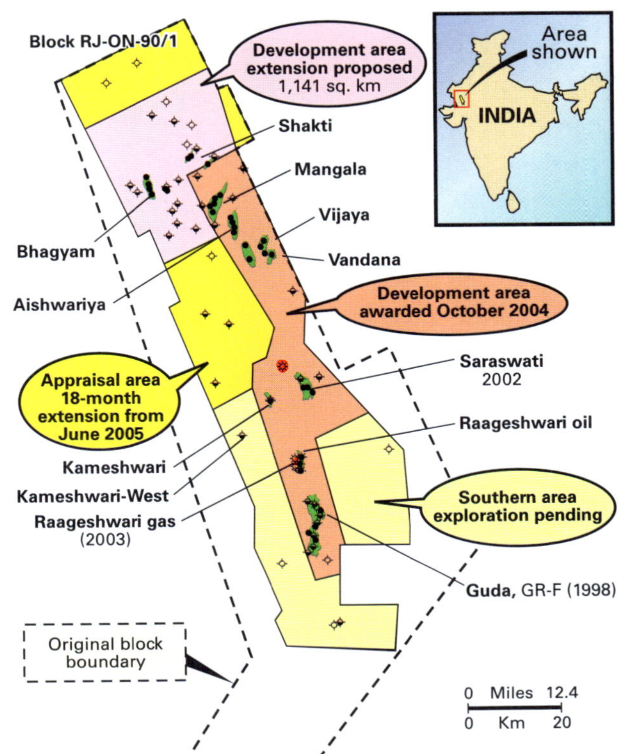

INDONESIA

Capital: Jakarta
Monetary unit: Rupiah
Refining capacity: 992,745 b/cd
Oil production: 892,000 b/d
Oil reserves: 4.300 billion bbl
Gas reserves: 97.780 tcf

Background

Indonesia's state-run PT Pertamina has had difficulty in recent years producing and delivering sufficient oil and gas to meet domestic needs.

The government announced in early 2006 that the country produced 955,000 b/d of oil and condensate in 2005, missing its target by 120,000 b/d. And in April, Pertamina said it faced a gas shortage of at least 48 MMcfd in West Java and had to trim deliveries to some large industrial customers. The company reported a 1 million-bbl increase in fuel oil imports in April over March.

In October 2005 the country's LPG production was reported to have declined by 20% due to problems in refineries in Balikpapan and Cilacap. Pertamina planned to import 5,000–6,000 tonnes of LPG from Singapore, Malaysia, and Thailand in early 2006.

Looking longer-term, the finance director of PT Medco Energi Internasional said it was unlikely that the country could raise its crude output to the planned level of 1.3 million b/d by 2009, due to a lack of exploration and development.

The economy also has been struggling to keep pace with the cost of increased oil imports. Indonesia consumes 1.35 million b/d of oil and imports about 390,000 b/d.

Pending pipeline projects and field developments should help improve the situation, including the impending resolution of questions about operation of the Cepu Block.

Cepu Block operation

Most of 2006 saw continued negotiations regarding operation of the Cepu Block. The oil and natural gas project straddles East Java and Central Java.

The Banyu Urip field—believed to be the largest of six on Cepu Block—has estimated reserves of more than 250 million bbl, making it one of the most significant oil discoveries in Indonesia since the mid-1990s.

ExxonMobil Oil Indonesia Inc. and state-run PT Pertamina are partners in a joint development agreement, each with a 45% share. The Central Java and East Java administrations hold the other 10%. However, the firms disagreed on the duration and type of operating arrangement.

MAJOR SOUTHEAST ASIA DISCOVERIES

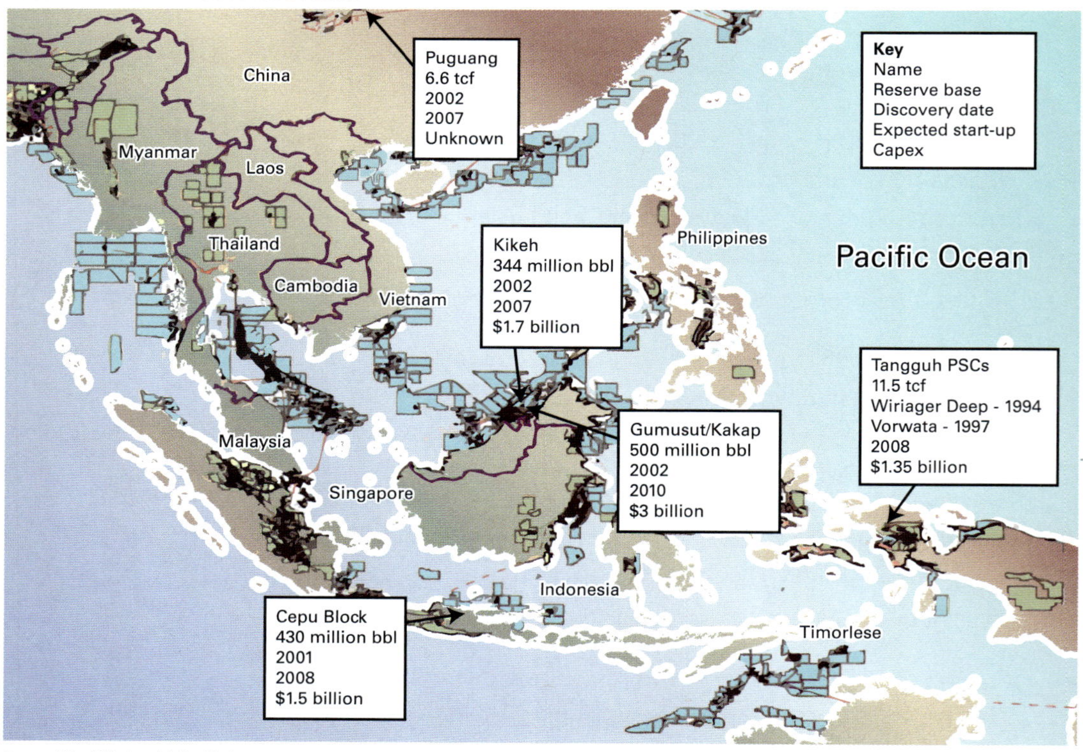

Source: Wood Mackenzie's Pathfinder

Southeast Asia perspective

Analysts with Wood Mackenzie (Edinburgh, Scotland) noted that exploration and appraisal drilling in Southeast Asia continued at relatively high levels through the first 10 months of 2006. Governments awarded 31 exploration licenses, with the highest levels in China and the Philippines.

China made seven awards, six in the South China Sea. The Philippines made seven service contract awards; Indonesia made five awards; Malaysia two, Thailand three, Brunei two, and Vietnam three.

About 230 exploration and appraisal wells were drilled across the region through October 2006, about the same overall activity level as in 2005. Year-on-year well counts increased in Indonesia, Thailand, and Vietnam but dropped in Malaysia and Myanmar.

Southeast Asia license awards, 2000-06

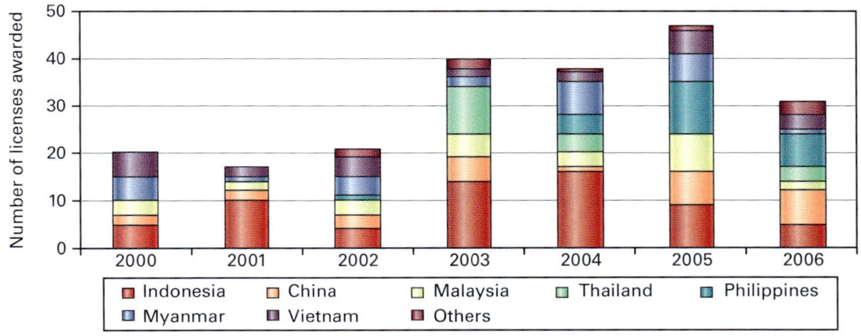

Source: Wood Mackenzie's Pathfinder

Southeast Asia E & A* wells drilled, 1995-2006

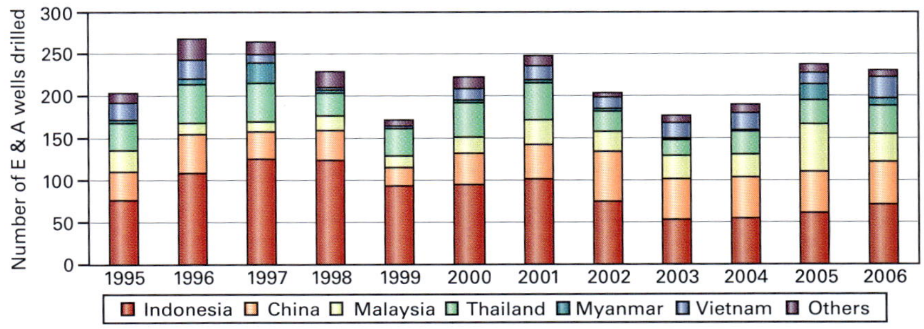

*Exploration and appraisal.
Source: Wood Mackenzie's Pathfinder

About 30 discoveries were made across Southeast Asia through October 2006, more than half of them in Indonesia and Malaysia together. Exploration efforts remained focused on proved provinces, including the Malay basin, the Gulf of Thailand, Sarawak, Java, Sumatra, and the Bohai Gulf.

Exploration through October 2006 resulted in the addition of about 300 million bbl of oil and 10 tcf of gas reserves. On an oil-equivalent basis, China added the largest volumes, due to the Liwan discovery by Husky Oil Ltd. (Calgary) on Block 29/26. This was the first discovery in China's deep water. The company made a preliminary reserves estimate of 4–6 tcf.

Myanmar increased its reserves significantly with the 3-tcf Mya discovery on Daewoo's Block A3 off Northwest Myanmar. The field will be developed alongside Shwe and Shwe Phyu fields, with combined reserves estimated at 8.6 tcf.

Discovered reserves by country, 2006 (BOE)

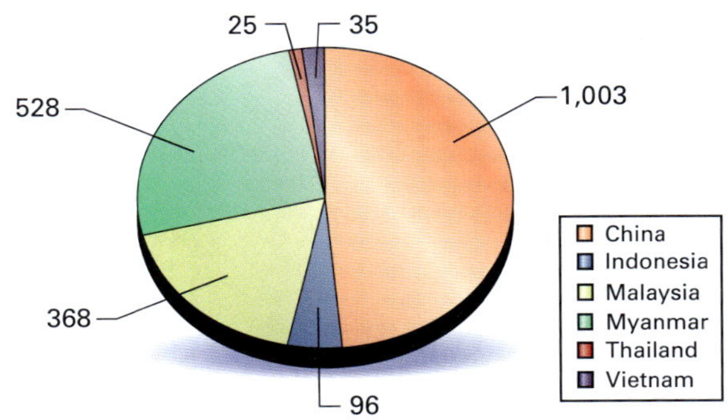

Source: Wood Mackenzie's Pathfinder

In Malaysia, Shell Malaysia and its partners discovered Pisagan field on Block G, off Sabah. Petronas also had a discovery in the PC4 field in unlicensed acreage off Sarawak. Initial oil and gas reserves estimates for these discoveries are 100 million bbl and 1 tcf, respectively.

A working plan was approved mid-year by Indonesia's upstream oil and gas regulatory body, BPMIGAS. State media reported that ExxonMobil was talking with drilling contractors, aiming to conclude negotiations by mid-2007.

In September, ExxonMobil said it hoped to start limited production by 2008.

Exploration and Development

Barossa-1 well

A joint venture of ConocoPhillips (Houston) and Santos Ltd. (Adelaide) confirmed the presence of natural gas in another field in the eastern Timor Sea that will provide supply to a second LNG train at its Darwin liquefaction plant.

Strong flow was recorded from the Barossa-1 well on Permit NT/P69, about 295 km (183 miles) northeast of Darwin and near the maritime boundary with Indonesia. The reservoir lies about 4,250 m (13,950 ft) subsea in 233 m (765 ft) of water

The main test flowed gas at 30 MMcfd and condensate at 7–9 bbl/MMcf of gas, constrained by rig equipment. The one drawback is the 16% carbon dioxide content of the gas.

Tunu field extension

Total SA confirmed an extension of Tunu gas field's southern zone with a gas discovery between Tunu and Peciko fields off East Kalimantan. The Tunu Great South-1 well, in very shallow water 8 km (5 miles) southwest of Tunu field's southernmost platform, encountered numerous gas reservoirs on Mahakam block.

Total E&P Indonesia was planning a production test and an appraisal-well program. The company said the extension could begin production by 2009.

Sebuku PSC

Pearl Energy Ltd. (Singapore) reported that an appraisal well on the 8,773 sq-km (3,387 sq-mile) Sebuku PSC encountered gas in Oligocene Upper Berai limestone, 2.6 km (1.6 miles) northwest of an earlier discovery well in the Makassar Straits off southern East Kalimantan.

Operator Pearl Energy said the Makassar Straits-2 well was drilled to TD 1,675 m (5,500 ft). It found Upper Berai to have 43 m (140 ft) of gross thickness but only 3 m (10 ft) of gas-bearing reservoir-quality rock.

Natuna gas block

The Indonesian Cabinet must decide what to do about gas production from the Natuna D-Alpha Block in the South China Sea. That conclusion followed action by the Energy and Mineral Resources Ministry to terminate the development contract awarded earlier to ExxonMobil Oil Indonesia Inc.

The ministry said ExxonMobil failed to submit a plan for developing the block and selling the gas. Work on the offshore project east of Natuna in the South China Sea reportedly has been delayed for several years due to lack of funds. It has natural gas reserves of 222 tcf.

ExxonMobil said it believes it has until 2009 to begin production under terms of the company's PSC with Pertamina. The company said that late in 2005 it was proceeding with a plan to start producing from the block by 2014.

Oil shipped from Tiaka field

PT Medco Energi Internasional shipped the first consignment of crude oil from Tiaka field off central Sulawesi. The company sent a tanker loaded with 75,000 bbl of crude to PT Pertamina's refinery at Plaju, South Sumatra.

Pertamina operates Tiaka field, which lies in 53 m (175 ft) of water on the Senoro Toili Block off Toili in Tolo Bay. The field produces 1,850 b/d of oil. PT Medco aims to eventually raise output to 5,000 b/d.

The company plans to drill three more production wells in early 2007.

Drilling and Production

Maleo field

Santos Ltd. (Adelaide) began natural gas production from Maleo field off eastern Java under an arrangement with Indonesia's state-owned PT Perusahaan Gas Negara (PGN). Santos has agreed to sell 243 bcf of gas to PGN for 8–12 years.

Production will flow from a new six-well wellhead platform through a 7.4-km (4.6-mile) spur line to the existing East Java Gas Pipeline, then into the Surabaya and Gresik grid. The field contains proved and probable reserves of 240 bcf.

Southeast Sumatra Gas Project

China's CNOOC Ltd. and partners started production from their Southeast Sumatra Gas Project, 120 km (75 miles) off West Java in 30 m (98 ft) of water. Phase I will deliver 55 MMcfd from four wells.

Phase II should start gas delivery in early 2007.

Refining and Processing

Indonesia aims to double the country's crude output to 300,000 b/d by 2010 and to modernize several of its biggest refineries, inviting overseas partners to participate in $18 billion worth of projects designed to boost crude oil production and fuel refining.

Situbondo refinery

The East Java government said a consortium of Russian investors and state-owned PT Petrogas Jatim Utama will build a $3 billion refinery in the province. Contract signing was expected by year-end for the facility, to be built in Situbondo.

The refinery will produce diesel, lubrication oil, and gasoline. Initial capacity will be 150,000 b/d, eventually rising to 300,000 b/d.

LNG

Sulawesi facility

PT Pertamina, PT Medco Energi Internasional, and Mitsubishi Corp. (Tokyo) announced plans to build a 2-mtpy LNG plant at Senoro on Sulawesi. Construction on the $1 billion plant is projected to begin in 2007, with completion targeted for 2009.

Senoro gas field, which has proved reserves of 2.3 tcf, will provide gas to the plant.

Imports needed

PT Pertamina planned to import 600,000 tpy of LNG from Oman, along with possible purchases from Qatar, to help meet its export contracts with Japan, South Korea, and Taiwan.

PT Pertamina has a contract to supply 12 mtpy of LNG to Japan alone. The shortfall to all three countries amounted to some 75 cargoes or 5 mtpy, including nine cargoes from Arun in Aceh and 70 cargoes from Bontang in East Kalimantan.

Interest in biofuels grows

The Indonesian government is working to cut dependency on oil, from the current 60% of the total energy mix to 30% by 2025. Alternative energy sources, such as biodiesel, are expected to account for around 10% of the total, with the remainder made up by coal and natural gas.

In mid-2006, the government announced plans to build four pilot biodiesel plants, for use as models to promote the use of alternative fuels.

The plants are to be built in Kalimantan and Sumatra by 2007, designed to produce 6,000 tpy of biodiesel from crude palm oil and castor oil. The agricultural ministry was in discussion with 14 regions about establishing new plantations to provide raw materials for more biodiesel plants.

Indonesia also offered Malaysia's top energy companies incentives to invest in its developing bioenergy industry. A team was to be established to draw up regulations by 2008 that would ease the entry of foreign firms starting in 2009. Pertamina and PT PLN would be standby buyers of biofuel produced from future plants.

Pipelines and Transportation

Riau to North Sumatra

State-owned gas distributor PT Perusahaan Gas Negara was planning to invite bids for construction of a $600 million, 360-km (224-mile) gas pipeline to link Duri in Riau and Medan in North Sumatra.

The 200–300 MMscfd pipeline could come on stream by yearend 2007. It would connect with two other lines already under construction.

One will extend between Gresik in South Sumatra and Rawa Maju in West Java. The other will join Pagardewa in South Sumatra and Cilegon in Banten. The two lines will be able to move 950 MMscfd.

East Kalimantan to Central Java

PT Bakrie & Bros. (Jakarta) said it planned to proceed with construction of a 1,115-km (693-mile), $1.26 billion pipeline to carry 1 bcfd of natural gas from East Kalimantan to Central Java.

Some government officials have questioned the project's feasibility in view of possible new gas supplies from Cepu Block and declining output from fields in Kalimantan..

Construction has been scheduled to start in 2007 and end in 2009.

South Sumatra to West Java

The Asian Development Bank approved two loans to support development of Indonesia's energy infrastructure.

A $75 million loan went to PT Perusahaan Gas Negara (PGN) for one section of a 661-km (411-mile) natural gas pipeline from South Sumatra to West Java.

The $652.5 million pipeline is one of two PGN is building to address a gas shortage in West Java.

The second loan, of $350 million, will help develop the Tangguh LNG project in Irian Jaya Barat.

Cirebon to Bekasi

Plans were under review for a 220-km (137-mile), $200–300 million natural gas pipeline between Cirebon and Bekasi in West Java. The project is part of the government's plan for a national transmission and distribution network.

The proposed line would link to the 367-km (228-mile) line from Pagar Dewa, South Sumatra to Cilegon, West Java. Other gas lines being built or under study as part of the national system are:

- Gresik to Kalimantan: 1,219 km (757 miles)
- Cirebon to Semarang, Central Java: 290 km (180 miles)
- Semarang-to-Gresik, East Java: 390 km (242 miles)

Licensing

2007 round

Indonesia plans to offer 30 new oil and gas blocks for licensing in mid-2007, mostly in the deep waters of Papua and Nusa Tenggara.

As an incentive, the government said companies could return their exploration rights if they judge areas unpromising after two to three years of exploration.

2006 round

The government planned to announce in first quarter 2007 the results of 20 tenders held in August 2006. Blocks offered at that time are in Madang, South Mandar, Sageri, and South Sageri, all off the South Sulawesi coast; Enrekang, onshore in South Sulawesi; and Karama, Malunda, and Mandar, off West Sulawesi.

2005 round

In late 2006, Indonesia awarded production-sharing contracts to 18 companies who had been invited to bid on 41 oil and gas blocks onshore and offshore Sumatra, Kalimantan, Sulawesi, and Java.

Winners included ConocoPhillips and partner Statoil ASA (Kuma Block in western Sulawesi); and CNOOC and its partner PT Gregori Gas Perkasa (Batanghari Block in central Sumatra).

Earlier in the year, the government announced four other winners from the 2005 regular tender round. They included Petronas (Lampung II block off South Sumatra); ExxonMobil (Suramana Block in the Makassar Straits off Sulawesi); Husky Energy (East Bawean II Block in East Java); and Marathon/Talisman (Pasangkayu Block in the Makassar Straits).

JAPAN

Capital: Tokyo
Monetary unit: Yen
Refining capacity: 4,676,690 b/cd
Oil production: 15,000 b/d
Oil reserves: 58.5 million bbl
Gas reserves:1.4 tcf

Changing energy mix

Petroleum's share of Japan's primary energy requirements is expected to drop from 47% in 2004 to 37% by 2030, according to the Institute of Energy Economics, Japan (IEEJ).

In place of petroleum, the report said, use of low-carbon fuel will increase. Nuclear power's share will increase to 20% from 11% in the forecast period, and the natural gas share will rise 18% from 14%.

Iranian, Russian projects falter

Troubled projects in Russia and Iran in 2006 set back the Japanese government's push to secure international oil and gas supplies, reported FACTS Inc. (Honolulu).

The government announced in mid-2006 a goal to increase the share of oil developed and imported by Japanese companies to 40% of total imports by 2030 from 15% in recent years.

Russia said it might divert gas from the Sakhalin-1 project, in which the Japanese group Sodeco holds a 30% interest, to China. And there are concerns that the project might fall subject to the type of environmental resistance that has stalled Sakhalin-2.

In Iran, Japan's Inpex Holdings Inc. had a substantial stake in development of the giant Azadegan oil field before the government substantially cut the Inpex stake.

In the face of these setbacks, Japan has turned its attention to Indonesia and Australia.

Inpex said it plans to invest about ¥500 billion ($4.22 billion) to launch natural gas production in Indonesia by 2015. It will start drilling in early 2007 and plans to submit a production plan to the Indonesian government in 2008. The company aims to ship 3–5 mtpy of LNG to Japan and elsewhere.

Inpex was pursuing a similar project in an Australian field. There the firm is spending ¥700 billion, with plans to produce 6 mtpy of LNG starting in 2012.

These two fields could jointly supply 15–20% of Japan's LNG needs.

Exploring methane hydrates

In early 2006, researchers from the University of Tokyo and the Japan Agency for Marine-Earth Science and Technology (JAMSTEC) announced successful extraction of a deposit of methane hydrate not far below the seabed, 30 km (19 miles) north of Joetsu.

The substance is a mix of methane trapped in frozen water. Some reports say the energy held in each cubic meter of hydrate is equal to that contained in up to 180 cu m of natural gas.

The new cache was said to be the first discovered in Asia lying relatively close to the seabed. Most known deposits lie hundreds of meters below the ocean floor.

JAMSTEC found and extracted samples just 6–12 m (20–40 ft) below the ocean floor, in water 800–1,000 m (2,625–3,280 ft) deep.

Seabeds around Japan are now thought to contain 7 trillion cu m of methane hydrate, equal to 100 years of natural gas consumption in Japan at the current rate of use. The novel resource could help cut Japan's dependence on crude oil.

In late 2006, a production test was planned in Canada, involving Japan's Ministry of Economy, Trade and Industry, Japan Oil, Gas & Metals National Corp., and the Canadian government.

If the experiments go well, the Japanese interests may begin trial extractions of methane hydrate from the Pacific Ocean floor by fiscal 2009. The Japanese government has said it aims to start commercial extraction of methane hydrate in fiscal 2017.

CHINA
RUSSIA
NORTH KOREA
SOUTH KOREA
JAPAN
Sea of Japan
Pacific Ocean
HOKKAIDO
HONSHU
SHIKOKU
KYUSHU
Vladivostok
Nakhodka
Peter the Great Bay
Chongjin
42
Pohang
Ulsan
817
Pusan
10
Korean Channel
Tsushima Channel
Wakkanai
Soya
Katsu
Menas
Koetoi
Toyotomi
Kitatoyotomi
Teshio Fields
Iskikari Bay
Ishikari
Atsuta
Barato
Sapporo
Tokachi Fields
Azuma
Fureoi
Kinausu
Tomakomai
133
180
Muroran
Oshamambe
Mori
Hakodate
Tsugaru Strait
Aomori
Nichi-Hachimori
Higashi-Hachimori
Iwako
Sawame
Sakake
Akita Fields
Surukawa
Kurokawa
Asakawa
Funakawa
Akita
Morioka
Yabase
Yuri
Kamekawa
Shonai
Innai
Kisagata
Okuoguni
Yokuka
Chokaizan
Yamagata Fields
Fukura
Sakata
Nogouchi
Mogaami
Shonai
Yamagata
Amarume
145
Sendai
Sendai Bay
Iwafune
Aga
Niitsu
Shibata
Niigata
Niigata
Niigata-Plain
Higashi
Okawatsu
Mitsuke
Higashiyama
Echigo Fields
Nishiyama
Nagaoka
Kashiwazaki
Sekihara
Ojiya-Sanjo
Kubiki
Ubase
Toyama Bay
Gozu
Kubiki
Nadachi
Menji
Maki
Iwaki
Iwaki
Asakawa
Odagiri
Toyama
57
Kanazawa
Kashima
181
Choshi
Yokoshiba
Fukui
Wakasa Bay
Tokyo
Kawasaki
Chiba
Shirasato
Mobara
Yokohama
Goi
Negishi
Sodegaura
Chiba Fields
120
296
65
228
175
209
192
340
Kyoto
Nagoya
Shimizu
Shimizu
Yaizu
Sagara
Chita
152
Yokkaichi
147
222
Ise Bay
Himeyi
Hyogo
115
Osaka
Sakai
250
76
155
Wakayama
Kainan
Owase
160
50
Shimotsu
Kii Strait
Mizushima
Takamatsu
Hiroshima
Sakaide
114
Yamaguchi
127
110
Marifu
Iwakuni
114
Shimonoseki
Onoda
Ube
Tokuyama
Kikuma
Kudamatsu
Matsuyama
Yahata
Ehime
120
Fukuoka
Sasebo
Oita
130
Bungo Strait
Tosa Bay
Kashima
Nagasaki
Kumamoto
Miyazaki
Osumi Strait
Okinawa
90
100
Off Map
JAPAN
0
50
100 Mi.
0
100 Km.

Chubu, Toho LNG terminals

Japan's Chubu Electric Power Co. and Toho Gas Co. announced plans to expand their LNG facilities by constructing two natural gas pipelines and further developing the jointly owned Chita LNG import terminal in Aichi Prefecture.

The pipelines will connect the Chita terminal to Chubu's Kawagoe thermal power plant and to Toho's regasification plant. The pipelines are scheduled for completion in 2013.

By 2009, the firms plan to refit and expand the Chita terminal to receive tankers that can carry 200,000 cu m of LNG. Current maximum vessel size is 170,000 cu m.

Chubu Electric also plans to refit by 2010 another LNG terminal near the Kawagoe thermal power plant to accept tankers that carry as much as 200,000 cu m of LNG.

LNG could come primarily from Qatar Liquefied Gas Co., a main supplier to Chubu since 1996.

Pilot GTL plant

Six Japanese firms announced plans to jointly establish a research entity and build a test plant for producing gas-to-liquid (GTL) fuel with an eye toward potential international development.

The six are Nippon Oil Corp., Inpex Holdings Inc., Japan Petroleum Exploration Co., Cosmo Oil Co., Nippon Steel Engineering Co., and Chiyoda Corp.

The group said a research organization could be set up by early 2007. Other objectives are to finish building a test plant in Japan during 2008, and then to conduct joint research through 2010.

The test facility, to be built in Niigata Prefecture, would be designed to produce 500 b/d of synthetic oil.

Working with Japan Oil, Gas & Metals National Corp., the consortium also would study the potential for the construction of GTL plants overseas.

LPG operations alliance

Mitsui & Co., Sumitomo Corp., and Marubeni Corp. were negotiating to form an alliance for the joint procurement of liquid petroleum gas (LPG), operation of tankers to carry it, and eventual consolidation of their respective distribution and storage facilities.

The companies were said to view the alliance as a way of strengthening their operations. Competition is increasing among Japan's LPG wholesalers due to shrinkage of the wholesale market and a doubling of the cost of supplies from Middle East producers.

MALAYSIA

Capital: Kuala Lumpur
Monetary unit: Ringgit
Refining capacity: 544,832 b/cd
Oil production: 737,000 b/d
Oil reserves: 3.0 billion bbl
Gas reserves: 75.0 tcf

Exploration and Development

Petronas deepwater program

State-owned Petronas said that it anticipates finding approximately 8 billion boe in undiscovered reserves through a deepwater drilling program to be carried out through 2010.

The company was planning to drill 10 deepwater exploration wells in 2006 (nine off Sabah and one off Sarawak), another 10 in 2007, and 11 per year for 2008–2010.

Deepwater Sabah licenses

Petronas awarded Murphy Sabah Oil Co. Ltd. a deepwater license for Block P off Sabah, and an extension to the exploration term of a portion of Block K. Murphy is the operator and has working interests in Blocks K and P. The blocks lie in 915–2,750 m (3,000–9,000 ft) of water.

The Block K extension was granted for five years, while Block P carries a seven-year exploration term. On Block P, the partners will acquire and process 500 sq km (190 sq miles) of 3D seismic data and drill four wildcat wells.

Murphy Oil said that Block P lies on trend 110 km (68 miles) to the northeast of Block K, and that the extension of part of Block K centers on the company's Kikeh complex on the block's southern corner, where production is to begin in the second half of 2007.

MALAYSIA-INDONESIA
MYANMAR (BURMA)
THAILAND
LAOS
CAMBODIA
VIETNAM
MALAYA
MALAYSIA
SABAH
BRUNEI
SARAWAK
SINGAPORE
SUMATRA
KALIMANTAN (BORNEO)
INDONESIA
JAVA
MADURA
BALI
SUMBAWA
BANGKA
BELITUNG
KEPULAUAN LINGGA
NATUNA ISLAND (INDONESIA)
HAINAN DAO
ANDAMAN ISLANDS
NICOBAR ISLANDS
SIMEULUE
NIAS
SIBERUT
SIPORA
PAGAI UTARA
PAGAI SELATAN
ENGGANO
PALAWAN
Bay of Bengal
Gulf of Martaban
Andaman Sea
Indian Ocean
Gulf of Thailand
Gulf of Tonkin
South China Sea
Strait of Malacca
Java Sea
Sunda Strait
Strait of Makassar
Sulu
Noakhali
Patuakhali
Chittagong
Patiya
Shahbazpur
Mandalay
Ayadaw
Yenangyaung
Chauk/Lanywa
Mann
Htaukshabin
Padaukpin
Pyaye
Pyato
Prome
Myanaung
Shwepyitha
Yangon (Rangoon)
Mae Sot
Martaban
Yadana
Yetagun
Mae Fang
Mae Suhn
Chiang Mai
Nan
Vientiane
Loei
Uttaradit
Mae Sot
KDN-1
Sirikit
Pru Krathiarn
Bung Ya
Phu Horm
Nam Phong
Chonnabot
Khon Kaen
Sisaket
Sawng Neung
Saraburi
Kanchanaburi
Bangkok
Sri Racha
Rayong
Phnom Penh
Sihanoukville
Nang Nuan
Plamuk
Kapong
Pakarang
Pladang
Tantawan
Surat
Chaba
Dara
Erawan
Baanpot
Pailin
Platong
Platong S.
Trat
Sabun
Jakrawan
Gomin
Punan
Khanom
Si Thammarot
Arthit
Bongkot
17-B-1
Songkhla
Bua Ban
Pilong
Hanoi
Hai Phong
Tien Hai
Weizhou 11-1
Weizhou 10-3
Weizhou 6-1
Haik'ou
Jinfeng
Wenchang 9-2
Yacheng 13-1
Moaming
Macau
Hong Kong
Lufeng 13-1
Lufeng 15-1
Xijiang 24-3
Xijiang 30-2
Huizhou 21-1
Huizhou 26-1
Liuhua 4-1
Liuhua 11-1
Hue
Da Nang
Nha Trang
Ho Chi Minh City
Vung Tau
Long Xuyen
Sutu Den
Rang Dong
Bach Ho
Rong Doang
Dai Hung
Dua-1X
Bursa
Lawit
Jerneh
Sepat
Damar
Bintang
Inas
Noring
Bedong
Larut
Bubu
Bujang
Tangga
Lerek
Tujoh
Dulang
Tabu
Barat
Tapis
Kakap
Besar
Duyong
Belumut
Angsi
Sotong
N. Belut
W. Belut
Malong
Mamung
Kerisi
Belanak
Terebuk
Sepat
Sotong
Semangkok
Tembang
Keong
Udang
Buntal
Belida
Kijing
Kepiting
Malong
Natuna
Kutaradja
Julu Rayeu
Perlak
Serang Jaya
Pangkalan Brandan
Belawan
Meulabah
Medan
Pinang
Kerteh
Kuala Lumpur
Port Dickson
Sibolga
Dumai
Singapore
Batam
Pinang
Bangko
Buaya
Menggala
Mandar
Libo
Minik
Langgak
Duri
Mellbur
Minas
Zamrud
Topaz
Lirik Fields
Pekanbaru
N. Palai
S. Palai
Andan
Sago
Lirik
Molek
Binio
Bukittinggi
Pontianak
Kuching
Djambi
Setiti
Bedjubang
Tempino
Limau Fields
Sumpal
Suban
Burong
Bentayan
Dayung
Keban
Sungei Gerong
Palembang
Muraraenim Fields
Tanjung Laban
Benakat
T. Djimar
Sambar
Ogan Timur
Kampong Minyak
Taham
T. Tiga
Lapang
Prabukeras
Pagardewa
Telukbetung
Intan
Indri
Winduri
Sundari
Karmila
Krisna
Cinta
Nora
AV
AA
Rama
KN
Arjuna
HZE
Cilegon
Jakarta
Arimbi
Cirebon
Jatibarang
Randegan
Bandung
Semarang
Rembang
Ledok
Ngiobo
Petak
Camar
Candi
Cepu
Banyubang
Kawengan
Maleo
Kendangan
Gresik
Waron
Cilacap
Jogjakarta
Wonokromo
Singaradja
Raba
M1
M3
M4
F6
F28
Jintan
Seria
F27
Bijan
F14
B11
B12
B22
Samarang
Kecil
Iron Duke
Magpie
Glayzer
Kecil
E8
D12
J4
W. Bayan
C2
Temana
Patricia
Bintulu
Miri
Kuala Belait
Bandar Seri Begawan
Labuan
Kota Kinabalu
Kinarut
S.W. Emerald
Tembungo
Erb W.
Barton
S. Furious
St. Joseph
Kudat
Sandakan
Tawau
Mamburungan
Bunyu Nibung
Tarakan
Batuputih
Sangkulirang
Kerindingon
Panyilatan
Serang
Attaka
Tunu
Kerendan
Semberah
Badak
Sanga-Sanga
Pamaguan
Louise
Sanggur
Bangkerak
Balikpapan
Handil
Bekapi
Mutiara
Sepinggan
Wailawi
Kambatin
Tanjung
Tapian Timur
Warukin Tengah
Warukin Selatan
Palangkaraya
Bandjarmasin
Subic Bay
Calauit
Galoc
Octon
West Linapacan
Malampaya
Camago
Destacado
Nido
Matinloc
Cadlao
Sampaguita
0 50 100 150 200 250 Miles
0 100 200 300 400 Km
20°
15°
10°
5°
0°
5°
10°
95°
100°
105°
110°
115°

Pacific Ocean
Philippine Sea
PHILIPPINES
LUZON
MINDORO
PANAY
LEYTE
NEGROS
BOHOL
MINDANAO
Yungan
Pingtung
270
Subic Bay
Quezon City
Manila
Limay
180
Batangas
Tabango
153
Calauit
Galoc
Saddle Rock
Malajon
Linapacan
Matinloc
Cadlao
Tacloban
Bacolod
Cebu
Davao
Sulu Sea
Celebes Sea
Nibung
Manado
Batuputih
Soasiu
Makassar
Palu
Molucca Sea
SULAWESI (CELEBES)
Kendari
Ujung Pandang
Sorong
1-X TBM
Matoa
Salawati
Kasim
Walio
Klamono
Sele
Wiriagar
Manokwari
Teruk Berau C
Teruk Berau A
Mogo
Wasian
Wiriager Deep
Ubadari
Wos
Vorwata
Roabiba
Ofaweri
Steenkool
Fakfak
Bula
Ceram Sea
BURU
CERAM
Ambon
Banda Sea
ARU
TANIMBAR
Flores Sea
FLORES
SUMBA
Dili
Matai
Ossulari
Aliambata
TIMOR
Kupang
Roti
Bayu
Jabiru
Challis
Cassini
Puffin
Timor Sea
Arafura Sea
Darwin
AUSTRALIA
Bonaparte Gulf
Gulf of Carpentaria
IRIAN JAYA
Hollandia
Bismarck Sea
Wewak
PAPUA NEW GUINEA
Madanda
Mt. Hagen
Goroka
Lae
Elvala
Ketu
Juha
Hides
S.E. Mananda
S.E. Hedinia
Elevala
Agogo
Iagifu
2XGobe
4XGobe
S.E. Gobe
Bwata
Barikewa
Kuru
Omati
Puri
Kikori
Baimuru
Uramu
Lake Murray
Gulf of Papua
Pasca
Pandora-B1X
Pandora-1X
33
Port Moresby
Solomon Sea
Coral Sea
120°
125°
130°
135°
140°
145°
Samarang Kecil
Champion
Glayzer
Pelican
Iron Duke
SABAH
Petrel
Magpie
Osprey
9
Egret
Bandar Seri Begawan
Fairley
Fairley Baram
Ampa
S.W. Ampa
Baram
Tali
Seria
W. Lutong
Kuala Belait
BRUNEI
Rasau
SARAWAK

Block G oil, gas strike

Shell Malaysia and its partners announced a fourth discovery with the Pisagan-1A exploration well on deepwater Block G off northwest Sabah. The well was drilled to 1,465 m (4,800 ft) and encountered both oil and gas.

Since 2003, the group has made three other oil and gas discoveries off Sabah: Ubah-2, Malikai-1, and Gumusut-1. Partners with operator Shell Malaysia are ConocoPhillips (East Malaysia) and Petronas.

Kikeh field

In late 2006, Technip (Paris) completed topsides installation on the deepwater Kikeh spar, in Kikeh oil field. The field is in Block K, 120km (75 miles) northwest of the island of Labuan, offshore Sabah..

Technip called it the first spar hull application outside the Gulf of Mexico, in a field being developed in the deepest waters in Southeast Asia—1,330 m (4,365 ft). An FPSO vessel will receive production from the wells to be drilled from the spar. Production could begin in late 2007.

Kikeh has a recoverable reserve base of at least 400 million bbl.

Murphy Sabah Oil—a unit of Murphy Oil Corp. (El Dorado, AR, U.S.A.) operates Block K and the Kikeh field in partnership with Petronas Carigali.

Drilling and Production

First Cendor cargo

Petrofac Malaysia lifted its first cargo totaling 311,000 bbl from Cendor oil field on Block PM304 off Malaysia in late 2006.

The field's production was averaging 12,000 b/d of oil, with an established peak rate of 16,000 b/d from the H15 and H20 formations. The company began oil production of 3,500 b/d from one of the five wells in the field.

Petrofac also completed drilling of the seven development wells comprising the field's first phase. The company, in a partnership with Petronas, was reevaluating estimates of proven reserves.

CENDOR FIELD

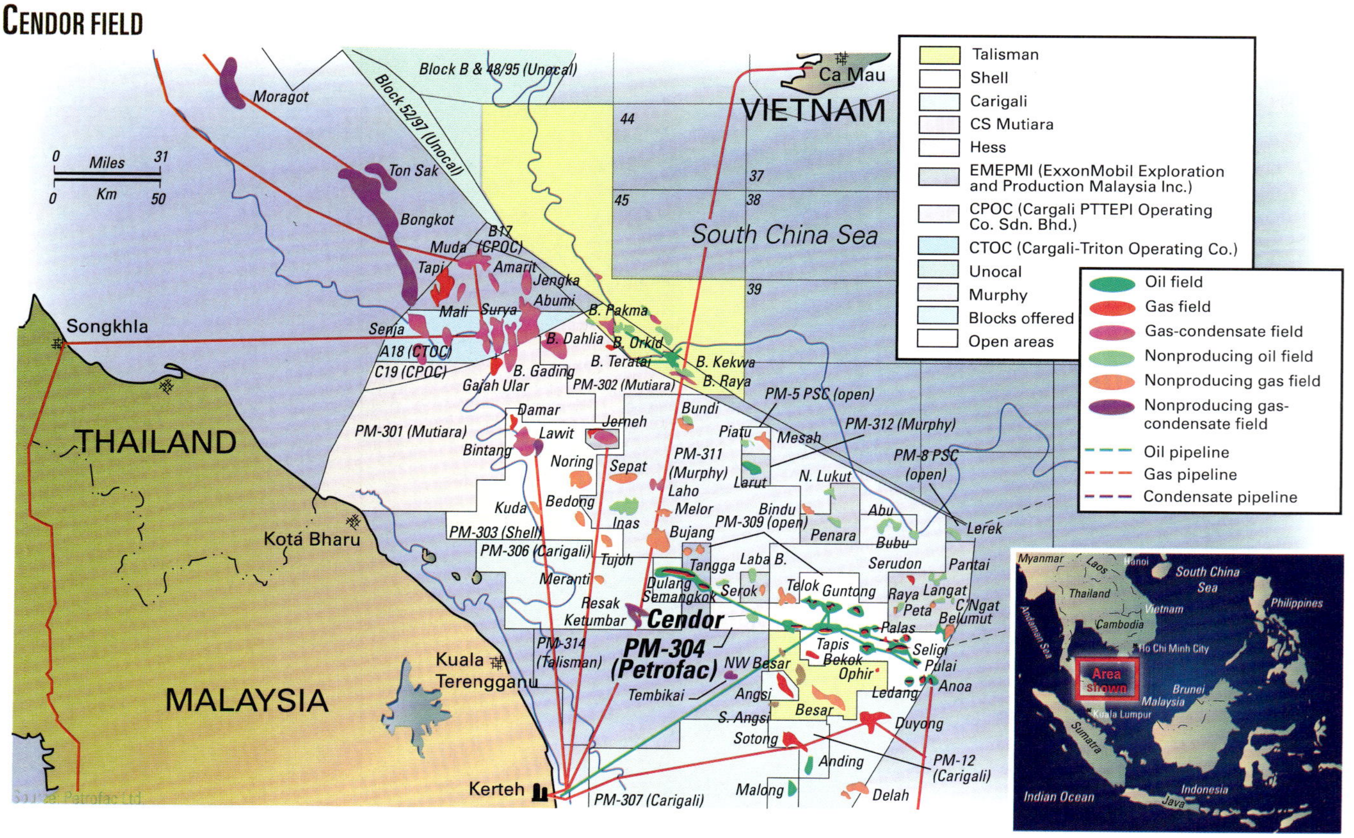

E8 field

Shell Malaysia delivered natural gas for the first time from E8 field off Sarawak. Production started in late July and flow was stabilizing toward a target flow of 600 MMscfd.

Sarawak Shell Bhd. operates E8, a key component of the E11 Hub Integrated Gas Project. Production from surrounding fields goes to the E11 complex for processing before being delivered onshore by pipeline to the Petronas LNG complex at Bintulu.

E8 is one of five fields developed to supply the LNG complex. The others are E11, F23, F6, and F13.

Refining

Group III base oil plant

Petronas began construction on a Group III lubricant base oil plant within its refinery complex in Melaka. The company calls it the first of its kind in Malaysia and Southeast Asia. It will have the capacity to produce 6,500 bpd of oils for use in low-viscosity automotive motor oil. The plant is to be completed in 2008.

LNG

LNG for Japanese firms

Malaysia LNG Sdn. Bhd. (MLNG) signed a confirmation of intent to supply Japan's Shikoku Electric Power Co. Inc. with 420,000 tpy of LNG for 15 years.

Delivery will begin in 2010, with an option for five additional years. The LNG will come from Petronas LNG's complex in Bintulu, Sarawak. The LNG will go to a receiving terminal in Sakaide, Japan, now under construction.

Malaysia LNG Tiga Sdn. Bhd. (MLNG Tiga) signed an agreement to sell Toho Gas Co. Ltd. 520,000 tpy of LNG for 20 years beginning in 2007. The LNG will come from the Bintulu LNG complex, and will be shipped to Toho's receiving terminal in Chita, Japan.

The Bintulu complex has production capacity of 23 mtpy of LNG from the 3-train, 8.1-mtpy MLNG plant; the 3-train, 7.8-mtpy MLNG Dua plant; and the 2-train, 6.8-mtpy MLNG Tiga plant.

MONGOLIA

Capital: Ulan Bator
Monetary unit: Tughrik
Refining capacity: 0
Oil production: 0
Oil reserves: 0
Gas reserves: 0

Production starts at Changbei field

Commercial gas production began in early 2007 at Changbei gas field in China, said Royal Dutch Shell PLC, field operator. The field is expected to plateau at 3 bcm/year in 2008 and deliver gas to Beijing, Tianjin, Shandong, and Hebei.

The field is in the Ordos basin of Shaanxi Province and Inner Mongolia Autonomous Region. PetroChina and Shell are jointly developing the field under a production-sharing contract. The drilling program is expected to involve about 50 horizontal and multilateral wells over 10 years.

Trans-Mongolia gas pipeline

Russian oil company TNK-BP Ltd. holds substantial stranded East Siberian gas reserves estimated at 1.4 tcm (49.3 tcf) in Kovykta field. For several years, the firm has been trying to secure government approval to build a pipeline to China and potentially beyond to South Korea.

KOVYKTA FIELD POTENTIAL GAS EXPORT ROUTES*

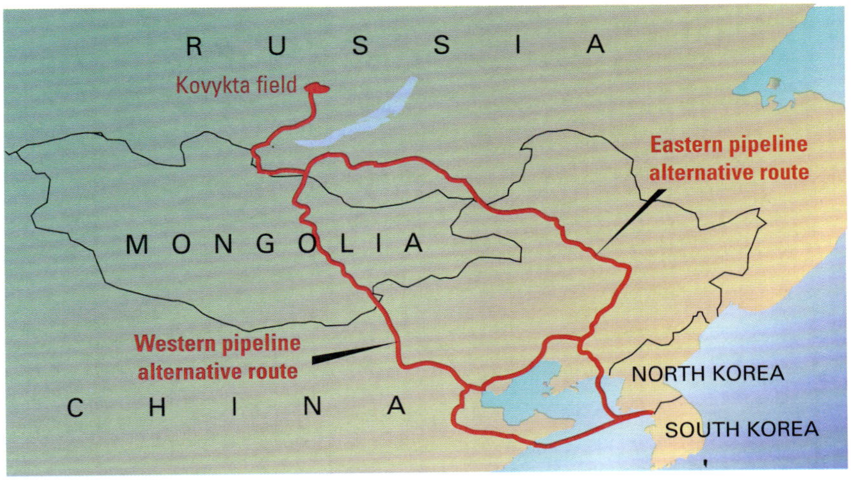

*Based on a TNK-BP prefeasibility study completed in 2001.
Source: David Wood & Associates

A pre-feasibility study has identified two potential routes for such a pipeline. The western route would extend 2,645 km (1,644 miles) across Mongolia to Beijing. The eastern option is a 4,485-km (2,787-mile) line extending across Ulan Ude and Chita to Beijing.

TNK-BP and Gazprom have discussed and disagreed over the preferred pipeline route for several years. In 2006 it was made clear that if such an export pipeline is to be built, it will be under Gazprom's operatorship and that work likely would not begin until 2012 or later.

MYANMAR

Capital: Pyinmana
Monetary unit: Kyat
Refining capacity: 57,000 b/cd
Oil production: 13,000 b/d
Oil reserves: 50.0 million bbl
Gas reserves: 10.0 tcf

Gas flows at Sein field

Total SA started production from Sein gas field in the Andaman Sea. Two development wells produce a combined 35 MMcfd of dry gas that is routed to Yadana field facilities. The wells are on Blocks M-5 and M-6.

Rakhine gas finds assessed

Independent consultants Gaffney, Cline & Associates said 3.3 tcf is the best estimate of recoverable gas from two discoveries in the Rakhine basin off Northwest Myanmar by a group led by Daewoo International Corp. (Seoul).

The appraised 2004 Shwe and 2005 Shwe Phyu discoveries on Block A-1 are the basin's first substantial gas discoveries.

Two more prospects are identified on A-3, Daewoo said, and one or two exploratory wells will be drilled, likely in 2007.

Best estimates for Shwe field are 2.9 tcf recoverable from 3.4 tcf in place; and for Shwe Phyu 0.4 tcf recoverable from 0.5 tcf in place.

Gas strike on A-3 block

South Korea's Daewoo International Corp. and its partners made a natural gas discovery on A-3 Block. The Mya-1 well cut a 32-m (105-ft) gas column and flowed 57.6 MMcfd of gas on test.

Mya-1 could be another giant gas find, with an estimated 1.5 tcf of gas reserves in-place, Daewoo said. The discovery adds to the success at Shwe and Shwe Phyu gas fields on adjacent A-1 Block.

Partners with Daewoo on A-1 and A-3 Blocks are India's Oil & Natural Gas Corp., Gas Authority of India Ltd., and Korea Gas Corp.

Development options being considered are pipeline exports to India, China or Thailand, as well as LNG and compressed natural gas projects.

Yetagun gas field

Myanmar agreed to step up natural gas production from Yetagun field to increase supplies to neighboring Thailand, according to the Ministry of Energy. Myanmar provides about 20% of Thailand's gas supply.

Thailand reportedly wants to buy 100 MMcfd of gas in addition to the 400 MMcfd of gas it pipes from the offshore reserves in the Andaman Sea. That gas comes mainly from two fields: Yetagun and Yadana.

RAKHINE BASIN GAS DISCOVERY

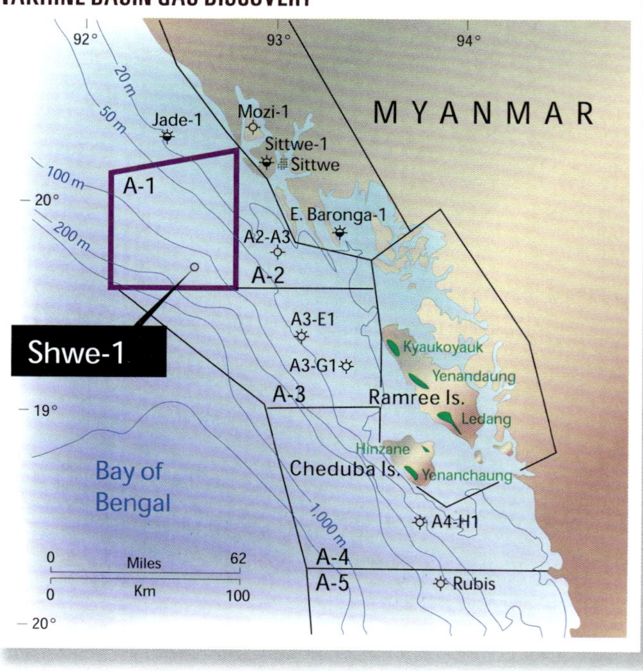

NEW ZEALAND

Capital: Wellington
Monetary unit: Dollar
Refining capacity: 103,550 b/cd
Oil production: 14,000 b/d
Oil reserves: 53.0 million bbl
Gas reserves: 900 bcf

Second Cheal B well

Austral Pacific Energy Ltd. (Wellington) achieved a TD of 1,866 m (6,122 ft) with its Cheal B2 well, the second of four wells to be drilled from the Cheal B site. The Cheal oil field is in the Taranaki basin.

This well targeted and intersected the late Miocene Mount Messenger sands over a gross interval of 1,709–1,735 m (5,607–5,692 ft). Wireline logs indicate these sands are oil bearing.

Drilling of the Cheal B3 well was slated to begin by year-end 2006. The four wells are expected to produce a combined 1,900 b/d by mid-2007.

Maari platform

A consortium led by OMV New Zealand Ltd. signed a contract with Clough Ltd (Perth) for a contract to build and install the Maari oil field wellhead platform in the Taranaki basin. The field lies in 100 m (328 ft) of water about 80 km (50 miles) off the North Island.

Five producing wells are planned, tied back to an FPSO vessel, as well as a shuttle tanker export arrangement. Total development cost is estimated at $360 million.

Production is expected to reach 35,000 b/d, starting in first quarter 2008. Reserves are estimated at 50 million bbl.

Westland basin

A group led by McKenzie Petroleum (Wellington) offered farmouts on three areas in 336 sq-km (130 sq-mile) PEP 38521 in the Westland basin south of Greymouth on the northwestern South Island.

Area A contains most of the Brunner anticline, where the 1943 SFL-1 exploration well flowed 6.5 MMcfd of 97% methane from the Oligocene Omotumoto sandstone member of the Kaiata formation at less than 275 m (900 ft).

Area B contains the southern part of the Brunner anticline as well as other prospects and leads. Block C is the deep oil prospective Grey Valley half-graben with the Pounamu prospect and other leads.

IEA inventory target

The energy ministry announced that, effective January 1, 2007, New Zealand would meet its International Energy Agency (IEA) obligation to hold oil stocks representing 90 days of net oil imports.

New Zealand's oil stocks had dipped as low as 60 days of net oil imports as a result of increasing demand and declining domestic oil production.

The government arranged contracts that provide options for New Zealand to buy petroleum and diesel in case of an IEA-declared emergency. The contracts cover petroleum and diesel stored in Australia, the Netherlands, and the UK from BP PLC, Royal Dutch Shell PLC, and Total SA.

Bilateral arrangements with Australia, the UK, and the Netherlands to enable these stocks to count toward New Zealand's IEA obligations.

Gas imports pondered

Several New Zealand companies were considering the importation of natural gas to make up the shortfall in indigenous supplies. One concern is the expected decline in production from long-running mainstay Maui field in offshore Taranaki basin.

Local gas transmission and distribution company Vector Ltd. was studying bringing in compressed natural gas from Papua New Guinea with a fleet of tankers across the Tasman Sea.

The company has joined Papua New Guinea firm Oil Search Ltd. to investigate this potential trade. One avenue could be the shipment of CNG from Gladstone in Queensland if the Papua New Guinea-Australia pipeline is ever completed.

Also, the nation's two largest gas users, Contact Energy Ltd. and Genesis Energy, were studying the feasibility of importing LNG, most likely from Australia. Both firms have gas-fired power stations in New Zealand and require secure generating fuel supplies beyond 2010.

PAKISTAN

Capital: Islamabad
Monetary unit: Rupee
Refining capacity: 268,975 b/cd
Oil production: 63,500 b/d
Oil reserves: 289.202 million bbl
Gas reserves: 28.00 tcf

Exploration and Development

Margalla blocks

MOL Pakistan Oil and Gas BV, a subsidiary of Hungary's MOL Group, announced plans to explore the Margalla and Margalla North blocks under concession agreements with the Pakistani government.

The blocks are in Punjab, North West Frontier Province (NWFP), and Islamabad. The Margalla Block covers 1,387 sq km (535 sq miles) and the Margalla North Block, 1,562 sq km (603 sq miles).

Seismic in Sindh Province

The joint venture of Jura Energy Corp.(Calgary) with local partner, Petroleum Exploration (Pvt.) Ltd., signed a seismic acquisition contract with the Oil & Gas Engineering Co. of the Sichuan Petroleum Administration (SPA) to explore for oil and gas in Sindh Province.

SPA will acquire 600 line-km of 2D seismic data on five exploration licenses—Mirpur Mathelo, Kandra (Lower Goru), Salam, Badin IV North, and Badin IV South—and on the Kandra (SML) development lease. All blocks are in the Middle and Lower Indus gas basin.

Discovery in Northwest Frontier Province

Pakistan's state-owned Oil & Gas Development Company (OGDC) reported discovery of oil and gas with its exploration well drilled 175 km (109 miles) west of Islamabad in the North West Frontier Province on Block 3371-10 (Kohat) in Zone I. The well has tested at 4,000 b/d of oil and 12 MMcfd of natural gas.

Licensing

Block awards

To attract investment into the oil and gas sector and boost Pakistan's economy by displacing imported oil with indigenous oil and gas, the government of Pakistan granted numerous exploration licenses in 2006. The awards generally require geological and geophysical studies, evaluation of seismic data, and drilling of exploratory wells.

- To BP for Arabian Sea Blocks U, V, and W (Blocks 2266-9, 2265-4, and 2264-1) in the Indus Delta. The blocks collectively encompass an area of more than 21,000 sq km (8,100 sq miles).
- To state-owned OGDC for Blocks 2769-14 (Tegani) in Zone III, 2769-15 (Thal) in Zone III, and 2567-11 (Tano Beg) in Zone II. Tagani is in Shikarpur, Jacobabad, and Sukkur districts. Thal is in Khairpur, Sukkur, and Ghotki districts of Sindh Province. Thano Beg is in Lasbella, Dadu, and Karachi districts of Baluchistan and Sindh provinces.
- To Techno Petroleum (Pvt.) Ltd. (TPPL) for exploration of Block No. 2770.3 (Islamgarh) in the Bahawalpur and Rahim Yar Khan districts of Punjab province in Zone-III. The block covers 230 sq km (89 sq miles).
- To Petroleum Exploration (Pvt.) Ltd. (PEL) for exploration on Blocks No. 2468-6 (Badin-IV North) and No. 2468 (Badin-IV South). The blocks cover 1,246 sq km (480 sq miles) and 1,265 sq km (488 sq miles), respectively. Badin-IV North is in Badin, Thatta, and Hyderabad districts of Sindh province, and Badin IV South is in Badin and Thatta.
- To state-owned exploration and production firm Pakistan Petroleum Ltd. (PPL) for Block 2969-8 (Barkhan), covering 2,104 sq km (812 sq miles) in the Dera Bugti, Kohlu, Musa Khail Bazar, Barkhan, and Loralai districts of Baluchistan Province (Zone II).
- To Eni Pakistan Ltd. for the Mithi (2470-3), Rajar (2470-2), Thar (2569-2), and Umarkot (2469-8) blocks in the Tarparkar, Umerkot, and Sanghar districts of Sindh Province.
- To a joint venture of government-owned Government Holdings (GHPL) and Pakistan Petroleum Ltd (PPL) for the Eastern Offshore Indus-C Block offshore Pakistan in the Arabian Sea. The license covers the 2,494 sq km Block 2366-7.

LNG

Karachi LNG terminal

A consortium of Dana Gas, Single Buoy Moorings (SBM), and Granada Group signed a memorandum of understanding to develop a 3.5-mtpy LNG terminal at Karachi's Port Qasim.

Dana Gas, based in the Arab Emirate of Sharjah, said it plans to develop a network of LNG terminals, mainly in the Middle East and North Africa. Initial targets are Pakistan, Lebanon, and Kuwait.

Under a cooperation agreement with SBM, Dana would focus on LNG marketing and SBM on the supply and operation of LNG floating storage and regasification terminals.

Refining

Hub refinery upgrade

Bosicor Refinery Ltd. (Karachi) signed a contract with Born Heaters Canada ULC for a 35,000 b/d crude distillation unit-fired furnace heater for its 18,000 b/d refinery at Hub in Baluchistan Province. The unit is part of an upgrade that will increase the refinery's capacity to 30,000 b/d.

Pipelines

Baluchistan bombings

Militant rebels continued attacks against natural gas pipelines, compressor stations and other facilities in the province of Baluchistan as part of an ongoing campaign for autonomy and a share in the revenues from oil and gas produced in the impoverished province.

PAPUA NEW GUINEA

Capital: Port Moresby
Monetary unit: Kina
Refining capacity: 32,500 b/cd
Oil production: 56,500 b/d
Oil reserves: 240.0 million bbl
Gas reserves: 12.2 tcf

PNG-Australia gas pipeline dead

Plans for a major pipeline to carry natural gas from the southern highlands of Papua New Guinea to Australia were abandoned. See the Australia section of this document for details.

South East Mananda field on stream

Oil Search Ltd. (Sydney) and its partners started production from the South East Mananda oil field in the southern highlands. The mid-year production target was 7,000 b/d.

Developers had to overcome a number of technical and commercial challenges, including installation of 16 km (10 miles) of parallel oil, gas-lift, and chemical pipelines in a remote, high-rainfall area.

Oil Search built an innovative 470-m (1,540-ft) cable suspension bridge to carry the bundle of pipelines across 400-m (1,312-ft) deep Hegigio Gorge.

On the other side of the gorge, the oil pipeline connects with the Agogo field facilities, which feed in to the Kutubu central production hub and the head of a pipeline that runs to a shipping terminal in the Gulf of Papua.

Gas find in Papuan basin

InterOil Corp. (Toronto) announced a gas discovery at its Elk-1 well in the onshore eastern Papuan basin. Test data indicated an open flow potential of 28.5 MMcfd of gas. Reserves are estimated at 3 tcf.

In a related development, Natural Gas Development Co. (NGDC)—owned by InterOil, Merrill Lynch Commodities, and Clarion Finanz AG—signed a memorandum of understanding with the government of Papua New Guinea to develop a gas processing plant and gas liquefaction facility.

InterOil owns the 32,500-b/d Papua New Guinea refinery in Port Moresby and conducts onshore exploration in the country.

The LNG proposal envisions a plant of 5 mtpy capacity. First LNG delivery is 2011 is possible.

PPL 235 area

Austral Pacific Energy Ltd. (Wellington, New Zealand) and Rift Oil PLC were planning feasibility studies to assess commercialization options for their recent gas discovery on the Douglas prospect in the PPL 235 area, which covers 2,910 sq km (1,124 sq miles).

The Douglas-1 well reached TD of 1,978 m (6,490 ft), and wireline logs indicated that the well intersected two gas-bearing columns. Seismic mapping of the Douglas structure indicated a 50-sq km (19 sq-mile) closure. Further seismic evaluation was slated to begin in 2006.

GTL project with Kufpec

Kuwait Foreign Petroleum Exploration Co. (Kufpec) signed a joint development agreement with Syntroleum Corp. (Tulsa, OK, U.S.A.) for participation in the development of a 50,000 b/d gas-to-liquids plant in Papua New Guinea. The facility will produce sulfur-free diesel fuel and other petroleum products. Financing and construction schedules were being developed.

Licensing round

In September, Papua New Guinea opened its new licensing round of 15 onshore and offshore blocks in Singapore. This offering was the first large-scale release of petroleum exploration blocks in the country's history.

The offshore areas, which cover 107,000 ha in the Gulf of Papua, were drawn up with the assistance of newly acquired seismic data from the Fugro Group. The area is regarded as frontier despite the discovery of Pasca and Pandora gas fields in the gulf during the 1980s. Even so, little exploration data are available.

Thirty companies expressed interest in the offering.

PHILIPPINES

Capital: Manila
Monetary unit: Peso
Refining capacity: 333,000 b/cd
Oil production: 16,000 b/d
Oil reserves: 138.5 million bbl
Gas reserves: 3.48 tcf

Cebu Island

Forum Energy PLC (London) declared Libertad gas field commercial and planned to start production in 2006. Gas from Libertad, in SC 40 on Cebu Island in the Visayan basin, is expected to generate as much as 2 MW of electricity following a $2.5 million development program.

Libertad was the first field to be declared commercial in the Philippines since Shell's Malampaya gas field came on stream in 2001.

Developing Calauit

NorAsian Energy Ltd. and Vital Resources Corp. (Calgary) were working to develop the Calauit oil discovery in the Northwest Palawan-Mindoro basin and then use the revenue to pursue other promising targets in the area.

Calauit-1B well, 160 km (100 miles) north of Palawan Island in area SC-50, is calculated to have 8.9 million bbl of recoverable oil. Vital plans to reenter Calauit-1B by early 2007 and drill a 500-m (1,640-ft) lateral at the top of the 154-m thick (505-ft) oil column.

The plan for SC-51 in the East Visayas basin called for seismic in 2006 and drilling in 2007. The area has the undeveloped Villaba-1 discovery with a 19-m (62-ft) gas column on the Leyte Block and several prospects on the Cebu Straits Block.

The plan for SC 55 in the Southwest Palawan basin calls for seismic evaluation in 2006 and drilling in 2008. Several promising structures have been identified.

SC-41

Tap Oil Ltd. (Perth) took 85% equity in service contract (SC) 41 in Philippines waters off northeastern Borneo. The 2006–07 work program calls for gathering of more than 300 sq km (116 sq miles) of 3D seismic data, in water 800–1,200 m (2,625–3,940 ft) deep in the Sulu Sea.

SC-55

A unit of Ottoman Energy Ltd. (Sydney) let a contract to Veritas Geophysical (Asia Pacific) to gather 400 km (248 miles) of 2D seismic data over the giant Marantao Reef prospect. The prospect is on SC-55 off southwestern Palawan Island.

Ottoman said SC-55 is in the center of a more than 1,000-km long (621-mile) oil and gas fairway from fields off Brunei and Sarawak to Ottoman's undeveloped Calauit field north of Palawan. Play types in SC-55 include a string of reefs on trend with Marantao and deepwater Miocene Pagasa turbidite sand targets.

Biodiesel plant

Shomar International Trading Corp. (Japan) announced plans to build facilities in the Philippines to produce biodiesel fuel from the jathropa plant.

At the planned site in Hermosa, some 5,000 hectares (12,350 acres) of idle land will be planted with jathropa. The company will seek an additional 70,000 hectares (173,000 acres) to maximize production.

Also, the Philippine government announced that Saudi Aramco agreed to build a jathropa processing plant and an ethanol distillery in Mindanao. It would secure feedstock from idle military camps that are now jathropa plantations.

SINGAPORE

Capital: Singapore
Monetary unit: Dollar
Refining capacity: 1,336,600 b/cd
Oil production: 0
Oil reserves: 0
Gas reserves: 0

Shell ethylene cracker

Shell Eastern Petroleum (Pte.) Ltd. started work on significant upgrades to its Eastern Petrochemicals Complex in Singapore. The resulting integrated refinery and petrochemicals operation will incorporate modifications and additions to the 458,000 b/cd Pulau Bukom refinery and a new 750,000-tpy monoethylene glycol (MEG) plant on Jurong Island.

A key element in the project will be a new world-scale 800,000-tpy ethylene cracker. Shell Eastern awarded a contract to ABB Lummus Global and partner Toyo Engineering Corp. for the cracker, covering engineering, procurement, and construction management. The cracker will use ABB Lummus's proprietary liquid feedstock steam cracking technology.

The MEG plant and the cracker will be integrated with the Bukom refinery, and the cracker will supply products to the MEG plant. Construction on the ethylene cracker was slated to begin in late 2006, with start-up anticipated in 2010.

ExxonMobil ethylene cracker

ExxonMobil Asia Pacific Pte. gave the go-ahead to Foster Wheeler Ltd. and partner WorleyParsons to begin FEED work on selected downstream units and associated plant infrastructure for the Singapore Parallel Train (SPT) project.

The project includes the possible construction of a new ethylene cracker that ExxonMobil describes as "world-scale." It would be integrated with an ethylene plant that will have capacity exceeding 900,000 tpy after completion of a 75,000-tpy expansion in late 2006.

The chemical facility operates alongside ExxonMobil's 605,000 b/cd Singapore refinery.

Along with the new ethylene cracker, the parallel train under study would include new polyethylene, polypropylene, and specialty elastomer plants, an aromatics extraction unit, and oxoalcohol expansion.

Storage project

PetroChina agreed to take a 35% stake in an oil storage facility being built in Singapore. Hin Leong Trading (Pte.) Ltd. owns the rest. The $455 million Universal Terminal on Jurong Island has a design capacity of 14.5 million bbl and is expected to be operational by the end of 2007. It will be the first independent oil terminal in the region with berths for two very large crude carriers.

SOUTH KOREA

Capital: Seoul
Monetary unit: Won
Refining capacity: 2,576,500 b/cd
Oil production: 0
Oil reserves: 0
Gas reserves: 0

Exploration and Development

Donghae 6-1 area

Korea National Oil Corp. made a second gas discovery in the Donghae 6-1 area in the Sea of Japan. The discovery, 75 km (47 miles) southeast of Ulsan, is smaller than the Donghae 1 discovery but close enough to be developed with that earlier find. The state-owned firm said a pipeline could be built to connect the two fields. Evaluation of the new gas find began in mid-2006.

Refining

Upgrades at Ulsan refinery

SK Corp. awarded Haldor Topsoe AS (Lyngby , Denmark) a contract for a 110-MMscfd hydrogen plant at its 817,000-b/cd refinery at Ulsan, South Korea. The hydrogen plant will be part of a new resid hydrodesulfurization unit able to process multiple feedstocks. The project will use Topsoe's steam reformer technology.

SK Corp. also awarded a contract to Shaw Stone & Webster (Stoughton, MA, U.S.A.) for design, procurement, and project management and installation of a 60,000 b/sd FCC unit at the refinery.

LNG

LNG supply gap looms

According to analysts at Poten & Partners, Inc. (New York), South Korea likely will face a significant LNG supply gap starting in 2008.

Consumption in Korea has been rising at nearly 10%/year since 2000, reaching 23 million tonnes in 2005. At the same time, Korea Gas (Kogas; Gyeongiu) has scrambled every year to fill the winter shortfall with expensive cargoes purchased on spot terms.

Concern is growing that Indonesia, Korea's second largest supplier, might be unable to meet its contractual obligations over the next several years. In 2005, Indonesia supplied Korea with 5.6 million tonnes, equivalent to one-quarter of the nation's total imports.

However, South Korea's supply outlook becomes particularly precarious in 2008, when short-term deals totaling about 2.8 mtpy with Qatar's RasGas and Malaysia LNG's Tiga end, as does another 2.3-mtpy contract with Pertamina.

The loss of these supplies will be offset, said Poten, by start-up of new long-term supplies from Yemen LNG, MLNG Tiga, and Russia's Sakhalin II, totaling 5.5 million tpy. But these purchases, which begin in 2008 and 2009, leave Kogas virtually no room to cover any demand growth.

Poten forecasts further supply reductions for 2010, when another 2.5 million tpy in medium-term sales from MLNG Tiga and Australia's North West Shelf cease.

Based on its review of existing and expected official projections from the Korea Energy Economics Institute (KEEI), MOCIE, and Kogas, Poten expects that the supply gap could be nearly 6 million tonnes in 2008, rising to 8–10 million tonnes in 2010.

Several private industrial companies import LNG for their own use, and some regional generators under Korea Electric Power Co. have lobbied MOCIE for approval to secure their own supplies. But MOCIE has put off letting Kogas or private importers make further commitments.

In Poten's view, securing 5.5–6.0 million tpy of supply by 2008 might no longer be possible. Kogas may have no choice but to continue relying on spot product. And even if it succeeds in lining up an additional 40 cargoes/year, Poten estimates that Kogas will still be short 3.3 million tonnes in 2008 and potentially as much as 7.8 million tonnes in 2010.

Korea LNG demand forecast

	Kogas/ Mocie	KEEI	Delta
	Million tonnes		
2005	21.3	22.9	1.6
2006	22.6	25.4	2.8
2007	24.0	27.2	3.2
2008	24.0	27.5	3.5
2009	23.8	29.3	5.5
2010	25.3	30.0	4.7

Sources: Kogas, Mocie, KEEI

Korea spot purchases

	2004	2005	2006
	Cargoes, no.		
Jan.-Mar.	15	5	*18
Apr.-June	3	3	—
July-Sept.	5	3	—
Oct.-Dec.	15	10	—
Total	**38**	**21**	**18**

*Estimated.
Source: Poten & Partners

Korea faces large supply gap

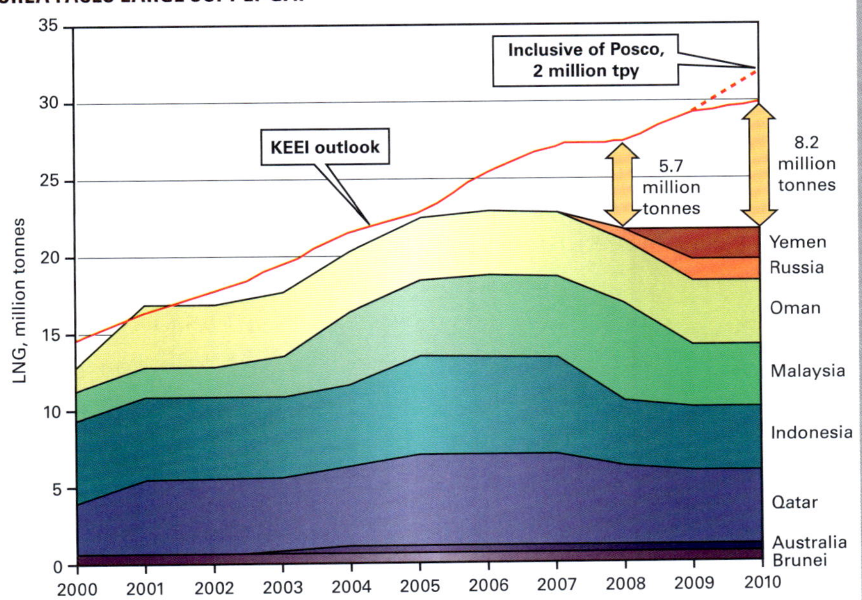

Source: Poten & Partners, KEEI

LNG shipbuilding leader

South Korean shipbuilders led the global market in construction of LNG carriers in 2005, according to the Korea Shipbuilders' Association.

It said South Korean companies won 79% of market share, with orders for 33 LNG carriers from a worldwide total of 42 carriers. That marks the third consecutive year that South Korean yards have led the global LNG market, after winning 73% market share in 2004 and 80% in 2003.

The association projected that the world's top three shipbuilders—Hyundai Heavy Industries Co., Daewoo Shipbuilding & Marine Engineering Co., and Samsung Heavy Industries Co.—would win orders for about 40 LNG carriers in 2006.

Two LNG vessels for BG

BG Group PLC (London) signed an agreement with Samsung Heavy Industries Co. Ltd. (Seoul) for the delivery of two newbuild dual-fuel diesel-electric (DFDE) LNG vessels.

The 170,000-cu m capacity carriers will incorporate hull modifications and DFDE propulsion technology that is expected to improve operating efficiency and reduce air emissions. The ships will use the GTT Mark III membrane cargo containment system. Both are scheduled for delivery in 2010.

BG also has four more vessels, each with 145,000 cu m of capacity, under construction at Samsung's South Korea shipyard, scheduled for delivery in 2006 and 2008.

TAIWAN

Capital: Taipei
Monetary unit: New Taiwan dollar. .
Refining capacity: 1,290,000 b/cd
Oil production: 1,000 b/d
Oil reserves: 2.38 million bbl
Gas reserves: 297 bcf

First offshore field

Taiwan's state-run Chinese Petroleum Corp. (CPC) announced plans to start production in 2009–2010 from the country's first offshore field, a gas-condensate deposit in the Taiwan Strait west of Kaohsiung.

The F structure gas field 100 km (62 miles) offshore in 120 m (394 ft) of water is believed to contain 240 bcf of recoverable gas plus an undisclosed volume of condensate.

CPC said it plans to drill 10 wells. A platform will be connected by subsea pipeline to processing facilities at Yungan and will sustain commercial exploitation for up to 10 years.

Refinery, chemical complex planned

Kuokuang Petrochemical Technology Corp. (KPTC) announced plans to spend as much as $12.46 billion to build a refinery and petrochemical complex in Yunlin County.

Facilities will include a 300,000 b/d refinery, a naphtha cracker capable of producing 1.2 million tpy of ethylene, an aromatics complex with output of 800,000 tpy of paraxylene, 23 plants for middle and downstream operations, 14 cogeneration plants, and 13 docks. Construction was expected to begin in late 2006, with completion scheduled for 2014.

KPTC is a joint venture of several leading Taiwan petrochemical companies.

THAILAND

Capital: Bangkok
Monetary unit: Baht
Refining capacity: 729,100 b/cd
Oil production: 212,000 b/d
Oil reserves: 290.0 million bbl
Gas reserves: 14.750 tcf

Exploration and Development

PTTEP E&P budget

Thailand's state-owned PTT Exploration & Production (PTTEP) announced plans to spend $5.8 billion over five years to expand drilling and exploration in Thailand as well as in Indonesia, Malaysia, Myanmar, and Vietnam.

PTTEP officials said the company aims to boost production from 179,000 boe/d in 2006 to 238,400 boe/d by 2010.

Part of the funding would aim to further develop a natural gas field in a Malaysia-Thailand joint development area (JDA) in the Gulf of Thailand. PTTEP said the current plan is to start gas production in late 2008 at 270 MMcfd.

Regarding overseas investments, PTTEP plans to expand three projects currently under exploration in Myanmar, where it already has two gas fields.

In Vietnam, it seeks two more projects in addition to existing exploration in two fields.

In Indonesia, PTTEP plans to increase its oil reserves with the Merangin project in South Sumatra.

JDA discussion continues

Thailand and Cambodia failed to resolve their disagreement about terms of production-sharing from their overlapping territorial claims in the Gulf of Thailand, despite high-level negotiations in August.

The matter has stalled development of energy resources by both nations from the 30,000-sq km (11,580-sq mile) area in the gulf's upper half, which is believed to be gas-prone.

In 2001, both countries agreed to adopt a Joint Development Area (JDA) scheme without resolving exact maritime boundaries. Their model was the Thailand-Malaysia JDA in the southern Gulf of Thailand, where production is split equally.

The original Thai-Cambodian JDA terms also called for a 50-50 split of resources arising from the middle portion of the JDA, with differing sharing ratios on areas flanking the area.

However, Phnom Penh authorities continued to press for a 60-40 ratio, according to Thai officials.

Blocks 15 and 16

Total SA and partners reported a natural gas find on the Bongkot concession 600 km (373 miles) south of Bangkok. Three gas discoveries were made at exploration wells Ton Chan-1X, Ton Chan-2X, and Ton Rang-2X, drilled on Blocks 15 and 16 in the Gulf of Thailand.

The two Ton Chan area wells are on Block 16, about 15 km (7.5 miles) southeast of the Bongkot Central Complex (BCC).

Ton Chan-1X—the first exploration well drilled in eight years on the Bongkot concession—was drilled to 3,442 m TD (11,290 ft). It found gas-bearing reservoirs with a total of 143 m (470 ft) of net pay thickness. Ton Chan-2X found gas with a total of 44 m (145 ft) of net pay thickness.

Ton Rang-2X, on Block 15, some 20 km (66 miles) northwest of the BCC, found gas-bearing sand of 72 m (236 ft) thickness.

A development plan was under study, with production tentatively planned for 2009.

The discoveries will increase production and prolong the life of Bongkot, which is producing 600 MMcfd of gas and 19,259 b/d of condensate.

Fang block

Niko Resources Ltd. (Calgary) took its first interest in Thailand, a 50% equity in the Fang production and exploration block 740 km (460 miles) north of Bangkok, near the borders with Myanmar and Laos. Thailand's Defense Energy Department holds the block, where 103 million bbl of oil in place have been discovered in shallow, marginal fields.

The joint venture is to further develop Mae Soon field. It is the largest field discovered in the area, with an estimated 44 million bbl of oil in place. Production likely would be trucked south to a pipeline at Sirikit oil field.

The joint venture was acquiring a 3D seismic survey and planning to drill 10 exploration wells starting in late 2006.

Licensing

Nineteenth bid round

Eleven companies submitted applications for exploration rights to 15 blocks—8 onshore and 7 offshore—in Thailand's 19th bidding round. The government had offered 82 blocks.

Occidental Petroleum Corp. bid on Block G6/48 in the central Gulf of Thailand, where several gas fields are on production, according to the Thai Department of Mineral Fuels.

And Adani Port Infrastructure Ltd. India submitted an application for onshore Blocks L25/48, which covers 3,962 sq km (1,530 sq miles) in the central plains as well as 3,975-sq km (1,535-sq mile) Block L39/48 in the northeast.

Drilling and Production

Phu Horm gas flows

Amerada Hess (Thailand) Ltd. reported the start of production from Phu Horm gas field on Block E5N in Udon Thani Province, northeastern Thailand. The field's gas reserves are estimated at 550 bcf plus condensate capable of yielding 500 b/d of production.

Gross initial production from two wells was about 60 MMcfd. When additional wells are brought into production in 2007, deliveries are expected to increase to as much as 108 MMcfd.

Other partners with operator Amerada Hess are Apico LLC, PTT Exploration & Production PLC (PTTEP), and ExxonMobil Exploration & Production Khorat Inc.

A sales agreement calls for more than 500 bcf of gas from the field to be supplied to a power station at Nam Phong over 15 years.

Arthit field

To meet increasing domestic demand, PTTEP announced plans to boost production by as much as 45% from its Arthit gas field in the Gulf of Thailand in the first three years of operation. The increase amounts to an additional 120–150 MMcfd of gas, in addition to the earlier scheduled rate of 330 MMcfd.

The production increase is facilitated by an FPSO vessel at the field, 230 km (143 miles) off the southern Thai province of Songkhla.

Following a series of delays, PTTEP has rescheduled Arthit production start-up to first quarter 2008.

Petrochemicals

Aromatics complex

Aromatics (Thailand) PLC (ATC) said it was teaming with Rayong Refinery Co. to build a second aromatics complex in eastern Rayong Province. Originally, ATC had planned the complex alone.

The $1.07 billion project includes an aromatics complex and a reforming facility. Total production capacity will be 2.88 mtpy and output will include 616,000 tpy of paraxylene and 363,000 tpy of benzene. Commercial operations are slated to start third quarter 2008.

VIETNAM

Capital: Hanoi
Monetary unit: Dong
Refining capacity: 0
Oil production: 350,000 b/d
Oil reserves: 600.0 million bbl
Gas reserves: 6.8 tcf

Exploration and Production

Block 16-1

A new appraisal well in Te Giac Trang on Block 16-1 off Vietnam flowed oil at high rates on a drillstem test. The TGT-5X well, drilled to a TD of 3,405 m (11,170 ft), gauged 7,000 b/d, said Thailand's PTT Exploration & Production PLC, which holds part interest in the block.

A second drillstem test recorded an initial flow rate of 7,300 b/d. Other appraisal wells on the same structure tested at flows as high as 9,908 b/d (from TGT-3X).

Venezuela joint projects

Venezuela and Vietnam signed a petroleum cooperation agreement to assess joint development of oil and gas assets and building a refinery and storage facilities in Vietnam, said Petroleos de Venezuela SA.

PdVSA said it would study supplying state-owned Petrovietnam with oil. The agreement specifically mentioned heavy oil from Venezuela's Orinoco belt.

Ca Ngu Vang field commercial

Hoan Vu Joint Operating Co., operator of Block 09-2 in the offshore Cuu Long basin, reported that Ca Ngu Vang field is commercial. SOCO International PLC (London)—whose SOCO Vietnam Ltd. holds an interest in the block—said it expected more commercial fields to follow in the oil-rich exploration province.

Block 124

Pogo Producing Co. (Houston) and Keeper Resources Inc. (Calgary) signed an exploration and production-sharing contract with state-owned Petrovietnam for Block 124, covering 1.48 million acres (2,310 sq miles) in the Phu Khanh basin northeast of Ho Chi Minh City.

The two firms will acquire 850 sq km of 3D seismic data and drill two wells during the first three years of a seven-year exploratory phase. Pogo said existing 2D seismic data indicate several leads in Miocene and Paleozoic rocks.

Pipelines

Pipeline will support fertilizer project

Vietnam-Russia oil and gas joint venture company Vietsovpetro let a contract to a unit of J. Ray McDermott SA to engineer and install a 298-km (185-mile) natural gas pipeline as part of the Gas Power Fertilizer Coordination Project. Vietnam state-owned PetroVietnam operates GPFC.

J. Ray said the PM3-Ca Mau gas pipeline will originate at the Bunga Raya B platform on Block PM3 in the Commercial Arrangement Area of the Gulf of Thailand and extend to a landfall station in Ca Mau province in Vietnam.

Refining

Nhon Hoi refinery

The Hong Kong General Chamber of Commerce signed an agreement with Vietnam's Binh Dinh province for the construction of a $450 million refinery in the Nhon Hoi economic zone.

The project is expected to be completed in 2010, with a refining capacity of 2 mtpy of oil, increasing to 8–10 mtpy during 2016–2020.

Vietnam exports substantial volumes of crude oil, but a lack of refineries has required the country to import all refined products.

In August 2005, state oil firm Petrovietnam reported that major work contracts were in place for construction of a 140,000 b/d refinery—the country's first—at Dung Quat in the central province of Quang Ngai. To further increase refining capacity, PetroVietnam plans to build another at Nghi Son.

Key Stats

Section 1:

Future energy supply

IEA: Fossil fuels important for decades

Fossil fuels will dominate the energy market through 2030 as oil's share slips to 33% from 35%, according to the *World Energy Outlook 2006* of the International Energy Agency (IEA; Paris).

The report said reliance on fossil fuels prevails both in a reference scenario, where current policies remain the same, and in an alternative policy scenario, where governments act strongly to create "an energy future which is clean, clever, and competitive," but at a lower level.

In the reference scenario, oil demand would grow by 1.3%/year during 2005–30—broadly in line with gross domestic product—averaging 1.7% in 2005–2015 and 1.1% in 2015–2030. It would reach 99 million b/d in 2015 and 116 million b/d in 2030, up from 84 million b/d in 2005.

Oil supply is increasingly dominated by a small number of major producers where oil resources are concentrated. The share of global energy supply provided by the Organization of Petroleum Exporting Countries grows to 48% by 2030 from 40% in 2006 and 42% in 2015. Conventional crude production from non-OPEC countries peaks by the middle of the next decade, but natural gas liquids production continues to grow.

Conventional oil accounts for the lion's share of the increased oil supply over the Outlook period, but nonconventional

OFFSHORE EUROPE'S OIL PRODUCTION HISTORY, OUTLOOK

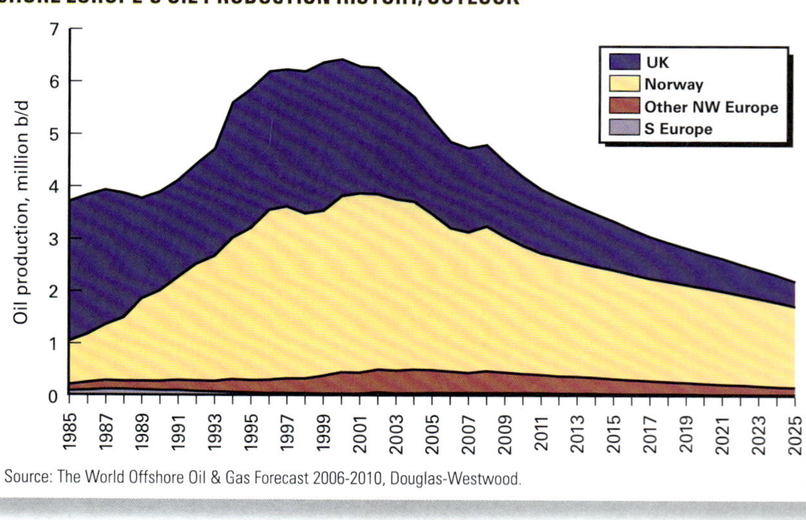

Source: The World Offshore Oil & Gas Forecast 2006-2010, Douglas-Westwood.

Oil production: an 'undulating plateau'

The "peak oil" theory, stipulating that world oil production will soon peak and sharply decline, is flawed, said Cambridge Energy Research Associates (CERA).

Instead of a peak, CERA said, production is more likely to trace an "undulating plateau" that will last for a decade or more beyond 2030. During later decades of the plateau period, unconventional production of heavy oil, gas-related liquids (condensate and natural gas liquids), gas-to-liquids, and coal-to-liquids will supplement traditional liquids supply.

CERA contends that the often-cited model developed by M. King Hubbert, which patterns production as a bell curve, fails to recognize that recoverable reserve estimates evolve with time and are subject to significant change. The Hubbert model also underplays the impact of technological advances.

The company also challenges the reliance of the peak-oil theory on assertions that new discoveries aren't replacing production. It calls the argument "incomplete," saying it ignores the role of development and economic considerations that can encourage companies and governments to favor development over exploration. CERA contends that aboveground factors—such as geopolitics, conflict, economics, and technology—will dictate future oil supply, not reservoir constraints.

Offshore Europe's gas production history, outlook

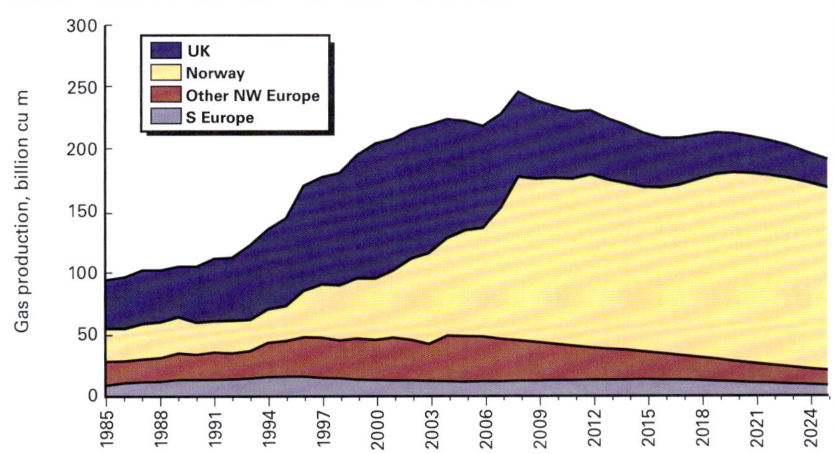

Source: The World Offshore Oil & Gas Forecast 2006-2010, Douglas-Westwood.

resources—mainly oil sands in Canada—and, to a lesser extent, gas-to-liquids plants, play a growing role. Canadian oil sands are projected to triple to 3 million b/d by 2015 and reach 5 million b/d by 2030.

The oil industry needs to invest $164 billion/year from 2005–2030—a total of $4.3 trillion (in 2005 dollars). Three-quarters of this will go to the upstream, indicates IEA, adding that upstream investment needs are more sensitive to changes in decline rates at producing fields than to the demand growth rate of oil.

Offshore Europe outlook

According to Douglas-Westwood energy analysts (Canterbury, UK), offshore Europe is the world's largest offshore oil and gas producing province. With nearly 450 offshore fields developed to date, the region is responsible for approximately 24% of the global offshore total.

Despite being mature with oil production having reached its peak in 2000, the province's prospects again seem bright in a future of rising oil prices.

Global oil, gas outlook bright; U.S. gas production nears plateau

Henry R. Linden, Max McGraw Professor of Energy and Power Engineering and management at the Illinois Institute of Technology reviewed current data and suggested that global crude oil, natural gas liquids (NGL), and Alberta oil sands production will plateau at above 118 million b/d and productive capacity at 123 million b/d in 2031.

This is far later than current predictions, even without the likely development of tar sands and other bitumens and extra-heavy crude.

There are no reliable data available of the kind needed to estimate the time and quantity of peak gas production, said Linden, but the global natural gas outlook is also quite bright.

He pointed to January 1, 2006, proved reserves of 6,112 tcf, and total 2003 global consumption of 95.5 tcf rising to 182 tcf in 2030, according to the U.S. EIA Reference Case in its International Energy Outlook 2006. This is equivalent to a roughly 40-year gas reserve life.

Linden called this outlook especially positive, as global gas exploration and resources development have not been nearly as intensive as for crude oil.

In the United States, however, he noted that current gas production of 19 tcf/year is already near its projected plateau, and a decline in domestic production seems inevitable unless there is a much more aggressive leasing policy, both off and onshore, especially for unconventional natural gas resources.

A key conclusion is the continued important role of research to develop new technology for finding, producing, and processing oil and gas from challenging formations.

Giant oil discoveries still to be found

The world continues to discover giant oil fields, said a report by the Centre for Global Energy Studies (CGES; London). The report contradicts key views held by "peak oil" advocates—that the industry has not been finding enough oil, has been relying on too few aging oil fields, and has not been discovering new giants.

CGES compiled figures that show 17 giant oil discoveries in the 1950s, 29 in the 1960s, 24 in the 1970s, 15 in the 1980s, and 11 in the 1990s. Four found in recent decades are supergiants.

Nevertheless, it said that 20% of global oil production comes from the world's largest 14 fields and that the large fields found in the 1990s produce only one-tenth of the oil produced by 36 giant fields discovered more than 40 years ago.

The four relatively recent supergiant discoveries are Cantarell off Mexico; Tengiz and Kashagan in Kazakhstan; and Azadegan in Iran.

A CGES list of large oil fields set to come on stream in 2007–20 excludes the emerging prospects of the Gulf of Mexico Lower Tertiary Trend. Twelve confirmed Lower Tertiary discoveries are thought to contain 3–15 billion bbl of recoverable reserves.

CGES said that large fields discovered since 1976 are expected to have supplied 60% of the 19.5 million b/d of incremental demand in 1976–2010.

Estimated World Oil Resources,[1] 1995-2025

Region	Proved reserves	Reserve growth	Undis-covered	Total
	Billion bbl			
OECD[2]				
US	21.4	76.0	83.0	180.4
Canada	178.8	12.5	32.6	223.9
Mexico	12.9	25.6	45.8	84.3
OECD Europe	15.1	20.0	35.9	71.0
Japan	0.1	0.1	0.3	0.5
Australia-New Zealand	1.5	2.7	5.9	10.1
Non-OECD				
Russia	60.0	106.2	115.3	281.5
Other Europe-Eurasia	19.1	32.3	55.6	107.0
China	18.3	19.6	14.6	52.5
India	5.8	3.8	6.8	16.4
Other Asia	10.3	14.6	23.9	48.8
Middle East	743.4	252.5	269.2	1,265.1
Africa	102.6	73.5	124.7	300.8
Central and South America	103.4	90.8	125.3	19.5
Total world	**1,292.5**	**730.2**	**938.9**	**2,961.6**
OPEC[3]	901.7	395.6	400.5	1,697.8
Non-OPEC	390.9	334.6	538.4	1,263.9

[1]Reserves include crude oil (including lease condensate) and natural gas plant liquids. [2]Organization for European Cooperation & Development countries. [3]Organization of Petroleum Exporting Countries.
Sources: International Energy Outlook 2006, Energy Information Administration (2006), US Geological Survey, World Petroleum Assessment 2000 (http://pubs.usgs.gov/dds/dds-060/), Proved reserves as of Jan. 1, 2006 (OGJ, Dec. 19, 2005, p. 24)

Global Liquids Productive Capacity vs. Production

Date	Production	Productive capacity
	Million b/d	
2003	79.6	82.3
2010	91.6	94.3
2015	98.3	101.6
2020	104.1	107.6
2025	110.7	114.9
2030	118.0	123.3

Source: Energy Information Administration, International Energy Outlook 2006

Conventional Fossil Fuel Reserves, Resources

Fuel	Proved reserves	Technically recoverable resources estimate, current upper bound
Natural gas	$6,112 \times 10^{12}$ cu ft $6,265 \times 10^{15}$ btu	$20,000 \times 10^{12}$ cu ft $20,500 \times 10^{15}$ btu
Conventional crude oil and condensate, Canadian oil sands, natural gas liquids (NGL)*	$1,544 \times 10^{9}$ bbl $8,492 \times 10^{15}$ btu	$3,600 \times 10^{9}$ bbl $19,800 \times 10^{15}$ btu
Coal and lignite	1.001×10^{12} short tons $21,021 \times 10^{15}$ btu	7×10^{12} short tons $147,000 \times 10^{15}$ btu
Total	**$35,778 \times 10^{15}$ btu**	**$187,300 \times 10^{15}$ btu**

*Proved global crude oil reserves, including Canadian oil sands, as of Jan. 1, 2006, are 1,292.5 billion bbl. There are no global data on proved reserves of natural gas liquids. The estimate of 251.7 billion bbl reflects the US ratio of 7.928 billion bbl of proved NGL reserves to 192.5 tcf of proved reserves of dry natural gas as of Dec. 31, 2004. See Reference 21.
Sources: OGJ, Dec. 19, 2005, p. 20; Energy Information Administration's International Energy Outlook 2006, EIA's Annual Report 2004

Traditional US gas resources excluding proved reserves

Resource category-depth Area, Dec. 31, 2004	Probable resource +	Possible resource +	Speculative resource =	Total potential resource
	Mean values, bcf			
Lower 48				
Onshore	154,755	244,893	182,959	582,958
Offshore	15,177	59,954	97,285	172,418
Total Lower 48	**169,930**	**304,848**	**280,285**	**756,494**
Alaska				
Onshore	31,717	22,300	40,417	94,432
Offshore	5,142	19,499	74,788	99,366
Total Alaska	**36,856**	**41,816**	**115,129**	**193,831**
Total traditional	**206,765**	**346,208**	**396,628**	**949,957**
Coalbed gas*	**17,570**	**56,780**	**94,948**	**169,298**
Grand total US	**224,335**	**402,988**	**491,576**	**1,119,255**

*Aggregated coalbed gas estimates are arithmetically added to subtotal of means to derive grand total.
Note: Totals are subject to rounding and slight differences due to statistical aggregation of distributions.
Source: Potential Gas Committee report December 2004, Potential Gas Agency, Colorado School of Mines September 2005

Projected US unconventional gas production

Lower 48	2002	2025
	– Production, tcf/year –	
Total tight sands gas	3.877	6.041
Rocky Mountain region	(1.502)	(3.998)
Total coalbed methane	1.502	1.992
Rocky Mountain region	(1.320)	(1.611)
Devonian shale gas	0.600	1.132
Total	**5.979**	**9.165**

Source: Energy Information Administration

US unconventional resources accessibility

	Unconventional resources, tcf
Officially inaccessible	23.44
Inaccessible due to development constraints	83.72
Accessible with lease stipulations	47.51
Accessible under standard lease terms	172.92
Total	**327.58**

Source: Energy Information Administration

1990s discoveries

Discovery year	Field	Peak output, 1,000 b/d
1991	Vankorskoye (Russia)	280
1992	Norne (Norway)	150
1993	Darkhovin (Iran)	200
1993	Cupiagua (Colombia)	200
1994	Ourhoud (Algeria)	230
1994	Caratinga (Brazil)	200
1996	Bonga (Nigeria)	250
1995	Hassi Berkine S. (Algeria)	230
1996	Girassol (Angola)	250
1997	Dalia (Angola)	230
1998	Kizomba A (Angola)[1]	200
1998	Kizomba B (Angola)[2]	250
1999	Agbami (Nigeria)	230
1999	Azadegan (Iran)	400
1999	Grane (Norway)	210
Total		**3,510**

[1]The Kizomba A project combines Chocalho and Hungo fields. [2]The Kizomba B project combines Dikanza and Kissanje fields.
Source: CGES et al.

Expected field start-ups

Year of peak output	Field	Peak output, 1,000 b/d
2007	Buzzard (UK)	180
2007	Roncador (Brazil)	180
2008	Atlantis (US)	200
2008	Greater Plutonio (Angola)[1]	200
2008	Kizomba C (Angola)[2]	200
2008	Tahiti (US)	125
2009	Thunder Horse (US)	250
2009	Valhall redevelopment (Norway)	150
2009	Urugua (Brazil)	150
2009	Kizomba D (Angola)[3]	125
2009	Frade (Brazil)	100
2010	Ku Maloob Zaap (Mexico)	800
2010	Yadavaran (Iran)	300
2010	Vankorskoye (Russia)	280
2013	Verkhnechonsk (Russia)	200
2020	Kashagan (Kazakhstan)	1,200
Total		**4,640**

[1]Greater Plutonio project combines Galio, Cromio, Paladio, Plutonio, Cobalto, and Platina fields.
[2]Kizomba C project combines Mondo, Saxi, and Batuque fields. [3]Kizomba D project combines output from other Block 15 discoveries.
Source: CGES et al.

Section 2:

Reserves and production

Oil, gas reserves increase slightly

Worldwide crude oil and gas reserves rose in 2006 as oil production posted a small increase from a year earlier, according to *Oil & Gas Journal*.

Oil reserves held by members of the Organization of Petroleum Exporting Countries moved up slightly from the previous survey, but OPEC's estimated gas reserves were down a bit. The organization controls 68.5% of the world's proved oil reserves and 51% of total proved gas reserves.

Worldwide crude and condensate production in 2006 year averaged 72.39 million b/d, according to *OGJ* estimates. This was up from 72.26 million b/d in 2005.

The region with the largest gain in crude output for 2005 was Eastern Europe and the former Soviet Union, driven by oil production in Kazakhstan, Russia, and Azerbaijan.

Oil production up, reserves down for big companies

The world's top 100 oil and gas firms based outside the United States recorded a modest gain in oil production during 2005. Collective production by the top 20 totaled 17.6 billion bbl, up about 2% from 2004. The top 20 held a combined total of 970.1 billion bbl in oil reserves, down about 1% from 982.6 billion bbl at yearend 2004.

Among 138 U.S.-based companies, liquids production moved up almost 5% worldwide in 2005 and climbed nearly 7% in the United States. Liquids reserves grew 7.2% worldwide and 8.6% domestically. A surge in drilling was behind the increase.

Gas production, reserves down

U.S. gas production was affected by Gulf of Mexico hurricanes, dipping to 8.36 tcf in 2005 from 8.49 tcf in 2004. Worldwide, annual gas production slid to a collective 13.87 tcf from 13.96 tcf. The group's gas reserves declined 4% worldwide to 157.7 tcf but surged almost 10% domestically to 106.1 tcf.

RUSSIAN PRODUCTION

	1,000 b/d
2005	9,190
2004	8,887
2003	8,216
2002	7,405
2001	6,781
2000	6,325
1999	5,930
1998	5,919
1997	5,914
1996	5,844

A DECADE OF RESERVES CHANGES

Jan. 1	World (1,000 bbl)	OPEC (1,000 bbl)	World gas, bcf
2007	1,317,447,415	902,343,000	6,182,692
2006	1,293,344,534	901,659,000	6,101,158
2005	1,277,701,992	885,188,000	6,040,208
2004	1,265,811,583	869,521,000	6,068,302
2003	1,212,880,852	819,007,000	5,501,424
2002	1,031,100,681	818,842,000	5,451,332
2001	1,028,457,585	814,398,710	5,278,484
2000	1,016,041,221	802,479,710	5,146,207
1999	1,034,264,678	800,479,710	5,144,753
1998	1,019,545,664	797,134,589	5,086,469

Source: OGJ Worldwide Production Reports

OGJ100: OIL PRODUCTION AND RESERVE LEADERS

Rank	Company	Production, million bbl
1	Saudi Arabian Oil Co.	3,321.5
2	National Iranian Oil Co.	1,416.2
3	Petroleos Mexicanos	1,372.0
4	Petroleos de Venezuela SA	1,127.9
5	BP PLC	935.1
6	Kuwait Petroleum Corp.	885.9
7	Nigerian National Petroleum Corp.	878.6
8	PetroChina Co. Ltd.	822.9
9	Abu Dhabi National Oil Co.	748.3
10	Royal Dutch Shell PLC	729.0
11	Petroleo Brasileiro SA	674.2
12	OAO Lukoil	664.3
13	Iraq National Oil Co.	660.7
14	National Oil Corp. (Libya)	598.6
15	Total SA.	591.7
16	OAO Rosneft	544.3
17	Sonatrach	492.8
18	Sonangol	447.1
19	ENI	404.0
20	Qatar Petroleum Corp.	279.2
	Total	**17,594.0**

Rank	Company	Reserves, million bbl
1	Saudi Arabian Oil Co.	259,800.0
2	National Iranian Oil Co.	125,800.0
3	Iraq National Oil Co.	115,000.0
4	Kuwait Petroleum Corp.	99,000.0
5	Abu Dhabi National Oil Co.	92,200.0
6	Petroleos de Venezuela SA	77,200.0
7	National Oil Corp. (Libya)	39,000.0
8	Nigerian National Petroleum Corp.	35,255.0
9	OAO Lukoil	16,114.0
10	Qatar Petroleum Corp.	15,207.0
11	OAO Rosneft	14,900.0
12	Petroleos Mexicanos	13,671.0
13	Sonatrach	11,800.0
14	PetroChina Co. Ltd.	11,536.2
15	Petroleo Brasileiro SA	11,364.8
16	Petronas	7,600.0
17	BP PLC	7,161.0
18	Total SA.	6,592.0
19	Petroleum Development Oman LLC	5,506.0
20	Sonangol	5,412.0
	Total	**970,119.0**

Top 20 in liquids production

Rank	Company	US liquids production, million bbl
1	Occidental Petroleum Corp.	202.0
2	Chevron Corp.	165.0
3	ConocoPhillips	145.0
4	ExxonMobil Corp.	136.0
5	Devon Energy Corp.	43.0
6	Anadarko Petroleum Corp.	37.0
7	Burlington Resources Inc.	33.5
8	Kerr-McGee Corp.	33.0
9	Marathon Oil Corp.	28.0
10	Apache Corp.	26.9
11	Hess Corp.	20.0
12	Plains Exploration & Production Co.	18.7
13	XTO Energy Inc.	18.1
14	Pioneer Natural Resources Co.	16.3
15	Dominion Exploration & Production	14.7
16	EOG Resources Inc.	10.2
17	Pogo Producing Co.	9.6
18	Noble Energy Inc.	9.5
19	Murphy Oil Corp.	9.4
20	Forest Oil Corp.	9.3
	Total	**985.2**

Rank	Company	Worldwide liquids production, million bbl
1	ExxonMobil Corp.	782.0
2	Chevron Corp.	609.0
3	ConocoPhillips	449.0
4	Occidental Petroleum Corp.	252.0
5	Apache Corp.	89.0
6	Hess Corp.	89.0
7	Devon Energy Corp.	88.0
8	Anadarko Petroleum Corp.	72.0
9	Marathon Oil Corp.	70.0
10	Kerr-McGee Corp.	59.0
11	Burlington Resources Inc.	58.3
12	Murphy Oil Corp.	33.1
13	Pioneer Natural Resources Co.	23.8
14	Noble Energy Inc.	20.8
15	Plains Exploration & Production Co.	18.7
16	XTO Energy Inc.	18.1
17	Dominion Exploration & Production	15.6
18	EOG Resources Inc.	13.2
19	Pogo Producing Co.	10.9
20	Forest Oil Corp.	10.6
	Total	**2,782.0**

Top 20 in liquids reserves

Rank	Company	US liquids reserves, million bbl
1	Occidental Petroleum Corp.	2,338.0
2	ExxonMobil Corp.	2,113.0
3	ConocoPhillips	1,929.0
4	Chevron Corp.	1,831.0
5	Anadarko Petroleum Corp.	708.0
6	Burlington Resources Inc.	516.6
7	Apache Corp.	461.6
8	Pioneer Natural Resources Co.	385.8
9	Devon Energy Corp.	370.0
10	Plains Exploration & Production Co.	356.3
11	Kerr-McGee Corp.	322.0
12	XTO Energy Inc.	256.1
13	Whiting Petroleum Corp.	199.2
14	Dominion Exploration & Production	198.6
15	Marathon Oil Corp.	189.0
16	Noble Energy Inc.	151.7
17	Encore Acquisition Co.	148.4
18	Hess Corp.	124.0
19	Energen Resources Corp.	106.9
20	Denbury Resources Inc.	106.2
	Total	**12,811.3**

Rank	Company	Worldwide liquids reserves, million bbl
1	Chevron Corp.	8,000.0
2	ExxonMobil Corp.	7,813.0
3	ConocoPhillips	6,168.0
4	Occidental Petroleum Corp.	3,478.0
5	Anadarko Petroleum Corp.	1,130.0
6	Apache Corp.	975.9
7	Devon Energy Corp.	895.0
8	Marathon Oil Corp.	704.0
9	Hess Corp.	692.0
10	Burlington Resources Inc.	662.7
11	Pioneer Natural Resources Co.	429.0
12	Kerr-McGee Corp.	362.0
13	Plains Exploration & Production Co.	356.3
14	Noble Energy Inc.	290.8
15	XTO Energy Inc.	256.1
16	Dominion Exploration & Production	217.7
17	Whiting Petroleum Corp.	199.2
18	Murphy Oil Corp.	182.0
19	Encore Acquisition Co.	148.4
20	Pogo Producing Co.	144.0
	Total	**33,104.2**

Top 20 in gas production

Rank	Company	US gas production, bcf
1	ExxonMobil Corp.	764.0
2	Chevron Corp.	604.0
3	ConocoPhillips	593.0
4	Devon Energy Corp.	555.0
5	Chesapeake Energy Corp.	422.4
6	Anadarko Petroleum Corp.	414.0
7	XTO Energy Inc.	377.1
8	Kerr-McGee Corp.	351.0
9	Burlington Resources Inc.	347.0
10	Dominion Exploration & Production	275.0
11	EOG Resources Inc.	267.4
12	Williams Cos. Inc.	224.0
13	Apache Corp.	218.1
14	Marathon Oil Corp.	211.0
15	El Paso Corp.	207.0
16	Pioneer Natural Resources Co.	197.4
17	Newfield Exploration Co.	183.2
18	Noble Energy Inc.	125.5
19	Houston Exploration Co.	105.8
20	Cimarex Energy Co.	100.3
	Total	**6,542.2**

Rank	Company	Worldwide gas production, bcf
1	ExxonMobil Corp.	2,819.0
2	Chevron Corp.	1,553.0
3	ConocoPhillips	1,355.0
4	Devon Energy Corp.	827.0
5	Burlington Resources Inc.	695.0
6	Anadarko Petroleum Corp.	516.0
7	Apache Corp.	461.3
8	EOG Resources Inc.	449.5
9	Chesapeake Energy Corp.	422.4
10	XTO Energy Inc.	377.1
11	Kerr-McGee Corp.	374.0
12	Marathon Oil Corp.	324.0
13	Dominion Exploration & Production	290.0
14	Pioneer Natural Resources Co.	263.1
15	Williams Cos. Inc.	224.0
16	El Paso Corp.	223.0
17	Hess Corp.	214.0
18	Noble Energy Inc.	185.5
19	Newfield Exploration Co.	183.4
20	Occidental Petroleum Corp.	165.0
	Total	**11,921.2**

Top 20 in Gas Reserves

Rank	Company	US gas reserves, bcf
1	ExxonMobil Corp.	13,692.0
2	ConocoPhillips	7,586.0
3	Chesapeake Energy Corp.	6,900.8
4	Anadarko Petroleum Corp.	6,578.0
5	XTO Energy Inc.	6,085.6
6	Burlington Resources Inc.	5,275.0
7	Devon Energy Corp.	5,164.0
8	Dominion Exploration & Production	4,856.0
9	Chevron Corp.	4,428.0
10	Kerr-McGee Corp.	3,633.0
11	Williams Cos. Inc.	3,382.0
12	EOG Resources Inc.	2,948.1
13	Pioneer Natural Resources Co.	2,750.9
14	Apache Corp.	2,566.2
15	Ultra Petroleum Corp.	1,900.2
16	El Paso Corp.	1,831.0
17	Noble Energy Inc.	1,641.1
18	Occidental Petroleum Corp.	1,636.0
19	Newfield Exploration Co.	1,327.2
20	Questar Corp.	1,324.8
	Total	**85,505.9**

Rank	Company	Worldwide gas reserves, bcf
1	ExxonMobil Corp.	33,355.0
2	ConocoPhillips	19,061.0
3	Burlington Resources Inc.	8,508.0
4	Anadarko Petroleum Corp.	7,910.0
5	Devon Energy Corp.	7,296.0
6	Chesapeake Energy Corp.	6,900.8
7	Apache Corp.	6,848.0
8	XTO Energy Inc.	6,085.6
9	EOG Resources Inc.	5,557.4
10	Dominion Exploration & Production	4,962.0
11	Kerr-McGee Corp.	3,633.0
12	Marathon Oil Corp.	3,547.0
13	Williams Cos. Inc.	3,382.0
14	Pioneer Natural Resources Co.	3,346.1
15	Noble Energy Inc.	3,091.2
16	Chevron Corp.	2,968.0
17	Hess Corp.	2,406.0
18	Equitable Supply	2,359.2
19	Occidental Petroleum Corp.	2,127.0
20	Ultra Petroleum Corp.	1,900.2
	Total	**135,243.5**

GOM reservoir properties may indicate reserve size

Much can be learned about potential reserves by examining the properties of reservoirs in the Gulf of Mexico—properties such as depth of production, thickness of pay zones, productive areal extent, porosity, formation pressure, and water saturation.

Frederick Haeberle, an experienced consulting geologist in Ohio, conducted such an evaluation of published data regarding proved reserves for oil pools in the northern gulf.

North American gas shale, CBM contributions

According to Warlick International global energy consultancy (Houston), at the end of 2005, production from gas shale basins accounted for about 8% of U.S. total gas production and about 2% of total Canadian production. There are more than 39,500 producing gas shale wells in the United States.

Share of Oil and Gas Reserves by Depth Interval

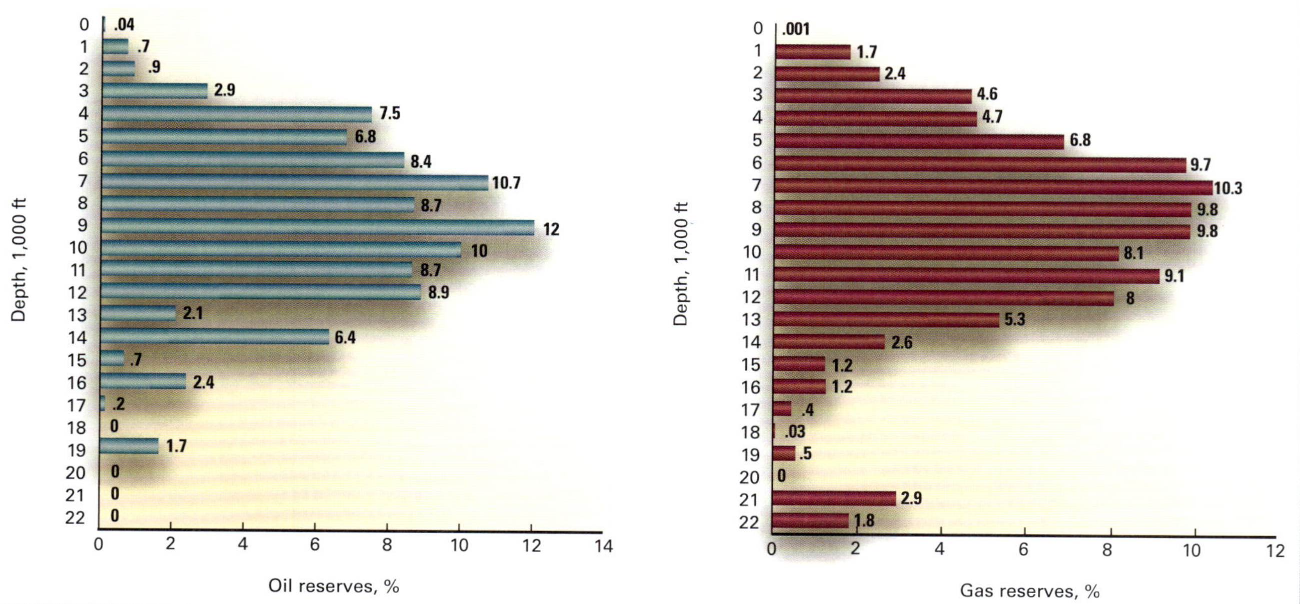

Share of Oil and Gas Reserves by Pay Thickness

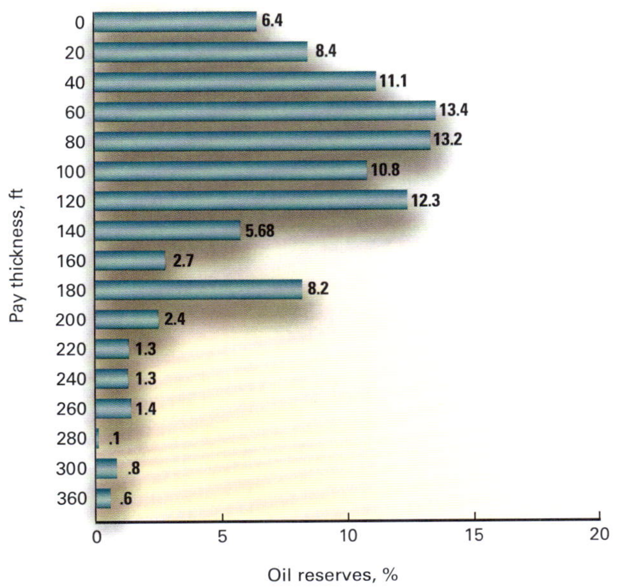

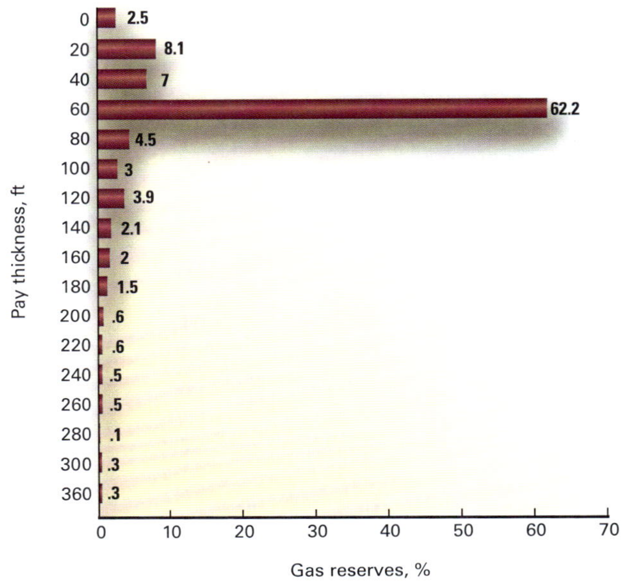

Share of Oil and Gas Reserves by Areal Extent of Reservoir

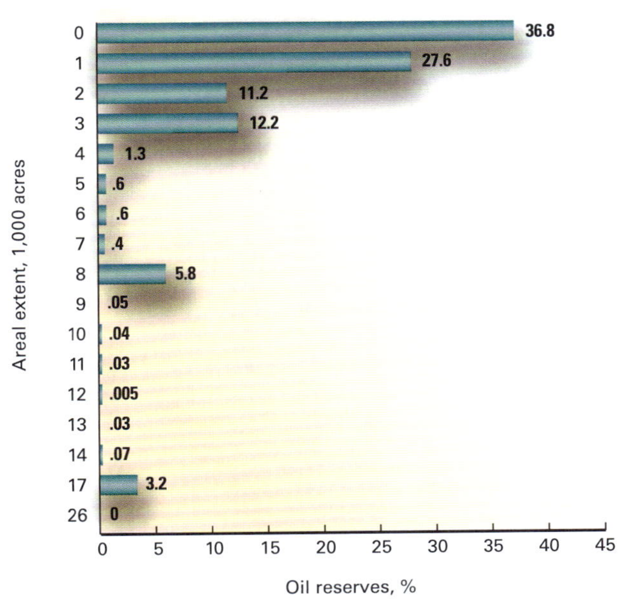

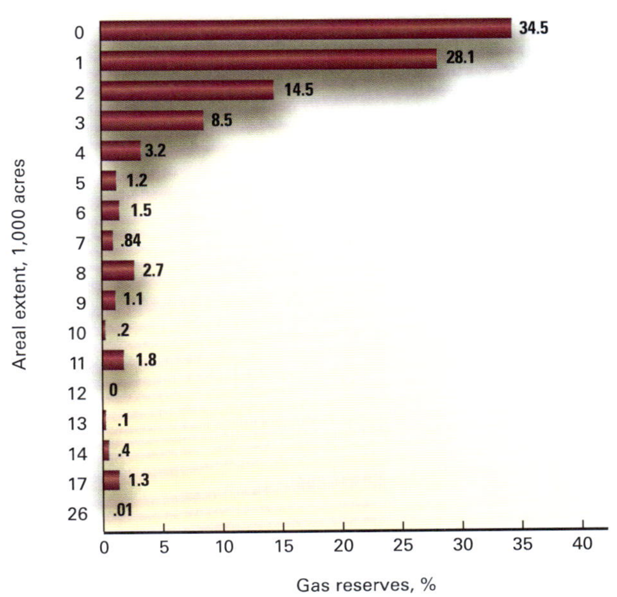

Share of oil and gas reserves by reservoir porosity

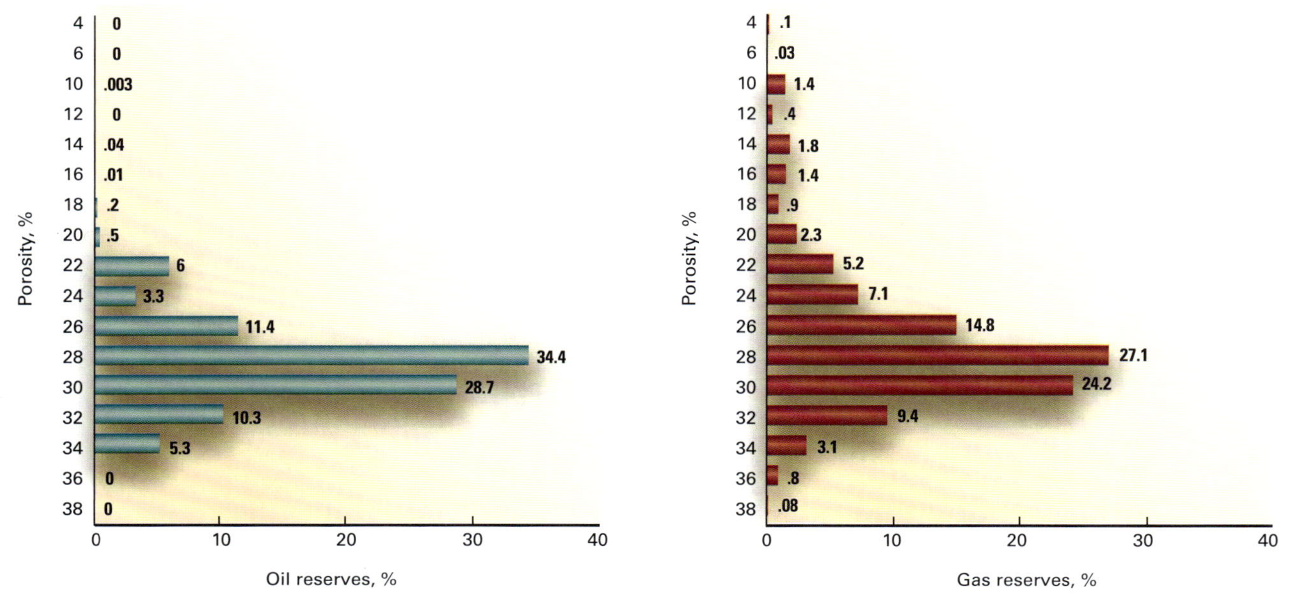

Share of oil and gas reserves by formation pressure

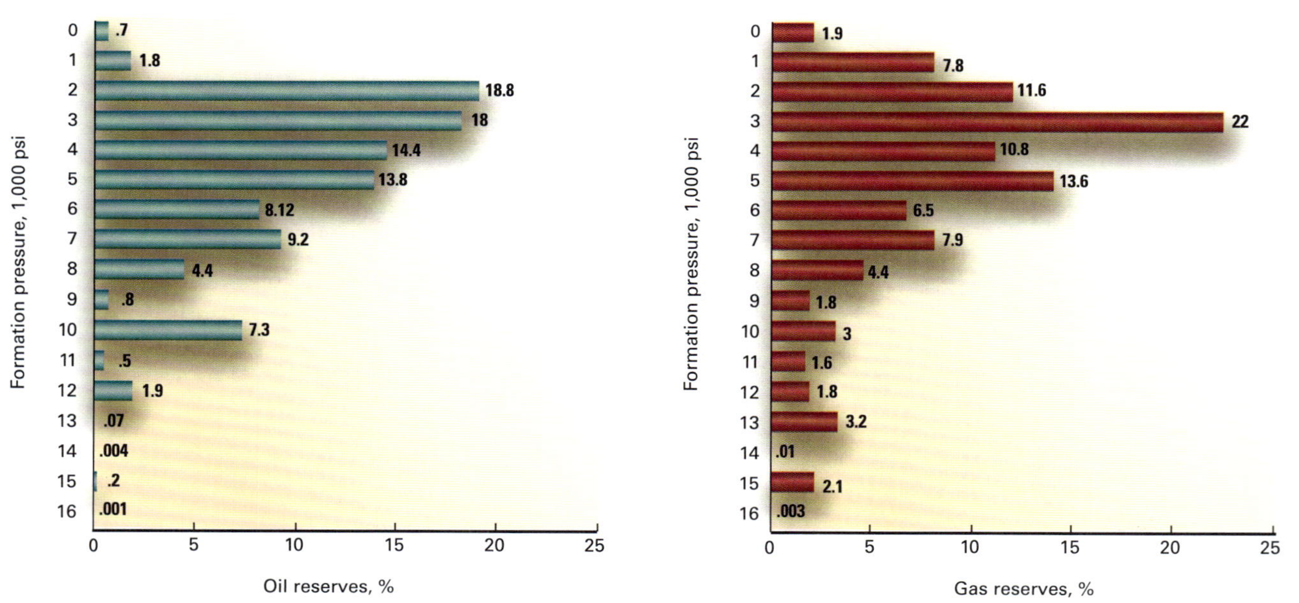

Share of Oil and Gas Reserves by Formation Water Saturation

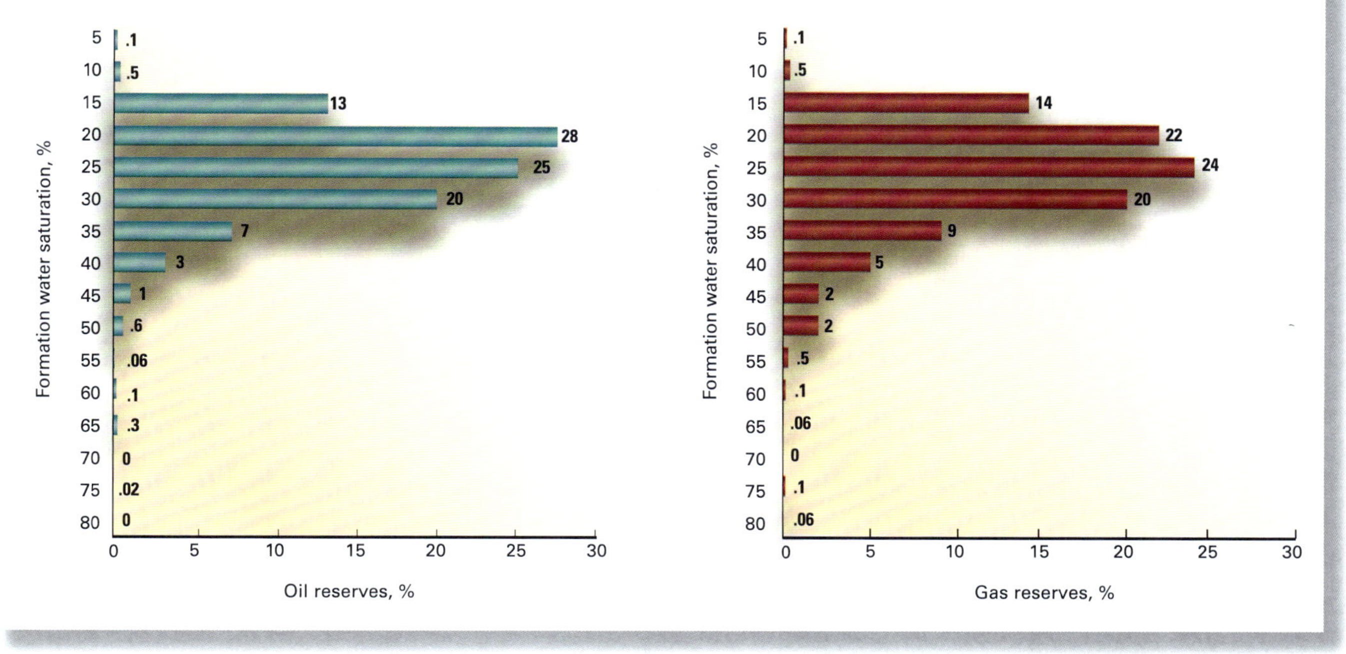

Significant US Gas Shale Basins

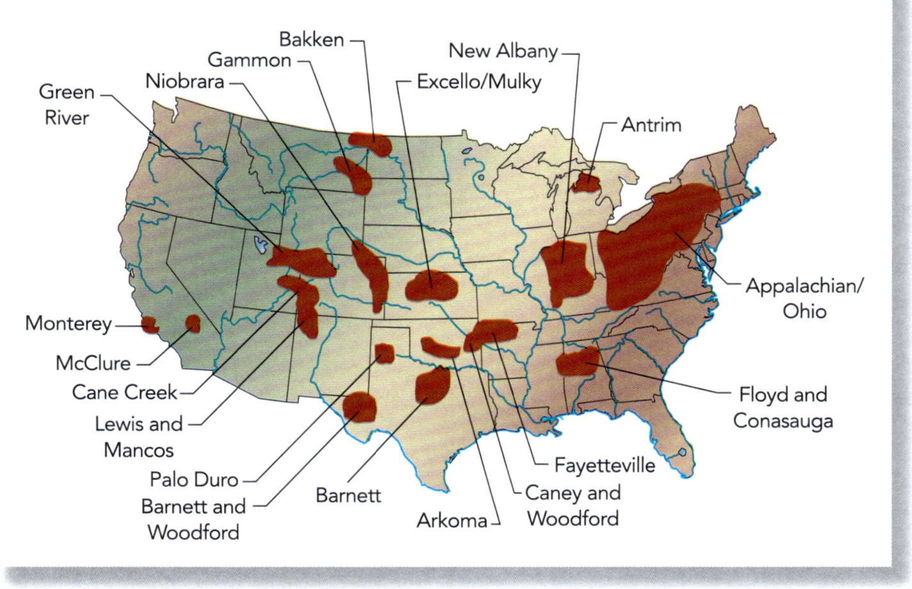

Coalbed methane (CBM) accounted for another 8% of U.S. production and about 2% of Canada's total gas production. North America has 16 significant coalbed methane areas.

Warlick expected gas shale drilling to increase in 2006 over 2005 due to big increases in the Fayetteville, Woodford/Caney shales and a steady uptrend in the Barnett shale.

CBM drilling also could increase because of activity in Canada, the Cherokee Platform area of Kansas/Oklahoma, and the Powder River and San Juan basins.

Leading gas shale areas in North America

Area/Basin*	Current status
Barnett	Producing over 1.3 tcf in 2005 with about 640 new wells, this is a huge multi-county collection of horizontal completions and intelligent fracturing, essentially an ongoing "technology lab" for future gas shale development. Devon accounts for ~ 60% of production followed by EOG, XTO, Chesapeake, Range Resources, and others
Antrim	Overall production in 2005 was 144 bcf. Predictable, uncomplicated drilling (about 350 new wells last year) this is a mature development with about 30 E&P players led by DTE, Dominion, Quicksilver, and Terra
Appalachian	The Appalachian/Ohio has most of its activity in KY, VA, WV, OH and PA. Producing over 48 bcf last year this historic play is being developed by active E&P players like Chesapeake, NGAS, Geo Maxx, and others drilling around 300 wells in the area in 2005
Arkoma including Woodford and Caney	Few gas shale wells were drilled in this established area during 2005 but production exceeded 21 bcf. The play has similarities to the Barnett. The new pacesetter is Newfield Exploration with big plans for the Woodford Shale
Western Canadian Sedimentary basin	Not known for gas shale development (CBM has dominated the area for years), estimated production from shales may have reached 4 - 5 bcf last year. Area has attractive shale prospects for development, should pick up in 2007 - 2008
Fayetteville	A growth area for sure, Fayetteville Shale production in 2005 was ~ 5 bcf. But more will be coming with Southwestern Energy working on fast-cycle development using horizontal and unique fracturing technologies. Others like Shell, Chesapeake, and XTO are entering; new wells drilled this year will quadruple wells drilled in 2005
New Albany	The New Albany Shale of the Illinois basin produced around 2 bcf in 2005. This is a measured gas shale area with development led by independents like Cabot, Harken, and others, mostly in Illinois with Kentucky and Indiana following.

Source: Warlick International
* appearing in order of area annual production

Significant US CBM basins

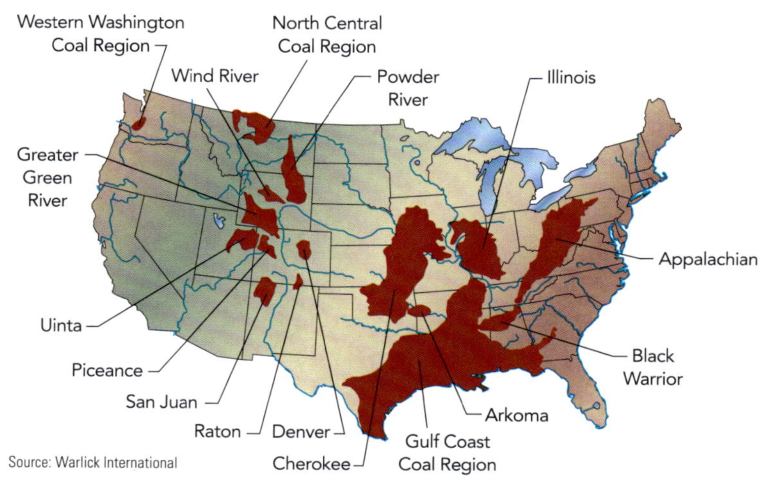

Source: Warlick International

Leading coalbed methane areas in North America

Basin/Area*	Overview
San Juan	PThe most prolific CBM basin in North America with just under 1 tcf in production and almost 700 wells drilled in 2005. ConocoPhillips/Burlington, BP, and Devon are the leaders here
Powder River	Last year produced ~ 340 bcf. This is the busiest CBM drilling play in North America with almost 3,000 wells drilled in 2005. Williams, Anadarko (including Western Gas Resources), Huber, and Devon are production leaders.
Western Canadian sedimentary basin	Produced over 135 bcf last year, CBM drilling has ramped up significantly with over 1,900 new wells in 2005. Continued commitment by large players like Apache Canada, EnCana, and Nexen is driving development.
Black Warrior	North Americca's longest-producing CBM area with steady production year to year (127 bcf in 2005). E&P companies include Dominion Resources, El Paso, Black Warrior Methane, Energen
Appalachian (central)	Producing over 70 bcf in 2005 this area has attractive prospects with good potential for efficient operators. CNX well ahead of others in CBM development, Cabot has long-time presence, Range Resources a new force
Arkoma	Good area with established infrastructure and low-cost horizontal completions. El Paso, Williams, and small independents were busy drilling ~ 250 new CBM wells in 2005. Production that year exceeded 20 bcf.
Cherokee platform	Simple CBM completions with minimal dewatering, largely in SE Kansas with some activity in NE Okla., production was about 16 bcf last year. Successful payout can occur in 12 months. Quest Resources dominates the area.

Source: Warlick International
* appearing in order of area annual production

Section 3:

Drilling and exploration

U.S. drilling activity high

Drilling by 138 U.S.-based companies resulted in 18,009 net wells in the United States in 2005, up 24% from 2004, according to *Oil & Gas Journal*. Meanwhile, capital and exploration expenditures grew to $85.9 billion from a 2004 level of $64 billion.

Mid-2006 figures from the American Petroleum Institute showed that in 2005, the total number of exploratory and development wells drilled in the United States climbed 15% from 2004. Of the 44,755 wells drilled in 2005, 13,047 were oil wells, 26,973 were gas wells, and 9% were exploratory wells.

Dynamic drilling market

Drillers in 2006 were working at or near capacity in most markets, said *OGJ*. Many operators continued to pursue ambitious drilling programs despite recent fluctuations in natural gas prices. Utilization rates for land and marine rigs remained high, even with new builds augmenting the fleets.

Around the world, drilling was strong and activity has increased over the past four years. The worldwide rig count reached 3,155 in July, a 14% year-over-year increase. The activity was driven by increasing demand in the face of strong oil and natural gas prices.

TOP 20 IN SPENDING AND US NET WELLS DRILLED

Rank	Company	Capital, exploratory spending, $1,000
1	ExxonMobil Corp.	13,839,000
2	ConocoPhillips	11,620,000
3	Chevron Corp.	8,701,000
4	Devon Energy Corp.	4,090,000
5	Apache Corp.	3,715,856
6	Anadarko Petroleum Corp.	3,408,000
7	Marathon Oil Corp.	2,890,000
8	Burlington Resources Inc.	2,469,000
9	Occidental Petroleum Corp.	2,423,000
10	Chesapeake Energy Corp.	2,371,854
11	Hess Corp.	2,341,000
12	Kerr-McGee Corp.	1,920,000
13	EOG Resources Inc.	1,724,763
14	El Paso Corp.	1,718,000
15	Dominion Exploration & Production	1,690,000
16	XTO Energy Inc.	1,304,000
17	Murphy Oil Corp.	1,246,242
18	Pioneer Natural Resources Co.	1,123,297
19	Whiting Petroleum Corp.	1,096,495
20	Newfield Exploration Co.	1,047,000
	Total	**70,738,507**

Rank	Company	US net wells drilled
1	Chevron Corp.	986.0
2	Dominion Exploration & Production	955.0
3	Williams Cos. Inc.	867.0
4	Chesapeake Energy Corp.	816.0
5	Devon Energy Corp.	815.6
6	XTO Energy Inc.	637.5
7	Range Resources Corp.	593.5
8	EOG Resources Inc.	593.0
9	ExxonMobil Corp.	563.0
10	Pioneer Natural Resources Co.	553.5
11	Noble Energy Inc.	529.4
12	Western Gas Resources	522.0
13	Occidental Petroleum Corp.	496.1
14	Kerr-McGee Corp.	485.1
15	Newfield Exploration Co.	436.7
16	Burlington Resources Inc.	423.5
17	Anadarko Petroleum Corp.	391.0
18	El Paso Corp.	371.0
19	ConocoPhillips	370.0
20	Quicksilver Resources Inc.	359.2
	Total	**11,764.1**

WORLD RIG ACTIVITY

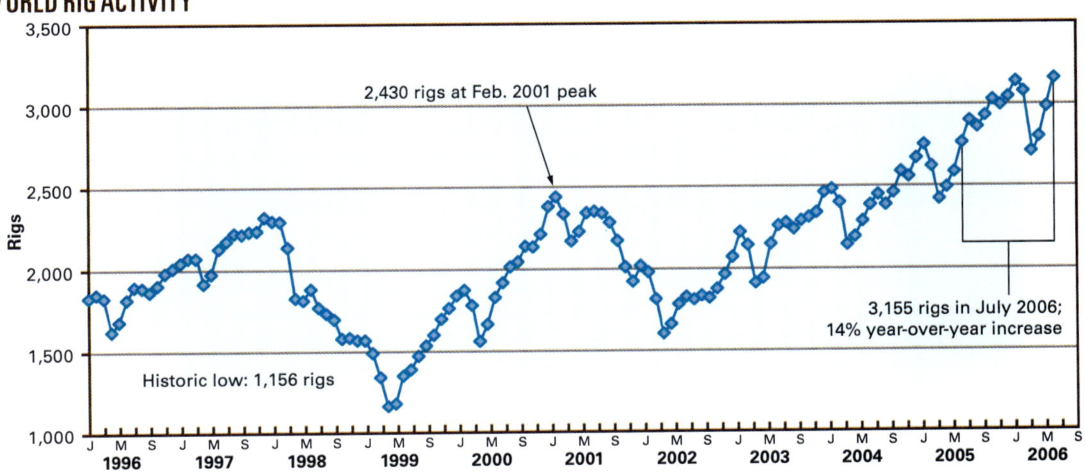

Source: Baker Hughes Inc.

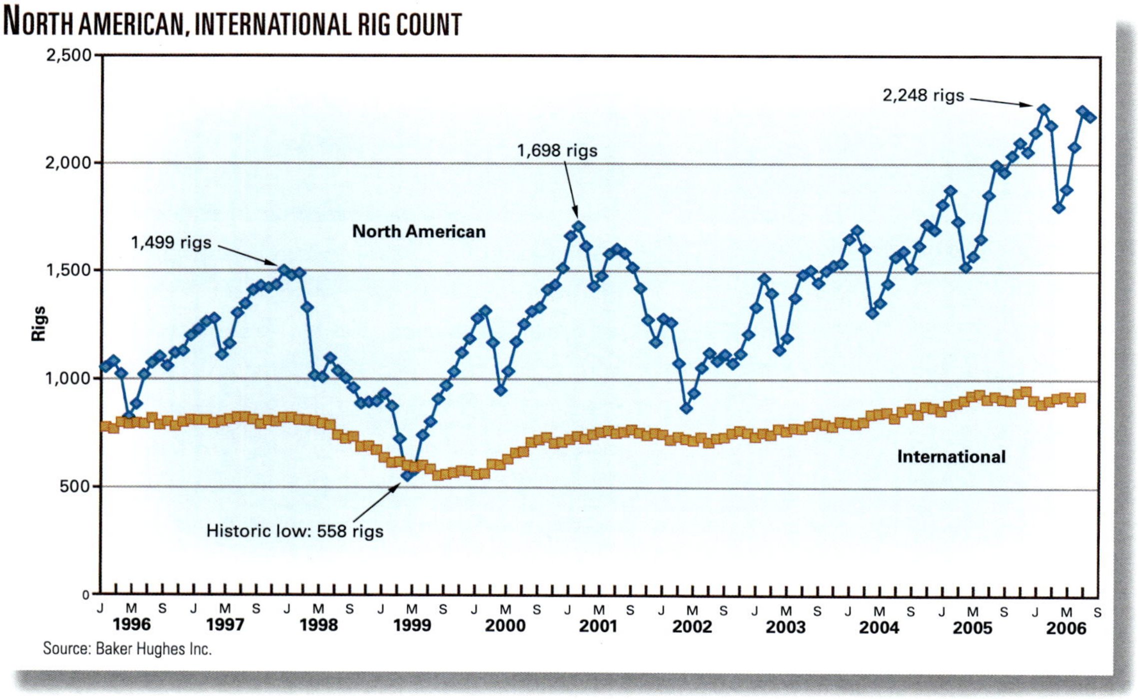

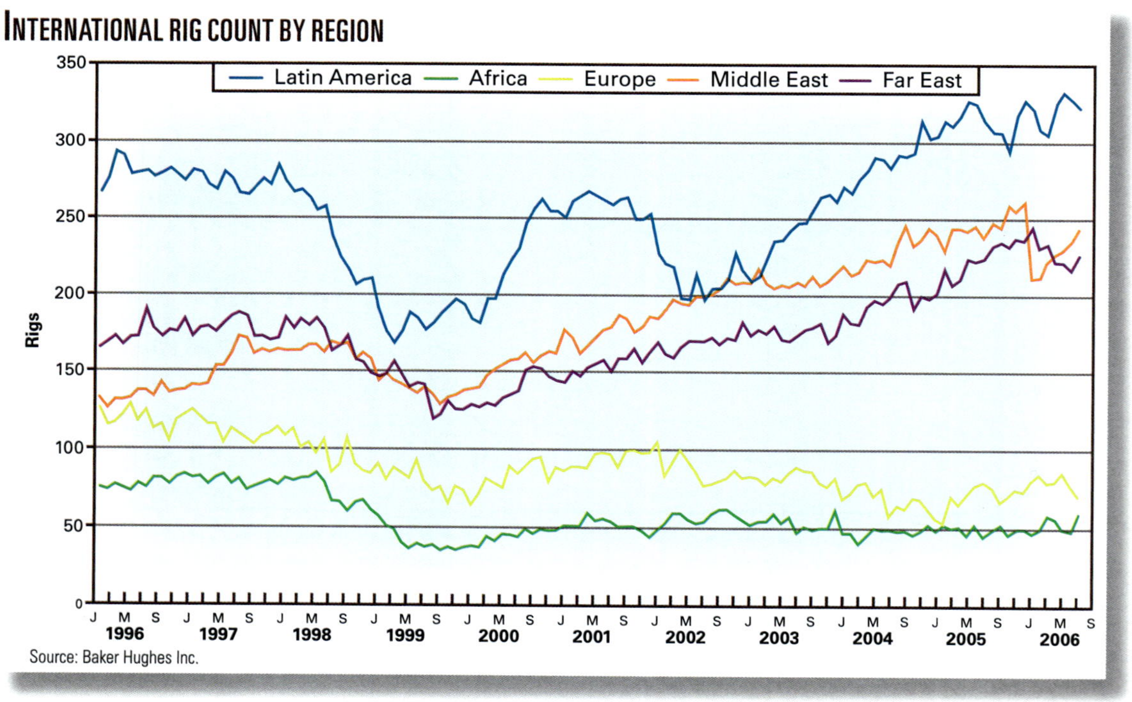

Baker Hughes Inc.'s international rig count (excluding North America) reached a new record high of 948 rigs in December 2005 but dropped back to 921 in July 2006. The North American rig count reached a new high of 2,248 rigs in February 2006 before dropping to 2,234 in July. This continued its upward, seasonally cyclical trend.

Rig activity has been steadily increasing in Latin America, the Middle East, and Asia Pacific since early 2002. All three reached historic highs in 2006: Latin America (333 rigs, May 2006); Middle East (262 rigs, December 2005); and Asia Pacific (244 rigs, January 2006).

SCORE INDEX EXCEEDS BENCHMARK

GlobalSantaFe Corp.'s Summary of Current Offshore Rig Economics (SCORE) reached and exceeded the 100 benchmark in most categories in 2006, for the first time since 1981:

- Jack ups jumped to 102 from 91 in November-December 2005 and reached a high of 119 in June 2006. Heavy-duty jack up day rates dropped 22% from June-July 2006.
- Semisubmersibles jumped to 101 from 99 in September-October 2005 and reached a high of 140 in July. This was a 10.8% increase month-to-month and nearly 67% year-over-year.
- North Sea jumped to 101 from 95 from September-October 2005 and reached a high of 143 in July, up 6% from June. The increase was due primarily to jack up contracts.
- Gulf of Mexico jumped to 101 from 96 in October-November 2005 and reached a high of 119 in June. Lower day rates caused the SCORE to drop 6.7% from June-July 2006.
- West Africa jumped to 105 from 98 in December 2005-January 2006 and reached a high of 144 in July, up 14.1% because of semisub contracts.
- The overall worldwide SCORE jumped to 103 from 96 in November-December 2005 and reached close to 124 in July 2006, a year-over-year increase of nearly 59%.

The greatest increase in rig activity in 2006 was in the Middle East, where 242 rigs were operating in July, up from 210 rigs in January. National oil companies seemed anxious to shore up production, with many major fields in steady decline. Saudi Arabia led drilling and was absorbing dozens of newbuild rigs. The kingdom said it plans to increase development drilling by 61%.

U.S. drilling hits 20-year high

In early August, Baker Hughes Inc. said that U.S. drilling activity reached a new 20-year high of 1,728 operating rigs. The top 50 operators drilled about 21,706 wells

TOP 50 OPERATORS FOR 2006

Rank[1]	Company	Well starts[2]	Wells, %	Footage	Footage, %	Wells with PTD[3]	Average footage
1	Aera Energy LLC	667	3.1	—	—	—	—
2	Chesapeake Operating Inc.	592	2.7	6,189,686	4.0	588	10,527
3	XTO Energy Inc.	545	2.5	5,535,508	3.5	542	10,213
4	Devon Energy Production Co. LP	481	2.2	4,398,731	2.8	480	9,164
5	EOG Resources Inc.	481	2.2	4,084,569	2.6	473	8,635
6	Encana Oil & Gas (USA) Inc.	470	2.2	4,786,465	3.1	470	10,184
7	Dominion Oklahoma Texas E&P Inc.	446	2.1	3,810,336	2.4	446	8,543
8	Burlington Resources Oil & Gas Co. LP	404	1.9	3,862,662	2.5	388	9,955
9	Chevron USA Inc.	320	1.5	1,399,297	0.9	158	8,856
10	Atlas America Inc.	293	1.3	1,399,293	0.9	291	4,809
11	Williams Production RMT Co.	284	1.3	2,149,257	1.4	284	7,568
12	BP America Production Co.	263	1.2	2,522,050	1.6	252	10,008
13	Apache Corp.	252	1.2	2,078,169	1.3	214	9,711
14	Pioneer Natural Resources USA Inc.	228	1.1	2,168,684	1.4	223	9,725
15	Noble Energy Production Inc.	212	1.0	1,982,569	1.3	211	9,396
16	Great Lakes Energy Partners LLC	202	0.9	972,768	0.6	194	5,014
17	Whiting Oil & Gas Corp.	190	0.9	1,029,568	0.7	181	5,688
18	ConocoPhillips Co.	174	0.8	1,450,756	0.9	168	8,635
19	Dominion E&P Inc.	169	0.8	1,080,760	0.7	161	6,713
20	Plains E&P Co.	169	0.8	6,949	—	2	3,475
21	Anadarko Petroleum Corp.	151	0.7	1,761,181	1.1	150	11,741
22	Kerr-McGee Rocky Mountain Corp.	151	0.7	1,165,576	0.7	151	7,719
23	Cabot Oil & Gas Production Corp.	148	0.7	1,145,640	0.7	146	7,847
24	Occidental of Elk Hills Inc.	148	0.7	562,457	0.4	97	5,799
25	Westport Oil & Gas Co. LP	140	0.6	1,346,851	0.9	137	9,831
26	Houston Exploration Co., The	134	0.6	1,169,250	0.7	133	8,791
27	OXY Permian Ltd.	127	0.6	809,000	0.5	127	6,370
28	Berry Petroleum Co.	126	0.6	255,251	0.2	34	7,507
29	BP Exploration (Alaska) Inc.	125	0.6	345,769	0.2	40	8,644
30	ExxonMobil Development Co.	122	0.6	757,788	0.5	84	9,021
31	Newfield Production Co.	119	0.5	734,480	0.5	119	6,172
32	Samson Lone Star LP	119	0.5	1,400,915	0.9	118	11,872
33	Chesapeake Appalachia LLC	114	0.5	390,572	0.2	85	4,595
34	Mariner Energy Inc.	114	0.5	931,584	0.6	103	9,045
35	Cimarex Energy Co. of Colorado	112	0.5	921,810	0.6	108	8,535
36	Questar E&P Co.	109	0.5	1,347,111	0.9	108	12,473
37	Anadarko E&P Co. LP	107	0.5	983,068	0.6	107	9,188
38	Petroleum Development Corp.	103	0.5	852,461	0.5	103	8,276
39	KCS Resources Inc.	99	0.5	958,725	0.6	99	9,684
40	Range Production Co.	99	0.5	732,350	0.5	98	7,473
41	OXY USA Inc.	97	0.4	322,360	0.2	65	4,959
42	Texas Keystone Inc.	95	0.4	364,000	0.2	90	4,044
43	Marathon Oil Co.	94	0.4	921,156	0.6	86	10,711
44	El Paso E&P Co. LP	93	0.4	1,014,980	0.6	91	11,154
45	Energen Resources Corp.	89	0.4	452,844	0.3	89	5,088
46	Latigo Petroleum Texas LP	88	0.4	609,425	0.4	88	6,925
47	Encore Operating LP	81	0.4	750,330	0.5	79	9,498
48	Seneca Resources Corp.	80	0.4	64,150	—	19	[illegible]
49	SEECO Inc.	79	0.4	386,461	0.2	77	[illegible]
50	Continental Resources Inc.	78	0.4	518,222	0.3	[illegible]	[illegible]
	Total for 2006	**21,706**	**87%**	**156,463,289**	**100%**	[illegible]	[illegible]

[1]Rank according to number of wells drilled. [2]Wells deeper than 2,500 ft. [3]Wells that have a proposed total depth.
Source: RigData; data from Jan. 1-Aug. 17, 2006

this year (through Aug. 17) deeper than 2,500 ft, totaling about 156 million ft.

The North American land rig market was near capacity in August. For the week ending Aug. 11, the Baker Hughes rig count listed 1,611 land rigs drilling in the United States. There were 18 rigs drilling in inland waters and 99 rigs drilling offshore. Baker Hughes data does not include rigs in transit, testing, rigging up, or rigging down.

Newbuilds, offshore

About 100 new mobile offshore drilling units (MODUs) were under construction worldwide. As of August 2006, there were 59 jack ups, 29 semisubmersibles, nine drillships, and three drilling tenders under construction.

Three of the semisub contracts were announced in June-July, augmenting the 26 under contract for about $13 billion. Fewer than one-third of these will be delivered in 2006 (6 rigs) and 2007 (24 rigs). Most of the new rigs (44) will hit the market in 2008, followed by another 22 in 2009.

US ONSHORE AND OFFSHORE DRILLING ACTIVITY

	Baker Hughes's rig count				RigData's working rigs				
	4-week avg.		Week, as of		4-week avg.		Week, as of		Wait-ing to
State, district	2005	2006	8/04/06	8/11/06	2005	2006	8/04/06	8/11/06	spud
Alabama	2	4	4	4	5	7	7	6	—
Offshore	—	1	1	1	—	1	1	1	—
Total	**2**	**5**	**5**	**5**	**5**	**8**	**8**	**7**	**—**
Alaska	9	8	8	8	12	15	14	15	5
Arizona	—	—	—	—	1	1	1	1	—
Arkansas	13	26	26	26	19	30	28	30	—
California	23	29	29	29	43	48	49	48	—
Offshore	4	5	5	5	—	—	—	—	—
Total	**27**	**34**	**34**	**34**	**43**	**48**	**49**	**48**	**—**
Colorado	72	92	93	92	81	91	89	94	1
Florida	1	—	—	—	1	—	—	—	—
Offshore	1	—	—	—	—	—	—	—	—
Total	**2**	**—**	**—**	**—**	**1**	**—**	**—**	**—**	**—**
Hawaii	1	—	—	—	—	—	—	—	—
Illinois	—	—	—	—	4	7	6	6	—
Indiana	—	—	—	—	1	3	3	3	—
Kansas	6	15	13	16	46	48	46	48	2
Kentucky	4	6	6	6	10	12	10	11	3
Louisiana									
North, land	51	56	57	52	61	59	61	58	3
South, land	31	35	34	36	45	50	48	52	3
Offshore	93	80	81	79	82	69	69	68	2
Inland waters	26	15	14	14	35	36	37	35	—
Total	**201**	**186**	**186**	**181**	**223**	**214**	**215**	**213**	**8**
Maryland	—	—	—	—	82	—	—	—	—
Michigan	3	4	4	3	11	9	8	8	1
Mississippi	15	12	13	13	16	21	22	22	2
Montana	24	21	22	21	32	30	30	29	4
North Dakota	24	31	32	32	27	38	37	38	—
Nebraska	—	—	—	—	3	2	2	2	—
Nevada	3	2	2	2	2	1	1	1	—
New Mexico	83	92	88	99	91	94	96	98	1
New York	4	7	7	7	5	5	4	5	—
Ohio	9	6	6	5	11	14	16	14	2
Oklahoma	153	192	194	195	196	225	227	229	11
Oregon	—	—	—	—	1	—	—	—	—
Pennsylvania	19	17	17	16	41	36	35	36	1
South Dakota	2	1	1	1	1	2	2	2	—
Tennessee	—	—	—	—	—	2	2	2	—
Texas									
District 1	15	20	18	20	26	19	19	17	1
2	47	25	25	27	52	40	40	41	9
3	82	63	62	58	67	68	67	67	7
4	72	78	75	80	80	101	101	103	5
5	72	141	141	149	112	136	134	143	4
6	91	113	110	115	99	121	116	125	11
7B	19	47	47	45	30	63	68	61	5
7C	39	40	40	41	46	62	64	62	13
8	65	100	100	101	91	112	108	115	6
8A	24	28	28	25	34	36	35	39	2
9	30	33	31	31	51	39	37	38	5
10	62	73	75	76	65	83	84	81	3
Inland waters	—	3	3	4	2	6	5	5	—
Offshore	7	12	12	14	15	17	18	16	—
Total	**625**	**776**	**767**	**786**	**770**	**903**	**896**	**913**	**71**
Utah	27	44	45	44	32	46	47	47	1
Virginia	4	—	—	—	2	—	—	—	—
Washington	—	1	1	1	—	1	1	1	—
West Virginia	15	28	28	28	25	33	32	33	3
Wyoming	77	105	107	107	128	149	150	148	4
US deep water	—	—	—	—	40	49	50	51	3
Total	**1,424**	**1,711**	**1,705**	**1,728**	**1,962**	**2,137**	**2,127**	**2,155**	**123**

Source: RigData; a summary of data presented in the Aug. 11, 2006, edition of "LOCATION & OPERATORS" report

2006 UTILIZATION OF TOP 25 US DRILLING CONTRACTORS—ONSHORE AND OFFSHORE*

Rank	Company	Jan.	Feb.	Mar.	Apr.	May	June	July	Aug.	Fleet	Rigs used throughout the year
1	Patterson-UTI Drilling Co. LP	290	291	294	293	296	297	301	304	310	307
2	Nabors Drilling USA LP	253	253	253	254	255	253	255	257	293	264
3	Grey Wolf Drilling Co. LP	110	111	110	111	111	111	112	112	112	113
4	Helmerich & Payne IDC	96	96	97	100	103	107	108	111	117	113
5	Unit Drilling Co.	104	104	104	103	103	101	97	78	79	106
6	Pioneer Drilling Co.	52	54	54	54	55	55	58	59	59	59
7	Bronco Drilling Co. Inc.	40	40	41	40	43	44	47	47	47	47
8	Ensign US Drilling Inc.	45	44	45	44	45	46	45	44	47	46
9	Unit Texas Drilling LLC	6	6	7	9	11	17	35	36	36	36
10	Union Drilling Inc.	32	32	32	30	31	31	32	28	35	35
11	Cactus Drilling Co. LLC	27	28	29	28	30	31	33	33	33	33
12	The Offshore Drilling Co.	27	27	28	28	28	29	30	30	54	32
13	Thornton Drilling Co.	15	15	15	15	27	16	16	16	16	29
14	Nomac Drilling Corp.	19	19	21	22	23	25	27	26	26	28
15	Gene D. Yost & Son Inc.	9	9	11	10	12	13	22	14	15	27
16	Cyclone Drilling Inc.	23	24	24	25	25	25	25	25	25	25
17	Lariat Services Inc.	20	20	20	20	22	21	24	24	31	24
18	Diamond Offshore Drilling Inc.	23	23	22	21	20	21	22	20	21	23
19	Big Dog Drilling Co.	20	20	19	19	20	21	21	21	22	22
20	Capstar Drilling LP	19	20	20	20	20	20	21	20	20	21
21	Nabors Offshore Corp.	11	16	17	18	18	18	16	12	41	20
22	Rowan Drilling Co. Inc.	17	17	17	17	17	18	19	20	19	20
23	Cheyenne Drilling LP	17	17	18	18	18	18	19	18	19	19
24	Felderhoff Bros. Drilling LLC	14	14	15	15	16	17	19	19	19	19
25	ENSCO Offshore Co. Inc.	15	15	16	17	18	18	18	17	17	18
	Active rigs	**1,304**	**1,315**	**1,329**	**1,331**	**1,367**	**1,373**	**1,422**	**1,391**	**1,513**	**1,486**

*Some rigs that worked earlier in the year may no longer be marketed.
Source: RigData; based on data Jan. 1-Aug. 17, 2006

New semisub orders exceeded $13 billion

The market for semisubmersible drilling rigs continued to improve in mid-2006. Day rates and utilization levels increased, leading the push to build fleet capacity with new rigs and major upgrades.

Drilling contractors had made contracts worth more than $10 billion to build 26 new semisubmersible drilling rigs, up from just four under construction in mid-2005.

Forecast sees ramp up in floating production system orders

A March 2006 report from the International Maritime Associates Inc. (IMA; Washington, DC) predicted a ramp-up in floating production system, 2006–2010.

MODUs* UNDER CONSTRUCTION

	Jack ups	Semisub-mersibles	Ships	Tenders	Total
US Gulf of Mexico	12	—	—	—	12
Europe	—	—	1	—	1
Middle East	4	4	—	—	8
Southeast Asia	38	12	—	3	53
Far East	4	12	8	—	24
Caspian Sea	—	1	—	—	1
Russian Arctic	1	—	—	—	1
Total	**59**	**29**	**9**	**3**	**100**
MODUs under construction by delivery date					
2006	4	1	—	1	6
2007	21	1	1	1	24
2008	26	12	5	1	44
2009	8	12	2	—	22
Undetermined	—	3	1	—	4
Total	**59**	**29**	**9**	**3**	**100**
MODUs under construction with no drilling contract					
2006	2	—	—	—	2
2007	16	1	—	—	17
2008	24	10	3	1	38
2009	7	4	1	—	12
Undetermined	—	1	1	—	2
Total	**49**	**16**	**5**	**1**	**71**

*MODU—mobile offshore drilling unit.
Sources: ODS—Petrodata Inc., Offshore Rig Locator, Aug. 22, 2006

Semisubmersibles under construction, 2006-11

Rig name	Owner	Design	Shipyard, country	Delivery date	Cost, $ million [shipyard]
				2006 (2)	
Iran Alborz	National Iranian Oil Co. (NIOC)	Fourth-gen., Gotaverken GVA 4000-M	Sadra shipyard at Nekkairan, Caspian Sea, Iran		226
West Berani	Eastern Drilling ASA (Smedvig, Seadrill)	Sixth-gen., Moss Maritime Keppel SSDT-3600	Keppel FELS Ltd.. Singapore	Fourth-quarter 2006	—
				2007	
West E-drill	Eastern Drilling ASA (Smedvig, Seadrill)	Sixth-gen., Moss Maritime SSDT	Samsung Heavy Industries Co. Ltd., South Korea	Fourth-quarter 2007	550
				2008 (11)	
Aker TBN-1	Aker Drilling AS	Sixth gen., Aker H-6e	Aker Kvaerner Stord, Norway and Dubai Drydocks, UAE	February 2008	627
SeaDrill TBN-1	SeaDrill Ltd.	Sixth gen., Gotaverken GVA 7500-N	Daewoo Shipbuilding & Marine Engineering Co. (DSME), South Korea	March 2008	496.5
SeaDrill TBN-2	SeaDrill	Sixth gen., F&G ExD	SembCorp Marine, Jurong shipyard,Singapore	June 2008	486.5 [390]
ENSCO 8500	ENSCO International Inc.	Sixth gen., Ensco 8500	Keppel FELS Ltd.	June 2008	[312]
ORS Pioneer	Offshore Rig Services ASA	Sixth gen., Global Maritime GM-4000	Yantai Raffles, China	Second-quarter 2008	—
Maersk TBN-1	Maersk Contractors	Sixth-gen., DSS-21	KFELS	August 2008	[237]
Aker TBN-2	Aker	Aker H-6e	Stord and Dubai Drydocks	October 2008	627
Odfjell TBN-1	Odfjell Drilling AS	Sixth gen., GVA 7500	DSME, Okpo yard	October 2008	—
ORS TBN-1	Offshore Rig Services	GM-4000	Yantai Raffles	Fourth-quarter 2008	—
TBN	Frigstad Offshore Drilling Ltd.	Frigstad D90	DSME	Fourth-quarter 2008	500
Noble Bingo 9000-4	Noble Drilling	Fourth gen., Bingo 9000	DSD Shipping AS, Dalian shipyard, China	2008	—
				2009 (9)	
GSF Development Driller III	GlobalSantaFe Corp.	MSC DSS 51	KFELS	March 2009	590
Noble Danny Adkins	Noble Drilling	Fifth gen., Bingo 9000	Dalian	March 2009	—
Bingo 9000 Hull 4	Noble Drilling	Fifth gen., Bingo 9000	Dalian	2009	—
Maersk TBN-2	Maersk	MSC DSS-21	KFELS	April 2009	[235]
ENSCO 8501	ENSCO	Ensco 8500	KFELS	June 2009	[338]
Frigstad TBN-1	Frigstad Offshore	Frigstad D90	DSME	Second-quarter 2009	500
SeaDrill TBN-3	SeaDrill	F&G ExD	SembCorp, Jurong	First-half 2009	486.5
Sevan Deepsea Driller	Sevan Drilling Pte. Ltd.	Sevan stabilized platform (SSP)	Hantong shipyard, China Jurong, Singapore	First-half 2009	[390]
PetroMena TBN-1	PetroMena AS	Sixth gen., F&G ExD	SembCorp, Jurong	December 2009	400
				2010 (3)	
Maersk TBN-3	Maersk	MSC DSS-21	Keppel FELS	2010	262
SeaDrill TBN-4	SeaDrill	F&G ExD	SembCorp	Early 2010	—
TBN	Larsen Oil & Gas AS	F&G ExD	SembCorp, Jurong	January 2010	462 [398]
				2011	
PetroMena TBN-2	PetroMena AS	Sixth gen., F&G ExD	SembCorp, Jurong	2011	—
				Total funds committed	**$7.3 billion +**

Sources: Colton Co., Rigzone, R.S. Platou Offshore, ODS-Petrodata, various owner, operator, and shipyard press releases.

The report forecasts the number of units and the capital expenditures needed, and includes floating production, storage, and offloading (FPSO) vessels; tension leg platforms (TLPs); semisubmersible production units; and production spars.

As of mid-2006, there were 179 floating production systems operating worldwide. The existing fleet consisted of 61% FPSOs, 21% production semis, 10% TLPs, and 8% spars. Also, 82 floating storage vessels (without production capability) were in operation, four of which store LPG.

Increase in Floater Installations

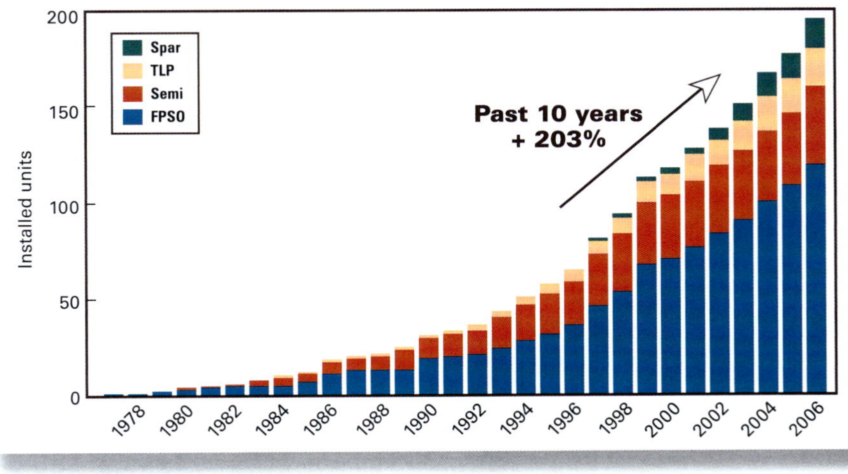

More than 60 new jack ups being built

Without accounting for likely attrition of older units, the world's supply of jack ups will expand more than 16% by 2009, adding 61 jack ups now under construction in addition to three delivered earlier in 2006, said *O&GJ* in October.

Nearly two-thirds (39) are being built in Southeast Asia. Twelve jack ups are being built on the U.S. Gulf coast. Yards in China and United Arab Emirates are each building four, a yard in Indonesia is building two, and India and Russia are each building one.

Most of these newbuilds will come to market in 2007 (21 rigs) and 2008 (27 rigs).

Jack-up rigs under construction, 2006-09

Rig name	Owner	Design	Shipyard, country	Delivery date	Cost, $ million
				2006 (4)	
Gulf 004	Gulf Drilling International	KFELS Mod V B	Keppel FELS (KFELS), Singapore	September	153
Hank Boswell	Rowan Cos. Inc.	LeTourneau Tarzan class	LeTourneau, USA	October	153
Deep Driller 003	DDI Holding AS	KFELS Mod V Super B Class	KFELS, Singapore	December	153
WilCraft	Awilco	KFELS Mod V B	KFELS, Singapore	December	153
				2007 (21)	
Soehanah	Apexindo	Baker Marine Pacific Class 375	PPL Shipyard Pte Ltd., Singapore	January	134
PV Drilling 001	PetroVietnam	KFELS Mod V B	KFELS, Singapore	March	144
Petrojack 001	A.P. Møller	Baker Marine Pacific Class 375	Jurong Shipyard Pte Ltd., Singapore	March	175
ENSCO 108	ENSCO International Inc.	KFELS Mod V B Bigfoot	KFELS, Singapore	March	134
Offshore Courageous	Scorpion Offshore	Letourneau Super 116	Keppel AmFELS, USA	April	160
WilSuperior	Awilco	Baker Marine Pacific Class 375	PPL Shipyard, Singapore	May	134
Panuco	Perforadora Central	Letourneau Super 116E	LeTourneau, USA	May	131
Deep Driller 004	DDI Holding AS	Baker Marine Pacific Class 375	PPL Shipyard, Singapore	June	144
Deep Driller 005	DDI Holding AS	KFELS Mod V Super B Class	KFELS, Singapore	June	140
West Prospero	SeaDrill Invest II	KFELS Mod V B	KFELS, Singapore	June	163
Arcticheskaya	Gazprom	CDB Corrall 6500/10-30	Zvezdochka Shipyard, Russia	July	130
Noble Roger Lewis	Noble Corp.	Friede & Goldman (F&G) JU-2000E	Dalian Shipbuilding Industry Co., China	August	129
West Atlas	Stella	KFELS Mod V B	KFELS, Singapore	September	147
JP Bussell	Rowan	LeTourneau Tarzan class	Signal International, USA	October	126
Offshore Defender	Scorpion Offshore	LeTourneau Super 116	Keppel AmFELS, USA	October	120
West Triton	Caleb	Baker Marine Pacific Class 375	PPL Shipyard, Singapore	October	133
Maersk JU Tbn001	A.P. Møller	Marine Structure Consultants (MSC) CJ50-X100MC	KFELS, Singapore	November	145
Thule Energy	Thule Drilling	F&G Super M2	QGM, UAE	November	140
Wilforce	Awilco	Baker Marine Pacific Class 375	PPL Shipyard, Singapore	December	140
COSL 942	China Oilfield Services Ltd.	F&G JU-2000E	Dalian Shipbuilding, China	December	150
Gulf 005	Gulf Drilling International	KFELS Mod V B	KFELS, Singapore	December	150
				2008 (27)	
Petrojack 002	Petrojack	Baker Marine Pacific Class 375	Jurong Shipyard, Singapore	January	171
Ocean Scepter	Diamond Offshore Drilling Inc.	KFELS Mod V Super B Class	Keppel AmFELS, USA	February	117
Ocean Shield	Diamond Offshore	KFELS Mod V Super B Class	KFELS, Singapore	February	100
Noble Hans Deul	Noble	F&G JU-2000E	Dalian Shipbuilding, China	February	182
Hakuryu 010	Japan Drilling	Baker Marine Pacific Class 375	PPL Shipyard, Singapore	February	-
Thule Force	Thule Drilling	F&G Super M2	QGM, UAE	March	90
Offshore Resolute	Scorpion Offshore	LeTourneau Super 116	Keppel AmFELS, USA	March	130
WilBoss	Awilco	KFELS Mod V B	KFELS, Singapore	April	180
Maersk JU TBN2	A.P. Møller	MSC CJ50-X100MC	KFELS, Singapore	May	130
Aban JU TBN1	Aban Offshore	Baker Marine Pacific Class 375	PPL Shipyard, Singapore	June	184
Wilseeker	Awilco	Baker Marine Pacific Class 375	PPL Shipyard, Singapore	June	180
Mosvold JU TBN1	Mosvold Drilling Ltd.	F&G Super M2	Maritime Industrial Services (MIS), UAE	June	143
Rowan JU TBN1	Rowan	LeTourneau Class 240-C	LeTourneau, USA	June	143
West Ariel	SeaDrill Ltd	KFELS Mod V B	KFELS, Singapore	June	153
Offshore Vigilant	Scorpion Offshore	LeTourneau Super 116	Keppel AmFELS, USA	June	156
Petrojack 003	A.P. Møller	Baker Marine Pacific Class 375	Jurong Shipyard, Singapore	July	190
Deep Driller 006	DDI Holding AS	KFELS Mod V Super B Class	KFELS, Singapore	August	133
Deep Driller 007	DDI Holding	Baker Marine Pacific Class 375	PPL Shipyard, Singapore	September	140
Standard JU TBN1	Standard Drilling	MSC CJ-46-X100D	Labroy Shipbuilding and Engineering, Indonesia	September	110
Atwood Aurora	Atwood	LeTourneau Super 116E	Keppel AmFELS, USA	September	100
Mosvold JU T2	Mosvold Drilling	F&G Super M2	MIS, UAE	October	125
Maersk JU TBN3	A.P. Møller	MSC CJ50-X100MC	KFELS, Singapore	November	165
Great Eastern JU TBN 2	Great Eastern	—	Bharati Shipyard, India	November	165
Offshore Intrepid	Scorpion Offshore	LeTourneau Super 116	Keppel AmFELS, USA	November	148
Jindal JU Tbn1	Discovery Hydrocarbons	KFELS Mod V B	KFELS, Singapore	December	156
Petrojack 004	JackInvest I Pte	Baker Marine Pacific Class 375	Jurong Shipyard, Singapore	December	143
Jindal JU Tbn2	Jindal Pipes	KFELS Mod V B	KFELS, Singapore	December	144
				2009 (8)	
Rowan JU Tbn2	Rowan	LeTourneau Class 240-C	LeTourneau, USA	February	148
Deep Driller 008	DDI Holding AS	KFELS Mod V Super B Class	KFELS, Singapore	March	121
Mercator JU Tbn1	Mercator Lines	KFELS Mod V B	KFELS, Singapore	March	132
Noble Scott Marks	Noble	F&G JU-2000E	Dalian Shipbuilding, China	March	146
Standard JU Tbn2	Standard Drilling	MSC CJ-46-X100D	Labroy Shipbuilding and Engineering, Indonesia	April	146
Maersk JU Tbn004	A.P. Møller	MSC CJ50-X100MC	KFELS, Singapore	May	129
Wilstrike	Awilco	KFELS Mod V B	KFELS, Singapore	May	126
Great Eastern JU Tbn 1	Great Eastern	KFELS Mod V B	KFELS, Singapore	October	126

Total funds committed, $7.344 billion

Source: ODS-Petrodata

So. America Drilling, July 2006*

	Land	Offshore	Total
	Rigs		
Argentina	84	—	**84**
Bolivia	3	—	**3**
Brazil	13	15	**28**
Chile	—	—	**—**
Colombia	21	—	**21**
Ecuador	11	—	**11**
Peru	4	1	**5**
Venezuela	71	17	**88**
Total	**207**	**33**	**240**

*Data from Baker Hughes Inc. international rig count.

South America drilling pace picking up

Companies were rewriting contracts and developing new working relationships to accelerate drilling in 2006 in South America, *OGJ* said. Governmental shifts to regain sovereignty in Venezuela, Ecuador, and Bolivia affected drilling plans, patterns, and fleets.

Applying advanced technology

Across the world, the oil and gas industry continued to develop and apply advanced technology to find, develop and produce hydrocarbon resources for growing markets.

Troll Field Quad Lateral Well Completion

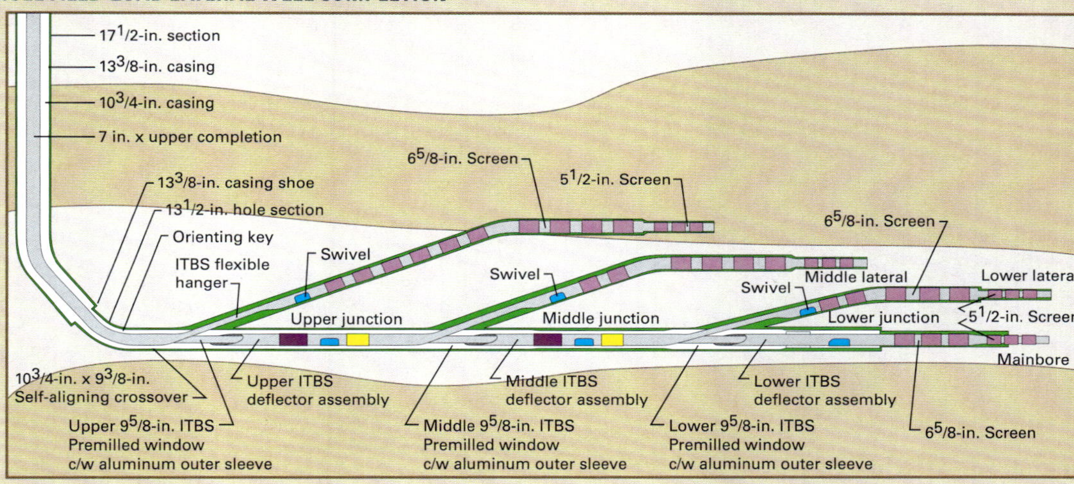

Troll field "starfish"

The Troll field lies beneath the Norwegian North Sea, about 80 km (50 miles) west of Bergen at a water depth of about 335 m (1,100 ft), covering 700 sq km (270 sq miles).

Norsk Hydro Produksjon AS and its service partners reported successful use of several advanced technologies in the Troll field, which lies beneath the Norwegian North Sea, about 80 km (50 miles) west of Bergen at a water depth of about 335 m (1,100 ft).

Troll West Starfish Well with Five Laterals

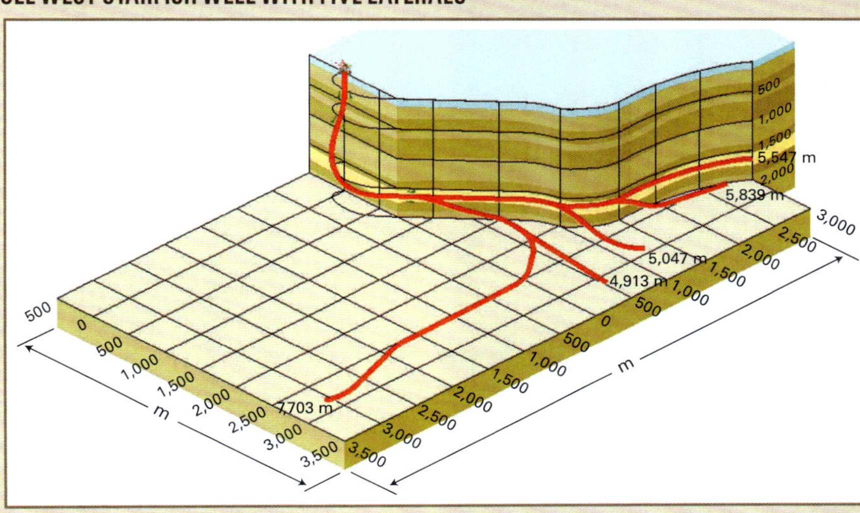

One highlight in the Troll West gas province was the construction of a "starfish" multilateral well with five extended-reach horizontal sections drilled from a single slot. Using technology aggressively, Hydro has produced more than 1 billion bbl from what was once judged an uneconomical oil field because of its thin productive reservoir sands.

Microhole coiled-tubing drilling

A specially designed hybrid "microhole" coiled tubing rig drilled 25 test wells in 2006 in the intractable Niobrara gas formation in western Kansas and eastern Colorado. The effort delivered cost savings of 25–35% per well drilled compared with conventional drilling equipment, said the U.S. Dept. of Energy (DOE), the research sponsor.

About 1 tcf of shallow gas that had been bypassed by conventional drilling has become economic as a result.

The commercial Niobrara drilling program—in which 3,000-ft wells were drilled in as little as 19 hr, from rig move-in to move-out—followed earlier DOE-funded research by Gas Technology Institute (GTI; Des Plaines, IL, U.S.A.).

GTI and its partners adapted a conventional coiled tubing rig for drilling exploratory and development wells with ultrasmall diameters.

GTI's project aimed to develop tools for drilling boreholes with diameters of 1.75–4.5 in, as well as related downhole microinstrumentation, using small, easily transportable coiled-tubing drilling rigs.

Producers have only recently begun to use coiled tubing rigs in grassroots exploratory and development areas, largely in Alaska, Canada, and other higher-cost operating areas.

DOE said the microhole coiled tubing drilling technology could be applied to bypassed resources in thousands of U.S. hydrocarbon reservoirs, tapping a volume of bypassed oil shallower than 5,000 ft that exceeds 215 billion bbl.

Intelliserv Inductive Coupler

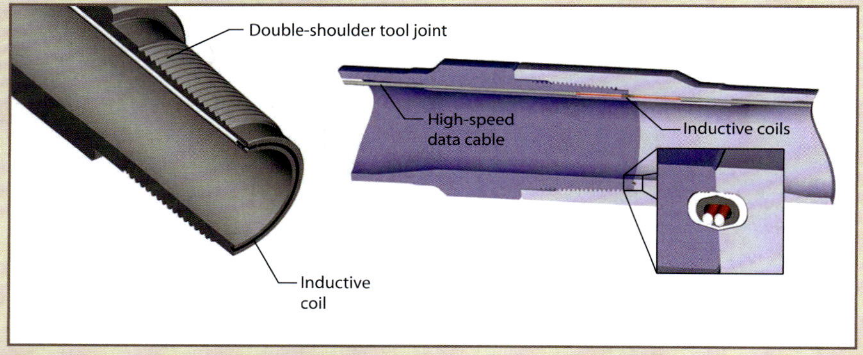

The CSEM Process

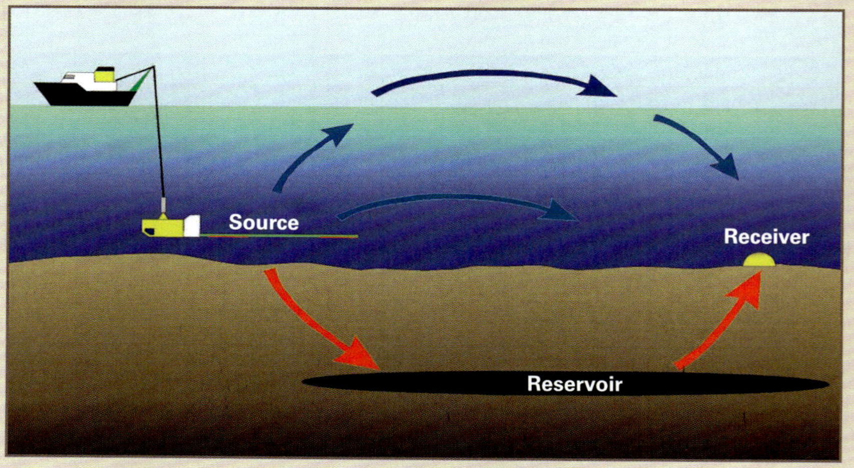

CSEM imaging

Controlled Source Electromagnetic (CSEM) imaging can help distinguish between hydrocarbon fluids and water; a distinction that seismic data alone cannot make.

The technology exploits the large variation in resistivity of a formation when conductive saline water is replaced by hydrocarbon. This variation can be mapped at a local or regional scale by studying its effect on an electromagnetic field transmitted through the sea floor, giving considerable insight into the subsurface geology.

A high-powered electromagnetic source is towed through the water transmitting a low-frequency signal to the seabed, and electromagnetic receivers placed on the seabed detect the resulting signal at offsets of up to 10–15 km (6–9 miles).

Data are interpreted in one, two, and three dimensions to provide geoelectric cross-sections and volumes, and hence to map complex reservoir structures in subsurface structures anywhere from a few meters to several kilometers depth.

A 2005 trial in the UK North Sea demonstrated that CSEM data could be collected in water shallower than previously thought possible. Commercial surveys have now been completed in 70 m (230 ft) of water, and even shallower surveys are planned.

Wired-pipe telemetry system

Baker Hughes INTEQ (Houston) and IntelliServ, Inc. (Provo, UT, U.S.A.) demonstrated an innovative system for gathering data while drilling at much higher rates than conventional approaches.

The system uses is based on wired-pipe telemetry. WPT provides data transmission at 57,000 bits/sec compared to 3 bits/sec using conventional mud-pulse telemetry (MPT).

A high-speed, bidirectional drillstring WPT system was developed by IntelliServ Inc., a joint venture between Grant Prideco Inc. and Novatek Engineering Inc. with partial funding from the U.S. Dept. of Energy.

In the WPT system, each joint in the drillstring contains an inductive coupler that allows the data-carrying signal to "jump" each tool joint without the need for physical contact between conductors.

Within each joint of pipe, a stainless steel tube in the pipe bore contains a shielded, coaxial data cable. Two interface components, one on the surface and one downhole, translate data to and from the IntelliServ network.

Testing showed that a wired drillstring can be used in all conventional operations in which a standard drillstring is used, including cementing and slick-line operations.

In 2005–2006, Baker Hughes INTEQ successfully connected both its downhole and surface MWD networks to the IntelliServ network and ran MWD services in commercial wells for BP America in the Arkoma basin.

IntelliServ, INTEQ Networks

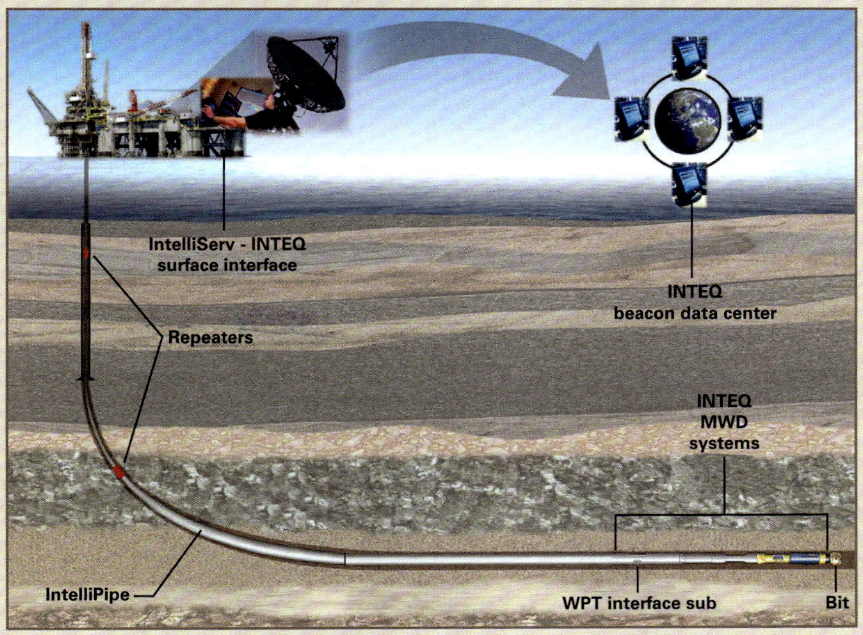

Rocket Rig

A new-design, fast moving, self-elevating land rig debuted in Oklahoma in September. The triple rig can be erected in about 20 min and drill to 18,000 ft.

Cactus Drilling Co. LLC finished Rig. 136, the first of four Rocket Rigs currently under construction, and was planning to complete and deliver the remaining three by early 2007.

The innovative system uses hydraulic winches to raise the mast and the entire drill floor assembly with equipment, all in one single shot. The drill floor stays horizontal throughout the procedure.

The new automated and self-elevating Rocket Rig (left). The bull line attached to the hydraulic raising system (right) is used to erect the rig. (Both images by Ann Sherman Photography, provided by Cactus Drilling Co. LLC.)

Marginal U.S. wells still important

A report by the U.S. Interstate Oil and Gas Compact Commission (IOGCC) estimated that marginal wells—producing 10 b/d or fewer of oil, or 60 Mcfd or fewer of gas—remain an important element of U.S. oil and gas production operations. These wells in 2005 accounted for about 17% of oil and 9% of onshore U.S. natural gas production.

OPGCC said that these wells in 2005 produced more than 321 million bbl of crude and 1.76 tcf of natural gas, with average daily production reaching its highest in 10 years.

ERD projects target 40,000 ft and beyond

Operators have used extended-reach drilling around the world since 1995. New ERD projects are driving the development of innovative technologies and changing operating practices.

C02 injection gains momentum

Oil & Gas Journal's biennial enhanced oil recovery (EOR) survey showed that the industry continues to increase the number of carbon-dioxide-injection EOR projects, while steam injection is producing less oil in California and more in the heavy oil and bitumen regions of Western Canada.

US MARGINAL OIL WELLS

Year	Wells active	Oil produced Million bbl/year	Oil produced B/d/well	Wells P&A'd
1996	428,842	323.5	2.06	16,674
1997	420,674	323.5	2.11	15,172
1998	406,380	316.9	2.14	13,912
1999	410,680	315.5	2.10	11,227
2000	411,629	325.9	2.16	10,718
2001	403,459	316.1	2.15	12,234
2002	402,072	323.8	2.21	13,635
2003	393,463	313.7	2.18	14,300
2004	397,362	310.9	2.14	11,977
2005	401,072	321.8	2.20	11,058

US MARGINAL GAS WELLS

Year	Wells active	Gas produced Bcf/year	Gas produced Mcfd/well	Wells P&A'd
1996	168,702	987	16.0	4,671
1997	189,756	1,042	15.0	4,661
1998	199,745	1,104	15.2	4,203
1999	207,766	1,139	15.3	3,546
2000	223,222	1,259	15.4	3,534
2001	234,507	1,354	15.8	3,600
2002	245,961	1,418	15.8	3,870
2003	260,563	1,478	15.5	3,883
2004	271,856	1,539	15.5	4,129
2005	288,898	1,760	16.8	4,517

SELECTED ERD MILESTONES

Operator	Well(s)	Field	Location	Date	Measured depth, ft	Horizontal displacement, ft	ERD ratio*
Exxon Neftegas Ltd. (ENL), for the Sakhalin-1 consortium	Chayvo	Chayvo	Northeast Sakhalin Island, Sea of Okhotsk, Russia	2003-2005	30,758	27,621 30,335	3.59 3.90
Statoil ASA	34/8-A-6 AHT2	Visund	North Sea	2004-2005	29,797	24,911	
Shell Offshore. Inc	A-9	Ram Powell	Viosca Knoll Blocks 956-> 913, Gulf of Mexico	2004	25,400		2.19
China National Offshore Oil Corp. Ltd. (CNOOC)	C02ST1, A4ERW3, and B3ERW4	Liuhua 11-1	South China Sea Block 24/09	2003-2004			4.58
Hibernia Management and Development Co.	B-16 36 (OPA1)	Hibernia Block A	Grand Banks, Newfoundland	2003	30,698	23,728	2.36
BP Amoco PLC and CNOOC	A2ERW1 and D5ERW2	Liuhua 11-1	South China Sea Block 24/09	2002			3.89
RWE-DEA AG	Dieksand 5	Mittelplate	Germany	2000-2001	29,511	26,725	
BP Exploration & Production Inc.			Gulf of Mexico	2000	25,709	22,056	2.61
BP Exploration PLC	M-16Z	Wytch Farm	Dorset, England	1999	37,001	35,196	6.89
ConocoPhillips, Royal Dutch/ Shell, and CNOOC	A-14 (first of 13 ERD wells)	Xijiang 24-3	South China Sea	1997	30,308	26,446	
BP	M-5	Wytch Farm	Dorset, England	1995	28,592	26,361	5.43

*Extended-reach drilling ratio of measured depth to vertical depth; generally greater than 2.
Data reported by operators.

The PT Caltex-operated Duri field on Sumatra Island in Indonesia remains the largest single EOR project in the world, producing about 220,000 b/d.

Petróleos Mexicanos (Pemex) has successfully used immiscible nitrogen injection in the Cantarell field, increasing oil production to more than 2 million b/d in 2004 from 1 million b/d in 1996.

Contracts for five new drillships

Drilling contractors awarded contracts worth $2.5 billion to two Korean shipyards to build five new drillships. Rapidly escalating contract rates suggest that the new capacity will be easily absorbed as the new ships enter the market in 2007–2009.

US EOR Production

	1986	1988	1990	1992	1994	1996	1998	2000	2002	2004	2006
Thermal											
Steam	468,692	455,484	444,137	454,009	415,801	419,349	439,010	417,675	365,717	340,253	286,668
Combustion in situ	10,272	6,525	6,090	4,702	2,520	4,485	4,760	2,781	2,384	1,901	13,260
Hot water	705	2,896	3,985	1,980	250	250	2,200	306	3,360	3,360	1,776
Total thermal	**479,669**	**464,905**	**454,212**	**460,691**	**418,571**	**424,084**	**445,970**	**417,675**	**371,461**	**345,514**	**301,704**
Chemical											
Micellar-polymer	1,403	1,509	617	254	64	0	0	0			
Polymer	15,313	20,992	11,219	1,940	1,828	139	139	1,598			
Caustic/alkaline	185										
Surfactant			20					60	60	60	
Total chemical	**16,901**	**22,501**	**11,856**	**2,194**	**1,892**	**139**	**139**	**1,658**	**60**	**60**	**0**
Gas											
Hydrocarbon miscible/ immiscible	33,767	25,935	55,386	113,072	99,693	96,263	102,053	124,500	95,300	97,300	95,800
CO_2 miscible	28,440	64,192	95,591	144,973	161,486	170,715	179,024	189,493	187,410	205,775	234,420
CO_2 immiscible	1,349	420	95	95				66	66	102	2,698
Nitrogen	18,510	19,050	22,260	22,580	23,050	28,017	28,117	14,700	14,700	14,700	14,700
Flue gas (miscible and immiscible)	26,150	21,400	17,300	11,000	—	—	—	—			
Other				6,300	4,400	4,350	4,350	0	0	0	0
Total gas	**108,216**	**130,997**	**190,632**	**298,020**	**288,629**	**299,345**	**313,544**	**328,759**	**297,476**	**317,877**	**347,618**
Other											
Carbonated waterflood											
Microbial				2	2	0	0	0	0	0	0
Total other	**0**	**0**	**0**	**2**	**2**	**0**	**0**	**0**	**0**	**0**	**0**
Grand total	**604,786**	**618,403**	**656,700**	**760,907**	**709,094**	**723,568**	**759,653**	**748,092**	**668,997**	**663,451**	**649,322**

Active US EOR Projects

	1986	1988	1990	1992	1994	1996	1998	2000	2002	2004	2006
Thermal											
Steam	181	133	137	119	109	105	92	86	55	46	40
Combustion in situ	17	9	8	8	5	8	7	5	6	7	12
Hot water	3	10	9	6	2	2	1	1	4	3	3
Total thermal	**201**	**152**	**154**	**133**	**116**	**115**	**100**	**92**	**65**	**56**	**55**
Chemical											
Micellar-polymer	20	9	5	3	2						
Polymer	178	111	42	44	27	11	10	10	4	4	0
Caustic/alkaline	8	4	2	2	1	1	1				
Surfactant	8	4	2	2	1	1	1				
Total chemical	**206**	**124**	**50**	**49**	**30**	**12**	**11**	**10**	**4**	**4**	**0**
Gas											
Hydrocarbon miscible/ immiscible	26	22	23	25	15	14	11	6	7	8	13
CO_2 miscible	38	49	52	52	54	60	66	63	66	70	80
CO_2 immiscible	28	8	4	2	1	1		1	1	1	2
Nitrogen	9	9	9	7	8	9	10	4	4	4	3
Flue gas (miscible and immiscible)	3	2	3	2							
Other				1	1						
Total gas	**104**	**90**	**91**	**89**	**79**	**84**	**87**	**74**	**78**	**83**	**97**
Other											
Microbial	1			2	1	1	1				
Total other	**1**	**0**	**0**	**2**	**1**	**1**	**1**	**0**	**0**	**0**	**0**
Grand total	**512**	**366**	**295**	**273**	**226**	**212**	**199**	**176**	**147**	**143**	**153**

In addition to the five newbuilds, eight other drillships were undergoing shipyard modifications in early 2006. Two more are cold-stacked, and the majority (28) were drilling under contract.

The utilization rate for the worldwide fleet of 33 drillships was about 82% for the period October 2005–January 2006.

According to Clarkson PLC (London), as of February 2006, South Korean ship builders led the world, accounting for 35% of the global industries' orders:

- Hyundai Heavy Industries Co. Ltd.: 10.82 million compensated gross tons (CGT)
- Daewoo Shipbuilding & Marine Engineering: 7.82 million CGT
- Samsung Heavy Idustries Co. Ltd.: 7.44 million CGT.

Deepwater rig market drivers

Expanding activity around the globe is expected to drive demand for deepwater rig fleets, said OGJ.

- Lease expirations in the Gulf of Mexico
- Drilling successes in Brazil and the influx of international operators there.
- Several new development programs in West Africa
- Rapid expansion of exploration offshore India
- Continuing work in Malaysia, Indonesia, and Australia.

Additional capacity likely will be needed for frontier areas such as deepwater Mexico, the Orphan basin in Eastern Canada, East Africa, Pakistan, and the South China Sea.

Special rigs—such as the Stena DrillMAX ships—will be needed to explore harsh environment and environmentally sensitive areas such as the Barents Sea.

Cantarell production

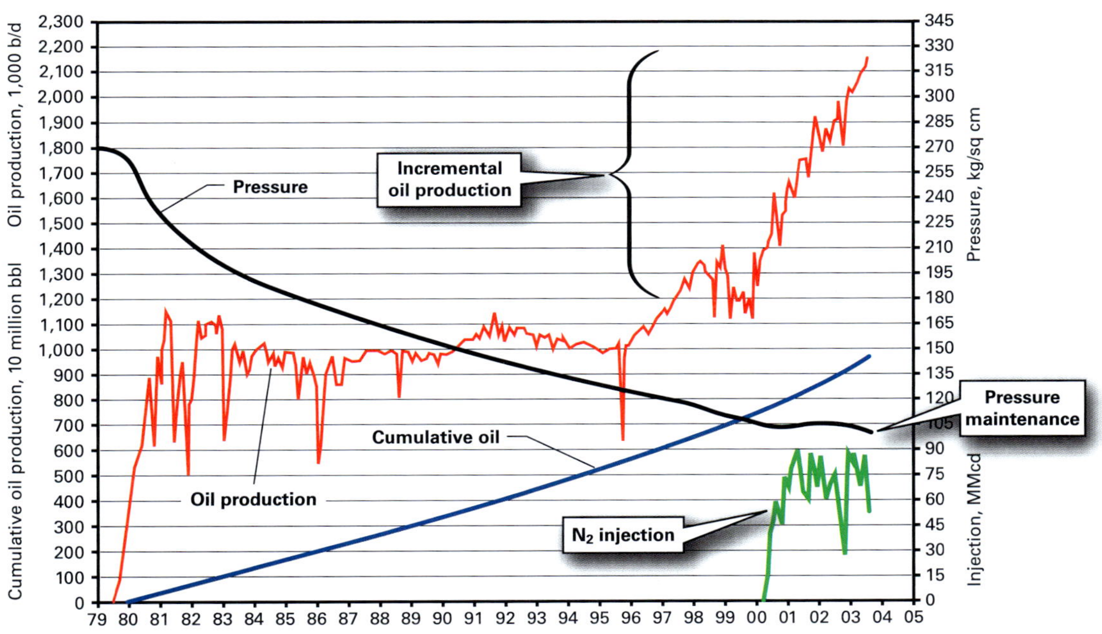

Production	1996	2003
Oil, 1,000 b/d	1,074	2,096
Associated gas, MMcfd	443	770

Drillships under construction, 2006-09

Ship	Owner	Water, drilling depth, ft	Shipyard, country	Delivery date	Cost, $ million (shipyard)
				2007	
Stena DrillMAX 1	Stena Drilling Ltd.	10,000; 36,000	Samsung Heavy Industries Co. Ltd., Geoje Shipyard, South Korea	December 2007	600 (526)
				2008	
Stena DrillMAX 2	Stena Drilling	10,000; 36,000	Samsung	June 2008	600 (550)
Mosvold Drillship TBN 1	SeaDrill Ltd.(Mosvold Drilling Ltd.)	10,000	Samsung	June 2008	(453)
Mosvold Drillship TBN 2	SeaDrill (Mosvold)	10,000	Samsung	December 2008	(464)
				2009	
Discoverer Clear Leader	Transocean Inc.	12,000; 40,000	Daewoo Shipbuilding & Marine Engineering, Okp'o, South Korea	September 2008 or first quarter 2009	650 (470)
Total: 5 ships				**Total shipyard contracts**	**$2.463**

Source: Various owner, operator, and shipyard press releases

Drillships undergoing modification, April 2006

Ship	Owner	Shipyard, country	Year built	Completion date	Cost, $ million
Aban Abraham	Aban Loyd Chiles Offshore Ltd.	Singapore	1976	Mid-2007	180-200
Gazprom I	OAO Gazprom	Malta Shipyards	1998	—	—
GSF Explorer	GlobalSanta Fe Corp.	US	1998	Mid-April	—
Frontier Discoverer	Frontier Drilling USA Inc.	Pan-United Pte. Ltd. shipyard, Singapore	1976	Early 2007	~65
Neptune Discoverer	Neptune Marine Drilling	Pan-United	1977	May 2006	—
Neptune Explorer	Neptune	Pan-United, Batam, Indonesia	1973	Second-half 2006	—
Deep Venture	Premium Drilling Inc.	Puerto Belgrano, Argentina	1981	Third-quarter 2006	290
Songa Saturn	Songa Offshore AS	Tzula shipyard,Turkey	1983	Third-quarter 2006	70
Total: 8 ships under repair					

Source: Rigzone.com; various owner, operator, and shipyard press releases

Section 4:

Investment and markets

Upstream work paces capital spending surge

Worldwide oil and gas industry planned sharp increases in capital expenditures for 2006, especially for upstream work, said *OGJ*.

High oil and gas prices were supporting the investment needed to boost worldwide production capacity. In the United States and outside North America, upstream spending was projected to grow by more than 10% in 2006.

Less money likely would be invested in U.S. downstream operations than during 2005, however, as pipeline construction and refinery upgrades decline.

Oil and gas industry capital expenditures in Canada were expected to reach $35.8 billion in 2006, up 7%. Exploration, drilling, and production outlays would likely account for $26.2 billion of the total. The rest would be allocated to all other industry activity, including oil sands development and pipeline construction.

Outside the United States Canada, construction of LNG facilities and petrochemical complexes were expected to boost investment already lifted by exploration and production. E&P spending in developing countries and high-risk areas was expected to grow in 2006, according to spending plans of many of the largest U.S.-based oil and gas firms.

WHERE FUNDS WILL GO FOR 2006 US PROJECTS

	2006, million $	Change 2006-2005, %	2005, million $	Change 2005-2004, %	2004, million $
Exploration-production					
Drilling-exploration	88,902	11.4	79,791	13.8	70,086
Production	16,891	11.4	15,160	13.8	13,316
OCS lease bonus	1,100	62.0	679	27.2	534
Subtotal	**106,894**	**11.8**	**95,631**	**13.9**	**83,936**
Other					
Refining	6,500	–9.3	7,170	10.0	6,520
Petrochemicals	675	–6.9	725	0.0	725
Marketing	2,225	9.9	2,025	–2.2	2,070
Crude and products pipelines	64	–92.2	828	0.5	824
Natural gas pipelines	2,427	–31.0	3,517	4.4	3,369
Other transportation	850	11.8	760	7.0	710
Mining, other energy	1,100	10.0	1,000	11.1	900
Miscellaneous	3,400	3.0	3,300	–1.5	3,350
Subtotal	**17,241**	**–10.8**	**19,325**	**4.6**	**18,468**
Total	**124,135**	**8.0**	**114,956**	**12.3**	**102,404**

CANADIAN SPENDING PLANS*

	2006, million $	Change 2006-2005, %	2005, million $	Change 2005-2004, %	2004, million $
Exploration-production					
Drilling-exploration	18,260	5.0	17,390.0	12.0	15,528
Production	7,960	5.0	7,578.0	12.0	6,766
Subtotal	**26,220**	**5.0**	**24,968.0**	**12.0**	**22,294**
Other					
Refining	400	5.3	380.0	7.0	355
Petrochemicals	210	–8.7	230.0	–2.1	235
Marketing	635	10.2	576.0	20.0	480
Crude and products pipelines	528	–20.0	659.8	724.7	80
Natural gas pipelines	5	386.4	1.0	–99.5	219
Other transportation	240	9.1	220.0	6.8	206
Mining, other energy	7,000	16.7	6,000.0	16.8	5,138
Miscellaneous	550	14.6	480.0	–8.6	525
Subtotal	**9,567**	**11.9**	**8,546.8**	**18.1**	**7,238**
Total	**35,787**	**6.8**	**33,514.8**	**13.5**	**29,532**

*US dollars.

US DEPARTMENT OF ENERGY'S ANNUAL ENERGY OUTLOOK

Year projected	Reference case crude price projections,[1] $2005/bbl 2010	2020	WTI[2] year average	
2001	24.50	25.70	2000	34.00
2002	26.20	27.70	2001	28.40
2003	26.30	27.90	2002	28.10
2004	26.00	28.00	2003	32.80
2005	26.40	30.10	2004	42.60
2006	45.20	46.25	2005	56.60

[1]Refiner acquisition cost of imported crude adjusted to $2005, using the GDP deflator.
[2]West Texas Intermediate benchmark crude.
Source: US Department of Energy Annual Energy Outlook, Petroleum Industry Research Foundation Inc.

Crude oil price projections challenge producers

High oil and gas prices and record profits are generating complex challenges for producers, said a report from the Petroleum Industry Research Foundation Inc. (Pirinc; New York).

The organization pointed to the U.S. Dept. of Energy's 2005 year-end Annual Energy Outlook (AEO) that forecasts crude price projections for 2010 and 2020.

"The evolution of prices corresponds broadly to the shifts in the long-term price structure of the futures market over the same period," said Pirinc. Though changes were small during 2001–2004, the projection at yearend 2005 has substantial price variations. Projected prices are well below current near-term levels, but "since [early 2004], the futures market has been signaling increasing confidence that high prices will persist," Pirinc said.

Section 5:

Trade and tankers

Oil trade movements

Internationally traded volumes of oil totaled about 49.9 million b/d (2,462 mtpy) in 2005, said BP PLC.

For detailed data, please refer to Tables 6 and 7 in the Statistical Tables section.

Gas trade movements

Internationally traded volumes of gas totaled about 722 bcm in 2005. Gas moved by pipeline accounted for 74 % (533 bcm), while gas traded as LNG represented the other 24% (189 bcm).

For detailed data, please refer to Table 15 in the Statistical Tables section.

Design approved for CNG carriers

The American Bureau of Shipping issued full-class design approval for compressed natural gas (CNG) carriers to Sea NG Corp. (Calgary). The approval is the first granted to CNG carriers, ABS said.

Sea NG's proprietary Coselle system uses small-diameter, high-strength steel pipe coiled into a carousel. Each Coselle will comprise about 10 miles of 6-in. pipe for storing and transporting CNG.

Sea NG is expected to let contracts for construction of three Coselle CNG carriers by early 2007. These first vessels, which will carry the notation "A1 CNG Carrier," will be designed to transport gas 320–3,200 km (200–2,000 miles). They are intended to service projects in the Caribbean or Mediterranean seas.

The vessels will be 120 m (394 ft) long and have the capacity to carry 50 MMscf of gas in 16 Coselles (160 miles of pipe).

China seeks oil security with new tanker fleet

Gabe Collins, research fellow at the U.S. Naval War College (Newport, RI) shared his analysis of China's plans to build a national tanker fleet.

SINO-US SHARE OF TOTAL GLOBAL OIL DEMAND

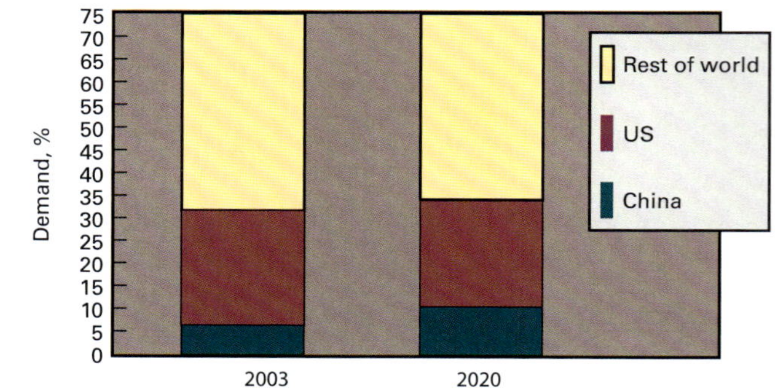

LICENSED CHINESE OIL SHIPPERS

State-owned[1]	Private
• Cosco (Dalian) Shipyard Co. Ltd. (3 VLCCs[2])	• Dalian Haichang Group (1 VLCC, 3 on order)
• China Shipping (Group) Co. (2 VLCCs)	• China Merchants Group (7 VLCCs)
• China Changjiang National Shipping Group (1 VLCC, 10 on order[3])	
• Sinotrans Ltd. (4 ULCCs)	
• Hebei Ocean Shipping Co. (3 VLCCs)	

[1]State-owned shippers control almost two thirds of Chinese VLCCs, and this share is expected to grow by 2010.
[2]Very large crude carriers. [3]Most VLCCs on order are due to be delivered by 2010.

CHINESE SHIPYARDS' PROPORTION OF GLOBAL TANKER ORDERS

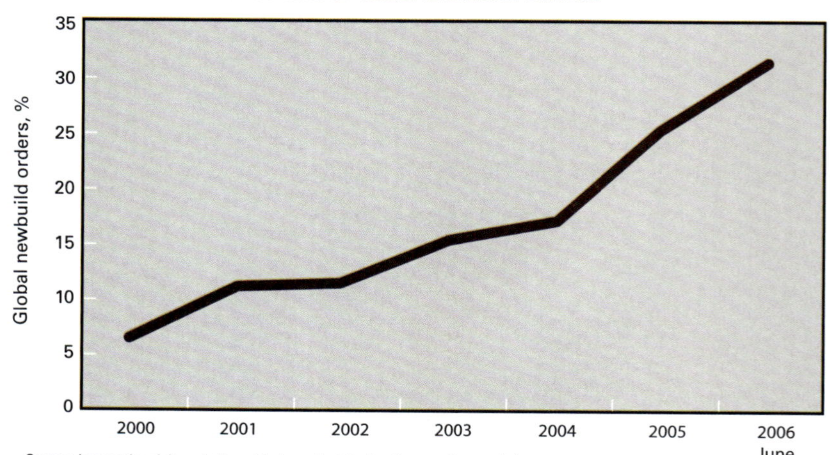

Source: International Association of Independent Tanker Owners (Intertanko)

His views, summarized here, are his own, and do not reflect the official views and policies of the U.S. Navy or any other U.S. government entity.

China aims to haul nearly three-fourths of its oil imports within the next 15 years with the new vessels, apparently driven by a complex mix of commercial and national security factors.

Paradoxically, Collins said the buildup of a state-controlled, Chinese-flagged tanker fleet may actually increase that country's vulnerability to energy supply interdiction.

China's percentage of world oil consumption is set to more than double in the next 15 years, and combined U.S. and Chinese oil requirements will represent 35% of total global demand by 2020.

By 2010, China intends to transport 40–50% of its oil imports in government-owned tankers. By 2020, it hopes to carry 60–70%. Chinese analysts predict that by 2010 the country will need as many as 40 very large crude carriers (VLCCs) and a fleet capable of transporting more than 700 million bbl/year.

China currently owns 21 VLCCs that cumulatively account for about 30% of its tanker tonnage. Most of the remaining vessels are small or old tankers better suited for the coastal trade than for international oil carriage.

China is one of the world's leading tanker builders. The Chinese shipyards' share of new global tanker contracts has grown at an average annual rate of roughly 30% from 2003 to 2006.

Two primary national security factors drive Beijing's tanker fleet construction: insufficient overland oil supplies and the fear that terrorists, pirates, or the U.S. Navy could bottleneck its sea lanes. Overland oil imports from Russia and Kazakhstan likely will continue to grow but will not solve China's dependence on seaborne oil.

China also worries about its present inability to militarily protect the sea lines of communication and choke points through which much of its oil imports flow, especially the Hormuz and Malacca straits.

A clear indicator that China's tanker fleet expansion is at least partially driven by commercial interests is that a significant number of VLCCs are being built in China for foreign operators. This fact also appears to be a vote of confidence in China's shipbuilding abilities.

Seven Chinese companies now possess government-issued crude oil shipping licenses, and five are state-controlled. Like China's state oil producers, the tanker companies compete among themselves yet are collectively expected to act in the Chinese national interest.

CHINA'S OVERLAND VS. MARITIME IMPORTS

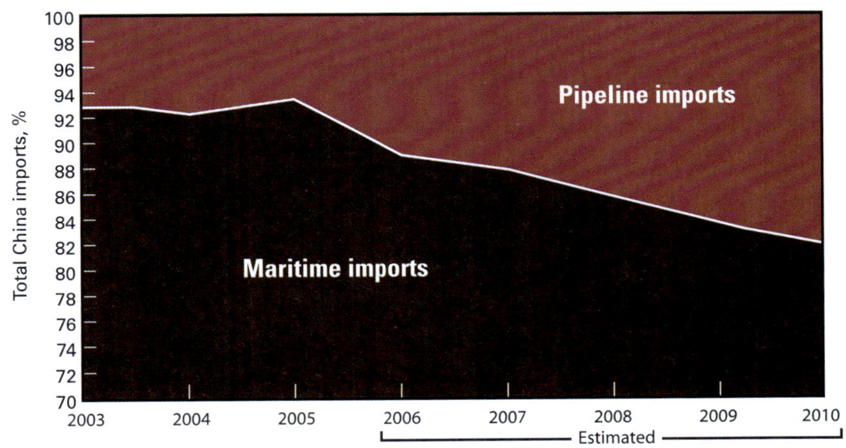

Sources: People's Daily Energy Intelligence, Asia Pulse, Prime-Tass

GLOBAL CRUDE OIL SHIPMENT CHOKE POINTS

Potential choke point	Volume shipped, million b/d
Bab al-Mandab	3.3
Bosporus	2.0
Hormuz Straits	17.0
Malacca Straits	*11.7
Panama (pipeline + canal)	0.6
Suez (pipeline + canal)	3.8
Total choke-point volumes	**38.4**
Percentage of total seaborne oil flow	75%

*At least 2.2 million b/d to China, or two thirds of Chinese oil imports.
Source: Energy Information Administration

Some Chinese analysts have outlined potential structures for a national tanker fleet.

Global energy security can improve, concluded Collins, if China overcomes its mercantilist inclinations and relies on world oil market to fulfill its oil and oil transportation needs.

He said there may be value in allowing China to join the International Energy Agency membership, advising the country as it strives for a secure, sustainable energy future.

In the meantime, he said, China's growing oil tanker fleet stands to substantially affect the global oil transportation sector.

Chinese foreign tanker construction deals*

Company	Customer	No. of tankers	Deal value, million $
China State Shipbuilding Corp. (CSSC) and China Shipbuilding Industry Corp. (CSIC)	PDVSA (Venezuela)	18	1,300
CSIC (Bohai Shipbuilding Heavy Industry)	BW Group (Norway)	2 VLCCs	240 (est.)
CSSC	Frontline (Norway)	4 VLCCs	440
Cosco (Dalian) Shipyard Co. Ltd.	Frontline	2 VLCCs	220 (est.)
Cosco, Kawasaki Heavy Industries	Sonatrach (Algeria)	1 VLCC	120

*Selected deals. Total value of non-Chinese deals = $2.4 billion.

Possible Chinese tanker fleet structures

Solution	Advantages to China	Disadvantages
Chinese oil importers and tanker operators reach exclusive long-term lease agreements[1]	Keeps entire oil supply chain under Chinese control. Reduces net capital outflow from China.	International tanker companies may offer more professional and reliable tanker services
Create a tanker shipping pool and reach a long-term agreement with oil importers	A shipping pool would increase shippers' negotiating power and reduce financial and operating risks[2]	Establishing a national shipping cartel could raise World Trade Organization compliance issues

[1]"Why Doesn't China have a National Oil Tanker Fleet?" Water Transportation Digest: Shipping Survey, September 2005, pp. 24-25. [2]Xiguang, Luo, and Zoujun, Liu, "The Step of Creating a United Chinese Oil Shipping Enterprise," Water Shipping Management (Shuiyun Guanli), Vol. 27, No. 10, October 2005, pp. 9-11.

Shell: Growth in gas trade poses challenges

At the World Gas Conference in June 2006 (Amsterdam), Royal Dutch Shell PLC shared its perspectives on a range of issues facing the gas industry.

- Global gas consumption will grow by more than 2%/year, exceeding projections by the International Energy Agency for overall energy demand growth of about 1.5%/year.
- LNG demand will increase by as much as 10%/year through 2020. The company's target for its own LNG business is 14%/year growth from 2004 to 2009.
- Cost competitiveness is driving LNG expansion. Engineering, procurement, and construction for the recently completed Qalhat LNG train in Oman cost less than $150/tonne/year—"probably the lowest ever."
- Gas-to-liquids (GTL) technology will see wider use. (Shell began construction on its Pearl GTL plant in Qatar in third quarter 2006.) The IEA projects that GTL "could account for 2% of world oil supply by 2030."
- High gas prices could prompt power investors to consider alternative fuels.
- Larger LNG trains and carriers will continue cutting the cost of moving gas over great distances.
- Key technical challenges include development of coal and shale gas and construction of long-distance pipelines in harsh or subsea environments.
- Political challenges include gaining assurance from governments of a "transparent, predictable and secure framework" that allows customers and "suppliers to invest and customers to make commitments."

Decline in tanker rates seen

Tanker rates are expected to decline in 2007–08 because the worldwide tanker fleet will grow faster than originally assumed, said Simmons & Co. International (Houston).

In the short term, tanker rate support will come from Nigerian crude oil supply disruptions and from changing fuel specifications in the United States. Outages in Nigeria bode well for relative strength in very large crude carrier (VLCC) rates, the company said, as Middle East tonnage replaces West African volumes (primarily a Suezmax trade).

China's entry into the tanker-building market is expected to reduce asset prices through 2007–2008, with an expected decline of 15%/year for two years. Tanker orders will slow as existing yards "grow wary at emerging Chinese capacity," said Simmons.

Oil shipping rates expected to fall

Freight rates for oil tankers will decline further in 2006 due to excess shipping capacity, several analysts told delegates at an industry forum in early 2006.

Clarksons Research Services said rates will decline in 2006 as owners spent $20.7 billion on new ships in 2005, marking the third successive year of heavy ordering.

The company said supply is set to continue to grow by 6% over the 2007–2009 period, but oil trade is expected to grow by just 1.5 million b/d/year until 2010—about 500,000 b/d less than required to meet the fleet additions. In 2006 alone, Clarksons saw 12.1 million dwt of new ships over the demand balance.

RS Platou said the market was declining, given that the peak of fourth quarter 2005 was much lower than the peak of fourth quarter 2004. The company said freight rates would decline in 2006–2007 as additional oil production and refining capacity would lead to demand growth of just 2–4%, while fleet growth would be higher at 5–6%.

Section 6:

LNG—Industry outlook

CERA: LNG growth exceeds expectations

In a market growing faster than had been expected, LNG will meet 15% of global gas demand by 2012, according to an analysis by Cambridge Energy Research Associates (CERA).

CERA now expects the LNG industry to double in six to seven years from 2004. The conclusion came from their report, "Progress in the Face of Adversity."

More than 25 mtpy of liquefaction capacity has been commissioned since October 2005, adding 18% to global capacity. For 2006, LNG trade was projected to grow to 159 mtpy, up 12% from 2005, with U.S. imports rising by almost 25%.

Based on projects currently under construction alone, global LNG supply capacity will rise 60% by 2012 or earlier. Almost half of the global construction is planned for Ras Laffan Industrial City, Qatar.

LNG as a pipeline alternative

According to *Oil & Gas Journal*, the Russian-Ukrainian pipeline-based natural gas dispute of early 2006 is likely to boost European LNG terminal projects to the degree they provide access to alternative supplies.

European LNG terminal projects, outside Italy

Country	Terminal	Capacity, million tpy	Start-up	Ownership
UK	Canvey Island —Essex	4.0	2010	Calor, LNG Japan Corp
Netherlands	Rotterdam	4.4	end of 2009	Petroplus International BV
Netherlands	Eemshaven	4.4	—	NV Nederlandse Gasunie, Royal Vopak NV
Netherlands	Rotterdam	—	—	ConocoPhillips, Essent Energie BV
Germany	Wilhelmshaven	7.4	2010	E.on Rurhgas AG
Poland	Gdansk	3.7	—	PGNiG SA
Croatia	Krk	5.9	—	Adria LNG

Asian LNG imports supplies, January-September 2005

Source	Alaska	Australia	Brunei	Indonesia	Malaysia	Abu Dhabi	Qatar	Oman	Egypt
	%								
Importer									
Japan	2.1	17.8	10.9	25.0	22.8	8.6	11.0	1.8	
Korea		2.0	1.6	24.8	22.9	0.4	27.5	20.2	0.7
Taiwan		7.9		50.6	38.0			3.4	
India		1.9					96.2	1.9	

Poland and Germany in particular are likely to push for the construction of LNG terminals to reduce their dependence on Russian natural gas, despite the higher costs involved.

LNG SHARE OF NATURAL GAS PORTFOLIO

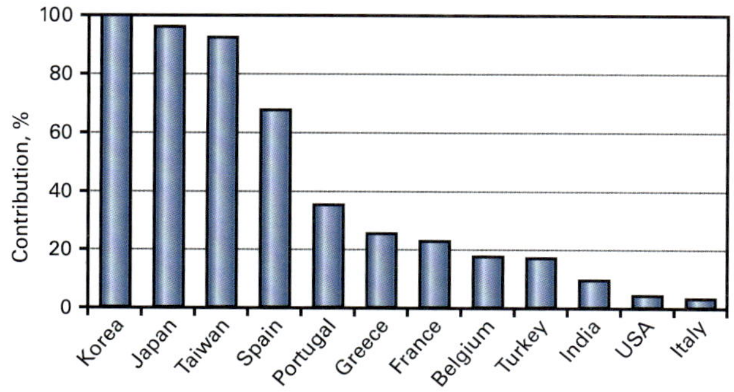

Source: BP Statistical Review of World Energy, 2005.

NORTH AMERICAN IMPORT CAPACITY*

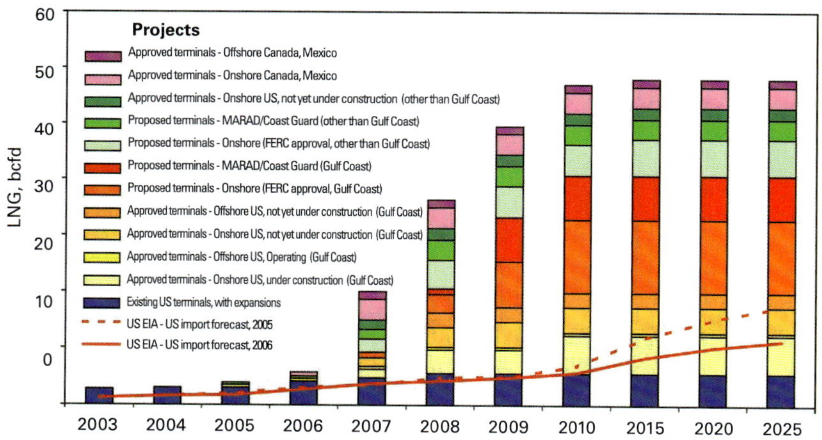

*CEE outlook. Based on agency pre-filings, approvals, and industry information as of Dec. 31, 2005.

REGIONAL LNG PRODUCTION

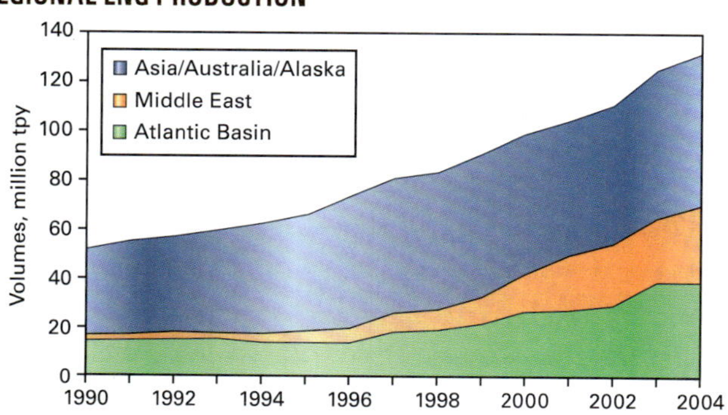

Source: CEE/UT - Austin

The dispute will likely have a long-term influence on supply strategies, going beyond European frontiers and encouraging construction of LNG receiving terminals. It could also provide an opportunity to increase LNG flows from Asia to Europe.

The Asia-Pacific region is generally less at risk than Europe to supply disruptions because of the dominant role played by LNG in the region.

North American receiving capacity

The United States has long been able to satisfy natural gas demand largely with domestic production, supplemented in recent years by pipeline imports from Canada (16% of delivered supply) and, more recently, by LNG imports (2–3%).

The Center for Energy Economics (CEE), part of the Bureau of Economic Geology at the University of Texas-Austin, said interest in LNG is growing.

The resurgence is driven by maturing U.S. and Canadian natural gas fields, burgeoning demand growth in Mexico, and underinvestment in oil and gas exploration and production in that country.

Several new LNG import terminals are under development, and three of the four existing land-based terminals are expanding.

As of early 2006, about 50 new projects throughout North America had been proposed, planned, were moving through the regulatory process, or were beginning to be built, said CEE.

It appears that more than enough LNG import capacity to meet projected demand is in place or on the drawing board.

Customers of existing projects in North America have enjoyed a rich array of supply sources from both the Atlantic and Pacific basins.

The total liquefaction capacity of current operating trains worldwide totaled about 151 mtpy as of mid-2005, compared to 132 mtpy produced in 2004.

Gazprom LNG plans

OAO Gazprom plans to become a major LNG supplier and is building its LNG expertise, particularly in North American markets, according to the company's affiliate, Gazprom Marketing & Trading Ltd. (GM&T; London). The company plans to partner with a number of major LNG players.

Gazprom supplies gas by pipeline to 21 countries in Western Europe, and in 2005 sold one LNG cargo—resulting from a swap with Gaz de France—in the United States. The company views LNG as a way to more easily commercialize some of Russia's abundant gas reserves.

GM&T said it intends "to participate in all elements of the value chain," from production and liquefaction to shipment and regasification.

In one planned LNG project, Gazprom would liquefy gas in Russia from Shtokman gas and condensate field awaiting development in the Barents Sea. The field is too remote for the gas to be sold by pipeline. Plans call for development in three phases of 15 mtpy each. Start-up of the first phase is planned for 2012.

Another project is Baltic LNG, which would liquefy—in a 3- to 5-mtpy plant near St. Petersburg—700 MMcfd of the 55 bcfd Gazprom transports through the Unified Gas System.

The primary target market would be North America, although there would be access to the UK or Western Europe. Baltic LNG gas could start moving to markets in 2009–2010.

Global LNG liquefaction capacity, June 2005

Region	Operating	Under construction	Planned	Total
	Million tpy			
Pacific Basin	70.8	20.8	75.8	167.4
Middle East	33.6	28.7	67.7	130.0
Atlantic Basin	46.3	32.3	101.0	179.6
Total	**150.7**	**81.8**	**244.5**	**477.0**

Source: CEE/UT-Austin

Nigerian gas exports

Project	Capacity, million tpy
NLNG Trains 1, 2, 3	10.0
NLNG Trains 4, 5, 6	12.0
NLNG Trains 7, 8	16.0
Brass LNG	10.0
OK LNG Trains 1, 2, 3, 4	20.0
ExxonMobil	4.8
Total	**72.8**

Source: CEE/UT-Austin

Looking at worldwide LNG supply and demand, a reasonably good balance is projected through 2020, assuming that some of the planned capacity additions come to fruition toward the end of that time frame.

LNG projects in Nigeria

Nigeria is poised to become a key LNG exporting country. Several existing and new LNG projects by Brass LNG and others, taken together, will contribute more than 70 million tonnes to worldwide LNG supply.

The country's 176 tcf of proved reserves of natural gas are estimated to be more than adequate to all announced LNG plants (more than 72 million tpy).

Construction-cost challenge

A late-2005 study by the New York-based shipping consultant Poten & Partners Inc. (New York) sees problems ahead for the LNG industry, particularly related to construction costs.

The study points out that before 2004, technology and better economies of scale consistently pushed liquefaction plant construction costs down—to near $200/tonne of capacity in some instances.

But some recent engineering, procurement, and construction (EPC) contracts were hovering near $350/tonne of capacity. Despite the undoubted economies of scale of the newer mega-trains, other factors are driving up their unit construction costs as well.

The study cites cost data from contracts awarded between mid-2004 and the end of 2005 for such projects as RasGas III, Yemen LNG, Tangguh, Sakhalin II and Snøhvit. The study conceded that the problems for the last two are probably more project-specific.

Worldwide LNG supply-demand projection*

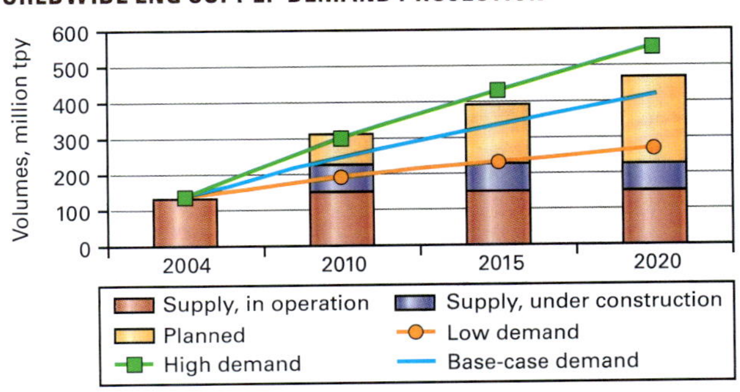

*As of June 2005.
Source: CEE/UT - Austin

The main culprit has been sharp cost increases since 2003 for many raw materials used in LNG projects, notably steel, nickel (used in cryogenic and stainless grades of steel), and cement.

The price of nickel hit more than $17,000/ton midway through 2005, from $7,000/ton in early 2003—then fell back to "only" $12,000/ton by year-end 2005.

The study also pointed to rising labor costs as contractors vie for the skilled workers needed for LNG projects. The demand for talent will grow as some EPC contractors now envision working on as many as 10 LNG trains that are scheduling completion by 2009.

The Poten study points to a forecast in early 2005 by engineering consultants Merlin Associates that said that LNG inflation through 2010 would run between 7.5% and 10%/year. Developers will need to decide whether to eat production-cost increases to preserve markets or pass them along and risk demand destruction.

Nigeria Projects

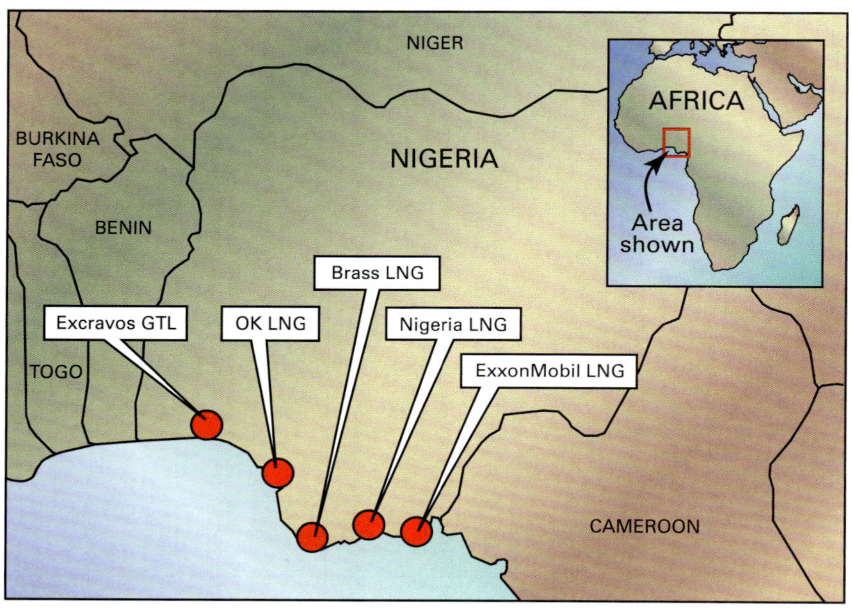

Source: CEE/UT - Austin

Section 7:

LNG—Liquefaction and receiving

Liquefaction plants

Just under 195 mtpy of liquefaction capacity was being built or planned as of mid-2006, according to *LNG Observer*.

New liquefaction construction

Leena Koottungal, Survey Editor, Oil & Gas Journal

Country	Operator	Location	Capacity, million tpy	Status	Completion	Contractor	Notes
Algeria	Repsol YPF/ Gas Natural/ Sonatrach	Arzew	5.2	Planning	2009		New.
	Sonatrach	Skikda	4.5	Engineering	2010	JGC/KBR/SNC-Lavalin/Black & Veatch	New. $700 million. Replaces plant destroyed in 2004.
Angola	Angola LNG Ltd.	Soyo	5.0	Engineering	2009	Bechtel/KBR/JGC/Technip	One train. $2 billion. JV Chevron Corp., BP PLC, Exxonmobil Corp., Total SA.
Australia	Gorgon LNG	Barrow Island	5.0	Engineering	2009	KBR/JGC/Clough/Hatch/ JGC	New. Two trains. JV of Chevron Corp. (57.1%), Shell (28.6%), ExxonMobil (14.3%).
	Woodside Energy Ltd.	Withnell Bay	4.5	Under constr.	2008	Foster Wheeler/Worley Parsons/CB&I	Expansion. Train 5. $2 billion.
Egypt	Egyptian LNG	Idku	3.3	Under constr.	2007		Expansion. Train 3.
Equatorial Guinea	Marathon Oil	Bioko Island	3.4	Under constr.	2008	Bechtel	New. $1.4 billion. Export to BG, Lake Charles, La.
	Marathon Oil	Bioko Island	4.4	Engineering		Bechtel	Expansion. Train 2. Decision to proceed will be made in 2007. Marathon (60%), Sonagas (25%), Mitsui (8.5%), Marubeni (6.5%).
Indonesia	BP Tangguh	Berau Bay, Papua	7.6	Under constr.	July 2008	KBR/JGC Corp./Wood Group Indonesia	New. Two trains. $1.4 billion.
Libya	National Oil Corp.	Marsa Al-Brega	3.2	Planning	2008		Expansion. $400 million. Shell and NOC.
Nigeria	Brass LNG Ltd.	Brass Terminal	10.0	Engineering	2009	Bechtel	New. $3 billion. NNPC, ConocoPhillips, Total SA, ENI.
	Nigeria LNG Ltd.	Bonny Island	4.0	Under constr.	2007	KBR/JGC Corp./Technip/ Snamprogetti SPA	Expansion. Train 6.
	Nigeria LNG Ltd.	Bonny Island	8.0	Planning	2010		Expansion. Train 7.
	Nigeria LNG Ltd.	Bonny Island	8.0	Planning			Expansion. Train 8.
	Olokola LNG	West Niger Delta	20.0	Planning	2010	KBR	New. Four trains. Chevron (18.5%), BG (13.5%), NNPC (49.5%), Shell (18.5%).

New liquefaction construction (cont.)

Country	Operator	Location	Capacity, million tpy	Status	Completion	Contractor	Notes
Norway	Statoil ASA	Snøhvit, Hammerfest, Melkøya Island	4.1	Under constr.	2007	Linde AG/Statoil	New.
Peru	Peru LNG Co.	Pampa Melchorita	4.4	Planning	2009	KBR	New. $1.6 billion. Sales gas to Repsol YPF. Hunt Oil.
Qatar	QatarGas II	Ras Laffan	15.6	Under constr.	2007-08	Chiyoda Corp./Technip	New. Two trains: 7.8 million tpy/train. To UK. Qatar Petroleum/ExxonMobil Corp.
	QatarGas III	Ras Laffan	7.8	Engineering	2009	Chiyoda Corp./Technip	New. $4 billion. Qatar Petroleum (68.5%), ConocoPhillips (30%), Mitsui (1.5%).
	QatarGas IV	Ras Laffan	7.8	Engineering	2010-11	Chiyoda Corp./Technip	New. Qatar Petroleum (70%), Shell (30%).
	RasGas II	Ras Laffan	4.7	Under constr.	2007	Chiyoda Corp./Snamprogetti SPA	Expansion. Train 5 to export to Europe.
	RasGas III	Ras Laffan	15.6	Under constr.	2008-09	Chiyoda Corp./Technip/ Snamprogetti SPA	Expansion. Two trains: 7.8 million tpy/train. Train 6: Aug. 2008; Train 7: Aug 2009. Train 6 to export to US. Qatar Petroleum/ExxonMobil.
Russia	Repsol/Anadarko/ Tambeineftegaz	Yamal Peninsula	10	Planning	2010		New.
	Sakhalin Energy Investment Co. Ltd.	Prigorodnoye, Sakhalin	9.6	Under constr.	2007-08	Chiyoda Corp./Toyo Engineering	New. Two trains: 4.8 million tpy/train. Train 1: July 2007; Train 2: Feb. 2008. Export to South Korea, Japan, among others. Consortium of Shell, Mitsui, Mitsubishi.
Trinidad & Tobago	Atlantic LNG Ltd.	Point Fortin	3.0	Planning	2009		New. Train 5.
	Atlantic LNG Ltd.	Point Fortin	3.0	Planning	2010		New. Train 6.
Venezuela	PDVSA	Gran Mariscal de Ayacucho, Sucre	4.7	Planning	2010		New. $2.7 billion.
	PDVSA	Jose, Anzoategui	2.1	Planning	2007		New. $600 million.
Yemen	Total SA Yemen LNG	Bal Haf	6.2	Under Constr.	2009	Technip/Bechtel/KBR/JGC	New. Two trains. Export to South Korea, France, Belgium.

LNG receiving terminals

As of December 15, 2006, 15 facilities for importing LNG were under construction worldwide, according to *LNG Observer.*

World's first fixed offshore LNG terminal passes milestone

Construction of the storage tanks for the world's first fixed offshore LNG terminal was completed in late 2006, and the tanks were loaded out by Hyundai Heavy Industries (HHI) in early November at Algeciras, Spain.

Each tank weighs 4,800 tonnes, measures 33 m wide, 28 m high, and 155 m long (108 by 90 by 508 ft), and is capable of storing 250,000 cu m (8.83 MMcf) of LNG. The overall project is more than 50% complete and will commence operations in 2008.

The terminal will consist of a concrete gravity-based structure containing inert ballasting material and placed on the seabed about 15 km (9 miles) off Italy in water about 30 m (100 ft) deep, according to Adriatic LNG. The two LNG storage tanks are made of 9% nickel steel and will be contained inside the GBS. The tanks are protected with high- resistance concrete double walls with inert material between the two walls.

LNG terminals: Newly operating in 2006 or under construction*

Terminal name, location	Owner	Start up	Receiving capacity, million tpy (expansion in parentheses)	Sources	EPC contractor
Americas					
Bear Head, NS, Canada	Anadarko	Delayed	7.5		CB&I
Canaport LNG, St. John, NB	Irving Oil, Repsol	2008	7.5		SNC-Lavalin, Saipem
Cameron LNG, Hackberry, La.	Sempra, ENI	2008	11.8		Aker Kvaerner, IHI
Freeport LNG, Freeport, Tex.	Freeport LNG Investments 45%, Cheniere Energy 30%, Dow Chemical 15%, Contango Oil & Gas 10%	2008	11.8		Technip, Saipem, Zachry
Golden Pass, Sabine Pass, Tex.	ExxonMobil 15%, ConocoPhillips 15%, QP 70%	2009	15.6	Qatar?	CB&I
Cheniere LNG, Sabine Pass, Tex.	Cheniere; capacity reserved by Total and Chevron	2008	20.4		Bechtel
Energía Costa Azul, México	Sempra. Shell has half of capacity.	2008	7.7	Indonesia, Russia	Techint SA, Black & Veatch, Mitsubishi, Vinci
Quintero, Chile	ENAP, BG	Late 2008	7.5		NA
Asia					
Fujian, China	CNOOC, Fujian Investment	2008-09	2.5 (2.5)	Indonesia	Sofregaz-Technigaz
Zhong Ximentang Island, Zhejiang Province, China	Shanghai LNG Co. Ltd.	2009	3		
Taichung, Taiwan	Chinese Petroleum Corp.	2009	1.7		IHI, Toa, CTCI
Dabhol, India	Ratnagiri Gas & Power	2007	2.1	Qatar (RasGas)	Whessoe, Punj Lloyd, Aker Kvaerner
Europe					
Fos-sur-Mer, France	Gaz de France	Mid-2007	6	Egypt, Nigeria, Algeria	Saipem, Sofregaz
Isola di Porto Levante, Italy (offshore)	QP, ExxonMobil, Edison	End 2007	5.8	Qatar (RasGas)	Aker Kvaerner
South Hook LNG, Milford Haven, UK	QP 67.5% , ExxonMobil 24.15%, Total 8.35%	2007	7.8 (15.6)	Qatar (Qatargas II)	CB&I
Dragon LNG, Milford Haven, UK	BG 50%, Petronas 30%, Petroplus 20%	2007	4.4 (6.6)	Trinidad, Egypt	Whessoe Oil & Gas Ltd. Volker Stevin

*As of Dec. 15, 2006. Source: GTI, Des Plaines, Ill.

Massive tanks for Adriatic LNG roll out at their fabrication base in Algeciras, Spain, late in 2006. (Photograph from Hyundai Heavy Industries.)

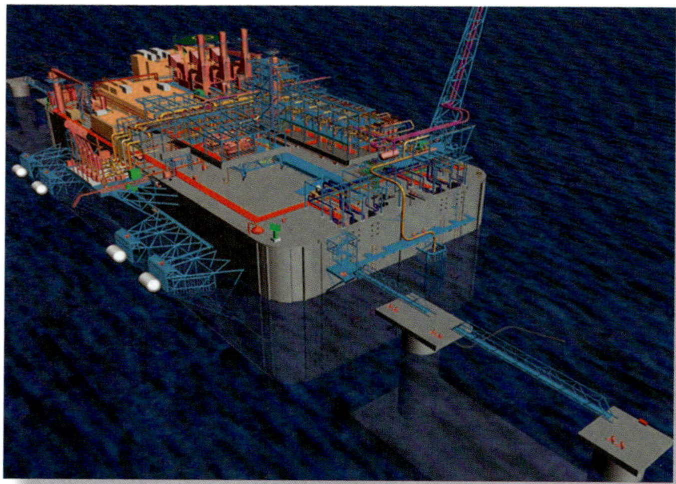

GBS STRUCTURE

Future LNG regasification capacity

Science Applications International Corp. (SAIC; San Diego, CA, U.S.A.) estimated expected LNG regasification capacity additions through 2012, using their World LNG (WLNG) database model.

Atlantic Coast America (including Mexico) shows the biggest rise in annual import capacity to more than 3 tcf/year by 2012. Most of the capacity likely will be added through new Gulf Coast terminals, although SAIC expects that one terminal could still be built to serve Florida and the U.S. Northeast through Canada.

CUMULATIVE LNG REGASIFICATION CAPACITY ADDITIONS

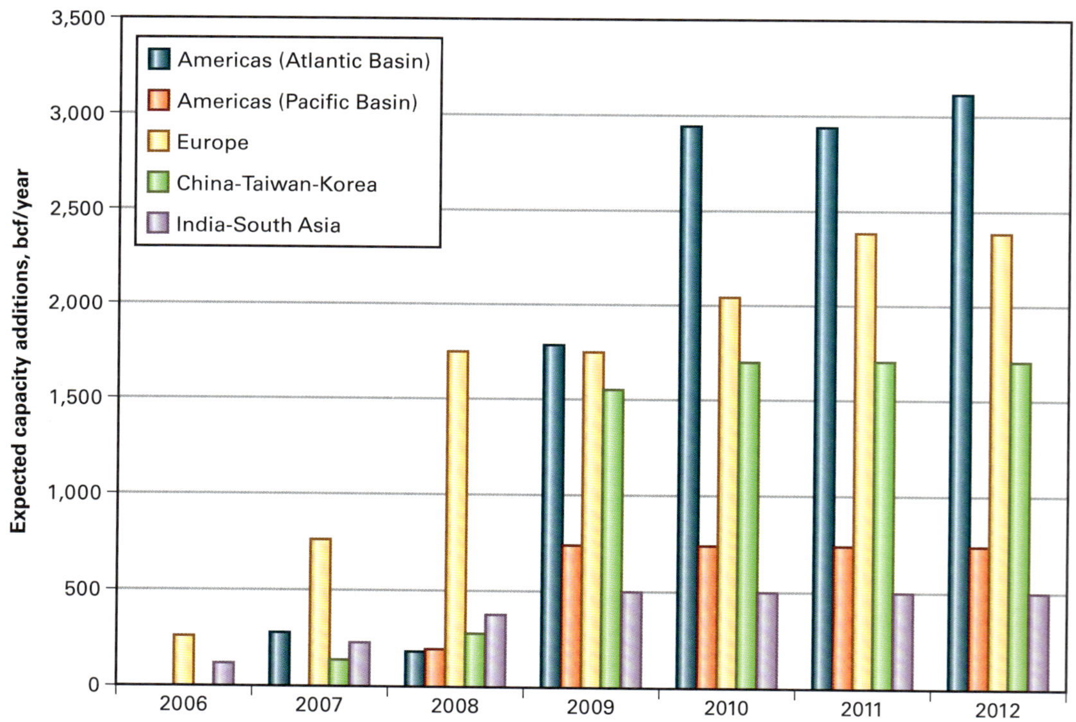

Source: SAIC's WLNG model

Section 8:

LNG—Carrier fleet

Carrier status

More than 130 new LNG tankers were under construction or on order as of January 1, 2007, and due for delivery through 2010, said *LNG Observer*.

Vessels on order by shipyard

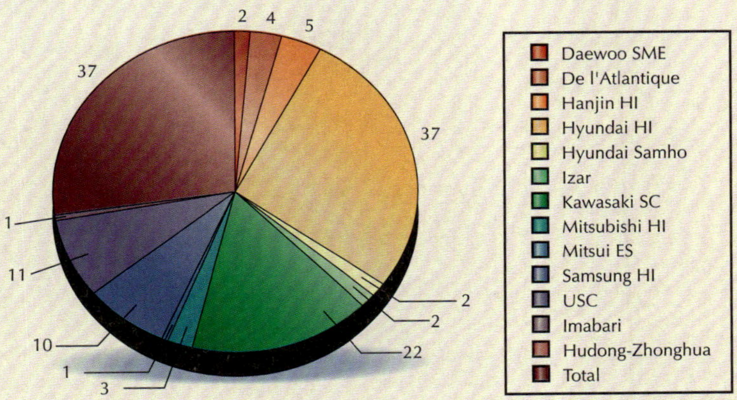

Orderbook by containment system

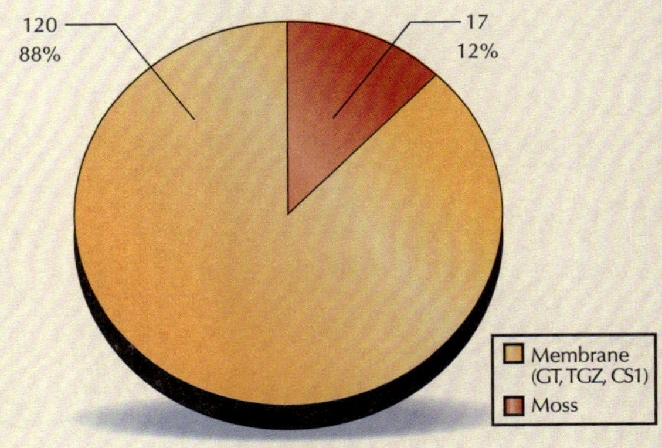

LNG tankers: present and future*

Fleet by containment system

	Total	On order	Existing
Worms/Gaz de France	1	—	1
Esso	3	—	3
IHI SPB Prismatic	2	—	2
GT	126	49	77
TGZ	106	68	38
Moss	111	16	95
CS1	3	2	1
Total	352	135	217
Summary			
Others	6	—	6
Membrane (GT,TGZ,CS1)	235	120	115
Moss	111	17	94
Total	**352**	**135**	**217**

Chartering status including orderbook

Status	Total	On order	Delivered
On charter	327	113	214
Open	25	22	3

Propulsion system

	Total	Existing	On order
ST	276	214	62
DFDE	35	1	34
SSDR	37	—	37
Total	**348**	**217**	**131**

Fleet numbers

		Count
Existing fleet	Delivered	217
Under construction	On order	135
Total		**352**

Delivery schedule

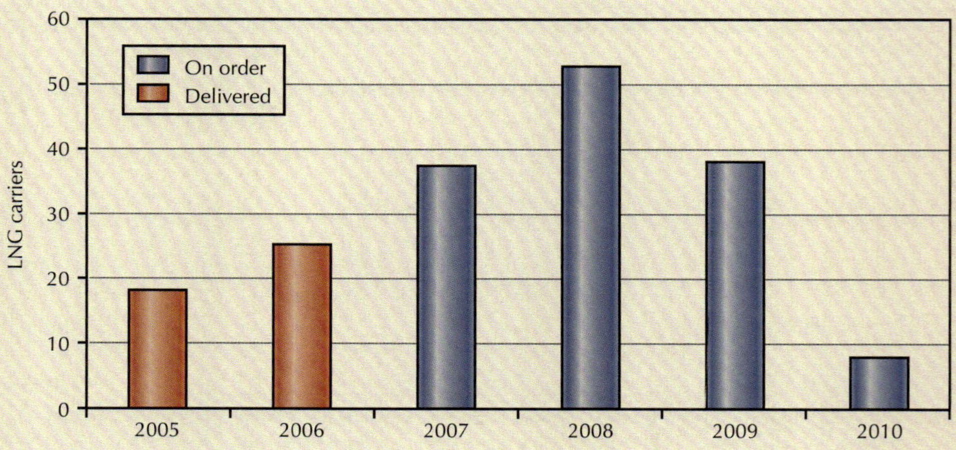

LNG tankers: present and future* (cont.)

Orders placed

Year	2002	2003	2004	2005	2006
	17	15	70	40	35

Vessels on order by region

Korea	Japan	Europe	China	Total
100	27	3	5	135
74%	20%	2%	4%	

Delivered fleet by containment system

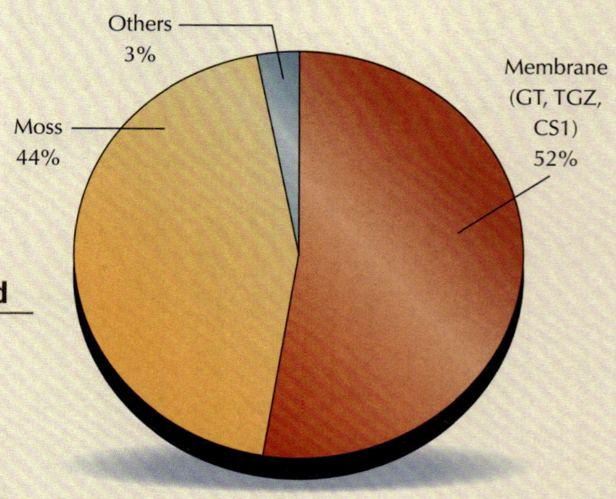

Vessels on order by shipyard

Daewoo SME	36
De l'Atlantique	2
Hanjin HI	2
Hyundai HI	22
Hyundai Samho	3
Izar	1
Kawasaki SC	9
Mitsubishi HI	11
Mitsui ES	1
Samsung HI	37
USC	2
Imabari	4
Hudong-Zhonghua	5
Total	**135**

Delivery schedule

	On order	Delivered
2003	—	15
2004	—	21
2005	—	18
2006	1	24
2007	37	—
2008	53	—
2009	38	—
2010	8	—
Total	**135**	**80**

*As of Jan. 1, 2007. Source: LNG Shipping Solutions, London

Gaz de France launches large LNG vessel

Gaz de France launched the 154,500-cu-m Provalys LNG carrier, one of the world's largest, at Montoir-de-Bretagne. The 190 m by 43.5 m (623 by 143 ft) vessel is equipped to load LNG in arctic conditions. Gaz de France is both owner and charterer.

Turning sideways in ice conditions allows an icebreaker designed by Aker Arctic Technology to create a sufficiently wide shipping channel with a single vessel. (Photos from Aker Arctic Technology Inc.)

The "double-acting" LNG tanger (DAT) has a regularly shaped bow plus a stern modified for icebreaking service.

New LNG tanker designs for Arctic service

Operation of LNG vessels in heavy ice is likely to be critical to the success of projects proposed for Russia's Arctic coast and the Baltic Sea, said Poten & Partners, Inc. (New York).

Such operation will impose far more rigorous requirements on LNG vessels than anything experienced to date. New approaches to ice-class ship design and icebreaker operations are now in the works.

One alternative under development by Aker Arctic Technology (Finland) is an oblique-shaped icebreaker that would turn sideways in ice conditions to create a sufficiently wide shipping channel with a single vessel.

Another concept under development at Aker Arctic is a "double-acting" LNG tanker (DAT). It has a regularly shaped bow for efficient movement in the open sea, but the stern is modified to allow icebreaking as the vessel sails astern.

The DAT design has been applied to oil tankers for Arctic conditions but remains a conceptual design for LNG transport.

The DAT propulsion system functions as a bow propeller when the ship is running astern in icebreaking conditions, creating downward pressure on ice. Other features include stronger gearboxes for screw shaft drives and diesel-electric propulsion to provide high-torque, low-rpm power.

New-design ships will require significant changes from standard LNG vessel design, extensive use of low-temperature steels, and significant winterization modifications.

LNG carrier to become storage, regas unit

Golar LNG Ltd. (Hamilton, Bermuda) awarded a $90 million (Sing.) contract to Keppel Shipyard Ltd. (Singapore) to convert an existing LNG carrier into a floating LNG storage and regasification unit (FSRU), a first conversion of this type, Golar said.

Saipem SPA unit Moss Maritime (Norway) will perform design and engineering for the project.

The FSRU will have a steel monohull with LNG tanks in the middle, a regasification plant in the forward section, and crew quarters with control room and utility machinery aft. The vessel, scheduled for completion by mid-2007, will be capable of sending 2.75 bscm/year of natural gas ashore through a high-pressure subsea pipeline.

First commercial LNG transshipment

Exmar NV, Antwerp, completed the first commercial ship-to-ship (STS) transfer of LNG in the Gulf of Mexico. Until recently, STS transfer of LNG was only performed between LNG carriers in emergency situations.

A total of 20,650 cu m of LNG was transferred from the LNGRV Excelsior, a 130,800 cu m vessel built in 2005, to the Excalibur, a 138,000 cu m vessel built in 2002. Exmar operates these vessels along with Excelerate Energy LLC. The process does not require the construction of an offshore platform.

LNG shipbuilding leader: South Korea

South Korean shipbuilders led the global market in construction of LNG carriers in 2005, according to the Korea Shipbuilders' Association.

South Korean companies won 79% of market share, with orders for 33 LNG carriers from a worldwide total of 42 carriers. South Korean yards won 73% market share in 2004 and 80% in 2003.

The association projected that South Korea's shipbuilding industry will continue to control more than 70% of the market in 2006. It predicted that the world's top three shipbuilders—Hyundai Heavy Industries Co., Daewoo Shipbuilding & Marine Engineering Co., and Samsung Heavy Industries Co.—would win orders for about 40 LNG carriers in 2006.

Section 9:

LNG—Risks and opportunities

Safety of LNG shipping

LNG shipping has neither taken a human life nor precipitated a major LNG release in 47 years, according the Center for Energy Economics of the University of Texas, Sugar Land, TX.

Through 2005, said the Center, there had been "more than 33,000 LNG carrier trips, covering more than 60 million miles... without major accidents or safety problems, either in port or on the high seas."

During 2005, one LNG carrier entered Tokyo harbor every 20 hours and Boston harbor nearly every week without incident.

According to ABSG Consulting (Houston), about 40 LNG receiving terminals currently operate worldwide, as well as 17 LNG liquefaction export plants, and more than 240 peak-shaving and LNG storage facilities.

Since the first transocean LNG shipment in 1959, incidents (such as explosions or fires) at LNG production plants and import terminals have been extremely rare. Of three major accidents resulting in loss of life, none resulted from an explosive release of LNG or a spreading cloud of vaporizing LNG.

LNG fire training gets a boost

Located at the Texas Engineering Extension Service (TEEX) at Texas A&M University is one of the world's largest LNG-fire training facilities.

In 2004, BP PLC approached TEEX with a proposal to upgrade its 20-year-old LNG program. BP also initiated support for LNG vapor-cloud-behavior research in Texas A&M's department of chemical engineering.

TEEX officials told *OGJ* that, with BP's support and advice, the center built a larger LNG-fire training facility (for real-life, full-scale fire training). Completed in 2006 at its Brayton Fire Training Field in College Station, TEEX calls the field the "largest live-fueled fire training facility in the world."

Between September 2005 and August 2006, TEEX ran nine training programs with about 20 students in each class. Crews and companies from 45 countries have been trained at TEEX over the years, and TEEX also has set up cooperative learning centers in Riyadh and Qatar.

Risk-based decision making for LNG terminal siting

The terrorist attacks of Sept. 11, 2001, led to a renewed focus on LNG security.

For about a month after those attacks, authorities stopped shipments of LNG to the Everett, MA, terminal. In parallel, from October through December 2001, the U.S. Coast Guard (USCG) studied what safety and security measures should be applied to allow reactivation of imports to the Cove Point, MD, LNG terminal.

The U.S. Federal Energy Regulatory Commission (FERC) and the U.S. Dept. of Energy (DOE) sponsored separate studies in 2004 regarding the consequences of LNG releases onto water.

In late 2004, FERC began implementing the methodology approaches from the consequence study it sponsored. FERC also revised its regulations to require LNG terminal applicants to pre-file a notice of their intention to file for LNG terminal approval.

The USCG also implemented various security-assessment approaches based on Maritime Transportation Security Act (MTSA) regulations, what was done for previous terminals, and local community inputs.

LNG Security Requirements Chronology

Date	Event
Prior to Sept. 11, 2001	49 CFR 193, 33 CFR 127, and NFPA 59A requirements apply.
September-October 2001	USCG interrupts LNG shipments to Everett, Mass., terminal.
December 2001	Risk assessment of Cove Point, Md., terminal; LNG import activities resume.
November 2002	Maritime Transportation Security Act of 2002 passed by Congress.
July 2003	MTSA security regulations (33 CFR subchapter H) published.
May 2004	Federal Energy Regulatory Commission issues ABS Consulting's consequence analysis report of LNG releases on water.
June 2004	FERC issues revised consequence analysis methodology by ABS Consulting, as well as a response to industry comments.
July 2004	Effective date of MTSA security regulations.
December 2004	DOE issues Sandia National Laboratories LNG releases on water report.
2004-05	Various LNG import terminals approved under captain of the port letter of recommendation requirement (based on 33 CFR 127).
June 2005	Waterway suitability assessment guidance (NVIC 05-05) published by USCG.
October 2005	NFPA 59A revised to include changes in security requirements.

Sample WSA Time Line

Months*	Applicant, WSA team	USCG	Keys to success
9 to 7	Collect port and LNG project information.	Provide area maritime security assessment and plan information	Early coordination with USCG.
	Meet with local USCG representatives to discuss WSA.	Provide input regarding local concerns and maritime activities.	Building on contacts the proposed LNG project has previously established within the local community.
	Establish initial stakeholder contacts for WSA purposes (i.e., collect data and concerns).		Being inclusive in collecting potential impacts and concerns, while staying within the safety and security focus of the WSA.
	Draft preliminary WSA report.		Drafting a report that speaks to nonexperts regarding LNG and shipping.
Prefile with FERC: Submit letter of intent and preliminary WSA report to USCG			
6 to 4	Select participants and schedule WSA team meeting for scenario review.	Start letter of recommendation process and review of preliminary WSA.	Including representation from appropriate state, municipal, port, maritime, law enforcement, and emergency response personnel without allowing WSA working meetings to grow to unmanageable size.
	Address USCG comments on preliminary WSA report.		Providing meetings for collecting concerns separate from WSA risk assessment meetings.
	Prepare draft scenarios and other materials for WSA scenario review.		Preparing effective informational materials on LNG and the project for use with WSA participants.
	Collect inputs for safety and security resource estimation efforts.		
4 to 3	Hold facilitated scenario review-risk assessment meetings.	Participate in WSA meetings to provide oversight.	Recognizing that the meetings serve many purposes beyond getting the WSA risk effort completed (e.g., information exchange, building trust, developing real understanding of potential issues).
2 to 1	Prepare follow-up WSA report.		Documenting WSA process results in a manner to help the project sponsors, USCG, and the local stakeholders make risk-informed decisions.
File with FERC: Submit follow-up WSA report to USCG			
During EIS process		Review, validate follow-up WSA report with area maritime security committee.	

*Months before project application is filed with FERC.

Interchangeability

The issue of natural gas interchangeability continues to challenge LNG producers, end-users, and regulators. At the 2006 World Gas Conference, the American Gas Association reported on efforts of the Gas Interchangeability Task Group to develop recommendations and interim guidelines.

These were presented to the U.S. Federal Energy Regulatory Commission in mid-2006.

The group also identified several research priorities that it said should be acted on by the U.S. Dept. of Energy, including the following:

- Lead federally sponsored research on interchangeability issues
- Implement projects on gas turbines used in power generation and industrial applications
- Conduct analyses to forecast future gas compositions available in the United States and identify the gas specifications used by other countries

Using lessons learned in those activities, USCG developed guidance for an analysis called a Waterway Suitability Assessment (WSA).

A preliminary WSA report to the USCG must be submitted when the applicant submits pre-filing documentation to FERC. That document must outline how the applicant is going to perform the safety and security risk assessments required of the marine activity associated with the project.

After review, the applicant must then provide a completed "Follow-on WSA Report," incorporating a thorough risk, threat, and vulnerability analysis. That analysis is similar to the security assessments required under MTSA regulations.

Section 10:

Refining and products

Global refining capacity up slightly in 2006

The worldwide refining industry added a slight amount of crude distillation capacity in 2006. For the fifth year in a row, worldwide capacity reached a record high. The number of refineries remained relatively stable.

A total capacity of nearly 85.2 million b/cd in 658 refineries was reported as of January 1, 2007—an increase of less than 52,000 b/cd, due mainly to capacity creep—unannounced minor expansions. The increase is the smallest net gain in refining capacity since 2001, when refining capacity decreased.

Some country-specific data points:

- Asia-Pacific showed the largest increase in refining capacity, up 105,000 b/cd or 0.5%.
- North America, Eastern Europe, and the Middle East showed smaller changes, increasing 27,000 b/cd, 28,000 b/cd, and 4,300 b/cd, respectively.
- U.S. capacity increased more than 146,000 b/cd.
- Africa, Western Europe, and South America lost capacity: 18,250 b/cd, 81,500 b/cd, and 8,000 b/cd, respectively.
- This year's survey lists four fewer refineries than last year's survey reflecting corrections and refineries that had shut down in previous years.

Worldwide refining

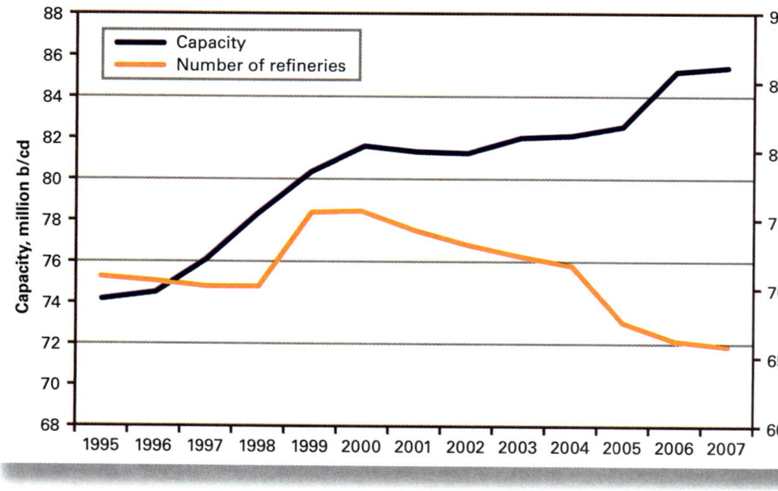

This year's survey lists no new refineries. The only closure reported was the 2,000 b/cd Norilsk, Russia, refinery. Two other refineries were removed—one in Ontario, Canada, and one in Poland.

For information on how to view or download the text version of the *OGJ Worldwide Refining Report 2006* tables from the *OGJ* Web site, send an email to orcinfo@pennwell.com.

U.S. motor fuel markets

The U.S. motor fuel market is being transformed by higher prices, tightening environmental requirements, changing demographics, growing world demand for oil, and expanding fuel options, said Cambridge Energy Research Associates (CERA; Cambridge, MA, U.S.A.).

U.S. gasoline pump prices rose from an average $1.59/gal in 2003 to $2.30/gal in 2005 and averaged $2.61/gal through mid-November 2006. Gasoline started at $2.30 in January 2006, hit a high of $3.00/gal in July, and averaged $2.22 in November.

How the World's Largest Refiners Rank

Rank: Jan. 1, 2007	Rank: Jan. 1, 2006	Company	Crude capacity, b/cd[1]
1	1	ExxonMobil Corp.	5,706,000
2	2	Royal Dutch Shell PLC	5,181,000
3	4	Sinopec	3,611,000
4	3	BP PLC	3,474,000
5	8	ConocoPhillips	2,927,000
6	5	Valero Energy Corp.	2,834,000
7	6	Petroleos de Venezuela SA	2,678,000
8	7	Total SA	2,668,000
9	9	China National Petroleum Corp.	2,440,000
10	10	Saudi Aramco	2,369,000
11	11	Chevron Corp.[2]	2,072,000
12	12	Petroleo Brasileiro SA	1,997,000
13	13	Petroleos Mexicanos	1,706,000
14	14	National Iranian Oil Co.	1,451,000
15	17	OAO Lukoil	1,217,000
16	15	OAO Yukos	1,182,000
17	16	Nippon Oil Co. Ltd.	1,157,000
18	18	Repsol YPF SA	1,105,000
19	19	Kuwait National Petroleum Co.	1,085,000
20	20	Pertamina	993,000
21	21	Marathon Oil Corp.	974,000
22	22	Agip Petroli SpA	906,000
23	23	Sunoco Inc.	880,000
24	24	SK Corp.	817,000
25	25	Indian Oil Corp. Ltd.	777,000

[1]Includes partial interests in refineries not wholly owned by the company. [2]Includes holdings in Caltex.

World's Largest Refineries

	Company	Location	Crude capacity, b/cd
1	Paraguana Refining Center	Cardon/Judibana, Falcon, Venezuela	940,000
2	SK Corp.	Ulsan, South Korea	817,000
3	Reliance Industries Ltd.	Jamnagar, India	660,000
4	LG-Caltex	Yosu, South Korea	650,000
5	ExxonMobil Refining & Supply Co.	Jurong/Pulau Ayer Chawan, Singapore	605,000
6	ExxonMobil Refining & Supply Co.	Baytown, Tex.	563,000
7	Saudi Arabian Oil Co. (Saudi Aramco)	Ras Tanura, Saudi Arabia	550,000
8	Formosa Petrochemical Co.	Mailiao, Taiwan	520,000
8	S-Oil Corp.	Onsan, South Korea	520,000
10	ExxonMobil Refining & Supply Co.	Baton Rouge, La.	503,000
11	Hovensa LLC	St. Croix, Virgin Islands	500,000
12	Shell Eastern Petroleum (Pte.) Ltd.	Pulau Bukom, Singapore	458,000
13	BP PLC	Texas City, Tex.	446,500
14	Kuwait National Petroleum Co.	Mina Al-Ahmadi, Kuwait	442,700
15	OAO Yukos	Angarsk, Russia	440,700
16	Citgo Petroleum Corp.	Lake Charles, La.	440,000
17	Shell Nederland Raffinaderij BV	Pernis, Netherlands	406,000
18	Sinopec	Zhenhai, China	403,000
19	Saudi Arabian Oil Co. (Saudi Aramco)	Rabigh, Saudi Arabia	400,000
20	Saudi Aramco-Mobil	Yanbu, Saudi Arabia	400,000

Regional Look at Worldwide Refining Operations

Region	No. of refineries	Crude distillation (b/cd)	Vacuum distillation (b/cd)	Catalytic cracking (b/cd)	Catalytic reforming (b/cd)	Catalytic hydrocracking (b/cd)	Catalytic hydrotreating (b/cd)	Coke, tonnes/day
Africa	45	3,212,112	510,144	205,085	462,174	62,208	886,649	1,841
Asia	156	22,310,291	4,179,246	2,682,659	1,984,118	793,468	8,673,206	20,200
Eastern Europe	91	10,272,594	3,759,573	928,944	1,496,038	318,354	4,267,106	12,570
Middle East	42	7,038,115	1,975,715	361,150	651,897	588,621	2,055,363	3,300
North America	156	20,853,787	9,091,606	6,622,194	4,217,628	1,759,180	15,651,468	128,795
South America	66	6,602,703	2,845,795	1,309,337	401,325	132,400	1,899,451	24,640
Western Europe	102	14,889,771	5,915,811	2,259,002	2,155,500	1,033,013	9,851,401	11,494
Total	**658**	**85,179,373**	**28,277,890**	**14,368,371**	**11,368,680**	**4,687,244**	**43,284,644**	**202,840**

Companies with 200,000+ b/cd refining capacity in Asia, the US, Western Europe

Rank	Company	No. of refineries	Crude capacity b/cd[1]
Asia[2]			
1	Sinopec	26	3,611,000
2	China National Petroleum Corp.	24	2,425,000
3	Royal Dutch Shell PLC	13	1,375,463
4	ExxonMobil Corp.	10	1,332,469
5	Nippon Oil Co. Ltd.	6	1,157,000
6	Pertamina	8	992,745
7	SK Corp.	1	817,000
8	Indian Oil Co. Ltd.	10	787,290
9	Chinese Petroleum Corp.	3	770,000
10	Reliance Petroleum Ltd.	1	660,000
11	LG-Caltex Corp.	1	[3]650,000
12	Chevron Corp.	6	648,093
13	Tonen/General Sekiyu Seisei KK	4	[4]629,375
14	Idemitsu Kosan Co. Ltd.	4	608,000
15	Hyundai Oil Refinery Co.	3	589,500
16	Cosmo Oil Co. Ltd.	4	565,250
17	S-Oil Corp.	1	[5]520,000
17	Formosa Petrochemical Co.	1	520,000
19	BP PLC	4	332,338
20	Hindustan Petroleum Corp. Ltd.	2	296,250
21	Saudi Aramco	2	254,000
22	Showa Yokkaichi Sekiyu Co. Ltd.	1	[6]222,000
US			
1	Valero Energy Corp.	15	2,320,160
2	ConocoPhillips	13	2,222,200
3	ExxonMobil Corp.	7	1,956,000
4	BP PLC	6	1,475,150
5	Royal Dutch Shell PLC	9	[7]1,084,550
6	Marathon Oil Corp.	7	974,000
7	Chevron Corp.	5	909,000
8	Sunoco Inc.	5	880,000
9	Petroleos de Venezuela SA	4	[8]849,400
10	Flint Hills Resources (Koch Industries)	3	773,775
11	Motiva Enterprises LLC[9]	3	730,000
12	Tesoro Corp.	6	558,000
13	Saudi Aramco	3	[10]372,500
14	Lyondell Chemical Co.	1	282,600
15	Total SA	1	231,452
Western Europe[11]			
1	Total SA	16	2,329,145
2	Royal Dutch Shell PLC	14	1,902,275
3	ExxonMobil Corp.	10	1,724,856
4	BP PLC	10	978,124
5	AgipPetroli SPA	10	876,210
6	Repsol YPF SA	5	709,200
7	ConocoPhillips	4	616,625
8	Turkish Petroleum Refineries Corp.	4	613,275
9	Compania Espanola de Petroles SA (CEPSA)	3	427,000
10	Ineos Group Holdings Inc.	2	402,800
11	OMV AG	3	398,635
12	ERG Group	4	396,214
13	Chevron Corp.	2	327,800
14	Preem Raffinaderi AB	2	316,000
15	Hellenic Petroleum SA	3	313,000
16	Galp Energia SA	2	304,172
17	Statoil AS	3	303,000
18	Neste Oil	6	[12]302,300
19	Saras SPA	1	300,000
19	Petroplus International NV	3	300,000
21	Petroleos de Venezuela SA	8	294,528

[1]Includes partial interest in refineries not wholly owned by the company. [2]Asia includes Australia, Bangladesh, Brunei, China (and Taiwan), India, Indonesia, Japan, Malaysia, Myanmar, New Zealand, North Korea, Pakistan, Papua New Guinea, the Philippines, Singapore, South Korea, Sri Lanka, and Thailand. [3]Includes Caltex's 50% stake. [4]Includes ExxonMobil Corp.'s 50% stake. [5]Includes Saudi Aramco's 35% stake. [6]Includes Royal Dutch Shell's 50% stake. [7]Includes Shell's stakes in Motiva and its 50% stake in the Deer Park, Tex., refinery. [8]Consists of PDVSA's ownership of Citgo and its 50% stake in the ExxonMobil Chalmette, La., refinery. [9]50/50 joint venture between Shell and Saudi Aramco. [10]Consists of 50% stake in Motiva. [11]Western Europe includes Austria, Belgium, Denmark, Finland, France, Germany, Greece, Ireland, Italy, the Netherlands, Norway, Portugal, Spain, Sweden, Switzerland, Turkey, and the UK. [12]Includes 50% stake in AB Nynas refineries.

The fuel efficiency of the entire automobile fleet—new and old cars—averaged 22.1 mpg in 2001, then flattened out at 22.2 mpg by 2005, the last year for which complete data were available.

U.S. refinery capacity up

No new conventional refineries have been built in the United States since the 1970s. However, refiners have invested $5–7 billion/year to significantly increase refinery capacity at existing plants in recent years.

Modernizing, debottlenecking, adding capacity, and completing environmental upgrades have increased capacity to 17.4 million b/d in 2006 from 15.3 million b/d in 1996. That resulted in about the same effect as building 17 average-sized new refineries, CERA said.

Transition to ULSD, ethanol smooth for U.S. refiners

During the last half of 2006, the initial transition to ultralow-sulfur diesel (ULSD) and ethanol blending went smoothly for U.S. refiners, reported KBC Advanced Technologies (Houston). The Oct. 15 deadline for 15-ppm (wt) sulfur ULSD at the retail outlet was ahead met of schedule.

To keep engine systems clean, the U.S. government has mandated ULSD be sold at 80% of U.S. diesel pumps, with 100% of pumps to be ULSD by 2010.

Of the 111 refineries operating in the United States, 88 ULSD units were started up in 2006 and about 11 other refineries indicated that "no changes" were needed to produce ULSD.

The changeover to ULSD is part of a broader program to encourage the wider use of cleaner-burning renewable fuels. The renewable fuel standards incorporated in the Energy Policy Act of 2005 calls for an increase in U.S. ethanol use to 7.5 billion gal/year by 2012.

US AVERAGE GASOLINE PRICES, 1918-2006*

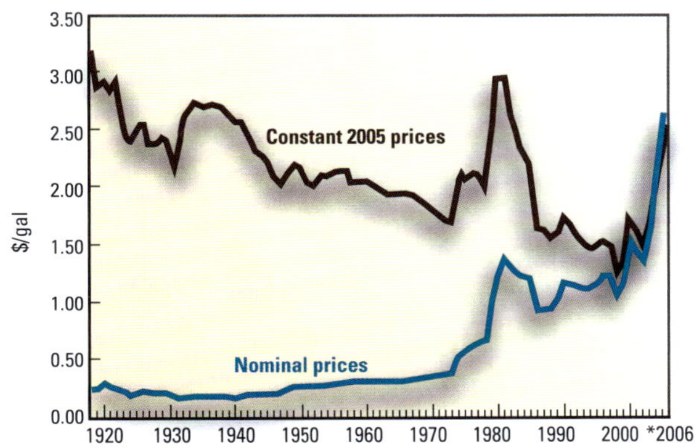

*Through November 2006.
Sources: Cambridge Energy Research Associates, American Petroleum Institute, US Department of Energy

US VEHICLE ON-ROAD FUEL EFFICIENCY*

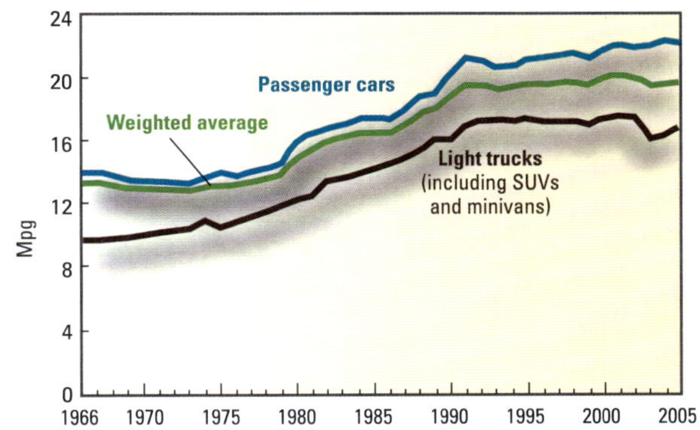

*Total vehicle fleet (cars and light-duty trucks).
Sources: Cambridge Energy Research Associates, Federal Highway Administration, US Department of Transportation, Highway Statistics (various years), Energy Information Administration

US OPERABLE REFINERIES, TOTAL CAPACITY

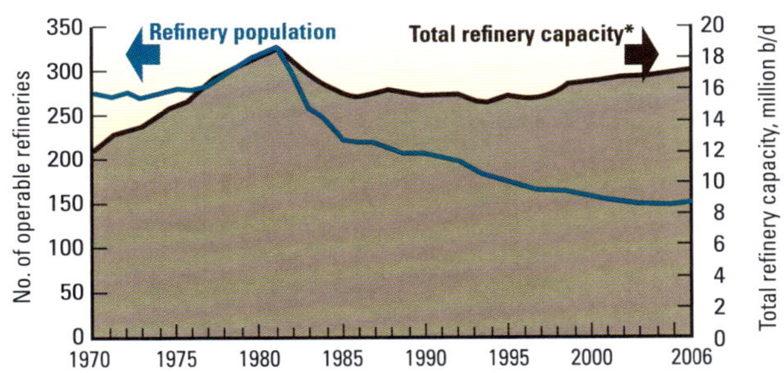

*Capacity as of Jan. 1 of each year.
Sources: Cambridge Energy Research Associates, Oil & Gas Journal, Energy Information Administration

Biofuels research

Chevron Technology Ventures LLC announced plans to provide as much as $25 million to scientists and engineers at the University of California-Davis to develop technology for converting cellulosic biomass into transportation fuels.

The five-year program seeks to develop commercially viable processes for producing transportation fuels from forest and agricultural residues, municipal solid waste, and other renewable resources.

Biofuels plant construction

Lurgi AG (Frankfurt, German) reported that it was engineering and building 31 biodiesel plants with total capacity of more than 4.2 mtpy.

The company said it received orders for seven plants in Argentina, Malaysia, France, and Indonesia. Capacities will range from 100,000–250,000 tpy, using rapeseed, soybean, or palm oil as feedstock.

Asia-Pacific refined product demand to rebound

After a strong 2004, demand growth for oil in the Asia-Pacific region slowed in 2005, said a 2006 study from FACTS Inc. (Honolulu). For 2006 and beyond, the study predicts stronger demand growth, especially in China, South Korea, Indonesia, Malaysia, and Vietnam.

FACTS reduced its 2005 preliminary estimate for the region's oil demand growth to 445,000 b/d.

China is the major driving force of the region's demand growth. Its oil demand growth accounts for about two-thirds of growth in the entire Asia-Pacific region.

ULSD PHASE-IN

Fuel	2006	2007	2008	2009	2010	2011	2012	2013	2014
	Sulfur level, ppm								
Highway		80% 15 ppm 20% 500 ppm			15	15	15	15	15
Nonroad		500	500	500	15	15	15	15	15
Locomotive, marine		500	500	500	500	500	15	15	15
Nonroad locomotive, marine with credits		High sulfur	High sulfur	High sulfur	500	500	500	500	15

Future growth of ULSD

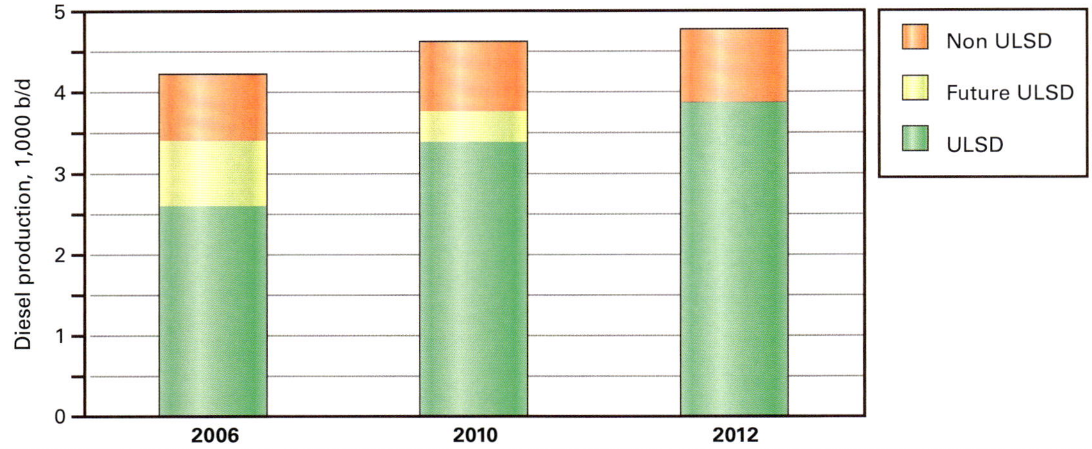

US renewable fuels requirement

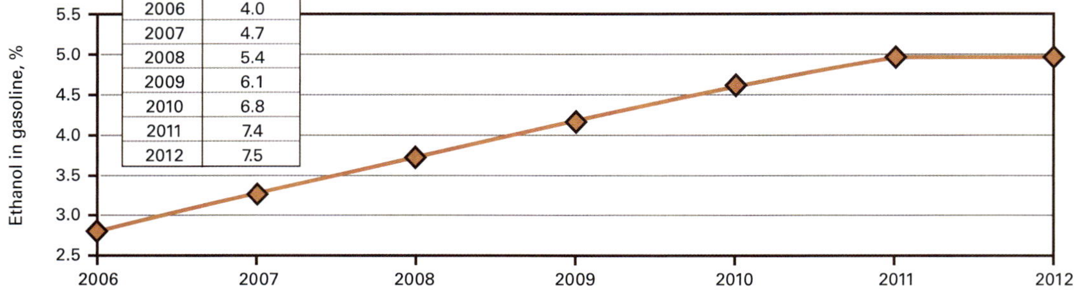

Regulated volume	
Year	billion gal
2006	4.0
2007	4.7
2008	5.4
2009	6.1
2010	6.8
2011	7.4
2012	7.5

Refined product demand

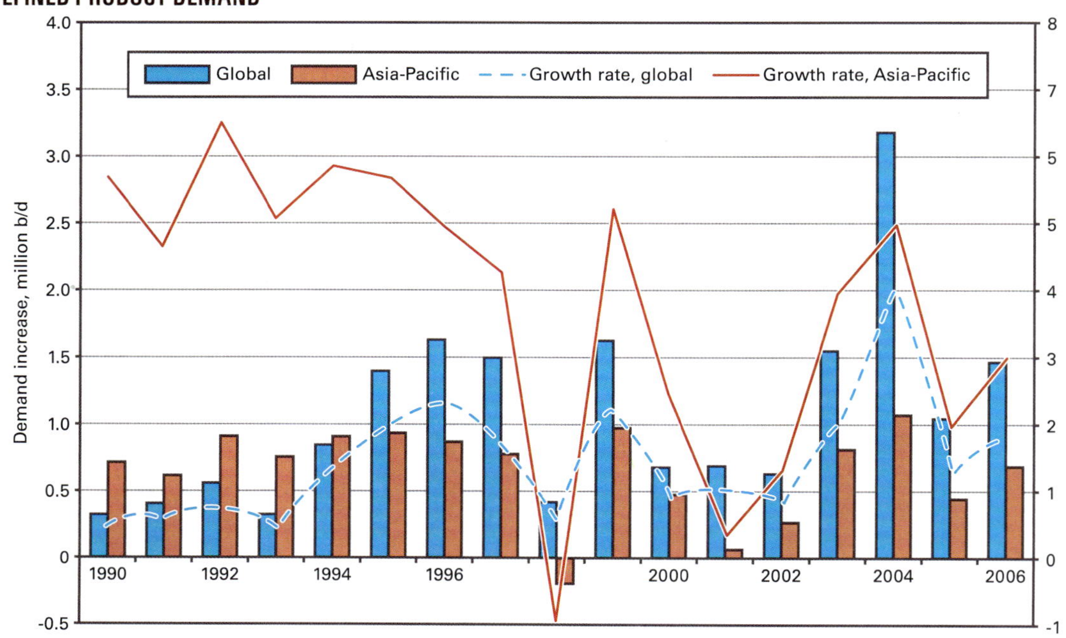

DEMAND GROWTH IN SELECTED COUNTRIES*

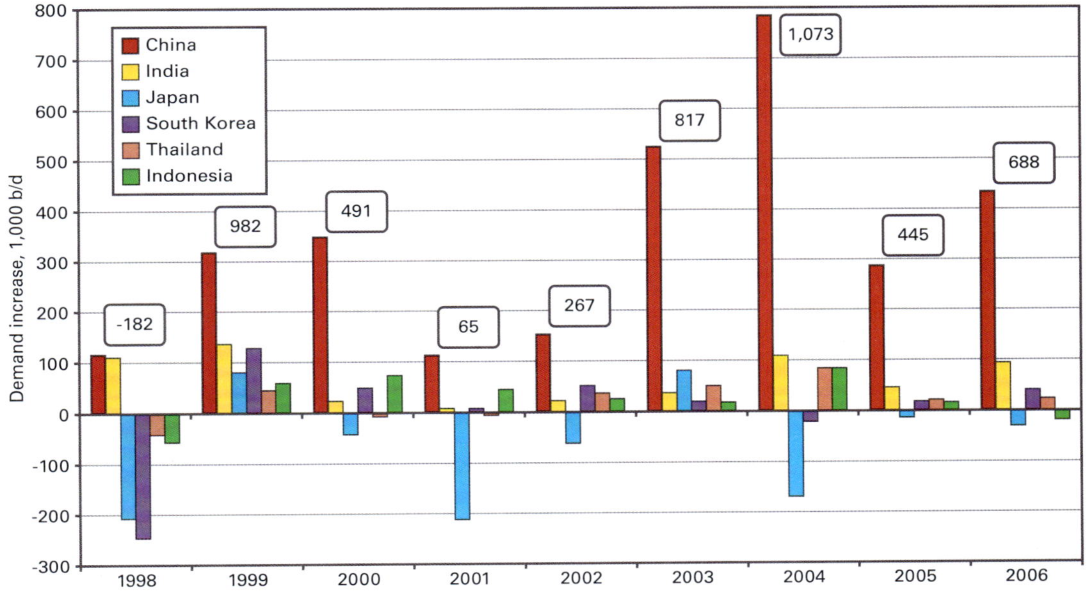

* The box indicates annual demand growth of the entire Asia-Pacific region

DEMAND BY PRODUCT

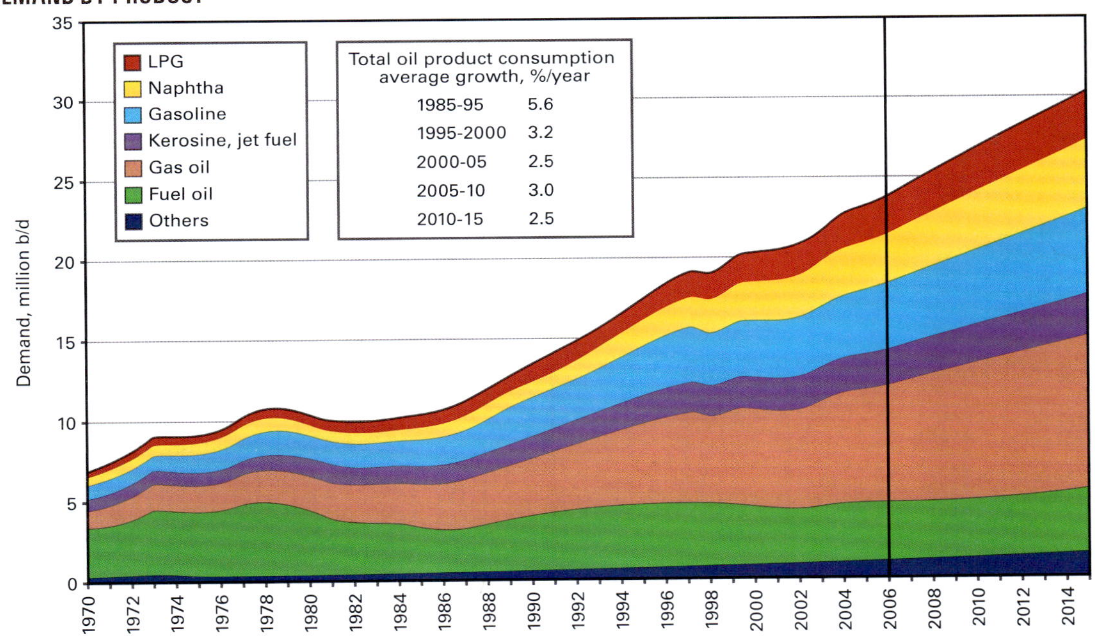

OPEC: Product demand outrunning refining capacity

OPEC projected investment needs through 2015 for refining capacity at $160 billion and for maintenance and replacement of lost capacity at $150 billion. The estimates exclude pipelines and terminals.

The group said "a more orchestrated effort [by consuming countries and international oil companies] is clearly required to ensure sufficient capacities are in place in the future."

Compared with 2005 levels, an OPEC reference case forecast that worldwide crude runs will need to climb 6.5 million b/d by 2010 and 13 million b/d by 2015.

OPEC projected that the world's refiners will need more than 13 million b/d of additional desulfurization capacity by 2015 compared with 2005 to satisfy ultralow-sulfur gasoline and ultralow-sulfur diesel requirements. Most new desulfurization units are projected for OECD regions.

WORLD REFINING INVESTMENTS*

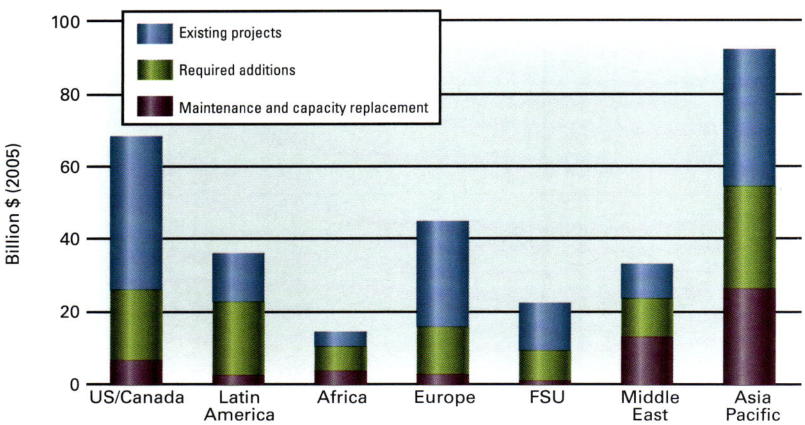

*2005-15 reference case
Source: OPEC Oil Outlook to 2025

WORLD DESULFURIZATION CAPACITY NEED, 2015 VS. 2005

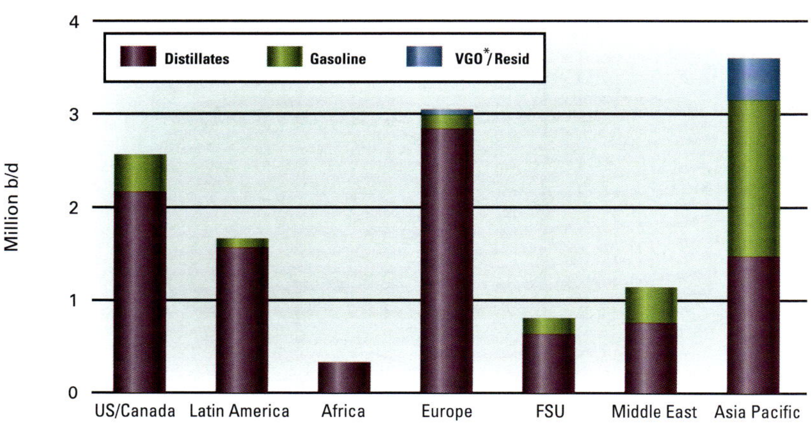

*Vacuum gas oil
Source: OPEC Oil Outlook to 2025

Section 11:

Gas processing and products

Major shift in gas processing, production in 2005

Declining natural gas production in the United States and Canada in 2005, combined with robust growth elsewhere in the world, tipped the balance of worldwide gas processing activity away from its historic center for the first time, *Oil & Gas Journal* noted.

The largest influence on gas processing in North America was the impact of Hurricanes Katrina and Rita on gas production and processing facilities along coast of the Gulf of Mexico in August, September and beyond. Much of the region's production and processing was shut in for the rest of the year.

Gas production also increased (2005 vs. 2004) outside North America:

- 2.8 tcf (10.4%) in Eastern Europe, including Russia and FSU states
- Nearly 2 tcf (24% in the Middle East
- More than 7 bcf (6.5%) in Asia-Pacific countries
- More than 433 bcf (9.5%) in Africa

Because of declining gas production, Canada and the United States combined in 2005 to produce just over 33% of the world's NGL, compared to more than 42% in 2002. For countries outside the United States and Canada, gas processing parameters included:

- Capacity: more than 126 MMcfd at the start of 2006 (vs. 115.5 bcfd in 2004)
- Throughput: average of more than 81 bcfd for 2005 (vs. nearly 74 bcfd in 2004)
- NGL production: more than 201 million gpd for 2005 (vs. more than 166.5 million gpd in 2004)

World's 20 Top Gas Producing Countries: 2005

Country	Production, bcf	Share, %
Russia/FSU	27,840.0	29.5
US	19,017.0	20.1
Canada	5,999.2	6.4
UK	3,269.0	3.5
Algeria	3,180.0	3.4
Netherlands	2,442.3	2.6
Norway	2,998.1	3.2
Iran	3,200.0	3.4
Indonesia	2,266.0	2.4
Saudi Arabia	2,294.0	2.4
Mexico	1,758.6	1.9
Argentina	1,549.1	1.6
Malaysia	1,561.0	1.7
UAE	1,614.0	1.7
China	1,783.1	1.9
Pakistan	1,365.0	1.4
Australia	1,307.4	1.4
Venezuela	1,000.0	1.1
Qatar	1,354.0	1.4
India	937.8	1.0
Top 20 total	**86,735.6**	**91.9**
Other	7,678.5	8.1
Total	**94,414.1**	**100.0**

Source: OGJ, Mar. 13, 2006, p. 68.

On Jan. 1, 2006, U.S. gas processing capacity stood at nearly 70 bcfd, essentially flat compared with 2004. Throughput in 2005 averaged more than 46 bcfd, down from 2004; and NGL production, more than 76 million gpd (more than 1.8 million b/d), again essentially flat compared with 2005.

Worldwide Gas Plant Construction*

Country	2006	2005
Africa	2	4
Asia-Pacific	5	5
Canada	-	-
Eastern Europe	3	-
Middle East	7	5
Latin America	4	2
US	4	2
Western Europe	9	1
Total	**34**	**19**

*Includes new plants, expansions, and modifications scheduled for completion in the year listed. 2006 data are from OGJ, Apr. 24, 2006. 2005 data are from OGJ, Apr. 25, 2005.

World's Major Gas Reserves

Country	Est. proved reserves, tcf	Share, %
Russia	1,680.0	27.8
Iran	971.2	16.1
Qatar	910.5	15.1
Saudi Arabia	241.3	4.0
Abu Dhabi	198.5	3.3
US	192.5	3.2
Nigeria	184.7	3.1
Algeria	160.5	2.7
Venezuela	151.4	2.5
Iraq	112.0	1.9
Indonesia	97.8	1.6
Australia	27.6	0.5
Malaysia	75.0	1.2
Norway	84.3	1.4
Turkmenistan	71.0	1.2
Uzbekistan	66.2	1.1
Kazakhstan	65.0	1.1
Netherlands	62.0	1.0
Egypt	58.5	1.0
Canada	56.6	0.9
Kuwait	55.5	0.9
China	53.3	0.9
Libya	52.7	0.9
Ukraine	39.6	0.7
Oman	29.3	0.5
Argentina	18.9	0.3
UK	18.6	0.3
Subtotal	**5,734.5**	**94.9**
Others	305.7	5.1
Total	**6,040.2**	**100.0**

Source: OGJ, Dec. 19, 2005, p. 24.

Worldwide Gas Processing Roundup

Region	[1]2004	[2]2005	Change	Change, %
US				
Gas capacity, MMcfd	70,277.1	69,815.4	–461.7	-0.7
Gas throughput, MMcfd	46,776.0	45,685.3	–1,090.7	–2.4
NGL production, 1,000 gpd	76,203.0	75,609.5	–593.5	–0.8
Proved reserves, bcf	189,044.0	192,513.0	3,469.0	1.8
CANADA				
Gas capacity, MMcfd	52,502.9	56,218.5	3,715.6	6.6
Gas throughput, MMcfd	28,050.9	32,364.8	4,313.9	13.3
NGL production, 1,000 gpd	28,597.5	25,966.8	–2,630.7	–10.1
Proved reserves, bcf	56,600.0	56,577.0	–23.0	–
WESTERN EUROPE				
Gas capacity, MMcfd	24,375.0	24,375.0	–	–
Gas throughput, MMcfd	9,812.2	9,812.2	–	–
NGL production, 1,000 gpd	9,834.6	9,834.6	–	–
Proved reserves, bcf	180,447.0	186,555.0	6,108.0	3.3
EASTERN EUROPE				
Gas capacity, MMcfd	2,796.0	2,796.0	–	–
Gas throughput, MMcfd	1,426.5	1,426.5	–	–
NGL production, 1,000 gpd	10,142.4	10,142.4	–	–
Proved reserves, bcf	1,966,730.0	1,966,790.0	60.0	–
LATIN AMERICA				
Gas capacity, MMcfd	18,648.2	18,448.2	–200.0	–1.1
Gas throughput, MMcfd	13,635.5	13,635.5	–	–
NGL production, 1,000 gpd	28,043.5	28,043.5	–	–
Proved reserves, bcf	265,370.0	266,823.0	1,453.0	0.5
MIDDLE EAST				
Gas capacity, MMcfd	29,112.0	27,917.0	–1,195.0	–4.3
Gas throughput, MMcfd	20,565.7	19,733.0	–832.7	–4.2
NGL production, 1,000 gpd	74,789.9	72,161.1	-2,628.8	–3.6
Proved reserves, bcf	2,522,125.0	2,565,400.0	43,275.0	1.7
AFRICA				
Gas capacity, MMcfd	17,254.2	17,254.2	–	–
Gas throughput, MMcfd	9,642.4	9,642.4	–	–
NGL production, 1,000 gpd	16,777.3	16,777.3	–	–
Proved reserves, bcf	476,509.0	485,841.0	9,332.0	1.9
ASIA-PACIFIC				
Gas capacity, MMcfd	23,259.2	25,004.2	1,745.0	7.0
Gas throughput, MMcfd	18,907.8	20,253.5	1,345.7	6.6
NGL production, 1,000 gpd	26,963.9	25,741.8	–1,222.1	–4.7
Proved reserves, bcf	444,383.0	391,645.0	–52,738.0	–13.5
TOTALS-excl. US				
Gas capacity, MMcfd	167,947.5	172,013.1	4,065.6	2.4
Gas throughput, MMcfd	102,041.0	106,867.9	4,826.9	4.5
NGL production, 1,000 gpd	195,149.1	188,264.8	–6884.3	–3.7
Proved reserves, bcf	5,912,164.0	5,919,631.0	7,467.0	0.1
TOTALS-incl. US				
Gas capacity, MMcfd	238,224.6	241,828.5	3,603.9	1.5
Gas throughput, MMcfd	148,817.0	152,553.2	3,736.2	2.4
NGL production, 1,000 gpd	271,352.1	263,874.3	–7,477.8	–2.8
Proved reserves, bcf	6,101,208.0	6,112,144.0	10,936.0	0.2

[1] Proved gas reserve totals for 2004 are as of Jan.1, 2005 (OGJ, Dec. 20, 2004, p.22). [2] Proved gas reserve totals for 2005 are as of Jan.1, 2006 (OGJ, Dec. 19, 2005, p. 24).

Worldwide Gas Processing

Leena Koottungal, Survey Editor

Capacities as of Jan. 1, 2006, and average production

Country	No. plants	MMcfd Gas capacity	MMcfd Gas throughput	Production, 1,000 gpd (average based on the past 12 months) Ethane	Propane	Iso-butane	Normal or unsplit butane	LPG mix	Raw NGL mix	Debut. natural gasoline	Other	Total products
Algeria	4	5,335.2	4,268.2	–	–	–	–	632.8	–	–	5,449.2	6,082.0
Angola	2	200.0	137.0	–	–	–	–	–	199.0	–	–	199.0
Argentina	14	4,885.3	3,076.0	712.0	654.7	–	414.0	–	7.0	493.2	–	2,280.9
Australia	5	4,384.5	3,261.5	2,500.0	710.0	–	1,034.9	–	1,875.0	345.0	4,418.4	10,883.3
Austria	2	86.0	80.5	–	–	–	–	–	–	58.6	76.0	134.6
Azerbaijan	1	–	–	–	–	–	–	–	–	–	–	–
Bahrain	1	280.0	260.0	–	118.4	112.5	–	–	–	204.9	–	435.8
Bangladesh	4	140.0	83.4	–	–	–	–	–	–	–	41.9	41.9
Bolivia	12	882.8	622.4	–	–	–	–	171.0	238.0	132.4	383.0	924.4
Brazil	15	1,083.4	370.3	–	–	–	–	669.8	101.6	108.0	1.6	881.0
Brunei	2	1,182.0	967.4	52.2	9.1	–	–	–	–	290.0	–	351.3
Canada	940	56,218.5	32,364.8	7,167.5	1,611.9	–	1,011.4	207.1	12,207.9	3,345.8	12.5	25,564.1
Chile	2	477.0	381.6	–	39.6	12.9	14.3	–	185.0	19.3	–	271.1
Colombia	7	333.0	247.3	76.5	58.0	–	63.0	88.8	27.0	63.0	–	376.3
Denmark	1	933.0	680.0	–	–	–	–	–	–	–	–	–
Ecuador	2	41.0	20.5	–	–	–	–	97.1	–	37.3	–	134.4
Egypt	19	4,510.0	2,133.0	–	–	–	–	1,379.6	–	–	1,671.3	3,050.9
Equatorial Guinea	1	870.0	170.0	–	–	–	–	102.8	–	–	36.7	139.5
France	1	570.0	563.0	–	139.0	–	132.0	–	–	270.0	–	541.0
Hungary	8	1,137.4	461.0	–	71.0	38.8	40.7	139.8	–	71.3	228.3	589.9
India	11	3,691.9	2,945.1	–	227.8	–	–	3,939.7	768.2	714.3	884.5	6,534.5
Indonesia	13	7,972.3	7,089.7	–	625.1	–	304.4	305.3	1,521.9	–	1,965.1	4,721.8
Iran	15	5,867.0	2,946.0	–	–	–	–	–	3,671.6	–	1,227.4	4,899.0
Iraq	1	2,081.0	1,550.0	–	–	–	–	7.3	–	2.2	–	9.5
Italy	20	5,175.0	1,495.0	–	–	–	–	–	–	25.8	–	25.8
Kazakhstan	4	155.0	10.0	–	–	–	–	296.0	35.9	–	0.6	332.5
Kuwait	3	1,680.0	1,011.0	2,308.1	2,833.1	–	2,057.7	8,595.8	–	1,155.7	–	16,950.4
Libya	9	4,897.0	2,567.0	–	273.0	294.0	126.0	37.8	1,749.0	567.0	551.0	3,597.8
Malaysia	5	4,250.0	3,400.0	–	23.2	–	28.2	–	–	57.0	–	108.4
Mexico	8	5,079.2	3,963.2	–	–	–	–	–	13,605.2	–	–	13,605.2
Mozambique	1	–	–	–	–	–	–	–	–	–	–	–
Myanmar	1	24.0	–	–	17.0	26.2	–	–	–	4.7	–	47.9
Netherlands	2	725.0	315.0	–	–	–	–	–	–	–	230.5	230.5
New Zealand	3	944.5	497.7	–	2.6	–	5.7	273.4	–	17.3	802.3	1,101.3
Nigeria	4	1,268.0	228.0	–	–	–	–	–	3,415.1	–	84.0	3,499.1
Norway	4	3,150.0	530.8	–	2,596.7	586.7	1,452.7	–	175.4	1,389.7	–	6,201.2
Oman	8	826.0	309.4	–	–	–	–	–	202.0	43.0	–	245.0
Pakistan	12	1,195.0	858.7	–	–	–	–	53.8	64.9	–	9.1	127.8
Peru	4	545.7	298.2	–	–	–	–	14.9	1,399.0	–	12.8	1,426.7
Poland	1	55.0	–	–	–	–	–	33.6	–	–	16.8	50.4
Qatar	2	1,010.0	220.0	1,397.0	1,681.0	–	1,103.0	–	1,618.7	122.0	–	5,921.7
Russia	24	1,405.6	925.5	563.7	–	–	709.7	6,559.9	–	1,056.8	237.8	9,127.9
Saudi Arabia	10	10,040.0	8,496.0	–	–	–	–	–	24,234.0	–	12,810.0	37,044.0
Serbia and Montenegro	1	43.0	30.0	–	11.3	6.0	8.4	–	–	16.0	–	41.7
Syria	1	450.0	–	–	–	–	–	507.0	–	–	–	507.0
Thailand	5	1,070.0	1,030.0	1.6	–	–	–	557.0	49.1	84.0	–	691.7
Trinidad & Tobago	3	1,394.0	1,101.0	–	290.3	88.6	99.3	–	–	195.0	–	673.2
Tunisia	1	174.0	139.2	–	–	–	–	–	–	–	209.0	209.0
Ukraine	1	–	–	–	–	–	–	–	–	–	–	–
United Arab Emirates	9	3,923.0	3,295.6	–	792.5	–	612.0	–	374.9	–	7,199.2	8,978.6
United Kingdom	11	13,736.0	6,147.9	–	–	–	–	–	1,093.0	1,572.5	36.0	2,701.5
United States	566	69,815.4	45,685.3	12,155.4	10,689.0	1,976.2	3,967.1	226.6	39,450.1	4,303.5	2,841.8	75,609.5
Uzbekistan	2	–	–	–	–	–	–	–	–	–	–	–
Venezuela	14	3,927.0	3,555.0	–	–	–	–	–	7,312.9	–	–	7,312.9
Vietnam	1	150.0	120.0	–	281.0	–	188.0	495.0	–	–	167.9	1,131.9
Yemen	4	1,760.0	1,645.0	–	–	–	–	903.0	676.2	–	–	1,579.2
Total, excl. US	**1,261**	**172,213.3**	**106,867.9**	**14,778.6**	**13,066.3**	**1,165.7**	**9,405.4**	**26,068.3**	**76,807.5**	**12,461.7**	**38,762.9**	**192,516.4**
World total	**1,827**	**242,028.7**	**152,553.2**	**26,934.0**	**23,755.3**	**3,141.9**	**13,372.5**	**26,294.9**	**116,257.6**	**16,765.2**	**41,604.7**	**268,126.1**
Canada												
Alberta	880	49,568.5	27,260.2	7,091.9	933.2	–	528.5	93.2	10,792.6	2,699.4	–	22,138.8
British Columbia	34	5,670.2	4,535.6	–	637.8	–	470.6	–	863.0	618.7	–	2,590.1
Northwest Territories	2	219.9	28.3	–	–	–	–	–	–	–	–	–
Nova Scotia	2	600.0	413.0	–	–	–	–	–	515.0	–	–	515.0
Ontario	1	–	–	–	–	–	–	–	–	–	–	–
Saskatchewan	21	159.9	127.7	75.6	40.9	–	12.3	113.9	37.3	27.7	12.5	320.2
Total Canada	**940**	**56,218.5**	**32,364.8**	**7,167.5**	**1,611.9**	**--**	**1,011.4**	**207.1**	**12,207.9**	**3,345.8**	**12.5**	**25,564.1**
United States												
Alabama	13	1,633.0	449.3	–	120.0	–	73.4	–	195.5	50.2	123.4	562.5
Alaska	5	9,525.0	9,298.0	–	1.2	–	–	–	168.4	–	0.6	170.2
Arkansas	5	873.8	512.4	–	–	–	–	–	27.1	–	–	27.1
California	31	1,124.2	847.6	–	407.8	56.1	247.5	56.2	79.9	264.0	12.1	1,123.6
Colorado	41	1,441.7	1,201.9	417.2	475.3	57.3	177.1	8.7	877.6	168.7	170.2	2,352.1
Florida	1	84.0	9.0	–	–	–	–	–	41.0	–	–	41.0
Illinois	1	2,100.0	1,426.0	1,440.0	655.0	127.0	159.0	–	2,940.0	75.0	–	5,396.0
Kansas	13	2,908.5	1,429.2	69.2	449.7	62.6	145.5	–	2,019.6	133.5	260.4	3,140.5
Kentucky	4	120.0	106.1	–	–	–	–	–	153.5	–	–	153.5
Louisiana	72	18,876.3	10,050.2	1,931.4	1,482.4	449.2	375.4	–	6,861.0	605.9	382.4	12,087.6
Michigan	22	1,549.4	625.7	–	5.6	–	3.0	2.6	476.5	1.5	9.8	499.0
Mississippi	4	1,303.4	816.1					–	1,130.2		–	1,130.2
Montana	5	15.1	10.0	–	25.2	–	15.7	–	3.6	15.2	–	59.7
Nebraska	1	10.0	8.0	–	–	–	–	–	2.0	–	–	2.0
New Mexico	27	3,289.0	2,521.2	2,149.4	1,305.6	210.7	501.7	–	1,427.2	567.6	318.6	6,480.8
North Dakota	8	302.0	190.9	–	360.2	37.9	183.1	–	22.3	159.7	–	763.2
Ohio	4	25.0	10.0	0.1	2.8	0.8	1.9	–	8.2	–	2.3	16.1
Oklahoma	56	3,388.5	2,295.3	610.9	663.2	105.1	239.9	56.9	4,884.3	268.3	255.9	7,084.5
Pennsylvania	8	43.0	32.5	–	26.0	–	14.8	–	4.1	7.8	6.5	59.2
Tennessee	2	8.0	1.5	–	–	–	–	–	–	–	–	–
Texas	181	14,765.3	10,224.5	4,722.2	3,611.2	725.8	1,284.5	60.2	15,802.5	1,512.4	827.1	28,545.8
Utah	14	531.0	243.2	–	21.6	–	24.0	–	253.0	12.1	9.8	320.5
West Virginia	8	585.0	304.0	250.9	176.0	30.3	86.2	–	378.0	65.5	–	986.9
Wisconsin	1	–	–	–	–	–	–	–	–	–	–	–
Wyoming	39	5,314.2	3,073.3	564.1	900.2	113.4	434.4	42.0	1,694.6	396.0	462.8	4,607.5
Total US	**566**	**69,815.4**	**45,685.3**	**12,155.4**	**10,689.0**	**1,976.2**	**3,967.1**	**226.6**	**39,450.1**	**4,303.5**	**2,841.8**	**75,609.5**

Source of Alberta data: Alberta Energy and Utilities Board, data series ST50, ST13A. Source of British Columbia data: Government of British Columbia, Ministry of Energy and Mines, www.em.gov.ba.ca. Source for Saskatchewan, Canada data: Saskatchewan Industry and Resources, tel. (306) 787-2596.

US GAS PROCESSING TRENDS

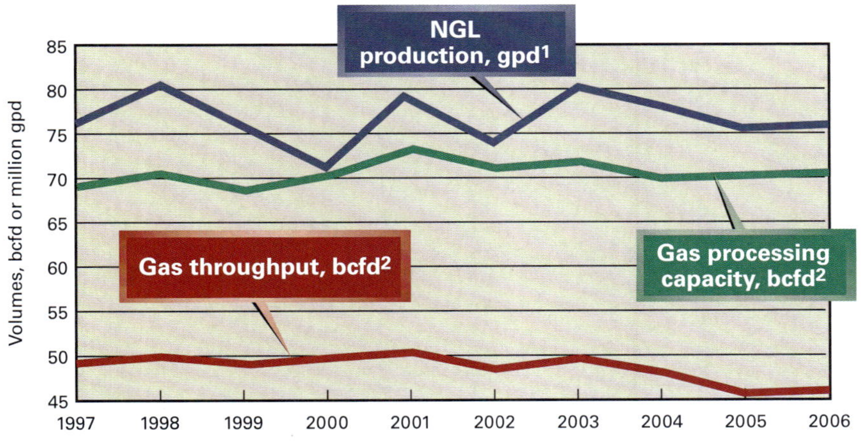

[1]Based on a 12-month average for previous year. [2]As of Jan. 1 of each year.

US LPG DIFFERENTIAL, PRICE TRENDS: 2005

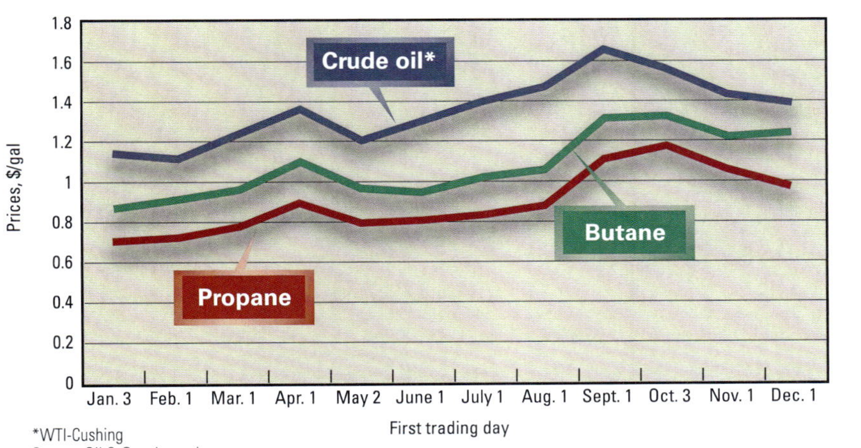

*WTI-Cushing
Source: Oil & Gas Journal

WORLD LPG DEMAND

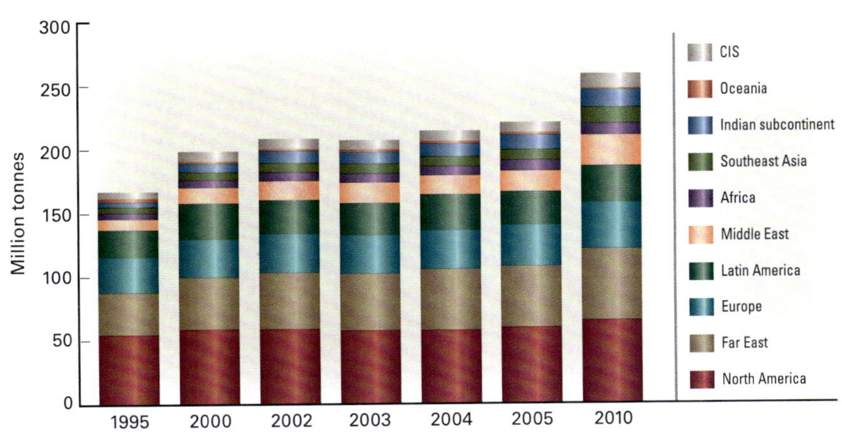

The historically normal price relationship between LPG and crude oil continued throughout 2005. In North America, natural gas processing margins were unusually volatile during 2005, as natural gas prices remained relatively high during the first half of the year. After two massive hurricanes in the Gulf of Mexico in late 2005, gas processing margins soared to record highs then declined at year end.

World LPG production may outpace demand

Expansion of the global LPG industry has been faster than most other energy markets and will accelerate during the next few years, according to Purvin & Gertz Inc. (Houston).

The global market for LPG increased to slightly less that 220 million tonnes in 2005 from only 168 million tonnes in 1995—an overall increase of 31% or about 3.1%/year. Virtually all end-use sectors contributed to this demand growth.

Through the second half of this decade, the LPG market will become supply-driven by many new production projects that will be started up around the world.

Global supplies of LPG rose to about 220 million tonnes in 2005 from 168 million tonnes in 1995, or about 3.1%/year. Purvin & Gertz expects supplies will reach about 258 million tpy by 2010.

Natural gas processing continues to be the largest supply source of LPG supply, accounting for nearly 60% of total worldwide production during the last decade. Refineries accounted for nearly all of the rest.

In the next few years, however, a growing volume of LPG is likely to be recovered from LNG liquefaction or regasification operations.

World LPG supply

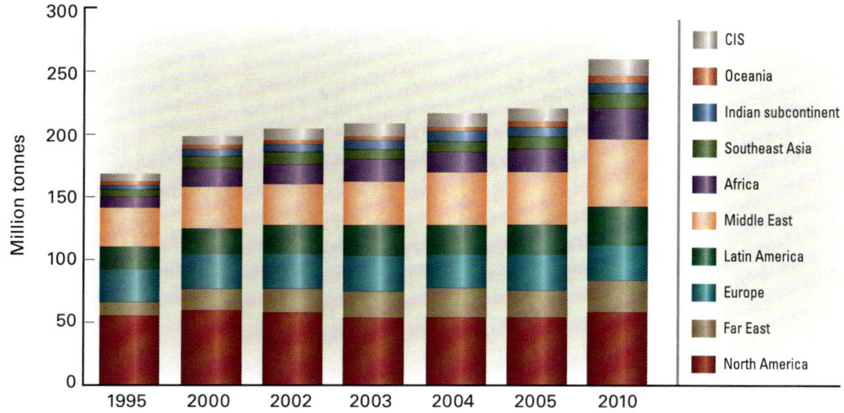

Regional LPG trade

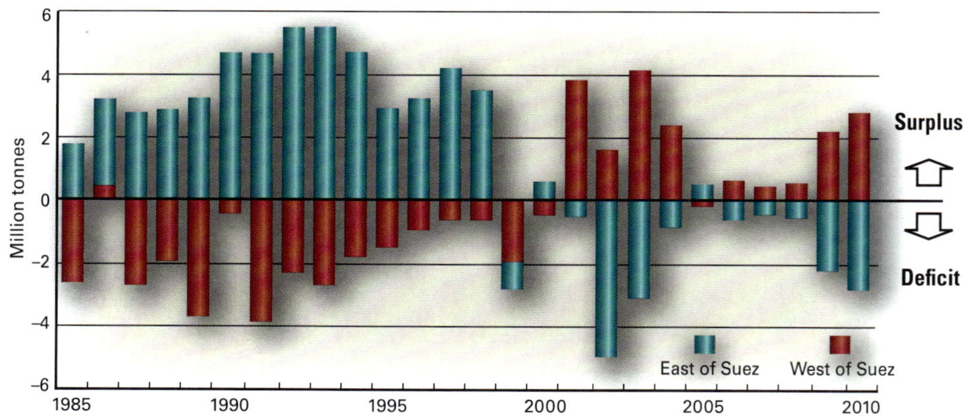

A shift in global LPG trade patterns occurred in the late 1990s. Between 1995 and 2000, LPG demand in the East of Suez region increased by about 17.3 mtpy. During this same period, supplies in the region increased by only 15 mtpy. That meant the overall LPG supply-demand balance East of Suez shifted from a surplus of supply to a deficit. Conversely, West of Suez experienced an equal but opposite shift in its LPG supply-demand balance.

These regional changes affected world LPG trade patterns. The Far East continues to be the largest LPG importing region in the world. Additional consumption of LPG to produce petrochemicals in some of the exporting countries in the Middle East, however, has reduced the amount of LPG available for export.

Therefore, in recent years the Far East has had to import LPG from increasingly diverse sources to meet regional demand—including from the West, North Africa and even the North Sea.

The United States continues to be the primary swing destination or supplier to the global LPG market. It can store large volumes of LPG in times of excess global supply and can also supply LPG from storage

Gas condensate capacity to increase substantially in Middle East, Asia

Condensate is defined broadly as an NGL that precipitates out of, or is stripped from, produced natural gas. Unlike LPG, it remains a liquid without specialized infrastructure and in many ways can be considered a sweet, light crude equivalent.

Condensate production in the Middle East and Asia will more than double through 2008 and increase by an additional 60% by 2013, according to a study by FACTS Inc. (Honolulu) and Asia Pacific Energy Consulting (Seattle).

The role of condensate splitters will continue to change during the next few years. The Middle East will begin using splitters to maximize the production of transportation fuels. And Asian countries will use them to satisfy a shortage of light-end hydrocarbons.

Condensate production, base case

	Current study 2001	2002	2003	2008	2013	2002 study 2001	2006	2011
	Production, 1,000 b/d							
Middle East								
Saudi Arabia	215	235	308	690	1,330	175	533	720
Iran	95	168	182	608	972	95	387	765
UAE	388	424	452	624	607	386	526	486
Qatar	187	180	200	818	1,254	187	407	542
Iraq	5	5	3	95	150	5	9	62
Oman	5	7	12	16	165	5	150	195
Yemen	—	—	—	—	40	—	—	30
Total	**895**	**1,019**	**1,157**	**2,851**	**4,518**	**853**	**2,012**	**2,800**
South Asia								
India	57	52	50	96	116	45	91	124
Pakistan	4	5	6	12	16	4	7	8
Bangladesh	5	4	4	7	11	5	6	21
Total	**66**	**61**	**60**	**115**	**143**	**54**	**104**	**153**
Southeast Asia								
Indonesia	108	125	114	150	233	100	146	197
Malaysia	94	95	98	108	139	94	160	186
Thailand	58	62	61	96	141	57	38	64
Brunei	14	16	17	17	23	14	16	25
Vietnam	4	5	7	20	41	7	19	33
Myanmar	7	10	16	23	31	7	15	16
Philippines	6	18	22	24	24	6	18	23
Total	**291**	**331**	**335**	**438**	**632**	**285**	**412**	**544**
Northeast Asia								
China	50	55	59	149	214	49	132	188
Russia	—	—	—	8	190	5	15	166
Total	**50**	**55**	**59**	**157**	**404**	**54**	**147**	**354**
Australasia								
Australia	129	145	137	209	263	120	206	178
New Zealand	24	21	18	27	12	25	24	16
Papua New Guinea	3	3	3	3	6	3	2	7
Total	**156**	**169**	**158**	**239**	**281**	**148**	**232**	**201**
East of Suez	1,458	1,635	1,769	3,800	5,978	1,394	2,907	4,052

According to the study, the East of Suez countries in 2013 will use 4.3 million b/d of condensate, up from 1.3 million b/d in 2003. Importing countries will use 405,000 b/d of condensate, about the same as in 2003. Total demand will therefore be nearly 4.7 million b/d in 2013.

Again, substantial increases in Persian Gulf countries account for most of the additional demand.

Condensate demand, base case

	Current study					2002 study		
	2001	2002	2003	2008	2013	2001	2006	2011
	Demand, 1,000 b/d							
Middle East								
Saudi Arabia	102	153	218	673	1,055	65	260	265
Iran	75	82	89	395	775	80	145	250
UAE	375	430	435	545	655	293	450	530
Qatar	80	97	87	232	357	75	172	182
Iraq	5	5	3	35	140	5	9	52
Oman	—	—	3	8	14	—	5	20
Yemen	2	2	2	3	17	—	—	15
Total	**639**	**769**	**837**	**1,891**	**3,013**	**518**	**1,041**	**1,314**
South Asia								
India	56	52	50	79	125	43	69	99
Pakistan	4	5	6	12	16	4	7	5
Bangladesh	5	4	4	4	5	5	6	9
Total	**65**	**61**	**60**	**95**	**146**	**52**	**82**	**113**
Southeast Asia								
Indonesia	93	39	50	191	221	73	125	187
Malaysia	83	90	93	105	115	83	110	112
Thailand	73	91	118	234	337	57	105	108
Brunei	—	—	—	—	—	—	—	—
Vietnam	4	5	7	20	30	7	13	21
Myanmar	—	—	—	—	—	3	6	6
Philippines	16	8	7	11	13	16	6	9
Total	**269**	**233**	**275**	**561**	**716**	**239**	**365**	**443**
Northeast Asia								
China	50	55	59	224	270	49	126	183
Russia	—	—	—	8	89	5	8	51
Total	**50**	**55**	**59**	**232**	**359**	**54**	**134**	**234**
Australasia								
Australia	26	36	39	27	31	30	38	44
New Zealand	21	18	15	24	4	25	23	14
Papua New Guinea	3	3	3	3	5	3	2	7
Total	**50**	**57**	**57**	**54**	**40**	**58**	**63**	**65**
East of Suez	**1,073**	**1,175**	**1,288**	**2,833**	**4,274**	**921**	**1,685**	**2,169**
Importing countries								
Japan	104	104	133	117	129	104	145	151
South Korea	149	170	169	151	166	149	194	207
Taiwan	69	31	13	13	16	69	86	103
Singapore	55	75	89	82	94	55	76	88
Total	**377**	**380**	**404**	**363**	**405**	**377**	**501**	**549**
Total demand	**1,450**	**1,555**	**1,692**	**3,196**	**4,679**	**1,298**	**2,186**	**2,718**

Propane production 2006: Short of full recovery

From January–September 2006, U.S. gas plants enjoyed record high profit margins, said Petral Worldwide Inc. (PW; Houston). Propane prices in all regional markets were strong enough to encourage refineries to maximize sales into the merchant market.

However, gas-plant production—even adjusted for post-hurricane problems—and net production from refineries failed to reach historic full-recovery volumes.

US Gas-Plant Propane Production

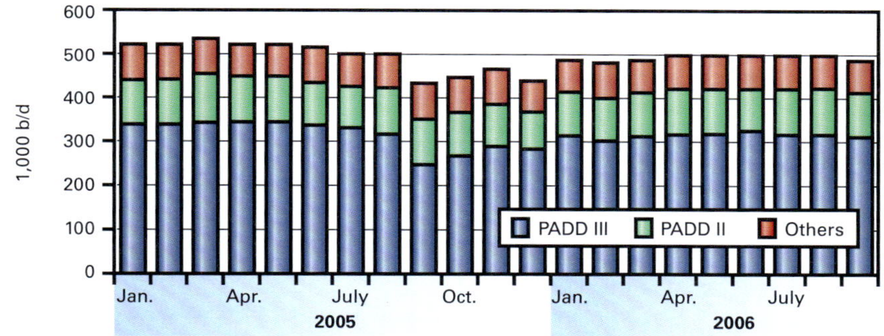

Source: US Energy Information Administration

Total US Propane Production

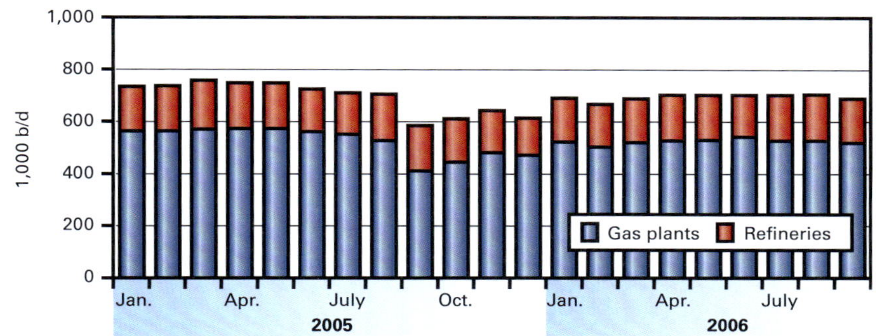

Source: US Energy Information Administration.

US Propane Inventory

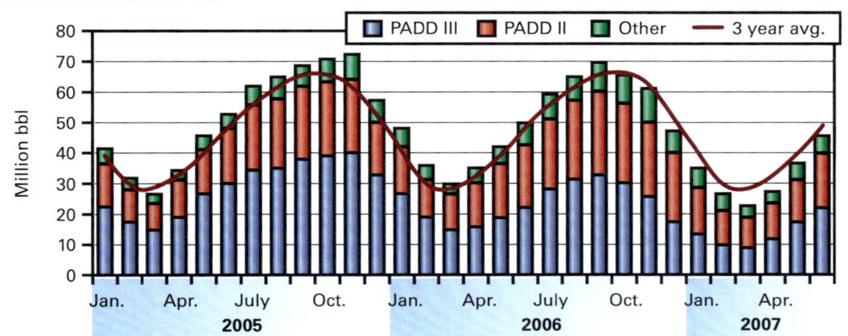

Source: US Energy Information Administration.

Section 12:

Petrochemicals outlook

Global ethylene market and outlook

The global ethylene industry added 4.4 mtpy of capacity in 2005, an increase from the 2004 addition rate of 2.1 mtpy, according to an *Oil & Gas Journal* survey published in March 2006.

Worldwide capacity as of January 1, 2006, was 117.3 mtpy, an increase from 112.9 mtpy a year earlier. This was an increase of nearly 4%.

Nearly 3 mtpy of capacity started up in four new plants. No capacity was removed because of plants being idled or shut down. About 1.4 mtpy of additional capacity resulted from expansions at existing sites, capacity creep, and debottlenecking.

In a ranking of the world's largest ethylene complexes, Nova Chemical Corp.'s Joffre plant in Alberta, Canada, retains the top spot. There is one newcomer—the Chocolate Bayou, Texas, plant of Ineos Olefins & Polymers USA, previously owned by BP PLC.

A few others moved positions due to changing capacities: ChevronPhillips Chemical Co.'s Sweeney, Texas, complex (decrease); Dow Chemical Co.'s Terneuzen, Netherlands, plant (increase); and Dow's Freeport, Texas, plant (increase).

Formosa Plastics Corp.'s Point Comfort, Texas, plant fell off the list.

Top 10 Ethylene Complexes*

	Company	Location	Capacity, tpy
1	Nova Chemicals Corp.	Joffre, Alta., Canada	2,812,000
2	Arabian Petrochemical Co.	Jubail, Saudi Arabia	2,250,000
3	ExxonMobil Chemical Co.	Baytown, Tex.	2,197,000
4	ChevronPhillips Chemical Co.	Sweeny, Tex.	1,868,000
5	Dow Chemical Co.	Terneuzen, Netherlands	1,800,000
6	Ineos Olefins & Polymers	Chocolate Bayou, Tex.	1,752,000
7	Equistar Chemicals LP	Channelview, Tex.	1,750,000
8	Yanbu Petrochemical Co.	Yanbu, Saudi Arabia	1,705,000
9	Dow Chemical Co.	Freeport, Tex.	1,640,000
10	Shell Chemicals Ltd.	Norco, La.	1,556,000

*As of Jan. 1, 2006.

On a list of the top 10 owners of ethylene capacity worldwide, there were only two changes in 2005. The first is the appearance of Ineos, previously noted. The other is the appearance of BASF in tenth place, due to the start-up of the BASF-YPC plant in China, of which it owns 50%. That move knocked Nova Chemicals Corp. off the list.

Global Ethylene Capacity Additions

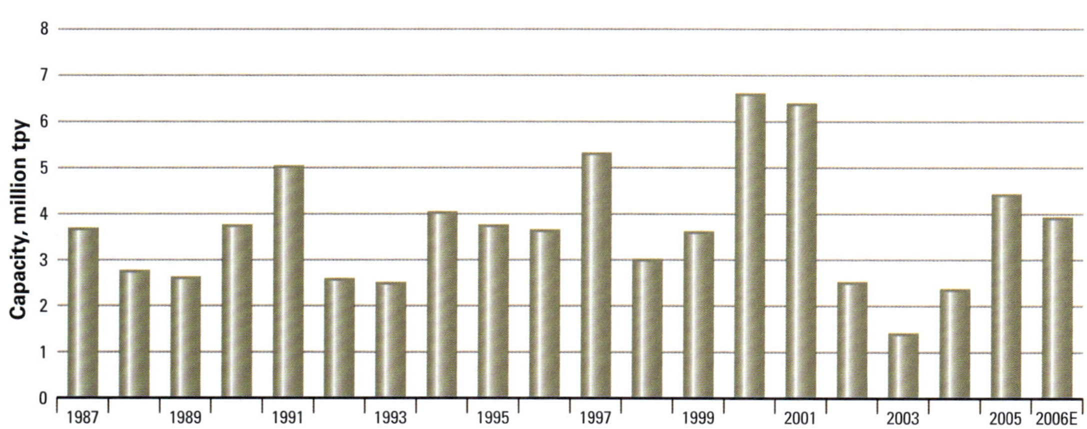

The Asia-Pacific region added 1.5 mtpy and the Middle East/Africa region added more than 1.1 mtpy of capacity. Other regions also showed substantial increases.

On a percentage basis, South America and the Middle East showed the most significant growth.

A country-by-country analysis shows China had the largest increase in ethylene production capacity—1.65 mtpy due to two new plants. Saudi Arabia had the second largest increase. Poland and the United States both increased capacity more than 300,000 tpy.

According to the latest *OGJ* construction data, nearly 3.9 mtpy of new capacity were slated to come on stream in 2006, and more than 13 mtpy involving 14 projects, in 2008. Seven of the projects list capacities of more than 1 mtpy each. Almost all are in the Middle East.

Most future growth in ethylene capacity will occur in the Middle East and Asia due to feedstock advantages. North America and Western Europe will experience relatively flat capacity growth. North American capacity growth will continue to be limited due to high natural gas costs.

Top 10 ethylene producers[1]

			Capacity, tpy	
	Company	No. of sites	Of entire complexes	With only company partial interests
1	Dow Chemical Co.	14	13,155,000	10,369,500
2	ExxonMobil Corp.	15	11,460,000	8,327,000
3	Royal Dutch Shell PLC	10	8,935,000	6,816,000
4	Saudi Basic Industries Corp.	6	8,095,000	6,298,000
5	Ineos	8	6,546,000	5,091,000
6	Lyondell Chemical Co. (Equistar Chemicals LP)	6	4,880,000	4,880,000
7	Sinopec	9	4,375,000	4,075,000
8	Chevron Phillips Chemical Co. LP[2]	4	3,956,000	3,701,000
9	Total AS	9	5,523,000	3,327,000
10	BASF AG	7	4,932,600	3,100,000

[1]As of Jan. 1, 2006. [2]Ownership: ChevronTexaco Corp. 50%, ConocoPhillips 50%.

Regional capacity breakdown

	Ethylene capacity, tpy		Change	
	Jan. 1, 2006	Jan. 1, 2005	tpy	%
Asia-Pacific	31,597,000	30,095,000	1,502,000	4.99
Eastern Europe	8,452,000	8,137,000	315,000	3.87
Middle East, Africa	12,357,000	11,217,000	1,140,000	10.16
North America	35,542,600	35,114,000	428,600	1.22
South America	4,938,500	4,385,500	553,000	12.61
Western Europe	24,443,000	23,957,000	486,000	2.03
Total capacity	117,330,100	112,905,500	4,424,600	3.92

National ethylene capacities

Country	Jan. 1, 2006 Ethylene capacity, tpy	Jan. 1, 2005 Ethylene capacity, tpy	Change, tpy
Algeria	133,000	133,000	0
Argentina	838,500	808,500	30,000
Australia	532,000	504,000	28,000
Austria	500,000	345,000	155,000
Azerbaijan	330,000	330,000	0
Belarus	193,000	193,000	0
Belgium	2,180,000	2,180,000	0
Brazil	3,355,000	2,835,000	520,000
Bulgaria	400,000	400,000	0
Canada	5,418,000	5,407,000	11,000
Chile	45,000	42,000	3,000
China	6,988,000	5,338,000	1,650,000
China, Taiwan	2,421,000	2,421,000	0
Colombia	100,000	100,000	0
Croatia	90,000	90,000	0
Czech Republic	485,000	510,000	–25,000
Egypt	300,000	300,000	0
Finland	330,000	330,000	0
France	3,383,000	3,373,000	10,000
Germany	5,567,000	5,516,000	51,000
Greece	20,000	20,000	0
Hungary	620,000	620,000	0
India	2,475,000	2,463,000	12,000
Indonesia	520,000	510,000	10,000
Iran	1,214,000	1,214,000	0
Israel	200,000	200,000	0
Italy	2,170,000	2,170,000	0
Japan	7,300,000	7,507,000	–207,000
Kazakhstan	130,000	130,000	0
Kuwait	800,000	800,000	0
Libya	350,000	350,000	0
Malaysia	1,649,000	1,630,000	19,000
Mexico	1,384,000	1,384,000	0
Netherlands	3,950,000	3,915,000	35,000
Nigeria	300,000	300,000	0
North Korea	60,000	60,000	0
Norway	550,000	450,000	100,000
Poland	700,000	360,000	340,000
Portugal	330,000	330,000	0
Qatar	1,030,000	1,030,000	0
Romania	844,000	844,000	0
Russia	3,670,000	3,670,000	0
Saudi Arabia	6,845,000	5,845,000	1,000,000
Serbia and Montenegro	200,000	200,000	0
Singapore	1,940,000	1,940,000	0
Slovakia	200,000	200,000	0
South Africa	585,000	445,000	140,000
South Korea	5,440,000	5,450,000	–10,000
Spain	1,430,000	1,430,000	0
Sweden	625,000	610,000	15,000
Switzerland	33,000	33,000	0
Thailand	2,272,000	2,272,000	0
Turkey	520,000	400,000	120,000
Ukraine	450,000	450,000	0
UAE	600,000	600,000	0
United Kingdom	2,855,000	2,855,000	0
United States	28,740,600	28,323,000	417,600
Uzbekistan	140,000	140,000	0
Venezuela	600,000	600,000	0
Total	**117,330,100**	**112,905,500**	**4,424,600**

Ethylene Expansions, 2006-10

Location	Company	2006	2007	2008	2009	2010
		Ethylene capacity, tpy				
Antwerp, Belgium	BASF Antwerpen NV		280,000			
Santo Andre, Sao Paulo, Brazil	Petroquimica Uniao SA			200,000		
Fujian Province, China	Fujian Petrochemical Co. Ltd.			800,000		
Lanzhou, China	Lanzhou Petrochemical Co.			360,000		
Dushanzi, Xinjang, China	PetroChina Dushanzi Petrochemical Co.			1,000,000		
Maoming, Guangdong, China	Sinopec					320,000
Ningbo, Zhejiang, China	Zhenhai Refining & Chemical Co. Ltd.					1,000,000
Litvinov, Czech Republic	Chemopetrol AS	50,000				
Gelsenkirchen, Germany	Ineos		90,000			
Bandar Imam, Iran	Arvand Petrochemical Co.		1,000,000			
Assaluyeh Bushehr, Iran	Arya Sasol Polymer Co.	1,000,000				
Ilam, Iran	Ilam Petrochemical Co.			318,000		
Assaluyeh Bushehr, Iran	Jam Petrochemical Co.	1,320,000				
Kharg Island, Iran	Kharg Petrochemical Co.			500,000		
Bandar Imam, Iran	Marun Petrochemical Co.	1,100,000				
Bandar Assaluyeh, Iran	Petrochemical Industries Dev. Mgmt. Co.			1,200,000		
Bandar Assaluyeh, Iran	Petrochemical Industries Dev. Mgmt. Co.				1,900,000	
Haifa, Israel	Carmel Olefins Ltd.	50,000				
Shuaiba, Kuwait	Petrochemical Industries Co., Dow Chemical Co.			850,000		
Geleen, Netherlands	Saudi Basic Industries Corp.				535,000	
Sohar, Oman	Dow Chemical Co.				850,000	
Mesaieed, Qatar	Qatar Chemical Co. Ltd.	195,000				
Mesaieed, Qatar	Qatar Petroleum Co., Honam Petrochemical Co.				900,000	
Ras Laffan, Qatar	Qatar Chemical Co. Ltd., ChevronTexaco Corp., Total AS			1,300,000		
Ras Laffan, Qatar	Qatar Petroleum Co., ExxonMobil Corp.			1,600,000		
Al-Jubail, Saudi Arabia	Chevron Phillips Chemical Co. LLC		200,000			
Al-Jubail, Saudi Arabia	Eastern Petrochemical Co.			1,300,000		
Rabigh, Saudi Arabia	Saudi Aramco, Sumitomo Chemical Co.			1,500,000		
Yanbu, Saudi Arabia	Saudi Basic Industries Corp.			1,300,000		
Singapore	ExxonMobil Corp.	75,000				
Yeosu, South Korea	LG Daesan Petrochemicals	100,000				
Rayong, Thailand	Siam Cement PLC, Dow Chemical Co.					900,000
Jose, Anzoategui, Venezuela	Pequiven			1,050,000		
Total		**3,890,000**	**1,570,000**	**13,278,000**	**4,185,000**	**2,220,000**

Source: Oil & Gas Journal construction survey

According to Chemical Market Associates, Inc. (CMAI; Houston), global ethylene production capacity will grow on average at 5.4%/year in 2005–2010, up from 3.5%/year in 2000–2005.

Ethane will be the most in-demand feedstock during 2005–2010. Its demand growth will average 7.1%/year in 2005–2010, up from 3.2%/year in 2000–2005.

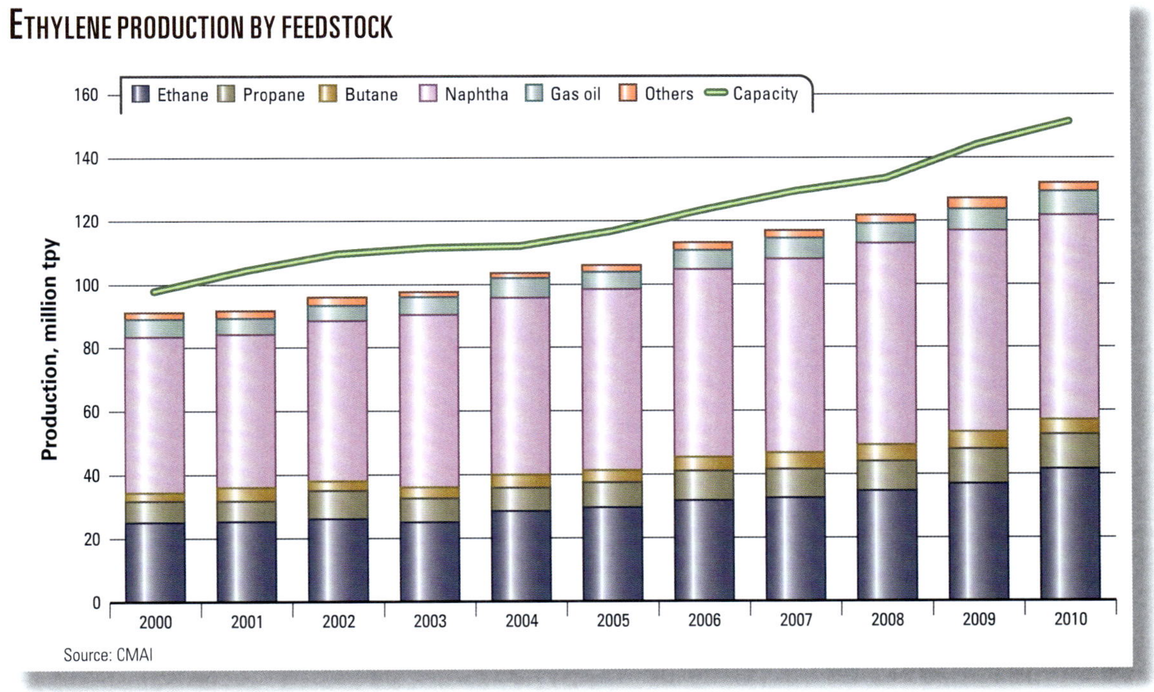

CMAI estimates that ethylene demand growth will average 4.3%/year during 2005–2010. Most growth will occur for ethylene oxide, which will increase 5.4%/year during 2005–2010, up from 3.7%/year during 2000–2005. Polyethylene will have the next-highest average growth of 4.6%/year in 2005–2010.

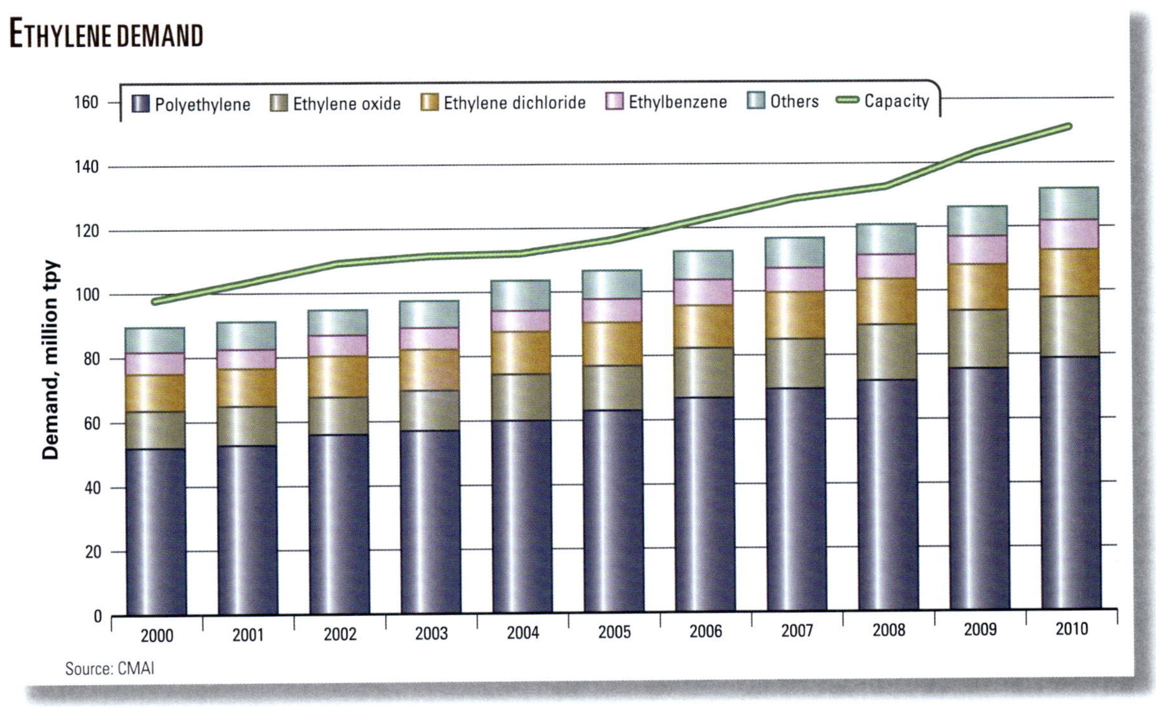

Middle East producers are projected to dominate the net equivalent ethylene trade in the future. All other regions will become net importers by 2009, according to CMAI.

ETHYLENE TRADE

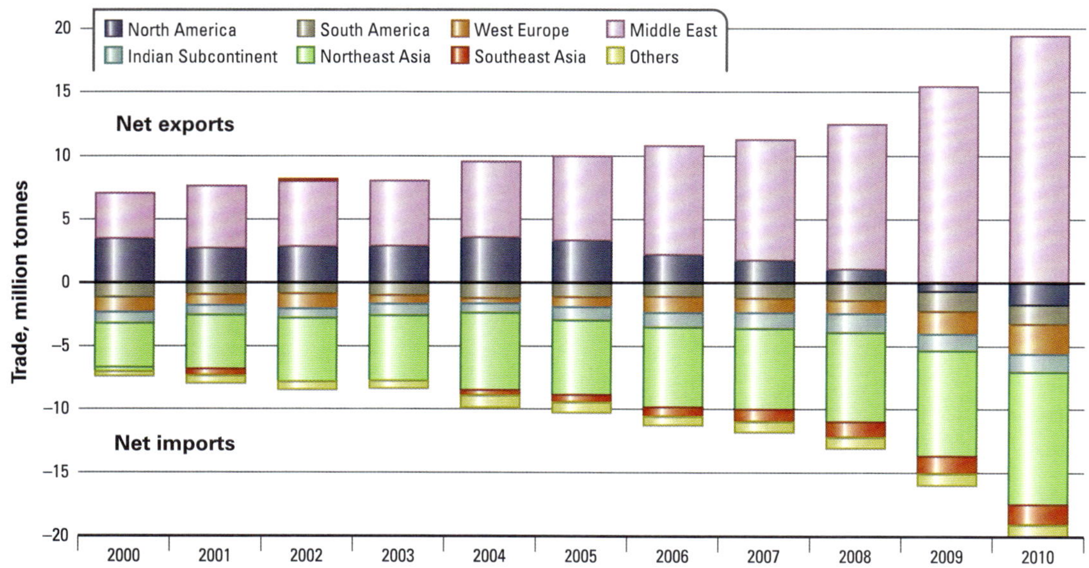

Source: CMAI

Ethylene prices, which have increased steadily since 2002 for U.S. producers, were expected to peak in 2006 but then remain between 40–50¢/lb through 2007.

US ETHYLENE PRICES

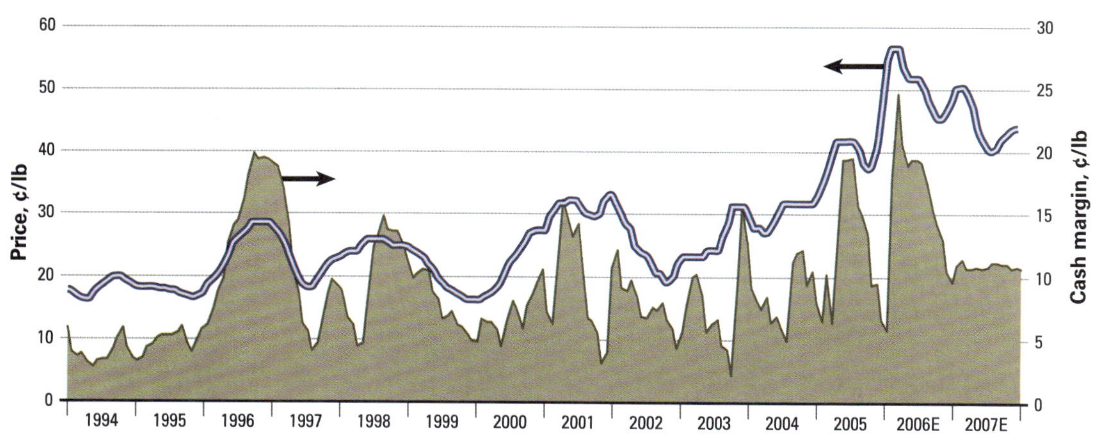

Source: Lehman Bros.

Saudi Arabia plans major petrochem push

Saudi Arabia aims to become a global hub for commodity petrochemical manufacturing by 2010.

The Saudi Arabian General Investment Authority (SAGIA) said that the country is working to increase its overall share of total world petrochemical output to 13% by 2010, compared to 7% in 2005.

Petrochemical production capacity increased to 36 mtpy in 2005 from 3.7 mtpy in 1985. SAGIA expects capacity to grow to 70 mtpy 2010.

Saudi Petrochemical Production

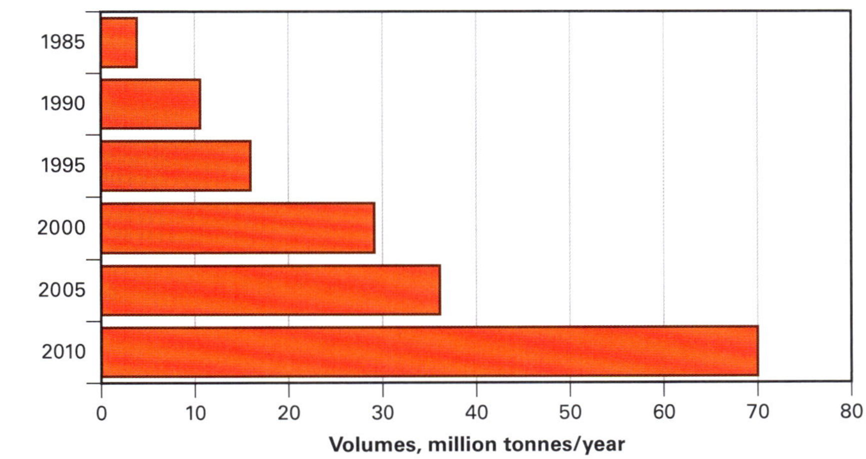

By 2010, Saudi Arabia aims to be one of the top four global producers for key commodity petrochemicals.

Saudi Production, 2004

Company	Basic petrochemicals: Methanol	Ethylene	Propylene	Benzene	Primary derivatives (1,000 tonnes/year): Methyl tertiary butyl ether	Polyethylene	Polypropylene	Mono-ethylene glycol	Ethylene dichloride	Styrene
Saudi Basic Industries Corp.	3,930	6,770	1,005	500	2,900	3,995	900	2,840	840	1,060
Tasnee Petrochemicals Co.	—	—	450	—	—	—	450	—	—	—
Saudi International Petrochemical Co.	950	—	—	—	—	—	—	—	—	—
Saudi Chevron Phillips Co.	—	—	—	550	—	—	—	—	—	—
Total	**4,880**	**6,770**	**1,450**	**1,050**	**2,900**	**3,995**	**1,350**	**2,840**	**840**	**1,060**

Saudi Arabia Petrochemical Production, 2010

Product	Capacity, million tonnes	Global ranking
Methanol	7.60	2
Ethylene	11.50	4
High-density polyethylene	3.85	3
Linear, low-density polyethylene	3.55	2
Low-density polyethylene	1.20	4
Mono-ethylene glycol	5.50	1
Propylene	5.19	4
Polypropylene	5.20	3
Styrene	2.25	2

Petrochemical producers in the country are planning to add 35 million tonnes of new capacity during 2004–2010, more than 17 million tonnes of it for basic petrochemicals.

Basic Petrochemical Projects

Petrochemical, companies building capacity	Capacity, 1,000 tonnes	Start-up dates
Methanol		
Saudi International Petrochemical Co., Tasnee Petrochemicals Co., Saudi Formaldehyde Chemical Company Ltd.	4,500	2007-08
Ethylene		
Saudi Basic Industries Corp. (YanSab, Sharq), Jubail Chevron Phillips Co., Sahara Petrochemicals Co./Tasnee Petrochemicals Co., Kayan Petrochemicals Co., Rabigh Refining & Petrochemical Co.	7,350	2007-09
Propylene		
Sahara Petrochemicals Co., National Polypropylene Co., Lujain Co., Jubail Chevron Phillips Co., Saudi Basic Industries Corp. (YanSab, Sharq), Kayan Petrochemicals Co., Rabigh Refining & Petrochemical Co.	3,920	2006-09
Benzene derivatives (styrene)		
Saudi Basic Industries Corp., Saudi Chevron Phillips Co.	1,215	2007
Total	**16,985**	

Planned capital investments in the Saudi petrochemical industry from 2004–2009 are projected to exceed $40 billion. There are 59 petrochemical projects in the pipeline, 74% of which are greenfield facilities.

Planned Projects, 2004-09

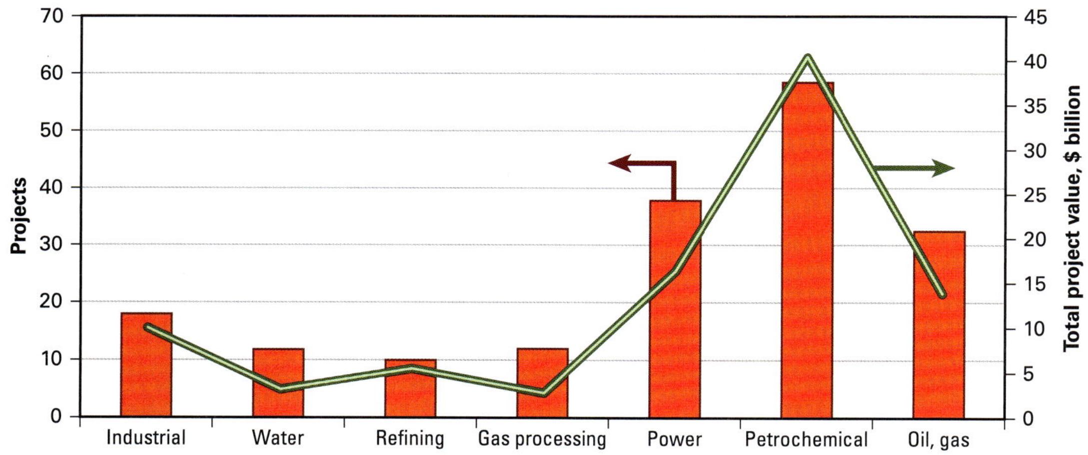

Source: Contax, 2005

Lost manufacturing slows U.S. petrochem growth

The U.S. petrochemical business is experiencing a lower level of growth due to lost manufacturing or "deindustrialization" trends, said Accenture Research (Houston).

In a study for the American Chemistry Council, Accenture quantified the impact of lost downstream manufacturing, in 17 selected industries, on the future chemical industry.

The study concluded that although domestic production of finished goods will still increase, imports will rise faster. This implies that U.S. manufacturers will lose market share and, therefore, chemicals manufacturers will lose the demand for chemicals associated with manufacturing these products.

High Gas Costs Lead to Shift in Ethylene Trade

High gas commodity prices in North America have accelerated a shift in investment strategy by the region's ethylene producers, according to Chemical Market Associates Inc. (CMA; Houston). Investment in new or expansion capacity in North America has moved to the Middle East.

CMA describes a "global assembly line," based on the fixed low costs of feedstock in the Middle East and on the world's largest market for durable and non-durables in North America.

This transition began in about 1995, when the United States was the world's largest exporter of ethylene to the world. By 2004, the United States was a net importer of ethylene-derived products. The Middle East had become the world's largest exporter.

CMA expects no expansions or new ethylene production capacity in North America for the foreseeable future.

CHEMISTRY USE LOSSES*

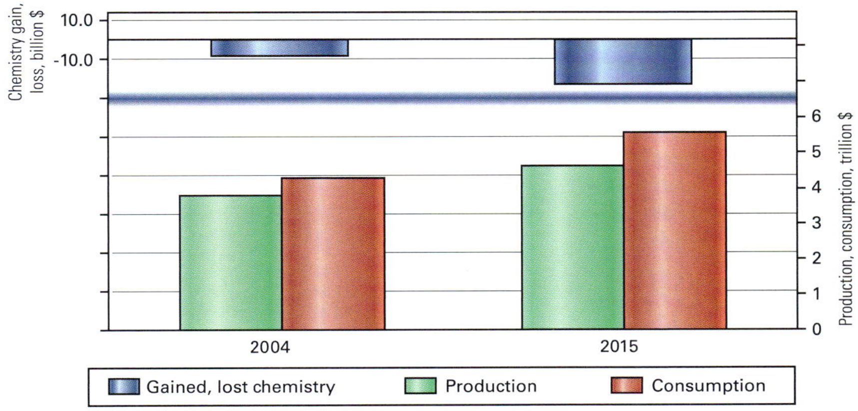

*Aggregate of 17 industries. 2005 constant dollars.

For example, by 2015, about 11 million tonnes of the top seven thermoplastics will be imported instead of produced domestically.

As of late 2006, no greenfield plants for producing ethylene—the most important base organic chemical—were scheduled for completion in the continental United States.

The study said that U.S.-based ethylene producers—uncertain about potential environmental legislation and energy policy—have announced construction plans for about 6 million tonnes of new ethylene capacity investment (equivalent to about $15 billion) during the next 10 years in other regions.

US ETHYLENE PRODUCERS' INVESTMENTS, 2005-15*

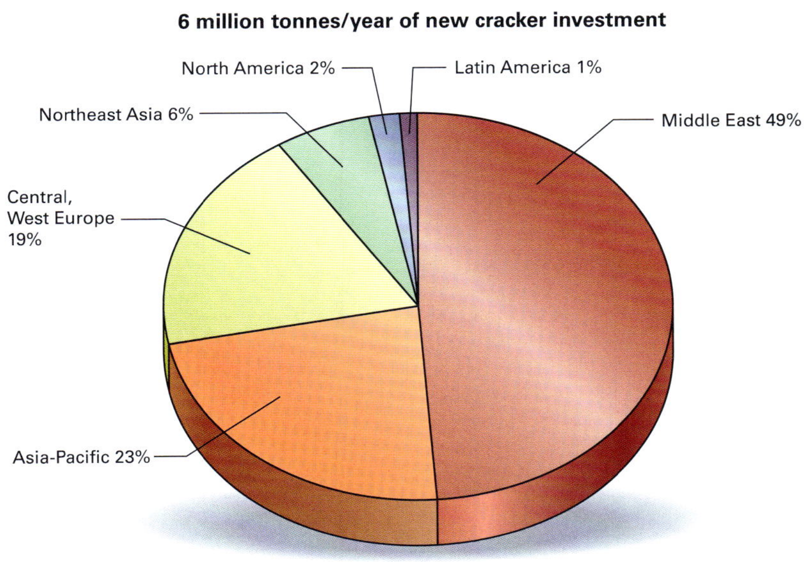

*Accenture Research analysis. Based on Parpinelli Tecnon data.

Table 1: World Reserves and Production

	ESTIMATED PROVED RESERVES				OIL PRODUCTION			
	Jan. 1, 2007		Jan. 1, 2006					
COUNTRY	Oil, 1,000 bbl	Gas, bcf	Oil, 1,000 bbl	Gas, bcf	Producing oil wells, Dec. 31, 2005	Estimated 2006, 1,000 b/d	Change from 2005, %	Actual 2005, 1,000 b/d
Canada	179,210,000	57,946	178,792,400	56,577	58,966	2,500.0	5.5	2368.8
US	21,757,000	204,385	21,371,000	192,513	507,928	5,135.0	–0.8	5,178.4
North America	200,967,000	262,331	200,163,400	249,090	566,894	7,635.0	1.2	7,547.2
Argentina	2,468,000	16,090	2,320,450	18,866	15,874	635.0	–1.6	645.4
Bolivia	440,000	24,000	440,500	24,000	346	45.0	8.2	41.6
Brazil	11,772,640	10,820	11,243,300	11,515	11,995	1,710.0	4.7	1,634.0
Chile	150,000	3,460	150,000	3,460	315	10.0	0	10.0
Colombia	1,453,000	3,996	1,542,000	4,040	7,644	530.0	0.7	526.1
Cuba	124,000	2,500	750,000	2,500	251	39.0	1.0	38.6
Ecuador	4,517,000	0	4,629,600	345	1,044	500.0	–6.0	532.0
Guatemala	83,070	0	526,000	109	20	16.5	–10.6	18.5
Mexico	12,352,000	14,557	12,882,200	15,985	3,052	3,273.0	–1.8	3,334.2
Peru	929,600	8,723	929,600	8,723	4,660	113.0	1.6	111.2
Trinidad & Tobago	728,300	18,770	990,000	25,880	3,821	150.0	3.8	144.5
Venezuela	80,012,000	152,380	79,729,000	151,395	15,669	2,563.0	–5.3	2,705.8
Other Latin America	120,552	6	113,500	5	600	14.8	13.9	13.0
Latin America	115,150,162	255,302	116,246,150	266,823	65,291	9,599.3	-1.6	9,754.9
Albania	198,130	30	198,130	30	1,338	7.0	15.3	6.1
Austria	50,000	570	62,000	530	905	17.4	0.6	17.3
Bulgaria	15,000	200	15,000	210	100	1.0	0	1.0
Croatia	74,320	1,052	69,144	1,000	864	16.8	–4.0	17.5
Czech Republic	15,000	140	15,000	140	0	4.7	–2.1	4.8
Denmark	1,277,000	2,542	1,328,000	2,786	217	342.0	–9.4	377.3
France	121,500	341	158,400	378	463	21.5	–0.8	21.7
Germany	367,000	9,000	367,200	9,076	1,006	69.0	–3.4	71.5
Greece	5,000	35	7,000	35	12	1.8	0	1.8
Hungary	20,180	286	102,480	1,210	875	16.5	–13.2	19.0
Ireland	0	350	0	350	0	0	0	0
Italy	600,000	5,800	621,700	8,000	207	105.0	–8.7	115.0
Lithuania	12,000	0	12,000	0	0	3.6	–16.3	4.3
Netherlands	100,000	50,000	106,000	62,000	203	23.0	–20.7	29.0
Norway	7,849,300	82,320	7,705,000	84,260	801	2,465.0	–8.6	2,697.6
Poland	96,375	5,820	96,375	5,820	512	17.0	–14.6	19.9
Romania	600,000	2,225	955,620	3,550	6,000	98.4	–1.6	100.0
Serbia	77,500	1,700	77,500	1,700	646	15.0	0	15.0
Slovakia	9,000	500	9,000	530	0	1.0	0	1.0
Spain	150,000	90	157,626	90	3,300	3.0	–9.9	3.3
Turkey	300,000	300	300,000	300	852	42.0	0	42.0
UK	3,875,000	17,000	4,029,480	18,750	1,360	1,480.0	–11.8	1,678.0
Europe	15,812,305	180,301	16,392,655	200,745	19,661	4,750.7	-9.4	5,243.1
Abu Dhabi	92,200,000	198,500	92,200,000	198,500	1,200	2,450.0	6.5	2,300.0
Bahrain	124,560	3,250	124,560	3,250	496	172.0	–1.4	174.4
Dubai	4,000,000	4,000	4,000,000	4,000	200	90.0	–11.8	102.0
Iran	136,270,000	974,000	132,460,000	971,150	1,120	3,850.0	–1.0	3,890.8
Iraq	115,000,000	112,000	115,000,000	111,950	1,685	1,915.0	5.8	1,810.0
Israel	1,960	1,275	2,000	1,375	6	0.1	0.0	0.1
Jordan	1,000	213	1,000	220	4	0	0	0
Kuwait	99,000,000	54,500	101,500,000	55,515	790	2,200.0	3.3	2,130.0
Neutral Zone	5,000,000	1,000	5,000,000	1,000	578	575.0	–0.9	580.0
Oman	5,500,000	30,000	5,506,000	29,280	2,298	738.0	–2.6	757.5
Qatar	15,207,000	910,500	15,207,000	910,520	421	820.0	2.7	798.3
Ras al Khaimah	100,000	1,200	100,000	1,200	7	0.7	–30.0	1.0
Saudi Arabia	259,800,000	239,500	264,310,000	241,340	1,560	8,990.0	–0.8	9,060.0
Sharjah	1,500,000	10,700	1,500,000	10,700	49	50.0	0	50.0
Syria	2,500,000	8,500	2,500,000	8,500	132	422.5	–8.2	460.0
Yemen	3,000,000	16,900	4,000,000	16,900	1,402	415.0	0.9	412.8
Middle East	739,204,520	2,565,538	743,410,560	2,565,400	11,948	22,688.3	0.7	22,526.9

Table 1: World Reserves and Production, cont'd.

COUNTRY	ESTIMATED PROVED RESERVES Jan. 1, 2007 Oil, 1,000 bbl	Jan. 1, 2007 Gas, bcf	Jan. 1, 2006 Oil, 1,000 bbl	Jan. 1, 2006 Gas, bcf	OIL PRODUCTION Producing oil wells, Dec. 31, 2005	Estimated 2006, 1,000 b/d	Change from 2005, %	Actual 2005, 1,000 b/d
Algeria	12,270,000	161,740	11,350,000	160,505	1,285	1,345.0	–0.5	1,351.7
Angola	8,000,000	2,000	5,412,000	1,620	1,068	1,385.0	11.9	1,238.1
Cameroon	400,000	3,900	400,000	3,900	255	87.5	6.1	82.5
Chad	1,500,000	0	1,500,000	0	250	180.0	0	180.0
Congo (Brazzaville)	1,600,000	3,200	1,505,900	3,200	460	240.0	0	240.0
Congo (Zaire)	180,000	35	187,000	35	150	20.0	0	20.0
Egypt	3,700,000	58,500	3,700,000	58,500	1,267	667.5	–4.0	695.0
Equitorial Guinea	1,100,000	1,300	12,000	1,300	38	320.0	–8.6	350.0
Gabon	2,000,000	1,000	2,499,000	1,200	393	232.0	–0.9	234.2
Ivory Coast	100,000	1,000	100,000	1,000	9	30.0	0	30.0
Libya	41,464,000	52,650	39,126,000	52,650	1,472	1,700.0	3.7	1,640.0
Mauritania	100,000	1,000	0	0	0	32.0	0	0
Morocco	988	58	1,069	60	7	0.2	0	0.2
Mozambique	0	4,500	0	4,500	0	0	0	0
Nigeria	36,220,000	181,900	35,876,000	184,660	2,379	2,220.0	–7.8	2,406.7
South Africa	15,000	0	15,680	1	28	20.0	–16.6	24.0
Sudan	5,000,000	3,000	563,000	3,000	9	295.0	1.7	290.0
Tanzania	0	230	0	800	0	0	0	0
Tunisia	400,000	2,300	307,560	2,750	239	65.0	–8.6	71.1
Other Africa	23,428	6,120	25,138	6,160	11	6.0	0	6.0
Africa	**114,073,416**	**484,433**	**102,580,347**	**485,841**	**9,320**	**8,845.2**	**-0.2**	**8,859.5**
Azerbaijan	7,000,000	30,000	7,000,000	30,000	130	625.0	41.0	443.3
Belarus	198,000	100	198,000	100	0	35.0	–1.4	35.5
Georgia	35,000	300	35,000	300	281	2.0	0	2.0
Kazakhstan	30,000,000	100,000	9,000,000	65,000	705	1,060.0	6.6	994.2
Kyrgyzstan	40,000	200	40,000	200	0	1.0	0	1.0
Russia	60,000,000	1,680,000	60,000,000	1,680,000	38,173	9,475.0	3.1	9,190.0
Tajikistan	12,000	200	12,000	200	0	0	0	0
Turkmenistan	600,000	100,000	546,000	71,000	2,460	165.0	–15.4	195.0
Ukraine	395,000	39,000	395,000	39,600	1,353	97.0	–1.0	98.0
Uzbekistan	594,000	65,000	594,000	66,200	2,190	105.0	–7.1	113.0
Former Soviet Union	**98,874,000**	**2,014,800**	**77,820,000**	**1,952,600**	**45,292**	**11,565.0**	**4.5**	**11,072.0**
Afghanistan	0	1,750	0	1,750	0	0	0	0
Australia	1,591,790	30,370	1,437,000	27,640	1,325	450.0	–5.3	471.5
Bangladesh	28,000	5,000	28,000	5,000	41	4.0	0	4.0
Brunei	1,100,000	13,800	1,350,000	13,800	779	197.0	5.62	186.5
China	16,000,000	80,000	18,250,000	53,325	71,542	3,700.0	2.0	3,627.1
India	5,624,640	37,960	5,847,840	38,880	3,674	682.0	3.5	659.1
Indonesia	4,300,000	97,780	4,301,000	97,786	8,331	892.0	–5.6	945.0
Japan	58,500	1,400	58,500	1,400	145	15.0	–4.0	15.6
Malaysia	3,000,000	75,000	3,000,000	75,000	788	737.0	–4.3	770.0
Myanmar	50,000	10,000	50,000	10,000	450	13.0	0	13.0
New Zealand	53,000	900	53,000	900	72	14.0	0.7	13.9
Pakistan	289,202	28,000	289,202	28,153	204	63.5	–1.6	64.5
Papua New Guinea	240,000	12,200	240,000	12,200	44	56.5	22.8	46.0
Philippines	138,500	3,480	138,500	3,960	11	16.0	4.6	15.3
Taiwan	2,380	297	2,380	297	71	1.0	0	1.0
Thailand	290,000	14,750	291,000	14,754	905	212.0	14.0	186.0
Vietnam	600,000	6,800	600,000	6,800	28	350.0	2.9	340.0
Asia-Pacific	**33,366,012**	**419,487**	**35,936,422**	**391,645**	**88,410**	**7,403.0**	**0.6**	**7,358.5**
Total World	**1,317,447,415**	**6,128,212**	**1,292,549,534**	**6,112,144**	**806,816**	**72,486.5**	**0.2**	**72,362.1**
Total OPEC	**902,343,000**	**3,152,350**	**901,659,000**	**3,152,871**	**36,746**	**29,660.7**	**–0.4**	**29,771.3**

Source: *Oil & Gas Journal* data as of Jan. 1, 2007

NOTES: All reserves figures are reported as proved reserves recoverable with current technology and prices except Former Soviet Union, some eastern European countries, and Canada's gas reserves; these figures include proved plus some probable. Some totals might not compute exactly due to rounding and other factors. Other Latin America includes Barbados and Suriname. Other Africa includes Benin, Ethiopia, Ghana, Namibia, Rwanda, and Somalia.

Table 2: World Oil Consumption

COUNTRY	2003	2004	2005	2003	2004	2005	Change 2005 over 2004	2005 share of total
	Million tonnes per year (mtpy)			Thousand barrels per day (1,000 b/y)			%	
North America								
US	912.3	948.8	944.6	20,033	20,732	20,655	-0.2	24.6
Canada	95.9	100.6	100.1	2,132	2,248	2,241	-0.2	2.6
Mexico	83.7	85.2	87.8	1,885	1,898	1,978	3.3	2.3
Total	1,091.8	1,134.6	1,132.6	24,050	24,877	24,875	0.1	29.5
Latin America								
Argentina	17.6	18.7	20.1	372	394	421	7.6	0.5
Brazil	82.0	81.9	83.6	1,785	1,776	1,819	2.4	2.2
Chile	10.5	11.3	11.9	229	244	257	5.5	0.3
Colombia	10.0	10.1	10.4	222	223	230	3.7	0.3
Ecuador	6.2	6.4	6.6	137	144	148	2.7	0.2
Peru	6.5	7.2	6.4	140	152	139	-9.6	0.2
Venezuela	22.0	24.2	25.4	479	525	553	5.3	0.7
Other Latin America	57.2	58.1	58.8	1,173	1,188	1,208	1.6	1.5
Total	212.0	217.9	223.3	4,537	4,647	4,776	2.8	5.8
Europe/Former Soviet Union								
Austria	14.1	13.8	14.2	293	285	294	3.4	0.4
Azerbaijan	4.3	4.6	5.1	86	92	103	11.3	0.1
Belarus	7.4	7.5	6.7	152	153	137	-10.4	0.2
Belgium/ Luxembourg	36.4	38.4	39.5	748	785	809	3.1	1.0
Bulgaria	4.5	4.7	5.0	98	102	109	6.7	0.1
Czech Republic	8.7	9.5	9.9	185	203	211	4.1	0.3
Denmark	9.2	9.1	9.1	193	189	189	0.2	0.2
Finland	11.4	10.6	11.0	239	224	233	4.2	0.3
France	93.1	94.0	93.1	1,965	1,978	1,961	-0.7	2.4
Germany	125.1	124.0	121.5	2,664	2,634	2,586	-1.7	3.2
Greece	19.6	21.3	20.9	404	435	429	-1.5	0.5
Hungary	6.1	6.3	7.0	132	136	151	10.6	0.2
Iceland	0.9	1.0	0.9	18	20	19	-2.4	--
Ireland	8.5	8.9	9.4	178	185	196	5.4	0.2
Italy	92.1	89.7	86.3	1,927	1,873	1,809	-3.5	2.2
Kazakhstan	8.8	9.0	10.0	183	188	208	11.4	0.3
Lithuania	2.4	2.6	2.7	51	54	57	4.0	0.1
Netherlands	44.1	46.2	49.6	962	1,003	1,071	7.7	1.3
Norway	9.9	9.6	9.8	219	210	213	1.8	0.3
Poland	19.9	21.1	21.9	435	460	478	4.1	0.6
Portugal	15.2	15.4	15.3	317	322	320	-0.5	0.4
Romania	9.4	10.9	11.3	199	230	240	4.5	0.3
Russia	124.7	128.5	130.0	2,645	2,714	2,753	1.4	3.4
Slovakia	3.3	3.2	3.5	71	68	73	9.6	0.1
Spain	75.5	77.6	78.8	1,559	1,593	1,618	1.7	2.1
Sweden	15.9	15.3	15.1	332	319	315	-1.2	0.4
Switzerland	12.1	12.0	12.2	259	258	262	2.1	0.3
Turkey	31.2	32.0	30.0	668	688	650	-5.9	0.8
Turkmenistan	4.2	4.6	4.9	95	103	110	6.8	0.1
Ukraine	13.5	13.9	13.9	286	293	294	0.4	0.4
United Kingdom	79.0	81.7	82.9	1,717	1,764	1,790	1.7	2.2
Uzbekistan	7.2	7.5	7.8	148	155	161	4.3	0.2
Other Europe/FSU	22.9	23.3	24.3	475	482	502	4.4	0.6
Total	940.8	957.6	963.3	19,903	20,195	20,350	0.9	25.1

Table 2: World Oil Consumption, cont'd.

COUNTRY	2003	2004	2005	2003	2004	2005	Change 2005 over 2004	2005 share of total
	Million tonnes per year (mtpy)			Thousand barrels per day (1,000 b/y)			%	
Middle East								
Iran	71.8	74.6	78.4	1,513	1,575	1,659	5.4	2.0
Kuwait	12.2	13.7	14.4	238	266	280	5.2	0.4
Qatar	3.0	3.3	3.8	77	84	98	17.1	0.1
Saudi Arabia	77.7	83.7	87.2	1,684	1,805	1,891	4.5	2.3
UAE	16.3	17.4	18.3	333	355	376	5.5	0.5
Other Middle East	67.4	68.1	69.2	1,394	1,407	1,436	1.9	1.8
Total	248.3	260.7	271.3	5,238	5,492	5,739	4.4	7.1
Africa								
Algeria	10.1	10.6	11.2	231	240	254	6.0	0.3
Egypt	25.9	26.8	29.2	550	567	616	9.3	0.8
South Africa	24.2	24.8	24.9	513	523	529	0.6	0.6
Other Africa	59.8	61.9	64.0	1,274	1,315	1,363	3.6	1.7
Total	120.1	124.2	129.3	2,568	2,646	2,763	4.4	3.4
Asia-Pacific								
Australia	38.3	38.8	39.7	851	856	884	2.7	1.0
Bangladesh	3.8	3.9	4.0	78	80	82	2.8	0.1
China	271.7	318.9	327.3	5,803	6,772	6,988	2.9	8.5
Hong Kong	13.0	15.3	13.8	269	314	285	-9.6	0.4
India	113.1	120.2	115.7	2,420	2,573	2,485	-3.5	3.0
Indonesia	53.9	54.7	55.3	1,132	1,150	1,168	1.4	1.4
Japan	248.9	241.4	244.2	5,455	5,286	5,360	1.4	6.4
Malaysia	22.2	22.8	22.0	480	493	477	-3.3	0.6
New Zealand	6.9	7.0	7.0	148	150	152	1.1	0.2
Pakistan	15.8	16.0	17.4	321	325	353	8.8	0.5
Philippines	15.5	15.8	14.7	330	336	314	-6.9	0.4
Singapore	33.9	38.1	42.2	668	748	826	11.0	1.1
South Korea	105.6	104.9	105.5	2,300	2,283	2,308	0.8	2.7
Taiwan	41.1	41.7	41.6	868	880	884	0.2	1.1
Thailand	40.0	44.0	45.6	836	913	946	4.0	1.2
Other Asia-Pacific	19.0	20.3	21.1	400	427	445	4.3	0.5
Total	1,042.6	1,103.6	1,116.9	22,359	23,586	23,957	1.5	29.1
Total World	**3,655.6**	**3,798.6**	**3,863.8**	**78,655**	**81,444**	**82,459**	**1.3**	**100.0**

Source: Adapted from *BP Statistical Review of World Energy 2006*.

NOTES: Some totals might not compute exactly due to rounding and other factors.
Growth rates are adjusted for leap years.
Differences between these world consumption figures and world production statistics are accounted for by stock changes, consumption of non-petroleum additives and substitute fuels, and unavoidable disparities in the definition, measurement or conversion of oil supply and demand data.

Table 3: World Petroleum Product Consumption

REGION/ PRODUCT	2000	2001	2002	2003	2004	2005	Change 2005 over 2004	2005 share of total
	Thousand barrels per day (1,000 b/d)						%	
US								
Light distillates	8,813	8,890	9,167	9,275	9,518	9,495	-0.2%	46.0%
Middle distillates	5,852	5,884	5,735	5,886	6,116	6,137	0.3%	29.7%
Fuel oil	893	794	686	763	859	907	5.7%	4.4%
Others	4,143	4,082	4,172	4,109	4,239	4,116	-2.9%	19.9%
Total	19,701	19,649	19,761	20,033	20,732	20,655	-0.4%	100.0%
North America (including US)								
Light distillates	10,106	10,211	10,523	10,675	10,979	10,970	-0.1%	44.1%
Middle distillates	6,811	6,812	6,655	6,861	7,133	7,188	0.8%	28.9%
Fuel oil	1,518	1,411	1,209	1,271	1,340	1,390	3.7%	5.6%
Others	5,086	5,137	5,278	5,243	5,425	5,326	-1.8%	21.4%
Total	23,522	23,571	23,665	24,050	24,877	24,875	0.0%	100.0%
South & Central America								
Light distillates	1,274	1,256	1,233	1215	1,184	1,227	3.7%	25.7%
Middle distillates	1,681	1,722	1,719	1723	1,825	1,872	2.6%	39.2%
Fuel oil	788	751	732	690	691	704	1.9%	14.7%
Others	918	1,011	1,003	909	947	973	2.8%	20.4%
Total	4,661	4,739	4,688	4,537	4,647	4,776	2.8%	100.0%
Europe								
Light distillates	4,194	4,085	4,026	3,923	3,870	3,758	-2.9%	22.9%
Middle distillates	6,734	6,991	6,935	7,138	7,360	7,572	2.9%	46.1%
Fuel oil	1,967	1,987	1,996	1,953	1,896	1,882	-0.7%	11.5%
Others	3,047	3,052	3,103	3,137	3,208	3,203	-0.2%	19.5%
Total	15,942	16,116	16,059	16,151	16,334	16,415	0.5%	100.0%
Former Soviet Union								
Light distillates	797	851	882	917	903	923	2.2%	23.4%
Middle distillates	1,008	1,021	1,048	1,087	1,113	1,146	3.0%	29.1%
Fuel oil	765	693	670	670	684	699	2.1%	17.8%
Others	1,053	1,062	1,066	1,078	1,161	1,168	0.6%	29.7%
Total	3,623	3,627	3,667	3,752	3,861	3,936	1.9%	100.0%
Middle East								
Light distillates	909	949	1,011	1,053	1,134	1,210	6.6%	21.1%
Middle distillates	1,565	1,619	1,667	1,715	1,751	1,816	3.7%	31.6%
Fuel oil	1,312	1,326	1,335	1,338	1,415	1,473	4.1%	25.7%
Others	948	960	1,034	1,132	1,192	1,240	4.1%	21.6%
Total	4,735	4,854	5,047	5,238	5,492	5,739	4.5%	100.0%
Africa								
Light distillates	577	587	590	599	613	631	2.8%	22.8%
Middle distillates	1,029	1,048	1,078	1,110	1,134	1,180	4.1%	42.7%
Fuel oil	457	433	415	421	444	477	7.5%	17.3%
Others	395	407	427	438	455	475	4.4%	17.2%
Total	2,458	2,475	2,511	2,568	2,646	2,763	4.4%	100.0%
China								
Light distillates	1,270	1,256	1,352	1,489	1,673	1,749	4.6%	25.0%
Middle distillates	1,590	1,669	1,791	1,945	2,285	2,425	6.1%	34.7%
Fuel oil	775	780	792	880	956	899	-6.0%	12.9%
Others	1,136	1,167	1,352	1,489	1,858	1,915	3.0%	27.4%
Total	4,772	4,872	5,288	5,803	6,772	6,988	3.2%	100.0%

Table 3: World Petroleum Product Consumption, cont'd.

REGION/ PRODUCT	2000	2001	2002	2003	2004	2005	Change 2005 over 2004	2005 share of total
	Thousand barrels per day (1,000 b/d)						%	
Japan								
Light distillates	1,735	1,720	1,759	1,788	1,793	1,819	1.5%	33.9%
Middle distillates	1,958	1,957	1,930	1,904	1,861	1,880	1.0%	35.1%
Fuel oil	804	690	649	770	672	698	3.9%	13.0%
Others	1,079	1,067	1,021	993	960	962	0.3%	18.0%
Total	5,577	5,435	5,359	5,455	5,286	5,360	1.4%	100.0%
Asia-Pacific (including China, Japan)								
Light distillates	5,543	5,650	5,927	6,154	6,494	6,600	1.6%	27.5%
Middle distillates	7,765	7,916	8,050	8,171	8,650	8,810	1.9%	36.8%
Fuel oil	3,574	3,413	3,339	3,462	3,486	3,526	1.1%	14.7%
Others	3,956	4,019	4,327	4,571	4,956	5,021	1.3%	21.0%
Total	20,839	20,998	21,644	22,359	23,586	23,957	1.6%	100.0%
World								
Light distillates	23,400	23,588	24,192	24,536	25,178	25,319	0.6%	30.7%
Middle distillates	26,593	27,130	27,153	27,804	28,966	29,584	2.1%	35.9%
Fuel oil	10,381	10,014	9,696	9,807	9,956	10,150	2.0%	12.3%
Others	15,405	15,648	16,239	16,508	17,344	17,406	0.4%	21.1%
Total	75,779	76,379	77,280	78,655	81,444	82,459	1.2%	100.0%
Total European Union 25								
Light distillates	3,858	3,757	3,688	3,599	3,525	3,428	-2.8%	23.2%
Middle distillates	6,146	6,373	6,316	6,495	6,687	6,859	2.6%	46.4%
Fuel oil	1,720	1,724	1,740	1,700	1,668	1,689	1.2%	11.4%
Others	2,678	2,698	2,728	2,753	2,806	2,797	-0.3%	19.0%
Total	14,402	14,553	14,471	14,546	14,687	14,772	0.6%	100.0%
OECD								
Light distillates	17,069	17,064	17,391	17,502	17,759	17,684	-0.4%	35.9%
Middle distillates	16,279	16,530	16,296	16,682	17,109	17,388	1.6%	35.3%
Fuel oil	4,642	4,416	4,175	4,309	4,218	4,266	1.1%	8.7%
Others	9,656	9,694	9,826	9,796	9,995	9,916	-0.8%	20.1%
Total	47,646	47,704	47,687	48,289	49,082	49,254	0.4%	100.0%
Other EMEs								
Light distillates	5,534	5,674	5,919	6,117	6,516	6,712	3.0%	22.9%
Middle distillates	9,306	9,578	9,809	10,035	10,744	11,050	2.8%	37.8%
Fuel oil	4,975	4,904	4,852	4,828	5,053	5,185	2.6%	17.7%
Others	4,696	4,891	5,347	5,634	6,188	6,323	2.2%	21.6%
Total	24,511	25,048	25,926	26,614	28,501	29,270	2.7%	100.0%

Source: Adapted from *BP Statistical Review of World Energy 2006*.

NOTES: Some totals might not compute exactly due to rounding and other factors.
Light distillates consists of aviation and motor gasolines and light distillate feedstock (LDF).
Middle distillates consists of jet and heating kerosines, and gas and diesel oils (including marine bunkers).
Fuel oil includes marine bunkers and crude oil used directly as fuel.
Others consists of refinery gas, LPGs, solvents, petroleum coke, lubricants, bitumen, wax, other refined products and refinery fuel and loss.

Table 4: World Oil Production

COUNTRY	1970	1980	1990	2000	2001	2002	2003	2004	Actual 2005	Est. 2006
	1,000 b/d									
North America										
Canada	1,263.6	1,412.0	1,508.0	2,035.0	2,052.4	2,212.9	2317.4	2,417.8	23,68.8	2,500.0
US	9,630.0	8,569.0	7,220.0	5,822.0	5,801.2	5,746.0	5,681.6	5,418.7	51,78.4	5,135.0
Total	10,893.6	9,981.0	8,728.0	7,857.0	7,853.6	7,958.9	7,999.0	7,836.5	7,547.2	7,635.0
Latin America										
Argentina	382.9	487.0	473.0	750.7	763.0	738.1	707.0	738.3	645.4	635.0
Bolivia	16.3	30.0	19.2	27.6	31.3	30.9	32.2	35.0	41.6	45.0
Brazil	160.5	182.0	633.0	1,128.0	1,302.5	1,495.0	1,535.7	1,479.2	1,634.0	1,710.0
Chile	34.6	29.0	20.3	7.0	7.0	10.0	10.0	10.0	10.0	10.0
Colombia	214.0	125.0	445.0	687.3	604.3	578.5	541.3	527.6	526.1	530.0
Cuba	--	--	15.0	44.0	40.3	40.0	40.0	40.0	38.6	39.0
Ecuador	4.1	222.0	287.0	401.1	407.1	392.5	417.8	525.5	532.0	500.0
Guatemala	0	5.0	4.0	20.7	21.1	23.6	23.9	20.2	18.5	16.5
Mexico	430.2	1,936.0	2,633.0	3,012.0	3,127.0	3,177.0	3370.4	3,383.3	3,334.2	3,273.0
Peru	72.2	191.0	132.0	98.0	93.1	92.8	87.3	88.6	111.2	113.0
Trinidad & Tobago	139.8	211.0	151.0	119.0	113.3	130.6	136.5	122.9	144.5	150.0
Venezuela	3,708.0	2,167.0	2,118.0	3,028.0	2,685.0	2,285.0	2,006.7	2,208.3	2,705.8	2,563.0
Other Latin America	--	--	--	--	--	--	12.8	12.3	13.0	14.8
Total	5,162.6	5,585.0	6,935.7	9,323.4	9,209.1	9,007.6	8,921.6	9,191.1	9,754.9	9,599.3
Europe										
Albania	--	--	--	5.9	6.0	6.2	6.0	6.1	6.1	7.0
Austria	53.3	29.0	24.0	19.1	18.9	18.6	18.2	17.1	17.3	17.4
Bulgaria	6.7	6.0	5.0	1.0	1.0	1.0	1.0	1.0	1.0	1.0
Croatia	--	--	--	23.0	21.7	20.7	20.7	19.1	17.5	16.8
Czech Republic	4.1	1.9	3.0	3.2	3.3	4.8	5.8	5 .9	4.8	4.7
Denmark	0	6.0	118.0	363.0	348.4	370.8	367.7	389.1	377.3	342.0
France	45.8	26.0	62.0	28.4	27.8	26.4	24.5	22.9	21.7	21.5
Germany	148.4	100.0	78.0	62.5	68.9	74.1	76.0	70.3	71.5	69.0
Greece	--	--	15.0	5.1	4.0	3.3	2.4	2.3	1.8	1.8
Hungary	38.7	40.6	40.0	25.8	23.6	22.0	23.7	22.5	19.0	16.5
Italy	26.3	39.0	97.0	90.8	65.2	84.4	95.8	115.0	115.0	105.0
Lithuania	--	--	--	--	--	--	--	--	4.3	3.6
Netherlands	35.8	25.0	71.0	29.3	27.0	43.7	45.8	44.0	29.0	23.0
Norway	0	528.0	1,574.0	3,212.0	3,244.6	3,148.8	3,065.0	2,955.1	2,697.6	2,465.0
Poland	8.5	6.6	3.0	14.3	17.4	16.6	17.6	20.1	19.9	17.0
Romania	269.6	230.9	160.0	122.9	120.0	116.1	115.2	103.0	100.0	98.4
Serbia	--	--	--	--	--	--	15.5	15.0	15.0	15.0
Slovakia	--	--	--	--	--	--	1.0	1.0	1.0	1.0
Spain	3.1	32.0	14.3	4.6	6.9	6.5	6.6	5.3	3.3	3.0
Turkey	67.9	44.0	70.0	56.0	48.4	46.9	44.4	42.0	42.0	42.0
UK	1.7	1619.0	1,860.0	2,513.7	2,330.8	2,305.8	2,103.2	1,888.3	1,678.0	1,480.0
Total	709.9	2,734.0	4,194.3	6,580.6	6,383.9	6,333.8	6,056.1	5,745.1	5,243.1	4,750.7
Middle East										
Bahrain	76.6	48.0	42.0	102.3	173.6	174.1	33.8	174.2	174.4	172.0
Iran	3,328.8	1,467.0	3,120.0	3,681.7	3,695.8	3,430.0	3,784.2	3,931.7	3,890.8	3,850.0
Iraq	1,548.6	2,638.0	2,083.0	2,566.7	2,355.0	2,014.2	1,328.3	2,002.5	1,810.0	1,915.0
Israel	--	--	--	--	--	--	0.1	0.1	0.1	0.1
Kuwait	2,734.5	1,382.0	1,080.0	1,765.0	1,715.0	1,600.0	1,870.0	2,050.0	2,130.0	2,200.0
Neutral Zone	500.6	540.0	315.0	630.0	565.0	538.0	610.0	600.0	580.0	575.0
Oman	332.4	283.0	658.0	933.3	964.2	904.2	828.3	764.2	757.5	738.0
Qatar	362.4	472.0	387.0	688.3	671.7	644.2	736.7	782.5	798.3	820.0
Saudi Arabia	3,548.9	9,630.0	6,215.0	7,995.0	7,695.0	6,850.0	8,480.0	8,750.0	9,060.0	8,990.0
Syria	83.1	165.0	385.0	522.5	518.3	508.3	527.8	503.3	460.0	422.5
UAE	693.8	1,709.0	2,101.0	2,230.5	2,153.5	1,873.5	2,287.2	2,353.7	2,453.0	2,590.7
Yemen	0	0	179.0	354.0	350.0	350.0	350.0	350.0	412.8	415.0
Total	13,280	18,335	16,566	21,469	20,857	18,886.6	20,836.4	22,262.2	22,526.9	22,688.3

Table 4: World Oil Production, cont'd.

COUNTRY	1970	1980	1990	2000	2001	2002	2003	2004	Actual 2005	Est. 2006
	1,000 b/d									
Africa										
Algeria	1,029.1	1,016.0	797.0	808.0	835.8	850.0	1,110.8	1,206.7	1,351.7	1,345.0
Angola	13.7	150.0	480.0	740.0	696.0	894.2	879.7	986.2	1,238.1	1,385.0
Cameroon	--	58.0	164.0	85.0	80.0	69.0	65.0	74.0	82.5	87.5
Chad	--	--	--	--	--	--	--	175.4	180.0	180.0
Congo (Brazzaville)	--	56.0	161.0	265.0	265.0	254.1	240.3	240.0	240.0	240.0
Congo (Zaire)	--	20.0	28.0	25.0	24.0	22.7	25.5	20.0	20.0	20.0
Egypt	327.3	596.0	873.0	810.0	760.0	751.3	750.1	709.2	695.0	667.5
Equatorial Guinea	--	--	--	167.5	183.6	200.0	241.6	320.0	350.0	320.0
Gabon	108.8	180.0	289.0	325.0	300.8	248.0	240.4	235.0	234.2	232.0
Ivory Coast	--	--	2.1	12.0	5.2	10.0	25.8	30.0	30.0	30.0
Libya	3,318.0	1,785.0	1,369.0	1,410.0	1,365.0	1,315.8	1,427.5	1,546.7	1,640.0	1,700.0
Morocco	0	0	0	0	0	0	0.2	0.2	0.2	0.2
Nigeria	1,083.3	2,057.0	1,808.0	2,030.0	2,083.3	1,945.0	2,145.0	2,342.5	2,406.7	2,220.0
South Africa	--	--	--	25.9	22.9	20.1	16.5	26.2	24.0	20.0
Sudan	--	--	--	185.0	200.0	200.0	255.0	286.7	290.0	295.0
Tunisia	87.7	110.0	93.0	75.9	69.6	72.2	65.8	69.7	71.1	65.0
Other Africa	--	--	--	--	--	--	6.0	6.0	6.0	6.0
Total	5,969.4	6,032.0	6,048.4	6,971.6	6,901.3	6,858.6	7,495.2	8,274.5	8,859.9	8,845.2
Former Soviet Union										
Azerbaijan	--	--	--	275.0	298.5	300.0	306.3	300.0	443.3	625.0
Belarus	--	--	--	35.0	35.0	35.0	35.0	37.0	35.5	35.0
Georgia	--	--	--	2.0	2.0	2.0	2.8	3.0	2.0	2.0
Kazakhstan	--	--	--	675.0	707.5	818.3	888.0	970.0	994.2	1,060.0
Kyrgyzstan	--	--	--	1.0	2.0	2.0	2.0	2.0	1.0	1.0
Russia	--	--	--	6,325.0	6,780.8	7,404.6	8,240.0	8,886.7	9,190.0	9,475.0
Turkmenistan	--	--	--	140.0	160.0	180.0	210.0	200.0	195.0	165.0
Ukraine	--	--	--	73.0	80.0	85.0	80.0	84.0	98.0	97.0
USSR	7,089.2	12,031.0	11,500.0	--	--	--	--	--	--	--
Uzbekistan	--	--	--	155.0	145.0	149.0	144.0	150.0	113.0	105.0
Total	7,089.2	12,031.0	11,500.0	7,896.2	8,210.8	8,975.9	9,908.1	10,632.7	11,072.0	11,565.0
Asia-Pacific										
Australia	175.6	379.0	582.0	700.0	632.6	622.6	518.2	440.4	471.5	450.0
Bangladesh	--	--	0.9	3.3	3.3	5.0	3.7	4.0	4.0	4.0
Brunei	148.0	230.0	143.0	177.9	180.5	188.4	193.4	189.2	186.5	197.0
China	500.0	2,119.0	2,755.0	3,237.3	3,296.7	3,405.7	3,422.9	3,485.2	3,627.1	3,700.0
India	138.6	182.0	679.0	646.0	643.8	661.8	669.4	687.7	659.1	682.0
Indonesia	853.6	1,576.0	1,274.0	1,267.3	1,214.2	1,116.3	1,013.3	971.7	945.0	892.0
Japan	16.3	10.0	10.5	12.7	13.1	12.5	14.2	14.3	15.6	15.0
Malaysia	--	288.0	605.0	690.0	744.2	770.0	800.8	859.2	770.0	737.0
Myanmar	16.4	30.0	13.0	8.0	8.0	10.0	13.0	13.0	13.0	13.0
New Zealand	--	7.0	39.0	36.0	34.0	30.0	23.9	21.0	13.9	14.0
Pakistan	9.9	10.0	60.0	53.3	59.7	60.9	61.1	62.1	64.5	63.5
Papua New Guinea	--	--	--	69.7	57.0	47.0	48.2	46.0	46.0	56.5
Philippines	--	15.0	5.0	1.1	7.1	14.4	14.8	12.4	15.3	16.0
Taiwan	1.8	5.0	2.5	0.6	0.8	0.9	0.8	0.9	1.0	1.0
Thailand	0.4	--	41.0	111.0	113.1	128.7	159.7	154.1	186.0	212.0
Vietnam	--	--	40.0	304.0	304.8	330.7	333.2	340.0	340.0	350.0
Total	1,860.6	4,851.0	6,249.9	7,318.2	7,312.7	7,404.9	7,290.6	7,301.3	7,358.5	7,403.0
Total World	**45,059**	**59,684**	**60,317**	**67,234**	**66,747**	**65,426.3**	**68,507.0**	**71,243.4**	**72,362.1**	**72,486.5**

Table 4: World Oil Production, cont'd.

COUNTRY	1970	1980	1990	2000	2001	2002	2003	2004	Actual 2005	Est. 2006
	Million tonnes per year (mtpy)									
North America										
Canada	62.9	70.3	75.1	101.4	102.2	110.2	115.4	120.4	118.0	124.5
US	479.6	426.7	359.6	289.5	288.9	286.1	282.9	269.8	257.9	255.7
Total	542.5	497.0	434.7	390.9	391.1	396.3	398.3	390.2	375.9	380.2
Latin America										
Argentina	19.1	24.3	23.6	37.4	38.0	36.8	35.2	36.8	32.1	31.6
Bolivia	0.8	1.5	1.0	1.4	1.6	1.5	1.6	1.7	2.1	2.2
Brazil	8.0	9.1	31.5	56.2	64.9	74.4	76.5	73.7	81.4	85.2
Chile	1.7	1.4	1.0	0.3	0.3	0.5	0.5	0.5	0.5	0.5
Colombia	10.7	6.2	22.2	34.2	30.1	28.8	27.0	26.3	26.2	26.4
Cuba	--	--	0.7	2.2	2.0	2.0	2.0	2.0	1.9	1.9
Ecuador	0.2	11.1	14.3	20.0	20.3	19.5	20.8	26.2	26.5	24.9
Guatemala	--	0.2	0.2	1.0	1.1	1.2	1.2	1.0	0.9	0.8
Mexico	21.4	96.4	131.1	150.0	155.7	158.2	167.8	168.5	166.0	163.0
Peru	3.6	9.5	6.6	4.9	4.6	4.6	4.3	4.4	5.5	5.6
Trinidad & Tobago	7.0	10.5	7.5	5.9	5.6	6.5	6.8	6.1	7.2	7.5
Venezuela	184.7	107.9	105.5	150.8	133.7	113.7	99.9	110.0	134.7	127.6
Other Latin America	--	--	--	--	--	--	0.6	0.6	0.6	0.7
Total	257.2	278.1	345.5	464.3	458.6	448.5	444.3	457.7	485.8	478.0
Europe										
Albania	--	--	--	0.3	0.3	0.3	0.3	0.3	0.3	0.3
Austria	2.7	1.4	1.2	1.0	0.9	0.9	0.9	0.9	0.9	0.9
Bulgaria	0.3	0.3	0.2	--	--	--	--	--	--	--
Croatia	--	--	--	1.1	1.1	1.0	1.0	1.0	0.9	0.8
Czech Republic	0.2	0.1	0.1	0.2	0.2	0.2	0.3	0.3	0.2	0.2
Denmark	--	0.3	5.9	18.1	17.4	18.5	18.3	19.4	18.8	17.0
France	2.3	1.3	3.1	1.4	1.4	1.3	1.2	1.1	1.1	1.1
Germany	7.4	5.0	3.9	3.1	3.4	3.7	3.8	3.5	3.6	3.4
Greece	--	--	0.7	0.3	0.2	0.2	0.1	0.1	0.1	0.1
Hungary	1.9	2.0	2.0	1.3	1.2	1.1	1.2	1.1	0.9	0.8
Italy	1.3	1.9	4.8	4.5	3.2	4.2	4.8	5.7	5.7	5.2
Lithuania	--	--	--	--	--	--	--	--	0.2	0.2
Netherlands	1.8	1.2	3.5	1.5	1.3	2.2	2.3	2.2	1.4	1.1
Norway	0	26.3	78.4	160.0	161.6	156.8	152.6	147.2	134.3	122.8
Poland	0.4	0.3	0.1	0.7	0.9	0.8	0.9	1.0	1.0	0.8
Romania	13.4	11.5	8.0	6.1	6.0	5.8	5.7	5.1	5.0	4.9
Serbia	--	--	--	--	--	--	0.8	0.7	0.7	0.7
Slovakia	--	--	--	--	--	--	--	--	--	--
Spain	0.2	1.6	0.7	0.2	0.3	0.3	0.3	0.3	0.2	0.1
Turkey	3.4	2.2	3.5	2.8	2.4	2.3	2.2	2.1	2.1	2.1
UK	0.1	80.6	92.6	125.2	116.1	114.8	104.7	94.0	83.6	73.7
Total	35.4	136.0	208.7	327.8	317.9	315.4	301.6	286.1	261.1	236.6
Middle East										
Bahrain	3.8	2.4	2.1	5.1	8.6	8.7	1.7	8.7	8.7	8.6
Iran	165.8	73.1	155.4	183.4	184.1	170.8	188.4	195.8	193.8	191.7
Iraq	77.1	131.4	103.7	127.8	117.3	100.3	66.1	99.7	90.1	95.4
Israel	--	--	--	--	--	--	--	--	--	--
Kuwait	136.2	68.8	53.8	87.9	85.4	79.7	93.1	102.1	106.1	109.6
Neutral Zone	24.9	26.9	15.7	31.4	28.1	26.8	30.4	29.9	28.9	28.6
Oman	16.6	14.1	32.8	46.5	48.0	45.0	41.2	38.1	37.7	36.8
Qatar	18.0	23.5	19.3	34.3	33.5	32.1	36.7	39.0	39.8	40.8
Saudi Arabia	176.7	479.6	309.5	398.2	383.2	341.1	422.3	435.7	451.2	447.7
Syria	4.1	8.2	19.2	26.0	25.8	25.3	26.3	25.1	22.9	21.0
UAE	34.6	85.1	104.6	111.1	107.2	93.3	113.9	117.2	122.2	129.0
Yemen	--	--	8.9	17.6	17.4	17.4	17.4	17.4	20.6	20.7
Total	661.3	913.1	825.0	1,069.2	1,038.7	940.5	1,037.7	1,108.6	1,121.8	1,129.9

Table 4: World Oil Production, cont'd.

COUNTRY	1970	1980	1990	2000	2001	2002	2003	2004	Actual 2005	Est. 2006
	Million tonnes per year (mtpy)									
Africa										
Algeria	51.2	50.6	39.7	40.2	41.6	42.3	55.3	60.1	67.3	67.0
Angola	0.7	7.5	23.9	36.8	34.7	44.5	43.8	49.1	61.7	69.0
Cameroon	--	2.9	8.2	4.2	4.0	3.4	3.2	3.7	4.1	4.4
Chad	--	--	--	--	--	--	--	8.7	9.0	9.0
Congo (Brazzaville)	--	2.8	8.0	13.2	13.2	12.7	12.0	12.0	12.0	12.0
Congo (Zaire)	--	1.0	1.4	1.2	1.2	1.1	1.3	1.0	1.0	1.0
Egypt	16.3	29.7	43.5	40.3	37.8	37.4	37.4	35.3	34.6	33.2
Equatorial Guinea	--	--	--	8.3	9.1	10.0	12.0	15.9	17.4	15.9
Gabon	5.4	9.0	13.4	16.2	15.0	12.3	12.0	11.7	11.7	11.6
Ivory Coast	--	--	0.1	0.6	0.3	0.5	1.3	1.5	1.5	1.5
Libya	165.2	88.9	68.2	70.2	68.0	65.5	71.1	77.0	81.7	84.7
Morocco	--	--	--	--	--	--	--	--	--	--
Nigeria	53.9	102.4	90.0	101.0	103.7	96.9	106.8	116.6	119.9	110.6
South Africa	--	--	--	1.3	1.1	1.0	0.8	1.3	1.2	1.0
Sudan	--	--	--	9.2	10.0	10.0	12.7	14.3	14.4	14.7
Tunisia	4.4	5.5	4.6	3.8	3.5	3.6	3.3	3.5	3.5	3.2
Other Africa	--	--	--	--	--	--	0.3	0.3	0.3	0.3
Total	297.2	300.5	301.2	347.2	343.7	341.5	373.3	412.0	441.2	440.5
Former Soviet Union										
Azerbaijan	--	--	--	13.7	14.9	14.9	15.3	14.9	22.1	31.1
Belarus	--	--	--	1.7	1.7	1.7	1.7	1.8	1.8	1.7
Georgia	--	--	--	0.1	0.1	0.1	0.1	0.1	0.1	0.1
Kazakhstan	--	--	--	33.6	35.2	40.7	44.2	48.3	49.5	52.8
Kyrgyzstan	--	--	--	0	0.1	0.1	0.1	0.1	0.1	0.1
Russia	--	--	--	315.0	337.7	368.7	410.4	442.5	457.7	471.9
Turkmenistan	--	--	--	7.0	8.0	9.0	10.5	10.0	9.7	8.2
Ukraine	--	--	--	3.6	4.0	4.2	4.0	4.2	4.9	4.8
USSR	353.0	599.1	572.7	--	--	--	--	--	--	--
Uzbekistan	--	--	--	7.7	7.2	7.4	7.2	7.5	5.6	5.2
Total	353.0	599.1	572.7	393.2	408.9	447.0	493.4	529.5	551.4	575.9
Asia-Pacific										
Australia	8.7	18.9	29.0	34.9	31.5	31.0	25.8	21.9	23.5	22.4
Bangladesh	--	--	--	0.2	0.2	0.2	0.2	0.2	0.2	0.2
Brunei	7.4	11.5	7.1	8.9	9.0	9.4	9.6	9.4	9.3	9.8
China	24.9	105.5	137.2	161.2	164.2	169.6	170.5	173.5	180.6	184.3
India	6.9	9.1	33.8	32.2	32.1	33.0	33.3	34.2	32.8	34.0
Indonesia	42.5	78.5	63.4	63.1	60.5	55.6	50.5	48.4	47.1	44.4
Japan	0.8	0.5	0.5	0.6	0.7	0.6	0.7	0.7	0.8	0.7
Malaysia	--	14.3	30.1	34.4	37.1	38.3	39.9	42.8	38.3	36.7
Myanmar	0.8	1.5	0.6	0.4	0.4	0.5	0.6	0.6	0.6	0.6
New Zealand	--	0.3	1.9	1.8	1.7	1.5	1.2	1.0	0.7	0.7
Pakistan	0.5	0.5	3.0	2.7	3.0	3.0	3.0	3.1	3.2	3.2
Papua New Guinea	--	--	--	3.5	2.8	2.3	2.4	2.3	2.3	2.8
Philippines	--	0.7	0.2	--	0.4	0.7	0.7	0.6	0.8	0.8
Taiwan	0.1	0.2	0.1	--	--	--	--	--	--	--
Thailand	--	--	2.0	5.5	5.6	6.4	8.0	7.7	9.3	10.6
Vietnam	--	--	2.0	15.1	15.2	16.5	16.6	16.9	16.9	17.4
Total	92.6	241.5	310.9	364.4	364.2	368.7	363.1	363.6	366.5	368.7
Total World	**2,243.9**	**2,972.0**	**3,003.4**	**3,348.2**	**3,324.0**	**3,257.9**	**3,411.6**	**3,547.6**	**3,603.6**	**3,609.8**

Source: *Oil & Gas Journal* data as of Jan. 1, 2007.

NOTES: Some totals might not compute exactly due to rounding and other factors.
Other Latin America includes Barbados, Suriname. Other Africa includes Benin, Ethiopia, Ghana, Namibia, Rwanda, and Somalia.
Slovakia production is included in Czech Republic's total before 2000.
USSR includes total of all Former Soviet Union countries before 2000.

Table 5: Historical Oil Reserves

	1-Jan											
	1980	1990	1998	1999	2000	2001	2002	2003	2004	2005	2006	2007
COUNTRY	Billion bbl											
North America												
Canada	6.4	6.1	4.8	4.9	4.9	4.7	4.9	180.0	178.9	178.8	178.8	179.2
US	26.4	25.9	22.0	22.5	21.0	21.8	22.0	22.4	22.7	21.9	21.4	21.6
Latin America												
Argentina	2.5	2.3	2.6	2.6	2.8	3.0	3.0	2.9	2.8	2.7	2.3	2.5
Brazil	1.3	2.8	4.8	7.1	7.4	8.1	8.5	8.3	8.5	10.6	11.2	11.8
Colombia	1.0	2.0	2.8	2.6	2.6	2.0	1.8	1.8	1.8	1.5	1.5	1.5
Ecuador	1.1	1.5	2.1	2.1	2.1	2.1	2.1	4.6	4.6	4.6	4.6	4.5
Mexico	44.0	56.4	40.0	47.8	28.4	28.3	26.9	12.6	15.7	14.6	12.9	12.3
Venezuela	17.9	58.5	71.7	72.6	72.6	76.9	77.7	77.8	77.8	77.2	79.7	80.0
Europe												
Norway	5.5	11.5	10.4	10.9	10.8	4.4	9.4	10.3	10.4	8.5	7.7	7.9
UK	14.8	4.2	5.0	5.2	5.2	5.0	4.9	4.7	4.7	4.5	4.0	3.9
Middle East												
Iran	57.5	92.8	93.0	89.7	89.7	89.7	89.7	89.7	125.8	125.8	132.5	136.3
Iraq	30.0	100.0	112.5	112.5	112.5	112.5	112.5	112.5	115.0	115.0	115.0	115.0
Kuwait	64.9	94.5	94.0	94.0	94.0	94.0	94.0	94.0	96.5	99.0	101.5	99.0
Neutral Zone	6.0	5.2	5.0	5.0	5.0	5.0	5.0	5.0	5.0	5.0	5.0	5.0
Oman	2.3	4.2	5.2	5.3	5.3	5.5	5.5	5.5	5.5	5.5	5.5	5.5
Qatar	3.6	4.5	3.7	3.7	3.7	13.2	15.2	15.2	15.2	15.2	15.2	15.2
Saudi Arabia	165.0	254.9	259.0	259.0	261.0	259.2	259.3	259.3	259.4	259.4	264.3	259.8
Syria	1.3	1.7	2.5	2.5	2.5	2.5	2.5	2.5	2.5	2.5	2.5	2.5
UAE	30.4	96.2	96.2	96.2	96.2	96.2	97.8	97.8	97.8	97.8	97.8	97.8
Yemen	--	3.0	4.0	4.0	4.0	4.0	4.0	4.0	4.0	4.0	4.0	3.0
Africa												
Algeria	8.2	9.2	9.2	9.2	9.2	9.2	9.2	9.2	11.3	11.8	11.4	12.3
Angola	1.2	2.0	5.4	5.4	5.4	5.4	5.4	5.4	5.4	5.4	5.4	8.0
Egypt	2.9	4.5	3.8	3.5	2.9	2.9	2.9	3.7	3.7	3.7	3.7	3.7
Gabon	--	0.7	2.5	2.5	2.5	2.5	2.5	2.5	2.5	2.5	2.5	2.0
Libya	23.0	22.8	29.5	29.5	29.5	29.5	29.5	29.5	36.0	39.0	39.1	41.5
Nigeria	16.7	16.0	16.8	22.5	22.5	22.5	24.0	24.0	25.0	35.3	35.9	36.2
Former Soviet Union												
Azerbaijan	--	--	--	--	--	1.2	1.2	7.0	7.0	7.0	7.0	7.0
Kazakhstan	--	--	--	--	--	5.4	5.4	9.0	9.0	9.0	9.0	30.0
Russia	--	--	--	--	--	48.6	48.6	60.0	60.0	60.0	60.0	60.0
USSR	63.0	24.0	59.1	59.1	59.0	--	--	--	--	--	--	--
Asia-Pacific												
Australia	2.4	1.7	1.8	2.9	2.9	2.9	3.5	3.5	3.5	1.5	1.4	1.6
Brunei	1.7	1.4	1.4	1.4	1.4	1.4	1.4	1.4	1.4	1.4	1.4	1.1
China	--	24.0	24.0	24.0	24.0	24.0	24.0	18.3	18.3	18.3	18.3	16.0
India	2.5	7.5	4.3	3.9	4.8	4.7	4.8	5.4	5.4	5.4	5.8	5.6
Indonesia	9.5	8.2	4.9	4.9	4.9	4.9	5.0	5.0	4.7	4.7	4.3	4.3
Malaysia	3.0	3.0	3.9	3.9	3.9	3.9	3.0	3.0	3.0	3.0	3.0	3.0
Total World	**648.5**	**1001.6**	**1,019.5**	**1,034.3**	**1,016.0**	**1,031.6**	**1,031.6**	**1,212.9**	**1,265.8**	**1,277.7**	**1,292.5**	**1,309.3**

Source: *Oil & Gas Journal* data as of Jan. 1, 2007.

NOTES: Some totals might not compute exactly due to rounding and other factors.
Canada's large increase in 2002–03 was due to including its oil sands.

Table 6: World Oil Imports and Exports

COUNTRY/REGION	2000	2001	2002	2003	2004	2005	Change 2005 over 2004	2005 share of total
	Thousand barrels per day (1,000 b/d)						%	
IMPORTS								
US	11,092	11,618	11,357	12,254	12,898	13,525	4.9	27.1
Europe	11,070	11,531	11,895	11,993	12,538	13,261	5.8	26.6
Japan	5,329	5,202	5,070	5,314	5,203	5,225	0.4	10.5
Rest of World	14,911	15,401	15,306	16,238	17,471	17,895	2.4	35.8
Total World	42,402	43,752	43,628	45,799	48,110	49,906	3.7	100.0
EXPORTS								
US	890	910	904	921	991	1,129	14.0	2.3
Canada	1,703	1,804	1,959	2,096	2,148	2,201	2.5	4.4
Mexico	1,814	1,882	1,966	2,115	2,070	2,065	-0.2	4.1
South & Central America	3,079	3,143	2,965	2,942	3,233	3,528	9.1	7.1
Europe	1,967	1,947	2,234	2,066	1,993	2,149	7.8	4.3
Former Soviet Union	4,273	4,679	5,370	6,003	6,440	7,076	9.9	14.2
Middle East	18,944	19,098	18,062	18,943	19,630	19,821	1.0	39.7
North Africa	2,732	2,724	2,620	2,715	2,917	3,070	5.2	6.2
West Africa	3,293	3,182	3,134	3,612	4,048	4,358	7.7	8.7
Asia-Pacific	2,767	2,879	2,863	3,025	3,009	2,967	-1.4	5.9
Rest of World	940	1,506	1,551	1,361	1,631	1,542	-5.5	3.1
Total World	**42,402**	**43,754**	**43,628**	**45,799**	**48,110**	**49,906**	**3.7**	**100.0**

Source: Adapted from *BP Statistical Review of World Energy 2006*.

NOTES: Some totals might not compute exactly due to rounding and other factors.
Rest of World includes unidentified trade.
Asia-Pacific excludes Japan.

Table 7: World Oil Trade Movements

2005 EXPORTS FROM	2005 IMPORTS TO											Total
	North America		Latin America				Asia-Pacific					
	US	Canada	Mexico	Other Latin America	Europe	Africa	Australasia	China	Japan	Other Asia-Pacific	Rest of World	
	1,000 b/d											
US	- -	154	211	323	242	15	- -	8	84	73	19	1,129
Canada	2,172	- -	2	4	17	- -	- -	- -	6	- -	- -	2,201
Mexico	1,647	34	- -	135	211	2	- -	- -	- -	32	4	2,065
Latin America	2,868	109	44	- -	309	21	- -	107	2	68	- -	3,528
Europe	1,100	444	50	48	- -	270	- -	12	6	125	94	2,149
FSU	473	- -	2	60	5,811	10	- -	398	47	72	202	7,076
Middle East	2,345	143	10	157	3,144	752	113	1,360	4,269	7,466	63	19,821
N. Africa	547	169	6	115	1,959	83	- -	64	2	109	12	3,070
W. Africa	1,943	40	- -	169	696	88	- -	574	60	765	18	4,358
E. & S. Africa	- -	- -	- -	- -	26	- -	- -	135	80	25	- -	266
Australasia	14	- -	- -	- -	- -	- -	- -	25	65	117	- -	222
China	32	2	- -	33	4	2	8	- -	47	289	8	427
Japan	- -	- -	- -	- -	8	- -	8	69	- -	21	- -	107
Other Asia-Pacific	170	4	2	6	128	15	545	626	211	301	10	2,318
Unidentified	214	111	- -	6	706	- -	39	6	44	44	- -	1,169
Total	13,525	1,210	328	1,056	13,261	1,258	722	3,384	5,225	9,507	431	49,906
	million tonnes											
US	- -	7.4	10.1	15.5	11.6	0.7	- -	0.4	4.0	3.5	0.9	54.1
Canada	107.1	- -	0.1	0.2	0.8	- -	- -	- -	0.3	- -	- -	108.5
Mexico	81.8	1.7	- -	6.7	10.5	0.1	- -	- -	- -	1.6	0.2	102.6
Latin America	140.9	5.3	2.1	- -	15.1	1.0	- -	5.3	0.1	3.3	- -	173.1
Europe	53.3	22.0	2.4	2.3	- -	12.9	- -	0.6	0.3	6.0	4.5	104.3
FSU	23.0	- -	0.1	3.0	287.0	0.5	- -	19.6	2.3	3.5	10.0	349.0
Middle East	116.5	7.1	0.5	7.8	156.1	37.2	5.6	67.4	211.7	369.2	3.0	982.1
N. Africa	26.7	8.4	0.3	5.7	97.0	4.1	0.2	3.2	0.1	5.4	0.6	151.7
W. Africa	96.5	2.0	- -	8.4	34.6	4.4	0.2	28.6	3.0	38.1	0.9	216.7
E. & S. Africa	- -	- -	- -	- -	1.3	- -	- -	6.7	4.0	1.2	- -	13.2
Australasia	0.7	- -	- -	- -	- -	- -	- -	1.2	3.2	5.8	- -	10.9
China	1.6	0.1	- -	1.6	0.2	0.1	0.4	- -	2.3	14.0	0.4	20.7
Japan	- -	- -	- -	- -	0.4	- -	0.4	3.3	- -	1.0	- -	5.1
Other Asia-Pacific	8.3	0.2	0.1	0.3	6.1	0.7	26.8	30.3	24.8	14.6	0.5	112.7
Unidentified	10.3	5.5	- -	0.3	34.3	- -	1.9	0.3	2.1	2.1	- -	56.8
Total	666.7	59.7	15.7	51.8	655.0	61.7	35.5	166.9	258.2	469.3	21.0	2,461.5

Source: Adapted from *BP Statistical Review of World Energy 2006*.

NOTES: Some totals might not compute exactly due to rounding and other factors.
Rest of World includes unidentified trade.
Unidentified includes changes in the quantity of oil in transit, movements not otherwise shown, unidentified military use, etc.
Asia-Pacific excludes Japan.

Table 8: World Oil Balance

	Avg. 2002	Avg. 2003	Avg. 2004	2005 1Q05	2005 2Q05	2005 3Q05	2005 4Q05	Avg. 2005
	Million b/d							
SUPPLY								
OECD								
US	9.00	8.80	8.70	8.78	8.84	7.95	7.74	8.32
Canada	2.95	3.11	3.14	3.01	3.06	3.02	3.28	3.09
Mexico	3.59	3.80	3.85	3.77	3.89	3.72	3.75	3.78
North Sea	6.21	5.99	5.62	5.46	5.22	4.95	5.05	5.17
Other OECD	1.66	1.55	1.51	1.49	1.57	1.55	1.51	1.53
Total OECD	23.42	23.24	22.80	22.50	22.58	21.20	21.32	21.89
Non-OECD								
OPEC								
OPEC Crude								
Algeria	1.31	1.61	1.68	1.76	1.78	1.82	1.83	1.80
Indonesia	1.25	1.16	1.10	1.08	1.07	1.06	1.05	1.07
Iran	3.44	3.74	4.00	4.07	4.13	4.21	4.13	4.14
Iraq	2.02	1.31	2.01	1.90	1.90	1.99	1.72	1.88
Kuwait	1.89	2.14	2.38	2.48	2.50	2.53	2.60	2.53
Libya	1.32	1.42	1.52	1.61	1.63	1.64	1.65	1.63
Nigeria	2.12	2.28	2.33	2.50	2.68	2.64	2.70	2.63
Qatar	0.68	0.72	0.78	0.84	0.84	0.84	0.84	0.84
Saudi Arabia	7.63	8.78	9.10	9.50	9.60	9.80	9.50	9.55
UAE	2.08	2.35	2.48	2.52	2.47	2.55	2.60	2.54
Venezuela	2.60	2.34	2.56	2.64	2.54	2.54	2.54	2.56
OPEC NGL	2.40	2.55	2.70	2.87	2.92	2.93	2.95	2.92
Refinery proc. gain	0.06	0.05	0.05	0.05	0.05	0.05	0.05	0.05
Total OPEC	28.99	30.56	32.85	33.99	34.25	34.55	34.30	34.27
Non-OPEC								
FSU	9.40	10.40	11.31	11.53	11.62	11.72	11.97	11.71
China	3.53	3.56	3.64	3.73	3.76	3.80	3.75	3.76
Other Non-OECD	11.61	11.80	12.40	12.49	12.83	13.19	13.20	12.93
Total Non-OPEC	24.54	25.76	27.35	27.75	28.27	28.71	28.92	28.40
Total Non-OECD	53.54	56.32	60.20	61.74	62.45	63.25	63.21	62.67
Total World Supply	**76.95**	**79.57**	**83.00**	**84.24**	**85.03**	**84.45**	**84.54**	**84.56**

Table 8: World Oil Balance, cont'd.

	Avg. 2002	Avg. 2003	Avg. 2004	2005 1Q05	2005 2Q05	2005 3Q05	2005 4Q05	Avg. 2005
	Million b/d							
DEMAND								
OECD								
OECD North America								
US	19.76	20.03	20.73	20.85	20.65	20.92	20.79	20.80
US Territories	0.32	0.34	0.36	0.35	0.37	0.32	0.37	0.36
Canada	2.08	2.21	2.30	2.36	2.24	2.25	2.23	2.27
Mexico	1.95	1.95	2.00	2.04	2.11	2.06	2.10	2.08
Total OECD North America	24.11	24.53	25.39	25.60	25.37	25.55	25.49	25.51
OECD Asia-Pacific								
Australia & New Zealand	1.03	1.04	1.04	1.04	1.06	1.04	1.10	1.06
Japan	5.30	5.42	5.29	6.00	4.94	5.03	5.46	5.35
South Korea	2.15	2.18	2.16	2.40	2.07	2.01	2.23	2.18
Total OECD Asia-Pacific	8.48	8.64	8.49	9.44	8.07	8.08	8.79	8.59
OECD Europe								
France	1.98	2.00	2.01	2.11	1.93	2.00	1.96	2.00
Germany	2.72	2.68	2.67	2.54	2.55	2.75	2.63	2.62
Italy	1.87	1.87	1.79	1.77	1.69	1.68	1.78	1.73
UK	1.73	1.76	1.80	1.84	1.79	1.82	1.84	1.82
Other OECD Europe	6.98	7.13	7.22	7.36	7.22	7.30	7.49	7.34
Total OECD Europe	15.28	15.44	15.48	15.62	15.17	15.56	15.69	15.51
Total OECD	47.88	48.60	49.37	50.66	48.59	49.20	49.97	49.60
Non-OECD								
FSU	3.84	3.93	4.11	4.30	3.81	4.04	4.60	4.19
Non-OECD Europe	0.67	0.66	0.69	0.74	0.69	0.64	0.69	0.69
China	5.16	5.58	6.40	6.62	6.89	6.93	7.14	6.90
Other Asia	7.75	8.00	8.39	8.34	8.71	8.43	9.06	8.64
Other Non-OECD	12.79	12.97	13.50	13.83	13.90	14.13	14.13	14.00
Total Non-OECD	30.20	31.14	33.08	33.84	34.01	34.18	35.63	34.42
Total World Demand	**78.08**	**79.74**	**82.45**	**84.50**	**82.60**	**83.38**	**85.60**	**84.02**
Supply/Demand Balance	**–1.13**	**–0.17**	**0.55**	**–0.26**	**2.43**	**1.07**	**–0.06**	**0.54**

Source: U.S. Energy Information Administration, www.eia.doe.gov/emeu/ipsr/balance.html, data as of Jan. 15, 2007.

NOTES: Some totals might not compute exactly due to rounding and other factors.
OECD = Organization for Economic Cooperation and Development.

Table 9: Summary of Operating Refineries Worldwide

COUNTRY	Number of refineries	Charge capacity, b/cd		
		Crude	Catalytic cracking	Catalytic reforming
Albania	2	26,300	0	3,500
Algeria	4	450,000	0	88,900
Angola	1	39,000	0	1,900
Argentina	10	624,575	140,910	56,400
Aruba	1	270,750	0	0
Australia	7	704,659	228,014	164,700
Austria	1	208,600	26,250	32,725
Azerbaijan	2	398,978	57,750	24,466
Bahrain	1	248,900	38,700	10,800
Bangladesh	1	33,000	0	1,800
Belarus	2	493,323	44,000	92,000
Belgium	4	790,629	132,251	104,017
Bolivia	2	41,200	0	12,100
Brazil	13	1,908,278	505,387	24,386
Brunei	1	8,600	0	5,700
Bulgaria	1	115,240	23,300	4,060
Cameroon	1	42,000	0	7,000
Canada	19	2,041,200	516,600	379,850
Chile	3	226,800	50,540	26,460
China	52	6,246,000	588,000	156,000
China, Taiwan	4	1,290,000	217,900	115,000
Colombia	5	285,850	90,000	0
Congo, Republic of	1	21,000	0	2,000
Costa Rica	1	24,000	0	1,200
Croatia	3	250,317	51,002	49,368
Cuba	4	301,400	14,700	20,000
Czech Republic	4	198,000	53,740	28,140
Denmark	2	176,400	0	21,500
Dominican Republic	2	48,300	0	7,560
Ecuador	3	176,000	18,000	12,800
Egypt	9	726,250	0	62,240
El Salvador	1	22,000	0	3,200
Eritrea	1	14,564	0	1,465
Finland	2	251,800	51,000	44,430
France	13	1,958,794	397,546	282,289
Gabon	1	17,300	0	1,400
Germany	15	2,417,418	359,719	409,087
Ghana	1	45,000	14,000	65,000
Greece	4	413,000	75,200	47,800
Hungary	2	161,000	24,000	29,600
India	17	2,255,540	303,305	41,673
Indonesia	8	992,745	101,450	92,970
Iran	9	1,451,000	35,000	164,700
Iraq	8	597,500	0	82,000
Ireland	1	71,000	0	11,000
Israel	2	220,000	49,500	26,500
Italy	17	2,337,229	310,500	287,014
Ivory Coast	1	65,200	0	12,827
Jamaica	1	36,000	0	3,700
Japan	31	4,676,690	881,500	710,600
Jordan	1	90,400	4,000	10,900
Kazakhstan	3	345,093	38,356	51,586
Kenya	1	85,500	0	8,100
Kuwait	3	889,200	36,000	46,620
Kyrgyzstan	1	10,000	0	0
Liberia	1	15,000	0	2,000
Libya	5	378,000	0	20,250
Lithuania	1	190,000	43,200	45,900
Macedonia	1	50,000	0	10,860
Malaysia	6	544,832	0	87,000
Martinique	1	17,000	0	3,019
Mexico	6	1,540,000	374,500	310,300
Morocco	2	154,901	5,040	24,359

Table 9: Summary of Operating Refineries Worldwide, cont'd.

COUNTRY	Number of refineries	Charge capacity, b/cd		
		Crude	Catalytic cracking	Catalytic reforming
Myanmar	3	57,000	0	0
Netherlands	6	1,211,886	104,800	150,473
Netherlands Antilles	1	320,000	50,000	20,000
New Zealand	1	103,550	0	27,000
Nicaragua	1	20,000	0	2,900
Nigeria	4	438,750	82,700	70,070
North Korea	2	71,000	0	7,300
Norway	2	310,000	54,000	37,600
Oman	1	85,000	0	16,000
Pakistan	5	268,975	0	25,715
Papua New Guinea	1	32,500	0	0
Paraguay	1	7,500	0	0
Peru	6	192,950	32,000	2,100
Philippines	3	333,000	15,000	52,900
Poland	4	496,500	31,100	88,500
Portugal	2	304,172	40,500	50,182
Puerto Rico	1	73,000	0	21,000
Qatar	1	200,000	60,000	29,400
Romania	10	516,557	109,478	61,763
Russia	40	5,338,961	330,817	745,735
Saudi Arabia	7	2,095,000	103,600	193,360
Senegal	1	27,000	0	2,000
Serbia & Montenegro	2	214,826	18,950	18,822
Sierra Leone	1	10,000	0	0
Singapore	3	1,336,600	69,000	129,000
Slovakia	1	115,000	18,000	21,000
Slovenia	1	13,500	0	0
South Africa	4	488,297	103,345	79,643
South Korea	6	2,576,500	187,000	235,870
Spain	9	1,271,500	191,300	197,200
Sri Lanka	1	50,000	0	5,300
Sudan	3	121,700	0	1,900
Suriname	1	7,000	0	0
Sweden	5	434,000	29,700	69,900
Switzerland	2	132,000	20,400	28,000
Syria	2	239,865	0	31,242
Tanzania	1	14,900	0	2,500
Thailand	4	729,100	91,490	125,590
Trinidad & Tobago	1	168,000	24,000	18,000
Tunisia	1	34,000	0	3,300
Turkey	6	714,275	28,935	65,662
Turkmenistan	2	236,970	15,151	52,540
Ukraine	6	879,759	70,100	144,711
United Arab Emirates	5	781,250	34,350	25,875
United Kingdom	11	1,887,068	436,885	316,621
United States	131	17,272,587	5,731,094	3,536,478
Uruguay	1	50,000	12,000	12,000
Uzbekistan	3	222,271	0	23,487
Venezuela	5	1,282,100	231,800	49,500
Virgin Islands	1	500,000	140,000	105,000
Yemen	2	140,000	0	14,500
Zambia	1	23,750	0	5,320
Total	658	85,179,374	14,368,355	11,377,680

Source: *Oil & Gas Journal* data as of Jan. 1, 2007.

NOTES: Some totals might not compute exactly due to rounding and other factors.
b/cd = barrels per calendar-day.

Table 10: World Refining Capacity

COUNTRY	1960	1970	1980	1990	2000	2001	2002	2003	2004	2005	2006	2007
	1,000 b/cd											
North America												
Canada	920	1,400	2,222	1,852	1,912	1,906	1,944	1,983	1,988	2,017	2,017	2,041
US	10,250	12,600	17,720	16,244	16,541	16,539	16,564	16,623	16,698	16,775	17,126	17,273
Total	11,170	14,000	19,942	18,096	18,453	18,445	18,508	18,606	18,686	18,792	19,143	19,314
Latin America												
Argentina	238	457	676	689	662	639	639	639	639	625	625	625
Bolivia	12	12	74	58	48	63	63	63	63	47	47	41
Brazil	156	502	1,205	1,397	1,783	1,918	1,756	1,865	1,914	1,920	1,908	1,908
Chile	24	91	139	147	205	205	205	205	205	227	227	227
Colombia	51	137	193	227	286	286	286	286	286	286	286	286
Cuba	87	93	160	160	301	301	301	301	301	301	301	301
Ecuador	7	33	86	145	176	176	176	176	176	176	176	176
Guatemala	--	21	16	16	20	22	16	16	16	0	0	0
Jamaica	--	--	--	--	--	--	--	--	--	--	36	36
Mexico	357	495	1,394	1,514	1,525	1,525	1,525	1,684	1,684	1,684	1,684	1,540
Peru	49	92	170	172	182	182	182	191	191	193	193	193
Trinidad & Tobago	182	430	456	300	160	160	160	160	165	165	175	168
Venezuela	886	1,313	1,446	1,201	1,239	1,282	1,282	1,282	1,282	1,282	1,282	1,282
Other Caribbean	--	--	--	--	--	--	--	--	1,125	1,045	1,224	1,229
Other Latin America	--	--	--	--	--	--	--	--	--	--	131	131
Total	2,841	5,114	8,620	7,315	7,973	8,202	7,942	6,868	8,047	7,951	8,295	8,143
Europe												
Albania	--	--	--	--	26	26	26	26	26	26	26	26
Austria	47	94	280	204	209	209	209	209	209	209	209	209
Belgium	171	669	1,064	614	719	768	791	791	793	803	858	791
Bulgaria	--	--	--	--	115	115	115	115	115	115	115	115
Croatia	--	--	--	--	236	253	293	260	250	250	250	250
Czech Republic	46	220	455	455	186	198	198	198	198	198	198	198
Denmark	1	186	214	187	135	176	176	176	176	176	176	176
Finland	25	174	336	241	200	200	239	252	252	252	252	252
France	769	2,317	3,385	1,820	1,902	1,895	1,896	1,903	1,951	1,951	1,979	1,959
Germany	573	2,359	2,986	1,507	2,275	2,259	2,259	2,267	2,289	2,323	2,344	2,417
Greece	30	102	431	385	383	407	407	407	401	401	413	413
Hungary	58	130	290	220	232	232	161	161	161	161	161	161
Ireland	40	55	56	56	71	71	71	71	71	71	71	71
Italy	73	2,963	4,131	2,804	2,341	2,359	2,283	2,300	2,313	2,321	2,324	2,337
Netherlands	359	1,362	1,828	1,381	1,188	1,204	1,206	1,207	1,222	1,228	1,222	1,212
Norway	2	123	264	295	358	305	310	310	310	310	310	310
Poland	20	156	385	385	382	382	382	350	350	350	467	497
Portugal	23	37	378	313	304	304	304	304	304	304	304	304
Romania	260	356	608	617	499	504	504	501	501	517	517	517
Slovakia	--	--	--	--	115	115	115	115	115	115	115	115
Spain	146	685	1,455	1,293	1,316	1,294	1,294	1,322	1,272	1,272	1,272	1,272
Sweden	48	230	458	428	427	423	424	424	424	434	434	434
Switzerland	--	106	137	132	132	132	132	132	132	132	132	132
Turkey	7	149	356	725	691	694	719	719	714	714	714	714
UK	963	2,301	2,527	1,831	1,785	1,771	1,784	1,789	1,817	1,825	1,877	1,887
Other Europe	0	0	0	0	0	0	0	0	0	0	278	279
Total	970	3,471	4,933	4,409	16,227	16,296	16,298	16,309	16,366	16,458	17,018	17,048
Middle East												
Bahrain	187	205	250	243	249	249	249	249	249	249	255	249
Iran	495	645	921	530	1,474	1,484	1,484	1,474	1,474	1,474	1,451	1,451
Iraq	56	104	168	319	348	418	418	418	588	598	598	598
Israel	87	107	195	180	220	220	220	220	220	220	220	220
Kuwait	270	569	645	819	865	764	773	889	889	889	889	889
Oman	--	--	--	--	--	--	85	85	85	85	85	85
Qatar	1	1	11	62	58	58	58	200	200	200	200	200
Saudi Arabia	189	377	487	1,484	1,710	1,745	1,745	1,745	1,745	1,745	2,095	2,095
Syria	--	59	223	244	242	242	242	240	240	240	240	240
UAE	--	--	14	180	429	444	515	514	514	514	781	781
Yemen	120	178	175	162	120	130	130	130	130	130	130	140
Other Middle East	--	--	--	--	--	--	--	--	--	--	90	90
Total	1,429	2,290	3,163	4,447	5,928	5,965	5,532	6,164	6,334	6,344	7,034	7,038

Table 10: World Refining Capacity, cont'd.

COUNTRY	1960	1970	1980	1990	2000	2001	2002	2003	2004	2005	2006	2007
	1,000 b/cd											
Africa												
Algeria	--	48	122	465	503	503	450	450	450	450	450	450
Angola	2	14	36	32	39	39	39	39	39	39	39	39
Cameroon	--	--	--	43	35	42	42	42	42	42	42	42
Congo (Republic of)	--	14	--	21	21	21	21	21	21	21	21	21
Congo (Zaire)	--	21	16	17	17	15	15	15	0	0	0	0
Egypt	87	175	234	489	578	726	726	726	726	726	726	726
Gabon	--	13	20	24	17	17	17	17	17	17	17	17
Ivory Coast	--	19	50	69	69	65	65	65	65	65	65	65
Libya	--	10	138	329	348	343	343	343	343	380	380	378
Madagascar	--	--	--	--	--	--	--	15	15	15	0	0
Morocco	2	35	72	155	157	157	155	155	155	155	155	155
Nigeria	--	46	160	433	439	439	439	439	439	439	439	439
South Africa	25	184	478	434	467	474	469	490	520	490	505	488
Sudan	--	21	26	25	82	122	122	122	122	122	122	122
Tanzania	--	--	--	--	--	--	--	--	--	--	15	15
Tunisia	--	23	34	34	34	34	34	34	34	34	34	34
Other Africa	--	--	--	--	--	--	--	--	--	--	221	222
Total	116	787	1,665	2,824	3,046	3,265	3,169	2,973	2,988	2,995	3,231	3,213
Former Soviet Union												
Azerbaijan	--	--	--	--	--	--	442	442	399	399	399	399
Belarus	--	--	--	--	--	--	493	493	493	493	493	493
Georgia	--	--	--	--	--	--	106	106	106	0	0	0
Kazakhstan	--	--	--	--	--	--	427	427	427	345	345	345
Kyrgyzstan	--	--	--	--	--	--	10	10	10	10	10	10
Lithuania	--	--	--	--	--	--	--	--	263	263	190	190
Russia	--	--	--	--	--	--	5,435	5,435	5,435	5,433	5,341	5,339
Turkmenistan	--	--	--	--	--	--	237	237	237	237	237	237
Ukraine	--	--	--	--	--	--	1,206	1,025	1,025	880	880	880
USSR	2,800	5,640	10,950	12,300	9,762	8,400	--	--	--	--	--	--
Uzbekistan	--	--	--	--	--	--	222	222	222	222	222	222
Total	2,800	5,640	10,950	12,300	9,762	8,400	8,578	8,397	8,617	8,282	8,117	8,115
Asia-Pacific												
Australia	237	594	725	675	812	847	848	848	755	755	702	705
Bangladesh	--	--	31	31	33	33	33	33	33	33	33	33
Brunei	--	--	--	--	--	--	9	9	9	9	9	9
China	98	300	1,600	2,200	4,347	4,347	4,528	4,528	4,528	4,650	6,246	6,246
India	112	446	557	1,080	1,858	2,113	2,135	2,135	2,135	2,255	2,255	2,256
Indonesia	274	268	528	714	993	993	993	993	993	993	993	993
Japan	640	3,140	5,509	4,198	4,998	4,962	4,786	4,767	4,703	4,707	4,672	4,677
Malaysia	--	--	--	--	--	--	515	516	516	545	545	545
Myanmar	9	26	26	32	32	32	32	57	57	57	57	57
New Zealand	--	66	74	95	98	106	106	106	104	104	104	104
Pakistan	6	107	98	121	143	239	239	234	269	269	269	269
Papua New Guinea	--	--	--	--	--	--	--	--	--	33	33	33
Philippines	22	182	253	284	401	420	420	420	333	333	333	333
Singapore	46	320	1,093	1,049	1,788	1,792	1,259	1,259	1,319	1,337	1,337	1,337
South Korea	--	180	601	867	2,540	2,560	2,560	2,560	2,544	2,577	2,577	2,577
Taiwan	28	119	425	570	770	920	920	920	920	1,220	1,220	1,290
Thailand	1	62	186	215	713	682	682	703	703	703	703	729
Other Asia-Pacific	--	--	--	--	--	--	--	--	--	--	121	121
Total	1,472	5,866	11,986	12,181	19,645	20,166	20,185	20,088	19,921	20,595	22,209	22,314
Total World	**24,239**	**48,982**	**79,514**	**74,535**	**81,556**	**81,252**	**81,166**	**81,878**	**82,055**	**82,409**	**85,043**	**85,185**

Table 10: World Refining Capacity, cont'd.

COUNTRY	1960	1970	1980	1990	2000	2001	2002	2003	2004	2005	2006	2007
	Million tonnes per year (mtpy)											
North America												
Canada	45.8	69.7	110.7	92.2	95.2	94.9	96.8	98.8	99.0	100.4	100.4	101.6
US	510.5	627.5	882.5	809.0	823.7	823.6	824.9	827.8	831.5	835.3	852.8	860.0
Total	556.3	697.2	993.2	901.2	918.9	918.5	921.7	926.6	930.5	935.7	953.2	961.6
Latin America												
Argentina	11.9	22.8	33.7	34.3	33.0	31.8	31.8	31.8	31.8	31.1	31.1	31.1
Bolivia	0.6	0.6	3.7	2.9	2.4	3.1	3.1	3.1	3.1	2.3	2.4	2.0
Brazil	7.8	25.0	60.0	69.6	88.8	95.5	88.9	92.9	95.3	95.6	95.0	95.0
Chile	1.2	4.5	6.9	7.3	10.2	10.2	10.2	10.2	10.2	11.3	11.3	11.3
Colombia	2.5	6.8	9.6	11.3	14.2	14.2	14.2	14.2	14.2	14.2	14.2	14.2
Cuba	4.3	4.6	8.0	8.0	15.0	15.0	15.0	15.0	15.0	15.0	15.0	15.0
Ecuador	0.3	1.6	4.3	7.2	8.8	8.8	8.8	8.8	8.8	8.8	8.8	8.8
Guatemala	--	1.0	0.8	0.8	1.0	1.1	0.8	0.8	0.8	0.0	0.0	0.0
Jamaica	--	--	--	--	--	--	--	--	--	--	1.8	1.8
Mexico	17.8	24.7	69.4	75.4	75.9	75.9	75.9	83.9	83.9	83.9	83.9	76.7
Peru	2.4	4.6	8.5	8.6	9.1	9.1	9.1	9.5	9.5	9.6	9.6	9.6
Trinidad & Tobago	9.1	21.4	22.7	14.9	8.0	8.0	8.0	8.0	8.2	8.2	8.7	8.4
Venezuela	44.1	65.4	72.0	59.8	61.7	63.8	63.8	63.8	63.8	63.8	63.8	63.8
Other Caribbean	--	--	--	--	--	--	--	--	56.0	52.0	61.0	61.2
Other Latin America	--	--	--	--	--	--	--	--	--	--	6.5	6.5
Total	141.5	254.5	429.2	364.0	396.9	408.0	395.2	342.0	400.6	395.8	413.1	405.4
Europe												
Albania	--	--	--	--	1.3	1.3	1.3	1.3	1.3	1.3	1.3	1.3
Austria	2.3	4.7	13.9	10.2	10.4	10.4	10.4	10.4	10.4	10.4	10.4	10.4
Belgium	8.5	33.3	53.0	30.6	35.8	38.2	39.4	39.4	39.5	40.0	42.7	39.4
Bulgaria	--	--	--	--	5.7	5.7	5.7	5.7	5.7	5.7	5.7	5.7
Croatia	--	--	--	--	11.8	12.6	14.6	12.9	12.4	12.4	12.5	12.5
Czech Republic	2.3	11.0	22.7	22.7	9.3	9.9	9.9	9.9	9.9	9.9	9.9	9.9
Denmark	--	9.3	10.7	9.3	6.7	8.8	8.8	8.8	8.8	8.8	8.8	8.8
Finland	1.2	8.7	16.7	12.0	10.0	10.0	11.9	12.4	12.4	12.4	12.5	12.5
France	38.3	115.4	168.6	90.6	94.7	94.4	94.4	94.8	97.2	97.2	98.6	97.5
Germany	28.5	117.5	148.7	75.0	113.3	112.5	112.5	112.9	114.0	115.7	116.7	120.3
Greece	1.5	5.1	21.5	19.2	19.1	20.2	20.2	20.2	20.0	20.0	20.6	20.6
Hungary	2.9	6.5	14.4	11.0	11.6	11.6	8.0	8.0	8.0	8.0	8.0	8.0
Ireland	2.0	2.7	2.8	2.8	3.5	3.5	3.5	3.5	3.5	3.5	3.5	3.5
Italy	38.5	147.6	205.7	139.6	116.6	117.5	113.7	114.5	115.2	115.6	115.7	116.4
Netherlands	17.9	67.8	91.0	68.8	59.2	60.0	60.1	60.1	60.8	61.1	60.8	60.3
Norway	0.1	6.1	13.1	14.7	17.8	15.2	15.4	15.4	15.4	15.4	15.4	15.4
Poland	1.0	7.8	19.2	19.2	19.0	19.0	19.0	17.4	17.4	17.4	23.2	24.7
Portugal	1.1	1.8	18.8	15.6	15.1	15.1	15.1	15.1	15.1	15.1	15.1	15.1
Romania	12.9	17.7	30.3	30.7	24.9	25.1	25.1	24.9	24.9	25.7	25.7	25.7
Slovakia	--	--	--	--	5.7	5.7	5.7	5.7	5.7	5.7	5.7	5.7
Spain	7.3	34.1	72.5	64.4	65.5	64.4	64.4	65.8	63.3	63.3	63.3	63.3
Sweden	2.4	11.5	22.8	21.3	21.3	21.1	21.1	21.1	21.1	21.6	21.6	21.6
Switzerland	-	5.3	6.8	6.6	6.6	6.6	6.6	6.6	6.6	6.6	6.6	6.6
Turkey	0.3	7.4	17.7	36.1	34.4	34.6	35.8	35.8	35.6	35.6	35.6	35.6
UK	48.0	114.6	125.8	91.2	88.9	88.2	88.8	89.1	90.5	90.9	93.5	94.0
Other Europe	--	--	--	--	0	0	0	0	0	0	13.9	13.9
Total	208.5	702.6	245.6	761.0	808.2	811.6	811.4	811.7	814.7	819.3	847.3	848.8
Middle East												
Bahrain	9.3	10.2	12.4	12.1	12.4	12.4	12.4	12.4	12.4	12.4	12.7	12.4
Iran	24.7	32.1	45.9	26.4	73.4	73.9	73.9	73.4	73.4	73.4	72.3	72.3
Iraq	2.8	5.2	8.4	15.9	17.3	20.8	20.8	20.8	29.3	29.8	29.8	29.8
Israel	4.3	5.3	9.7	9.0	11.0	11.0	11.0	11.0	11.0	11.0	11.0	11.0
Kuwait	13.5	28.4	32.1	40.8	43.1	38.0	38.5	44.3	44.3	44.3	44.3	44.3
Oman	--	--	--	--	--	--	4.2	4.2	4.2	4.2	4.2	4.2
Qatar	--	--	0.5	3.1	2.9	2.9	2.9	10.0	10.0	10.0	10.0	10.0
Saudi Arabia	9.4	18.8	24.3	73.9	85.2	86.9	86.9	86.9	86.9	86.9	104.3	104.3
Syria	--	2.9	11.1	12.2	12.1	12.1	12.1	12.0	12.0	12.0	11.9	11.9
UAE	--	--	0.7	9.0	21.4	22.1	25.6	25.6	25.6	25.6	38.9	38.9
Yemen	6.0	8.9	8.7	8.1	6.0	6.5	6.5	6.5	6.5	6.5	6.5	7.0
Other Middle East	--	--	--	--	--	--	--	--	--	--	4.5	4.5
Total	71.2	114.0	151.4	221.6	295.4	297.2	275.6	307.0	315.6	316.1	350.4	350.4

Table 10: World Refining Capacity, cont'd.

COUNTRY	1960	1970	1980	1990	2000	2001	2002	2003	2004	2005	2006	2007
	Million tonnes per year (mtpy)											
Africa												
Algeria	--	2.4	6.1	23.2	25.0	25.0	22.4	22.4	22.4	22.4	22.4	22.4
Angola	0.1	0.7	1.8	1.6	1.9	1.9	1.9	1.9	1.9	1.9	1.9	1.9
Cameroon	--	--	--	2.1	1.7	2.1	2.1	2.1	2.1	2.1	2.1	2.1
Congo (Brazzaville)	--	0.7	--	1.0	1.0	1.0	1.0	1.0	1.0	1.0	1.0	1.0
Congo (Zaire)	--	1.0	0.8	0.8	0.8	0.7	0.7	0.7	0.0	0.0	0.0	0.0
Egypt	4.3	8.7	11.7	24.4	28.8	36.2	36.2	36.2	36.2	36.2	36.2	36.2
Gabon	--	0.6	1.0	1.2	0.8	0.8	0.8	0.8	0.8	0.8	0.9	0.9
Ivory Coast	--	0.9	2.5	3.4	3.4	3.2	3.2	3.2	3.2	3.2	3.2	3.2
Libya	--	0.5	6.9	16.4	17.3	17.1	17.1	17.1	17.1	18.9	18.9	18.8
Madagascar	--	--	--	--	--	--	--	--	--	0.7	0.0	0.0
Morocco	0.1	1.7	3.6	7.7	7.8	7.8	7.7	7.7	7.7	7.7	7.7	7.7
Nigeria	--	2.3	8.0	21.6	21.9	21.9	21.9	21.9	21.9	21.9	21.8	21.8
South Africa	1.2	9.2	23.8	21.6	23.3	23.6	23.3	24.4	25.9	24.4	25.1	24.3
Sudan	--	1.0	1.3	1.2	4.1	6.1	6.1	6.1	6.1	6.1	6.1	6.1
Tanzania	--	--	--	--	--	--	--	--	--	--	0.7	0.7
Tunisia	--	1.1	1.7	1.7	1.7	1.7	1.7	1.7	1.7	1.7	1.7	1.7
Other Africa	--	--	--	--	--	--	--	--	--	--	11.0	11.1
Total	5.7	38.8	82.9	140.4	151.2	162.3	157.1	147.3	148.0	148.3	160.7	160.0
Former Soviet Union												
Azerbaijan	--	--	--	--	--	--	22.0	22.0	19.9	19.9	19.9	19.9
Belarus	--	--	--	--	--	--	24.6	24.6	24.6	24.6	24.6	24.6
Georgia	--	--	--	--	--	--	5.3	5.3	5.3	0.0	0.0	0.0
Kazakhstan	--	--	--	--	--	--	21.3	21.3	21.3	17.2	17.2	17.2
Kyrgyzstan							0.5	0.5	0.5	0.5	0.5	0.5
Lithuania	--	--	--	--	--	--	--	--	13.1	13.1	9.5	9.5
Russia	--	--	--	--	--	--	270.7	270.7	270.7	270.5	266.0	265.8
Turkmenistan	--	--	--	--	--	--	11.8	11.8	11.8	11.8	11.8	11.8
Ukraine	--	--	--	--	--	--	60.1	51.0	51.0	43.8	43.8	43.8
USSR	139.4	280.9	545.3	612.5	486.1	418.3	--	--	--	--	--	--
Uzbekistan	--	--	--	--	--	--	11.1	11.1	11.1	11.1	11.1	11.1
Total	139.4	280.9	545.3	612.5	486.1	418.3	427.4	418.2	429.3	412.5	404.4	404.2
Asia-Pacific												
Australia	11.8	29.6	36.1	33.6	40.4	42.2	42.2	42.2	37.6	37.6	34.9	35.1
Bangladesh	--	--	1.5	1.5	1.6	1.6	1.6	1.6	1.6	1.6	1.6	1.6
Brunei	--	--	--	--	--	--	0.4	0.4	0.4	0.4	0.4	0.4
China	4.9	14.9	79.7	109.6	216.5	216.5	225.5	225.5	225.5	231.5	311.0	311.0
India	5.6	22.2	27.7	53.8	92.5	105.2	106.3	106.3	106.3	112.3	112.3	112.3
Indonesia	13.6	13.3	26.3	35.6	49.5	49.5	49.5	49.5	49.5	49.5	49.4	49.4
Japan	31.9	156.4	274.3	209.1	248.9	247.1	238.3	237.4	234.2	234.4	232.6	232.9
Malaysia	--	--	--	--	--	--	25.6	25.7	25.7	27.1	27.1	27.1
Myanmar	0.4	1.3	1.3	1.6	1.6	1.6	1.6	2.8	2.8	2.8	2.8	2.8
New Zealand	--	3.3	3.7	4.7	4.9	5.3	5.3	5.3	5.2	5.2	5.2	5.2
Pakistan	0.3	5.3	4.9	6.0	7.1	11.9	11.9	11.7	13.4	13.4	13.4	13.4
Papua New Guinea	--	--	--	--	--	--	--	--	--	1.6	1.6	1.6
Philippines	1.1	9.1	12.6	14.1	20.0	20.9	20.9	20.9	16.6	16.6	16.6	16.6
Singapore	2.3	15.9	54.4	52.2	89.0	89.3	62.7	62.7	65.7	66.6	66.6	66.6
South Korea	--	9.0	29.9	43.2	126.5	127.5	127.5	127.5	126.7	128.3	128.3	128.3
Taiwan	1.4	5.9	21.2	28.4	38.3	45.8	45.8	45.8	45.8	60.8	60.8	64.2
Thailand	--	3.1	9.3	10.7	35.5	34.0	34.0	35.0	35.0	35.0	35.0	36.3
Other Asia-Pacific	--	--	--	--	--	--	--	--	--	--	6.0	6.0
Total	73.3	292.1	596.9	606.6	978.2	1,004.3	1,005.1	1,000.4	992.0	1,025.4	1,105.6	1,111.0
Total World	**1,206.9**	**2,438.8**	**3,959.5**	**3,711.5**	**4,060.8**	**4,045.7**	**4,041.4**	**4,077.5**	**4,086.0**	**4,103.6**	**4,234.7**	**4,241.4**

Source: *Oil & Gas Journal* data as of Jan. 1, 2007.

NOTES: Some totals might not compute exactly due to rounding and other factors.
b/cd = barrels per calendar-day.
Other Caribbean includes Aruba, Dominican Republic, Martinique, Netherlands Antilles, Puerto Rico, and Virgin Islands.
Other Latin America includes Costa Rica, El Salvador, Nicaragua, Paraguay, Suriname, and Uruguay.
Other Europe includes Macedonia, Serbia & Montenegro, and Slovenia.
Other Middle East includes Jordan.
Other Africa includes Eritrea, Ghana, Kenya, Liberia, Senegal, Sierra Leone, and Zambia.
USSR includes total of all Former Soviet Union countries before 2002.
Brunei and Malaysia were totaled with Singapore before 2002.
Other Asia-Pacific includes North Korea and Sri Lanka.

Table 11: World Refinery Throughput

REGION	2000	2001	2002	2003	2004	2005	Change 2005 over 2004	2005 share of total
	Thousand barrels per day (1,000 b/d)						%	
US	5,430	5,646	5,695	5,867	5,790	5,958	2.9	8.1
Canada	2,233	2,463	2,251	2,200	2,286	2,424	6.0	3.3
Mexico	851	847	855	823	820	752	-8.3	1.0
South & Central America	4,218	4,215	4,395	4,823	5,382	5,748	6.8	7.8
Europe & Eurasia	4,145	4,107	3,986	4,118	4,037	4,136	2.5	5.6
Middle East	8,914	9,053	8,827	9,300	9,957	10,139	1.8	13.7
Africa	5,430	5,646	5,695	5,867	5,790	5,958	2.9	8.1
Australasia	2,233	2,463	2,251	2,200	2,286	2,424	6.0	3.3
China	851	847	855	823	820	752	-8.3	1.0
Japan	4,218	4,215	4,395	4,823	5,382	5,748	6.8	7.8
Other Asia-Pacific	4,145	4,107	3,986	4,118	4,037	4,136	2.5	5.6
Total World	**68,670**	**69,513**	**68,668**	**70,748**	**73,055**	**73,999**	**1.3**	**100.0**

Source: Adapted from *BP Statistical Review of World Energy 2006*.

NOTES: Some totals might not compute exactly due to rounding and other factors.

Table 12: World Refining Margins

MARKET	1996	1997	1998	1999	2000	2001	2002	2003	2004	Dec. 2005	Dec. 2006	2005–2006 Change, %
	$/bbl											
US Gulf Coast												
West Texas sour	3.37	3.83	3.31	2.15	5.54	2.91	3.81	3.83	5.67	12.05	9.69	-19.6
Composite US gulf refinery	--	--	--	--	--	--	4.54	4.16	7.87	13.42	10.33	-23.0
Arabian light	3.63	4.57	4.04	3.14	7.27	5.02	5.45	5.51	8.81	13.22	10.58	-20.0
Bonny light	2.41	2.90	2.83	1.74	6.40	0.97	1.54	2.89	5.27	7.74	1.87	-75.8
US PADD II Region												
Chicago (WTI)	3.74	3.78	3.17	2.47	4.60	2.52	4.30	3.18	3.31	7.90	7.58	-4.1
US East Coast												
NY Harbor (Arabian medium)	2.89	3.89	3.62	2.50	6.38	4.15	3.38	3.45	5.90	9.14	4.48	-51.0
East Coast composite-RFG	--	--	--	--	--	2.88	4.33	4.15	6.53	10.06	6.42	-36.2
US West Coast												
Los Angeles (ANS)	5.13	5.58	5.60	6.55	12.48	4.33	3.57	6.95	10.03	8.60	19.56	127.4
NW Europe												
Rotterdam (Brent)	2.78	3.26	3.26	2.19	4.71	0.88	0.76	2.21	0.52	0.65	-0.57	-187.7
Mediterranean												
Italy (Urals)	1.82	2.24	2.73	1.67	3.28	0.05	0.74	1.88	7.38	5.60	7.34	31.1
Far East												
Singapore (Dubai)	2.57	2.50	1.61	0.78	3.89	0.63	1.98	3.43	4.24	1.46	-1.39	-195.2

Source: Jacobs Consultancy, Inc., in *Oil & Gas Journal*

NOTES: WTI = West Texas Intermediate.
ANS = Alaska North Slope.

Table 13: World Natural Gas Consumption

COUNTRY	2003	2004	2005	2003	2004	2005	Change 2005 over 2004	2005 share of total
	Billion cubic meters (bcm)			bcfd			%	
North America								
US	643.1	645.0	633.5	62.2	62.2	61.3	-1.5	23.0
Canada	92.2	92.7	91.4	8.9	8.9	8.8	-1.1	3.3
Mexico	45.8	48.6	49.6	4.4	4.7	4.8	2.3	1.8
Total	781.1	786.3	774.5	75.6	75.9	74.9	-1.2	28.2
South & Central America								
Argentina	34.6	37.9	40.6	3.3	3.7	3.9	7.4	1.5
Brazil	15.9	19.0	20.2	1.5	1.8	2.0	6.7	0.7
Chile	7.1	8.3	7.6	0.7	0.8	0.7	-8.2	0.3
Colombia	6.0	6.3	6.8	0.6	0.6	0.7	7.5	0.2
Ecuador	0.1	0.2	0.2	0.0	0.0	0.0	0.0	0.0
Peru	0.5	0.9	1.6	0.1	0.1	0.2	84.2	0.1
Venezuela	25.2	28.1	28.9	2.4	2.7	2.8	3.2	1.1
Other S/C America	15.9	17.1	18.3	1.5	1.7	1.8	7.1	0.7
Total	105.3	117.7	124.1	10.2	11.4	12.0	5.7	4.5
Europe and Eurasia								
Austria	9.4	9.5	10.0	0.9	0.9	1.0	5.7	0.4
Azerbaijan	8.0	8.6	8.8	0.8	0.8	0.9	3.0	0.3
Belarus	16.3	18.5	18.9	1.6	1.8	1.8	2.8	0.7
Belgium & Luxembourg	16.0	16.5	16.8	1.6	1.6	1.6	2.3	0.6
Czech Republic	2.8	2.9	3.2	0.8	0.8	0.8	-1.1	0.3
Denmark	8.7	8.7	8.5	0.5	0.5	0.5	-3.7	0.2
Finland	5.2	5.2	5.0	0.4	0.4	0.4	-7.1	0.1
France	4.5	4.3	4.0	4.2	4.3	4.4	1.3	1.6
Germany	43.3	44.5	45.0	8.3	8.3	8.3	0.3	3.1
Greece	85.5	85.9	85.9	0.2	0.2	0.2	2.3	0.1
Hungary	2.3	2.5	2.5	1.3	1.3	1.3	3.6	0.5
Ireland	4.1	4.1	4.1	0.4	0.4	0.4	-4.6	0.1
Italy	70.9	73.6	79.0	6.9	7.1	7.6	7.7	2.9
Kazakhstan	13.3	15.4	17.8	1.3	1.5	1.7	15.8	0.6
Lithuania	3.1	3.1	3.2	0.3	0.3	0.3	5.1	0.1
Netherlands	40.3	41.1	39.5	3.9	4.0	3.8	-3.6	1.4
Norway	4.3	4.6	4.5	0.4	0.4	0.4	-2.1	0.2
Poland	11.2	13.1	13.6	1.1	1.3	1.3	4.1	0.5
Portugal	3.0	3.1	3.0	0.3	0.3	0.3	-0.5	0.1
Romania	18.3	17.5	17.3	1.8	1.7	1.7	-0.7	0.6
Russian Federation	392.9	401.9	405.1	38.0	38.8	39.2	1.1	14.7
Slovakia	6.3	6.1	5.9	0.6	0.6	0.6	-3.1	0.2
Spain	23.6	27.4	32.3	2.3	2.6	3.1	18.2	1.2
Sweden	0.8	0.8	0.8	0.1	0.1	0.1	-4.5	0.0
Switzerland	2.9	3.0	3.1	0.3	0.3	0.3	2.9	0.1
Turkey	20.9	22.1	27.4	2.0	2.1	2.6	24.1	1.0
Turkmenistan	14.6	15.5	16.6	1.4	1.5	1.6	7.5	0.6

Table 13: World Natural Gas Consumption, cont'd.

COUNTRY	2003	2004	2005	2003	2004	2005	Change 2005 over 2004	2005 share of total
	Billion cubic meters (bcm)			bcfd			%	
Ukraine	68.0	72.9	72.9	6.6	7.0	7.0	0.3	2.6
United Kingdom	95.3	97.0	94.6	9.2	9.4	9.2	-2.2	3.4
Uzbekistan	47.2	44.8	44.0	4.6	4.3	4.3	-1.4	1.6
Other Europe/Eurasia	14.2	14.4	15.3	1.4	1.4	1.5	7.0	0.6
Total	1,070.6	1,101.2	1,121.9	103.6	106.3	108.6	2.2	40.8
Middle East								
Iran	82.9	86.5	88.5	8.0	8.4	8.6	2.5	3.2
Kuwait	9.1	9.7	9.7	0.9	0.9	0.9	0.3	0.4
Qatar	12.2	14.9	15.9	1.2	1.4	1.5	6.8	0.6
Saudi Arabia	60.1	65.7	69.5	5.8	6.3	6.7	6.1	2.5
United Arab Emirates	37.9	40.2	40.4	3.7	3.9	3.9	0.8	1.5
Other Middle East	23.9	25.3	27.0	2.3	2.4	2.6	7.1	1.0
Total	226.1	242.3	251.0	21.9	23.4	24.3	3.9	9.1
Africa								
Algeria	21.4	22.0	24.1	2.1	2.1	2.3	9.8	0.9
Egypt	24.6	26.2	25.5	2.4	2.5	2.5	-2.6	0.9
Other Africa	19.2	20.4	21.6	1.9	2.0	2.1	6.3	0.8
Total	65.2	68.6	71.2	6.3	6.6	6.9	4.0	2.6
Asia-Pacific								
Australia	26.1	25.3	25.7	2.5	2.4	2.5	1.6	0.9
Bangladesh	12.3	13.3	14.2	1.2	1.3	1.4	7.1	0.5
China	33.2	39.0	47.0	3.2	3.8	4.5	20.8	1.7
China Hong Kong SAR	1.5	2.2	2.2	0.1	0.2	0.2	-1.3	0.1
India	29.9	32.7	36.6	2.9	3.2	3.5	12.2	1.3
Indonesia	33.4	36.9	39.4	3.2	3.6	3.8	7.2	1.4
Japan	82.6	78.7	81.1	8.0	7.6	7.8	3.3	2.9
Malaysia	31.8	33.9	34.9	3.1	3.3	3.4	3.4	1.3
New Zealand	4.1	3.7	3.6	0.4	0.4	0.3	-3.3	0.1
Pakistan	23.2	26.9	29.9	2.2	2.6	2.9	11.5	1.1
Philippines	2.7	2.4	3.0	0.3	0.2	0.3	27.6	0.1
Singapore	5.3	6.6	6.5	0.5	0.6	0.6	-1.2	0.2
South Korea	26.9	31.5	33.3	2.6	3.0	3.2	6.1	1.2
Taiwan	8.7	10.2	10.7	0.8	1.0	1.0	5.1	0.4
Thailand	26.3	27.4	29.9	2.5	2.6	2.9	9.5	1.1
Other Asia-Pacific	5.6	7.8	8.9	0.5	0.8	0.9	14.4	0.3
Total	353.8	378.5	406.9	34.2	36.5	39.4	7.8	14.8
Total World	**2,601.9**	**2,694.7**	**2,749.6**	**251.7**	**260.0**	**266.0**	**2.3**	**100.0**

Source: Adapted from *BP Statistical Review of World Energy 2006*.

NOTE: Some totals might not compute exactly due to rounding and other factors.

Table 14: World Natural Gas Production

COUNTRY	2003	2004	2005	2003	2004	2005	Change 2005 over 2004	2005 share of total
	Billion cubic meters (bcm)			bcfd			%	
North America								
US	551.4	539.4	525.7	52.3	51.2	49.9	-2.3	19.0
Canada	182.7	183.6	185.5	17.7	17.7	17.9	1.3	6.7
Mexico	36.4	37.4	39.5	3.5	3.6	3.8	6.0	1.4
Total	770.5	760.4	750.6	73.5	72.6	71.7	-1.0	27.2
South & Central America								
Argentina	41.0	44.9	45.6	4.0	4.3	4.4	1.9	1.7
Bolivia	6.4	8.5	10.4	0.6	0.8	1.0	23.2	0.4
Brazil	10.0	11.0	11.4	1.0	1.1	1.1	3.1	0.4
Colombia	6.1	6.4	6.8	0.6	0.6	0.7	7.0	0.2
Trinidad & Tobago	24.7	28.1	29.0	2.4	2.7	2.8	3.5	1.0
Venezuela	25.2	28.1	28.9	2.4	2.7	2.8	3.2	1.0
Other S/C America	2.2	2.8	3.5	0.2	0.3	0.3	26.3	0.1
Total	115.7	129.7	135.6	11.2	12.5	13.1	4.8	4.9
Europe and Eurasia								
Azerbaijan	4.8	4.7	5.3	0.5	0.4	0.5	13.9	0.2
Denmark	8.0	9.4	10.4	0.8	0.9	1.0	11.1	0.4
Germany	17.7	16.4	15.8	1.7	1.6	1.5	-3.2	0.6
Italy	13.7	13.0	12.0	1.3	1.3	1.2	-7.3	0.4
Kazakhstan	12.9	20.6	23.5	1.3	2.0	2.3	14.2	0.9
Netherlands	58.4	68.8	62.9	5.6	6.6	6.1	-8.4	2.3
Norway	73.1	78.5	85.0	7.1	7.6	8.2	8.6	3.1
Poland	4.0	4.4	4.3	0.4	0.4	0.4	-0.8	0.2
Romania	13.0	12.8	12.9	1.3	1.2	1.3	1.3	0.5
Russian Federation	578.6	591.0	598.0	56.0	57.0	57.9	1.5	21.6
Turkmenistan	55.1	54.6	58.8	5.3	5.3	5.7	7.9	2.1
Ukraine	17.7	19.1	18.8	1.7	1.8	1.8	-1.2	0.7
United Kingdom	102.9	96.0	88.0	10.0	9.3	8.5	-8.1	3.2
Uzbekistan	53.6	55.8	55.7	5.2	5.4	5.4	0.0	2.0
Other Europe/Eurasia	10.7	11.0	9.8	1.0	1.1	0.9	-10.7	0.4
Total	1,024.4	1,055.9	1,061.1	99.1	101.9	102.7	0.8	38.4
Middle East								
Bahrain	9.6	9.8	9.9	0.9	0.9	1.0	1.3	0.4
Iran	81.5	84.9	87.0	7.9	8.2	8.4	2.8	3.1
Kuwait	9.1	9.7	9.7	0.9	0.9	0.9	0.3	0.4
Oman	16.5	17.2	17.5	1.6	1.7	1.7	2.0	0.6
Qatar	31.4	39.2	43.5	3.0	3.8	4.2	11.4	1.6
Saudi Arabia	60.1	65.7	69.5	5.8	6.3	6.7	6.1	2.5
Syria	5.2	5.3	5.4	0.5	0.5	0.5	3.0	0.2
United Arab Emirates	44.8	46.3	46.6	4.3	4.5	4.5	0.9	1.7
Other Middle East	1.8	2.5	3.4	0.2	0.2	0.3	39.2	0.1
Total	259.9	280.4	292.5	25.1	27.1	28.3	4.6	10.6
Africa								
Algeria	82.8	82.0	87.8	8.0	7.9	8.5	7.3	3.2
Egypt	25.0	26.9	34.7	2.4	2.6	3.4	29.4	1.3
Libya	5.8	6.5	11.7	0.6	0.6	1.1	79.5	0.4
Nigeria	19.2	21.8	21.8	1.9	2.1	2.1	0.3	0.8
Other Africa	6.9	7.0	7.0	0.7	0.7	0.7	0.1	0.3
Total	139.7	144.3	163.0	13.5	13.9	15.8	13.3	5.9
Asia-Pacific								
Australia	33.2	35.3	37.1	3.2	3.4	3.6	5.5	1.3
Bangladesh	12.3	13.3	14.2	1.2	1.3	1.4	7.1	0.5
Brunei	12.4	12.2	12.0	1.2	1.2	1.2	-1.5	0.4
China	35.0	41.0	50.0	3.4	4.0	4.8	22.2	1.8
India	29.9	30.1	30.4	2.9	2.9	2.9	1.3	1.1
Indonesia	72.8	75.4	76.0	7.0	7.3	7.4	1.1	2.8
Malaysia	51.8	53.9	59.9	5.0	5.2	5.8	11.6	2.2
Myanmar	9.6	10.2	13.0	0.9	1.0	1.3	27.8	0.5
New Zealand	4.3	3.8	3.7	0.4	0.4	0.4	-3.3	0.1
Pakistan	23.2	26.9	29.9	2.2	2.6	2.9	11.5	1.1
Thailand	19.6	20.3	21.4	1.9	2.0	2.1	6.0	0.8
Vietnam	2.4	4.2	5.2	0.2	0.4	0.5	24.1	0.2
Other Asia-Pacific	6.7	6.5	7.3	0.7	0.6	0.7	12.2	0.3
Total	313.1	333.0	360.1	30.3	32.1	34.8	8.4	13.0
Total World	**2,623.3**	**2,703.8**	**2,763.0**	**252.8**	**260.1**	**266.4**	**2.5**	**100.0**

Source: Adapted from *BP Statistical Review of World Energy 2006.*

NOTE: Some totals might not compute exactly due to rounding and other factors.

Table 15a: World Natural Gas Trade Movements: Pipelines

2005 Pipeline Exports To	2005 Pipeline Exports From									
	North America		Latin America			Europe/Former Soviet Union				
	US	Canada	Mexico	Argentina	Bolivia	Belgium	Denmark	Germany	Netherlands	Norway
	Billion cubic meters (bcm)									
North America										
US	-	104.18	0.03	-	-	-	-	-	-	-
Canada	10.14	-	-	-	-	-	-	-	-	-
Mexico	10.14	-	-	-	-	-	-	-	-	-
South & Central America										
Argentina	-	-	-	-	1.74	-	-	-	-	-
Brazil	-	-	-	0.20	8.63	-	-	-	-	-
Chile	-	-	-	6.50	-	-	-	-	-	-
Other S/C America	-	-	-	0.12	-	-	-	-	-	-
Europe and Former Soviet Union										
Austria	-	-	-	-	-	-	-	1.10	-	0.78
Belgium	-	-	-	-	-	-	-	1.20	7.95	8.50
Bulgaria	-	-	-	-	-	-	-	-	-	-
Croatia	-	-	-	-	-	-	-	-	-	-
Czech Republic	-	-	-	-	-	-	-	-	-	2.35
Finland	-	-	-	-	-	-	-	-	-	-
France	-	-	-	-	-	1.90	-	0.10	8.30	14.20
Germany	-	-	-	-	-	-	2.28	-	21.30	26.30
Greece	-	-	-	-	-	-	-	-	-	-
Hungary	-	-	-	-	-	-	-	0.83	-	-
Ireland	-	-	-	-	-	-	-	-	-	-
Italy	-	-	-	-	-	-	-	2.50	8.00	6.90
Latvia	-	-	-	-	-	-	-	-	-	-
Lithuania	-	-	-	-	-	-	-	-	-	-
Luxembourg	-	-	-	-	-	0.70	-	0.70	-	-
Netherlands	-	-	-	-	-	-	2.13	4.50	-	6.16
Poland	-	-	-	-	-	-	-	0.39	-	0.54
Portugal	-	-	-	-	-	-	-	-	-	-
Romania	-	-	-	-	-	-	-	1.30	-	-
Serbia	-	-	-	-	-	-	-	-	-	-
Slovakia	-	-	-	-	-	-	-	-	-	-
Slovenia	-	-	-	-	-	-	-	-	-	-
Spain	-	-	-	-	-	-	-	-	-	2.10
Sweden	-	-	-	-	-	-	0.88	0.15	-	-
Switzerland	-	-	-	-	-	-	-	1.20	0.90	-
Turkey	-	-	-	-	-	-	-	-	-	-
UK	-	-	-	-	-	1.80	-	1.00	0.30	11.55
Other Europe/FSU	-	-	-	-	-	-	-	0.20	-	0.08
Middle East										
Iran	-	-	-	-	-	-	-	-	-	-
Jordan	-	-	-	-	-	-	-	-	-	-
UAE	-	-	-	-	-	-	-	-	-	-
Africa										
Tunisia	-	-	-	-	-	-	-	-	-	-
Asia-Pacific										
Singapore	-	-	-	-	-	-	-	-	-	-
Thailand	-	-	-	-	-	-	-	-	-	-
Total Exports	20.28	104.18	0.03	6.82	10.37	4.40	5.29	15.17	46.75	79.46

Table 15a: World Natural Gas Trade Movements: Pipelines, cont'd

2005 Pipeline Exports To	2005 Pipeline Exports From						
	Europe/Former Soviet Union					Middle East	
	UK	Russia	Turkmenistan	Ukraine	Other Europe & Eurasia	Iran	Oman
	Billion cubic meters (bcm)						
North America							
US	-	-	-	-	-	-	-
Canada	-	-	-	-	-	-	-
Mexico	-	-	-	-	-	-	-
South & Central America							
Argentina	-	-	-	-	-	-	-
Brazil	-	-	-	-	-	-	-
Chile	-	-	-	-	-	-	-
Other S/C America	-	-	-	-	-	-	-
Europe and Former Soviet Union							
Austria	-	6.80	-	-	-	-	-
Belgium	0.97	0.30	-	-	-	-	-
Bulgaria	-	2.85	-	-	-	-	-
Croatia	-	1.13	-	-	0.04	-	-
Czech Republic	-	7.13	-	-	-	-	-
Finland	-	4.20	-	-	-	-	-
France	0.20	11.50	-	-	-	-	-
Germany	3.08	36.54	-	1.20	-	-	-
Greece	-	2.40	-	-	-	-	-
Hungary	-	8.32	-	0.30	1.37	-	-
Ireland	3.05	-	-	-	-	-	-
Italy	0.54	23.33	-	-	-	-	-
Latvia	-	1.75	-	-	-	-	-
Lithuania	-	2.93	-	-	-	-	-
Luxembourg	-	-	-	-	-	-	-
Netherlands	1.82	2.97	-	-	-	-	-
Poland	-	6.40	1.10	-	1.78	-	-
Portugal	-	-	-	-	-	-	-
Romania	-	3.95	-	1.00	-	-	-
Serbia	-	2.15	-	-	-	-	-
Slovakia	-	6.40	-	-	-	-	-
Slovenia	-	0.56	-	-	0.10	-	-
Spain	-	-	-	-	-	-	-
Sweden	-	-	-	-	-	-	-
Switzerland	-	0.37	-	-	0.37	-	-
Turkey	-	17.83	-	-	-	4.32	-
UK	-	-	-	-	-	-	-
Other Europe/FSU	-	1.47	-	-	-	-	-
Middle East							
Iran	-	-	5.80	-	-	-	-
Jordan	-	-	-	-	-	-	-
UAE	-	-	-	-	-	-	1.40
Africa							
Tunisia	-	-	-	-	-	-	-
Asia-Pacific							
Singapore	-	-	-	-	-	-	-
Thailand	-	-	-	-	-	-	-
Total Exports	9.66	151.28	6.90	2.50	3.66	4.32	1.40

Table 15a: World Natural Gas Trade Movements: Pipelines, cont'd

2005 Pipeline Exports To	2005 Pipeline Exports From						
	Africa			Asia-Pacific			
	Algeria	Egypt	Libya	Indonesia	Malaysia	Myanmar	Total
North America							
US	-	-	-	-	-	-	104.21
Canada	-	-	-	-	-	-	10.14
Mexico	-	-	-	-	-	-	10.14
South & Central America							
Argentina	-	-	-	-	-	-	1.74
Brazil	-	-	-	-	-	-	8.83
Chile	-	-	-	-	-	-	6.50
Other S/C America	-	-	-	-	-	-	0.12
Europe and Former Soviet Union							
Austria	-	-	-	-	-	-	8.68
Belgium	-	-	-	-	-	-	18.92
Bulgaria	-	-	-	-	-	-	2.85
Croatia	-	-	-	-	-	-	1.17
Czech Republic	-	-	-	-	-	-	9.48
Finland	-	-	-	-	-	-	4.20
France	-	-	-	-	-	-	36.20
Germany	-	-	-	-	-	-	90.70
Greece	-	-	-	-	-	-	2.40
Hungary	-	-	-	-	-	-	10.82
Ireland	-	-	-	-	-	-	3.05
Italy	25.23	-	4.49	-	-	-	70.99
Latvia	-	-	-	-	-	-	1.75
Lithuania	-	-	-	-	-	-	2.93
Luxembourg	-	-	-	-	-	-	1.40
Netherlands	-	-	-	-	-	-	17.58
Poland	-	-	-	-	-	-	10.21
Portugal	2.62	-	-	-	-	-	2.62
Romania	-	-	-	-	-	-	6.25
Serbia	-	-	-	-	-	-	2.15
Slovakia	-	-	-	-	-	-	6.40
Slovenia	0.44	-	-	-	-	-	1.10
Spain	9.49	-	-	-	-	-	11.59
Sweden	-	-	-	-	-	-	1.03
Switzerland	-	-	-	-	-	-	2.84
Turkey	-	-	-	-	-	-	22.15
UK	-	-	-	-	-	-	14.65
Other Europe/ FSU	-	-	-	-	-	-	1.75
Middle East							
Iran	-	-	-	-	-	-	5.80
Jordan	-	1.10	-	-	-	-	1.10
UAE	-	-	-	-	-	-	1.40
Africa							
Tunisia	1.30	-	-	-	-	-	1.30
Asia-Pacific							
Singapore	-	-	-	4.83	1.78	-	6.61
Thailand	-	-	-	-	-	8.90	8.90
Total Exports	39.08	1.10	4.49	4.83	1.78	8.90	532.65

Source: Adapted from *BP Statistical Review of World Energy 2006.*

NOTES: Flows are on a contractual basis and may not correspond to physical gas flows in all cases.
Data excludes trade within the Former Soviet Union.

Table 15b: World Natural Gas Trade Movements: LNG

2005 LNG Exports To	2005 LNG Exports From													
	Americas		Middle East			Africa				Asia-Pacific				Total
	US	Trinidad	Oman	Qatar	UAE	Algeria	Egypt	Libya	Nigeria	Australia	Brunei	Indonesia	Malaysia	
	Billion cubic meters (bcm)													
Americas														
US	-	12.44	0.07	0.08	-	2.75	2.05	-	0.23	-	-	-	0.25	17.87
Dominican Republic	-	0.25	-	-	-	-	-	-	-	-	-	-	-	0.25
Puerto Rico	-	0.67	-	-	-	-	-	-	-	-	-	-	-	0.67
Europe														
Belgium	-	0.08	-	-	-	2.90	-	-	-	-	-	-	-	2.98
France	-	-	0.08	-	-	7.50	1.05	-	4.20	-	-	-	-	12.83
Greece	-	-	-	-	-	0.46	-	-	-	-	-	-	-	0.46
Italy	-	-	-	-	-	2.50	-	-	-	-	-	-	-	2.50
Portugal	-	-	-	-	-	-	-	-	1.58	-	-	-	-	1.58
Spain	-	0.50	1.65	4.56	0.31	5.19	3.53	0.87	5.00	0.08	-	-	0.16	21.85
Turkey	-	-	-	-	-	3.85	-	-	1.03	-	-	-	-	4.88
UK	-	0.07	-	-	-	0.45	-	-	-	-	-	-	-	0.52
Asia-Pacific														
India	-	-	0.08	5.80	-	-	-	-	-	0.16	-	-	-	6.04
Japan	1.84	-	1.25	8.35	6.75	0.08	-	-	-	13.05	8.35	19.00	17.65	76.32
South Korea	-	-	5.93	8.31	0.08	-	0.30	-	-	1.16	0.80	7.51	6.36	30.45
Taiwan	-	-	0.16	-	-	-	-	-	-	0.40	-	4.95	4.10	9.61
Total	**1.84**	**14.01**	**9.22**	**27.10**	**7.14**	**25.68**	**6.93**	**0.87**	**12.04**	**14.85**	**9.15**	**31.46**	**28.52**	**188.81**

Source: Adapted from *BP Statistical Review of World Energy 2006*.

Table 16: Crude Oil Prices

YEAR	US refiner acquisition cost			US landed cost of imports							US f.o.b. cost	
	Domestic	Imported	Composite	Total	Canada	Mexico	UK	Nigeria	Saudi Arabia	Venezuela	OPEC	Non-OPEC
	$/bbl											
1985	26.66	26.99	26.75	26.67	25.71	25.63	28.36	28.96	24.72	24.43	25.67	25.96
1986	14.82	14.00	14.55	13.49	13.43	12.17	14.63	15.29	12.84	11.52	12.21	12.87
1987	17.76	18.13	17.90	17.65	17.04	16.69	18.78	19.32	16.81	15.76	16.43	16.99
1988	14.74	14.56	14.67	14.08	13.50	12.58	15.82	15.88	13.37	13.66	13.43	13.05
1989	17.87	18.08	17.97	17.68	16.81	16.35	18.74	19.19	17.34	16.78	17.06	16.72
1990	22.59	21.76	22.22	21.13	20.48	19.64	22.65	23.33	21.82	20.31	20.40	20.32
1991	19.33	18.70	19.06	18.02	17.16	15.89	21.37	21.39	17.22	15.92	16.99	16.77
1992	18.63	18.20	18.43	17.75	17.04	15.60	20.63	20.78	17.48	15.13	16.87	16.66
1993	16.67	16.14	16.41	15.72	15.27	14.11	17.92	18.73	15.40	13.39	14.78	14.65
1994	15.67	15.51	15.59	15.18	14.83	14.09	16.64	17.21	15.11	13.12	14.00	14.34
1995	17.33	17.14	17.23	16.78	16.65	16.19	17.91	18.25	16.84	14.81	15.36	16.02
1996	20.77	20.64	20.71	20.31	19.94	19.64	20.88	21.95	20.49	18.59	18.94	19.65
1997	19.61	18.53	19.04	18.11	17.63	17.30	20.64	20.64	17.52	16.35	16.26	17.51
1998	13.18	12.04	12.52	11.84	11.62	11.04	13.55	14.14	11.16	10.16	10.20	11.21
1999	17.90	17.26	17.51	17.23	17.54	16.12	18.26	17.63	17.48	15.58	15.90	16.84
2000	29.11	27.70	28.26	27.53	26.69	26.03	29.26	30.04	26.58	26.05	25.56	26.77
2001	24.33	22.00	22.95	21.82	20.72	19.37	25.32	26.55	20.98	19.81	19.73	21.04
2002	24.65	23.71	24.10	23.91	22.98	22.09	26.35	26.45	24.77	21.93	22.18	22.93
2003	29.82	27.71	28.53	27.69	26.76	25.48	30.62	31.07	27.50	25.70	25.36	26.21
2004	38.97	35.90	36.98	36.07	34.51	32.25	39.28	40.95	37.11	33.79	33.95	33.58
2005	52.93	48.85	50.23	49.15	44.98	43.46	55.32	57.24	50.02	47.84	49.60	45.79

Source: U.S. Energy Information Administration: *Annual Energy Review*, August 2005 (www.eia.doe.gov/emeu/aer/petro.html); and *Monthly Energy Review*, Dec. 2006 (www.eia.doe.gov/emeu/mer/contents.html)

NOTE: All prices are US annual average (nominal dollars).

Table 17: Petroleum Product Prices

YEAR	Retail unleaded motor gasoline $/gallon (nominal)			Refiner resale (wholesale) Cents/gallon								
	Regular	Premium	All	Home Heating Oil	Motor Gasoline	Aviation Gasoline	Kerosene Jet Fuel	Kerosene	No. 2 Fuel Oil	No. 2 Diesel Fuel	Propane	Residual Fuel Oil
1985	1.20	1.34	1.20	77.4	83.5	113.0	79.4	87.4	77.6	77.2	39.8	57.7
1986	0.93	1.09	0.93	47.0	53.1	91.2	49.5	60.6	48.6	45.2	29.0	30.5
1987	0.95	1.09	0.96	53.1	58.9	85.9	53.8	59.2	52.7	53.4	25.2	38.5
1988	0.95	1.11	0.96	47.3	57.7	85.0	49.5	54.9	47.3	47.3	24.0	30.0
1989	1.02	1.20	1.06	56.6	65.4	95.0	58.3	66.9	56.5	56.7	24.7	36.0
1990	1.16	1.35	1.22	69.5	78.6	106.3	77.3	83.9	69.7	69.4	38.6	41.3
1991	1.14	1.32	1.20	61.8	69.9	100.1	65.0	72.2	62.2	61.5	34.9	31.4
1992	1.13	1.32	1.19	58.5	67.7	99.1	60.5	63.2	57.9	59.1	32.8	30.8
1993	1.11	1.30	1.17	55.9	62.6	96.5	57.7	60.4	54.4	57.0	35.1	29.3
1994	1.11	1.31	1.17	52.2	59.9	93.3	53.4	61.8	50.6	52.9	32.4	31.7
1995	1.15	1.34	1.21	53.0	62.6	97.5	53.9	58.0	51.1	53.8	34.4	36.3
1996	1.23	1.41	1.29	65.3	71.3	105.5	64.6	71.4	63.9	65.9	46.1	42.0
1997	1.23	1.42	1.29	60.2	70.0	106.5	61.3	65.3	59.0	60.6	41.6	38.7
1998	1.06	1.25	1.12	43.9	52.6	91.2	45.0	46.5	42.2	44.4	28.8	28.0
1999	1.17	1.36	1.22	53.6	64.5	100.7	53.3	55.0	49.3	54.6	34.2	35.4
2000	1.51	1.69	1.56	89.6	96.3	133.0	88.0	96.9	88.6	89.8	59.5	56.6
2001	1.46	1.66	1.53	77.9	88.6	125.6	76.3	82.1	75.6	78.4	54.0	47.6
2002	1.36	1.56	1.44	71.8	82.8	114.6	71.6	75.2	69.4	72.4	43.1	53.0
2003	1.59	1.78	1.64	88.2	100.2	128.8	87.1	95.5	88.1	88.3	60.7	66.1
2004	1.88	2.07	1.92	117.8	128.8	162.7	120.8	127.1	112.5	118.7	75.1	68.1
2005	2.30	2.49	2.34	172.8	167.2	208.4	172.8	179.0	162.8	174.5	91.8	95.0

Source: U.S. Energy Information Administration, *Annual Energy Review*, August 2005 (www.eia.doe.gov/emeu/aer/petro.html).

NOTE: All prices are U.S. annual average (nominal dollars).

Table 18: Comparative Energy Prices

YEAR	Natural gas delivered to consumers, $/1000 cu ft				Retail electricity, cents/kWh			
	Residential	Commercial	Industrial	Electric Power	Residential	Commercial	Industrial	All
1985	6.12	5.50	3.95	3.55	7.39	7.27	4.97	6.44
1986	5.83	5.08	3.23	2.43	7.42	7.20	4.93	6.44
1987	5.54	4.77	2.94	2.32	7.45	7.08	4.77	6.37
1988	5.47	4.63	2.95	2.33	7.48	7.04	4.70	6.35
1989	5.64	4.74	2.96	2.43	7.65	7.20	4.72	6.45
1990	5.80	4.83	2.93	2.39	7.83	7.34	4.74	6.57
1991	5.82	4.81	2.69	2.18	8.04	7.53	4.83	6.75
1992	5.89	4.88	2.84	2.36	8.21	7.66	4.83	6.82
1993	6.16	5.22	3.07	2.61	8.32	7.74	4.85	6.93
1994	6.41	5.43	3.05	2.28	8.38	7.73	4.77	6.91
1995	6.06	5.05	2.71	2.02	8.40	7.69	4.66	6.89
1996	6.34	5.40	3.42	2.69	8.36	7.64	4.60	6.86
1997	6.94	5.80	3.59	2.78	8.43	7.59	4.53	6.85
1998	6.82	5.48	3.14	2.40	8.26	7.41	4.48	6.74
1999	6.69	5.33	3.10	2.62	8.16	7.26	4.43	6.64
2000	7.76	6.59	4.45	4.38	8.24	7.43	4.64	6.81
2001	9.63	8.43	5.24	4.61	8.58	7.92	5.05	7.29
2002	7.89	6.63	4.02	3.68	8.44	7.89	4.88	7.20
2003	9.63	8.40	5.89	5.57	8.72	8.03	5.11	7.44
2004	10.75	9.43	6.53	6.11	8.95	8.17	5.25	7.61
2005	12.84	11.59	8.56	8.48	9.45	8.67	5.73	8.14

Source: U.S. Energy Information Administration: http://tonto.eia.doe.gov/dnav/ng/ng_pri_sum_dcu_nus_a.htm; *Electric Power Monthly*: http://www.eia.doe.gov/fuelelectric.html

NOTE: All prices are U.S. annual average.

Table 19: Price History of Oil, Gas, and Gasoline

YEAR	Actual prices: Petroleum: Wellhead, $/bbl	Actual prices: Petroleum: Retail gasoline, $/gal	Actual prices: Natural gas: Wellhead, $/1000 cu ft	Actual prices: Natural gas: Residential, $/1000 cu ft	Producer price index	Inflation-adjusted prices (1982=100): Petroleum: Wellhead, $/bbl	Inflation-adjusted prices (1982=100): Petroleum: Retail gasoline, $/gal	Inflation-adjusted prices (1982=100): Natural gas: Wellhead, $/1000 cu ft	Inflation-adjusted prices (1982=100): Natural gas: Residential, $/1000 cu ft
1935	0.97	0.188	0.058	N/A	13.8	7.029	1.365	0.420	NA
1940	1.02	0.184	0.045	N/A	13.5	7.556	1.364	0.333	NA
1945	1.22	0.205	0.049	N/A	18.2	6.703	1.126	0.269	NA
1950	2.51	0.268	0.065	N/A	27.3	9.194	0.980	0.238	NA
1955	2.77	0.291	0.104	N/A	29.3	9.454	0.992	0.355	NA
1960	2.88	0.311	0.140	N/A	31.7	9.085	0.982	0.442	NA
1965	2.86	0.312	0.156	N/A	32.3	8.854	0.966	0.483	NA
1970	3.18	0.357	0.171	1.09	36.9	8.618	0.967	0.463	2.954
1975	7.67	0.567	0.445	1.71	58.4	12.945	0.971	0.762	2.928
1980	21.59	1.191	1.59	3.68	89.7	24.069	1.328	1.773	4.103
1981	31.77	1.311	1.98	4.29	98.0	32.418	1.338	2.020	4.378
1982	28.52	1.296	2.46	5.17	100.0	28.520	1.296	2.460	5.170
1983	26.19	1.241	2.59	6.06	101.3	25.879	1.226	2.559	5.982
1984	25.88	1.212	2.66	6.12	103.7	24.981	1.170	2.568	5.902
1985	24.09	1.202	2.51	6.12	103.2	23.366	1.166	2.435	5.930
1986	12.51	0.927	1.94	5.83	100.2	12.498	0.926	1.938	5.818
1987	15.40	0.948	1.67	5.54	102.8	14.981	0.922	1.625	5.389
1988	12.58	0.946	1.69	5.47	106.9	11.768	0.885	1.581	5.117
1989	15.86	1.021	1.69	5.64	112.2	14.135	0.910	1.506	5.027
1990	20.03	1.164	1.71	5.80	116.3	17.223	1.001	1.470	4.987
1991	16.54	1.140	1.64	5.82	116.5	14.197	0.979	1.408	4.996
1992	15.99	1.127	1.74	5.89	117.2	13.643	0.962	1.485	5.026
1993	14.25	1.108	2.04	6.16	118.9	11.985	0.932	1.716	5.181
1994	13.19	1.112	1.85	6.41	120.5	10.955	0.924	1.537	5.320
1995	14.62	1.147	1.55	6.06	124.8	11.724	0.920	1.243	4.856
1996	18.46	1.231	2.17	6.34	127.7	14.456	0.964	1.699	4.965
1997	17.23	1.234	2.32	6.94	127.6	13.503	0.967	1.818	5.439
1998	10.87	1.059	1.96	6.82	124.4	8.738	0.851	1.559	5.482
1999	15.56	1.165	2.19	6.69	125.5	12.398	0.928	1.657	5.331
2000	26.72	1.510	3.68	7.76	132.7	20.143	1.138	2.773	5.848
2001	21.84	1.461	4.00	9.63	134.2	16.274	1.089	3.070	7.183
2002	22.51	1.441	2.95	7.89	131.1	17.170	1.099	2.250	6.018
2003	27.56	1.638	4.88	9.63	138.1	19.957	1.186	3.533	6.973
2004	36.77	1.923	5.46	10.75	146.7	25.065	1.311	3.742	7.321
2005	50.28	2.338	7.51	12.82	157.4	31.944	1.485	4.771	8.145

Sources: U.S. Energy Information Administration: *Annual Energy Review (2005)*, www.eia.doe.gov/emeu/aer/petro.html;
Monthly Energy Report (Dec. 2006), http://www.eia.doe.gov/emeu/mer/contents.html;
Petroleum Navigator, http:// tonto.eia.doe.gov/dnav/pet/pet_pri_top.asp;
Capital District Regional Planning Commission, www.cdrpc.org/CPI_PPI.html.

NOTE: All prices are US annual average.

Table 20: International Rig Count

REGION	2000	2001	2002	2003	Dec. 2004			Dec. 2005			Dec. 2006		
					Land	Off.	Total	Land	Off.	Total	Land	Off.	Total
North America													
Canada	344	266	277	332	436	4	440	571	4	575	452	4	456
US	918	1,000	862	1,126	1,140	106	1,246	1,393	77	1,470	1,634	84	1,718
Subtotal	1,262	1,266	1,139	1,458	1,576	110	1,686	1,964	81	2,045	2,086	88	2,17
Latin America													
Argentina	57	62	56	60	73	1	74	80	1	81	81	0	81
Bolivia	11	6	5	5	4	0	4	5	0	5	3	0	3
Brazil	23	27	26	24	9	17	26	11	17	28	14	19	33
Chile	1	1	1	0	0	0	0	0	0	0	1	0	1
Colombia	14	16	10	11	13	0	13	21	0	21	24	0	24
Ecuador	7	12	8	7	12	0	12	14	0	14	12	0	12
Mexico	44	63	80	107	75	28	103	67	27	94	57	27	84
Peru	4	3	1	2	2	0	2	3	0	3	5	1	6
Trinidad	4	5	2	3	0	3	3	1	4	5	3	4	7
Venezuela	63	60	30	45	55	14	69	60	15	75	58	18	76
Other	--	--	--	--	--	--	--	1	--	1	2	0	2
Subtotal	228	255	219	264	243	63	306	263	64	327	260	69	329
Europe													
Croatia	0	0	3	3	3	0	3	2	1	3	1	0	1
Denmark	3	7	4	4	0	3	3	0	2	2	0	0	0
France	1	0	0	1	1	0	1	1	0	1	0	0	0
Germany	3	5	4	2	2	0	2	3	0	3	4	0	4
Hungary	1	1	2	2	2	0	2	2	0	2	3	0	3
Italy	8	7	5	7	4	1	5	2	1	3	4	1	5
Netherlands	4	8	5	5	1	3	4	2	3	5	1	2	3
Norway	22	24	15	17	0	14	14	0	15	15	0	15	15
Poland	15	13	12	11	3	0	3	2	0	2	2	0	2
Romania	0	2	2	2	2	0	2	2	0	2	2	0	2
Turkey	6	6	4	4	4	0	4	4	1	5	4	0	4
UK	17	23	23	21	1	15	16	1	26	27	1	21	22
Other	3	3	4	4	4	0	4	4	0	4	4	0	4
Subtotal	83	99	83	83	27	36	63	25	49	74	26	39	65
Middle East													
Iran*	27	33	34	39	33	7	40	32	6	38	0	0	0
Jordan	--	--	--	--	--	--	--	1	0	1	1	0	1
Kuwait	12	6	7	8	12	0	12	13	0	13	16	0	16
Oman	24	27	32	37	36	0	36	36	0	36	41	0	41
Qatar	6	14	9	8	1	10	11	1	13	14	2	7	9
Saudi Arabia	25	31	33	31	31	3	34	38	6	44	66	9	75
Syria	14	19	24	24	23	0	23	23	0	23	25	0	25
UAE	13	18	16	16	10	4	14	9	6	15	10	5	15
Yemen	6	7	11	7	12	0	12	14	0	14	15	0	15
Other	1	4	3	1	5	0	5	2	0	2	2	0	2
Subtotal	128	159	169	171	163	24	187	169	31	200	178	21	199

Table 20: International Rig Count, cont'd.

REGION	2000	2001	2002	2003	Dec. 2004 Land	Dec. 2004 Off.	Dec. 2004 Total	Dec. 2005 Land	Dec. 2005 Off.	Dec. 2005 Total	Dec. 2006 Land	Dec. 2006 Off.	Dec. 2006 Total
Africa													
Algeria	15	20	22	20	20	0	20	21	0	21	25	0	25
Angola	6	4	2	7	0	3	3	0	3	3	0	4	4
Congo (Brazzaville)	3	1	1	4	1	1	2	1	0	1	1	0	1
Egypt	18	24	24	27	18	13	31	26	6	32	31	8	39
Gabon	2	1	2	3	1	1	2	2	1	3	2	1	3
Libya	7	4	10	11	9	1	10	9	0	9	12	1	13
Nigeria	8	9	12	9	2	9	11	1	8	9	2	9	11
South Africa	1	1	0	1	0	0	0	0	0	0	0	0	0
Sudan*	--	--	5	7	17	0	17	19	0	19	0	0	0
Tunisia	1	2	3	2	2	0	2	0	0	0	3	2	5
Other	3	3	4	5	2	1	3	2	1	3	2	3	5
Subtotal	64	69	85	96	72	29	101	81	19	100	78	28	106
Asia-Pacific													
Australia	10	10	10	13	7	7	14	10	4	14	11	11	22
Brunei	2	2	3	2	0	2	2	0	2	2	1	3	4
China	9	8	9	10	0	12	12	0	17	17	0	17	17
India	49	50	58	62	49	27	76	60	23	83	54	31	85
Indonesia	32	49	53	42	27	24	51	38	22	60	35	18	53
Japan	6	10	9	4	2	0	2	3	0	3	1	0	1
Malaysia	7	11	15	14	0	11	11	0	12	12	0	14	14
Myanmar	9	6	7	10	8	1	9	7	2	9	8	2	10
New Zealand	1	2	1	2	3	0	3	6	0	6	4	2	6
Pakistan	10	5	13	12	12	0	12	11	0	11	17	0	17
Papua New Guinea	0	1	0	1	1	0	1	3	0	3	1	0	1
Philippines	0	1	2	2	3	0	3	2	0	2	1	0	1
Thailand	6	8	7	5	1	5	6	1	8	9	3	8	11
Vietnam	8	7	9	9	1	5	6	0	10	10	0	6	6
Other	1	0	1	1	2	2	4	5	1	6	2	2	4
Subtotal	150	170	197	189	116	96	212	146	101	247	138	114	252
Total World	1,913	2,018	1,892	2,261	2,197	358	2,555	2,648	345	2,993	2,766	359	3,125

Source: Baker Hughes Inc. in *Oil & Gas Journal*.

NOTES: US land count includes inland water rigs.
China count includes offshore rigs only.
No Iraq rig report received since the Iraq Invasion of Kuwait.
* Iran and Sudan rig counts discontinued in Jan. 2006.

Information Sources

National Oil Companies and Energy Ministries

A

Albania
Albpetrol ShA
c/o Geological Institute for Oil & Gas
Fieri, ALBANIA
Telephone: 355 423-4542
Fax: 355 423-4204

National Petroleum Agency
Agjensia Kombetare e Hidrokarbureve
Rruga "Duressit," No. 83
Tirana, ALBANIA
Telephone/Fax: 355 423-1034

Algeria
Ministry of Energy & Mining
B.P 677 AlgerGare
80 Av. Ghermoul
Algiers, ALGERIA
Telephone: 213 21 65-22-22
Fax: 213 21 65-19-04
E-mail: info@mem-algeria.org
Web site: www.mem-algeria.org

Sonatrach
80 Av. Ghermoul
Algiers, ALGERIA
Telephone: 213 21 54-80-11; -70-00
Fax: 213 21 54-77-00
E-mail: sonatrach@sonatrach.dz
Web site: www.sonatrach-dz.com

Angola
Sonangol (Sociedad Nacional de Combustiveis de Angola)
PO Box 1316
Luanda, ANGOLA
Telephone: 244-2-33-4448
Fax: 244-2-39-1782
Web site: www.sonangol.co.ao

Argentina
Secretary of Energy
Av. Paseo Colón 171
Capital Federal – CP (C1063ACB)
ARGENTINA
Telephone: 54-11-4349-5000
E-mail: energia@minplan.gov.ar
Web site: http://energia3.mecon.gov.ar/

Repsol YPF SA
Av. Roque Sáenz Peña 777
CP 1364 Buenos Aires, ARGENTINA
Telephone: 54 11 43 29 20 00
Fax: 54 11 43 29 20 01
Web site: www.repsolypf.com

Armenia
Republic of Armenia
Ministry of Energy
Government Building 2
Republic Square
Yerevan 375010, ARMENIA
Telephone: 3742 521964
Fax: 3742 151687
E-mail: minenegy@minenergy.am
Web site: www.minenegy.am

Australia
Commonwealth of Australia
Department of Industry,
Tourism and Resources
Resources and Energy
GPO Box 9839
Canberra ACT 2601, AUSTRALIA
Telephone: 61 2 6213 6000
Fax: 61 2 6213 7000
E-mail: inquiries@industry.gov.au
Websites: www.australia.gov.au
www.industry.gov.au

Australia, New South Wales
Department of Energy, Utilities and Sustainability
Level 17, 227 Elizabeth St.
GPO Box 3889
Sydney, New South Wales 2001, AUSTRALIA
Telephone: 02 8281 7777
Facsimile: 02 8281 7799
E-mail: information@deus.nsw.gov.au
Web site: www.deus.nsw.gov.au

Australia, Northern Territory
Department of Minerals and Energy
Telephone: 618 8999 5511
Websites: www.nt.gov.au
www.minerals.nt.gov.au/

Australia, Queensland
Department of Natural Resources and Mines
Mineral House
41 George St.
Brisbane, Queensland 4000, AUSTRALIA
GPO Box 2454
Brisbane Q 4001, AUSTRALIA
Telephone: 07 3896 3111
Web site: www.nrm.qld.gov.au

Australia, South Australia
Department of Primary Industries and Resources
Minerals, Petroleum and Energy
GPO Box 1671
Adelaide, South Australia5001, AUSTRALIA
Telephone: 61 8 8463 4154
Fax: 61 8 8463 4155
E-mail: pirsa.minerals@saugov.sa.gov.au
Web site: www.pir.sa.gov.au

Australia, Tasmania
Department of Infrastructure, Energy and Resources
GPO Box 936
Hobart, Tasmania 7001, AUSTRALIA
Telephone: 1300 135 513
E-mail: info@dier.tas.gov.au
Web site: www.dier.tas.gov.au

Australia, Victoria
Department of Infrastructure
GPO Box 2797
Melbourne, Victoria 3001, AUSTRALIA
Telephone: 61 03 9655 6666
Fax: 61 03 9655 6752
Web site: www.doi.vic.gov.au

Australia, Western Australia
Office of Energy
Level 9
Governor Stirling Tower
197 St Georges Terrace
Perth, Western Australia 6000, AUSTRALIA
Telephone: 61 8 9420 5600
Fax: 61 8 9420 5700
E-mail: enquiries@energy.wa.gov.au
Web site: www.energy.wa.gov.au

Austria
Federal Ministry of Economics and Labour
A-1010 Wien, Stubenring 1, AUSTRIA
Telephone: 43/1/711 00-0
E-mail: service@bmwa.gv.at
Web site: www.bmwa.gv.at

OMV Aktiengesellschaft
Otto-Wagner-Platz 5
A-1090 Wien, AUSTRIA
Ph.: 43 (1) 40 440-0
Fax: 43 (1) 40 440-20091
E-mail: info.austria@omv.com
Web site: www.omv.com

Azerbaijan
AIOC (Azerbaijan International Operating Co.)
Villa Petrolea, 2 Neftchilar Prospect (Bayil)
Baku, AZERBAIJAN
Telephone: (994-12) 97-90-00; 97-97-54
Fax: 97-96-02, 97-97-37
E-mail: bayatltf@bp.com
Web site: www.azer.com

Socar (State Oil Co. of Azerbaijan)
73 Neftchilar Prospect
Baku, AZERBAIJAN
Telephone: (994-12) 493-2312, 492-0745
Fax: (994-12) 493-6492
Web site: www.azer.com

B

Bahrain

Bahrain National Gas Co. (Banagas)
P.O. Box 29099, BAHRAIN
Telephone: 973 17 756222
Fax: 973 17 756991
Web site: www.banagas.com.bh

Bahrain National Oil Co.
P.O. Box 25504
Awali, BAHRAIN
Telephone: 973 75-46-66
Fax: 973 75-32-03

Bangladesh

Petrobangla (Bangladesh Oil, Gas & Mineral Corp.)
Petrocentre
3 Kawranbazar
Dhaka – 1215, BANGLADESH
Fax: 880 2 9120224
Web site: www.petrobangla.org

Belize

Ministries of Natural Resrouces and Environment
Power Lane, Belmopan, Belize
Telephone: 08-22817
Fax: 08-23317
Web site: www.belize.gov.bz/legislative_houseofrep.html

Bolivia

Vice-Ministry of Energy and Hydrocarbons
Av. Mariscal Santa Cruz eaq. Oruro
Edificio Palacio de Comunicaciones, Piso 12
La Paz, BOLIVIA
Telephone: 591 2 374-050
Fax: 591 2 392-758
E-mail: enerhid@ceibo.entelnet.bo

Yacimientos Petroliferos Fiscales Bolivianos (YPFB)
Box 401
La Paz, BOLIVIA
Telephone: 591 2 374-468
Fax: 591 2 374-469

Brazil

ANP (Agência Nacional do Petróleo)
Web site: www.anp.gov.br

Ministry of Mines and Energy
Web site: www.mme.gov.br

Petrobras (Petroleo Brasilerio SA)
Avenida República do Chile 65, sala 401 E
20031-912 Rio de Janeiro, BRAZIL
Telephone: (+55 21) 3224-1510
Fax: (+55 21) 3224-6055
Web site: www.petrobras.com.br

Brunei

Office of the Prime Minister
Petroleum Unit
Bahirah Building, Jalan Menteri Besar
Darussalam BB3910, BRUNEI
Telephone: 673-2-387 102
Fax: 673-2-383 004
E-mail: brupet@brunet.bn
Web site: www.petroleum-unit.gov.bn

Bulgaria

Committee of Energy
8 Triaditza Str.
Sofia 1040, BULGARIA
Telephone: +359 2 92 63 636
Fax: +359 2 980 76 30; 988 14 43
E-mail: pressall@doe.bg
pressdir@doe.bg
Web site: www.doe.bg

C

Cameroon

Societe Nationale des Hydrocarbures (SNH)
Box 955
Yaounde, CAMEROON
Telephone: 237 20-19-10; -98-64
Fax: 237 20-46-51

Canada

National Energy Board
444 Seventh Avenue SW
Calgary, Alberta
T2P 0X8 CANADA
Telephone: 403-292-4800
1-800-899-1265
Fax: 403-292-5503
1-877-288-8803
E-mail: info@neb-one.gc.ca
Web site: www.neb-one.gc.ca

Natural Resources Canada
Energy Sector
Ottawa, CANADA
Telephone: 613-995-0947
Fax: 613-992-8738
Web site: www.nrcan-rncan.gc.ca

Canada, Alberta

Alberta Department of Energy
North Petroleum Plaza
7th floor, 9945-108 St
Edmonton, Alberta
T5K 2G6 CANADA
Telephone: 780-427-7425
Fax: 780-422-8731
Web site: www.energy.gov.ab.ca

Alberta Department of Energy
300, 801-6 Avenue SW
Calgary, Alberta
T2P 3W2 CANADA
Telephone: 403-297-8955
Web site: www.energy.gov.ab.ca

Alberta Energy and Utilities Board
Calgary Office
640 - 5 Avenue SW
Calgary, Alberta
T2P 3G4 CANADA
Telephone: 403-297-8311
Web site: www.eub.gov.ab.ca

Canada, British Columbia

Ministry of Energy, Mines and Petroleum Resources
Public Affairs Bureau
PO Box 9324, Stn Prov Govt
Victoria, British Columbia
V8W 9N3 CANADA
Phone: 604-660-2421
E-mail: EnquiryBC@gov.bc.ca
Web site: www.gov.bc.ca/bvprd/bc/home.do

Canada, Manitoba

Department of Industry, Economic Development & Mines
Petroleum Branch
The International Business Centre
1000 - 259 Portage Avenue
Winnipeg, Manitoba
R3B 3P4 CANADA
Tel. (204) 945-2475
Fax: (204) 945-3977
E-mail: mgi@gov.mb.ca
Websites: www.gov.mb.ca
www.gov.mb.ca/iedm/petroleum/

Canada, New Brunswick

Carleton Place
Room: 830
P.O. Box 6000
Fredericton, New Brunswick
E3B 5H1 CANADA
Telephone: 506-453-5399
Fax: 506-444-5405
Web site: www.gnb.ca/0085/index-e.asp

Canada, Newfoundland & Labrador

Department of Natural Resources
P.O. Box 8700
St. John's, Newfoundland
A1B 4J6 CANADA
Telephone: 709-729-2920
Fax: 709-729-0059; -2076
Websites: www.gov.nf.ca
www.nr.gov.nl.ca/nr/

Canada, Northwest Territories

Environment and Natural Resources Department
PO Box 1320
Yellowknife, Northwest Territories
X1A 2L9 CANADA
Telephone: 867-669-2355
Fax: 867-873-0596
Web site: www.gov.nt.ca

Canada, Nova Scotia

Department of Energy
Bank of Montreal Building, Suite 400
5151 George St.
P.O. Box 2664
Halifax, Nova Scotia
B3J 3P7 CANADA
Telephone: 902-424-4575
Fax: 902-424-0528
Web site: www.gov.ns.ca/energy

Canada, Nunavut
Department of Sustainable Development
Iqaluit, Nunavut, CANADA
Telephone: 867-975-5925
Fax: 867-975-5980
Web site: www.gov.nu.ca/Nunavut/English/departments/DSD

Canada, Ontario
Ministry of Energy
900 Bay Street, 4th Floor
Hearst Block
Toronto, Ontario
M7A 2E1 CANADA
Telephone: 1-888-668-4636
E-mail: write2us@energy.gov.on.ca
Web site: www.energy.gov.on.ca

Canada, Prince Edward Island
Energy and Minerals Division
Dept. of Environment, Energy, and Forestry
P.O. Box 2000
Charlottetown, Prince Edward Island
CIA 7N8 Canada
Telephone: 902-368-4000
E-mail: island@gov.pe.ca
Web site: www.gov.pe.ca/enveng/eam-info/index.php3

Canada, Quebec
Department of Natural Resources
5700, 4e Avenue Ouest, B302
Charlesbourg, Québec
G1H 6R1 CANADA
Telephone: 418-627-8600
1-866-248-6939
E-mail: hydrocarbures@mrnfp.gouv.qc.ca
Web site: www.mrn.gouv.qc.ca/english/energy

Canada, Saskatchewan
Department of Industry and Resources
2103 - 11th Avenue
Regina, Saskatchewan
S4P 3V7 CANADA
Telephone: 306-787-4765
1-866-727-5427
Fax: 306-787-8447
Web site: www.ir.gov.sk.ca

Canada, Yukon Territory
Energy, Mines and Resources
Box 2703
Whitehorse, Yukon
Y1A 2C6 CANADA
Telephone: 867-667-3130
Fax: 867-667-8601
E-mail: oilandgas@gov.yk.ca
Web site: www.emr.gov.yk.ca

Chile
Enap (Empresa Nacional del Petróleo)
Todos los derechos reservados
Av. Vitacura 2736 piso 10 Las Condes
Santiago, CHILE
Telephone: 56-2 280 3000
Fax: 56 2 280 3199

China
CNOOC Ltd. (China National Offshore Oil Corp. Ltd.)
Jingxin Bldg., Dongsanhuan Bei Rd.
Box 4705
Beijing 100027, PEOPLE'S REPUBLIC OF CHINA
Telephone: 86 10 6466-9001; -3696
Fax: 86 10 6466-2994; -9007

CNPC (China National Petroleum Corp.)
6, Liupukang Jie, Xicheng District
Beijing, 100724, PEOPLE'S REPUBLIC OF CHINA
Telephone: +86-10-6209-4114
Fax: +86-10-6209-4806

Sinopec Corp. (China Petroleum & Chemical Corp.)
No.A6 Hui xin East St.
Chaoyang District
Beijing 100029, CHINA
Web site: http://english.sinopec.com/index.jsp

Colombia
Ecopetrol SA (Empresa Colombiana del Petroles)
Edificio Principal Cr 13 No. 36 – 24
Bogota, COLOMBIA
Telephone: 57 1 234-4444; -4000
Fax: 57 1 234-4099
E-mail: prozo@infantas.ecp.com
Web site: www.ecopetrol.com.co

Ministerio de Minas y Energia de Colombia
Centro Administrative Nacional
Avenida El Dorado
Bogota, COLOMBIA
Telephone: 57 1 44-520; -525
Web site: www.minminas.gov.co

Congo (Brazzaville)

General Direction of Hydrocarbons
Ministry of Mines and Energy - S.P. 2120
Brazzaville, CONGO
Telephone: 242 83-12-81; -58-73; -59-74
Fax: 242 83-62-43

Hydro-Congo
B.P. 2008
Pointe Noire, Brazzaville, CONGO
Telephone: 242 81-35-60; -40-23
Fax: 242 83-12-38

Congo (former Zaire)

Petrozaire
B.P. 7617
Kinshasa/Gombe, CONGO ZAIRE
Telephone: 243 12 20-344; 492

Croatia

INA-Industrija nafte, dd
Avenija Veceslava Holjevca 10, p.p. 555
10002 Zagreb, CROATIA
Telephone: 00385 1 6450-000
E-mail: ina@ina.hr
Web site: www.ina.hr

Cuba

Comercial Cupet SA
km 37.5 Via Blanca
Caribe, Sta. Cruz del Norte
Havana 32900, CUBA
Telephone: 53 7 62-40-11
Fax: 53 7 33-34-60; -80-27; -30-72

Ministerio de la Industria Basica
Empedarado 113
Havana, CUBA
Web site: www.cubagob.cu

Czech Republic

Urad Vlady CR
Nabrezi Edvarda Benese 4
Mala Strana, PSC 118 01
Praha 1,CZECH REPUBLIC
Telephone: 22400 2335; 2309
Fax: 22400 2502
E-mail: sedlackova.silvie@vlada.zc
Web site: www.vlada.cz

D

Denmark

Ministry of the Environment
Frontlinien
Rentemestervej 8
2400 Copenhagen NV DENMARK
Ph.: + 45 70 12 02 11
E-mail: mim@mim.dk
frontlinien@frontlinien.dk
Web site: www.mim.dk

E

Ecuador

Petroecuador
Alpallana E8-86 y
Av. 6 de Diciembre
P.O. Box 17-11-5007
ECUADOR
Telephone: (593-2) 256 3060; 1589; 0525
Fax: (593-2) 250 3571
Web site: www.petroecuador.com.ec

Ministerio de Energía y Minas
Av. Orellana N26-220 y Juan León Mera (esquina) Edificio MOP
PBX (593-2) 2550-018 / 2550-041
Quito, ECUADOR
Web site: www.menergia.gov.ec

Olade (Organizacion LatinoAmericana de Energia)
Dir. Av Mariscal Antonio Jose de Sucre N 58-63 y
Fernandez Salvador Edif. Olade
Sector San Carlos
Quito, ECUADOR
Telephone: (593 2) 2598-122, -280; 2597-995; 2599-489
Fax: (593-2) 2531-691
E-mail: olade@olade.org.ec
Web site: www.olade.org.ec

Egypt

Arab Republic of Egypt
Ministry of Petroleum
1 (A) Ahmed El Zomor str.
8th District Nasr City
Cairo, EGYPT
Telephone: 00 202 670-64-35; -05
Fax: 00 202 670-64-10
Web site: www.petroleum.gov.eg

Egyptian General Petroleum Corp. (EGPC)
Palestine Street part 4, New Maadi
Cairo, EGYPT
Telephone: 706 53 45
Fax: 706 53 59
E-mail: info@egpc.com.eg
Web site: www.egpc.com.eg

Egyptian Natural Gas Holding Co. (Egas)
85C Nasr Road
Nasr City, Cairo, EGYPT
PO Box 8064
Nasr City 11371, EGYPT
Telephone: 40558 45; -46; -47; -48
Fax: 40558 76
E-mail: egas@egas.com.eg
Web site: www.egas.com.eg

Equatorial Guinea
Ministry of Mines, Industry and Energy
David Shaw, Technical Advisor
C/12 de Octubre
Malabo, EQUATORIAL GUINEA
Telephone: +240 09 3567; 27 7502
Fax: +240 09 3353
E-mail: d.shaw@ecqc.com
Web site: www.equatorialoil.com

F

Faroe Islands
Faroese Earth and Energy Directorate (Jardfeingi)
Brekkutún 1
Postmoga 3059
FO-110 Tórshavn, FAROE ISLANDS
Telephone: 35 70 70
Fax: 35 70 01
Email: jardfeingi@jardfeingi.fo
Web site: www.jardfeinfi.fo

Finland
Ministry of Trade and Industry
P.O. Box 32
FI-00023 GOVERNMENT
Aleksanterinkatu 4, FI-00170
Helsinki, FINLAND
Telephone: +358 9 160 01, 578 11
Fax: +358 9 1606 3666, 5786 3666
E-mail: kirjaamo@ktm.fi
Web site: www.ktm.fi

France
Ministry of the Economy, Finance and Industry
139, rue de Bercy 75012
Paris, FRANCE
Web site: www.minefi.gouv.fr

G

Gabon
Ministry of Mines, Energy, Oil and Hydraulic Resources
PO Box 2199
Libreville, GABON
Telephone: 241 76-10-71
Fax: 241 72-49-90

Petrogab (Societe Nationale Petroliere Gabonaise)
Box 564
Libreville, GABON

Greece
Hellenic Petroleum S.A.
17th km., Athens-Corinth National Rd.
GR-193 00 Aspropyrgos
Attikí, GREECE
Telephone: +30-210-55-33-000
Fax: +30-210-55-39-298
Web site: www.hellenic-petroleum.gr

Ministry of Development
Energy and Natural Resources
119 Mesogion Ave.
101 92 Athens, GREECE
Telephone: 30-210-6969 863
E-mail: Grammatia@ypan.gr
Web site: www.ypan.gr/fysikoi_poroi/diarthrosi_eng.htm

Guatemala
Ministerio de Energia y Minas
Diagonal 17 29-78, Zona 11
Las Charcas, GUATEMALA
Telephone: (502) 2477-0382, -0901
Fax: (502) 2476-8506
E-mails: informatica@mem.gob.gt
relapubli@mem.gob.gt
Web site: www.mem.gob.gt

H

Hungary
MOL plc (Hungarian Oil & Gas)
Október huszonharmadika u. 18
H-1117 Budapest, HUNGARY
Phe: +36-20-9665-665
Fax: +36-1-464-0794
Web site: www.mol.hu

I

India
Ministry of Petroleum & Natural Gas
Hindustan Times House
18-20 Kasturba Gandhi Marg.
New Delhi 110 001 INDIA
Telephone: 91 11 335-2650; -2617
Fax: 91 11 331-7081; 335-2649
E-mail: dspdi.png@sb.nic.in
Web site: www.petroleum.nic.in

ONGC (Oil & Natural Gas Corp. Ltd.)
Tel Bhavan
Dehradun, 248 003 INDIA
Telephone: 0135-275-9561-67; -2161-65
Web site: www.ongcindia.com

Indonesia
Pertamina (Perusahaan Pertambangan
Minyak Dan Gas Bumi Negara)
Jalan Merdeka Timur No. 1 A
Jakarta-10110, INDONESIA
Telephone:(62) (21) 3815-111
Fax. (62) (21) 384-6865; -3882
Web site: www.pertamina.com

Iran
National Iranian Oil Co.
Hafez Crossing, Taleghani Avenue
Tehran, IRAN
Telephone: +98-21-6615-2929
Fax: +98-21-6615-3886
E-mail: managing-director@nioc.com
Web site: www.nioc.com

Iraq
Ministry of Oil
Oil Complex Building
Port Saeed St.
Baghdad, IRAQ
Telephone: (946-1) 8177-000, -050, -051;
international (946-1) 2863-103
Fax: (946-1) 8869-432; international (946-1) 2869-432
E-mails: oil@uruklink.net
oilcompo@uruklink.net
oilministry@yahoo.com
Web site: www.iraqigovernment.org

State Oil Marketing Organization (SOMO)
Telephone: 873 763 705 018; 019
Fax: 873 763 705 020; 964-1-7742 797

Ireland
Department of Communications,
Marine and Natural Resources
29-31 Adelaide Rd.
Dublin 2, IRELAND
Telephone: +353-1-678 2000
Fax: +353-1-678 2449
Web site: www.marine.gov.ie

Israel
Israel Government Ministries
Web site: www.gov.il/firstGov/topNavEng/
Engoffices/EngMinistries

Israel National Oil Co. Ltd.
Beit Gibor, 6 Kaufman St.
Tel-Aviv 68012, ISRAEL
PO Box 50199
Tel-Aviv 61500, ISRAEL
Telephone: 972 3 514-2020
Fax: 972 3 514-2061

Italy
Ente Nazionale Idrocarburi SpA (Eni SpA)
Piazzale Mattei, 1
00144 Rome, ITALY
Telephone: 39 06 59821
Fax: 39 06 5992 2141
Web site: www.eni.it

Ministero dello Sviluppo Economico
Direzione Generale dell'Energia e delle Risorse Minerarie
Via Molise 2
00187 Roma, ITALY
Telephone: 06 4705 2638
E-mail: urp@sviluppoeconomico.gov.it
Web site: www.minidustria.it

Ivory Coast

Director of the Office of the Minister
Ministere de Mines
Box V 50
Abidjan, COTE D'IVOIRE
Telephone: 225 21-50-03
Fax: 225 21-53-20

Petroci (Societe Nationale d'Operations
Petrolieres de la Cote d'Ivoire)
Box V 194
Abidjan, COTE D'IVOIRE
Telephone: 225 20-25-00; -10
Fax: 225 21-68-24

J

Japan

Japan Ministry of Economy, Trade and Industry
1-3-1 Kasumigaseki, Chiyoda-ku
Tokyo 100-8901 JAPAN
Telephone: +81-3-3501-1619
Fax: +81-3-3501-6942
E-mail: webmail@meti.go.jp
Web site: www.meti.go.jp/english/index.html

Japan National Oil Corp. (JNOC)
Fukoku Seimei Bldg.
2-2-2 Uchisaiwai-cho, Chiyoda-ku
Tokyo 100-8511 JAPAN
Telephone: 81 3 3597-7522
Fax: 81 3 3591-0172

K

Kazakhstan

Kazmunaigaz
65 Kazibek Bi Str., Office 405-406
Almaty 480091, KAZAKHSTAN
Telephone: +7 (3272) 98-00-20
Fax: +7 (3272) 72-29-28
E-mail: info@o2.kz
Web site: www.o2.kz

Ministry of Fuel & Energy
Bakenbay-Batyj St. 142
480091 Almaty, KAZAKHSTAN
Telephone: 7 3272 62-64-10; -60-80
Fax: 7 3272 62-66-30

Kuwait

Kuwait Oil Co.
P.O. Box 9758 Ahmadi
61008 Ahmadi, KUWAIT
Telephone: 00965-398 9111
Fax: 00965-398 3661
E-mail: kocinfo@kockw.com
Web site: www.kockw.com

Kuwait Petroleum Co.
P.O. Box 26565
Postal code 13126
Safat, KUWAIT
Telephone: +965 2455 455
Fax: +965 2411 780; 2467 159
E-mail: info@kpc.com.kw
Web site: www.kpc.com.kw

Ministry of Oil
P.O.Box 5077
13051, KUWAIT
E-mail: alnaft@moo.gov.kw
Web site: www.moo.gov.kw

L

Libya

General Department of Technical Affairs
Ministry of Petroleum
PO Box 256
Tripoli, LIBYA

National Oil Corp.
Bashir Saadawi St.
Tripoli, LIBYA
Phone: +218-333-7141
Fax: +218-333-1930
E-mail: info@www.noclibya.com
Web site: www.noclibya.com

M

Malaysia

Petronas (Petroliam Nasional Berhad)
Tower 1, Petronas Twin Towers
Kuala Lumpur City Centre
Kuala Lumpur 50088 MALAYSIA
Telephone: +603 2051-5000
Fax: +603 2026-5050, -5055
E-mail: webmaster@petronas.com.my
Web site: www.petronas.com.my

Mexico

Pemex (Petroleos Mexicanos)
Marina Nacional 329, Col. Huasteca
C.P. 11311
México D.F., MEXICO
Telephone: +52-55 1944-2500
Web site: www.pemex.com

Mongolia

Mongol Petroleum Co.
Uildverchnii Street 37
Ulaanbaatar, MONGOLIA
Telephone: 976 1 38-34-83
Fax: 976 1 33-11-76

Mineral Resources and Petroleum Authority of Mongolia
Builders' Square 3, State Property
Building #5, West Leaf, 1st Building of MPRAM
Ulaanbaatar 211239 Post Office 38 MONGOLIA
Telephone: 976-11-327180, -263701
Fax: 976-11-327180, -310370
E-mail: mram@mram.mn
Web site: www.mrpam.gov.mn

Myanmar

Myanmar Ministry of Energy
Director General, Energy Planning Department
No. 23, Pyay Road, Lanmadaw Township
Yangon, MYANMAR
Telephone: 95-1-222-545; -220-010
Fax: 95-1-222-960
E-mail: myanmoe@mptmail.net.mm
Web site: www.energy.gov.mm

N

Netherlands

SenterNovem
Juliana van Stolberglaan 3
Postbus 93144
2509 AC Den Haag, NETHERLANDS
Telephone: +31 70 3735000
Fax: +31 70 3735100
E-mail: info@senternovem.nl
Web site: www.senternovem.nl/senternovem

New Zealand

Ministry of Economic Development
New Zealand Crown Minerals
PO Box 1473
Wellington, NEW ZEALAND
Telephone: +64 3 962 6179
Fax: +64 4 499 0968
E-mail: crown.minerals@med.govt.nz
Web site: www.crownminerals.govt.nz

Nicaragua

Nicaraguan Institute of Energy (INE)
Petronic Building, 2nd Floor
Post Box 3226
Managua, NICARAGUA
Telephone: 505-222-7527
Fax: 505-228-2049
Web site: www.ine.gob.ni

Nigeria

Ministry of Petroleum Resources
Federal Secretariat, 1st & 2nd floor
Shehu Shagari Way
P.M.B. 449
Abuja, NIGERIA
Telephone: 234-9-523-0763
Fax: 234-9-523-7332
Web site: http://www.nigeria.gov.ng

Nigerian National Petroleum Corp. (NNPC)
NNPC Towers, Herbert Macaulay Way,
Central Business District
Garki Abuja, NIGERIA
Telephone: +234-9-523-9141
Fax: +234-9-234-0029

Norway

Ministry of Petroleum and Energy
P.O. Box 8148 Dep
0033 Oslo, NORWAY
Telephone: +47 22 24 62 43
E-mail: postmottak@oed.dep.no
Web site: odin.dep.no/oed/engelsk

Statoil (Den norske stats oljeselskap AS)
Forusbeen 50
N-4035 Stavanger, NORWAY
Telephone: +47 51 99 00 00
Fax: +47 51 99 00 50
E-mail: statoil@statoil.com
Web site: www.statoil.com

O

Oman

Ministry of Oil & Gas
Director General of Petroleum and Gas Affairs
Sultanate of Oman
P.O. Box 600
P.C. 113 Muscat, OMAN
Telephone: 968 2460 3222
Fax: 968 2469 3770
E-mail: webmaster@omanet.om
Web site: www.omanet.om/english/government/ministers.asp?cat=gov

PDO (Petroleum Development Oman)
Muscat 113 OMAN
Telephone: +968 678 111
Fax: +968 677 106
E-mail: external-affairs@pdo.co.om
Web site: www.pdo.co.om

P

Pakistan

Ministry of Petroleum and Natural Resources
3rd Floor, A Block, Pak Secretariat
Islamabad, PAKISTAN
Telephone: +92 51-920-8233
Fax: +92 51-920-5437
E-mail: info@mpnr.gov.pk
Web site: www.mpnr.gov.pk

Papua New Guinea

Department of Petroleum and Energy
P O Box 1993
Port Moresby, PAPUA NEW GUINEA
Telephone: +675 322 4200
Fax: +675 322 4222
E-mail: Bernard_Pawih@petroleum.gov.pg
Web site: www.petroleum.gov.pg

Peru

Ministerio de Energia & Minas
Av. Las Artes Sur 260 San Borja
Lima, PERU
Telephone: (511) 4750065
Web site: www.minem.gob.pe

Petroperu (Petroleos del Peru SA)
Av. Paseo de la Republica 3361
Lima 27, PERU
Telephone: 511 211-7800; 641-5000
Web site: www.petroperu.com.pe/Main.asp

Philippines

Department of Environment and Natural Resources
Visayas Ave., Diliman
1100 Quezon City, PHILIPPINES
Telephone: 63-2-929-6626; 928-1178
Web site: www.gov.ph

Philippine National Oil Co.
PNOC Bldg.
7901 Makati Ave.
Makati, Metro Manila 1200, PHILIPPINES
Telephone: 63-2-85-90-61; -85-88-56; -88-75-51
Fax: 63-2-815-3094; -810-6728

Poland

Ministry of Environmental Protection,
Natural Resources, and Forestry
Ul. Wawelska 52/56
00-920 Warsaw, POLAND
Telephone: 48 22 57-92-900
Web site: www.mos.gov.pl

PGNiG (Polskie Górnictwo Naftowe i Gazownictwo SA)
ul. Krucza 6/14
00-537 Warszawa, POLAND
Ph.: +48 (0-22) 583 50 00, 79 00
Fax: +48 (0-22) 691 82 73
E-mail: pr@pgnig.pl
Web site: www.pgnig.com.pl

Portugal

Galp Energia SA
Edifício Galp Energia
Rua Tomás da Fonseca
1600-209 Lisboa, PORTUGAL
Telephone: (+351) 21 724 25 00
Fax: (+351) 21 724 29 65
E-mail: galp@galpenergia.com
Web site: www.galpenergia.com

Q

Qatar

Ministry of Foreign Affairs
Web site: http://english.mofa.gov.qa

Qatar Petroleum
P.O. Box 3212
Doha, QATAR
Telephone: (974) 440 2000
Fax: (974) 483 1125
Web site: www.qp.com.qa

R

Romania

Ministry of Economy and Finance
1, Victoriei Square, District 1
Bucharest 010366, ROMANIA
Telephone: (4021) 314 34 00
Fax: (4021) 230 36 60
Web site: www.gov.ro/engleza

National Agency for Mineral Resources
36-38, Mendeleev St., Sector 1
Bucharest 010366, ROMANIA
Phone: (4021) 313 22 04
Fax: (4021) 210 74 40

Petrom SA
239 Calea Dorobantilor, District 1
Bucharest 010567 ROMANIA
Telephone: +40-21-4060040
Fax: +40-21-4060420
Web site: www.petrom.com/engleza/petrom_en.htm

Russia

Gazprom
16 Nametkina St., 117997
Moscow, V-420, GSP-7, RUSSIA
Telephone: +7 (495) 719-30-01
Fax: +7 (495) 719-83-33, -35
E-mail: gazprom@gazprom.com
Web site: www.gazprom.com

OAO Rosneft Oil Co.
26/1 Sofiyskaya Embankment, 1
Moscow, GSP-8 115998, RUSSIA
Phone: +7-095-777-44-22
Fax: +7-095-777-44-44
E-mail: postman@rosneft.ru
Web site: www.rosneft.ru

S

Saudi Arabia

Saudi Aramco (Saudi Arabian Oil Co.)
P.O. Box 5000
Dhahran 31311, SAUDI ARABIA
Telephone: 966-3 872-0115
Fax: 966-3 873-8190
E-mail: webmaster@saudiaramco.com.sa
Web site: www.saudiaramco.com

South Africa

Department of Minerals and Energy
Pretoria, SOUTH AFRICA
Telephone: 27 12 317-9121
Fax: 27 12 322-0810
Web site: www.dme.gov.za

Petroleum Agency SA
PO Box 1174
Parow 7499, SOUTH AFRICA
Telephone: + 27 21 938 3500
Fax: + 27 21 938 3520
E-mail: info@petroleumagencysa.com
Web site: www.petroleumagencysa.com

South Korea

Korea Gas Corp. (Kogas)
215 Jeongja-dong
Bundang-gu, Seongnam
Gyeonggi-do, 463-754, KOREA
Telephone: (+82)31-710-0070-2
Fax: (+82)31-710-0079
E-mail: kogasmaster@kogas.or.kr
Web site: www.kogas.or.kr

Korea National Oil Corp. (KNOC)
Telephone: 82-31-380-2114
Fax : 82-31-387-9321
Web site: www.knoc.co.kr

Spain

Comisión Nacional de Energía
Dirección de Relaciones Externas y Documentación
c/ Alcalá, 47
28014 Madrid, SPAIN
Telephone: 91 432 96 00
Fax: 91 577 62 18
E-mail: dre@cne.es
Web site: www.cne.es

Enagás SA
Paseo de los Olmos, 19
28005 Madrid, SPAIN
Telephone: (+34) 902-443-700
E-mail: contacta@enagas.es
Web site: www.enagas.es

Repsol YPF SA
Paseo de la Castellana, 278-280
28046 Madrid, SPAIN
Telephone: 34 91 348 80 00
Fax: 34 91 348 09 77
Web site: www.repsolypf.com

Sweden

Ministry of Enterprise, Energy and Communications
Phone: +46 8 405 10 00
Address: Jakobsgatan 26, 103 33 Stockholm, SWEDEN
Web site: www.sweden.gov.se/sb/d/2067

Syria

Ministry of Petroleum and Mineral Resources
Al-Adawi Str. Al-Khateeb Katte
P.O.Box 31483
Damascus, SYRIA
Telephone: 00963-11-445-1624; -5972
Fax: 00963-11-4463942
E-mail: mopmr@net.sy
Web site: www.mopmr-sy.org

Syrian Petroleum Co.
Al-Mutanabbi St.
P.O.Box: 2849
Damascus, SYRIA
Telephone: 963 11 222-7007; -4272
Fax: 963 11 222-5648; 221-0953
E-mails: spccom2Dir@net.sy
spc@spc-sy.com
Web site: www.spc-sy.com

T

Taiwan

Chinese Petroleum Corp.
No. 3 Sungren Rd., Shinyi-Chiu
Taipei 11010, TAIWAN
Telephone: 886-2-8789-8989
Fax: 886-2-8789-9000
E-mail: ir@cpc.com.tw
Web site: www.cpc.com.tw

Tanzania

Ministry of Energy and Minerals
Permanent Secretary
Sokoine/Mkwepu St.
P.O. Box 2000/9152
Dar es Salaam, TANZANIA
Telephone: +255 22 211-7153-9; -2793
Fax: +255 22 211-6719
E-mail: madini@africaonline.co.tz
Web site: www.tanzania.go.tz/government

Thailand

Ministry of Energy
Rama VI Rd.
Bangkok 10400, THAILAND
Telephone: 66 2 511-0215; -0749
Fax: 66 2 245-9855
Web site: www.thaigov.go.th

PTT plc (formerly Petroleum Authority of Thailand)
555 Vibhavadi Rangsit Rd., Chatuchak
Bangkok 10900, THAILAND
Ph. 0-2537-2000
Fax. 0-2537-3498-9
E-mail: corporate@pttplc.com
Web site: www.pttplc.com

Trinidad & Tobago

Ministry of Energy and Energy Industries
Level 9, Riverside Plaza
Besson Street
Port of Spain, TRINIDAD
Telephone: (868) 623-6708
Fax: (868) 625-0306
E-mail: cblades@energy.gov.tt
Web site: www.energy.gov.tt

National Gas Co. of Trinidad and Tobago Ltd.
Orinoco Drive
Point Lisas Industrial Estate
P.O Box 1127
Port of Spain, TRINIDAD
Telephone: (868) 679-4662, -4680
Fax: (868) 679-2384
E-mail: ngc@ngc.co.tt
Web site: www.ngc.co.tt

Petroleum Co. of Trinidad & Tobago (Petrotrin)
Cor. Queen's Park West & Cipriani Blvd.
Port of Spain, TRINIDAD
Telephone: 868- 625-5240; -5239
Fax: 868-624-6461
E-mail: petrochair@petrotrin.com
Web site: www.petrotrin.com

Tunisia

Etap (Entreprise Tunisienne d'Activites Petrolieres)
27 bis, Ave. Khereddine Pacha
1002 Tunis, TUNISIA
Telephone: (216) 71-782-288
Fax: (216) 71-780-123; -786-141
E-mail: dexprom@etap.com.tn
Web site: www.etap.com.tn

Turkey

BOTAS Petroleum Pipeline Corp.
Bilkent Plaza A, II Blok, Bilkent
06530 Ankara, TURKEY
Telephone: 90-312 297 20 00
Fax: 90-312 266-07-33; -07-34
Web site: www.botas.gov.tr

TPAO Turkish Petroleum Corp.
Mustafa Kemal Mahallesi 2
Cadde No. 86
06520 Ankara, TURKEY
Telephone: 90 312 286-91 00
Fax: 90 312 286-90-00, -01
E-mail: tpaocc@petrol.tpao.gov.tr
Web site: www.tpao.gov.tr

Turkmenistan

Ministry of Energy and Industry
6 Nurberdy Pomma St.
Ashgabat 744000, TURKMENISTAN
Telephone: 9 (9312) 35-49-21; 51-08-82
Fax: 9 (9312) 29-06-82

U

Ukraine

Ukrtransnafta
60 Artema St.
Kyiv 04050, UKRAINE
Telephone: 044 246-80-07
Fax: 044 246-83-77

State Committee on Geology and
Utilization of Mineral Resources
Volodymyrska 34
252601, Kyiv 34, UKRAINE
Telephone: (044) 228-60-51; -32-43
Fax: (044) 228-60-51
E-mail: admin@brama.com
Web site: www.brama.com/ua-geology

United Arab Emirates

Adnoc (Abu Dhabi National Oil Co.)
P.O. Box : 898
Abu Dhabi, UNITED ARAB EMIRATES
Tel : (9712) 6020000
Fax: (9712) 6023389
Web site: www.adnoc.com

Ministry of Energy
P.O. Box 59
Abu Dhabi, UNITED ARAB EMIRATES
Telephone: 02 6262288; 02 6126500
Fax: 02 6272291; 02 6260037
E-mail: mopmr@uae.gov.ae
Web sites: www.uae.gov.ae
www.government.ae/gov/en/gov/federal/moew.jsp

United Kingdom

Department of Trade and Industry
Energy Division
Oil and Gas Directorate
1 Victoria St.
London SW1H 0ET, UNITED KINGDOM
Telephone: 020 7215 5000
Web site: www.dti.gov.uk/energy

Uzbekistan

Republic of Uzbekistan
Telephone: 998-71-137-4941
E-mail: info@uzinfocom.uz
Web site: www.gov.uz/

Uzneftegaz
Amir Temir Ulitsa 66
700084 Tashkent, UZBEKISTAN
Telephone: 371 2 35-83-23
Fax: 371 2 35-83-37

V

Venezuela

Ministerio de Energía y Petróleo
Edificio Petroleos de Venezuela, Torre Oeste
Av. Libertador con Av. Empalme, Urb. La Campiña
Parroquia El Recreo, Distrito Capital
Caracas, VENEZUELA
Telephone: 0212-708.75.81
Fax: 0212-708.75.98
Web site: www.mem.gov.ve

PdVSA (Petroleos de Venezuela)
Edificio Petróleos de Venezuela
Avenida Libertador, La Campiña, Apartado 169
Caracas, 1010-A, VENEZUELA
Telephone: +58-212-708-4111
Fax: +58-212-708-4661
Web site: www.pdvsa.com

Vietnam

Ministry of Petroleum & Gas
193/6A Nam Ky Khoi Nghia
Telephone: 080-48137, -48354
Fax: 080-48299
Ho Chi Minh City 3, VIET NAM

PetroVietnam (Vietnam Oil and Gas Corp.)
22 Ngo Quyen Street
Hoan Kiem District
Hanoi, VIET NAM
Telephone: 84-4-8252526
Fax: 84-4-8265942
E-mail: webmaster@hn.pv.com.vn
Web site: www.petrovietnam.com.vn

Y

Yemen

Ministry of Petroleum and Mineral Resources
Sanaa, YEMEN
Telephone: 967 2 263-246
Fax: 967 2 268-949
Web site: www.yemen.gov.ye

Yemen Petroleum Co.
YPC Bldg.
Box 3360
Al Hudaydah, YEMEN

GLOSSARY

Abbreviations and Acronyms

A

ACG	Azeri-Chirag-Guneshli (Azerbaijan oil fields)
Adma Opco	Abu Dhabi Marine Operations Co.
Adnoc	Abu Dhabi National Oil Co.
AED	Alberta Economic Development
AEP	American Electric Power Co.
A.G.A.	American Gas Association
Agip KCO	Agip Kazakhstan North Caspian Operating Co., formerly OKIOC
AGPA	Alaska Gasline Port Authority
AIOC	Azerbaijan International Operating Co.
altfuels	alternative fuels including wind power and solar energy
AMBO	Albanian Macedonian Bulgarian Oil Co.
AMPCO	Atlantic Methanol Production Co. LLC
ANP	Agencia Nacional de Petroleo, or National Petroleum Agency (Brazil)
ANWR	Arctic National Wildlife Refuge (Alaska, US)
AOSP	Alberta Oil Sands Project; Alberta Oil Sands Pipeline
APG	Aboriginal Pipeline Group (Inuvik, NT)
API	American Petroleum Institute
Aramco	(Saudi) Arabian Oil Co.
ASEAN	Association of Southeast Asian Nations
ASPO	Association for the Study of Peak Oil
ATHEER	Abu Dhabi Gas Co. (United Arab Emirates)

B

BALAK	national oil and gas agency's Indonesian-language acronym
Banagas	Bahrain National Gas Co.
Banco	Bahrain National Oil Co.
Bapco	Bahrain Petroleum Co.
bbl	barrels
b/cd	barrels per calendar day
bcf	billion cubic feet
bcfd	billion cubic feet per day
bcfe	billion cubic feet equivalent
bcm	billion cubic meters
bcmd	billion cubic meters per day
bcme	billion cubic meters equivalent
b/d	barrels per day
bbl-d	barrels daily
BEDB	Brunei Economic Development Board
BGL	Bhagyanagar Gas Ltd. (India)
b/h	barrels per hour
BLM	Bureau of Land Management (US)
BNA	Bundesnetzagentur (Germany)
BNDES	National Economic & Social Development Bank (Brazil)
bo/d	barrels of oil per day
boe	barrels of oil equivalent
boe/d	barrels of oil equivalent per day
BOOT	build, own, operate, and transfer
BOOTS	Bulk Oil Offshore Transfer System (US)
BOTAS	Turkey government natural gas and pipeline company
bscf	billion standard cubic feet
bscfd	billion standard cubic feet per day
bscm	billion standard cubic meters
b/sd	barrels per stream day
BTC	Baku-Tbilisi-Ceyhan oil pipeline (Azerbaijan-Georgia-Turkey)
Btu	British thermal unit
bw/d	barrels of water per day

C

CAA	commercial arrangement area (Malaysia/Viet Nam)
CAC	Central Asia-Center (Former Soviet Union pipeline system)
CAGC	Canadian Association of Geophysical Contractors
CAPP	Canadian Association of Petroleum Producers
Caricom	Caribbean Community
CBM	coalbed methane
CBNG	coalbed natural gas
CBOT	Chicago Board of Trade
CD	compact disk
CEO	chief executive officer
CEPA	Canadian Energy Pipeline Association
Cepsa	Compania Espanola de Petroleos SA (Spain)
CERA	Cambridge Energy Research Associates (CERA)
CERI	Canadian Energy Research Institute
cf	cubic feet
cfd	cubic feet per day
CFE	Comision Federal de Electricidad (Federal Electricity Commission, Mexico)
cfed	cubic feet of equivalent per day
CFTC	Commodity Futures Trading Commission (US)
CGES	Centre for Global Energy Studies (London)
CHDP	cricondentherm hydrocarbon dew point
Chinaoil	China National United Oil Co.
CICESE	Centro de Investigación Científica y de Educación Superior de Enseñada (Mexico)
CIS	Commonwealth of Independent States (Former Soviet Union)

CITIC	China International Trust & Investment Corporation
CLJOC	Cuu Long Joint Operating Co. (Viet Nam)
C-NLOPB	Canada-Newfoundland Labrador Offshore Petroleum Board
CNG	compressed natural gas
CNOOC	China National Offshore Oil Corporation
CNPC	China National Petroleum Corporation
CNR	Canadian Natural Resources Ltd.
CO	carbon monoxide
CO_2	carbon dioxide
COS	China Oilfield Services Ltd.
cP	centi-Poise (viscosity measurement)
CPC	Caspian Pipeline Consortium
CPC	Chinese Petroleum Corp. (Taiwan)
CPOT	Constanta-Pancevo-Omisalj-Trieste pipeline (Romania-Croatia-Italy)
CPU	central processing unit
CPUC	California Public Utilities Commission
CRE	Energy Regulatory Commission (Mexico)
CRP	Paraguana Refining Complex (Venezuela)
CSO	Coflexip Stena Offshore
CTL	coal to liquids
cu m	cubic meters (metres)
CUCBM	China United Coalbed Methane Corp. Ltd.
CV	core venture (Saudi Arabia gas initiative)

D

DDCV	deep-draft caisson vessel
DEPA	Public Gas Co. (Greece)
DIGP	Dauphin Island Gathering Partners
dilbit	diluted bitumen (heavy crude oil and bitumen blended with condensate diluent)
DOE	Department of Energy
DONG	Dansk Olie ag Naturgas AS (Denmark)
DUC	Dansk Undergrunds Consortium (Denmark)
dwt	deadweight tons

E

E&D	exploration and development
E&P	exploration and production
EBRVs	Energy Bridge regasification vessels
EC	European Commission
ECO	Economic Cooperation Organization (Central Asia)
EdF	Electricité de France
EDP	Electricidade de Portugal SA
EGAS	Egyptian Natural Gas Holding Co.
EGPC	Egyptian General Petroleum Corp.
EIA	Energy Information Administration (US)
EIA	environmental impact assessment
ELNG	Egyptian LNG joint venture
ENAP	Empresa Nacional de Petroleo (Chile)
Enarsa	Energía Argentina SA
Enel	Ente Nazionale Energia Elettrica SpA (Italy)
Eni	Ente Nazionale Idrocarburi SpA (Italy)
EPA	Environmental Protection Agency
EPC	engineering, procurement, and construction
EPCA	Energy Policy & Conservation Act (US)
EPDC	European Pipeline Development Co. BV
EPSA	exploration and production-sharing agreement
ERC	Energy Regulatory Commission (France)
ESP	electrical submersible pump
ETAP	Enterprises Tuniseienne d'Activites Petrolieres
ETLP	extended tension leg platform
EU	European Union
EUB	Energy and Utilities Board (Alberta, Canada)
EVE	Basque Energy Authority (Spain)
EWT	extended well test
Ex-Im Bank	Export-Import Bank of the US

F

FAWAG	foam-assisted water alternating gas
FCC	fluid catalytic cracking
FCCU	fluid catalytic cracking unit
FDI	foreign direct investment
FEED	front-end engineering design
FERC	Federal Energy Regulatory Commission
FNA	Frigg Norwegian Association (Norway)
fob	free on board
FPC	Formosa Petrochemical Corp.
fph	feet per hour
FPSO	floating production, storage, and off-loading
FPU	floating production units
FSRU	floating storage and regasification unit
FSU	Former Soviet Union
ft	foot
FTC	Federal Trade Commission
FTIR	Fourier transform infrared

G

GAIL	Gas Authority of India Ltd.
GASCO	Abu Dhabi Gas Industries Ltd. (United Arab Emirates)
GBS	gravity-based structure
GC	Green Canyon (US Gulf of Mexico)
GCC	Gulf Cooperation Council
GCDS	Gasoducto Cruz del Sur (Argentina-Uruguay)

GdF	Gaz de France SA
GDP	gross domestic product
GEA	Gas Energy Adria
GFU	Gas Negotiating Committee (Norway)
GNA	Gasoducto del Noreste Argentino (Bolivia-Argentina)
GNPOC	Greater Nile Petroleum Operating Co. (Sudan)
GOM	Gulf of Mexico
GOR	gas-oil ratio
GOSP	gas-oil separation plant
gpd	gallons per day
GSCP	Georgia Strait Crossing Project (Canada-US)
GTI	Gas Technology Institute
GTL	gas-to-liquids
GTLB	GTL Bolivia SA
GWh	gigawatt-hours

H

H&P	Helmerich & Payne Inc.
HCU	Hydrocarbon Unit of the Bangladesh Energy and Mineral Resources Division
HDPE	high-density polyethylene
HDS	hydroskimming
HOA	heads-of-agreement
hp	horsepower
HP	Hellenic Petroleum (Greece)
HPCL	Hindustan Petroleum Corp Ltd. (India)
HPHT	high-pressure, high-temperature
HSE	health, safety, and environment

I

IADB	Inter-American Development Bank
IADC	International Association of Drilling Contractors
IAEA	International Atomic Energy Agency (Vienna)
ICO	Interconexion Centro-Occidente pipeline (Venezuela)
ID	inside diameter
IDB	Inter-American Development Bank
IEA	International Energy Agency
IEC	Israel Electric Co.
IET	integrated exploration technology
IFP	Institut Français du Pétrole (France)
IGAT	Iran Gas Trunkline
IGCC	integrated gasification combined cycle
ILNGA	International LNG Alliance (US)
IMF	International Monetary Fund
IMO	International Maritime Organization
INA	Industrija nafte dd (Croatia)
INAC	Indian and Northern Affairs Canada
INGAA	Interstate Natural Gas Association of America
INOC	Iraq National Oil Co.
IOC	Indian Oil Corp.
IOC	international oil company
IPAA	Independent Petroleum Association of America
IPO	initial public offering
IRNA	Iranian news agency
ISAC	Information Sharing & Analysis Center (US)
IUA	International Unitization Agreement

J

Japex	Japan Petroleum Exploration Corp.
JDA	Joint Development Area (Malaysia-Thailand, Australia-East Timor)
JIPA	joint investment production activity
JNOC	Japan National Oil Company
JPSC	Joint Pipeline Security Commission (Azerbaijan, Georgia, and Turkey)
JV	joint venture

K

KGOC	Kuwait Gulf Oil Co.
KIO	Karachaganak Integrated Organization (consortium)
KKPL	Kenai Kachemak pipeline (Alaska)
km	kilometer
KNOC	Korean National Oil Corporation
Kogas	Korea Gas Corporation
KPC	Kuwait Petroleum Corp.
KVA	kilo-volt-amperes
KZM	Ku-Zaap-Maloob complex (Mexico)

L

LAB	linear alkylbenzene
LION	large increase in octane number
LLDPE	linear low-density polyethylene
LNG	liquefied natural gas
LOGGS	Lincolnshire offshore gas gathering system (UK)
Logic	leading oil and gas industry competitiveness (UK)
LPG	liquefied petroleum gas
LWD	logging-while-drilling

M

m	meter
M&A	merger and acquisition
M&NP	Maritimes and Northeast Pipeline (Canada-US)
Marad	Maritime Administration (US)
MC	Mississippi Canyon (US Gulf of Mexico)
Mcf	thousand cubic feet
MD	measured depth
MEG	monoethylene glycol
MFL	magnetic flux leakage
MHP	Market Hub Partners (US)
MIBEL	Mercado Ibérico de Electricidade (Portugal, Spain)
MIDOR	Middle East Oil Refinery Ltd.
Migas	Directorate General of Oil and Gas (Indonesia)
MMBtu	million British thermal units
MMcf	million cubic feet
MMcfd	million cubic feet per day
MMcfe	million cubic feet of equivalent
MMcfed	million cubic feet of equivalent per day
MMcm	million cubic meters
MMcmd	million cubic meters per day
MMS	Minerals Management Service
MODU	mobile offshore drilling unit
MOL plc	Hungarian Oil and Gas Co.
MSC	multiple-service contract
MTBE	methyl tertiary butyl ether
mtoe	million tonnes of oil equivalent
mtpy	million tonnes per year
MW	megawatt
MWD	measurement-while-drilling
MWh	megawatt-hour

N

NAM	Nederlandse Aardolie Maatschappij BV (Netherlands)
NARUC	National Association of Regulatory Utility Commissioners (US)
NCMA	North Coast Marine Area (Trinidad & Tobago)
NEGP	North European Gas Pipeline (Russia-Germany)
NELP IV	New Exploration Licensing Policy IV (India)
Nemed	North East Mediterranean deepwater concession (Egypt)
NETA	New Electricity Trading Arrangements (UK)
NGO	non-governmental organization
NGSA	Natural Gas Supply Association (US)
NIMBY	not in my back yard
NIOC	National Iranian Oil Company
NLNG	Nigeria Liquefied Natural Gas Co.
NNCL	Bangladesh Business Development Corp. Ltd.
NNPC	Nigerian National Petroleum Co.
NO_x	nitrogen oxides
NOIA	Newfoundland Ocean Industries Association
NPD	Norwegian Petroleum Directorate
NPPAP	New Players Petroleum Association of the Philippines
NPR-A	National Petroleum Reserve-Alaska
NPRA	National Petrochemical & Refiners Association (US)
NTN	Nova Transportadora do Nordeste SA (Brazil)
NTS	Nova Transportadora do Sudeste SA (Brazil)
NWS	North West Shelf (Australia)
NYBOT	New York Board of Trade
Nymex	New York Mercantile Exchange
NYSE	New York Stock Exchange

O

OCA	Overlapping Claims Area (Thailand-Cambodia)
OCP	Oleoducto de Crudos Pesados Ecuador SA (consortium and pipeline)
OCTG	oil country tubular goods
OD	outer diameter
OECD	Organization for Economic Cooperation and Development
OFAC	Office of Foreign Assets Control (US Department of Treasury)
OGDC	Oil & Gas Development Co. Ltd. (Pakistan)
OGIP	original gas in place
OGJ	*Oil & Gas Journal*
OKIOC	Offshore Kazakhstan International Operating Company, re-named Agip KCO
OLNGC	Oman Liquefied Natural Gas Co.
OML	oil mining lease (Nigeria)
Omnis	Office of National Mines and Strategic Industries (Madagascar)
ONAREP	Office National de Recherches et d'Explorations Pétrolières (Morocco)
ONGC	Oil & Natural Gas Corporation (India)
Onhym	National Office of Hydrocarbons and Mining (Morocco)
OPEC	Organization of Petroleum Exporting Countries
OPIC	Overseas Private Investment Corp. (US)
OPL	oil prospecting license (Nigeria)
OPS	Office of Pipeline Safety (US)
OSA	operating service agreement (Venezuela)

P

PADD	Petroleum Administration for Defense District (US)
PALs	Plantwide Applicability Limits (US)
Parco	Pak-Arab Refinery Co. Ltd.
Parsep	Proyecto de Asistencia para La Reglamentación del Sector Energético del Perú
PCO-W	Iraq Project and Contracting Office (Washington, DC), US Army
PCR	production change rate
PDET	Oil Outflow and Treatment Plan (Campos basin, Brazil)
PDO	Petroleum Development Oman LLC
PDO	plan for development and operation (Norway)
PdVSA	Petróleos de Venezuela SA
PEGASO	Excellence in Environmental Management & Operational Safety Program (Brazil)
Pemex	Petroleos Mexicanos
PEP	Petroleum Exploration Permit (New Zealand)
Pertamina	Perusahaan Pertambangan Minyak Dan Gas Bumi Negara (Indonesia)
PET	polyethylene terephthalate
Petrobangla	Bangladesh Oil, Gas & Mineral Corp.
Petrobras	Petroleo Brasileiro SA
Petroecuador	Petroleos del Ecuador
Petrogal	Petroleos de Portugal EP
Petronas	Petroliam Nasional Sdn. Berhad (Malaysia)
Petroperu	Petroleos del Peru SA
PetroSA	Petroleum Oil and Gas Corp. of South Africa
Petrotrin	Petroleum Co. of Trinidad and Tobago Ltd.
PetroVietnam	Vietnam Oil & Gas Co.
PG&E	Pacific Gas & Electric Co.
PGC	Potential Gas Committee (US)
PGC	Process Gas Consumers (US)
PGN	PT Perusahaan Gas Negara (Persero) Tbk (Indonesia)
PGNiG	Polskie Gomictwo Naftowe I Gazownictow SA (Polish Oil & Gas Co.)
PIC	Petrochemical Industries Co. (Kuwait)
PIO	plan for installation and operation (Norway)
PIRINC	Petroleum Industry Research Foundation Inc.
PJ	petajoule (approx. = 1 trillion Btu)
PKN	Polski Koncern Naftowy (Poland)
PL	Production License (Norway)
PLEM	pipeline end manifold
PMP	Petroleum Mining Permit (New Zealand)
PNSC	Pakistan National Shipping Corporation
Poas	Petrol Ofisis (Turkey)
PoG	Perseus-over-Goodwyn (Australia)
POGC	Polish Oil and Gas Co.
PPL	petroleum prospecting license (New Zealand)
PSA	production-sharing agreement (or arrangement)
PSA	pressure swing adsorption
PSC	production-sharing contract
PSO	Pakistan State Oil
PTEN	PT Exspan Nusantara (Indonesia)
PTT	polytrimethylene terephthalate
PTT plc	Thailand government oil and gas firm (formerly Petroleum Authority of Thailand)
PTTEP	PTT Exploration and Production plc (Thailand)

R

RasGas	Ras Laffan Liquefied Natural Gas Co. Ltd. (Qatar)
RFCC	residual fluid catalytic cracking
RFG	reformulated gasoline
RFR	right of first refusal
RIK	royalty-in-kind
ROSBOS	Rosetti Marino SPA-Bouygues Offshore joint venture
ROV	remotely operated vehicle
RR	railroad
RTO	regional transmission organization
RTP	rapid thermal conversion

S

Sabic	Saudi Basic Industries Corp.
SADC	South African Development Community
SAGD	steam-assisted gravity drainage
SAGE	Scottish Area Gas Evacuation
SAP	solvent-aided process
SAPREF	South African Petroleum Refineries (Pty) Ltd.
SARS	severe acute respiratory syndrome
scfd	standard cubic feet per day
SCO	synthetic crude oil
SCP	South Caucasus Pipeline
SDFI	state's direct financial interest (Norway), re-named Petoro
SEA	State Energy Administration (China)
SEEL	South-East European Line
semi	semisubmersible offshore production platform
Sinochem	China National Chemical Import & Export Company
SIR	Ivorian Refining Company (Ivory Coast)
SNG	synthetic natural gas
SNGPL	Sui Northern Gas Pipelines Ltd. (Pakistan)
SNH	National Hydrocarbon Corp. (Cameroon)
SNPC	Société Nationale des Pétroles du Congo (Brazzaville)
SO_2	sulfur dioxide
SO_x	sulfur oxides
SOCAR	State Oil Company of Azerbaijan Republic
SOEP	Sable Offshore Energy Project (Canada)

SOMO	State Oil Marketing Organization (Iraq)
Sonangol EP	Sociedade Nacionale de Combustiveis de Angola
SOTE	Trans-Ecuadorian pipeline's Spanish-language acronym
SPA	sale and purchase agreement
SPE	Society of Petroleum Engineers (US)
SPLA	Sudan People's Liberation Army
SPM	single-point mooring
SPP	Slovensky plynarensky priemysel (Slovakia)
sq km	square kilometer
sq mi	square mile
SPR	Strategic Petroleum Reserve
SRS	Schmierstoff Raffinerie Salzbergen (German refinery)
SRU	sulfur-recovery unit
SSGC	Sui Southern Gas Co. Ltd. (Pakistan)
SSP	Sevan Stabilized Platform (cylindrical monohull FPSO vessel design)
Statoil	Den norske stats oljeselskap AS (Norway)
STL	submerged turret loading
SVA	security vulnerability assessment (US)
synbit	syncrude-bitumen blend (raw bitumen diluted with synthetic crude oil)

T

TAC	technical assistance contract
TAG	Trans Austria Gasleitung pipeline
TAP	Trans-Alpine Pipeline
TAP	Turkmenistan-Afghanistan-Pakistan pipeline
TAWN	Tariki, Ahuroa, Waihapa, and Ngaere area (New Zealand)
tcf	trillion cubic feet
tcfe	trillion cubic feet equivalent
tcm	trillion cubic meters
tcme	trillion cubic meters equivalent
TCP	Trans-Caspian Pipeline
TD	total depth
TGS	Transportadora de Gas del Sur (South American pipeline operator)
TJ	terajoules
TMD	total measured depth
TNK	Tyumen Oil Co. (Russia)
toe	tons or tonnes of oil equivalent
TPAO	Turkiye Petrolleri Anonim Ortakligi (Turkey)
tpd	tons or tonnes per day
tpy	tons or tonnes per year
TRG	Rompetrol Group NV (Netherlands)
TSB	Transportadora Sulbrasileira de Gas pipeline (Brazil)
tscm	trillion standard cubic meters
Tupras	Turkish Petroleum Refineries Corp.
TVD	true vertical depth
TVDS	true vertical depth subsea
TVK	Tiszai Vegyi Kombinat Rt. (Hungary)
TW	terawatt
TWh	terawatt-hour

U

UAE	United Arab Emirates
UDS	Ultramar Diamond Shamrock Corp.
UES	Unified Energy Systems (Russia)
UKCS	UK Continental Shelf
UKOOA	UK Offshore Operators Association Ltd.
ULCC	ultra large crude carrier
UN	United Nations
Unipec	United International Petroleum & Chemicals Company Ltd. (China)
UOG	United Arab Emirates Offsets Group
URR	ultimate resource recovery
USACE	US Army Corps of Engineers
UXO	unexploded ordnance

V

Vapex	vapor extraction
VGO	vacuum gas oil
VLCC	very large crude carrier
VOCs	volatile organic compounds
VP	Vice President
VSS	vortex separation system

W

WAG	water-alternating-gas
WAGP	West Africa Gas Pipeline (Nigeria. Ghana, Togo, and Benin)
WDDM	West Delta Deep Marine (Egypt)
WLGP	Western Libyan Gas Project
WNTS	West Natuna Transportation System (Indonesia to Singapore)
Wocap	World Oil Production Capacity (model)
WOPP	White Oil Pipeline Project (Pakistan)
WTO	World Trade Organization

Z

ZIC	Zone d'Interet Commun (Angola/Congo-Brazzaville)

Glossary

North American Geographical Abbreviations

U.S. States

AL	Alabama
AK	Alaska
AS	Arkansas
AZ	Arizona
CA	California
CO	Colorado
CT	Connecticut
DC	District of Columbia
DE	Delaware
FL	Florida
GA	Georgia
HI	Hawaii
ID	Idaho
IL	Illinois
IN	Indiana
IA	Iowa
KS	Kansas
KY	Kentucky
LA	Louisiana
MA	Massachusetts
MD	Maryland
ME	Maine
MI	Michigan
MN	Minnesota
MO	Missouri
MS	Mississippi
MT	Montana
NC	North Carolina
ND	North Dakota
NE	Nebraska
NH	New Hampshire
NJ	New Jersey
NM	New Mexico
NV	Nevada
OH	Ohio
OK	Oklahoma
OR	Oregon
PA	Pennsylvania
RI	Rhode Island
SC	South Carolina
SD	South Dakota
TN	Tennessee
TX	Texas
UT	Utah
VA	Virginia
VT	Vermont
WA	Washington
WI	Wisconsin
WV	West Virginia
WY	Wyoming

Canada Provinces

AB	Alberta
BC	British Columbia
MB	Manitoba
NB	New Brunswick
NL	Newfoundland & Labrador
NS	Nova Scotia
NT	Northwest Territories
NU	Nunavut
ON	Ontario
PE	Prince Edward Island
QC	Quebec
SK	Saskatchewan
YT	Yukon Territory

Index

Advertisers

Contents

A

B

C

D

E

F

G

H

I

J

K

L

M

N

O

P

Q

R

S

T

U

V

W

X

Y

Z

Fold-out Map Lege

Use this legend for all IPE 2

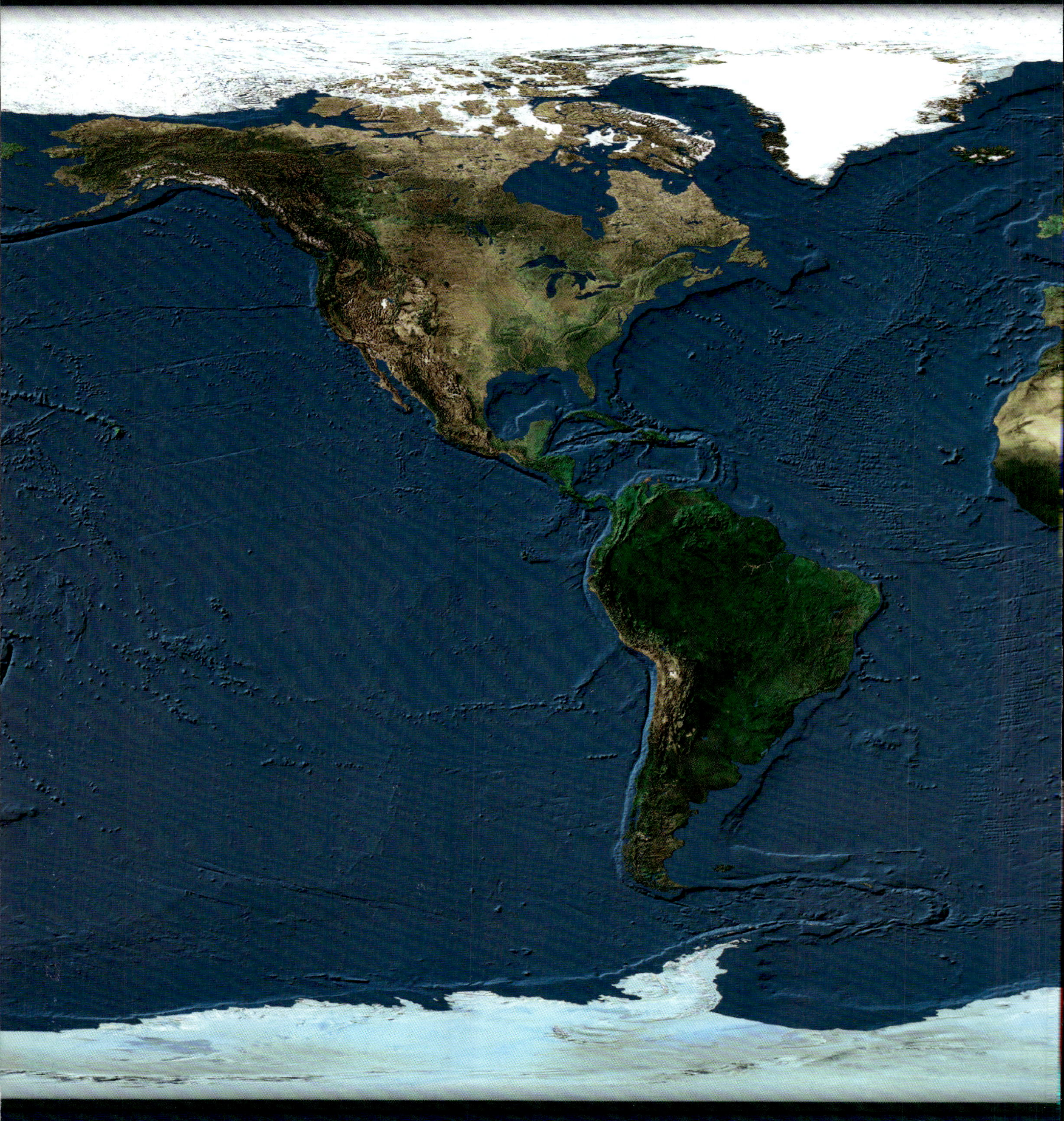